BETWEEN TWO WORLDS

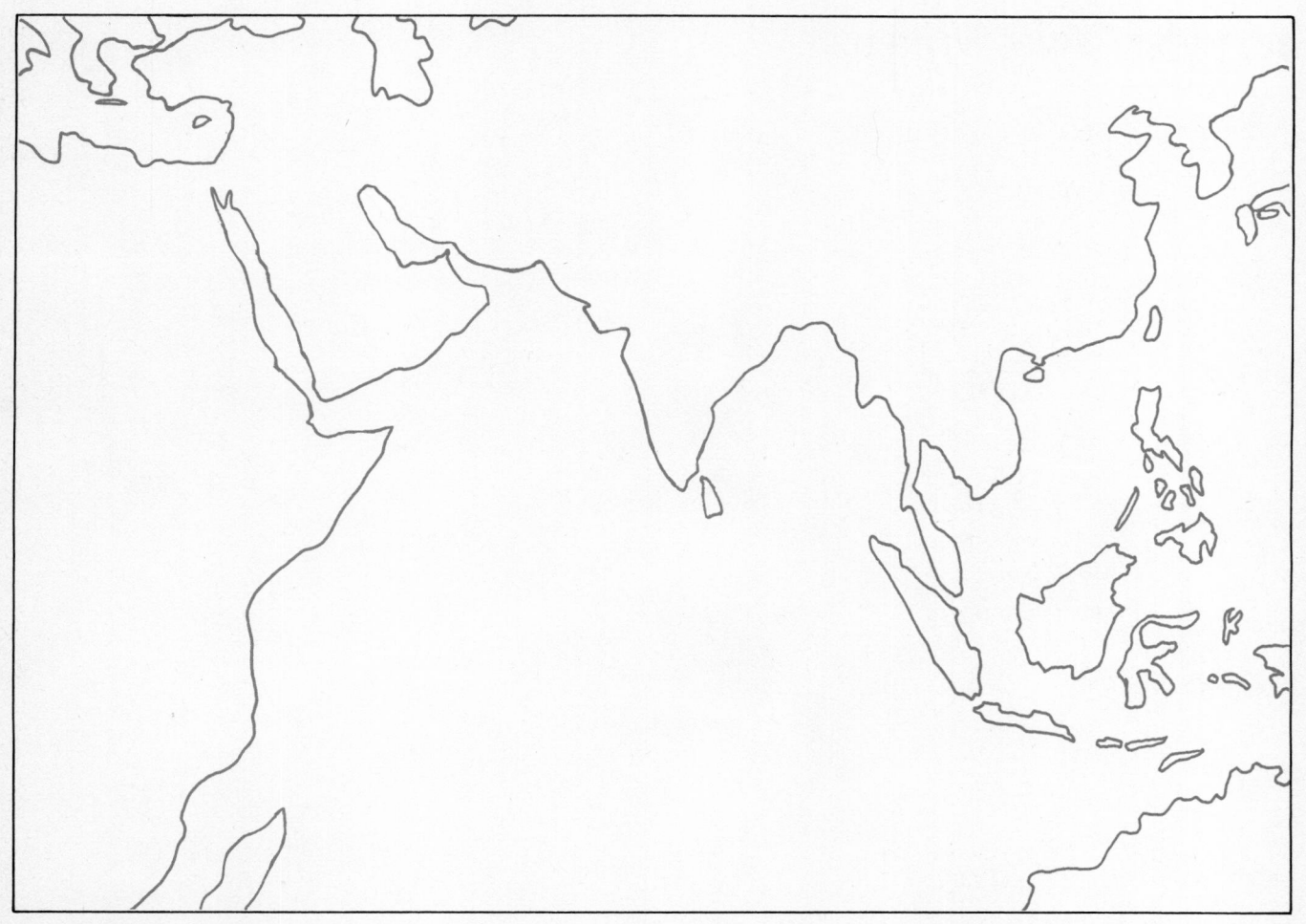

BETWEEN TWO WORLDS

An Introduction to Geography

SECOND EDITION

Robert A. Harper
UNIVERSITY OF MARYLAND

Theodore H. Schmudde
UNIVERSITY OF TENNESSEE

HOUGHTON MIFFLIN COMPANY   Boston

Dallas   Geneva, Illinois   Hopewell, New Jersey   Palo Alto   London

Cartography by Francis & Shaw, Inc., New York, New York; art on cover and on pp. ii, 1, 161, and 417 by ANCO, Boston.

Copyright © 1978 by Houghton Mifflin Company.

Printed in the U.S.A.

Library of Congress Catalog Card Number: 77-76418

ISBN: 0-395-25164-8

Contents

List of Figures

Preface

Between Two Worlds: An Introduction to Geography presents a geographical approach to the contemporary world. It is intended for the college student who may have little or no background in geography and is seeking a broadened understanding of the world. The approach and content of the book make it useful as a text for various introductory geography courses, such as world regional geography, world geography, introduction to geography, or human/cultural geography.

The title, *Between Two Worlds,* suggests our approach. A conceptual framework is developed for evaluating the multitude of places that exist on the surface of the earth, beginning in Chapter 1 with the contrast between "two worlds": the traditional one, shown in a village in India, and the modern one, illustrated by Washington, D.C. In traditional systems, each local group primarily depends on the possibilities of its local environment, whereas modern systems reach out to encompass resources and ideas from the entire world.

Different systems developed by human groups throughout the world can be seen as existing at various points along a continuum between the contrasting worlds of Chapter 1. The rest of Part One describes the models of the two contrasting systems, explains the continuum, and presents a tool, a "geographic equation" composed of environment, culture, technology, and spatial inter-action. With this framework, the student has a set of norms by which to compare and contrast places. Its basis is the idea that people in different places approach their environmental opportunities and obstacles from their own particular heritages and perspectives, and at the same time must function in the context of the economic, political, and technological realities of the world around them.

Parts Two and Three consist of detailed examinations of major world regions, largely in terms of the conceptual framework presented in Part One. Variations in the modern system are the focus of Part Two, with chapters devoted to the United States, Europe, the Soviet Union, Japan, and Canada. The concluding chapter of Part Two discusses the outreach of the modern system to less developed parts of the world. Part Three then turns to those regions of the world where the traditional system still plays an important role despite bridgeheads of modernization: Latin America, Black Africa, the Middle East–North Africa, and South and East Asia. In many parts of these regions, the two systems are found in close juxtaposition, and governments must deal with the problems of both systems simultaneously.

Throughout the text an effort has been made to stimulate the student's interest by incorporating photo essays; a large number of carefully developed maps and graphs; and forty-two boxes featuring thought-provoking themes, issues, and projects, such as "An Inventory of General Motors Corporation," "Australia: A Self-Study Project," "Brazil's Economic Growth in the 1970s," "The Sahel Drought," and "The Green Revolution." References are listed for the student who wants to pursue topics further. In this edition, a glossary has been provided to help the student master basic terms used in geography.

The second edition is a thorough revision, not only updated for contemporary developments and to provide recent data throughout the text, tables, and maps, but also improved along lines suggested by users of the first edition. The result is that we have made major changes while retaining what most users of the first edition have seen as the strength of the book: its conceptual framework. Much of our effort has been to sharpen that conceptual framework in Part One. The opening case studies of Washington, D.C., and the Indian village have been telescoped into a single chapter. The statement of the framework in Chapter 2 is now more explicit about the components of the geographic equation—environment, culture, technology, and spatial interaction—as they relate to one another in creating the geographical character of places and regions. Chapter 3, which deals with the physical environment, has been expanded, reorganized, and rewritten to amplify the significance and contributions of physical environments to human living patterns throughout the world. Chapter 4 takes up culture, and Chapter 5 offers a world overview of the pattern of population and the spatial variability of material circumstances under which people are living today.

Three major additions have been made to fill gaps in the regional coverage of the first edition. A chapter on Canada and a box on Australia have been added to Part Two. A chapter on the Middle East–North Africa has been added to Part Three. Also in Part Three, the coverage of Asia has been expanded to include all of South and East Asia, not only India and China. The book now covers all major nations or areas of the world. More geographic content has been incorporated, especially concepts, examples, and explanations, to clarify the relationship of the conceptual framework to these real world regions. More attention has been given to the physical environment of different parts of the world.

A set of twenty complimentary 35-mm slides made from illustrations in the text is available to adopters of the second edition.

We wish to thank all our colleagues who have stimulated us through personal discussion or shared their thoughts by correspondence. In particular, we wish to thank the following people who reviewed all or part of the manuscript for the second edition: Walter N. Duffett of Eastern Illinois University, Edward F. Bergman of Herbert Lehman College, John Edwin Coffman of The University of Houston, Martin B. Kaatz of Central Washington State College, J. Paul Herr of Indiana University, South Bend, Richard J. Parish of Cuyahoga Community College, Thomas F. Kelsey of the University of Pittsburgh, Vera L. Herman of Ohio State University, Ranjit Tirtha of Eastern Michigan University, E. D. Metevier of the State of Georgia, Governor's Office of Planning and Budget, Douglas L. Johnson of Clark University, and Melvin M. Vuk. We are also grateful to Roman Drazniowsky, map curator of the American Geographical Society, who reviewed our maps in preliminary form, and to Walter N. Duffett for his work in preparing the glossary. We wish to state, however, that none of these people are responsible for any inaccuracies or misstatements that may appear in the text.

Particular thanks are extended to those who contributed to the preparation of the manuscript, especially our typist, Donna Scripture, who did an excellent job of deciphering our scribblings.

R.A.H.
T.H.S.

Introduction

Geography is not just about the earth; it is also about the places that make up the earth and the people who live in those places. If the earth environment were the same in all places, and if people everywhere were alike, there would be no need for geography. But every place is unique, and the wide variety of cultures and ways of life throughout the world arouses our curiosity today as it has since before the beginnings of written history. Other places and people were the subject of Homer's *Odyssey* and the writings of the early Greek historian Herodotus. In the nineteenth century a similar curiosity about people and places gave rise to a number of geographic societies in Europe and the United States, organized mostly by interested amateurs in much the same way as local historical societies. Today continued interest in the diversity of places explains the popularity of *National Geographic* among magazine readers and the flourishing travel industry.

What should we learn about places and the people who live in them? How can we sort, and make understandable, the complexities of this planet? The aim of this book is to help you find answers to these questions.

There are many approaches to geography. One traditional aproach is to present a set of definitions and then to give information about people and places that reinforces these definitions. The result is usually an encyclopedic array of facts.

Other approaches work with theory about the locational and spatial aspects of people and places. They study geography in terms of locations and spaces on the earth, and they have made much use of quantitative methods and computers. Recently, many geographers have concerned themselves with psychological questions about how we see our environment and how our view of spatial phenomena may differ from the way we view other aspects of our surroundings. Ideas and theoretical concepts are important in geography, but they often fail to give us a picture of the real world of people and places.

This book focuses on the real world around us in terms of a particular framework of ideas. The conceptual framework is presented in the first two chapters of the book, through a pair of case studies (Chapter 1) followed by an explanatory chapter (Chapter 2). The remaining chapters in Part One deal with the geographic components of the framework in some detail in regional and global terms. In Parts Two and Three the overall framework for examining places and people is applied to major regions of the world; in Part Two, to those dominated by the modern interconnected system familiar to most readers of this book; and in Part Three, to those regions where a traditional lifestyle rooted in the local environment is a major part of the geographic picture.

The conceptual framework we use to give order to human life in the world has a systems perspective, in the sense that the objective is to understand the *workings* of human life at a place, rather than the component parts by themselves. In Chapter 2 we introduce what we call the *geographic equation,* which expresses the interrelationships between the geographic components of life at a place, and throughout the book we emphasize that the *interrelationships* are more significant than the individual components.

Our conceptual framework is a useful way of examining the geography of the world and its places and people. It is a way of looking at the world, not a rigid doctrine. We invite you to evaluate our frame of reference, to discuss its strengths and weaknesses, and to suggest your own modifications and improvements. Most important, you should test the conceptual framework and any ideas of your own against the data for regions of the world presented in Parts Two and Three. You are also encouraged to test the framework on any country or region not included in the book. You can use an encyclopedia and various data sources such as the United Nations *Statistical Yearbook* and its *Demographic Yearbook,* various atlases, and reference sources to provide data for your study. This book, then, is a starting point for your examination of the world today; it is not the last word about that world. You should consider the text as a "launch vehicle." How far you go will depend on your own interest.

Maps, tables, photographs, and inquiry boxes allow you to test ideas presented in the written text and to reach beyond it. Maps and tables are important parts of any geography book, because they often reveal a geographic idea more quickly and effectively than written words can. When reference is made to a map or table in the text, turn to it and study it carefully. Not only will maps and tables expand on the written material, but they will also provide tests of our conceptual framework and your understanding of it. The photographs have been selected not just to give you views of a place, but to help you become more sensitive to aspects of a place or its problems.

The inquiry boxes are of several sorts. Some expand on a point made in the text; others present data for further consideration; still others offer case studies. All are presented to involve you and to stimulate your thinking. They illustrate the kind of additional information that you can gather to add to your study of the world. They also exemplify the kinds of study you can do in the course of exploring and testing the conceptual framework of the book.

We have made a conscious effort to avoid the usual professional jargon and formal definitions. We emphasize the working concepts developed as part of the flow of ideas. As in any subject, however, the use of some terminology is unavoidable and even desirable because it makes for more efficient communication. Pay special attention to the words set in **boldface** type. To test your understanding of these terms, write out definitions of them; then check each definition against the one in the glossary at the back of the book.

Turn now to the case studies in Chapter 1. They illustrate two contrasting ways of life present in today's world, and two very different ways in which people make use of the earth environment. These two portraits will be referred to frequently throughout the book; they will be used as models to aid in understanding other people and places throughout the world.

PART ONE

The Geographic Problem:
Finding a Focus for the World

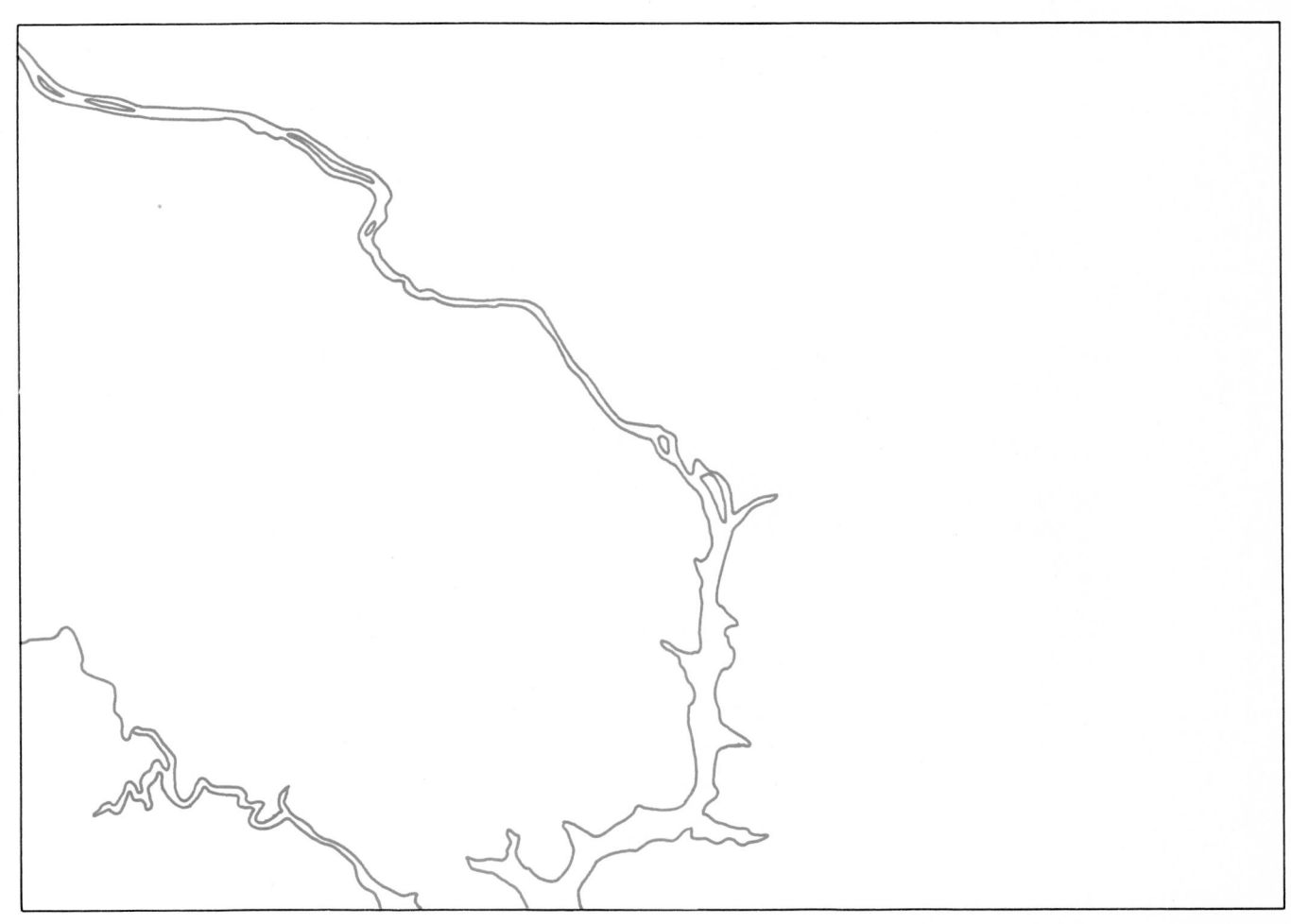

CHAPTER 1

Portraits of Two Worlds: Metropolitan Washington and a Village in India

The two places described in this chapter represent two very different ways of life. Washington, D.C., is a modern city with many connections to the rest of the world. Life in the village of Ramkheri is based on local traditions, and most villagers spend their lives within a short distance of their birthplace. In Chapter 2 these two places will be related to a conceptual framework that will be useful in understanding places throughout the world.

Above: The fictional village of Ramkheri may resemble this village in north central India. (United Nations)
Right: Takeoff at National Airport in Washington, D.C. (Paul Conklin)

Left: A typical Washington social gathering, at which politics and diplomacy dominate. (Paul Conklin)
Above: The craft of pottery is highly developed and of practical importance in many parts of the traditional world. (United Nations)

Most of us are aware that differences exist between places of the earth, yet we often unconsciously interpret the actions, lifestyles, and beliefs of other people in terms of our own place of living and our own way of life. Even those who have traveled to other places, or have the world brought to them by television, or hear and read extensively about the events and problems of the world, often have only a superficial understanding of other people's ways and environments. The two places described in this chapter are real illustrations of the striking differences that govern and mold the processes of living for people within this world. These two places represent many of the basic and essential differences that people must cope with in different places, and they provide a study in the contrasting roles that places and people can play in the wider world about them.

Few Washingtonians are likely to have the experience to understand the world in which the villager in India must live. It is also unlikely that an Indian villager can begin to imagine life in Washington. Nevertheless, places like Washington and the village in India are both parts of the contemporary world. Though each of these two places has its own unique characteristics, each can also stand as an example of many other, similar places. Living in Washington, and the functions of a city like Washington, have much in common with metropolitan living and functions elsewhere in the United States, or in Australia, Western Europe, the Soviet Union, or Japan. Similarly, although the experiences of a villager in India differ in detail, they have much in common with the experiences of Chinese, Egyptian, Indonesian, or perhaps Mexican villagers. Thus while these two examples seem worlds apart, they represent two different styles of life found throughout the world in many different environments and involving peoples of very different human groups.

As you read about these two places, look for characteristics that distinguish not just the place, but the distinctive style of human life. Think of the very different problems each place presents its inhabitants. Try to contrast the two places in terms of their use and consumption of the earth's resources. Consider also how the different resource uses affect the environment in different ways. The material existence of most Indian villagers appears harsh and without conveniences. Does metropolitan living as exemplified in Washington suggest ways of modifying the harshness of village life? Conversely, does village life have attractive aspects compared with some of the harsh realities of metropolitan living?

METROPOLITAN WASHINGTON: AN IMPORTANT PLACE IN THE INTERCONNECTED WORLD

Imagine Washington, D.C., on almost any weekday morning, winter or summer, between 7:30 and 9:00 A.M. The view from a traffic-reporting helicopter shows bumper-to-bumper traffic backed up as the cars try to cross the bridges over the lower Potomac River to enter downtown Washington. The mass of traffic coming in from the Virginia suburbs slows to a halt in the "mixing bowl" where the major freeways come together in the area of the Pentagon. This is just part of the multitude of autos, buses, and trucks bringing more than 400,000 workers into downtown Washington each workday morning.

Life in the teeming anthill of Washington calls for a daily migration of people—some of whom live many miles from their place of work—from their homes to their jobs and back home again. The bulk of this movement is into the heart of the city, but there are also traffic snarls on the Beltway that rims metropolitan Washington because many people go to jobs in the suburbs as well.

At the same time, more than half a million schoolchildren make their own daily migration on foot, by bicycle, by car, or in hundreds of schoolbuses. For most children this movement is only a few blocks or a few miles, and very few go to schools downtown. Years of planning and development by local school boards have created a system of schools located as near as possible to where the children live. But because of desegregation, some children are bused several miles. Since more than two out of every three people in the metropolitan area live in the suburbs outside the District of Columbia, school migration forms a pattern different from that of the people going to work.

Later in the morning still another, less concentrated movement will begin. Nonworking people will make their daily trips to the store, to the doctor, or to visit friends. Again, some of these trips will be on foot or by bus, but the majority will probably be by automobile. Like the schoolchildren, only a minority of these people will travel into downtown Washington. Most trips will be local, within one neighborhood of the city proper or within one suburban sector. Few Marylanders regularly go to suburban Virginia or even the District except to work; the same is true of Virginians with suburban Maryland.

The lower-income groups that occupy many areas in the District have fewer cars and less mobility than most Washingtonians. Thirty percent have no cars and entirely depend on public transportation. Their job opportunities, shopping possibilities, and social contacts are restricted to places in the metropolitan area with bus connections. Since the bus system is designed mostly to bring suburbanites and District residents into the central city, it does not help city residents find jobs in the suburbs or move easily from one sector of the city to another. Inner-city residents have complained bitterly about their relative inaccessibility to the job opportunities of the outer metropolitan area and about the expense of public transportation.

During the day, a massive fleet of trucks moves endlessly throughout the metropolitan area, delivering goods to offices, factories, stores, and even individual homes. The trucks pick up cleaning, trash, and special orders from those same places, and the workers on the trucks maintain the appliances, equipment, and public utilities essential to the operation of this complex metropolitan system.

On weekends the patterns change. There is an almost complete halt in the major migrations to work and school. In their place shopping patterns increase, and new flows appear to churches and recreational centers.

Most of the major movement in the metropolitan area occurs each weekday morning and evening as people travel to and from work. Otherwise, most Washingtonians—both suburbanites and inner-city residents—live in their own local "life space." They have little regular experience in other parts of the metropolitan area and have only a vague concept of the rest of the Washington area. Other neighborhoods and suburban communities are almost foreign territory.

Figure 1–1 shows the Washington of its residents. The daily life space of any individual or family occupies just a small fraction of the total metropolitan area. Each person encounters only a minute number of the almost 3 million inhabitants. Yet everyone identifies with Washington. Every family listens to Washington radio stations or watches Washington television news; many read the Washington daily papers, follow the Washington Redskins, and pay taxes to municipalities in the Washington area. Each family depends on others who live there and who usually are personally unknown to them. All residents partake of Washington's services; all suffer from the pollution and congestion that beset the whole metropolitan area.

As we can see, metropolitan Washington is more than a location measured by latitude and longitude, more than an environmental combination of atmosphere and land resources. It is a place where people live—and it is the living that focuses geographers' attention on the place. They are interested in understanding the how and why of human life at particular places on the earth. Each place has its unique combination of locational, environmental, and human conditions, and it is geographers' concern to understand how they interact.

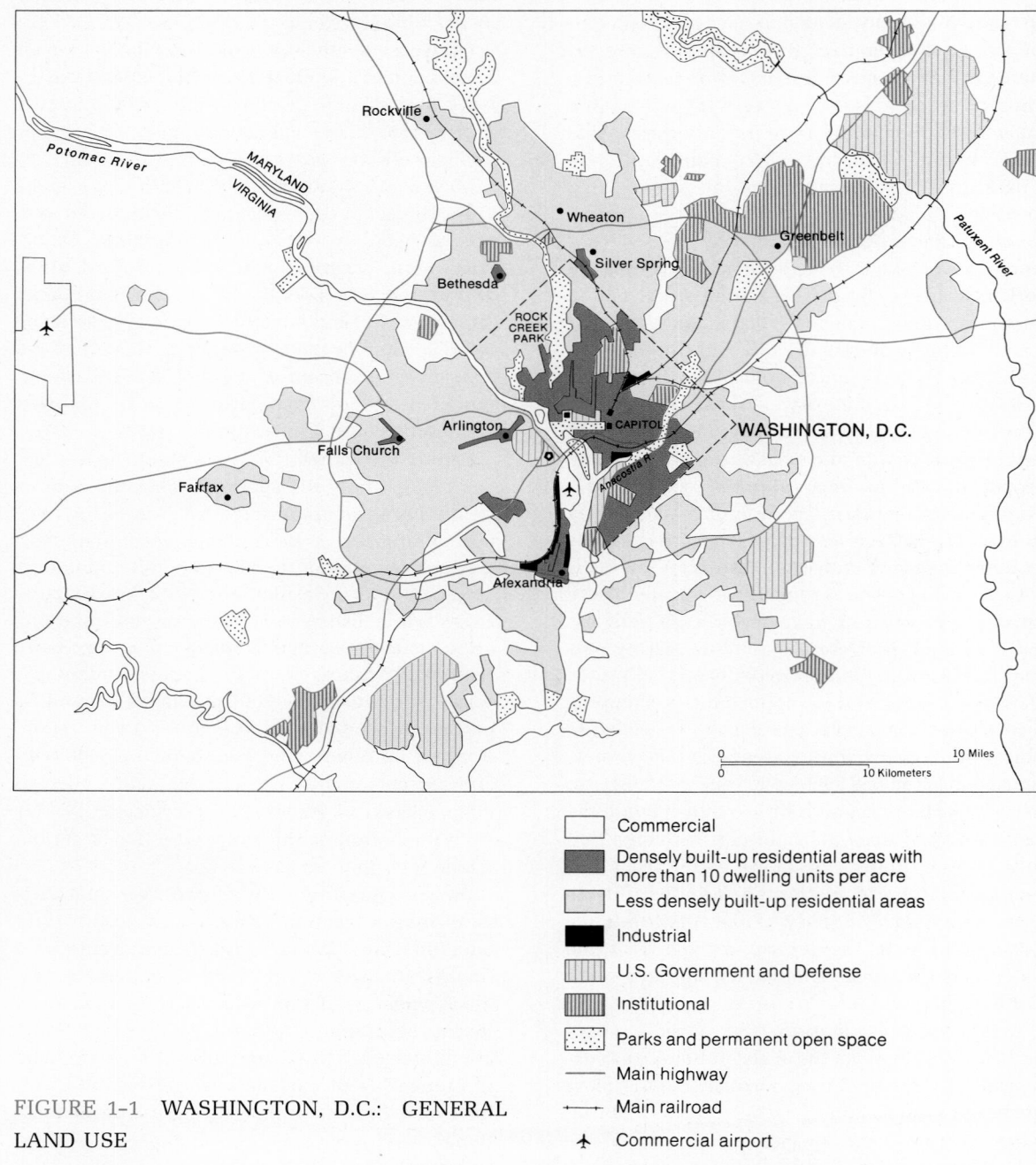

Commercial

Densely built-up residential areas with
more than 10 dwelling units per acre

Less densely built-up residential areas

Industrial

U.S. Government and Defense

Institutional

Parks and permanent open space

Main highway

Main railroad

Commercial airport

FIGURE 1-1 WASHINGTON, D.C.: GENERAL
LAND USE

What, then, should we know about Washington? What is significant? Are the figures on size, population, employment structure, and income important? Is Washington more important for its suburbs, for its inner-city communities and slums, or for its problems of pollution and race relations? Should we concentrate on Washington as the capital of the United States with influence over the farm program, the highway system, the use of wilderness areas, and the welfare of the poor? Or is its role as the capital of one of the world's strongest nations most significant?

WASHINGTON'S CONNECTIONS TO THE WIDER WORLD

Anyone who is familiar with the United States recognizes that Washington, as the capital of the country, probably ranks among the top metropolitan areas of the world in importance. This significance cannot be seen in the morning traffic jams or in the crowds at shopping centers and sports events, which are also found in other large cities. It is not the local movements within the metropolitan area that make Washington distinctive among cities; we must look at other evidence.

Washington is one of the world's busiest air centers. In a recent year more than 15.5 million passengers passed through the three public airports serving the metropolitan area (one of the three also serves the Baltimore metropolitan area). This means that the population coming and going between Washington and other places by air was more than five and one-half times the resident population. Of course, air travelers are just a portion of the people moving between Washington and other places. Many others travel by auto, bus, and railroad.

Even these movements of people into and out of Washington tell only part of the story of its connections with other places. The telephone company serving the Washington metropolitan area recently estimated that some 76 million long-distance telephone calls leave the area an-

nually. If we assume an equal number of incoming calls, the total would be about 152 million telephone calls per year—more than fifty times the total resident population of the metropolitan area. These figures do not include calls over permanently leased government lines, since the company does not monitor the number of calls on these lines. And there must be many other messages coming in and going out over other privately leased lines and by telegraph, cable, and microwave connections.

In light of these figures, we can see the resident population of Washington as just the upper part of a largely submerged iceberg of interaction. In other words, Washington's visible daily life is a very poor measure of what the city really does. The 3 million residents of the metropolitan area operate an enterprise involving many more people than those who actually live there.

The airline passengers include tourists, to be sure, but the majority are businesspeople and government officials on business. The long-distance telephone calls include many personal calls, but most are undertaken as part of governmental or private business. When one considers all these contacts by air, phone, and other means, the real basis of Washington's being comes into focus. The nation's capital exists on the basis of its interactions with the rest of the country and the rest of the world.

Washington does not operate as a self-contained entity. It cannot provide itself with all the required goods, services, and ideas of life. Rather, it lies within a much larger network that stretches into many different parts of the country and the world. The people of Washington depend on other parts of the world for food, clothing, and most other needs. Washingtonians buy foreign-made as well as Detroit-made cars. They drink Florida orange juice, eat California lettuce, and broil Middle Western beef. Wealthy Washingtonians wear the latest clothing designed in New York, Paris, and Rome. Other residents buy the same ready-to-wear clothes sold in cities and towns throughout the country.

TABLE 1-1 METROPOLITAN AREA POPULATIONS OF THE FIFTEEN LARGEST CAPITALS OF THE WORLD

Capital	Population (Millions)	Rank
Tokyo	21.6	1
London	10.7	2
Moscow	10.2	3
Mexico City	9.0	4
Paris	9.0	5
Buenos Aires	8.6	6
Peking	7.6	7
Cairo	6.6	8
Seoul	5.9	9
Djakarta	4.6	10
Delhi	4.5	11
Manila	4.4	12
Madrid	3.6	13
Lima	3.4	14
Washington	3.2	15

SOURCE: Edward Espenshade and Joel Morrison, eds., *Goode's World Atlas*, 14th ed., Rand McNally, Chicago, 1974.

TABLE 1-2 LARGEST METROPOLITAN AREAS OF THE UNITED STATES (*Estimates for Mid-1972*)

Metropolitan area	Population	Rank
New York	17,181,000	1
Los Angeles	10,231,000	2
Chicago	7,615,000	3
Philadelphia	5,642,000	4
Detroit	4,684,000	5
San Francisco–Oakland	4,585,000	6
Boston	3,918,000	7
Washington	3,015,000	8
Cleveland	2,921,000	9
Dallas–Fort Worth	2,499,000	10
Houston	2,402,000	11
St. Louis	2,371,000	12
Pittsburgh	2,334,000	13
Miami	2,223,000	14
Baltimore	2,140,000	15

SOURCE: U.S. Bureau of Census, *Current Population Reports*, Washington, D.C., 1976, Series p-25, No. 640.

Washington also performs important functions for people in other places. Many of the people who commute to work through Washington traffic are on their way to jobs in government offices that deal with problems of people, local governments, and business throughout the United States. Others work in agencies involved in international relations. Some of these workers are employed by business associations, lobbying groups, and professional societies representing interests of people and businesses in other parts of the country. Still others are foreign nationals who work for embassies or consulates of their governments or for international organizations.

WASHINGTON AS A METROPOLIS

Washington is the capital of one of the world's largest, most influential countries. In population, however, it does not stand out among either the world's capitals or the largest metropolitan areas of the United States. Table 1-1 shows that it ranks fifteenth among world capitals in population. It is smaller than Lima (Peru), Madrid (Spain), and Manila (Philippines), capitals of countries with only a fraction of the U.S. population. Among major metropolitan areas of the world it ranks fortieth, and even in the United States it ranks eighth (Table 1-2). The United

TABLE 1-3 EMPLOYMENT STRUCTURE OF MAJOR METROPOLITAN
AREAS OF THE UNITED STATES, 1970 *(percent)*

	Total Employment	Manufac- turing	Trade	Services	Transpor- tation and Utilities	Construc- tion	Finance, Insurance, Real Estate	Government
New York	4,861,000	20.9	20.9	20.5	7.8	3.5	10.6	15.8
Chicago	2,981,000	31.4	22.5	16.9	6.9	4.0	6.1	12.1
Los Angeles	2,897,000	28.2	22.3	18.9	6.0	3.8	5.9	14.5
Philadelphia	1,796,000	30.5	20.5	17.7	5.8	4.8	5.7	14.8
Detroit	1,483,000	37.6	19.9	14.8	5.3	3.5	4.6	14.3
Boston	1,291,000	21.5	22.7	24.9	5.9	4.0	7.3	13.7
San Francisco– Oakland	1,264,000	16.1	21.3	17.8	10.6	4.8	7.8	21.5
Washington	1,157,000	3.8	19.6	21.8	5.2	5.9	5.9	37.7
St. Louis	899,000	30.5	21.3	16.9	7.5	4.5	5.2	13.9
Pittsburgh	875,000	31.8	20.3	18.3	6.8	4.9	4.3	12.6
Cleveland	859,000	34.5	21.4	16.2	6.0	4.1	4.9	12.8
Baltimore	808,000	24.2	21.9	16.7	7.1	5.4	5.4	19.2

SOURCE: U.S. Bureau of Census, *Statistical Abstract of the United States, 1971*, 92nd ed., Washington, D.C., 1971, Sect. 33.

States is the most politically and economically powerful country in the world. Why is its capital not the greatest metropolitan center in the world? Let us look at the evidence.

A Special-Purpose Metropolis

Table 1–3 indicates one of Washington's distinctive characteristics as compared with the largest metropolitan areas of the country—those with populations of more than 2 million. The concentration of employment in one activity, government services, is obvious. Notice that government workers in Washington constitute 37.7 percent of all those employed there. This concentration in a single activity in a metropolitan area is matched only in Detroit, where 37.6 percent of employment is in manufacturing. If we think of Detroit as a specialized manufacturing center, then we must think of Washington as a specialized government center. Its distinctive businesses are the vast array of agencies of the federal government.

The concentration of 30 percent or more of all workers in one activity is not unique to Washington, as you can see from Table 1–3. Six of the other largest metropolitan areas have such concentrations; but in all six manufacturing is the most important activity. Thus Washington not only does not fit the pattern of a major metropolitan area in the United States; it also lacks some of the usual urban nongovernmental functions of other national capitals like London, Tokyo, Moscow, Paris, and even Mexico City and Buenos Aires. They are centers of business, finance, trade, manufacturing, and culture for their countries as well as being capitals. In the United

States, however, the major center for such activities is New York City, the largest metropolitan area.

Growth of the Capital

Washington was established specifically to be the seat of national government. Its location was selected primarily because of its proximity to both the geographical center and the center of population of the existing thirteen states from New Hampshire to Georgia. From the beginning, then, Washington's location with regard to other places has been most important. Through the years, Washington has continued to be a key center of transportation routes: first toll roads and sea routes, then a canal system and railroads, and now highways and air routes. Communications have been equally important. It has been necessary to develop systems of transportation and communication that can connect not only with points throughout the United States but also with capitals and large metropolitan centers throughout the world.

Washington began as a planned city. Maryland and Virginia ceded an area of ten miles by ten miles to the federal government, which at that time seemed more than adequate for future growth. The original plan covered an area only about three miles by six miles. Even after more than sixty years of growth, the principal means of commuting—horse, carriage, or walking—had not changed, and Washington had yet to reach the limits of the original design. This was the time of the two-story row house that still characterizes much of the District. Houses were tightly placed on narrow lots with no space between them.

By 1917, owing to streetcars and commuter railroads, the city of Washington had spread well beyond the original planned area but not beyond the District boundaries for the most part. Separate suburbs grew in both Maryland and Virginia, where towns had streetcar and commuter railroad connections to Washington. For the first time the District boundaries appeared inadequate for the functioning urban center. The automobile set the pattern for suburbia beyond the District, and today more than two of every three persons in the metropolitan area live in suburban Maryland or Virginia.

A single functioning metropolitan unit has thus sprawled across the original District boundaries. Metropolitan Washington now laps over into the political jurisdiction of two different states, four counties in Virginia and two in Maryland, and five incorporated cities in Virginia and twenty in Maryland.

Population Growth as a Measure of Governmental Influence

The growth of Washington as the capital city has reflected both the growth of the country and a greater role for the federal government in making the nation function effectively as a unit. In the period of nation building, Washington was a major center of population, ranking eighth among the country's cities in 1850 and ninth in 1880. But its single-purpose character was a limiting factor as business and industry developed throughout the country. By 1920 Washington had fallen to fourteenth place, and not until 1940 did its metropolitan population reach 1 million. Nine cities exceeded it in population at that time.

Since 1940 the metropolitan population of the United States has grown, and Washington has been one of the most rapidly expanding centers. This increase seems directly related to the growing importance of the federal government in the affairs of the country. Federal expenditures in 1940 amounted to only 9 percent of gross national product (GNP); by 1970 they were 30 to 35 percent. Moreover, the higher percentage represented a much larger sum, $313 billion in 1970 as compared with $9 billion in 1940. (The declining value of the dollar accounts for only a small part of this increase.) Although most of this money is actually spent elsewhere, the decisions concerning its allocation and administration are made in Washington. With 2,724,000 employees in 1974, the U.S. government is a huge enterprise, four times as large as General Motors, which is the world's largest private business employer. Between 1940 and 1974 federal employment in Washington more than doubled to 341,000 persons. In this period the total population of the metropolitan area increased two and one-half times.

WASHINGTON AS A FOCAL POINT FOR A WORLD BEYOND

It is common to look at a metropolitan area or any urban place in terms of the number of people who live there. We have seen that Washington is only eighth among the big metropolitan areas of the country. However, its impact on the nation is based less on the number of inhabitants than on the influence that its governmental and other decisions have on other areas. It is not so much what Washington does for Washingtonians as what it does for the population of the United States. This is difficult to measure, but we get some idea of Washington's ties with other areas by measuring its air-passenger traffic connections.

The Flow of People and Ideas

We can obtain one measure of Washington's direct contacts by examining airline schedules. In 1976 Washington airports had daily direct flights to 182 places in the United States. There was direct connection with all but 9 of the 100 largest metropolitan areas. But flights reached only 65 of the more than 100 places with populations between 100,000 and 500,000, six places with 50,000 to 100,000 populations, and only 46 of the hundreds of places with less than 50,000 people.

From this pattern we can deduce a series of different types of airline connections that Washington has with the rest of the United States. First is the frequent service to the major metropolitan centers of the country, such as New York and Los Angeles. Next are the frequent flights to smaller regional metropolitan centers in the South, Southwest, and Middle West. Third are

TABLE 1-4 CITIES WITH TWELVE OR MORE DIRECT FLIGHTS PER DAY TO WASHINGTON–BALTIMORE, 1976

City	Flights
New York–Newark	86
Philadelphia	48
Atlanta	45
Chicago	43
Boston	36
Norfolk	23
Miami	21
Salisbury, Maryland	21
Dallas	20
Pittsburgh	18
Baltimore	17
San Francisco	16
Detroit	15
Los Angeles	15
Cleveland	13
Minneapolis	13
Richmond	12

SOURCE: *Official Airline Guide, North American Edition* (semimonthly), The Reuben H. Donnelley Corporation, New York, New York.

the less frequent flights to metropolitan areas of 500,000 populations and less. Similar service connects with those smaller communities, many of them with less than 50,000 in population, within 200 miles of Washington or along the South Atlantic coast. Finally, there are scattered centers served by one or two flights per day.

The pattern of frequency of air service is quite different from that of flight connection. Only seventeen places within the United States were directly connected to Washington by twelve or more direct flights per day in 1976 (Table 1–4), yet flights to these seventeen places totaled over half of all flights from Washington. The New York

LIFE IN WASHINGTON

Socially and economically, Washington is a divided city. The more affluent and middle class, especially those with children, move to the suburbs to find open spaces and cleaner surroundings. Those in the lower middle class and on the bottom rung of the ladder are left in the less attractive and heavily congested pockets of the inner city, though of course there are exceptions. Many middle-class and affluent persons prefer to live in fashionable neighborhoods like Georgetown within the city limits. Conversely, some suburban towns, such as Arlington, have sections of lower middle class and the poor.

The McGrady family fits one of the more familiar metropolitan patterns. Their home is a large sprawling country house on five acres in a rural setting about twenty-five miles from the city. Seven members of the family plus an assortment of pets fill the house.

Dan McGrady works in Washington as Washington correspondent for a leading Middle Western newspaper. He has tried many ways to outwit the traffic jams he invariably runs into on his way to downtown Washington, but he can't beat the rush unless he leaves home before 6:30 A.M. or returns after 6 P.M. His wife, Laura McGrady, also a writer, regularly drives into Washington for interviews and research. She starts out later in the day so that the journey into the city goes somewhat more easily, accompanied by less frustration and headache for her.

The McGradys are deeply involved in many facets of Washington. Anything that happens in the metropolitan area not only gives them material for their typewriters but makes its impact on their daily life as well. Indeed they have much less interest in strictly local matters involving the township they live in than in the gossip and events that move the metropolitan area.

The Hamilton family lives on the fringe in Silver Spring, about eighteen miles from downtown Washington. Their colonial house, situated on a wooded one-acre plot, gives the family the kind of privacy it desires. Russ is a rising young telephone company executive whose office is but seven miles from the house. He is one of an increasing number of white-collar workers who are finding jobs in the suburbs. The Hamiltons have very little contact with the central city, but they view that as an asset. Russ feels a real sense of relief when he hears the reports of traffic snarls on the radio and knows it is a quick seven miles to his own suburban office.

For the most part, life has the mark of relative isolation and comparative ease. Distances mean little to a two-car family, although Betsy Hamilton is often irritated by her role as permanent chauffeur. She must use the car to get a loaf of bread, to cash a check, to take the kids to the dentist, or to visit friends. In good weather the kids can bicycle to school, but in winter she must drive them.

Like most suburban families, the Hamiltons go through Washington's downtown district only to show out-of-town friends the sights. All the suburbs have movies that show the same current features once found in the "big town," and the family has little interest in museums or theaters.

The local shopping mall about five miles from the house is the focus of shopping life. There the family can find a supermarket, a beauty parlor, a variety store, and a branch of the local bank. There is a larger shopping center twelve miles away with a choice of department stores and specialty shops for major purchases.

The Hamiltons' closest contact with downtown Washington is via television, but it has no more meaning for their personal lives than Washington news has for people in Denver. If Russ moves up the company path he may well be moved to another metropolitan area, and this potential mobility keeps them from feeling rooted to the larger community.

The life of the Richardson family in the inner city is neither as comfortable nor as glamorous. They live in one of Washington's poorest areas. Raised in the Deep South, Henry came to the capital with only ninth-grade education and experience as a farm worker. There is virtually no demand for his limited skills, and the best he can do is to find work as a janitor or helper. His wife works at night as a cleaning woman in one of the downtown office buildings to help make ends meet.

Home for the Richardsons is a four-room apartment in an old row house that was once fashionable but is now completely rundown. The house could be remodeled, but landlords in neighborhoods of this type make only absolutely essential repairs and have done nothing for years. Although the family manages to hold onto a ten-year-old car, both adult Richardsons use public transportation to travel to and from their jobs, and they now fear reports of a proposed ten-cent fare increase that will raise their monthly expenses by more than eight dollars.

Most of the Richardsons' life is confined to the neighborhood. Martha Richardson shops for the groceries at a store two blocks away, making daily trips to gather the family's supplies. The church and the children's school are another block-and-a-half distant. Young Henry, Judy, and Sam spend most of the time they are not in school playing on the sidewalks of the block or in the alley behind the house. If they want to try the public park a few streets away, they have to cross a busy thoroughfare crammed with buses and cars most of the day. Nor does the school playground offer an alternative; it is always badly overcrowded. Except for work, the Richardsons are anchored to the blocks around their apartment.

A few times each year, they venture out to the suburbs in their car to shop at discount stores for the clothing and other merchandise they need. What friends they have in Washington lead much the same circumscribed lives within the same neighborhood.

In many ways the Richardsons relate to their neighborhood as they might to a small town. They live in one of the most important world capitals, but the news on their local radio station seems totally divorced from their lives, and their only contact with the government is through local agencies such as the police or welfare departments. But Washington is a large city, not a small town, and the contrast between their lives and those of the more affluent citizens of Washington is apparent to them every time they turn on the television or take a bus through the other sections of the city. Like the poor in so many cities, they are confronted by the dichotomy between the cosmopolitan world that all the visitors to Washington see and the reality of their own poor neighborhood.

metropolitan area alone accounted for about 20 percent of all direct flights. Nine of the seventeen metropolitan areas with most service are among the country's twelve largest. Most other places with intensive service are regional metropolitan centers, like Atlanta, Minneapolis, and Cleveland. Miami is both a large metropolitan center and a famous resort area. Norfolk, Richmond, and Salisbury (Maryland) represent a different sort of connection: three important centers within the Washington regional area.

Surprisingly, Washington has few direct-flight international air connections. It has daily flights to only seven political capitals. The only major centers linked by more than one flight per day are Toronto, Montréal, London, and Mexico City. Some other cities, including resort areas of Mexico and Bermuda, also have direct flights. Flights to other places are on a less-than-daily basis, apparently showing the predominance of New York as the international airport for the eastern United States. Since New York is really very close to Washington, it serves as the dispatching point for most foreign-bound passengers from Washington.

Types of Transactions

The first order of connection between Washington and the rest of the world is through people and ideas, not goods. As a center of government, Washington makes decisions, not cars or television sets. Lawmakers come to Washington to establish the rules of corporate and personal conduct, business, and education for the country. Through indirect measures, such as price supports, and through direct investment, such as in irrigation projects and flood control facilities, they determine how the country will use its agricultural land and other resources. They also set the climate for business by means of tax structures and control over the country's financial structure. Government spending programs play a large part in the structure and population of cities and rural areas as well. Federally supported urban renewal has changed the appearance of inner cities, and government-supported freeways have allowed suburbs to sprawl. Government welfare programs have contributed to the migration of the poor to the cities, and government aid programs have been designed to support people in depressed areas such as Appalachia. The government controls interstate freight rates, issues television and radio licenses, charters banks, and supports schools. Its laws concerning segregation and civil rights have changed the country's cultural patterns.

As we have seen, the government is also the largest single purchaser of goods and services in the United States. In 1972 the Department of Defense alone spent $36 billion, and almost all of it was for contracts with business firms. It is not surprising that the major military contractors maintain Washington offices, and more and more research and development facilities are locating in the Washington area.

Washington and the Surrounding Region

Although it is the long-range national and international ties that are fundamental to Washington's function as the capital of the United States, the metropolitan area also has some ties to the region around it. Farmers in the local area produce milk for the large Washington market; people in surrounding Maryland and Virginia shop in downtown Washington or, more likely, the nearest suburban shopping center; young people go to local universities, particularly the state University of Maryland in College Park (in suburban Washington) and George Mason University, a state university in Virginia. Sports teams such as the Washington Redskins football team

and the Bullets basketball team have fans in the surrounding region. The large Washington hospitals provide specialized care not available in smaller local hospitals. Most of the surrounding area, however, as much as fifty miles away, is the commuting zone. The wealthy of Washington and many of the middle-class bureaucrats live in the beautiful rolling "hunt country" of the Virginia and Maryland piedmont, or even on the slopes of the Appalachians, or have homes along the shores of Chesapeake Bay and the large rivers that flow into it.

Only thirty-five miles from Washington, Baltimore, with its port and railroad focus, is the major wholesaling center serving the businesses in the small communities of Maryland and Virginia around Washington, and often the stores in Washington as well. Farmers on the eastern shore of Maryland, who specialize in poultry and vegetables, see Washington as just one part of the larger concentrated urban market called Megalopolis that extends northward to Boston and includes more than 30 million people.

Washington is, indeed, a part of Megalopolis, one of the most densely urbanized parts of the United States. This region of high population density contains several large cities, whose suburbs in many cases join to form a continuous urban zone. The cities of Megalopolis include New York, the largest metropolitan area; Philadelphia, the fourth; Boston, the seventh; Baltimore, the fifteenth; and a whole series of smaller metropolitan areas in addition to Washington. As Table 1–3 shows, the large metropolitan centers of Megalopolis are among the most important manufacturing centers in the country. More important, all have been major centers of business and culture for the country since colonial times. New York, Philadelphia, Baltimore, and Boston have been the most important ports of the country and the points of origin of the railroad systems reaching out to the rest of the country. At these nodes American business has taken root: giant banks, publishing companies, insurance companies, and trading firms. Here major advertising firms, fashion consultants, media experts, and artists set the styles and tastes followed by much of the rest of the country and the world.

Megalopolis is regarded as a huge metropolitan area with one giant metropolis merging into another. There is still open space, as anyone who has traveled Interstate 95 from Boston to Washington can attest. But the major movements of people, goods, and information of the country take place between the centers of Megalopolis. Notice from Table 1–4 how important airline connections are from Washington to New York, Philadelphia, and Boston.

RELATIONS WITH THE
LOCAL EARTH ENVIRONMENT

With about 3 million people, almost a third of whom have annual family incomes of $10,000 or more, the Washington metropolitan area is a great consumer of goods. It lies in an area with a long growing season but only moderately productive soils. Much of the land to the east of the city is low and sandy, and that to the west is hilly with only patches of good soil. There are no local deposits of oil or iron, and the coal of the Appalachians is more than 100 miles away. Clearly, the city is not supported industrially or agriculturally from its own locality.

Most goods used by Washington residents, like those of any metropolitan or urban place in the United States, are drawn from specialized producing areas spread over the country or the world. An elaborate system of transport, finance, and trade handles the movement of goods between the producing areas and Washington.

In one year food sales in the metropolitan area were more than $722 million, while the total value of all farm products raised there was less than 5 percent of that figure. In the same year nonfood retail sales were $2,744 million, while the total value of goods manufactured in the area was only 20 percent of the total.

Beyond providing the basic essentials for urban life—air, water, and land—the environment of the metropolitan area is not a major supplier of food and material goods for Washington residents. The water supply for most of the metropolitan area is drawn from rivers traversing the area, particularly from the Potomac and Patuxent rivers. These rivers flow from the higher Appalachian Mountains in the west so that even in this case, a large share of the city's needs is drawn from beyond its metropolitan area. Water storage and treatment facilities have been developed in the metropolitan area, but there is great difficulty expanding these fast enough to keep up with the growing population and its increasing use of water due to more air conditioning, extra bathrooms, and swimming pools. The Washington area has been forced to institute water rationing during recent periods of summer drought.

Waste disposal is an even greater problem. Most sewage is disposed of in the major rivers that flow into the lower Potomac and into Chesapeake Bay. Although much of it is treated, some is not, and there have been problems of pollution by detergents and phosphates.

Trash and garbage are burned or spread over waste disposal areas in sanitary landfills. Tighter and tighter restrictions are being placed on the burning of wastes that create air pollution; thus the demand for landfill increases greatly. But the problem of how to handle the ever-increasing sewage and garbage of the growing metropolitan area without despoiling some aspect of the local environment is not solved.

Metropolitan areas are designed to function on human, not environmental, terms. The weekday schedule of regular movement to and from work goes on every week of the year. Interruptions occur only when there is a particularly heavy rainfall or snowstorm, and then for only a few hours. Severe hot spells create overloads in electrical systems: the generators outside the metropolitan area are taxed by the tremendous demands of air conditioners, and breakdowns can result.

The need for Washington to reach out beyond its local land resources to support its large population can be illustrated by estimating the agricultural land required to feed the population.

1. Let us assume that all the people of the United States share equally in the food produced from the nation's cropland.

a. In 1970 there were about 203 million people in the United States and about 387 million acres in cropland.

b. By dividing the area from which most food products are derived by total population, we find it takes about 1.9 acres of cropland to supply each person.

2. The population of the Washington metropolitan area in 1970 was 2,861,000 people.

a. If we assume that each person in Washington requires the same amount of cropland to supply food as the average for the nation, a simple multiplication of 1.9 acres $\times$ 2,861,000 people shows that about 5,436,000 acres are needed to support Washington's population.

b. This is about equal to the total amount of cropland in Delaware, Maryland, Virginia, and West Virginia combined.

3. There are, however, 8,758,000 people living in these states in addition to those in metropolitan Washington. They too presumably require 1.9 acres of cropland for food. Thus an additional 16,640,000 acres of cropland is needed (1.9 $\times$ 8,758,000).

4. If our assumptions hold, such simple calculations show that the people in Washington, D.C., cannot be supported by the cropland in the nearby surrounding area. Clearly, metropolitan concentrations like Washington must reach out to major agricultural areas to provide for their food demands.

For contrast, compare the territorial outreach for food by Washington, D.C., to that for the Indian village of Ramkheri (see particularly the section entitled "Ramkheri Within Its Region," pages 27–28).

The exhausts of at least three-quarters of a million motor vehicles used in the metropolitan area produce an atmospheric haze almost every day. During periods of high pressure and low wind, there is no way for air to remove the pollutants being continuously added, and pockets of smog form. As with water use, the problem of smog comes not only from overloading humanly constructed distribution systems, but also from actually discharging more waste materials into the air than the atmosphere can readily absorb. Thus the technological ability and financial willingness of the people to clean up the environment become an important aspect of a large metropolitan area's relationship to its local environment.

THE WORLD OF WASHINGTONIANS

Today the built-up portions of the Washington metropolitan area spread out more than twenty miles from downtown Washington, far beyond the limits of the District of Columbia. Most recent development of the metropolitan area has been chaotic and piecemeal. No single water or sewer system has been established; various operating electricity companies serve different parts of the area; phone companies charge different rates; sales, property, income, and gasoline taxes vary from one area to another; and there are various school districts. Still, no serious discussions have been undertaken to integrate the political structure and the basic public services of the area. Planning agencies and elected officials focus attention on their own segments of the metropolitan area.

Today 38 percent of metropolitan Washingtonians live in suburban Maryland, 31 percent in Virginia, and 31 percent in the District. As in most large metropolitan areas in the United States, the division separates different populations: suburbanites generally have high incomes and city residents low incomes; suburbanites are predominantly white, District residents predominantly black.

Within the District of Columbia two different groups of people live in largely separate parts of the city. On the one hand, there is still a significant number of whites within the city. On the other hand, there are also great numbers of blacks. Other differences are more marked than that. The whites are generally young single people or childless couples, who work for the government or businesses in the center of the city, enjoying the active life of a major metropolitan center with its restaurants, nightclubs, museums, and other cultural attractions. Here too are elderly whites who have stayed on or perhaps have returned to city life after raising their families. Essentially missing from the white population are families. White families, faced with the problems associated with integrating public schools, have fled to the more attractive suburban areas to raise their children. In contrast, the black population of Washington still consists mostly of families, and the District's public schools are over 90 percent black.

In addition to the division into racial groups, Washington residents can also be grouped according to their participation in the decision making of government. Although governing the world's most affluent country is the business of Washington, the real management of government involves only a tiny fraction of its population. For most federal employees and most workers in embassies and associations dealing with government, work is an eight-hour day, a job typing, filing, and keeping accounts.

Decision making in the federal government is largely in the hands of Congress, the president and his cabinet, the Supreme Court, and a handful of military commanders—a total of less than a thousand persons in this metropolitan area of 3 million. This is the group that functions as the corporation board and the executive headquarters for the U.S. government. As such, it is also the focus of Washington life. The families of these leaders form the nucleus of Washington's society. They hold the endless cocktail parties, and it has been said that the social life of Washington is merely the after-sundown extension of the day's business—politics and government.

Yet this unrelenting single-mindedness of life in the inner circle of Washington attracts an important segment of the talent from communities throughout the country. The ambitious—both the young and those already successful in their own right—see Washington's inner circle as the opportunity to become a part of the power of U.S. government.

Thus despite its dependence on world connections, Washington itself is highly compartmentalized. It lacks a comprehensive local government and an overall identity with the metropolitan community as a whole. It comprises people who have no contact beyond their own tiny portions of the metropolitan area, and its residents often have little understanding of the way of life of other people in the many parts of the area. Government and business leaders within Washington are in regular contact with centers of commerce and politics throughout the United States and the world, but the majority of the Washington population does not see much beyond its own job, family, and neighborhood. Major contact with the outside world comes through the mass media—newspapers, magazines, radio, and television. And only a segment of such information is concerned with the metropolitan area or even originates within it. Thus Washington, as one of the world's key metropolitan centers, depends on constant contacts with the world beyond its local setting, but the daily territory of individual inhabitants includes primarily the residential neighborhood, nearby shopping, and the place of work.

FROM THE WIDE WORLD OF WASHINGTON TO THE LOCAL WORLD OF RAMKHERI

Urban dwellers in any metropolitan area of the United States—or of Canada, Europe, the Soviet Union, or Japan—can readily understand and identify with the description of life in Washington. However, more than half the world's population live under a system that contrasts sharply with Washington lifestyles.

Washington and its inhabitants are very much interdependent with other parts of the country and the world for goods, services, and management. This interdependence supports a complex economic and political system. The village in India described in the next section represents an alternative condition of living. The village as a whole has much less need for interaction with other places, and most of its inhabitants' roles in village life are not as specialized as they are for most people in Washington, D.C. As you read about the circumstances of living in a village in India, try to identify those differences in resource base, connections with other areas, and the interdependence of people that sharply distinguish the village from Washington.

THE VILLAGE OF RAMKHERI: A WORLD IN ITSELF[1]

Ramkheri is a small village of about 900 persons in central India, about 350 miles northeast of the

[1]The material in this section is largely based on Adrian C. Mayer, *Caste and Kinship in Central India*, University of California Press, Berkeley, 1960. *Ramkheri* is a fictitious name for a real village in this part of India.

FIGURE 1–2 INDIA, SHOWING THE LOCATION OF RAMKHERI

city of Bombay. It has neither a rail station nor an all-weather road. The nearest road passes three-quarters mile away, and entrance to the village is over a cart track in an old stream bed that is flooded in rainy weather.

The village itself is not very important even in Madhya Pradesh, the Indian state where it is located (Figure 1–2). Yet Ramkheri is important to us because it is representative of some 750,000 other villages in the world's second most populous country. It is an example of a way of life completely different from the one illustrated by Washington, D.C. Moreover, life in Ramkheri is much more typical of life for the majority of the world's population than is Washington.

In using Ramkheri as an example of a traditional village, it is important to remember that it is not typical of all villages, or even all villages in India. Every place has a distinctive character

TABLE 1-5 POPULATION AND EMPLOYMENT FOR TWO AREAS OF VILLAGES

| | Gujarat State Villages | | | West Bengal State Villages | | | | |
	A	B	Total	A	B	C	D	Total
Area (sq. km)	4.9	1.5	6.4	1.0	2.0	2.2	0.6	5.8
Number of houses	264	49	313	n.a.	n.a.	n.a.	n.a.	n.a.
Number of households	274	60	334	117	164	62	62	405
Population	1,135	265	1,400	414	784	165	273	1,636
Males	596	134	730	n.a.	n.a.	n.a.	n.a.	n.a.
Females	539	131	670	n.a.	n.a.	n.a.	n.a.	n.a.
Density (per sq. km)	232	177	219	414	392	75	455	282
Total agricultural employment								
Number	927	236	1,163	249	746	157	273	1,425
Percentage	81.6	89.1	83.1	60.0	96.4	93.9	100.0	87.1
Total nonagricultural population								
Number	208	29	237	165	38	8	0	211
Percentage	18.0	11.0	17.0	40.0	4.9	4.8	0	12.9

SOURCE: T. Fukutake, T. Ouch, and C. Nakane, *The Socio-Economic Structure of the Indian Village,* The Institute of Socio-Economic Affairs, Tokyo, 1964, pp. 5, 102–103.

based on the culture and traditions of the people and the nature of the terrain. It is only in the most general geographic terms that we refer to Ramkheri as being typical of other villages, but by examining that one place carefully we should get some insights into the nature of life in villages in all countries.

THE VILLAGE

Despite governmental pressure for change, life in Ramkheri goes on much as it has for many centuries. Families rise with the dawn, for in this agricultural community there is usually much to do. Since there is no electricity and fuel for lamps is expensive, the daylight hours are important to all. Essentially, village life takes place from dawn to dusk. When the sun goes down, most activity in Ramkheri comes to a halt.

As in Washington, there is a daily journey to and from work, but in Ramkheri the traffic is mostly along the footpaths leading from the village to the surrounding fields. Most of the workers in this village are farmers (Table 1–5), who walk from the village to their fields each morning and return each evening.

Ramkheri is supported by agriculture, and the village itself is mainly a place to live, not to work. Village life offers many advantages: the services of carpenter, blacksmith, barber, potter, basketweaver, and other artisans; the protection provided by the headmen; the pleasantries of talks with neighbors at night, of ceremonies; shrines essential for the practice of one's religion, so important to peace of mind and to crop production. One takes pride in being a part of a village, but the family is the basic unit of social and economic life.

The village consists mostly of one-story houses

made of mud reinforced by straw and cow dung. The poorer houses have palm-thatch roofs, and the better ones have roofs of locally made tiles or corrugated iron. The houses serve more as shelters, barns, and storehouses than as places to live. The few windows have no glass, only barred frames with wooden shutters. Houses have no chimneys. In the evening when food is cooked inside over open fires, it appears that the whole village has been set afire and left to smolder. Smoke can be seen oozing through cracks in the roofs.

There is little privacy in the houses not only because of the openness of their interior structure but also because they are closely spaced. A family's life goes on in close proximity to that of its neighbors, and villagers are involved in each other's affairs whether they want to be or not.

The two largest buildings near the village center were formerly the residence and the office of the landlord. One building is now the village school and the other is the headquarters of the Village Committee sponsored by the state government. Nearby is a small shrine. In the midst of these public buildings is the village square, a cleared, unpaved area strewn with carts and usually occupied by several cattle. Village festivals and meetings are held here; little else attracts people to the square. There is no cafe, no club, no clustering of shops and shopkeepers. The real life of the village takes place in the neighborhoods where people live. These neighborhoods are connected by streets and alleys, some wide enough for bullock carts and some so narrow that only two people can pass.

Because Ramkheri is off the main road, government officials and outsiders rarely visit it. Dewas, a town of more than 25,000 people, is seven miles away, and Indore, a major city of almost 400,000, is twenty-two miles farther, but it is a long journey to either place for people who must walk or go by bullock cart. Now some villagers have bicycles. They make the trip to sell their produce in the market at Dewas and to buy special items. In addition, young people from Ramkheri have taken jobs in Dewas and Indore. Thus the villagers have contacts with larger places. The larger centers benefit from the villagers in terms of the profits of trade with them and the taxes they pay, but seem unaware of them as people. Some people consider the villagers second-class citizens— hicks or rubes from the country. Government officials assigned to teach farming techniques, animal husbandry, or health in Ramkheri hope they will not have to remain long in such a small place.

THE LOCAL ENVIRONMENT

Although the villagers have contact with the world beyond their immediate area, the tiny space within a few miles of the village center provides the basic needs of the village families. The distance to neighboring villages is only two or three miles. Yet from slightly more than five square miles of the earth's surface must come basic food and material support for the 900 people who live there.

This area supplies most of the food eaten in the village, the sun-dried bricks and daub walls of the houses, the tile and thatch for roofs, and the dung-cake fuel used for heating and cooking. The chief sources of power are human muscles and bullocks and other local animals. Basic tools such as hoes, plows, threshing implements, and carts are locally made. Most of the manufactured products used in daily life are made by village carpenters, tailors, potters, and the blacksmith. Most of these artisans serve the farmers, who make up the majority of the population, and their major source of income is an annual payment of grain from each. In turn, the artisans provide basic services at no other charge to the farmers.

The Agricultural Base

The chief concern in Ramkheri is to provide food for the families of the village. Only secondary attention is paid to crops for sale outside the area, even though the market, Dewas, is just a few miles away.

The chief crop is a grain called sorghum. From it are made thin cakes of unleavened bread, the staple food of the village. These are the base for most meals. Mixtures of chili, leafy vegetables, and seasoning cooked in water are spooned over these cakes. So important is sorghum that the villagers who speak of "mother earth" also talk of "mother sorghum." A village saying is "just as our mother feeds us, so the earth feeds us; for without sorghum we could not live." Sorghum also has many ceremonial uses in village rites.

Other major crops contribute to the food supply. The edible seeds of pod-bearing plants such as peas, beans, and lentils are used in making the daily curry dishes. Wheat, much less important in the daily diet than sorghum, is a luxury for villagers. Because little land can be irrigated, rice is only a minor crop in this part of India.

In recent years the government has encouraged the sale of produce in Dewas. Cotton is now the second leading crop. Peanuts and sugar cane are both local necessities and sources of cash.

Most of the land is farmed by the landowners and their families. The average size of properties ranges from ten to twenty acres. Farms in this area are larger than the average Indian farm because of the drier nature of the local terrain, and this dryness results in lower production possibilities per acre. Rarely do the properties lie in a single unit; more commonly, a ten-acre farm may include several separate pieces. These farms represent about as much land as a family itself can operate. Even so, it is often necessary to hire additional labor because there are virtually no imports of energy fuels nor are there machines to use such energy. Thus a number of farmers hire villagers to work at contract labor, usually on a yearly basis for payment of some cash and a share of the harvest. Extra laborers needed at harvest time are paid a daily wage.

Changes with the Seasons

Life in an agricultural village such as Ramkheri varies with the season. Plowing, planting, and harvesting must fit into the climatic sequence of

this part of the world, and since farm work is of first importance in the village, the pattern of family and social activity follows the seasonal sequence, too.

The agricultural year starts in mid-April near the close of the hot, dry season. Religious farmers mark their bullocks with holy signs, and their wives make offerings in their homes to this new year. It is considered important to begin work on a day that is personally auspicious. From April through the middle of July is the first busy period of the agricultural year. This is the time to plant the crops that grow during the rainy summer monsoons. Since all work is accomplished by human and animal muscle and with very primitive implements, plowing, planting, and weeding take great physical effort and absorb all the daylight hours.

After mid-July until mid-September the workload lightens, partly because many days are rainy. The heavy soils cannot be worked when they are very wet, although the crops must be weeded when possible. Sorghum, planted in rows, can be weeded using a bullock-drawn cultivator; fields planted broadcast must be weeded by hand.

In mid-September when the rains cease and the crops are ripening, the workload is not heavy, but constant vigilance is needed in the fields to keep away bird and animal predators. During that season men live in small huts on stilts in the fields. Land left fallow during the first crop season is now plowed and harrowed in preparation for the cool-season crops.

Harvesting and threshing occur from October through December. Maize, a minor crop, is harvested first, then peanuts, pod-bearing plants, small amounts of rice and hay, and, in December, cotton and sorghum. At the same time newly harvested fields must be prepared for a second crop to be grown during the cooler season that extends from January through March. The final three months of the calendar year are the busiest of all, for there is no rain to hold up work and labor has to be expended on two different crop sequences at once. Everyone, even the artisans, is enlisted in this work.

By January work slackens. Only sugar cane is harvested at this time, and sugar making takes place right in the fields. Then in March the second-crop season is completed with the harvesting of wheat and pod-bearing plants. A slack period lasts until the start of the next agricultural year in April. This is the time of village festivals and family events, the time of betrothals and marriages and making visits and pilgrimages. Most other festivals occur between June and November.

Labor-Intensive Agriculture

To those of us in the modern world of machinery and inanimate energy, the amount of human effort and time that goes into traditional agriculture is unbelievable. It takes two months of steady work to prepare and plant a field smaller than a city block. The hard ground must be broken by a single-bladed wooden plow with a metal tip, the product of village artisans. Manure collected daily from the houses in the village has been placed in individual family compost pits on the edge of the village, but then it must be spread over the field, a task that can take two men with hoelike shovels, two baskets, a bullock team, and a cart as many as thirty days during the hot season when temperatures reach 100°F (38°C). The farmer with the bullock team might take another eight days to break up the plowed clods and level the field. Finally, several farmers can sow the field in a single day using a seed drill and the bullock team.

The harvests of sorghum at the beginning of the cool season and wheat at the start of the hot season require the efforts of almost every able-bodied person in the village. Wheat is harvested by hand with sickles. The task requires additional hired labor and takes several days. The grain, still on the stalk, must then be transported to the threshing floor, another time-consuming task.

Threshing requires from four days to a week.

The grain must be dried in the sun in the morning; then in the afternoon eight to ten bullocks or cows are driven round and round a stake in the center of the threshing floor until their hooves knock the grain off the stalks. The men thresh the grain, while the women guard the crop from stray cattle or goats and rake the threshed ears from the floor. Traditionally, threshing time had a festival air, when farm and village families worked and visited together. In recent years, however, individual farmers have developed threshing floors out in their own fields as a convenience, thus changing village social patterns as well as farming practices.

Agriculture in villages like Ramkheri generally does not produce a year-to-year surplus or profit for a farm family. Most food production goes to sustain the family diet; no family members are paid wages. Some of the crop must be sold to pay for seed and other farm needs; to provide clothing for the family; and to pay for birth, wedding, and funeral ceremonies. The prospects for a cash profit beyond bare essentials are poor, and the return for family labor and capital invested is very small.

ECONOMIC AND SOCIAL
DIMENSIONS OF CASTE

Life in Ramkheri has been altered by centuries of human habitation just as the people have been affected by the nature of the environment. The result is a remarkably sophisticated pattern of human relationships that form the economic, political, and social system of the village.

Like the rest of India, Ramkheri has an elaborate series of social castes based largely on occupation and ritualistic food habits. Hindus consider death and body wastes to be polluting, but all products of the cow to be pure. Vegetarians usually enjoy high social standing since, in theory, they live without taking life. Those who have not had to work with their hands—landlords, priests, and warriors—are of highest rank.

Farmers, around whom the whole village economy revolves, are at a middle level above the artisans and traders who depend on them. Lowest in the hierarchy are people who eat meat, kill animals, or touch polluted things such as dirty clothing. Leatherworkers, whose traditional duty involves removing dead animals from village streets and using leather to make sandals and drums, and sweepers, who have had to handle animal excrement, are at the bottom of the caste system. Stoneworkers and those who eat pork also are low in the scale.

Distinctions Among the Castes

Because there are various degrees of ritual purity associated with the hierarchy of castes, persons within a given caste are careful about their associations with persons of other castes. One's purity can be polluted by certain contacts with inferior castes.

There is great rigidity and formality in the relationships among people of different castes. The people of a given caste have grown into an almost extended-family relationship. Members owe each other respect, affection, and obligations. Such friendships are important because friends lend each other equipment and money, and work in each other's fields. In Ramkheri, as in most villages, each caste occupies a particular area, or neighborhood, in the village. Persons from the highest castes—those associated with leadership in the village—live near the center of town. Farmers and others of the middle castes live farther out, and the homes of the lowest tend to be on the fringe of the village.

The caste system produces a series of socially and ritually homogeneous groups ranked along a ladder of prestige. Each of the residential caste members has a definite economic, social, and ceremonial role within the village. Moreover, there are established practices to be followed in dealing with members of each caste. Caste relations guarantee economic cooperation outside the family circle. It is good business for farmers to be on favorable terms with basketweavers, carpen-

ters, and blacksmiths, who are sources of farm equipment. Farmers' good will and credit are important assets in survival, for farmers are rich only at harvest time when their needs are usually low; they may be very poor at sowing time when services are needed. Mistakes in dealing with members of other castes may cause hostility, and offended parties might withdraw their services.

Payment for services falls into three classes: those for which an annual fixed payment of grain is provided; those for which payment is made on each occasion that the service is used; and ceremonial duties, payment for which includes gifts at designated times. Artisans such as the carpenter, potter, and barber perform services in the first group. The second includes services of other craftspeople, such as the tailor, whose work is irregularly used, and the tobacco-curer, who buys from villagers and sells the products in the town market. The keepers of the two shops in town are commonly paid in amounts of grain that they, in turn, must sell to buy additional shop supplies. In the third group are the services of those supported by gifts, including the priest and the keepers of various village shrines. By custom, gifts are given when particular rites and ceremonies are performed. In addition, there are certain days when the priest is offered other gifts.

Breaks in the Occupational Hierarchy of Caste

Although an Indian is born into the caste of his or her parents and is not likely to rise above it socially, political and economic forces do allow a person of one caste to work at other things. There are landowner-cultivators in Ramkheri from more than two dozen castes. Persons of the warrior, goatherd, tobacco-curer, barber, and carpenter castes are, in fact, farmers. As a result, members of the farmer caste own a minority of the land around Ramkheri. In the same way persons of several other castes operate carpenter shops in the village. Some castes, such as cotton-carders, oil-pressers, weavers, and tobacco-curers, find that their services have largely been replaced by products of industry that can be

LIVES OF RICH AND POOR IN RAMKHERI

Under the traditional pattern the life of the richest and poorest people in Ramkheri is much the same. *Patra,* one of the headmen, lives in one of the largest houses of the village. Although there are perhaps twenty rooms in the maze of buildings in his compound, he and his immediate family occupy only two rooms. Other quarters house his brother's wife and children and his mother, who is in charge of the household by virtue of her age.

Patra rises at daylight like almost everyone else. As soon as most of the village is up, it is too noisy for anyone to sleep. Besides, Patra has a lot to do.

Like the other villagers, Patra's day starts with a bath carried out in Indian fashion by scrubbing and rinsing without removing his clothes. He wears a loosely wrapped garment that looks like a woman's full-length skirt; it is the standard garb of village men. When he has completed his bath, he wraps a fresh garment around himself and deftly removes the damp one. Then he prepares to eat breakfast.

The morning meal is usually large, consisting of fried sorghum cakes and a curried stew of vegetables topped off with a few cups of milky tea. But during the warm season he has a cold meal, prepared the night before, and washes it down with tea.

Most days Patra goes out into the fields to supervise the operation of the several hundred acres that he owns. He must oversee all stages of the crop from planting through har-vesting, threshing, and storing. He must check on the progress of his workers and deal with any problems that have arisen. It is not easy to get enough people to work for him, and he tries to get the most work he can out of his helpers.

In the dry season there is less to do. Patra may rest in the cool of his house or play cards on the temple porch. Much time is spent on his own front porch, hearing the problems of villagers. He may be called on to settle a dispute between a landlord and a tenant or between two villagers over divorce compensation. Some days he removes his village garb, changes into Western clothes, climbs on his bicycle, and rides to Dewas. At the headquarters of the government farm program he can order supplies, get information on crops, and perhaps place an order for special seed or fertilizers. He may also visit the many cloth and merchandise shops, or just stop for tea and sweets and conversation.

The evening meal comes just before going to bed. It is large and centers around hot dishes prepared that day. He eats alone. His wife eats after he finishes, and the children eat when it is convenient.

Patra's wife is rarely seen outside the confines of the house. She leaves only for latrine duties in the fields or to go visiting in other villages with her mother-in-law and to show off her young boys. She also regularly returns to her own village to visit her parents. She has numerous household duties, but the manage-ment of the house is still in the hands of her mother-in-law.

Patra's children live like other Ramkheri youngsters. Since there are few children of their social status, they play with the other village children even though none of the parents are ever pleased. They attend the three-grade village primary school. Patra can also afford to have his children tutored at home in the evening. They will later go on to secondary school elsewhere, and the boys will go to college if they are able.

Motiram, a farmer who owns three acres of land, lives in a family compound with the families of his two brothers. But their homes are just crude huts. He follows much the same daily routine as Patra, but his meals are more austere, he has simpler clothes and home furnishings, and he rarely travels to Dewas. His children get no tutoring and probably will go no further than the village school in their education.

Like Patra, Motiram begins his day at dawn with a morning bath. His bath, however, is taken at a village pond, and he uses no soap or scented oil.

Motiram works in the fields every day during the growing season. Most days he works for Patra, for he had to mortgage two of his three acres of land for money to marry off his eldest daughter. The remaining acre will not feed his family for the year, so he and his wife and two children must spend most of the day, six days a week, working for Patra. He makes twenty cents per day for this work, his wife makes ten cents, and each son makes seven cents. On Friday Patra gives his workers a holiday. This is the day Motiram must work in his own fields.

Motiram's meals vary greatly over the year. During the dry season there is little available. He cannot afford to buy vegetables, and his own supply could not be stored for use at this season. The sorghum cakes will be accompanied by a chili pepper or two. But during the best times of year he, too, will eat curried vegetables.

Motiram's leisure time is spent discussing crops with other villagers. Occasionally during the dry season he visits relatives in a distant village. The life of Motiram's wife is much less restrained than that of Patra's wife. Her daily duties bring her into contact with the other women of the village. They gather to gossip as they bathe in their corner of the pond and draw water for their families' daily needs. Of course, she works in the fields with her husband and children.

Motiram's children have the run of

the village when they are young, and in addition to field work the boys go to school. The girls, on the other hand, are taught womanly duties instead of going to school. All Motiram's children show the effects of their meager diet. They do not get either enough calories or proteins and vitamins, and they are susceptible to colds and childhood diseases. Motiram has lost three young children to either typhoid or dysentery.

---

obtained in the village shops or the town market. Some of the members of these castes have left the village altogether to become factory workers and teachers or to join police forces and the army. But they are still closely tied to their families and thus to their castes in Ramkheri.

Among the people who farm the land around Ramkheri, there is an important distinction between landowners, tenants, and farm laborers. Landowners are well recognized, since land is the fundamental source of wealth. As managers of their own property they have a high degree of independence. Tenants work parcels too small for effective individual operation or land owned by absentee landlords from Dewas or beyond. Most tenants are sharecroppers. Their shares depend on the proportion the landlord contributes. For a daily wage farm laborers are hired either on annual contracts to assist a farmer or when needed at harvest or planting time. Like landowner-cultivators, tenants and laborers come from a variety of castes. As a rule, they are more likely to come from the lower castes.

The Headmen, Symbols of Authority

The most respected people in the village are the three headmen. They are members of the Rajput caste, whose members traditionally have been

rulers, warriors, and landlords. In Ramkheri the Rajput headmen had been designees of the maharajah, who until recent times owned all the land in the village. As official representatives of the maharajah, the headmen were responsible for collecting the land taxes of the village and for maintaining law and order. They were both the symbols of political power in the village and the village representatives to higher authority.

The power of the maharajah is now gone; his land has been distributed among the villagers. A village council supported by the Indian government makes important decisions about local matters. Nevertheless, the headmen remain the most important people in the local hierarchy. They still collect the land taxes, and although they have no official power, much of their traditional authority remains. The village has no police force or court, so disputes have always been brought before the headmen for decisions. Like mayors, the headmen serve as hosts for the village when government officials and other dignitaries visit. They are the most important figures during village festivals. The position of headman may not continue much longer. As the children of the headmen become educated, they commonly leave the village. Once they have tasted life in the larger world, they often have little interest in returning to the village. They prefer to work in government service as officials or teachers.

THE BEGINNINGS OF CHANGE

Today much of the headmen's traditional power rests in the new Village Committee and an associated Comprehensive Committee. Both are elected at village meetings. The Village Committee, created by authority from the Indian government, has eight members, including two from a much smaller village nearby. The committee had begun to take over the collection of land taxes from the headmen and is now the official agent of the government in the village. As such, it is the local contact for new national social and economic programs aimed at producing change at the village level. The six members of the committee from Ramkheri are important people politically. Unlike the headmen, however, they do not hold arbitrary and continuing authority. A new election is held every five years. Committee members, surprisingly, come from various classes. Two members are from the Rajput caste, but neither is one of the village headmen. Each of the other four from Ramkheri is from a different caste.

The problems of change can be seen in Ramkheri's agriculture, too. The basic methods of food production have changed little in hundreds of years. Some farming practices among the small farms of Ramkheri are inefficient by modern standards. Farmers scatter wasteful quantities of unselected, untested seed. Young plants are lost to birds, insects, and other animals. Manure and compost lose much of their fertilizer value from exposure to sun and rain. Crop storage is poor and much is lost to rats, worms, and weevils.

It may be possible to increase productivity substantially by instituting better practices. New varieties of sorghum could double or triple production but only if insecticides, adequate grain storage facilities, chemical fertilizers, rat poison,

and good fences are also incorporated into the agricultural system of production. Change, therefore, becomes a formidable task; it also involves great risk. Farmers unskilled in the use of fertilizer may destroy their crops by applying it too liberally. An improperly mixed insecticide can ruin a crop. If seed is purchased through government channels rather than from traditional village sources, who will provide the loan for a daughter's wedding? Modern agricultural equipment would have to come from outside the village, thus jeopardizing the farmer's traditional ties with the blacksmith and carpenter—and such ties are more than economic. These artisans are neighbors and friends, but they also have religious functions that make their presence essential at births, marriages, and deaths. At the same time the blacksmith and carpenter hesitate to improve their products, for they have no assurance that their farmer-customers will accept the new implements. Besides, the new products are likely to cost more than farmers are willing to pay.

Change comes most easily to the larger landholders in the Ramkheri region. The larger acreages may yield a profit even at low prices. The wealthy can use chemical fertilizers or American-made moldboard plows so big that they must be pulled by three pairs of bullocks. Those who regularly visit the larger towns, such as Dewas, where government offices are located, have access to seed information and agricultural advice on new crops and techniques. Moreover, these changes favor relatively low-yield grain crops that occupy large acreages and require little labor. The moldboard plow requires only half as much labor as the single wooden plow; chemical fertilizers can be applied with much less labor than cow dung and compost. Adequate fertilizer, good seed, and healthy plants reduce the amount of labor needed for weeding, thinning, killing insects, and even for harvesting.

However, any change that would decrease the demand for labor on the lands of large landowners would be disastrous, since many villagers depend on the contract farm labor or harvest work. Even slight changes in farm technology could drastically reduce the need for laborers and further lower the living standard of the poorest farmers.

RAMKHERI WITHIN ITS REGION

Ramkheri has links with the world around it. With the concern of the central government for national economic and social progress to reach every village, outside pressures are increasing. Attempts to improve agriculture depend on the availability of new seed stocks and advice from agricultural experts. Farmers are being encouraged to use more fertilizer, and a program to develop wells for irrigation water depends on outside resources. Population growth is also influenced by outside pressures. Pesticides like DDT and vaccination programs result in reduced death rates. Governmental concern over continued high birth rates has resulted in attempts at better family planning through birth control.

Even so, it should be clear that the differences between Ramkheri and Washington, D.C., are more than differences in size or culture. Ramkheri's dependence on the environmental base is

VILLAGE HIERARCHY IN INDIA

In most of India four different levels of villages can be distinguished. At the first level are tiny hamlets of fewer than 300 people on poorer land, and since hamlets usually are found in poor environmental situations, everyone is poor. These villages are made up of people mostly from only one caste—often either farmers or shepherds. They have few artisans and must depend on larger centers for supporting services. Because of their poverty the hamlets are of little importance to the larger villages nearby and are largely ignored by regional government officials. They have little social life, few ceremonies, no theaters, and no fairs. No wealthy landowner lives there.

Villages with between 300 and 700 people form the second level. They have a range of castes and thus provide some supporting services for themselves. Such villages vary sharply. Some are supported by herding, others by different combinations of crops. Although there is a mixture of castes in each village, most people are middle-class farmers, and social life features parties, entertainments, and elaborate ceremonials. Commonly, one wealthy landlord has been the central figure in the community.

Medium-sized villages like Ramkheri—at the third level—contain 700 to 1,200 people. They may be the residence of several large landowners as well as many middle-class farmers and laborers. A large number of different castes are present. Such a village approaches self-sufficiency; its wide range of workers provides the basic goods and services, and it is the center of many fairs and ceremonies. A village of this size is the focus of attention for government officials and government programs. It has a good school and receives considerable government assistance.

The large village, which may have 6,000 or 7,000 people, is as much a central place for other villages as a unit in itself. The village at this level, the fourth, often occupies a key agricultural location and thus generates its own wealth, but it also serves other functions. It is a base of operation for minor government officials such as village-level workers with jurisdiction over smaller villages in the area. In fact, the large village may include a number of hamlets and smaller villages for administrative purposes. It often has the equivalent of a junior high school and is a market for surrounding communities.

fundamentally different from Washington's. The U.S. capital has vital ties with a myriad of other places, even other continents. Via these tielines come the basic food and other necessary resources. In the same way, Washington's influence spreads to all communities in the United States and to capitals in all countries of the world. In very small measure it even extends to rural India; decisions on foreign aid, for example, may affect life in an Indian village.

In contrast, Ramkheri's basic ties are with the environment within several miles of the village. From the land within sight of the village comes most of the food and other needs of the people. Within this limited sphere the essential cultural ties of the people of Ramkheri exist; here they feel comfortable. At the core of their lives is the farm family's dependence on the land they work.

The region in which Ramkheri is located is made up of numerous villages like Ramkheri that are essentially the residences of basically independent family units. Because some villages are in better environmental situations, or have developed more effective farm practices, or both, the villages vary in size from a handful of residences to several thousand families. Not all can produce the complete range of foodstuffs needed; not all can support a complete staff of specialists and technicians to produce things considered essential to economic and social life. Thus some exchange of goods and services between them takes place as different villages concentrate on the production of specialized goods and services. This interchange of goods is, however, secondary to the primary dependence of most families on their own farms within walking distance of their homes.

THE TWO WORLDS: RAMKHERI AND WASHINGTON

The description of Ramkheri has shown how the people of India and other traditional cultures have spread over the earth's surface. Each group utilizes the landscape within the limits of its own environment. Groups tend to be successful where the environment is suited to their traditional food-producing practices, and less successful in other environments. In every traditional culture the basic ties are to the land directly surrounding the community, and success and failure are determined by geographic and cultural factors.

India, like most other places, has a hierarchy of towns and cities that has emerged above the level of the villages. Dewas, only seven miles from Ramkheri, represents that "other world," which is so different and so impersonal. The road and railroad from Dewas lead to Indore, a metropolitan area of almost half a million people. Railroads and airlines in turn connect Indore to Hyderabad, Bombay, Calcutta, and to New Delhi, the capital of India. From these large centers India has air and communication links to Washington, London, and other capitals of the world.

The dichotomy between the life of people in Washington and those of Ramkheri is striking. They are extreme examples of two totally different ways of life. Most people in the world live in communities that lie somewhere between the cosmopolitan world of a political capital and an isolated village in a poor country. If we can understand the way in which the two extremes of the spectrum interact, it is then easier to understand the relationships among all the millions of places that lie between them.

More important, Washington and Ramkheri are examples of two different and contrasting systems that are major forces in today's world. The dichotomy between these two different ways of life lies behind many of the world's problems today. Moreover, issues such as the population explosion, the availability of world resources, and pollution are not simple questions having a single solution; they are very different matters in the two different systems. This point will become clearer as we investigate the various regions of the world. Before moving to a study of particular regions, however, let us set the examples of Washington and Ramkheri into the perspective of the overall framework that will be used throughout the book.

SELECTED REFERENCES

Alsop, Stewart J. O. *The Center: People and Power in Political Washington.* Harper & Row, New York, 1968. A discussion of Washington's first function, politics, by a noted columnist.

Beals, Alan R. *Gopalpur, A South Indian Village.* Holt, Rinehart and Winston, New York, 1962. An anthropologist's look at a village in another part of India.

Federal Writers Project. *Washington: City and Capital.* Government Printing Office, Washington, D.C., 1968. A revised version of a basic reference volume on Washington done during the Depression years.

Green, Constance M. *The Secret City.* Princeton University Press, Princeton, N.J. 1967. The most recent book by the woman who is Washington's personal historian.

Harper, Robert A., and Frank Ahnert. *Introduction to Metropolitan Washington.* Association of American Geographers, Washington, D.C., 1968. A look at Washington in terms of both its internal function and its role in the United States and the world.

Lewis, Oscar. *Village Life in Northern India.* University of Illinois Press, Urbana, 1958. An anthropologist's lucid view of village life in yet another part of India.

Mayer, Adrian C. *Caste and Kinship in Central India.* University of California Press, Berkeley, 1966. The full study of Ramkheri and its people upon which part of this chapter is based.

Nair, Kusum. *Blossoms in the Dust.* Praeger, New York, 1962. A novel in a village setting by an Indian author.

Randhawa, Tejinder Singh. "Progress in a Punjab Village," *Geographical Magazine,* August 1974, pp. 608–614. A description of life in a Punjab village and the changes that are occurring in traditional ways. The photographs of the village setting and activities of its people are especially useful.

Wiser, Charlotte V., and William H. Wiser. *Behind Mud Walls, 1930–1960.* University of California Press, Berkeley, 1963. One of the first studies of an Indian village to provide a comparison over time.

CHAPTER 2

A Framework for Understanding the Worlds of Humanity

The goal of geographic study is to see how a place "works." Our study will be aided by the use of a geographic equation, which is a way of describing the interaction of people and the environment at a place. The four variables in the equation are (1) the earth environment, (2) the human culture, (3) technology, and (4) spatial interaction. This equation will be applied in this chapter to Ramkheri and Washington, D.C., to develop models of two contrasting "worlds," or systems: the traditional, locally based system and the modern interconnected system. Most places in the world lie somewhere between the two models.

Left: South Royalton, Vermont, a town of about 400 population. (Owen Franken/Stock, Boston) *Above:* A town in northern Spain. (Klaus Francke/Peter Arnold)

Above: Sheepherding on the cold plains of Patagonia in southern Argentina. (Mike Andrews/Photo Trends) *Left*: Market activity along a canal in Bangkok (Krung Thep), the capital of Thailand. (United Nations)

The world we inhabit is tremendously varied—much more than the examples of Washington, D.C., and Ramkheri indicate. It is a world of ice caps and hot equatorial lands, of mountains and lowlands, of wet and dry climates, of crowded places and vast empty stretches. It is a world of many different peoples and cultures, social and political systems, human values, and ways of making a living. It is as real as a view through a window or a chat with a friend, yet it is as abstract as the idea that the earth is a sphere or the thought of the more than 4 billion people who inhabit it.

THE NEED FOR A CONCEPTUAL FRAMEWORK

With so many people of different cultures and with a seemingly infinite range of variations in environmental conditions from place to place, understanding the world is a complex task. The basic curiosity of people about this great variation from place to place over the earth led long ago to the study of geography and today continues to form the basis of interest in geography. Geography (Greek *geo,* earth, and *graph,* writing) is about particular people living in particular places on the earth, about how they live and use the earth environment, and about how they interact with people in other places. Geography deals with **spatial variables,** phenomena that vary from one section of space on earth to another.

"WHERE?" IN GEOGRAPHY

Since geography is concerned with people living in places, the first question it asks is "Where?" Where is the place on earth? Where is it in relation to other places? Many people, therefore, consider *where* to be the most important geographic question. They expect geography to pro-

vide a catalogue of places in the world and geographers to know all the countries, all the prominent mountains and rivers, all the capitals of countries and states. But *where* is the beginning of geographic study, not the end. Names of places are like the vocabulary of a language. A name tells the geographer where in the world a place is, and this information gives clues about what the people and the environment of the place are apt to be like. Knowing where the Sahara Desert is locates it among the environments of the world, places its inhabitants among the cultures of the world, and thus starts a process of geographic study.

Maps are basic tools of geographic study. Maps show *where* quickly and far more effectively than any written description. Moreover, particular distributions that characterize places —distributions of rainfall, of population, of industrial plants, for example—can be shown on maps, to provide insight into environmental conditions, culture, and technology. The many maps in this book are not just illustrations, but integral parts of the study of the earth being undertaken here. They should be studied for the insight they can provide into a place.

THE GEOGRAPHER'S PROBLEM

The geographer's problem is to find order among the almost infinite number of places and people, to bridge the gap between the concrete world of one's local environment and the abstractness of the globe. This book is organized around a particular conceptual framework for studying people in places. The case studies in Chapter 1 were chosen to introduce you to the framework by presenting two contrasting places with very different environments and lifestyles. Washington and Ramkheri are more than places with different ways of life; they represent two very different systems of human life and uses of the environment that are found in the world today. They provide two different **models** we can use to study people over the earth. Thus they are basic to the conceptual framework of the book, although in Chapter 1 they were presented without any direct reference to it. In this chapter the framework is explicitly presented, and in the following three chapters its components are examined in more detail. In the main part of the book the framework is used to examine the major regions of the world. It is important, therefore, to gain a working knowledge of the framework; it will be used again and again.

THE GEOGRAPHIC EQUATION

Geographic study involves the examination of the complex interrelationships between people and the earth environment. The object is to understand the results of those interrelationships: the way people *live* in places. Geography, like mathematics, is not concerned with simply understanding the components of a problem—the environment of a place or its culture. Rather, the goal of geographic study is to seek the solution to the problem of life at a place involving the complex interrelations of people and environment: to see how a place "works." Instead of looking at a place as exotic and unfathomable, geography seeks to understand the interactions of the components of the place that give it its character.

We can speak of a geographic "equation" that describes the complex interaction of people and environment at a place. In this equation there are two main variables, the **earth environment** and the human **culture.** But human cultures are multifaceted and complex in themselves, and thus we must single out two aspects of human culture—**technology** and **spatial interaction**—for special attention as additional variables in the geographic equation. We can define *spatial interaction* as the connections between the place being studied and other places in the world. The "solution" to the geographic equation is an understanding of these four variables and how they interact to make a place distinctive, with its own ways of life, its own ways of using the environment, and its own special character.

As with any equation, we must understand the nature of the variables. In our study of any particular place, we want to "solve" the equation for *each* of the four variables. We do so by focusing on each variable in turn while holding the other three steady. In this way we can understand the relative importance of each variable in the complex life process that the equation describes. This "solution" of the equation is not the same as the solution to an algebraic equation. It might be more accurate to say that the geographic equation *yields* a characterization of life at a place.

Although the geographer carefully studies the environment and the major aspects of human culture at a place, including technology and spatial interaction, such study is just a means to achieve an understanding of life at that place. In this text we shall examine the physical, cultural, and economic geography of places, but it is the interaction of these elements that gives character to places, and an understanding of that character is the purpose of our study. Thus we are following the ancient tradition that has been the essence of geographic inquiry since Herodotus's early writings, namely, regional geography. As we study today's world, where the modern system typified by Washington has an impact on all cultures and all environments, we shall need to pay particular attention to the technology and spatial interaction components of the geographic equation. Life in a place like Washington no longer depends on the local environmental conditions, and its population is not limited to the group of people born there. The national and international ties of the region must be considered as well.

THE EARTH ENVIRONMENT

The environment is an essential part of the geographic equation because all human life is lived at the surface of the earth within the envelope of air that is the atmosphere, and depends on the resources of the earth's crust.

Although we depend on the environment, modern technology can help us overcome many of the obstacles it presents. Human beings can live today any place on the earth—in deserts, on ice caps, in high mountains, even under the sea—if they are willing to pay the costs and if they are able to draw many of the necessary resources from other places on the earth. Modern technology, however, merely allows human beings to protect themselves against the problems of the earth's environment; it does not yet allow us to make fundamental changes in it. We have been trying, without much success, to "seed" clouds to increase rainfall on particular days when conditions are right. But we have not been able to produce major climatic changes to suit our will. Moreover, we can tunnel through mountains and build roads up mountainsides, but we cannot move the mountains to more desirable locations. We must take the environment at a place as given, and then discover ways to make the best use of it.

The environment at the earth's surface is really the interface between different zones: (1) the **atmosphere,** or gaseous envelope over the earth; (2) the **lithosphere,** the solid crust of the earth; (3) the **hydrosphere,** consisting of the oceans and other bodies of water; and (4) the **biosphere,** the living things on the earth. The atmosphere produces our climate. The lithosphere is the surface on which we live, with its soil and mineral resources. The hydrosphere, which accounts for about 70 percent of the earth's surface, contributes greatly to climatic conditions over the land, offers seafood and mineral resources, and acts as an important transportation path. The biosphere is a major source of food, building materials, and energy.

It is common in earth science to examine each of these environmental zones in detail, studying first climate, then earth processes, soils, and biological life in turn. But in this book we are concerned with the environment as part of the geographic equation. We shall look at climate, soils, and landforms, but we want to examine the environment in human terms to see the advantages or disadvantages to its human use. Thus we are interested in environmental systems that humans can use—systems that may involve elements from all the different environmental zones—rather than isolated facts about the environment.

We can identify three major aspects of the environment that are important to human life:

1. The land environment is a "growing machine" from which we draw food, other agricultural products, wood and other forest products, and energy.

2. The land environment is a treasure house of minerals: metallic minerals, mineral fuels, and nonmetallic minerals upon which the modern system depends.

3. The ocean environment is a separate growing environment and mineral treasure house that remains almost untapped. Fishers gather the animate food supply in much the same way that early hunters and gatherers used the land environment. We have scarcely approached the agricultural stage in our use of the ocean environment, breeding and cultivating sea creatures as we do the animals and plants of land. In the same way, we scarcely use the mineral treasures of the sea. Offshore oil drilling is the first major attempt to tap the mineral wealth under the seas.

The third aspect of the environment is not of major importance today; however, the other two are the basis of both traditional and modern societies.

The fundamental aspect of the environment as a growing machine and as a treasure house of minerals is that the processes of growth and mineral formation vary over the earth so that each location has a different resource base. Growing potential in the tropics is commonly twice as great as that in the midlatitudes, and that in the high latitudes is less than one-third the growing potential in the tropics. In Chapter 3 we shall identify high-growth (high-work), seasonal-growth (seasonal-work), and low-growth (low-work) environments, all of which have very different possibilities for agriculture and forestry. Moreover, the processes that have produced the world's mineral deposits are very different from those that have produced the earth as a growing machine. Thus rich mineral deposits are not commonly found in good agricultural areas, though sometimes the two occur close together, as in the U.S. Middle West. In addition, the rocks that are most likely to produce mineral fuels are formed by very different processes than those that produce rich metallic deposits. Oil-rich areas, therefore, such as the Middle Eastern countries or the Texas–Louisiana coast of the United States, are not likely also to have great deposits of iron ore or metallic minerals.

To understand the earth environment variable in the geographic equation, we must have a general understanding of growing processes and of the processes of rock formation. We should focus on the aspects of the process that produce some regularity of variation over the earth, rather than on the details of the processes themselves—details that would be important to the physical geographer. Our major concern is to see why and how growing potential varies over the earth—how it varies with sun energy (which largely depends on latitude) and with moisture availability within various growing areas. Our aim is to develop a "feel" for the variations in the earth environment from place to place, particularly in terms of growing potential and mineral resources.

PEOPLE AND CULTURE

The components of the environment include the materials that sustain life, but each natural resource is identified and interpreted by people. There have been no signs in nature saying: "This is corn; its ears should be roasted," "This is coal; light it, and it will provide warmth or energy to do work," or "This is pure water that can be drunk, but this is polluted water that should be avoided." Each human group has had to experiment and discover the resources it can use in various ways or to adapt ways learned from other groups.

At one time humanity had only stone tools. Then some peoples learned to use bronze and later others learned of iron and then steel. In addition to a history of the evolution of the use of these resources, there is also a geography of environmental uses today. All but a small number of peoples have the capability of using iron-bearing rock to make iron and even steel implements, but less than half the people of the world—only those with sufficient knowledge, capital, and a highly organized marketing and distribution system—can build steel mills and make modern machines from the steel.

The differences in people's perceptions of their environment, in value systems, and in ways of life are as great as the variations in the physical environment over the earth, and perhaps they are more important. In our studies of Ramkheri and Washington we saw great contrasts in ways of living—in social institutions, in values, and particularly in people's perceptions of the possibilities of their environment. The way a group of people in a particular part of the world lives is known as its culture, and the group is identified as a distinctive **culture group.** The particular culture group (or set of groups) occupying a place at a particular time is one of the fundamental components in the geographic interrelationships of that place, for this group develops the possibilities of the earth environment to support its lifestyle.

A culture group's ideas about how to live rest upon the traditions and values built up by many generations before them. Ideas on what to eat, what to wear, what is good and bad, come from the past. So do field patterns, streets and buildings, and political boundaries. The pattern of Ramkheri and its surrounding fields has changed little for centuries. Even Washington, despite massive building projects in the twentieth century, has streets that follow the routes established in the original city plan.

Technology

Among the contrasting cultural features of Ramkheri and Washington, probably the most prominent is the difference in the level of technological development. Technology is the stock of experience that allows a culture to develop tools and to control energy sources, and it determines the people's capabilities for manipulating their material existence. We are immediately struck by the tremendous amount of human effort expended for the small amount of agricultural product gained per person in Ramkheri. The great efforts of humans and animals there accomplish little when compared to places in the United States and other countries of the modern world, where for a century people have known how to harness electricity, natural gas, and other energy sources, and how to produce complicated labor-saving machines.

The separation of the world into high- and low-productivity areas, wealthy and poverty-stricken places, involves all human cultural values, desires, and social organizations, but it is first of all a matter of technology. Large metropolitan areas could not be supported today without the elaborate system of production, supply, and transportation that uses a technology far more advanced and sophisticated than Ramkheri's.

At any place, technological ideas and output accumulate. Each successive generation finds itself with a storehouse of technical know-how and a stock of facilities and tools with which to work. New investment, whether in tools or ideas, can build upon the supply inherited from the past. However, such accumulations include many products—goods or ideas—that are obsolete in addition to those that work successfully.

Ties Between Different Places

The final variable in the geographic equation is spatial interaction, the contacts of a given place with other places. As we saw in Chapter 1, Ramkheri is largely self-contained, with relatively few links to nearby towns and villages, the rest of India, and the world. Modern Washington, by contrast, has ties to all parts of the United States and to the other countries of the world. Moreover, its population depends on other regions for most of its food, clothing, and material goods. Washington, unlike most modern metropolitan areas, has almost no manufacturing production other than specialized printing and certain technological services required by the federal government.

The amount of transportation and communication between places depends on the other factors in the geographic equation. Actual distance is, of course, a factor, as are natural barriers such as mountains and deserts. The key factors, however, are the human culture and especially the technology it has developed. If a human group in one place wants to interact with a group in another place, and if it has modern technology and unlimited resources, no distance is too great, no mountain range too high, no desert too dry.

SEEKING THE SOLUTION TO THE GEOGRAPHIC EQUATION

The geographic equation, which describes the workings of human life at a particular place on the earth's surface, involves the interaction of two major factors—the earth environment and human culture—and of two important aspects of culture—namely, technology and interaction with other places and peoples on the earth. Of course, the four variables are not of equal weight in our equation. People are undoubtedly the most dynamic component. Groups of people—culture groups—have "discovered" the earth through time—in the physical sense of spreading over the earth's surface, and even more, in the sense of how to use the earth's environment. Natural resources are, after all, really only those environmental components that human groups have discovered to be valuable in developing their style of life. Human history and the development of technology can be viewed as humanity's increasing discovery of how to use the earth environment.

We have seen that although the people of Ramkheri make full use of their local environment as far as their know-how and the other limitations of their lifestyle allow, they are not getting the full potential from their environment. Ramkheri still depends in large part on the tools and skills developed by previous generations. With better seed, better insecticides and fertilizer, and knowledge of farm practices used elsewhere in the world, the Ramkheri environment could produce more. Places like Ramkheri indeed must produce more in order to meet the demands of a rapidly increasing population. Ramkheri must either produce more locally or develop specialties that can tie it into a distribution system using the environmental bases of many regions.

On the other hand, in certain places in today's world the living standard far exceeds the environmental potential of the local area to support population. Washington is a case in point. The environment of the Washington metropolitan area could not possibly produce food and other necessities for almost 3 million people. Using long-range connections, the people of Washington have reached out to tap other growing areas, not only in the United States, but elsewhere in the world. At the same time, Washington has access to the expertise and productive abilities of the most advanced peoples of the world—in the United States, in Europe, in Japan, and elsewhere. Today the long-range spatial interactions of places like Washington are more important than their ties to the local environmental base. (Of course, as we shall see in Chapter 3, the great concentration of people in this small area causes other problems: some environmental, such as pollution; some social, such as overcrowding and congestion.)

The geographic equation is in part like an economic formula, since economics measures the productivity of human life. But economic choices of what to eat and wear are more than matters of economics; they involve cultural values and aesthetic choices about what is acceptable and what is not. For example, people in similar earth environments have chosen to eat very different foods, and these choices affect their perception of the environment and what they choose to produce from it. Chinese and Americans occupy very similar physical environments, but the choice, made long ago, to eat rice or to eat wheat has produced very different relationships with the environment. Facets of history involving systems of political control, social structure, and religion have set powerful restrictions within which economic forces must operate. Moreover, economists may consider humans as rational "economic beings," but in fact political and cultural choices often override economic considerations, as we can see in the present-day conflicts in the Middle East.

TWO DIFFERENT HUMAN WORLDS

The geographic equation will help us describe life at any place on the earth, but it will give us a different solution for every place we study. Thus if we are to make order out of the cultural complexity of the world, we need a framework within which to place these many different solutions. In this book, instead of just examining one culture group or one country after another, we shall develop a broad framework for examining any place or culture.

Ramkheri and Washington illustrate that framework, for they represent almost the extremes of two cultural models or systems of human use of the world, two different ways in which people manipulate the geographic possibilities of their part of the world. These two models, together with the geographic equation, should give us the framework necessary to find order in the complexity of human life over the earth. Using this framework we can examine human life in different parts of the world, on different scales, and then we can compare the systems from one part of the world to another.

Ramkheri and Washington, as we have seen,

represent two very different systems of human life. Since the system illustrated by Ramkheri primarily depends on the use of the local environment by people whose view is limited by their local culture, and since this local system has been dominant over the earth through most of human life on the earth, we shall speak of it as the **traditional, locally based system.** The model represented by Washington is both new to the world (in historical terms) and in large part dependent on far-reaching connections to the resource base and other peoples; we shall therefore call it the **modern, long-range, interconnected system.** Let us now consider the general characteristics of each of these two systems.

THE TRADITIONAL, LOCALLY BASED
SYSTEM: A WORLD OF MOSAIC TILES

In the early days of human life on the earth, the capabilities of a group of people at any place in the world were extremely limited. Their ability to travel long distances was even more restricted. Transport overland was limited to walking or using camels, horses, or other beasts of burden. Waterways—rivers and oceans—offered the best highways, but boats were small and depended on human muscles or wind for propulsion. Contacts with other groups were therefore irregular and infrequent, and limited to groups in the same part of the world. There was traffic in ideas and particularly valuable commodities such as seeds, good-luck symbols, and medicinal products, but primitive people couldn't carry sufficient food or other necessities to support people in another group through trade. A transport system efficient enough to carry grain, for example, in quantities that would meet the needs of another group simply did not exist.

The attention of individuals and of groups was therefore centered on discovering how to live in their own little local corners of the world. People in hot, wet tropical areas had to learn how to live in a world of dense, lush vegetation where it was warm throughout the year. People in humid midlatitudes had to adjust their lives to hot summers with rapid plant growth and to winters when little or nothing grew, as well as protect themselves against the severe winter cold. Desert dwellers had to figure out how to live in an environment of little water. Groups in the polar lands had to work out solutions to living in a world of very slow, sparse plant growth.

Each group of humans living almost exclusively in its own local world, often limited to walking distance of its dwelling place, formed a culture group. In solving its problems, each culture group had to depend on the experiences of the group itself over a long period of time. Solutions to those problems—the discovery of what to eat, what to wear, what values were important, and how to obtain the essentials of life—had to come through the trials and errors of the individuals within the group. Exceptions occurred when new ideas were picked up through contact with neighboring groups or through journeys away from the local group.

Every culture group developed its own character and its own political and economic organization. Life from tribe to tribe, and from place to place, was distinct. The existence of different languages indicates this diversity. For example, more than 150 languages and hundreds of dialects are recognized in modern India. These linguistic differences reflect the many subtle cultural differences between different places in the Indian subcontinent and the isolation of villages like Ramkheri from the world beyond the horizon. Similar cultural variation based on isolation is found among tribal groups in Africa and among the native Indian populations of the Americas.

Some culture groups were more successful than others. Some had the benefit of bountiful environments; some were better organized or had more innovative leadership. Generally speaking, life was easier in warm, moist tropical environments where there was abundant plant growth throughout the year. There people needed little to wear and very simple shelter. Life was more difficult in the more severe growing environments of the deserts, midlatitudes, and polar lands.

Moreover, in any given environment particular groups appeared more successful than others, apparently through more efficient organization and through the development of innovations.

For the geographer, the earth consists of separate "mosaic tiles," each occupied by a different culture group. A mosaic, as you may know, is a picture or decorative design made by setting small colored pieces of tile in mortar. From a distance only the overall pattern can be discerned, but a closer view reveals the individual pieces of tile. Thus the geographer's mosaic tile view of the world, where each culture group occupies its own self-contained territory with its own boundaries, has given rise to the traditional view of territory as a single portion of the earth's surface. The modern extension of this idea is the **nation-state,** in which the group occupying the territory not only has claim over the part of the earth it occupies, but also expects to receive the loyalty of all persons living within that territory. On a political map each country appears as a separate entity (mosaic tile). Each government in the world sees itself as active on behalf of the people within its boundaries.

We can, however, take a broader view. Despite the many separate, local worlds of the different tribal groups in Africa, these native peoples share some basic similarities in lifestyle that distinguish them as a whole from, say, the native Indian populations in North and South America. American Indians also have similarities in lifestyle that set them apart from Middle Eastern peoples, or from the inhabitants of South and East Asia. To generalize, we may say that large sections of the world have developed broad cultural patterns distinct from those of other areas; these are called **culture realms.** In terms of our metaphor of a mosaic, culture realms are the broad patterns that become evident when a mosaic is viewed from a distance.

The Geographic Equation of the Traditional System

When we solve the geographic equation for different places in the world, we see that each place has its own unique solution, with different values for the four variables—earth environment, culture, technology, and spatial interaction—and different interrelationships between them. Nonetheless, for each of our two systems, the modern and the traditional, we can develop a generalized solution to the geographic equation, which we can apply to any place that seems to fit one of the two models and which we can modify as necessary. Let us develop a general solution to the geographic equation for the traditional, locally based system.

In this system the *earth environment* component of the geographic equation is restricted to the local environment, from which come the basic necessities of life—food supply, building materials, material for clothes, and basic tools. The *cultural* component of the equation is also limited, in this case to the local group, which sets the basic rules of life—what to wear, what style of house to build, what is right and wrong, what life is all about. The fundamental problem for each group is to live in its particular earth environment and use the knowledge that has evolved within its own culture. Analysis of different groups, of course, will reveal many variations in the fundamental model: different degrees of know-how, different group values, different earth environments.

The traditional world has a simple *technology* dominated by the use of tools powered by human muscles and animal power. Fire is the major means by which other sources of energy—wood, straw, and animal waste such as dung—are utilized. With a limited amount of available energy, productivity is low and a great deal of human effort produces remarkably little. Farmers expend a great deal of effort to produce small amounts. It takes artisans days to produce one pot or one rug. Also, the traditional system usually has no rapid means of transportation or communication. As a consequence, almost no important *spatial interaction* exists between places in the traditional world.

The traditional, locally based system is still the dominant life system in areas throughout much of the world today and for at least half the world's people. Some examples of extreme isolation exist. In our times a tribe of Eskimos was found that thought that no one else lived on earth and that all the earth was a cold and barren place in which the sea produced the most opportunities for food. Recently, tribes completely cut off from the outside world have been found in Brazil and the Philippines. Far more important, probably the majority of people in Africa and Asia, and large numbers in Latin America and the Middle East, still live in predominantly local worlds. They use manufactured goods and have contact with doctors and government agents from the outside world, but their needs are supplied primarily from local resources. Even in the United States—in Appalachia, in the rural South, and on Indian reservations—some human groups live basically this way.

The traditional model also affects the modern world in an important way because it underlies much political thinking. Since World War II there has been a strong worldwide feeling that individual peoples should be able to govern their own parts of the world. A primary tenet of the United Nations has been "self-determination of peoples." Dozens of countries have been created out of former empires so that separate culture groups can control their own affairs. Some of these countries have such small territories and so few people that they seem to have little chance of surviving economically. But in still other countries, minority culture groups cry out for land of their own: the French Canadians, the Welsh and Scots, the Bretons, the Tyroleans, the Somalis, and many more seek independence from the governments that control them.

Organization of the Traditional System

When we examine the traditional, locally based system in terms of its functional organization, we gain further insight into the way that it uses the environmental base, distributes the return from the land, and makes decisions about life. This approach allows us to see the overall system as a set of interacting subsystems: one for production, one for distribution and marketing, and one for managing the economy and making decisions important to local life.

Production **Production** in any society involves using the resources of the environment for food, shelter, clothing, transport, and communication—in short, all the essentials and luxuries that the group requires. The first step in the production system, then, is working to gather or produce products directly from the environmental base. This sort of production, called **primary production,** includes hunting and gathering, fishing, agriculture and herding, forestry, and mining. The second step, known as **secondary production,** is the further processing of the products of primary production into goods of greater value to the society. Secondary production includes all manufacture and construction.

In the traditional, locally based system most of the environment's resources are used directly or are further processed by the family or tribe. Food crops are often eaten raw. Fuel—wood or straw—is gathered directly from the forest or fields, and building materials—such as wood and stone—are gathered, not mined. Since their technology is limited, people in the traditional system can extract a large yield from the environment only at the expense of much effort and a great amount of time, as we saw in Ramkheri. The basic tasks of most people in traditional societies involve primary production. Thus more than half the work force in Ramkheri were identified as farmers. The importance of primary production is just as great when the traditional system is based on hunting or gathering, herding or fishing.

In Ramkheri the farmers could be readily identified because they spent most of their working effort on farming. But in more primitive societies the primary producers may hunt, fish, farm, and herd livestock, and even do some primitive mining. In others, distinctions are made between the people who farm and those who fish or hunt or raise livestock.

Manufacturing is also done in all traditional

societies: food must be cooked, buildings constructed, clothes made, and products necessary for work or other aspects of everyday life produced. In the simplest societies these tasks are done by the primary producers themselves and take only a fraction of their total working time. In Ramkheri many of the manufacturing tasks such as cooking and making clothing are still done within the family, but there are also artisans specializing in carpentry, basketmaking, and pottery. Still, only a handful of these specialists are needed to supply the basic manufacturing needs of Ramkheri's population beyond those satisfied by the farm families themselves.

In the traditional system many different services are provided, too. Hair has to be fashioned, ceremonial costumes must be prepared, midwives are needed, and the sick must be cared for and the dead buried. These tasks, however, can be carried out by the primary workers and their families. Ramkheri has few service workers; members of different castes have particular roles in village ceremonies, and families take care of most of their own service needs. Ramkheri has long had a priest to minister to the villagers, and now it has teachers and government officials as well.

Marketing and consumption In the traditional system much of the production is consumed by the family that produces it. Thus producers are consumers and little exchange goes on. However, even locally based systems commonly have marketplaces where goods are exchanged between one person as a producer and another as a consumer. The idea of "exchange," then, is a part of the **consumption** process. Much of the exchange within the traditional society is done by barter, the trading of one product or service for another. In Ramkheri even the artisans are customarily "paid" for their services in agricultural produce.

The exchange, in barter or money, depends not only on the willingness of producer and consumer to exchange, but also on their ability to do so. That, in turn, depends on the existence of a surplus to be sold or traded and of money or goods in kind with which to buy. In traditional systems, where production is low in comparison to modern output, there is less to be sold and buyers have little, either goods or money, with which to make purchases.

Decision making and management Even in the traditional, locally based system the production and distribution of the society must be managed. The tribal leaders or village officials, and the families and individuals, must decide what to make of their lives and their societies. In economic terms a culture group must decide what to produce, how to organize human and environmental resources to produce the product, how to carry out the actual production, and how to distribute what is produced. The process of **management** is carried out by traditional societies using mostly the guidelines of tradition.

To a large degree most people in the traditional system are producers, consumers, and decision-makers all at once. The individual family knows its needs, organizes itself to produce those needs, and consumes virtually all of what it produces. Economic decisions are limited by the narrow range of resources and by the small number of consumer-producers. The leaders of a tribe or the headmen in a village such as Ramkheri are likely to make major decisions that the group respects. Sometimes a member from each family will be involved in making a decision for the group, passing it on to the people, and seeing that it is carried out.

In a traditional system, production, consumption, and management occur within the same local area. People who manage the system and consume its products live in villages within walking distance of producing fields or productive herds. Thus the population distribution reflects to a high degree the more productive areas as judged by the culture at its particular stage in technology and within its frame of values. The more densely populated an area is, the more likely it is that the supporting land in the vicinity is productive.

The "Traditional" View of Life

In many traditional societies people live so close to the minimum conditions for survival that the risks of change are almost overwhelming. We have seen the conservatism of the Indian village, and it should be apparent that any farmers who introduce new farming methods are taking severe risks not only for themselves, but for their families as well. Thus in traditional society the emphasis is on the status quo, things as they are. The leaders of the society are the elders, and rules brought down from the past are revered. A premium is placed on conforming to accepted practices that have worked in the past, and decisions of the group leadership are rarely challenged.

Such a system allows little margin for error if physical suffering of human beings is to be avoided. Crops may be stored from harvest to harvest, but they are usually completely consumed in daily living. Commonly, little or no reserve accumulates from year to year to serve as a source of support if a harvest is bad. Even if there is a surplus, storage is difficult without modern facilities; much of a crop is lost to rodents or rot. Under these conditions the risk of change is tremendous. How can one afford to try something new—a different seed, a plow, a new fertilizer? What if it should fail?

Added to this dilemma of poor subsistence dwellers are their lack of education and a cultural structure built for preserving traditional ways. In such a society all accepted practices call for maintenance of the status quo. It is a brave person who attempts to make a change. Often it is only the wealthy, or those whose educational experiences extend beyond those of the local society, who can afford to take risks.

Technological development generally calls for a great deal of reorganization, too. It requires a change in institutions, in styles of living, and, most of all, in attitudes. Such changes are among the hardest to accomplish. Often the blinders of cultural heritage stand as greater obstacles than resource limitations.

THE MODERN INTERCONNECTED SYSTEM: A WORLD NETWORK

The traditional system does not enable us to understand Washington or any other major metropolitan centers of the world, or even most of the smaller cities and towns. Yet almost half the people of the world live in urban-centered societies. The majority of people in the United States and Canada, most European countries, the Soviet Union, Japan, and Australia live in metropolitan areas, and many of the world's great agricultural and mining areas are geared to supply those centers.

A Different Geographic Equation

In the modern, metropolitan-centered world the basic principles of environmental use are very different from those of the traditional, locally based system. None of the world's great cities or any of the industrial countries expects to support its population from local environmental resources. Rather, their stores and marketplaces are designed to operate on a continuing stock of goods gathered from different environments throughout the world. In turn, goods, money, and ideas from these centers move throughout the world.

In the same way, farmers in the major producing areas of the United States, in the wheat- and wool-producing districts of Australia, in coffee plantations in Latin America and Africa, and in other specialized agricultural areas do not expect to feed their families from their own land. Their plan is to mass-produce as much of a particular commodity as modern farming methods applied to the local environment will allow, and to buy food and other needs in local stores stocked with goods from the worldwide resource base.

The ideas that enable the modern system to operate are drawn from many places, too, not just from the local resource base. Modern electronic technology has been built on ideas contributed by scientists and engineers from all major European countries, the United States and Canada, the Soviet Union, Japan, and other modern countries. Modern fashion develops styles based on native costumes from the Pacific, Africa, Latin America, and the rest of the world. The doctors, scientists, politicians, musicians, and artists of the interconnected world system continuously interact with one another, discussing new ideas.

In the business world the **multinational corporation,** both industrial and financial, has developed to invest production capital anywhere in the world that promising resources are found. These corporations also develop distribution systems that reach markets wherever there are people willing to pay for their products and services. Yet the whole multinational complex is usually managed from a world headquarters operation in New York, London, or Tokyo with a few regional subcenters in places such as Hong Kong or Singapore, Cairo or Athens, Mexico City or Buenos Aires, Brussels or Zurich.

This whole system is made possible by an extensive worldwide network of modern transportation and communication that enables people in the centers of the modern system to interact with different parts of the world rapidly and cheaply and to move masses of goods from one place to another at low cost. Jet aircraft and long-distance telephones allow politicians and businesspeople to monitor worldwide interactions continuously. Modern supertankers and container ships, trains, trucks, and pipelines move masses of raw materials and finished manufactured goods by the extensive use of inanimate energy, usually obtained from fossil fuels such as petroleum and natural gas.

Origin and Spread of the Modern System

In the early stages of technology, caravan routes, rivers, and oceans offered alternatives to local self-sufficiency. The Chinese, Greeks, Phoenicians, Norse, Venetians, and Arabs all expanded their regular contacts well beyond their own culture groups. The Roman Empire was an early attempt at a controlled interconnected system, but it was limited by lack of rapid transportation and communication facilities. The Romans had to rely on goods and contacts that could be carried by human porters, animals, wagons, or sail-powered ships. Messages could not move faster than a horse and rider or the fastest ship.

The first effective example of a truly interconnected system was the British Empire. It connected producing areas on all continents, but it really organized only a small fraction of all the possible earth resources. Its great success did not occur until late in the nineteenth century as the result of British leadership in the new technology of the **Industrial Revolution.** It was based on breakthroughs in transportation and communication, steel-hulled, steam-powered ships and railroad trains, the telegraph, and the telephone. Other European countries that shared in the Industrial Revolution developed their own intercontinental networks on a smaller scale; the United States created its own, very different version of the interconnected system by organizing its varied resources into a functioning whole and by strengthening its ties with Europe.

Today other examples of the interconnected model on different scales can be seen in the Soviet Union and Japan. To an important degree, each country in the contemporary world is trying to fit into the worldwide modern system. Some produce just one or two commodities in modest amounts for export to the major industrial centers, but in the oil-rich countries of the Middle East–North Africa, governments have ambitions of becoming the centers of their own powerful systems like those of the present industrial countries. Today, then, just as there are many variations on the locally based model, there are numerous forms of the interconnected system.

Originally, the modern interconnected system reached out from Europe and the United States primarily for raw materials. The idea was to produce needed resources from elsewhere in the world as cheaply as possible. But the parts of Asia, Latin America, and Africa that supplied these resources were not seen as important consuming areas and were not expected to produce finished manufactured goods requiring high technology. Today's giant multinational corporations seek markets wherever they can be found and have also discovered that urban centers anyplace in the world have the capacity to produce modern industrial goods. Thus places such as Hong Kong, Singapore, Taiwan, Korea, Mexico, and Brazil contain centers of modern electronics and automobile manufacture not just for their home markets but for worldwide distribution. Products from these new manufacturing centers are sold in the United States and Europe. Moreover, as more and more workers in these places are drawn into the modern economy, they become consumers. Their wages may still be only a fraction of those in the United States or Western European countries, but they are high enough so that the people become consumers of bicycles and motorcycles, radios and televisions, home appliances and Western clothing.

Different Cultural Implications

Today, with modern transportation, each producing area in the modern system can supply customers in all other places throughout the system. At the same time, specialized producers expect to be able to consume the full range of goods and ideas from throughout the system. The major barriers to the movement of goods and ideas are commonly political, not technical.

The modern interconnected system presents a very different perspective of space than the traditional system. The traditional world is a mosaic of separate cultural units, each of which has its own individual distinctive ways and its own rights and privileges. Any communications that take place are from one mosaic unit to another. In the modern system the primary links are between metropolitan **nodes,** not between cultural mosaics. The nature of the new world is illustrated dramatically by modern jet air travel. The jet traveler moves from one metropolitan center to another, in the same country, the same continent, or anywhere in the world. In fact, jet travel is generally feasible only between the large metropolitan centers of the world. Not many cities of

fewer than 100,000 people have regularly scheduled multiflight jet service unless they happen to be cities in world-renowned resorts such as Bermuda, Jamaica, or Majorca.

For the jet traveler, life at one point in the system is very much like that at any other: the airplanes are the same, as are the airports, taxis, hotels, and restaurants. Money has to be exchanged and there is the bother of people speaking different languages, but most of the workers in the jet system—the flight attendants, the taxi drivers, the restaurant and hotel staffs—speak the basic language of the new international system, English.

Jet travel largely obscures the mass of land between the metropolitan points: planes fly too high for passengers to see much of the landscape between origin and destination. Commonly, travelers are carried into the clouds over one metropolitan area only to emerge out of them at a destination whose characteristics are mostly the same. The space between origin and destination is seen largely in a time dimension, not a space dimension. It is a necessary wait that must be endured by eating, drinking, watching movies, reading, or sleeping. Travelers arriving at the end of this time-void find it hard to realize that they are in another country or another continent.

Among politicians, wealthy tourists, and businesspeople who travel between the centers of the world airline system, a distinctive culture has emerged that is different from that of any single country. We speak of such people as members of the "jet set." Their ways may take most from the modern culture of the United States, but they have had inputs from all other cultures of the modern world. And for these people those ways have altered the traditional national culture into which they were born. There is much to the statement that members of the jet set from London, New York, Tokyo, Sydney, Buenos Aires, or Teheran have more in common with one another than they do with the ordinary citizens of their home countries. Consider the implications that this has for a world whose basic political unit—the nation-state—has been built on the traditional view of the common bonds between people of a given culture living within a single self-contained territory.

Organization of the Modern System

In the traditional, locally based system, production, consumption, and management all occur within the confined local area. In the modern interconnected system, however, each of these three subsystems is located in accordance with its own needs and thus has its own distinctive distribution pattern. Not only do the three occur in different places, but primary production and secondary production are even located separately.

Production   In the modern system, secondary production is far more important than primary production. It employs more workers and accounts for a greater value of output. As in the traditional system, primary production, by definition, must be carried out where the resources are. Thus farming, forestry, and herding still must spread over the landscape in order to produce sufficient quantities. Mining must be located where the mineral resources are. Because great quantities of inanimate energy are used, however, relatively few workers are now employed in primary production—in most modern countries less than a quarter of the work force; in the United States, less than 7 percent.

Secondary production generally is oriented to people, not to resources. Some manufacturing, such as lumbering and smelting and refining of low-grade metals, must be done at the site of the raw materials. Most, however, is concentrated in cities where abundant, varied labor can satisfy the need for specialized skills. Other advantages of cities are that they provide ready access to markets and that they are convenient places for the varied materials—some raw materials from different places, but mainly the products of other manufacturers in or accessible to the area—to be brought together.

Production in the modern system requires **specialization of labor.** Manufacturing processes are frequently so complex that they require a large number of different skills. It is also necessary to have many specialized services to support the modern production system. These services are provided by doctors, lawyers, morticians, hairdressers, florists, and a host of other workers. Most of these services are concentrated in large metropolitan centers where there is a large enough population to support such specialized activities. Because secondary production and services are urban-centered activities, in the modern society most of the labor force, and thus most of the population, is found in large metropolitan areas.

Marketing and consumption   In the modern system, with its lavish use of inanimate energy and complex machinery, many things are available for sale, and workers are paid comparatively high wages, thus enabling them to buy. Since most individuals specialize in just one occupation, the purchase of most of the needs of life is part of the system. Transactions are aided by a complex system of money and finance in which not only money, but also checks and credit cards, are presented. Moreover, the banking system allows credit purchases in anticipation of future payment.

Since most production and most of the population are centered in metropolitan areas and these centers are the major transport and communication nodes, consumption in the modern system is concentrated here, too. The object of major producers is to market their products wherever people are located. The U.S. economy has been based as much on mass marketing as mass production.

Decision making and management   As we have seen, decision making and management are essential in both the traditional and the modern systems, since even families and individuals must decide what to make of their lives and their societies. In the modern interconnected system much of the important decision making with major impact on the economy has been developed as a science, using masses of data manipulated by computers and directed by highly trained managers, the government officials and the corporate executives. Management is a specialized skill, and both business and government managers are among the most highly paid persons in the system.

With jet aircraft and modern communications, as we have seen, major multinational corporations can now manage worldwide operations from a single central headquarters. Modern international oil companies manage a system that involves production on all continents and distribution to worldwide markets.

Separate Networks for Each Subsystem

In the modern system, the three subsystems—production, consumption, and management—are often in widely separated locations. Products consumed by a city family come from producing areas throughout the world. A city such as Washington is thus important as a consuming center to producing areas throughout the world. As it grows, more supermarkets and shopping malls are established, more auto dealers and house builders open for business, and the production and distribution system adjusts to feed more to the market. But Washington is not a center of production; rather, it is one of the most specialized decision-making and management centers in the world. Less than 10 percent of its work force is engaged in production of goods, while almost 40 percent is involved in government, a management operation.

Washington is an exception to the general pattern. Since the modern interconnected system has risen to dominance in the last 200 years, most large metropolitan centers have been seen primarily as manufacturing and goods distribution centers. Manufacturing is the chief employer in most large cities in the United States, Europe, and Japan, and most of the manufacturing in modern industrial countries takes place in large cities. With the growth of the modern multinational corporation and international banks, however, much of the decision making has been removed from most manufacturing cities and concentrated in a few centers such as New York, Chicago, San Francisco, Detroit, and Houston in the United States; London, Paris, Amsterdam, and Milan in Europe; and Tokyo and Osaka in Japan. This is the result of the purchase and merger of local firms into large multinational organizations based in a few world centers. Now even metropolitan centers of a million or more people such as Atlanta, Cleveland, Buffalo, San Diego, Glasgow, Rotterdam, and Munich have very few large manufacturing operations with headquarters in their cities. Their large plants, employing 1,000 or even 10,000 persons, are controlled by management sometimes hundreds of miles away in their own country, or even thousands of miles away in another country.

Traditionally, individual culture groups and individual families within those groups made the basic decisions about their lives and the utilization of their local environment, but the concentration of decision making in large multinational business institutions in the modern interconnected system is changing that. More and more these businesses are making the big decisions concerning the utilization of the earth environment. They are emerging as the only institutions, other than governments, with the capital and technical competence to make the major decisions concerning new mines or large-scale manufacturing enterprises. They send the oil-drilling equipment out on offshore platforms in Europe, North America, the Middle East, and elsewhere in the world; they open new coal mines and metal-processing operations; they build new steel plants and auto assembly works or petrochemical plants; and they decide to close mines or factories that sometimes are the only support for small communities.

The multinational corporations are now making these decisions on a worldwide scale. In the past, in the traditional system, each region was evaluated by its own people, and its resources, rich or meager, were utilized at least on a small scale. The multinationals that now evaluate resources on a global scale are not trying to develop the resources of all regions. Rather, they are searching for the very few best producing areas—the world's major mineral resources, the best farming lands, the key locations for factories and headquarters. Thus many regions that have long been occupied by locally based systems are passed over by the global decision makers, for they lack resources of value on a global scale.

More and more the decisions about the use of the environment and daily lives that have traditionally been in the hands of local people are being made in headquarters far from the places where people live. The opening of a new mine or factory can mean a great deal to a locality, as can the closing of such an installation. But towns and now large cities are finding that these decisions, so important to local life, are being made in company board rooms miles, sometimes countries and continents, away.

Demands on the Environment

The modern system has made increasing demands on the environment, and the demands are accelerating rapidly. It is estimated that more minerals and timber have been extracted from the environment in the past 100 years than in all human history before. The United States, one of the largest modern industrial countries, has less than 6 percent of the population of the world. Yet, with its mass production and mass consumption, it is said to consume about one-third of earth resources at present. The modern system, which places a premium on transportation and communication equipment and consumer goods, has not only spread over North America, Europe, the Soviet Union, and Japan but has growing bridgeheads in all countries of the world. Thus the large multinational businesses seek new resources to meet the rapidly increasing demand for resource-using food and modern goods.

Whereas in the traditional system humans seemed at the mercy of the environment, it now appears that the environment is increasingly endangered by the works of humans. Using modern technology, humanity has the capacity to produce great changes in the environment. Rivers have been dammed to create lakes such as that behind the new high dam in Egypt or behind the hydroelectric complexes in the Soviet Union. In the Soviet Union the waters of rivers that flow to the Arctic Ocean are being diverted into the deserts of central Asia. Canals link ocean to ocean and provide ties between inland rivers and lakes and the sea. Artificial harbors have been created at ports, tunnels have been built through mountains, and whole hillsides have been modified for building sites. But more important, environmental problems have been created by the modern interconnected system. First are the demands that increasing use places on earth resources. Although most metallic minerals such as iron can be recycled by using scrap from obsolete equipment, mineral fuels are essentially nonrenewable. When coal, natural gas, or petroleum is burned as fuel, the burning changes it into gases that are released into the atmosphere and are not retrievable. Thus in the brief 200-year period of the modern system, we have made great inroads on the fossil fuels of the earth. Large deposits of coal still remain, but it is doubtful whether petroleum will be available for more than another century.

The most crucial impact of the modern interconnected system on the environment appears to be the contamination of the atmosphere, largely as a result of the burning of fuel such as oil and coal and the release of industrial gases into the atmosphere. Problems of smog have for a number of years been recognized in the large metropolitan areas of the United States as a result of the mass use of motor vehicles that burn gasoline. Now smog is developing in other cities of the world because of increasing use of motor vehicles there. More critical are changes in the general character of the atmosphere that makes the earth inhabitable. Scientists have detected changes in atmospheric temperature and in the character of the ozone layer, a crucial upper atmospheric layer. Some scientists hypothesize that the pollution of the atmosphere has already produced irreversible changes, and most agree that the atmosphere cannot continue to absorb large amounts of pollutants without serious consequences. Pollution of lakes and even the oceans has also reached serious levels in some places, and scientists call for greater control of industrial and municipal wastes.

Breakdowns in the System

Despite these problems, the modern interconnected system has for the present undoubtedly alleviated traditional environmental problems such as famine in the parts of the world in which this system is found. With multiple sources of supply, there is little danger of serious short-term food shortages in the modern world today; for example, when a crisis such as a flood or earthquake occurs, relief supplies can be airlifted in a matter of hours.

The modern interconnected system depends on a tremendously expensive and intricate **infrastructure** involving a complex system of transportation and utilities. Delivery of goods and messages to just one metropolitan area requires a bewildering complex of road, rail, ship, and air transport and an even more complicated system of communications. Massive amounts of energy must be moved throughout the delivery system, and producers, shippers, financiers, wholesalers, retailers, and purchasers are involved. Individual items will move over several different means of transportation and change hands from one business unit to another as many as a half-dozen times before reaching the consumer. The environmental dangers to such a system are not great, although a sudden snowstorm, flood, or windstorm can disrupt the infrastructure for hours, even days. Unusual weather conditions, such as the exceptionally cold winter of 1977 in most of the United States, can impair the functioning of the system for even longer periods.

The greatest problem of the complex modern system is a breakdown in the system due to equipment malfunctioning or human disruption. The whole northeastern United States was blacked out for hours on November 9, 1965, because of electrical equipment failure. People were caught in darkened skyscrapers whose elevators did not run. Lights to control traffic and streetlights did not function; hospitals were blacked out; electric trains stopped running. The most populous part of the United States quickly discovered how much its everyday operations depend on electricity. But strikes and other disruptions caused by people can be just as bad. In the winter of 1973–1974 the British Isles adopted a three-day workweek and did without electricity two days a week because of a prolonged and bitter strike by coal miners. The whole economy of the industrial world was shocked in 1973–1974 by the rapid price increase for petroleum instituted by the oil-exporting countries and a boycott on oil to countries sympathetic to the Israeli cause.

The Modern View of Life

The modern interconnected system has produced a psychological change. Whereas traditional systems protect the status quo from generation to generation, the modern system has developed a commitment to change. Change has been seen as progress during the advances of the last 200 years. Consequently, the society has come to assume change to be for the better until proved otherwise. Consumerism, particularly in the mass production–consumption society of the United States, has been built on changing fashion in clothes, appliances, and automobiles. The person or firm who can develop a new product or process and market it successfully is assured of wealth. Even education continuously presents new learning theories and teaching techniques.

Why is the modern world so committed to change? The crucial element seems to be the technology that has produced the modern world's material well-being. An affluent society has a large margin for taking risks. If the gamble fails, all is not lost; it can try again or attempt something else. In the modern business world, new products mean new markets, and new markets mean new profits. Since the emphasis is on improvement and new models, much of the profit is reinvested in research and the development of new products or new designs. The major operations of this type in Washington, D.C., for example, are the research and development programs funded by government grants for new military equipment and studies for social as well as technological change.

Much has been written about the psychological and sociological impact of "future shock" on the modern world. Throughout history up until the present century, ideas developed in the traditional system dominated even in the industrialized areas where the modern system had gained a foothold. Family, "home town," sovereign nation-state—such ideals were based on the continuity of an established lifestyle within the culture that would not change much from generation to generation. In such a world, the elderly, who had gained the most experience in the system, were looked to for leadership. The modern world of change undermines those values. Experience gained in the past may be a hindrance in dealing with change, and the young who are involved in such change give little credence to the older generation. In a world where a large portion of the population is nomadic, moving from job to job and from city to city, the concepts of home, community, and even family have much less meaning than they formerly did. Most large metropolitan areas have little feeling of community. Thus change brings insecurity to those whose values were established in a more traditional society.

LOOKING AT THE TWO SYSTEMS IN TODAY'S WORLD

The ideas presented in this chapter—the geographic equation and the two contrasting human systems, the traditional and the modern—form a useful framework for the study of the geography of a place and its people. The two human systems can be regarded as models of very different ways of solving different problems in environmental management and cultural development. Ramkheri and Washington may be seen as almost opposite extremes on a continuum of human life that ranges from the most traditional to the most modern. In the real world we would not expect to find the two models exactly as sketched here, but rather a variety of modifications of the two extremes. In each particular part of the world there will be variations in the components of the geographic equation that produce a unique system of life. Thus traditional life in Latin America is different from life in Black Africa or Asia; life in the Latin American highlands will show differences from that in the lowlands; life in rural Venezuela will be different from life in the Amazon Basin of Brazil. In the same way, the interconnected system that has evolved in the United States is different from that of Europe, the Soviet Union, Japan, or Canada.

Although the two different human systems offer a way of looking at the geography of places and the significance of those places in the world, we must remember that, in fact, human life in any place is likely to involve both systems. Part of our task as we examine regions of the world is to consider how these two basic systems relate to and affect each other at any place.

In some places in the world one system predominates; in other places the other system predominates, but there are aspects of both. Ramkheri is mostly traditional, but the modern interconnected system, particularly in the form of the Indian government, is present in Ramkheri daily, and traditional ways are changing. In Washington, which seems a prime example of the modern interconnected system, portions of the population—the elderly poor and the inner-city poor—do not fit into the modern system around them, although public assistance payments from the system may support them marginally. People in the United States were shocked in the early 1970s to find that an important segment of the rural population in the South not only had cash incomes of less than 200 dollars per year but also showed many of the symptoms of malnutrition found in poor traditional societies.

CHANGE FROM TRADITIONAL TO MODERN

It is important to remember that no place is locked into the traditional way of life. The modern interconnected world has evolved from the traditional system. Thus Western Europe, the United States and Canada, the Soviet Union, and Japan have their roots in traditional societies, but industrialization has allowed the modern system to replace old ways. The world today is in a transitional stage in which the traditional system is evolving into the modern interconnected system. It remains to be seen whether the modern system will continue to spread over the world until it encompasses all peoples. It is doubtful whether the earth environment has enough resources to support even the present population of the planet if all people were part of the mass-consumption modern system.

Nevertheless, places have evolved from traditional to modern ways, and the process continues in today's world. This transition has occurred, and is occurring, in different ways. The United States, the Soviet Union, and the European countries have maintained control over modernization themselves. They have used ideas about agriculture, manufacturing, management, and trade gained in other places, but they have retained control of development decisions. Where the modern system has reached out to Latin America, Africa, and most of Asia it has come from Europe, the United States, or another industrial area. Mines and plantations, as well as cities, ports, and railroads, have been established in those regions of the world as part of the modern system. But control over the system has remained largely in the hands of people and firms from the industrialized countries. African countries have recently gained political independence from colonial empires, but a key question in today's world remains: Can such countries develop their own economies or will they always be minor parts of larger systems centered in the industrialized world?

PROCEEDING FROM HERE

The components of the geographic equation are discussed in greater detail in the next two chapters. In Chapter 5 indexes of traditional and modern society are presented to give an idea of the global distribution of traditional and modern systems. In Part Two of this book we focus on the main centers of the modern interconnected system in the world: the United States, Canada, Europe, the Soviet Union, and Japan. In each center the modern interconnected system is overwhelmingly dominant, but while the technology factor in the geographic equation is much the same in each, the other components of the geographic equation—environment, culture, and spatial interaction—are different. The result is several different variations on the modern system.

In Part Three we examine the other major areas of the world—Latin America, Black Africa, the Middle East–North Africa, and South and East Asia with special attention to China and India. In these areas the traditional, locally based system that has existed since the dawn of history is still a dominant force. But all these areas have been drawn into the modern interconnected system as well. The centers of the modern system have reached out into these traditional areas—first through empires and then through trading systems that have not only tapped tropical agricultural areas and rich mineral resources but have organized mines and plantations linked by railroads and ports. Certain parts of some countries, such as Brazil, Mexico, Iran, and particularly China and India, have developed as important centers of the modern interconnected system with large modern metropolitan centers containing factories, office blocks, jet airports, and modern residential areas. All countries are involved in international diplomacy. Their capitals are nodes in a worldwide diplomatic network as well as an economic system.

SELECTED REFERENCES

Berry, Brian J. L., Edgar C. Conkling, and Michael D. Ray. *The Geography of Economic Systems.* Prentice-Hall, Englewood Cliffs, N.J., 1976. An excellent reference to augment the material in several chapters of this book. The text is well written and effectively supplemented with pertinent tables, charts, and maps. Part Five is especially appropriate as additional reading for this chapter.

Black, Cyril Edwin. *The Dynamics of Modernization, A Study in Comparative History.* Harper & Row, New York, 1966. Uses much the same framework proposed here, but in terms of political science and history. The focus is on the modernization process as it has evolved in different parts of the world.

Broek, J. O. M. *Geography: Its Scope and Purpose.* Charles Merrill, Columbus, 1965. A book for teachers by a distinguished geographer commissioned to prepare a statement on geography and its viewpoint.

de Souza, Anthony R., and Philip W. Porter. *The Underdevelopment and Modernization of the Third World,* Resource Paper No. 28. Association of American Geographers, Commission on College Geography, Washington, D.C., 1974. A review of diagnostic concepts and qualities about modernization and underdevelopment in the world. The discussion evaluates the descriptive and theoretical nature of underdevelopment and modernization. It is recommended for a deeper understanding of the current process of modernization and its application to the problems of underdevelopment.

English, Paul Ward. *World Regional Geography: A Question of Place.* Harper & Row, New York, 1977. The evolution of the world into its current developing and technological parts as the framework for regional treatments. Each country or region is approached through a summary of its historical development, an analysis of its geo-graphical setting, and a discussion of its major contemporary economic or social problems.

James, Preston. *All Possible Worlds, A History of Geographical Ideas.* Bobbs-Merrill, Indianapolis, 1972. The development of geographic thinking as interpreted by one of America's most esteemed geographers.

Keyfitz, Nathan. "World Resources and the World Middle Class." *Scientific American,* July 1976, pp. 28–35. Proposes a four-part world of capital-rich, resource-rich, developing, and poor nations or regions in contrast to the dichotomy presented in this chapter. The resource and pollution implications of continued expansion of affluent populations among the world's peoples are evaluated.

MacKaye, Benton. *The New Exploration.* Illini Books, Urbana, Ill., 1962. The conflict between the modern interconnected system and traditional life as seen by a regional planner in the 1920s.

Mesarovic, Mihajlo, and Eduard Pestel. *Mankind at the Turning Point,* The Second Report of the Club of Rome. E. P. Dutton, New York, 1974. A report using world regions as a basis for analysis. Its basic premises are that the various parts of the world are being integrated more and more into a world system; that the disparity in material welfare and population growth between regions threatens the world system; and that survival of the world system and its regional parts depends on a greater sharing of resources, capital, and technology. The emerging problems demand management approaches that are global in scope.

Philbrick, Allen. *Man's Human Domain.* John Wiley, New York, 1962. An introductory college text organized around the theme of the human organization of space.

CHAPTER 3

The Earth Environment

Our physical environment is a fundamental variable in the geographic equation. Mineral resources and the possibilities for vegetative growth at a place influence how people interact with the environment. In our study of environments as working systems, we shall find that solar energy and moisture availability are limiting factors. A general pattern of world climatic regions will emerge. We shall see how the different work patterns of the environment present different opportunities and problems to people at particular places, and also how people cause problems for the environment.

Flood damage in Pakistan: a washed-out road. (CARE photo)

Duststorm near a village on the outskirts of Delhi, India. (Lynn McLaren)

Above: Windmills as a source of energy in Spain. (Ampliaciones y Reproducciones MAS, Barcelona)   *Right:* A dried-up irrigation ditch in West Africa, the result of prolonged drought. (Max Hastings/Photo Trends)

In Chapter 2 we suggested that places can be viewed as being on a spectrum ranging from the traditional, locally based village to the modern interconnected city. This spectrum has developed over a long period as people have developed tools, technologies, and institutions to use the world's physical environments more and more effectively. The many different ways of life on earth today evolved in different physical environments, and they reflect in part the particular attributes of the physical environments that various peoples have experienced. For example, the knowledge, skills, and tools of desert dwellers are quite different from those of farmers in humid areas. But even in seemingly similar environments, different ways of living have evolved. Thus we may consider environments as working systems that people can exploit and manage in order to support themselves. We shall find that energy from the sun accompanied by the presence of water is a vital combination. Together, solar energy and water establish the conditions for plant growth, which is in turn the basis of agriculture, livestock grazing, and timber production. Plant growth and water resources are vital to human life.

Because the normal focus of human activity is on the surface of the earth, this interface of solids, gases, and water is most relevant to our discussion in this chapter. Although the deep interior of the earth and the outer atmosphere do affect conditions at the earth's surface, they are rarely part of direct human experience or the decisions of daily living. Obtaining adequate food and water from the earth's surface, however, has been and still is a basic problem for people wherever they live. In this chapter we therefore focus especially on aspects of the physical environment that relate significantly to the geography of agricultural and water resources over the world.

Extreme climatic conditions of aridity or persistent cold severely limit biological life and available water supplies, and most such areas with these conditions have sparse human populations. In these regions climate has an obvious and powerful influence on the local geography. For much of the land area of the world, however, climates are not so harsh or severe and their influence on food and water supplies not so obvious. Nevertheless, different climatic environments provide people with very different opportunities for food production and water resources. Knowledge of how climatic environments work is therefore particularly relevant to understanding how people manage agriculture and obtain their water supplies. This chapter presents a view of world climatic environments in terms that relate directly to biological activity and water availability. Understanding these climatic environments enables us to anticipate the plant-growing potential at particular places on the earth and to evaluate the potential for developing water resources.

Of course, biological productivity is not simply a matter of the climatic conditions alone, powerful as these influences might be. The character of soils, for example, can also have a significant effect on plants because soil is the immediate source from which plants obtain many of the nutrients essential to their growth. In addition, soil provides the moisture storage from which plants draw their water needs. As we shall see, soil conditions are strongly influenced by climate because the processes driven by the energy and moisture of climate modify the physical and chemical make-up of the surface of the earth's crust. This means that climatic environments and soils have much the same pattern of distribution over the world, though major exceptions do exist.

Surface configuration is another environmental factor that may influence the amount and kind of agricultural activity as well as the general population pattern over the world. The patterns of mountains and plains, in particular, appear to have a direct influence on the world pattern of population. This spatial association is discussed in the next chapter, but here let us look at some of the basic features of the earth's surface, especially landforms and mineral deposits.

THE EARTH ENVIRONMENT: SURFACE FEATURES

In Chapter 2 we described the surface of the earth as the interface between (1) the atmosphere, the gaseous envelope over the earth; (2) the lithosphere, the solid crust of the earth; (3) the hydrosphere, the water cycling between oceans, atmosphere, and land; and (4) the biosphere, the living organisms, plants and animals and the human population.

The relationships between the various environmental zones and the resulting complex environment at any place on the earth can be seen as the result of the combination of, and balance between, three different systems of forces:

1. Forces from within the earth itself are driven by energy from the earth's mass, such as radioactive decay and density differences. These forces, referred to as **diastrophic forces,** are most obvious in the action of volcanoes and earthquakes, but accomplish much more work in far slower, less violent ways. In fact, the earth's crust is now known to consist of large rigid plates bounded by faults and rifts where movements between the plates can occur. The San Andreas fault in California is one such slippage zone between plates. Gradually, over vast periods of time, the stresses on the crust and intrusions of molten materials from the deeper interior of the earth cause mountains to rise or broad areas of the crust to warp up or down. The persistent action of such forces has created the major structural features of the continents and ocean basins, has produced the foundation structures of mountains, and is responsible for the broad basins and rises on continents. Most of the rocks that make up the

crust of the earth, and the particular types of minerals in those rocks that people use as resources, have been produced or modified by chemical and physical reactions in the earth's crust and in the more pliable layer beneath it under varying conditions of heat and pressure.

2. Atmospheric and **hydrospheric forces** are powered by the continuing input of energy from the sun to the earth's surface zone. This sun-powered system produces the continuously varied condition of the atmosphere that creates our weather. It evaporates and precipitates water, thus creating the essential moisture cycle for plant growth and the basic environment that supports animals and human beings. The cycling of water involves movements of water, ice, and wind, which in turn cause changes in the earth's surface configuration. We easily recognize the changes in surface brought by high winds and floods, but like the diastrophic forces, other slower, more subtle changes—weathering and erosion—are continuously changing the form of the earth's surface in a less perceptible manner. Over long periods these subtle changes are capable of wearing away highlands and filling in troughs in the earth's crust.

The surface features of the earth are produced by the interaction of the diastrophic forces of change within the earth and the sculpting forces set off by climatic conditions. Internal earth forces tend to build up or increase elevations. Atmospheric and hydrospheric forces tend to tear down high areas and fill in low places.

3. Biological forces are derived from living organisms. Plants draw the necessities of life directly from the land and atmosphere around them; animals depend most directly on plants and the atmosphere for sustaining life. As all these living organisms thrive in a particular locality, they leave their mark on the environment. Though soil, for example, consists mostly of minerals, air, and water, its particular physical and chemical character is very much a product of the chemical and physical activities of plants and microorganisms and the organic matter that is derived from these organisms. Plant cover also protects the land surface from the direct impact of wind and rainfall, and reduces the erosive power of running water. Destruction of the plant cover, most often by people, can speed the effects of the sculpting forces of wind and water. Biological activity interacting with weathering and erosion over the long spans of geologic history, as well as in the present, has produced the variety of surface detail and soil conditions on the earth's surface.

Diastrophic activity, the weathering-and-erosion system, and biological activity have produced the distinctive environmental conditions at particular places. It is not the purpose of this chapter to explore the complexity of their interrelationships in detail. Rather, the aim is to get a picture of the major environmental variations over the world and a general insight into the reasons for these variations.

LANDFORMS: THE SETTING FOR LIFE

An elevation map of the world (Figure 3–1) can give a rough representation of gross features of the earth's surface. Such a map shows, for example, where the major mountain ranges are located, such as the Andes along the west coast of South America and the Himalayan Mountains in central Asia. Extensive plains are also detectable on a world elevation map, for example, the Gulf and coastal plains in the southeastern United States and the broad plains of Eastern Europe, mostly within the Soviet Union. The details of surface configuration, such as the ridges and valleys of a single mountain range or the pattern of stream valleys of a hilly or plains area, though important to the daily lives of people, cannot be shown at this scale. For our purposes, however, great detail is not as important as establishing a world overview of the pattern of major plains,

mountains, hills, and tableland areas. Though the distinction between plains, mountains, hills, and tablelands, or plateaus, is common in the experience of most people, a brief description of each may provide clarification.

A useful distinction is made between elevation and relief. **Elevation** is simply distance above a reference level, usually sea level. **Relief** is the difference between the highest and lowest points in a given land area, for example, between valley bottoms and ridge crests. In a region with low relief, there is relatively little difference in elevation between the high points and the low points. **Plains,** for example, consist mostly of land surfaces with gentle slopes and low relief. Most plains are at elevations close to sea level, as are the coastal plains of the United States and the extensive plains of the Soviet Union, but this is not always true. For example, the extensive plains of western Texas, Oklahoma, and Kansas are well over a half-mile above sea level. The dominance of nearly flat surfaces in plains areas makes them especially attractive for human uses such as agriculture or transportation routes.

Plateaus, or tablelands, have slope characteristics similar to those of plains—much flat land—but differ in that they have greater relief. The greater relief may come from deep valleys cut into these nearly flat, elevated surfaces, as is the case of the Grand Canyon cut into the Colorado Plateau. The relief may also be provided by an **escarpment,** a steep slope separating the higher surface from adjacent lowlands. Many of the plateaus of Brazil, Africa, and India have escarpments.

Hills and mountains have much in common, and both are distinguished from plains and plateaus by the lack of extensive areas of gentle slope. Much of their surface consists of steep slopes. **Hills** have more relief than plains, generally 300 to 1,000 ft (90 to 305 m). **Mountains** have greater relief than hills. From the perspective of human use, the lack of gentle slopes hampers people from using the hills and mountains for crop agriculture and large settlements, but the steep slopes and long vistas provide scenic qualities that encourage recreational use of these land surfaces.

ROCKS AND MINERALS

The crust of the earth, from which we get our metallic ores and fossil fuels, is made up of three major rock types:

1. Igneous rocks are formed from the cooling of molten material. Molten rock may pour out on the surface, as when volcanoes emit lava, but more often it is intruded into the crust and slowly cools there. Granite is a common igneous rock of the continents. Basalts, though found on continents, are the dominant igneous rocks of the crust under oceans.

2. Sedimentary rocks are, as the name indicates, composed of particles produced from any rock by the action of weathering and erosion. Examples are sandstone, limestone, and shale.

3. Metamorphic rocks have been subjected to strong modification by heat and pressure while in the earth's crust. For example, sandstone, a sedimentary rock composed of sand grains, is metamorphosed into quartzite by compressing and merging of the sand grains; granite, an igneous rock, is modified to become gneiss.

Of the three types of crustal rocks, igneous rocks have the least exposure at the surface of continents. Metamorphic rocks are more commonly found at the surface, and, of the three, sedimentary rocks have the greatest surface cover of most continents. Mountainous areas are usually composed of igneous and metamorphic rocks, but some, like parts of the Rocky Mountains in Canada and parts of the Alps, have developed on massive thicknesses of sedimentary rocks. Sedimentary rocks underlie most of the major plains areas of the world. Old metamorphic rocks are extensively exposed at the surface in the Canadian Shield and in the northern parts of Scandinavia and adjacent parts of the Soviet Union.

Because of the very different character of the three rock types, each is associated with particular types of minerals. **Fossil fuels**—coal, oil, and natural gas—are derived from the preservation and modification of organic materials. These may be preserved in sedimentary accumulations. Because of the great heat and pressure involved in the formation of igneous and metamorphic rocks, fossil fuels are absent from those two rock types. **Metallic minerals,** on the other hand, can be found in any of the rock types, generally combined with other substances in ores. Most metallic ores, however, are found in areas of igneous and metamorphic rocks where they have been concentrated from molten materials or by metamorphic processes.

It is rare, then, to find large deposits of metallic ores and mineral fuels in the same area of the world. The smelting and refining of metals, which calls for great quantities of fuel as well as ore rock, usually involves bulk transportation of either ores or fuels or both. In the United States, iron ore from metamorphic areas around Lake Superior, and coal and limestone found in sedimentary rocks of the Appalachians, are brought together at intermediate locations such as Chicago, Detroit, or Buffalo. In the Soviet Union similar large-scale transfers of ores and fuels take place.

ENVIRONMENTS AS WORKING SYSTEMS

The surface of the earth, as we have seen, is not just a passive platform upon which human life goes on. The earth environment can be imagined as a dynamic complex of machinery in which each machine performs a specialized part of the total work. We have seen something of the action of forces within the earth (diastrophic forces), of wind and water (atmospheric and hydrospheric forces), and of biological forces (the activities of

FIGURE 3-1 WORLD PATTERNS OF ELEVATIONS

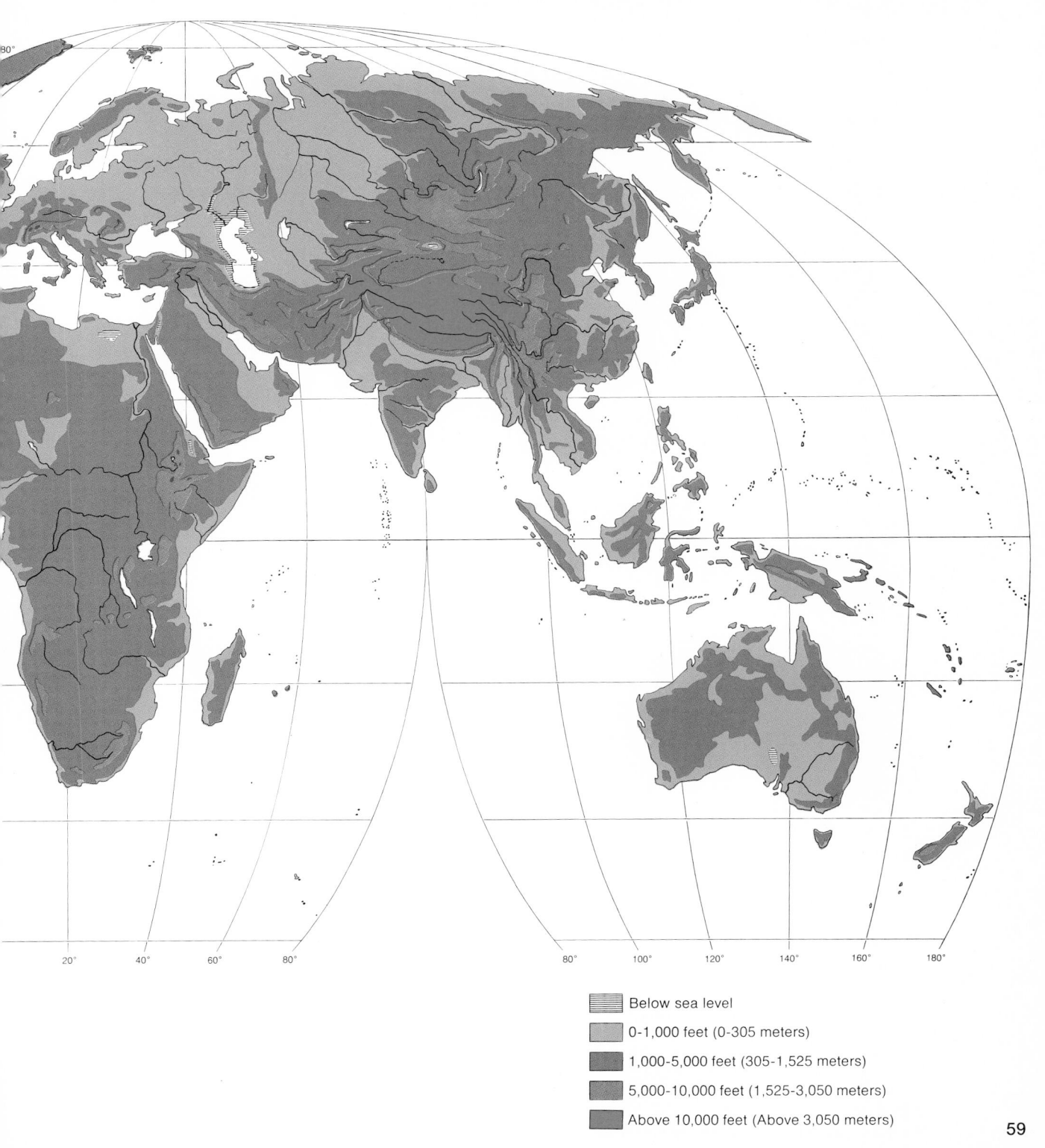

Below sea level

0-1,000 feet (0-305 meters)

1,000-5,000 feet (305-1,525 meters)

5,000-10,000 feet (1,525-3,050 meters)

Above 10,000 feet (Above 3,050 meters)

59

living things, from microorganisms in the soil to plants, animals, and humans). Our next task is to see how these environmental forces function as a working system.

Like all machinery, environmental processing requires energy. The principal source of energy of the earth environment is radiation from the sun. The amount of energy available varies from one part of the world to another, with latitude, and from season to season, and thus limits the work of the environment accordingly. The availability of water, without which very little environmental work can be accomplished, is also a limiting factor. Consider deserts, where the amount of solar energy can be high but where very little growth occurs. By contrast, where water is available as well as energy from the sun, biotic activity (animal and plant life), chemical activity, and the mechanical action of erosion all take place. The *combination* of energy and water is the driving force behind the "working machinery" of environments.

THE INTERACTION OF ENERGY AND WATER

Energy conditions at the earth's surface determine the potential limits to the amount of work that nature can accomplish in any particular environment. The actual work depends on the presence of water and how it is utilized. Energy causes water to evaporate from surfaces and to be transpired by plants. **Evaporation** of water from land surfaces occurs constantly as heat from the sun's rays converts water into vapor form. **Transpiration** is the process by which plants release water vapor into the atmosphere through the stomata (openings on leaf surfaces), which in turn causes them to take up water from soil through their roots. It is useful to combine evaporation and transpiration to obtain a single measure of the amount of water leaving the earth's surface: **evapotranspiration.** In a region of complete vegetative cover, normally more than 85 percent of evapotranspiration is accounted for by the transpiration of growing plants.

If we assume that there is unlimited water available for these processes in a particular environment, the maximum amount of evapotranspiration that can occur at that place directly depends on the amount of energy received from the sun. We refer to this maximum as **potential evapotranspiration** (PE). Most deserts are areas of high PE because of the large amount of solar radiation they receive. Deserts, however, are areas of low **actual evapotranspiration** (AE) because very little water is available to be evaporated or transpired.

Water that is not lost to evapotranspiration runs off the surface and contributes to erosion, transportation, and deposition of surface materials. Rather than contributing to growth and to the work of the environment as a biological system, this water participates in the movement of materials—soils and minerals, for example—over the earth's surface and in the reshaping of that surface. This work of **erosion** continuously changes surface configurations and is the source of materials deposited at the margins of the ocean basins. The rinsing action of water also modifies the composition of surface materials. As will be discussed later, these movements benefit human populations by removing wastes, supplying water, and regenerating soils.

THE WORLD HYDROLOGIC CYCLE:
AN OVERVIEW OF THE
ENVIRONMENTAL ENGINE

The world **hydrologic cycle** is a system that circulates water between land, atmosphere, and oceans and is powered by energy derived from the sun. It provides a very general overview of the work accomplished by energy and water at the earth's surface. The energy involved in the evaporation of water from both land and water surfaces furnishes the impetus for the work of the hydrologic cycle. For the world as a whole, as the values in Figure 3–2 indicate, more water evaporates from the oceans than falls back to them as precipitation. It follows, then, that more

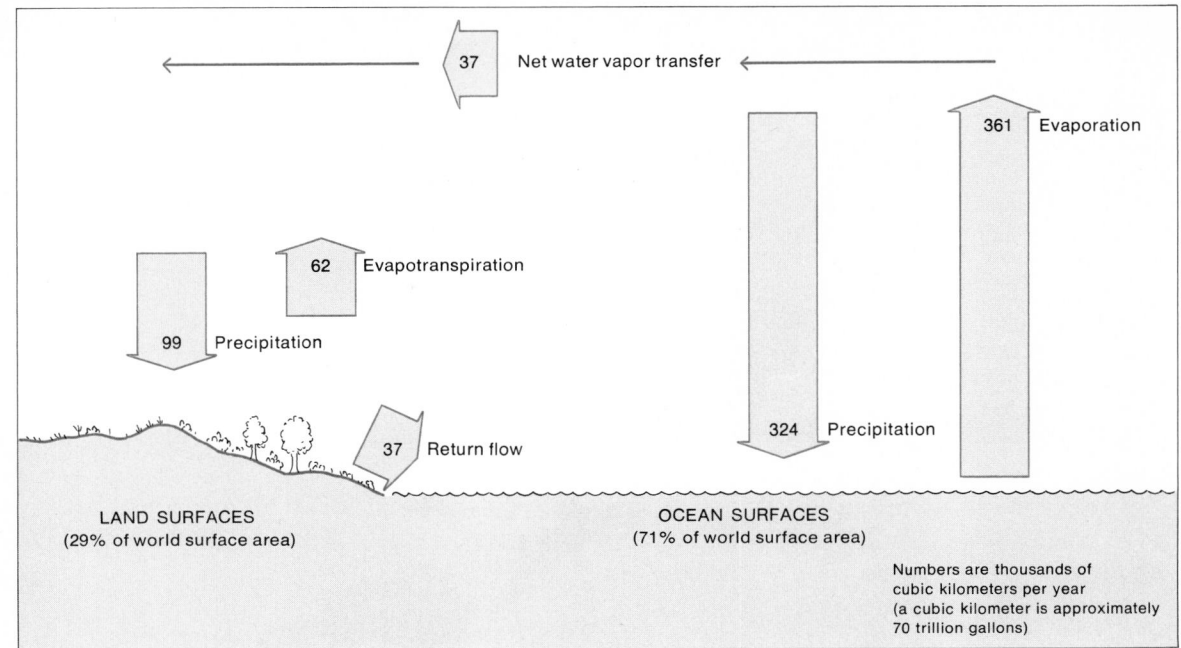

FIGURE 3-2 WORLD HYDROLOGIC CYCLE

water falls on land surfaces than is lost from them by evaporation. The additional precipitation on land surfaces from the atmosphere comes as a result of oceanic evaporation losses. The net transfer of water from oceans to continents is accomplished by movements of the atmosphere. The larger part of the precipitation received on land is soon returned to the atmosphere by evaporation and transpiration (evapotranspiration) by plants, while the smaller part not so lost becomes the surface and subsurface water of the land that runs off. Runoff returns to the oceans the water gained by continents from atmospheric circulation and thereby completes the world hydrologic balance.

The evaporation and transpiration of water transfers enormous amounts of energy and water into the atmosphere. When condensation of the water vapor occurs, energy and water for precip-itation are released into the atmosphere. The release of energy helps drive atmospheric circulation; the release of precipitation returns water to the earth's surface. Over the oceans precipitation only adds to their volumes, whereas precipitation on land surfaces performs work through the water that moves into the soil or across those surfaces in response to gravity.

Water reaching the land surface, therefore, goes to one of two possible destinations:

1. Some water goes to **soil-moisture storage.** It is stored in the small pores of the soil that retain it against the pull of gravity. The importance of water in soil storage is its availability to plants, whose physiological processes depend on the withdrawal of water from soil storage. It has been found that the growth of plants is strongly regulated by the amount of water they transpire. Thus water in soil storage and the energy to make

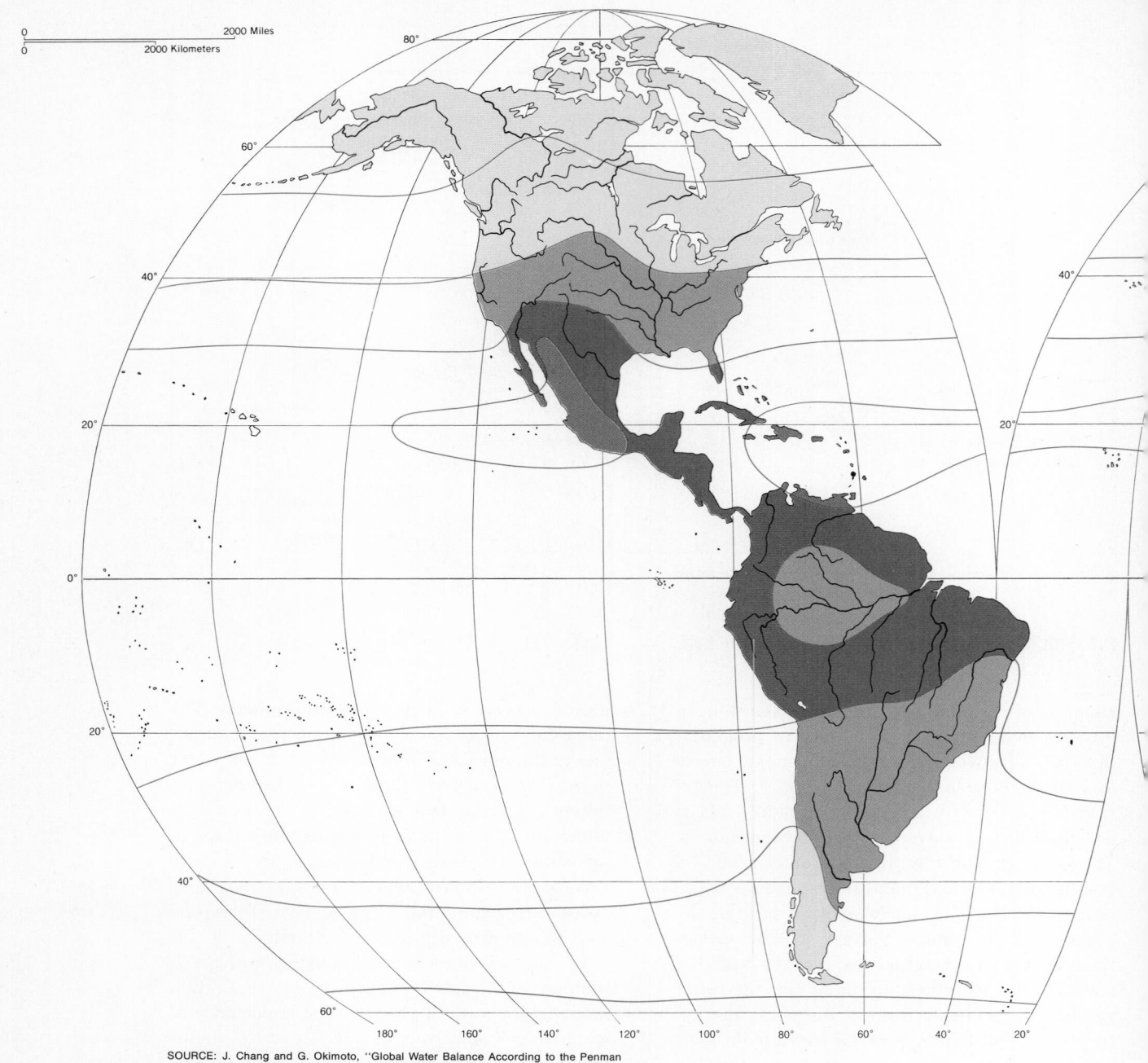

SOURCE: J. Chang and G. Okimoto, "Global Water Balance According to the Penman
Approach," *Geographical Analysis,* 2(1): 55–67, 1970. Reprinted by permission.

FIGURE 3–3 WORLD POTENTIAL EVAPOTRANSPIRATION

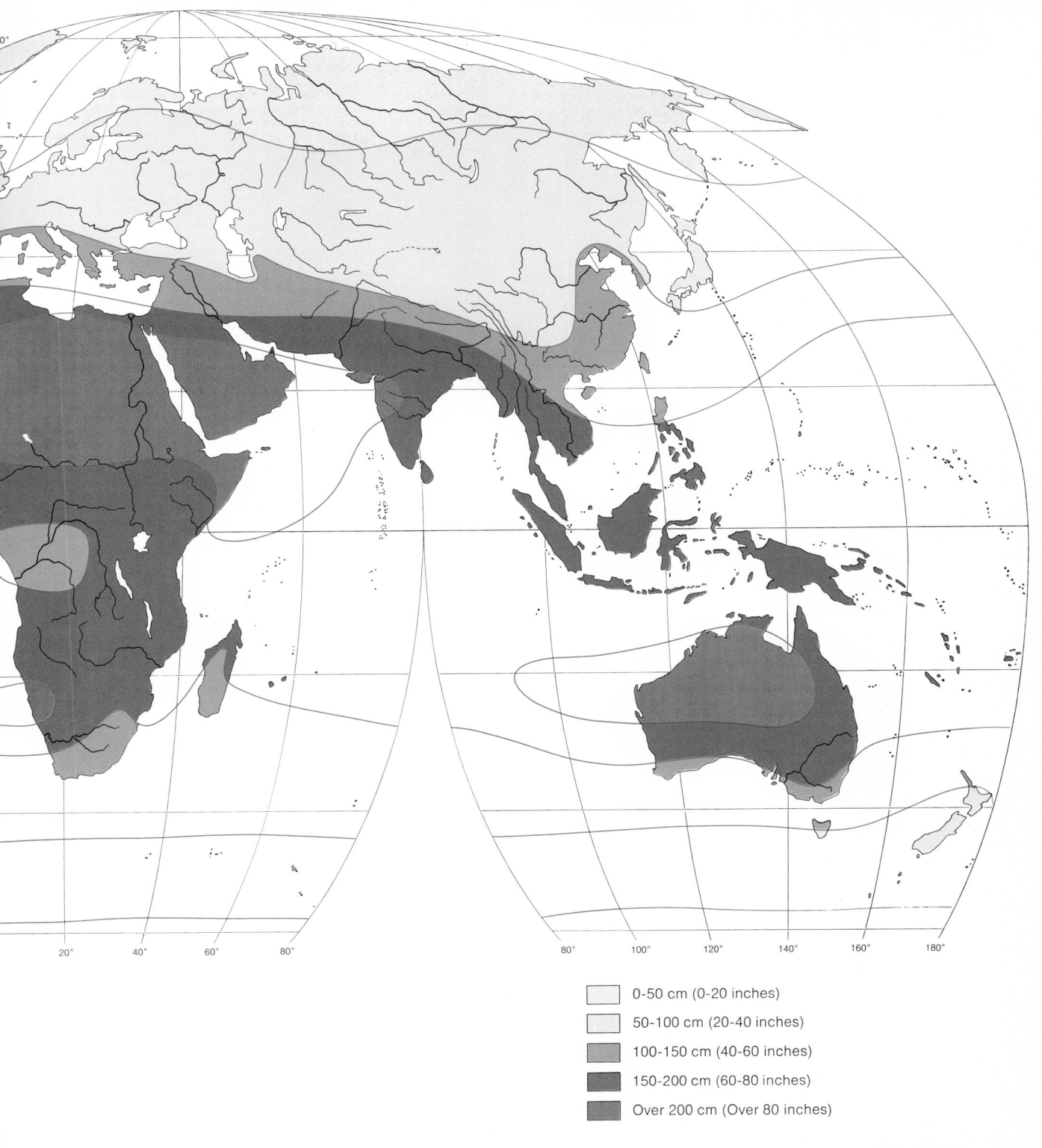

	0-50 cm (0-20 inches)
	50-100 cm (20-40 inches)
	100-150 cm (40-60 inches)
	150-200 cm (60-80 inches)
	Over 200 cm (Over 80 inches)

FIGURE 3-4   WORLD RAINFALL

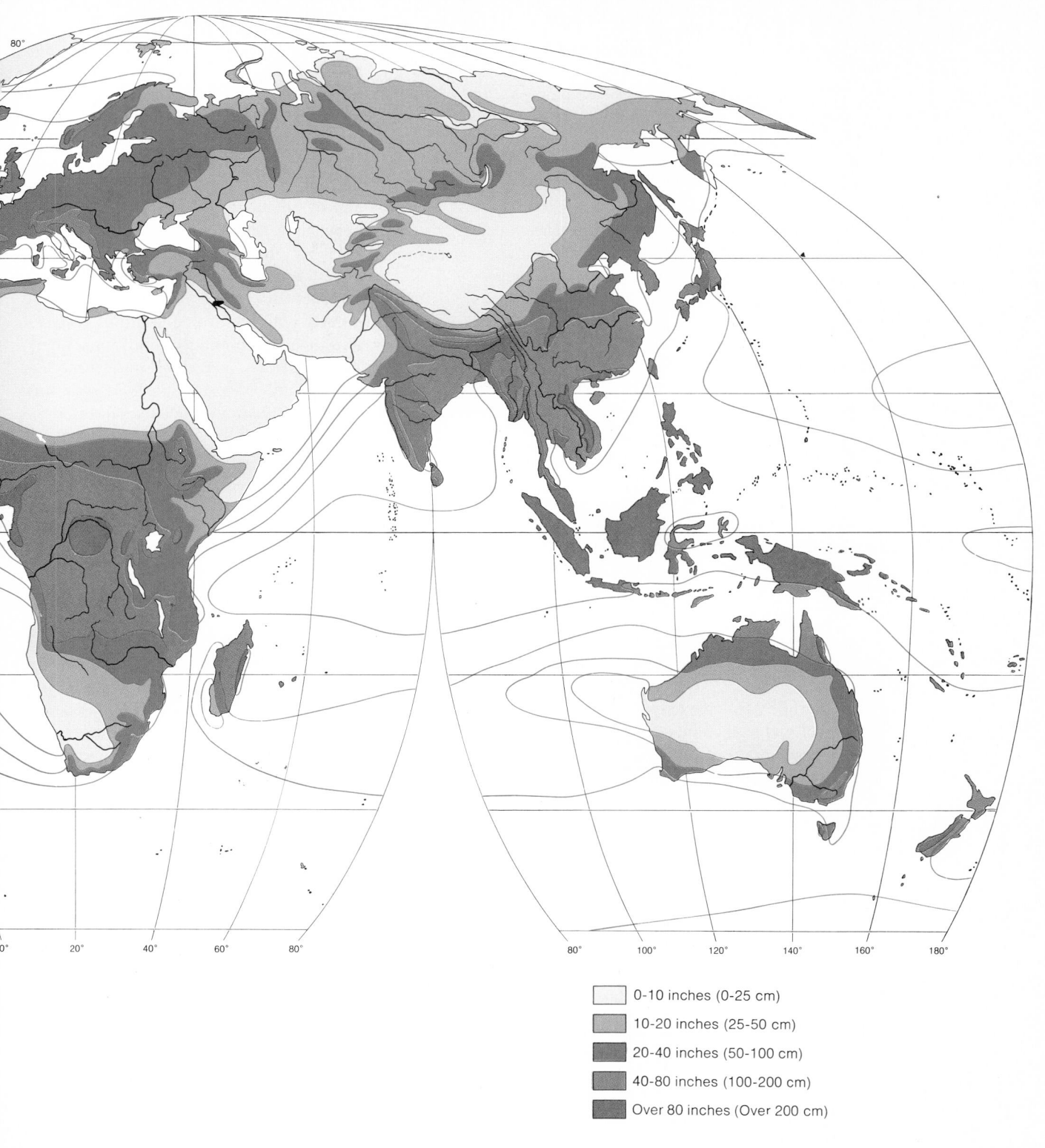

0-10 inches (0-25 cm)

10-20 inches (25-50 cm)

20-40 inches (50-100 cm)

40-80 inches (100-200 cm)

Over 80 inches (Over 200 cm)

plants transpire it are the primary factors in plant growth and productivity. While in the soil, water also functions as a solvent and is fundamental to the chemical work and changes that occur there.

2. Water that is in excess of the soil's capacity to store it or that moves rapidly across steep or bare surfaces, avoids soil storage and flows into streams, lakes, and ground-water reservoirs. This **runoff water** serves as our source of fresh water. Much of it is stored in underground deposits in porous rocks; there, as **ground water,** it becomes the source of water for household use through wells and springs. Water that does not go into ground-water reservoirs has energy to pick up and move materials as it flows across land surfaces. By eroding and depositing surface materials, it changes the surface configuration of the land. Runoff water that passes through the ground can dissolve materials and carry them away in solution. The work of the runoff component of the hydrologic cycle is found, therefore, in such processes as the sculpting of land surfaces and transporting of materials. A very important process of this type is **leaching,** in which the chemical constituents of the soil are dissolved and washed away by water. Direct benefits of runoff to human populations include removal of wastes, generation of power, and irrigation.

The world hydrologic cycle regulates the gross balance of water exchanges between oceans, the atmosphere, and land surfaces. But each particular place on earth has its own peculiar income and outgo of energy and water—the local **water balance**—which determines the amount and kinds of environmental work accomplished there.

MEASUREMENTS OF CLIMATIC CHARACTER

Data on the water balance, temperature, solar radiation, and other factors for a particular environment can be combined to describe the climate of that environment. Local measurements from a large number of places can also be combined to create broad climatic regions. The maps of potential evapotranspiration (Figure 3-3), rainfall (Figure 3-4), water surplus (Figure 3-5), and water deficit (Figure 3-6) will provide us with basic information for dividing the world into broad climatic environments. Let us study the data on these maps and see what generalizations they allow us to make.

The map of potential evapotranspiration (Figure 3-3) reveals a rather simple latitudinal pattern of relatively high values over a broad belt along the equator (from about 35° north latitude to 35° south latitude). The values of PE decline quite regularly with increasing latitude. High PE values, and hence high-energy conditions, dominate most of Africa, Australia, South America, and the southward-extending parts of Asia and North America. Substantial portions of the large land areas of the Northern Hemisphere have relatively low energy endowments. Notice, then, that in general PE varies with latitude. The tropics, with the highest energy totals, have the greatest potential for growing crops. The midlatitudes have about three-fourths as much energy, and the high latitudes less than half as much. To a large degree this variation reflects the increasing length of winter—the period of low PE—with latitude.

If we now look at the map of rainfall (Figure 3-4), we see that precipitation does not have a simple latitudinal pattern; many large areas with high PE have low precipitation. This is true for major portions of Africa, Australia, southwestern North America, and central and southwest Asia. These areas experience drought during nearly all months of the year. Because of the limited amount of water available in these environments, the actual evapotranspiration (AE) falls far short of the potential evapotranspiration (PE). The growing potential of these areas is extremely limited. The difference between the potential evapotranspiration and actual evapotranspiration is called the **water deficit:**

$$PE = AE + \text{Deficit}$$

or

$$\text{Deficit} = PE - AE$$

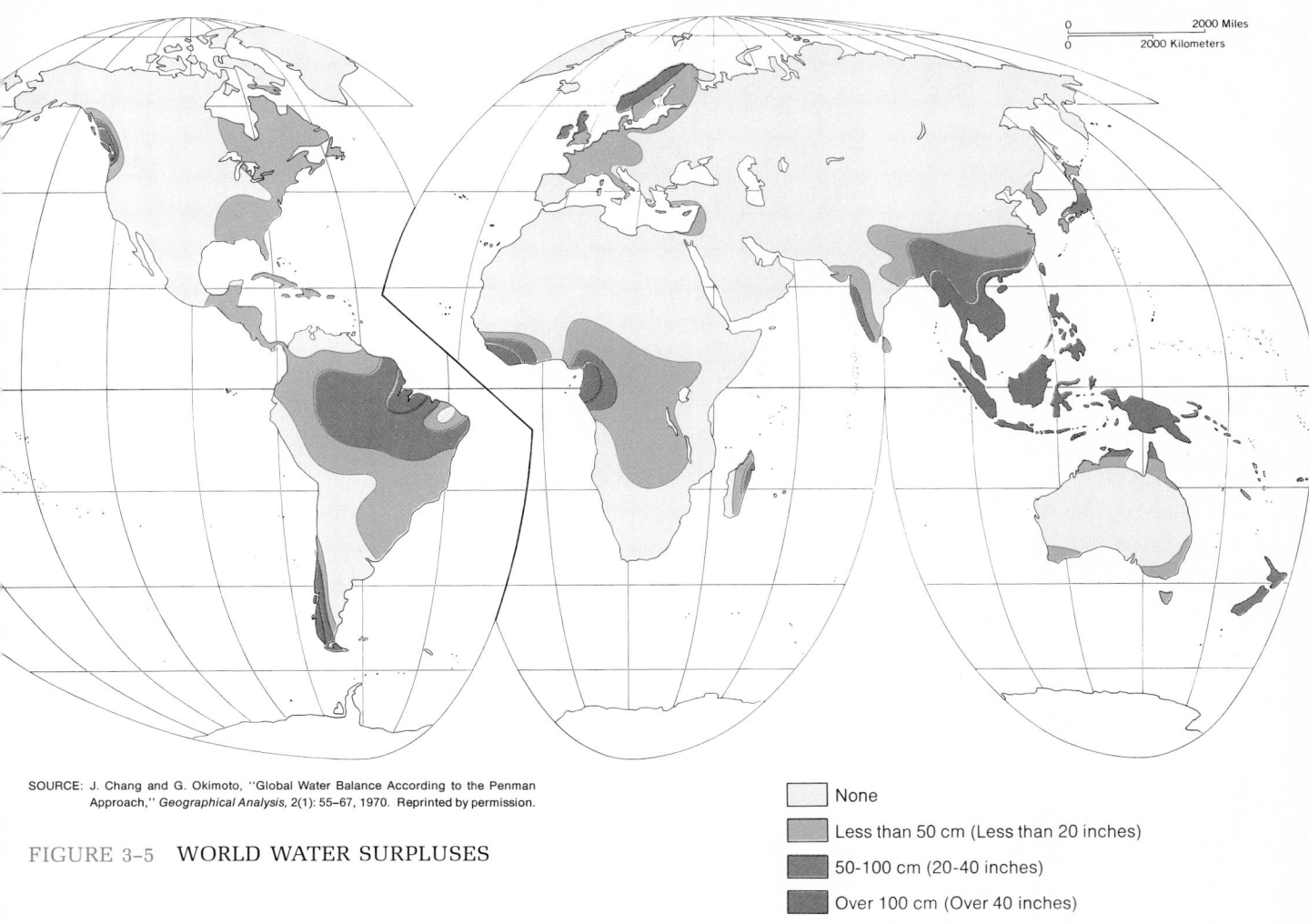

SOURCE: J. Chang and G. Okimoto, "Global Water Balance According to the Penman Approach," *Geographical Analysis*, 2(1): 55–67, 1970. Reprinted by permission.

FIGURE 3-5   WORLD WATER SURPLUSES

None

Less than 50 cm (Less than 20 inches)

50-100 cm (20-40 inches)

Over 100 cm (Over 40 inches)

Figure 3-4 shows high precipitation in most of South America, which also has relatively high PE. Under these conditions the actual evapotranspiration may equal the potential evapotranspiration. If there is more water available than can be used up by the processes of evaporation and transpiration, the excess will run off as a **water surplus:**

Precipitation = AE + Surplus

or

Surplus = Precipitation − AE

Figures 3–5 and 3–6 indicate the areas of water deficits and water surpluses throughout the world. Note the large deficits in most of Africa and Australia, the consequence of high potential evapotranspiration and low precipitation. In Figure 3–5 you see a high surplus in South America, especially in the Amazon Basin, an area of high PE and high precipitation. Surpluses also occur in equatorial Africa, South and East Asia, eastern North America, and Western Europe. A striking feature of the pattern of surpluses is their concentration in the tropics, where the potential

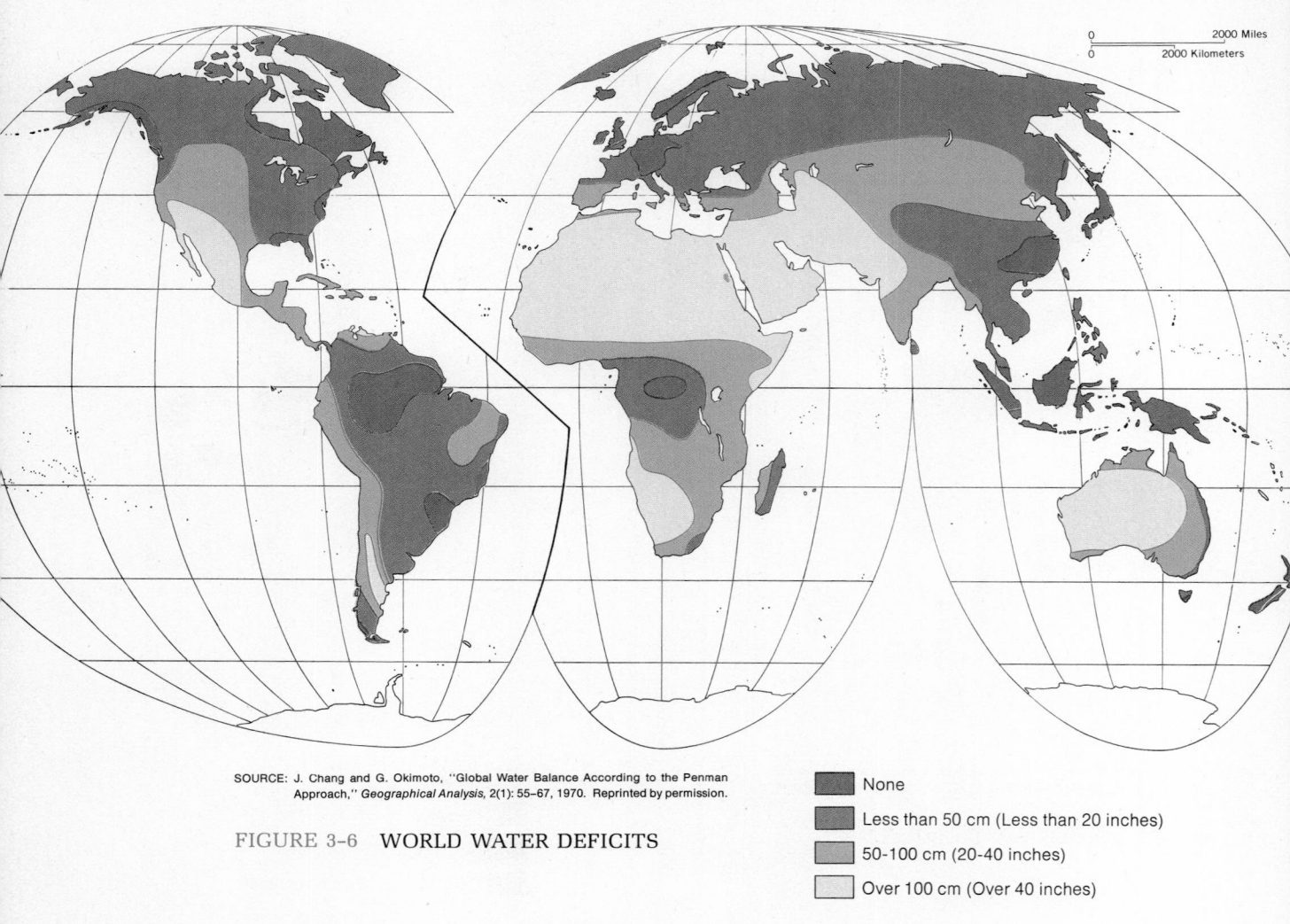

SOURCE: J. Chang and G. Okimoto, "Global Water Balance According to the Penman Approach," *Geographical Analysis*, 2(1): 55–67, 1970. Reprinted by permission.

FIGURE 3–6 WORLD WATER DEFICITS

■	None
■	Less than 50 cm (Less than 20 inches)
■	50-100 cm (20-40 inches)
▢	Over 100 cm (Over 40 inches)

for evapotranspiration is high. This means that the precipitation must be *very* heavy in these areas to offset the heavy losses of water by evapotranspiration.

Another way to generalize about the climatic environments of the world is shown in Table 3–1. This table presents estimated water balances for the continents. If you study the figures in the table, you will see that they agree with the above equations. Actual evapotranspiration and water deficit add up to the potential evapotranspiration

for each continent. Similarly, the total of surplus and actual evapotranspiration equals the precipitation. Note that *both* a water deficit and a water surplus exist for most parts of the world. This is not a contradiction, but simply a reflection of the fact that the deficit and the surplus for any particular environment occur at different times of the year. At one time of the year, precipitation is greater than the growing potential (PE), as in midlatitude winters; at times of high temperature, PE may be greater than the availa-

TABLE 3-1 ANNUAL WATER BALANCE OF THE CONTINENTS

(*In Centimeters*)

	Rainfall	Potential Evapo-transpiration	Actual Evapo-transpiration	Deficit	Surplus
Africa	74	188	65	123	9
Asia	54	104	43	61	11
Australia	49	167	41	126	8
Europe	60	70	47	23	13
North and Central America	70	120	63	57	7
United States	76	134	71	63	5
South America	151	146	111	35	40
All land	75	133	61	72	14

SOURCE: J. Chang and G. Okimoto, "Global Water Balance According to the Penman Approach," *Geographic Analysis*, 2 (January 1970): 55–67. Reprinted by permission.

ble precipitation. A deficit in one season cannot be erased by a later surplus, nor can a surplus be negated by a later deficit.

Note in Figure 3–6 and Table 3–1 the large areas of deficits throughout the world. We have already mentioned the deficits in large parts of Africa, southwestern North America, and central and southwest Asia. The great deserts of the world occur mostly between 40° north latitude and 40° south latitude and closely correspond to areas with deficits greater than 100 cm (40 in.). Above 40° latitude, deficits usually are not large, although short periods of drought do occur in the summer months. Can you explain why the summer should be the season in which droughts occur? Except for areas along west coasts, surpluses also are very small for these higher latitudes. The rather modest deficits of the eastern United States and Western Europe are worth noting.

The degree and pattern of water deficiency have important implications for agricultural land use because of the detrimental effect of water deficit on plant growth. Surpluses, on the other hand, offer the possibility of storing water for distribution during periods of drought. Yet the figures in Table 3–1 make it clear that except for South America, deficits exceed surpluses, often by a wide margin. Thus in most continents, even if elaborate water storage and transfer systems were developed, there would still not be enough precipitation to achieve the full potential evapo-transpiration.

WORLD CLIMATIC ENVIRONMENTS

For our purposes, the great variety of water balances over the earth can be reduced to three basic work budgets: low, high, and seasonal. Under seasonal work, two subtypes distinguish between those areas with a dormant season (a season with little or no work) and those without a dormant period but where the intensity and kind of work follow seasonal sequences. The generalized regional patterns of these climatic environments are shown in Figure 3–7.

SOURCE: Adapted from *Elements of Geography* by G. T. Trewartha, A. H. Robinson, and
E. H. Hammond. Copyright © 1967 by McGraw-Hill, Inc. Used with permission
of McGraw-Hill Book Company.

FIGURE 3–7 GENERALIZED CLIMATIC ENVIRONMENTS

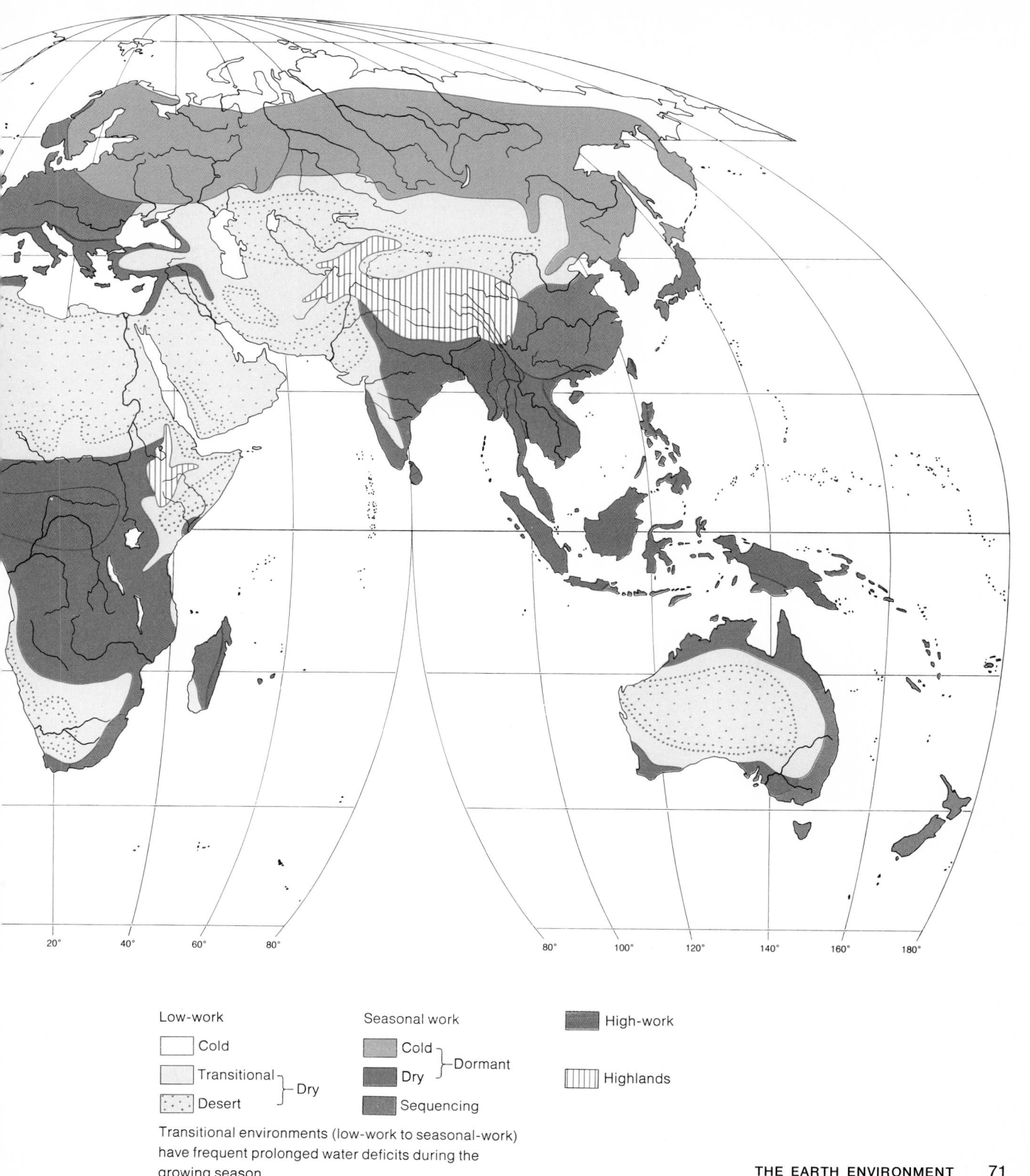

Low-work

- ☐ Cold
- ☐ Transitional ⎤
- ⦂ Desert ⎦ Dry

Seasonal work

- ☐ Cold ⎤
- ☐ Dry ⎦ Dormant
- ☐ Sequencing

☐ High-work

▥ Highlands

Transitional environments (low-work to seasonal-work) have frequent prolonged water deficits during the growing season.

LOW-WORK ENVIRONMENTS

Low-work environments are characterized by very short periods during which biological growth, chemical changes, leaching, or erosion are active.

In the cold climates, environmental work is confined to a short summer period when surface material thaws and temperatures are sufficient to activate vegetative growth. During this brief period, plants can carry out their reproductive and growth functions, the chemistry of the surface soil can be active, and movement of surface material may occur. Standing water and wet soils are widespread at such times because the subsurface is permanently frozen and impermeable to water movement. These conditions encourage mass movements of saturated surface materials (landslides) even on very gentle slopes. During most of the rest of the year the lack of energy (low temperatures) keeps moisture frozen, which in turn keeps all aspects of the environment inactive.

Dry climates are also characterized by very short periods when the environment can work. Since most areas of dry environment are in the middle and low latitudes, energy conditions are favorable for much if not all the year, and the infrequent precipitation events trigger environmental work. Following precipitation events, dormant vegetation can have a spurt of growth, seeds bring forth new vegetation, and limited chemical changes in the soil are possible. Precipitation, when it does occur in dry areas, is often intense and of short duration. The lack of vegetative cover leaves the ground surface exposed to erosion from runoff. The amount and size of material moved by individual storms in dry areas can be very impressive, even though such storms are very infrequent.

Neither the short summer of cold climates nor the infrequent moist periods of dry climates are attractive for human use of the land for food production or as a source of water supply. As Figure 3–7 shows, these environments cover ap-proximately two-fifths of the earth's land area—a substantial area that is unattractive for permanent settlement of large human populations.

HIGH-WORK ENVIRONMENTS

The opposite of low-work environments are those with high energy and moisture endowments throughout the year: **high-work environments.** Some areas within these environments have seasonal periodicities in their precipitation patterns, but times of inadequate moisture are short and the deficits are small compared to the moisture excesses (surpluses) that normally prevail. The persistently warm temperatures and ample water supply drive biological and chemical activity at high rates throughout the year. Biological decay is also very high and rapidly recycles organic matter on the surface of the soil. For most such areas precipitation exceeds evapotranspiration demand, leaving plenty of water to leach the soil and run off over the land surface. These environments, therefore, are source areas for large river systems such as those of the Congo and Amazon basins. High-work environments are limited to areas along the equator in portions of the Congo Basin of Africa, the Amazon Basin of South America, and the islands and peninsulas of Southeast Asia.

High-work environments would appear to be ideal for human use because of their high biological output and large water supplies. In fact, however, most of these areas are conspicuous for their sparse populations. The environmental conditions actually hinder agricultural activity. Without a cold season, growth is continuous and soil processes go on continuously. Thus the highly leached-out soils have poor chemical fertility, biological competition is high, and disease organisms and insects are many and varied. The sparse populations in most areas of high-work environments reflect these handicaps to agricultural production.

SEASONAL-WORK ENVIRONMENTS

Climatic environments that have at least one extended period of environmental work are called **seasonal-work environments.** For midlatitude areas seasonality limits all environmental work except for dry (low-work) areas. On the cold margins the work season is summer, when energy and moisture are at a maximum. Toward the drier margins the presence of moisture rather than temperature may be a more important determinant of the season of environmental work. For the warmer areas some type of environmental work goes on at all seasons, and seasons are distinguished by the type of environmental work rather than by the presence or absence of work. In tropical areas the season of work is determined by the timing of precipitation, since energy levels are adequate throughout the year.

Two distinctive types of seasonal-work environments are distinguished on the map (Figure 3–7). Some seasonal-work environments have a **dormant season,** during which little or no work occurs. In the higher latitudes this results from low temperatures, which freeze surplus water and make vegetation dormant. In subtropical and tropical latitudes the dry period causes soil moisture to be depleted, vegetation to go dormant, and water for runoff to be unavailable.

Other seasonal-work environments have **sequencing seasons,** during which the work goes on year round but the type of work differs from one part of the year to another. The characteristic sequence is described in detail in the box on pages 74–75. The essential distinction, however, is between the season of maximum biological and chemical work, which is usually the warmest part of the year, and the season of greatest leaching, runoff, and erosion, which normally occurs during winter and spring. These areas neither get too cold nor too dry to cause water and the local hydrologic cycle to cease work completely.

Areas of seasonal environments have most of the major concentrations of human population. In general, this is the consequence of an extended season of substantial biological activity that provides a basis for agricultural output. In addition, such environments provide seasonal surpluses of water that are available resources for human use. Because these seasonal environments have quite different work regimes, very different ways of utilizing them have been developed by different cultures. Even for environments of similar seasons of environmental work, different management schemes have been developed, as we shall see in later chapters.

NATURAL VEGETATION IN RELATION TO ENVIRONMENTAL WORK

Vegetation depends on the work of environments not only to provide the energy for photosynthesis but also to make nutrients available from soil and organic matter. Therefore one might expect the distribution of natural vegetation to resemble the pattern of world climatic environments. Studies of the productivity of vegetation have shown a positive relationship between productivity and actual evapotranspiration. The kind of vegetation, however, depends on other factors, especially moisture deficit. Most grasses and shrubs, for example, can tolerate extended drought by going dormant, whereas most trees cannot survive long periods of dryness. Consequently, grasses and shrubs are most prevalent in environments with extended dry seasons, such as the "transitional" environments of Figure 3–7, and trees tend to dominate in those environments where drought is uncommon during the growing season.

In latitudes where there are no great moisture deficits and where actual evapotranspiration is high during the growing season, trees are the dominant vegetation. Forests differ with latitude. Tropical forests, where actual evapotranspiration is high throughout the year, grow rapidly and produce the greatest amount of

ENVIRONMENTS AS SEQUENCES OF WORK

A **water-budget analysis** for a year using monthly data reveals the varying interplay of energy and water over the year. For example, low values for either potential evapotranspiration or precipitation for a season mean little environmental activity, whereas large values for both mean a great deal of activity. Environments of most places vary in intensities and particular sequences of work within their annual cycle. Therefore we can visualize an environment as a series of interrelated processes that vary in kind and intensity over the year, and not as a steady condition. From this perspective an environment gets its identity from the particular repetitive rhythms and changes that it goes through during the year. An analysis of the annual budget of one regional environment may be helpful in visualizing how energy, moisture, and soil-moisture storage interact. To illustrate, we shall describe the annual cycle of environmental activity representative of the sequencing seasonal environments shown in Figure 3–7.

The choice of a midlatitude environment for analysis is arbitrary, though relevant. Midlatitude areas experience annual cycles in their energy endowments, which allows us to study the effects of variable energy inputs and to clarify certain subtleties of the adequacy or inadequacy of moisture. For this analysis the water budget of Chicago (Figure 3–8 and Table 3–2) is used as a graphic summary, even though it is a borderline representative of the environment described.

The potential evapotranspiration curve shows that the distribution of energy for the year has one high period and one low period. The distribution of precipitation shows a summer maximum, but no dry period for the year. Precipitation for places within the sequencing seasonal environments, however, can be significantly different from Chicago in amount, timing, or both. For example, annual amounts of precipitation in other parts of the midlatitudes can range from 20 to over 60 in. (50 to 150 cm).

Winter

Let us begin our analysis at the low-energy season of the year, which is December or January in the Northern Hemisphere (June or July in the Southern Hemisphere). This is shown at the left of the graph in Figure 3–8.

With energy income (heat and light from the sun) at its lowest level, we can deduce that all processes directly dependent upon energy, such as material transformation or chemical processing, are also at their lowest levels of activity. All biotic activity (vegetative production in particular) is at its lowest level because most living matter is in a dormant state. We can also expect weathering of materials of the earth's surface to slow down considerably and the decay of organic matter to reach its lowest level. Evaporation of water and transpiration by plants are at minimal levels, and when the temperature drops below freezing these processes essentially cease. From this evidence it is apparent that the demand for moisture by evapotranspiration is at its lowest level.

These conditions give additional clues to other aspects of environmental work. For example, what happens to the income of moisture if the withdrawal by evapotranspiration is small? Under these circumstances might not surface runoff and soil leaching increase?

The surplus of moisture beyond evapotranspiration demand indicates that winter is the time when moisture is available for storage in the soil. However, it is also the time when the soil has reached its capacity for moisture storage. Therefore, once the soil is saturated, additional incoming moisture must either move deeper through the soil into the earth or run off over the surface. Water moving downward through the soil transports smaller soil particles physically or reduces them by chemical action. In the same way, water running over the soil erodes it physically. Translocational processes such as these can be active from the moment of soil saturation. However, during periods of freeze-up, leaching and surface erosion stop for lack of moving water. Thus we can clearly see that winter is a time of low overall environmental work.

Spring

With the approach of spring, energy and precipitation endowments normally increase. The atmosphere becomes warmer, which gives it greater capacity for water vapor storage and potential to bring about precipitation. Increases in energy levels also thaw frozen ground and release any moisture stored as snow and ice. Early spring, especially, is often the period of the year when the maximum moisture is available for leaching and erosion; it is also the time of maximum runoff and flood hazard. Nevertheless, spring is also normally the period of greatest fire hazard in midlatitude forest areas, because the absence of live vegetative cover permits a large proportion of solar radiation to reach the ground surface to warm and dry it. As a result, organic litter is dry and easily ignited.

A major change in environmental activity begins when energy levels are sufficient to stimulate renewed vegetative growth. Energy endowments by midspring usually have increased and accumulated to the point of being nearly double the winter lows. As a result, more and more moisture is lost to evapotranspiration, so that less and less is available for runoff and leaching. Soon evaporation and transpiration losses become so significant that some withdrawal of moisture from the soil may occur. Increasing soil temperatures accelerate chemical activity, which renews

the processes of mineral weathering and organic decay. This increased chemical activity necessarily accompanies increased biotic activity, which we see displayed dramatically in the leafing-out of plants. Vegetation in full leaf now insulates the ground surface from the direct effects of solar energy and moisture, and it drastically reduces the hazards of surface erosion.

As energy levels continue to rise, plant transpiration increases the demand on the moisture stored in the soil. The physiological demand for water from growing plants increases as photosynthesis increases. With this demand on soil moisture, some of the new moisture supply (precipitation) is entrapped in the soil, leaving less water as surplus for runoff. By late spring or early summer the volume of moisture withdrawn from the soil may approach the volume received from precipitation. This may be as much as 12 to 15 cm ($4\frac{1}{2}$ to 6 in.) of water per month. Probably, moisture receipt at this time of year will scarcely be in excess of what can go into soil-moisture storage, and moisture available for leaching and stream flow will approach zero. Under these circumstances there is little likelihood of erosion.

Summer

The steady rise in energy levels and biotic activity culminates in summer, when mineral weathering and organic decay (the **transformational processes**) are most active. This is also a time of decreased erosion, ground-

water recharge, stream flow, flood hazard, and leaching of solubles (the **translocational processes**). These conditions generally intensify through the summer until, by middle or late summer, the potential demand for moisture by evapotranspiration normally exceeds moisture income. Thus summer is often marked by frequent droughts, dry streams, low ground-water tables, and general water shortages, even though the actual amount of precipitation received at the surface may be at or near its peak. Moisture deficits in the soil slow down plant growth and, if they prevail for long, cause plants to go dormant and possibly die.

Fall

By late summer and early fall, soil moisture is at its minimum, and the capacity of the soil to store incoming precipitation is at a maximum. But, because the fall season is a time of decreasing energy levels compared to summer, the demands for moisture also decrease. Thus, during the fall, evapotranspiration demand drops below income from precipitation and the moisture in the soil is gradually replenished. For a month or two the opportunities for moisture to go into storage leave little if any surplus for runoff and ground-water recharge, thus minimizing such processes as leaching and erosion. Lowest stream flows usually occur at this time as a result of the combined effects of depleted ground-water stocks and the recharging of soil-moisture storage.

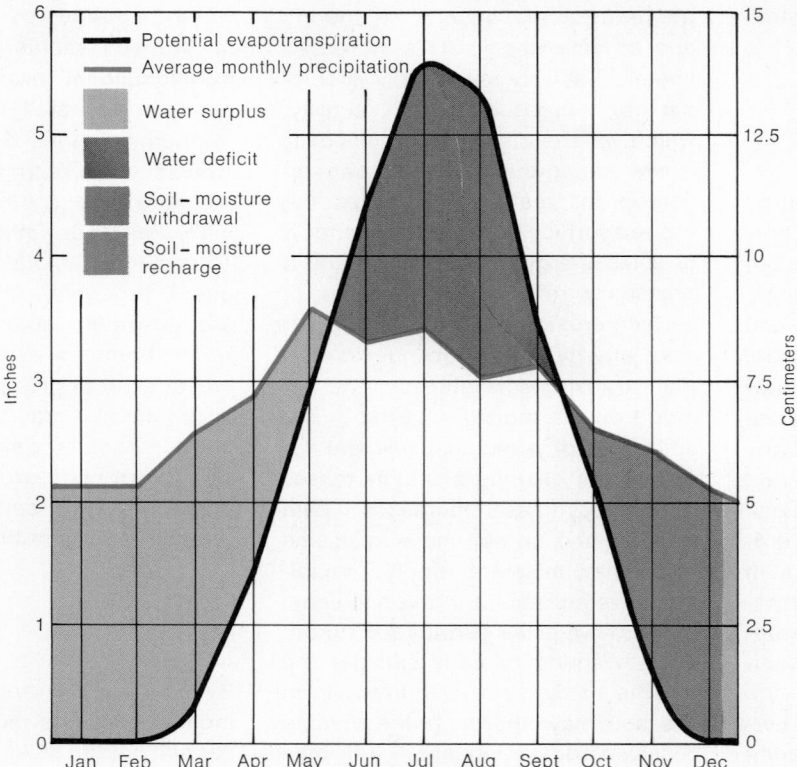

FIGURE 3–8   WATER BUDGET FOR CHICAGO

vegetative matter (see Table 3–3). They are composed of a great variety of trees, most of which
are evergreen. In the midlatitudes, forests are
made up of fewer kinds of trees, and most trees
are deciduous, that is, they lose their leaves during the winter season (see moist temperate forests
in Table 3–3). In the higher midlatitudes, forests
are still more uniform in species composition,
most trees are evergreen and coniferous (cone-
bearing, like pine and spruce), and growth is
quite slow (see boreal coniferous forests in
Table 3–3).

Grasslands are found in areas where there is
frequent moisture deficit during the growing season. Small trees, usually widely spaced, may also

be part of the vegetative landscape in these
areas. By going dormant temporarily or by con-
serving and storing water to maintain life until
the moisture supply is adequate again, grasses
and shrubs can adapt to seasonal moisture
deficits.

In the deserts, where there are persistent, severe moisture deficits, plant cover cannot be
maintained. Much bare rock and soil is in evidence. Only specialized plants adapted to brief
and infrequent periods of moisture are found
here. Shrubs with thick woody branches and
waxy leaves dominate, and grasses and flowering
plants adapted to rapid growth appear during the
brief period of moisture after a desert rain.

TABLE 3-2 A WATER BUDGET FOR CHICAGO, ILLINOIS

	Jan.	Feb.	Mar.	Apr.	May	June	July	Aug.	Sept.	Oct.	Nov.	Dec.
Temperature												
Celsius	−4	−3	2	8	14	20	23	22	19	12	5	−1
Fahrenheit	25	27	36	48	57	68	73	72	66	54	41	30
PE												
Inches	0	0	0.3	1.4	3.0	4.5	5.6	5.3	3.4	2.1	0.6	0
Centimeters	0	0	0.8	3.6	7.6	11.4	14.2	13.5	8.6	5.3	1.5	0
P												
Inches	2.1	2.1	2.6	2.9	3.6	3.3	3.4	3.0	3.1	2.6	2.4	2.1
Centimeters	5.3	5.3	6.5	7.4	9.1	8.4	8.6	7.6	7.9	6.5	6.1	5.3
AE												
Inches	0	0	0.3	1.4	3.0	4.5	5.6	3.6	3.1	2.1	0.6	0
Centimeters	0	0	0.8	3.6	7.6	11.4	14.2	9.1	7.9	5.3	1.5	0
D												
Inches	0	0	0	0	0	0	0	1.7	0.3	0	0	0
Centimeters	0	0	0	0	0	0	0	4.3	0.8	0	0	0
S												
Inches	2.1	2.1	2.3	1.5	0.6	0	0	0	0	0	0	0.4
Centimeters	5.3	5.3	5.7	3.8	1.5	0	0	0	0	0	0	1.0
SMS[a]												
Inches	4.0	4.0	4.0	4.0	4.0	2.8	0.6	0	0	0.5	2.3	4.0
Centimeters	10.0	10.0	10.0	10.0	10.0	7.0	1.4	0	0	1.2	5.6	10.0

[a] The storage values used a conversion of 2.5 cm = 1 in. in order to keep the assumed soil-storage values in round figures of 10 cm and 4 in. The calculated storage, however, is consistent with the budgetary accounting of the table in which the conversion was based upon 2.54 cm = 1 in. Hence, some inconsistency does occur between centimeters and inches for the months of June, July, October, and November.

KEY: PE = potential evapotranspiration. P = precipitation. AE = actual evapotranspiration. D = deficit. S = surplus. SMS = soil-moisture storage.

Polar areas have no moisture deficit, but actual evapotranspiration is so low and the growing season so short that plant growth is severely restricted. Such areas are called **tundra,** and the dominant vegetation is ground-hugging lichens with some grasses and flowering plants that appear during the short summer. Trees are few and stunted, and much of the surface appears to be barren of plants.

The data for gross primary productivity (Table 3–3) provide a useful measure of one dimension of environmental work (although they disregard other aspects of environmental work, such as erosional activity). Gross primary productivity is the total amount of vegetative mass produced annually, without reductions for losses by decay. Note that the world average is ten times the output of the low-work environments and about

TABLE 3-3 ESTIMATED GROSS PRIMARY PRODUCTIVITY FOR
SELECTED VEGETATIVE GROUPINGS

Vegetative grouping	Gross Primary Productivity (kcal/m²/yr)
Deserts and tundra	200
Grassland, pastures, and dry forests	2,500
Boreal coniferous forests	3,000
Moist temperate forests	8,000
Wet tropical and subtropical forests	20,000
Average for biosphere of the world	2,000

SOURCE: Eugene P. Odum, *Fundamentals of Ecology,* 3d ed., W. B. Saunders, Philadelphia, 1971, p. 81.

one-tenth the output of the high-work environments. In other words, the high-work environments have one hundred times more gross primary productivity than do the low-work environments.

SOIL AS A DIMENSION OF WORLD ENVIRONMENTS

The preceding discussion has emphasized that climatic energy and moisture provide the power for making physical processes at the earth's surface, including the processes of biological life, do their work. Keep in mind, however, that these processes work on materials and that biological life, of which we are a part and upon which we depend for our basic needs of food and many other products, consists of organisms structured out of minerals and elements drawn from their surroundings. The primary sources for the structural constituents of biological life are gases in the atmosphere, water, and minerals at the earth's surface. Since the gaseous composition of the atmosphere is relatively uniform worldwide, it has little direct bearing on the variable pattern of biological activity over the world. The pattern of water availability is determined by the pattern of climatic environments, reviewed in the previous section. The pattern of presence and availability of different minerals, particularly the mineral nutrients needed by plants, is related to that of the climatic environments. However, important differences can affect potential biological productivity, and we need to know them to understand the potentials and problems of agriculture in different parts of the world. This section presents a very broad overview of the world pattern of soil characteristics that are relevant to agricultural practices and problems. Of necessity it ignores much of the complexity and range of local soil differences. By drawing attention to a few important soil properties and then coupling their distribution to that of climatic potentials, we can obtain a useful world overview of the environmental side of agriculture—its opportunities and its problems.

THE IMPORTANCE OF SOIL FOR PLANT LIFE

Soil is important to biological existence because it is the immediate source from which plants draw most of the water they use and many of the essential **nutrients** that they need to build tissue and carry on physiological processes. The chemical elements that are essential to the survival of plants are called plant nutrients. Plants may differ somewhat in their requirements for essential nutrients, but all plants need relatively large quantities of seven elements, the **macronutrients**—oxygen, hydrogen, carbon, nitrogen, calcium, potassium, and phosphorus. In contrast, they normally need eight to ten other essential nutrients in much smaller quantities—the **micronutrients.**

Plants can obtain oxygen, hydrogen, and carbon directly from either the atmosphere or water. The inorganic minerals of the soil plus **organic debris** are the major sources for calcium, potassium, and phosphorus. Organic matter (often called humus when decayed) is the principal source from which plants obtain their nitrogen needs. Most plants cannot directly assimilate free nitrogen of the atmosphere, and because of its high solubility, mineralized nitrogen is not normally present in the mineral soil. Micronutrients are obtained primarily from the inorganic materials of the soil. Thus aside from oxygen, hydrogen, and carbon, essential plant nutrients are drawn mostly from the mineral and organic contents of the soil, and if these sources are poor in macronutrients, then plant growth and productivity are constrained.

When soils are used for agriculture, especially cropping and pasture, their natural stock of available nutrients declines and sooner or later productivity is reduced. As a rule, soils that are naturally rich in plant nutrients are easier to keep productive using appropriate **fertilization** practices than are those low in nutrient supplies. Therefore nutrient-poor soils often require quite different agricultural practices than do more fertile soils. In some places the limitations of soil nutrient conditions make repeated cropping difficult if not impossible in spite of fertilization. This is the case in some high-work environments, like the Amazon and Congo basins, where as a result of leaching the soil contains few nutrients.

THE INFLUENCE OF CLIMATIC ENVIRONMENTS ON SOIL CHARACTERISTICS

In general, the availability of the macronutrients calcium, potassium, and phosphorus is a function of the degree of weathering and leaching. The work of **weathering,** which encompasses the physical breakup and chemical decay of rocks and minerals, is closely related to the amount of actual evapotranspiration. As AE increases so do the rate and intensity of weathering, especially chemical weathering (decay). In areas of high year-round AE, such as the hot, wet tropics, soil weathering is at a maximum, whereas in the low-AE environments of desert and polar areas, soil weathering occurs very slowly. Soils under seasonal environments have intermediate degrees of weathering.

Chemical weathering, which involves breaking the molecular bonds of minerals, is most intense within the top few feet of soil, and it is within this rather shallow zone that most of the individual plant nutrients are released from their mineral structures. Weathering thereby makes plant nutrients available by releasing them from rock minerals and organic matter. As long as these nutrients remain bound in mineral and organic complexes, plants cannot utilize them.

When the nutrient elements are released they become vulnerable to solution and removal by water moving through the soil (leaching). The amount of leaching is a function of surplus water (water in excess of that stored in the soil or lost by evapotranspiration). Environments that have large year-round or seasonal surpluses make leaching very active. Soils under these conditions are usually poor in available plant nutrients.

Because organic matter is the primary source from which nitrogen is released, the availability of nitrogen to plants depends on organic decay (weathering). Since organic decay is a function of the AE level, areas with high AE have rapid organic matter decay and their soils are usually low in organic matter content. If leaching is also high, soils tend to be poor in available nitrogen.

Soils of the Seasonal Environments

Soils of seasonal environments have a combination of characteristics that have made them more suitable for agriculture than those of other climatic environments.

1. The rate of weathering is sufficient to release a continuous supply of nutrients from inorganic and organic materials but not so rapid as to quickly deplete the source materials for nutrients. Soils developed under low-work environments are hindered by too little weathering, and those under high-work environments have too drastic weathering.

2. Like weathering rates, leaching rates are sufficient to remove constituents from the soil, including plant nutrients, to keep the soil chemistry active but not so rapid as to completely deplete the supply of nutrients. Soils in the sequencing and tropical seasonal environments are often quite well leached of their nutrients and need fertilization.

3. The annual production and decay of organic material produce a continuing supply of nitrogen and, because the growing season is normally not a time of leaching, this nitrogen supply is available in the soil for plants to use.

4. Although their clay content can vary greatly, the soils of seasonal environments are likely to have chemically active clays that can retain positively charged plant nutrients. Soils of the sequencing and tropical varieties of seasonal environments as a rule have more clay content but are poorer in their nutrient-storing capacity. The lower clay content of soils under cooler or drier conditions is offset by their greater nutrient-storing capacity.

The most desirable soil conditions for growing most of the crops and grasses common to midlatitude agriculture are found in the subhumid margins of the seasonal environments. The soils of the wheat-growing areas of the Great Plains in North America, the pampa of Argentina, and the Ukraine in the Soviet Union exemplify such areas, which initially were naturally very rich in plant nutrients, have high organic matter contents (nitrogen supply), and have high capacities for storing nutrients. These ideal soil conditions reflect a combination of active weathering and slow leaching that conserves an abundant supply of available plant nutrients in the soil. The agricultural assets of soil in these areas are offset to a considerable extent, however, by relatively low precipitation and frequent droughts.

Soils of the Low- and High-Work Environments

The positive evidence for the more favorable characteristics of soils under seasonal environments can be emphasized by considering some of the drawbacks for agriculture of soils under low- and high-work environments.

1. Rates of weathering under low-work environments are very slow and create poor nutrient conditions for plants. In cold environments the release of available plant nutrients is slower than leaching losses, leaving soils infertile. In the dry environments weathering, though slow, is greater than leaching. Where little or no leaching occurs the products of weathering can accumulate to the point of being detrimental to plants. Most agricultural crops cannot tolerate an overabundance of some salts, particularly those containing sodium. In high-work environments rapid rates of weathering and very active leaching produce soils very poor in plant nutrients.

2. Nitrogen supplies are low in soils of both the low- and high-work environments because organic matter conditions are unfavorable. Although some organic matter accumulates on the soils of cold environments, its slow rate of decay and poor quality produce limited nitrogen. In dry environments there is very little organic matter and its rate of decay is exceedingly slow. By contrast, high-work environments produce a large amount of organic matter but rapid decay and leaching leave little nitrogen in the soil.

3. Soils of low-work environments are normally poor in clays. This deprives them of the chemically most active part of the soil and thus the capacity to retain nutrient ions from loss by leaching. Soils of high-work environments are normally dominated by clays, but they are types that for the most part are chemically inert and have very little capacity to retain nutrient ions. Thus in any of these environments the important functions of the clays are mostly missing.

In low-work environments, not only are energy and moisture conditions a hindrance to agriculture, but the soils have important deficiencies as well. In high-work environments the poor nutrient conditions and generally inert character of soils are major deterrents to crop agriculture.

Soils That Depart from the Pattern of Climatic Environments

Such factors as poor drainage, excessive erosion, recent deposition (alluvial soils), and initial materials impart special characteristics to the soil, such as coarseness of texture or exceptional richness of plant nutrients. Such soils are widespread and are of local significance within the broad soil regions described above. Seasonal-work environments, for example, have some areas of soils that are quite unsuited for agriculture because they are deficient in nutrients or water-holding capacity; some areas may even lack soil itself. But other areas of soils in low- and high-work environments do respond well when cropped. These are mostly alluvial soils in the dry environments

and soils derived from materials rich in nutrients in the high-work environments. For our world overview we will consider only the extensive areas of alluvial soils, soils with poor drainage, soils developed on nutrient-rich materials, and areas of little or no soil.

Areas of **alluvial soils** are often intensively used for agriculture. In fact, river floodplains and deltas are the prime agricultural lands in China, India, and peninsular Southeast Asia. Other important areas of alluvial soils are along the Tigris–Euphrates lowland, the deltas of the Rhine and Po in Europe, and the extensive alluvial lands along the lower Mississippi, the lower Colorado, and in the Central Valley of California. Here again the pattern suggests that alluvial landscapes of the midlatitudes have the most suitable agricultural soils.

Alluvial soils consist of deposits derived principally from the topsoil of upland areas. Usually, alluvial soils are also deep and flat-lying, attributes that are very attractive for agriculture. They are not always well endowed with plant nutrients, and flooding and poor drainage are liabilities that may hamper their use for agriculture.

Soils developed under conditions of poor **drainage** are very often rich in plant nutrients, which results from accumulation of organic matter and slow loss or perhaps even gain of soluble nutrients. If the drainage problems can be solved, these soils are excellent for agriculture. Much of central and northern Europe, and the area north of the Ohio and Missouri rivers in North America, have sub-

stantial areas of soils developed under poor drainage left from glacial deposition in the recent geologic past. Both locations now contain major areas of agriculture.

If surface materials are especially rich in the macronutrients, they will usually weather into soil abundantly endowed with these nutrients. Certain basaltic rocks of relatively recent volcanic origin have developed into soils rich in macronutrients. Where these occur in wet tropical areas, especially, they are conspicuous as unusually fertile areas compared to the much poorer surrounding soils. Parts of peninsular India, the island of Java in Indonesia, and patchy areas in eastern Africa are noted for these soils. In India and Java, particularly, these soils are intensively used for agriculture and support dense populations.

Areas of little or no soil cover are very widespread and can be mostly related to any one of three primary factors. Desert areas are especially barren of soil because there is so little weathering. Very steep slopes, especially in mountainous areas, also have thin or stony soils and much bare rock exposed by erosion. A third area with little or no soil cover occurs in northern North America and extreme northern Europe and Asia, where recent glaciation has removed most of the weathered surface materials and insufficient time has passed for soil to redevelop. These are also areas with environments that have long dormant seasons, making the rate of soil development slow.

In dry environments organic matter content of soils is extremely low and nitrogen deficiencies prevail because of the scarcity of live vegetation.

The **clay content** of soil also can be an important indication of potential agricultural capabilities. Clays are the smallest particles of soil and, along with organic matter, are where most of the chemical activity of the soil takes place. The chemical activity of clays results from their small size and capacity to attract and store positively charged ions (individual atoms) of plant nutrients. Organic matter has this same characteristic. Nutrients stored in this manner resist loss by leaching but are readily available for uptake by plant roots. High storage capacities are generally advantageous to agriculture because fertilizers are retained better by the soil.

Clay content can also indicate the soil's capacity to store moisture, especially for midlatitude soils. As a rule, as the clay content increases up to about half the volume of soil solids, so does the soil's capacity to hold water available for plants. As clay content increases beyond half the volume of soil solids, soil's water-holding capacity tends to decrease. High clay content impedes water movement, thus encouraging waterlogging, a process highly detrimental to plant growth.

The clay content of soil is related to weathering. The greater the rate and intensity of weathering, the higher the clay content will generally be. The degree of weathering also affects the nutrient-storing capacity of clays. As a general rule, the greater the degree of weathering, the lower the nutrient-storing capacity of the clay. The net result of these two relationships is that the clay content of midlatitude soils is less than that of soils in tropical areas, but the clays of tropical soils tend to be much less active chemically and have much lower capacities to store nutrients. Within the midlatitudes the clay content is greatest in soils of the warmest and most humid areas and decreases in drier and colder environments.

The work performed by climatic environments on the surface materials of the earth results in a soil cover that varies in quality with respect to conventional agricultural activity. An appropriate general statement about the variable quality of soil over the world is that soils of the seasonal environments, especially those of the midlatitudes, have been the most adaptable to agricultural use. Conversely, soils of high- and low-work environments have been the least adaptable to agricultural uses.

PEOPLE AS PART OF THE PHYSICAL ENVIRONMENT

During the first decades of this century, geographers placed great emphasis on the study of the physical environment as a basis for understanding the geography of the human world. According to a widely held view of the time, the differences in the human condition from one place to another over the world were thought to be the result of differences in the local climatic environment and other land resources. This concept placed the physical environment in a dominant or even determinant role with respect to people's material well-being and means of livelihood.

Such a view was perhaps understandable in the years before rapidly expanding technology made its impact. Until recently most people lived on the land as agriculturalists of one type or another. Management of the environment was almost entirely dependent on human and animal power (biological energy). Under these conditions the scope for human alteration of the environment was narrow and thus the local environment set the limits within which people had to work.

Over the past fifty years, however, advances made by human technology have shown that local land resources no longer need to be the only basis of progress. Modern technology based on machines and inanimate energy, such as energy from petroleum fuels, has offered new capabilities to exploit and to modify the physical environment. As a consequence, some people are awed by technological accomplishments and are persuaded that there are practically no environmental constraints—that manipulation and control of the environment through technology is unlimited. Such a position is questionable because of limited supplies of inanimate energy and the limitations imposed by our biological characteristics, which demand that the overall environment must not deviate substantially from its long-term norms in order for our species to survive.

Human use of the physical environment has given rise to several serious problems, some of which we shall discuss later in this chapter. In preparation for that discussion, let us see how the three basic types of working environments—low work, high work, and seasonal work—lend themselves to human management.

ENVIRONMENTS AND HUMAN MANAGEMENT

The different work patterns of environments at particular places present different opportunities and problems. Many aspects of human living involve responding to particular environmental conditions. This is perhaps most obvious in areas with traditional lifestyles and a near total dependence on local biological productivity to sustain life. The climatic environment also affects the management of resources in modernized societies, though more subtly.

Low-Work Environments

Because change occurs very slowly in low-work environments, they might be called the storage environments of the world. For example, during the Korean conflict, World War II aircraft stored in the Arizona desert were quickly reactivated because they had undergone little deterioration. Environments of this type tend to preserve human artifacts and thus offer great potential for the archaeologist. On the other hand, they are not places for us to dispose of our wastes because whatever we put there will be preserved, not decayed. Deserts might be reasonable places to leave nuclear wastes because they would not be

moved by environmental action there. In contrast, when nuclear wastes are dumped in the oceans, the circulation of contaminants is a constant hazard.

The low activity of these environments also means that they have limited capacity to recover quickly from disturbances. Plant growth will not re-cover an exposed area. Roads across the fragile alpine and arctic tundra and wagon trails across deserts leave their marks long after their last use. Undoubtedly, the recent exploration of the North Slope of Alaska for oil will leave an indelible imprint in that environment for generations to come, because the surface has been cut by truck and tractor routes, and patterns of vegetation have been disrupted.

Obviously, such environments could never sustain large populations if the people had to depend on local resources. Bringing in outside resources or reorganizing available ones can, however, provide opportunities for larger numbers of people to thrive there.

Irrigation represents one such reorganization of resources in dry areas. It depends on using water from natural streams or on schemes to relocate water. By adding water to the soil, the generally high potential energy of the environment is put to use by increasing evapotranspiration. Greater biological growth is usually the goal. Nevertheless, irrigation management is not a simple matter of putting water into the soil. The chemical reactions activated by the added moisture set in motion a sequence of changes in the soil. Weathering increases and alters the chemical and physical character of the soil.

A common practice in irrigation is to add more water than needed for evapotranspiration so that the accumulating products of weathering and erosion leach away. Often, however, this practice causes groundwater levels to rise and leads to further undesirable consequences. Because of these problems, irrigation has usually been applied not to desert soils, but to soils transported to, and deposited in, arid regions by streams from more humid areas. Alluvial lands such as floodplains, deltas, and alluvial fans in dry areas have been found the most suitable sites for irrigated agriculture. Their soils are more stable than desert soils under the moist conditions of irrigation and are the easiest lands to manage.

In spite of problems such as irrigation, current successes in some dry and cold environments suggest that these environments offer opportunities for the settling of urbanized populations. An example is the spectacular growth and development in the dry southwestern part of the United States. Vigorous efforts have also been made to develop the resources of the low-energy environments of arctic Alaska, Canada, Europe, and the Soviet Union. At the moment, such development is largely a matter of creating facilities for importing most of the things needed to support human existence. Iceland, however, has been able to feed its population by harnessing its thermal energy (hot springs and geysers) to grow agricultural produce, including bananas, in hothouses.

What problems can we foresee from large-scale settlement of these low-work environments? To make intelligent estimates, we need to know more about the effects of various manufactured wastes on the environment. The strong seasonal fluctuations of environmental activity also influence the design of buildings and create specific problems for road and runway maintenance. The controversy associated with the construction of the oil pipeline in Alaska emphasizes the uncertainties of our knowledge in the use of this type of environment.

High-Work Environments

Since a lack of biological resources is a constraint to human settlement in cold and dry environments, it would seem reasonable that an abundance of biological resources in environments with high energy and moisture endowments (humid tropical areas) would attract human settlement. In most such environments, however, population densities are low or even sparse. Does this mean that humanity has overlooked an obvious opportunity, or do these environments pose major problems for sustaining human populations?

It is apparent from the pattern of population in tropical areas (see Figure 5–4) that the higher densities are largely limited to areas with seasonal, not year-round, precipitation. Where moisture income equals or exceeds potential evapotranspiration for all or nearly all months of the year, population densities are low. The Amazon and Congo basins are outstanding examples of this. The vegetative growth in these areas is large, yet in general people have not successfully used this seemingly rich environment. Grandiose efforts, even when backed by modern technology, have often resulted in equally grand failures. Is there any logical reason for the problems in these high-work environments?

Some researchers have laid most of the blame on poor soils. The high-work performance of the environment has caused very active chemical decay of the surface material, and the surpluses of moisture have leached most of the released solubles away. Thus the potential and actual stock of nutrients for plants have been removed from much of the soil in these areas. How, then, can the rich forest cover exist? The answer is in the continuous recycling of those nutrients tied up in standing and decaying vegetation. Essentially, the vegetation lives off its own debris **(detritus).** Dead residue of vegetation is rapidly broken down into basic constituent minerals under high-energy and high-moisture conditions, and as rapidly as these constituents are released, the living vegetation takes them up again. As a result, the standing vegetation is the source of nutrients. The highly weathered soils play only a minor role in supplying plant nutrients under these circumstances.

In this situation standing vegetation and its residue, not soil, are the primary sources of nutrients to growing plants. When vegetation is removed in the tropics, the most available stock and source of plant nutrients in the plant-soil system is also removed. Thus when clearings are made for agriculture, the stock of nutrients in the vegetation is released by burning and decay; but because of harvesting and rapid leaching by the environment, these nutrients are soon lost from the soil and productivity of crops drops. Usually within two or three years a clearing is exhausted of plant nutrients and is abandoned.

A possible solution might be to add nutrients obtained from other sources such as fertilizers. Although this procedure is perhaps technically feasible, it poses special management problems. Highly weathered soils are unable to store the mineral nutrients in the form provided by fertilizers and needed by plants. Leaching removes fertilizer nearly as fast as it is added. By comparison, the less weathered soils of the midlatitudes have much greater capacities for retaining nutrients. Under the high-work conditions of the tropics a logical plan might be to add a little fertilizer every few days instead of larger amounts once a season or every few years, as is done in the midlatitudes. Frequent application of fertilizer, however, is costly and may not be practical.

Management of nutrient supplies for crop agriculture is not the only problem in high-work environments. Insects, diseases, and other plants flourish under these warm, moist conditions. Natural tropical vegetation is noted for its great diversity and variety, a circumstance that acts as a natural check to the spread of insects and diseases with their own special habitats. However, fields of a single crop, which are easy to manage, also make the spread of diseases and pests easy. Unwanted plants also spring up rapidly, competing for nutrients and sunlight. Each of these factors compounds the problem of increasing food output by conventional agriculture.

Another possible reason for sparse human habitation in the humid tropics is that decay and production are continuously occurring at approximately equal rates. There is no dormant period. In order to acquire food from plants, people must quickly gather the constant flow of products before it is consumed by other organisms or chemical decay. These circumstances are in clear contrast to the orderly and manageable cycles of production and decay of the seasonal environments, where most of the world's food production occurs.

Seasonal-Work Environments

Compared to year-round, high-work environments, the management of midlatitude environments has proved easier for agriculture. In environments with seasonal sequences there is a growth phase in spring and summer, a dormant or storage phase in winter, and a decay phase during the following spring and summer. Little can happen during the winter storage phase, which is why decay is delayed to the next year. Such a sequence also means that products of the growing season mature more or less simultaneously, usually in midsummer or by late summer or early fall before the dormant season begins.

Such a timetable is convenient because plant production becomes available at a time when competition from other organisms or chemical decay is at a minimum. Most grains, vegetables, and many fruits are easily kept over the winter even without elaborate technology. Meat, too, can be stored, although not with equal success. Thus people get to consume much of this production during the winter without competition from decay. This gives them first access to the energy stored in the production phase. Waste of consumption is then recycled by the environment during the following growing season when decay processes are active again.

Moreover, the sequence of environmental work makes management easier. People can adapt their activities to those of the season: a time to plant, a time to cultivate and protect crops, a time to harvest, and a time to rest and prepare for the next cycle. In the high-work environments, though productivity per unit area may be greater than in the seasonal midlatitudes, there is no sequence to environmental and biological activity and the cultivator's activities overlap in time.

In this context it might be suggested that seasonal environments offer people an easier setting for managing food production than do the high-work environments. An intriguing idea is that the timing and the amount of environmental work may limit the potential for certain forms of human management, particularly agriculture. This idea suggests that resource management schemes that have worked well in one environment may be quite inappropriate for another; we may therefore be making a fundamental error by applying midlatitude agricultural practices in tropical environments.

HUMAN SPATIAL ORGANIZATION AND ENVIRONMENTAL WORK

In this section we shall consider the two models developed in Chapter 2—the modern interconnected system and the traditional, locally based society—as they relate to the environmental processes analyzed earlier in this chapter. For each model we shall explore how the human spatial organization is related to the work and seasonality of environments.

In general, traditional societies, like Ramkheri, are largely confined to exploiting and modifying the existing physical environment. Modern societies, on the other hand, go beyond their natural biological environment and use other chemical and physical processes. They also reach out beyond local areas for material resources of other environments wherever they may be found. That is why modern societies can produce so much more material wealth than traditional societies, but it is also why they have such a large capacity to pollute the world's environments and why they are rapidly depleting the world's most useful resources.

Traditional Society and the Management of Local Environments

Since a traditional society depends on the biological productivity of its own local environment, it would appear that the greater the natural productivity of the environment, the better off the local population will be. Undoubtedly, environments of low biological productivity have less potential to support populations of traditional societies than do environments of high productivity. This fact is reflected in the relatively low population densities of areas with low biological productivity. However, the material well-being of people in traditional societies does not solely depend on the environmental base, because people can use different approaches to their local environment to yield different returns. We can gain a better understanding of environments as life-supporting systems by inspecting the distinctive people-environment relationships. The most common ways in which traditional societies manage their local environment include gathering, hunting, herding, and crop cultivation.

Gatherers, hunters, and herders The few societies in the contemporary world that still depend on gathering, hunting, and fishing scarcely manage their environment at all. Their members are shrewd scavengers and opportunists, gathering whatever the wild plant and animal communities produce at any given time and bringing down whatever their skills and weapons allow them to kill. These people survive by their practical knowledge of their local resources. They must know the seasonalities of different foods and the local habitats and migratory habits of animals.

Though gathering and hunting peoples are often considered primitive, their understanding of the local wild environment is hardly simple. Eskimos and Bushmen are contemporary representatives of hunting and gathering societies who have learned to exploit expertly the limited food resources of low-work environments.

Societies based on herding are less vulnerable to starvation because they maintain stocks of food and other necessities in the form of tended animal herds. However, they still must provide adequate forage for their animals and they must know where and when feed and water will be available. In environments of low or seasonal productivity, herders learn to move from place to place in order to harvest the local natural vegetation at appropriate seasons. Typically, they move to moist, mountain pasturelands in summer when lowlands are parched, and then go down into lowlands, where temperatures are warmer, to take advantage of fall, and possibly winter, plant growth. Some herders practice food storage, such as drying hay for use as feed in the winter.

Cultivators Cultivators, by far the most numerous group in traditional societies today, have developed practices that combine managing the type of plants grown with some modification of the local natural environment. Management involves reducing the competition from unwanted plants by clearing them away by means of fire and cutting, and then carefully tending the desired plants. Even the chosen crop is thinned so that the most desirable plants are favored.

Cultivators also have learned that the land and climate of most natural environments have deficiencies that limit crop productivity. The most common techniques for modifying these limitations of the environment are the use of manures to return nutrients to the soil and the development of irrigation works to add water to the land in dry areas or dry seasons.

The food crops grown by the processes of plant selectivity and land modification depend on people to protect and husband them. Otherwise, most could not survive the threats from other plants, diseases, and insects. Many could not reproduce themselves. But by domesticating plants and applying some relatively simple technology to their management, cultivators have gained a degree of control over their food supply not possible for gatherers, hunters, or herders. As a rule, cultivating societies have denser populations and higher standards of living than do other traditional societies.

Modern Spatial Organizations and Environments

Perhaps the most obvious and outstanding attribute of the modern world is the concentration of people in metropolitan centers. Viewed in the context of the work potential of the local environment, how can so many people sustain themselves in so little space? Obviously, they obtain very little of their food or other material needs locally.

In the modern spatial organization, this apparent detachment from the local environment is made possible by transportation links and trade arrangements with other producing areas. For example, by establishing links with a variety of places with growing seasons at different times, it is possible to obtain a supply of fresh produce year round, or at least to modify the seasonal rhythms of food supply. Similarly, this outreach of the modern interconnected system gives it a form of immunity from local failures and food shortages. The immunity is reinforced by the technology of food storage, in which overproduction in good years can be kept available for use at other times. Thus the food habits of modernized societies are no longer as seasonally dependent as those of people who rely on their local environment for foodstuffs.

Modern technology has also made possible greater management control over agricultural productivity. For example, phosphorus, calcium, and nitrogen can be added as fertilizers to the soil. Such developments give modern societies much wider prospects for exploiting their physical environments. In addition to the basic resources for food production, modern technology has developed the capability of using a wide spectrum of the earth's physical and chemical processes. This is evident in the wide variety of materials and products available today that did not exist a century or even a few decades ago.

Clearly, modern metropolitan populations depend on efficient transportation and communications for their resource needs. But does the spatial organization make metropolitan dwellers independent of their local environments? The answer to this question may become clear after we explore the effect of the local environment on other aspects of our daily lives.

THE GEOGRAPHY OF YOUR FOOD SUPPLY

To investigate the extent to which your diet involves foods that are seasonal in production, even though they are available to you year round, list seasonal food items and the months when they are harvested. Also, list foods that you know are not grown in your general area, and identify the environments appropriate for their production. Locate these environments on a map, and then, using whatever additional information you are able to find, develop a geography of your food supply. If others in the class have done this, you might compare results. No doubt there will be conflicting opinions. To what extent are you able to resolve these conflicts?

You might also identify foods that need special preparation to preserve them and those that can be stored easily. How do drying, salting, and pickling retard decay? What environmental processes do they arrest?

PROBLEMS CAUSED BY PEOPLE

People must sustain their biological and material existence from the raw products and the energy of the physical environment. Human life ultimately rests on continuous consumption of the materials gained from the earth's physical environments. People clearly cannot live independent of the physical earth. The physical environments we know today are not only used by people, however; they also carry the influence of people. Hence for our purposes, human populations and their activities can be viewed as integral parts of the environment.

Changes people make in physical environments are usually piecemeal and designed to attain specific, limited goals. Initially, each change may seem to modify only the local environment, but with the increasing use of technology, the variety and magnitude of such changes have increased to the point where modifications of environments go well beyond local limits. A recent example is the localized use of low concentrations of pesticides such as DDT. These concentrations have spread, via the processes of the environment, to all parts of the earth's surface. DDT, like many manufactured pesticides, resists breakdown in nature, and once it gets into the food chain, it can become highly concentrated in some animals. Some of these animals live far from the areas where DDT is actually used. DDT contamination of animal food sources may prove disastrous for people, too.

Neither the stability nor the wide diffusion of these pesticides was anticipated by the people using them, which indicates that human modifications of environments for human purposes often cause new conditions that are unanticipated and that can become undesirable because they are not compatible with the biological requisites of human life. This seems to be the crux of current concern over environmental quality.

THE CONSEQUENCES OF TECHNOLOGY

We must be concerned about what is happening to the earth's environments as we apply our technological power to create more wealth. But the use we make of environments and their resources depends more on human ingenuity and technology than on the environment itself. It is therefore important, when we study the human-environment relationship, to be aware of mutual **feedbacks** between people and the environment.

People make material and other demands on the environmental resource base for their physical existence. They manipulate environments in order to survive. In recent times their manipulations have often been on a large scale: long-distance water diversions, chemical farming, mineral exploitation, stripping and reshaping of the surface, and outpouring of large quantities of combustion wastes. People have also brought about less grand but very widespread alterations to vegetation, soils, and water quality. All such changes to environments affect the well-being and continued existence of all humans. Since we cannot escape from our built-in biological dependence on a compatible physical environment, we must make sure that our prowess in exploiting our physical resources does not change the overall world environment very much from long-standing norms.

The interdependence of environment and people becomes more apparent with every increase in material abundance. Greater exploitation of the earth's resources causes depletions and shortages, waste disposal problems, and increasing hazards to health from water pollution and air pollution. Until recently we have been ignorant of, or tended to overlook, the problems raised by our pursuits of greater material wealth through use of the physical environment. A specific example of what happens from ignorance of such problems can be seen in a study of our use of inanimate energy.

The material affluence of modernized societies is directly related to the lavish use of inanimate energy, as is discussed more fully in Chapter 4. At present the most widely used sources of energy are derived from fossil fuels such as coal, oil, and natural gas. The earth's stocks of these fuels are limited, yet the demand for them continues to rise. We expect that alternative sources can be found. However, most of the alternative sources being considered are going to be very costly to find and develop. The costs in energy used in developing these sources may in some cases be almost as much as the value of the energy extracted. Also, extraction and development activity can cause potentially serious changes to local and regional environments.

An especially good example of the problem is **oil shale.** Shale is a fine-grained sedimentary rock composed of silt and clay. In some areas—notably the Green River formation in Wyoming—it contains organic material that can be crushed and heated to release energy-rich chemical compounds; this type of shale is known as oil shale. It appears now that the costs in energy necessary to exploit oil shales may equal or exceed the energy that can be recovered from them. In addition, the stripping of land, water loss and decline in water quality, and pollution from processing the shale will drastically alter, perhaps devastate, the physical environments of the mined areas.

It is becoming ever more apparent that hasty implementation of new technologies, such as oil shale exploitation, can create more problems than it solves. Piecemeal approaches by modernized societies to the problems of energy, mineral resources, and pollution have relied on technological "solutions" that may cause more damage than benefits through long-term degradation of the environment and of the welfare of people. Thus we need to increase our understanding of how environments work and to anticipate the long-term consequences of our resource management schemes.

LOCAL ENVIRONMENTS AS DISPOSAL SYSTEMS

How do cities get rid of their smoke, noxious gases, excess heat, and other pollutants? Most depend on the atmosphere to disperse these wastes. Because cities are so compact, however, there is a limit to the amount of waste that can be scattered without changing the local characteristics of the atmosphere or threatening people's health.

Another problem of congested areas is sewage disposal. The output of sewage systems is usually dumped into local streams in the hope that it will be carried away. The stream offers a low-cost removal system for the city dumping the waste, but cities downstream incur purification costs if they depend on the same stream for their water supply. Also lost to all along the stream are the fishing, swimming, and enjoyment of beauty that go along with clean natural streams.

Solid wastes also cause problems in congested areas. Much solid waste is **biodegradable** and decomposes rather readily at the soil surface, where it is then easily washed away or incorporated into the soil. However, the amount of biodegradable material that the local environment can process is limited. Moreover, most of the solid wastes of cities are made of materials that resist decay. The rates of decay of these materials are usually much slower than the rates at which they are being dumped. It is therefore no surprise that cities are finding it difficult to dispose of the wastes produced by affluent urban living.

These examples point to a basic fact: concentration of population at a location means that wastes concentrate there, too. The enormous capabilities of modern societies to produce and consume material wealth, especially in the metropolitan centers, has created a flow of raw materials in one direction, to the locations of human concentrations. Such a flow depletes the source areas by the activities of mining, lumbering, erosion, and soil tillage, and makes the cities into dumps for wastes from manufacturing operations, energy consumption, and consumer junk. Can such a system, geared to providing a greater and greater abundance of material wealth, be sustained without the development of an adequate means for handling wastes? Probably not, because the inadequate disposal efforts of most U.S. cities are even now a growing financial and environmental problem for them. Cities are going to have to restructure waste disposal, and in doing so they might well emulate the biological system. This requires reuse of wastes as raw materials. Domestic sewage, for example, is a source of plant nutrients, and much of the solid waste could be reused as a substitute for new raw materials from mines and forests.

Although modern societies have now come to recognize the need to maintain environmental quality, most efforts so far have involved the application of still more technology. Yet, as we have seen, it is the untested application of new technology that is in part responsible for degrading environmental quality. Must modern societies use less or more technology in order to manage their pollution and waste disposal problems? Perhaps we should reconsider our material desires and reorder our needs as a first step toward reducing environmental changes.

SELECTED REFERENCES

The Biosphere, A Scientific American Book. W. H. Freeman, San Francisco, 1970. A series of articles addressed to the cycles of the biosphere and human dependence and use of the processes of the biosphere.

Carter, Douglas B., Theodore H. Schmudde, and David M. Sharpe. *The Interface as a Working Environment: A Purpose for Physical Geography,* Technical Paper No. 7. Association of American Geographers, Commission on College Geography, Washington, D.C., 1972. Develops the case for approaching the environment at the earth's surface as a working system. Examples of how this working system is regulated by energy and moisture and how it interrelates with human activities are explored.

Chorley, Richard J., ed. *Introduction to Geographical Hydrology.* Barnes & Noble, New York, 1971. A collection of articles dealing with specific aspects of water resources and human management of them.

Commoner, Barry. *The Closing Circle.* Alfred A. Knopf, New York, 1971. Evaluates the role of the technology of modern production systems in environmental pollution. The case is presented in nontechnical language that is easy for a reader to follow.

Darling, F. Fraser, and John P. Milton, eds. *Further Environments of North America.* Natural History Press, Garden City, N.Y., 1966. Presents papers (some very specific, some general) and discussion given at a conference of future environments of North America sponsored by the Conservation Foundation. The papers treat a very broad range of topics and issues concerning environments and society. Collectively, the material touches on a wide spectrum of problems concerning people and our use of environments.

Energy and Power, A Scientific American Book. W. H. Freeman, San Francisco, 1971. A series of articles on the flows and uses of energy in the universe, on earth, and by different societies and technologies.

Foin, Theodore C., Jr. *Ecological Systems and the Environment.* Houghton Mifflin, Boston, 1976. In Chapters 1 to 3, discusses environment in terms of ecological systems and includes useful supplementary reading to the material of this chapter. Chapters 6 to 14 are also relevant in their consideration of environmental stresses caused by resource utilization and waste disposal from growing populations and by the increasing affluence of our advanced technologies.

Man and the Ecosphere, Readings from *Scientific American* with commentaries by Paul R. Ehrlich, John P. Holdren, and Richard W. Holm. W. H. Freeman, San Francisco, 1971. A collection of articles covering a wide range of subjects. Some primarily concern aspects of environment and some are mostly concerned with questions of human management of resources.

Manners, Ian R., and Marvin W. Mikesell, eds. *Perspectives on Environment.* Publication No. 13, Association of American Geographers, Commission on College Geography, Washington, D.C., 1974. A selection of review papers that represent both breadth and depth of environmental problems of interest and concern to geographers. Articles 1, 2, 3, 4, 7, 8, 11, and 12 are especially appropriate to the ideas developed in this chapter. The extensive bibliographies ac-

companying each article offer a useful guide to additional reading and study.

Miller, David H. *A Survey Course: The Energy and Mass Budget at the Surface of the Earth.* Publication No. 7, Association of American Geographers, Commission on College Geography. Washington, D.C., 1968. Illustrates that the principles of mass and energy exchange and transformation can be used in understanding the functioning of urban, rural, and wild land parts of the earth's surface. Each of the eight units consists of a brief introductory overview or frame of reference and a comprehensive annotated bibliography. An excellent reference source for problems and research concerning the environment and human involvement with it.

Miller, G. Tyler, Jr., *Living in the Environment: Concepts, Problems, and Alternatives.* Wadsworth, Belmont, Calif., 1975. Develops basic concepts of the physical environment and its operation, such as the laws of thermodynamics and the structure and function of ecosystems, in a manner that articulates with the problems of population, resources, and pollution. Chapters 2, 3, 4, 5, 6, and 9 are especially relevant to the ideas developed in this chapter. Twenty enrichment studies offer a wide assortment of in-depth looks at specific problems of humanity and our physical environment. The book effectively integrates the environmental problems confronting humankind.

Trewartha, G. T., A. H. Robinson, E. H. Hammond, and L. H. Horn. *Fundamentals of Physical Geography,* 3rd ed. McGraw-Hill, New York, 1977. A concise, basic introduction to the elements of climate, landforms, vegetation, and soil that make up the physical system of environments at the earth's surface.

THE EARTH ENVIRONMENT AS A WORKING SYSTEM

Farmer in India using a plow drawn by bullocks. (CARE photo)

Above: Eskimos of St. Lawrence Island in the Bering Sea hunting walrus. (George Gerster, Rapho/Photo Researchers, Inc.)   *Below:* A segment of the Trans-Alaska pipeline. (Ledree/Syzma)

The contrasts between the modern and traditional systems are sharply displayed in these photographs showing human use of harsh environments. Modern uses often depend on extensions of the infrastructure of the modern interconnected system into these environments by such means as pipelines or hydroelectric plants. Resources are thus transferred to more populous regions, where people can benefit from these resources without having to endure the local environmental limitations. Very few people in modern societies rely directly on these environments for their necessities of food, clothing, and shelter. People of traditional societies who live in harsh environments, however, have had to make a way of life by extracting the necessities of life from these environments. Traditional peoples have developed ingenious hunting skills, nomadic foraging strategies, and special practices of cultivation to survive. They must deal directly with the severe limitations of these environments in their everyday lives.

Above: A group of nomadic herders in Afghanistan. (United Nations/H. K. Lall)

Right: Plowing after a drought in Mali. (George Gerster, Rapho/Photo Researchers, Inc.)
Below: Hydroelectric power station at the Aswan High Dam in Egypt. (Novosti from Sovfoto)

Left: A wagonload of vegetable produce in Indonesia. (Burt Glinn/Magnum Photos, Inc.)

Above: Banana harvest on a plantation in Panama. (United Fruit)   *Left:* Citrus groves in Florida, the southern part of which has a semitropical, high-work environment. (Florida Development Commission)

High-work environments yield a great diversity of biological output. Traditional societies, through their special local adaptations, have probably made better use of this biological diversity than have the people of the modern interconnected system. One of the common modern approaches in utilizing these environments is to concentrate on specialty products, employing some form of single-crop agriculture. Pests and competition from other plants are a constant and serious threat to agricultural practices of this type, and such commercial ventures are quite risky in tropical environments. Where urban concentrations develop in these environments, a modern infrastructure must be provided, sometimes with the aid of international organizations.

Above: Clearing a section of forest for agricultural use in Bolivia. (Cornell Capa/Magnum Photos, Inc.) *Below left:* Dusting with pesticides in Guinea. (FAO photo by A. Tessors) *Below right:* Expansion of the water distribution system in Accra, the capital city of Ghana, with aid from the International Development Association. (International Development Association)

As is now known, manipulation of water supply to overcome water deficiencies of the natural environment was a well-established practice in the early centers of civilization in China, India, and the Fertile Crescent. The design of irrigation works has not changed dramatically since these early times. One of the impressive aspects of irrigation, especially then but also now, is the tremendous human investment put into the building of canals and dikes that distribute water or maintain appropriate water conditions in the fields. These photographs portray the importance of human works in managing the environments of irrigated landscapes. They also convey a sense of the continuity of social cooperation and discipline that the practice of irrigation imposes. Thus, irrigation undoubtedly has had a significant influence on the culture of people in areas that require irrigation.

Above: A traditional form of irrigation in the Nile Valley. (FAO photo by Patrick Morin)　*Below left:* Irrigation ditch in Pakistan. More modern irrigation systems have been developed and are being expanded in both Egypt and Pakistan. (Charles Harbutt/Magnum Photos, Inc.)　*Below right:* Construction of an irrigation canal in the Imperial Valley of California. (George Gerster, Rapho/Photo Researchers, Inc.)

Water and land transportation networks form an interlocking system in the Netherlands. (Mastboom Vliegbedrijf / Rotterdam)

Below: These rows of cactus in Tunisia have been planted to prevent further erosion. (FAO photo by F. Botts)   *Right:* Terraced rice fields near Bandung in Indonesia. (United Nations)

CHAPTER 4

Human Cultures

Culture is the human component of the geographic equation. Ideas and traditions from the past, religious beliefs, and political customs are important influences on the way of life of any human group. Technology, however, is perhaps the most important difference between villagers in India and the people of Washington, D.C. Modern technology, the product of the Industrial Revolution, is based on the lavish use of inanimate energy, as opposed to human and animal muscle power. Besides increasing productivity, modern technology makes possible a greater range of interactions between places than can occur in a traditional village.

Above: A country restaurant on the Île d'Orléan near Québec City in Canada. (John Launois/Black Star) *Below:* Spectators at a parade in Abu Dhabi, United Arab Emirates. (Azzi/Magnum Photos, Inc.)

Above: Meeting of the Council of the Wise in a village in the Kirghiz Soviet Socialist Republic. (Tass from Sovfoto)   *Right:* Urban street scene in India, where the cow is considered sacred according to the Hindu religion. (Marc Riboud/Magnum Photos, Inc.)

As we look at the human component of the geographic equation for different places in the world, we must remember that almost everywhere the modern interconnected system has evolved from the locally based system. At the present time, however, the transformation from old to new is not complete. Indeed, the modern interconnected system may never encompass the whole world.

The interconnected system rests on vestiges of traditional physical structures and remnants of ideas from the past. The ideas and traditions often pose greater obstacles to change than do the structures. The destruction of the centers of many Western European cities, with their 500-year-old buildings and street patterns, during World War II seemed to present an excellent opportunity to rebuild them to fit modern needs. Although the buildings were gone, the titles to property and traditions remained. Thus while buildings were of new architectural design, in almost all cases the rebuilt city center repeated the pattern of the past with little regard for the movement of modern street traffic and the need for auto parking. Culture, religion, and the present political pattern of separate states are fundamental facets of the twentieth century inherited from the traditional, locally based world of the past.

The modern interconnected world is a very recent development. Important interaction between peoples on different continents dates largely from the European Age of Exploration. This age began with Columbus's voyages at the end of the fifteenth century, but explorations did not become extensive until development of steel-hulled, steam-powered ships in the nineteenth century. Efficient trains, planes, pipelines, telegraph, telephone, and television have developed since 1880. Before that people depended mainly on animal power, wind, or their own muscles, all of which are inefficient and slow. For most people in the industrial world, the interconnected system is a product of the twentieth century.

THE RESULTS OF THE ISOLATION OF TRADITIONAL SOCIETY

Anthropologists present evidence of the existence of separate human groups on different continents at least 40,000 years ago. The physical isolation of these groups meant that each interbred within its own population. Consequently, in the view of most anthropologists, the earth was occupied not by one human population, but by several almost completely separate populations. This is the basis of the idea of **race,** whereby we distinguish groups by physical appearance, primarily skin color. As Figure 4–1 roughly indicates, we identify particular races with distinctive parts of the earth. Frequent long-range interconnections over the world are mostly phenomena of only the past several hundred years, and most people in one part of the world trace their ancestry back through time within their own region. The movement of Europeans into the Americas, South Africa, and Australia; the migration of Asians to the Americas and Africa; and the forced removal of blacks from Africa to the Americas have largely come in the past 200 years.

Many of the characteristics we associate with peoples of particular races are not the result of differences in their physical appearance, but rather of differences in ways of life, or culture. A group decides what to eat, what clothing to wear, what is "right" and "wrong" on social, political, and ethical matters. Not all these decisions are made at one time. Neither are they unchanging once established. Each group communicates its way of life and its values from one generation to the next. Each evolves its own system of living, indeed its own way of perceiving the environment. Each has its own culture.

BASES OF CULTURAL DIFFERENCES

Variations in ways of life from one culture group to another may have been affected by a number of variables. The particular earth environment

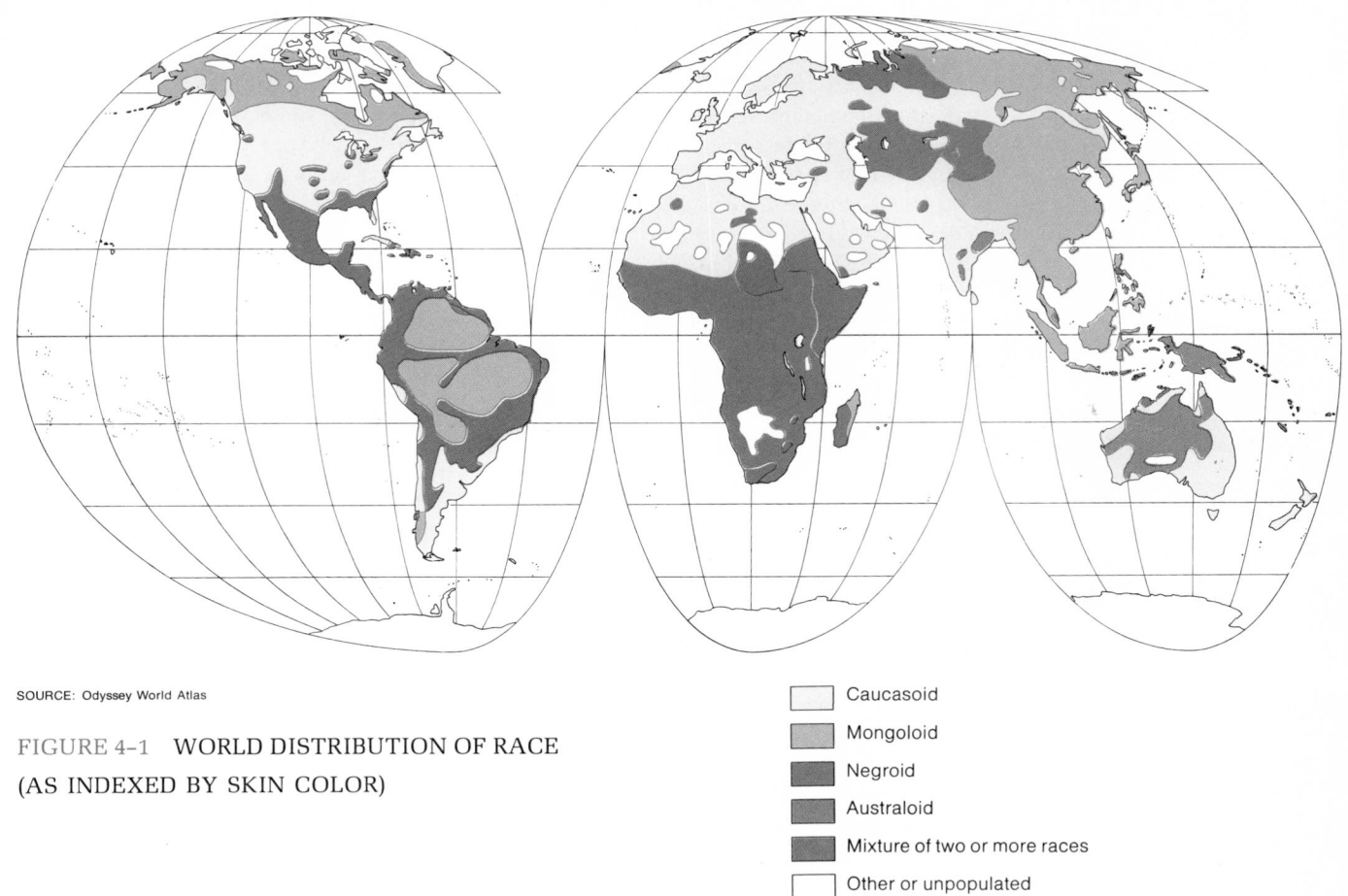

FIGURE 4-1   WORLD DISTRIBUTION OF RACE
(AS INDEXED BY SKIN COLOR)

Caucasoid

Mongoloid

Negroid

Australoid

Mixture of two or more races

Other or unpopulated

presents its own set of obstacles and possibilities. For example, midlatitude Europe and tropical Africa present very different problems of survival. From time to time particular groups have also been inspired by their creative leaders. Aristotle, Jesus, Julius Caesar, Luther, Marx, and Darwin all have had tremendous impact on European culture. But it has been within the group, rather than the individual, that the accepted patterns of life and the "conventional wisdom" about how life should be lived has gradually evolved.

Some individuals within the group may not accept its conventional wisdom and demand change. The group may adjust its ways to suit protesters within the community, or it may reject the proposed changes. Cultures may change as the result of individual inventiveness, but only to the degree that the group accepts new ideas.

ADVANTAGES OF CULTURE

It is generally recognized that the culture group has been one of humanity's great assets. Drawing on the abilities of particular individuals, the group provides not only the frame of reference for everyone within it at a particular time but also a continuity of life from generation to generation. The individual learns the lessons of living from the group; he or she does not have to start from the beginning. Each generation has a basic body of knowledge about life with which to start. It can then add to the sum of previous experience. In this way cultural continuity extends from one generation to another, but it is also modified by the additions of each generation.

LIABILITIES OF CULTURE

Culture can be a great human asset, but it is also in some ways a liability. A culture's preconceived way of living greatly limits its possibilities for using any environment. People develop persistent convictions about what is right or wrong, what can or cannot be done. These cultural **mind-sets** often present obstacles to the development of the earth that are as great as, or greater than, environmental limitations. An outsider often can quickly see a group's cultural "blinders." It is not hard for Washingtonians to think of solutions to some of Ramkheri's problems. It is less easy for Washingtonians to see the mind-sets affecting Washington.

An example of cultural mind-set within the United States is in order. The U.S. Middle West has one of the world's finest environments for agriculture. There Americans have developed one of the most productive agricultural systems humanity has ever known. Yet, although the Middle Western environment permits the production of more than 200 different crops, most of the land is used to raise less than half a dozen: corn, soybeans, wheat, oats, and hay. Why? Because these are the crops our culture uses as a food base for eating and for feeding livestock.

In recent years, the very productiveness of Middle Western agriculture has made obvious the limitations of this cultural outlook. We have produced so much of some of these crops that we have had great surpluses. To prevent a glut on the market that would drive prices down and hurt the farmers, the government established a program for limiting the acreage that they could plant of a particular crop. In return, the government guaranteed farmers a base price for what they did produce.

When too much wheat was produced, wheat acreage was limited by the government. Farmers had to decide what to plant in former wheat fields. Faced with such a problem, they planted more of the other established crops—soybeans, corn, and hay. The land could be used to produce dozens of other crops, but the farmers did not even consider most possibilities. Moreover, they probably could not have found a market for them anyway, since eating millet or using sunflower seed is not generally acceptable in our culture. Limitations are not only self-imposed by the farmer; the culture that has a narrow perspective of what is good to eat imposes them, too.

LANGUAGE, RELIGION, AND POLITICAL STATES: FURTHER RESULTS OF ISOLATION

Today, when daily contacts throughout the world seem to make worldwide communication essential, some 3,000 different languages are in use. Thirty of them are spoken by at least 20 million people. In Europe alone about 120 different languages are in current use. The many language differences pose communication problems for the United Nations and other international bodies. Many people are trained in foreign languages, but at the United Nations it is still necessary to provide simultaneous translation for the five major languages. The business of the world is usually conducted in the languages of the major industrial powers, but these are the native languages of only a tiny minority of the world's people.

The presence of so many languages in the modern world stands as a measure of diversity. Each group evolved its own word designations

WORLD LANGUAGE GROUPS

Language, the basic medium of communication, is probably the most distinctive and important cultural index. A given language indicates the limits of easy communication within a group.

The subdivisions of language within any cultural group present further insight into the group's cultural divisions; they show the importance of tribal or national divisions on a more local level. Even the map in Figure 4–2 does not indicate individual languages. Notice that it lumps almost all the languages of Europe, most of India, and the greater part of North and South America in one group. All Indian languages in Latin America and the Canadian-Alaskan area are included together. The large number of different languages in the world—about 3,000—prohibits rendering them on a world map of the scale used in this book.

Is there a difference between the patterns of languages of the predominantly modern regions of the world and those of regions with dominantly traditional ways of living? Why does Europe have a variety of languages whereas Anglo-America, Latin America, and the Soviet Union are each dominated by a single language group?

for the same physical phenomena or the same ideas. The multitude of different languages in the world is also an indication of the separateness of different groups in the past. Language is a key factor used by anthropologists to distinguish separate culture groups.

If humanity were emerging in today's modern interconnected world without a past history of separateness, George Bernard Shaw's dream of a single worldwide language might be possible. But because of the diversity of human history, language has become a cherished cultural value. Second languages are taught in schools, but one's mother tongue continues to be the primary one.

As the need for international communication has increased, interest in individual local languages has risen simultaneously. Each culture group has shown interest in its past and considers language an important aspect of that past. India, for example, rejected English as its national language when freed from the United Kingdom and sought a language of its own. Moreover, the political subdivisions of India were drawn largely to distinguish individual language groups. Newly emerging countries have encouraged the teaching of the languages of their minority groups, and some are developing written languages for the first time. Even minorities within modern industrial countries have sought roots in their own local languages. Welsh is taught over British television, and Gaelic has been reintroduced in Ireland. In the United States blacks with renewed interest in their African heritage are learning Swahili.

RELIGION: VARIATIONS IN VIEWS OF THE MEANING OF LIFE

The views held by various culture groups about the meaning of life are fundamental to each group. Each geographic area has its own religious beliefs to answer the basic questions of human existence. Religions have formed the basis for different rules of conduct, for distinctive perspectives of life, for speculation about the place of the individual in the world, and even for particular uses of the earth environment. These are a major factor inhibiting communication between cultures.

Figure 4–3 shows the distribution of religious groups throughout the world. Animism, which

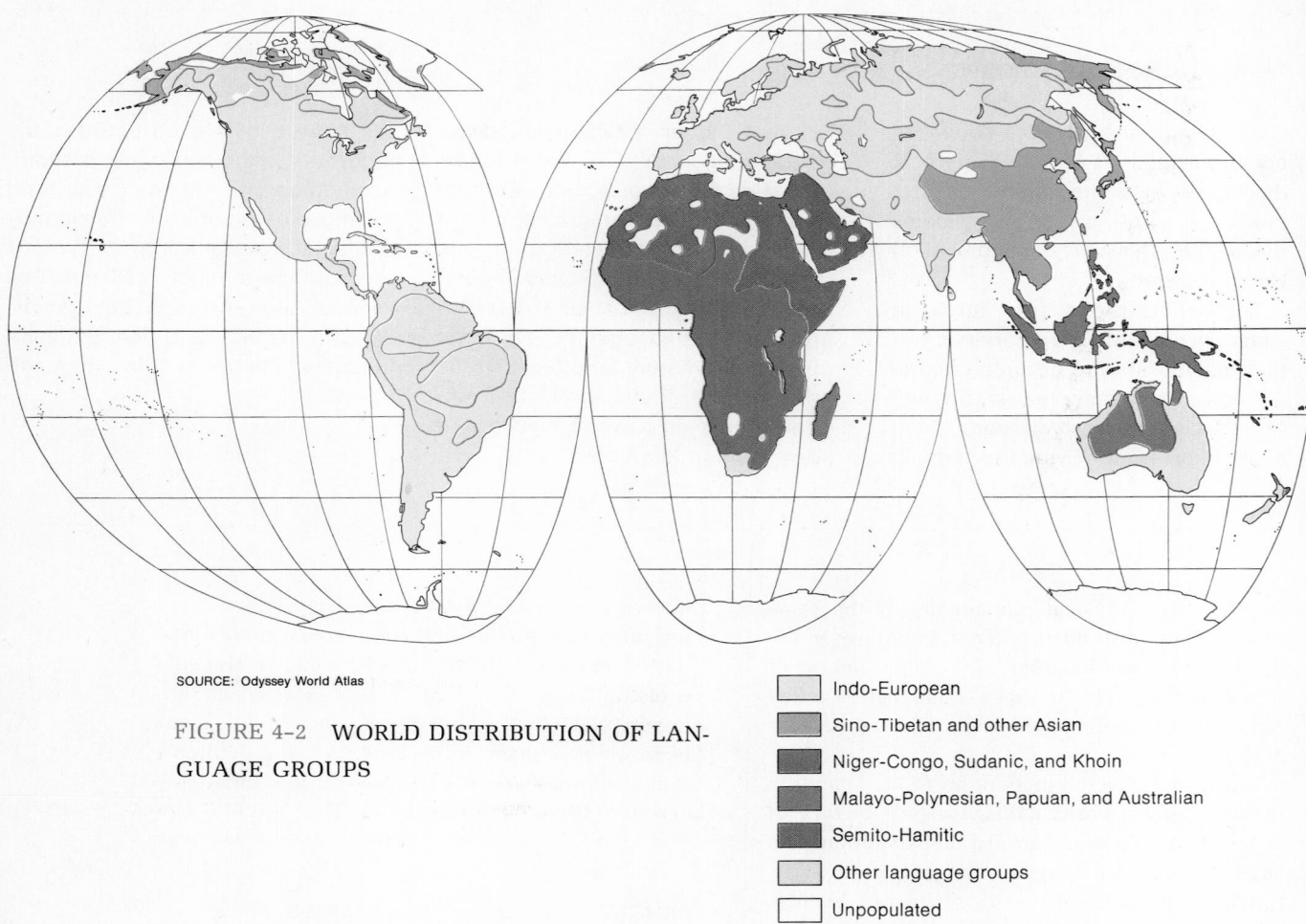

SOURCE: Odyssey World Atlas

FIGURE 4-2   WORLD DISTRIBUTION OF LAN-
GUAGE GROUPS

Indo-European

Sino-Tibetan and other Asian

Niger-Congo, Sudanic, and Khoin

Malayo-Polynesian, Papuan, and Australian

Semito-Hamitic

Other language groups

Unpopulated

attaches spiritual power to earth, atmosphere, and all living things, has been dominant among the peoples of Africa, the polar areas, and the Pacific islands. In South Asia, Hinduism has dominated. In East Asia, the religions have been Buddhism, Confucianism, Taoism, and, in Japan, Shinto. Buddhism embraces the idea of the submergence of the individual to a nonworldly state of infinite being. In Confucianism people are seen as related to the supreme deity, Heaven, through their ethical actions on earth. Disasters

are viewed as the result of the chastisement of Heaven for misdeeds, particularly of the emperor, the "Son of Heaven." Shinto stresses the descent of the Japanese people and islands from the gods. An East Asian is likely to hold a combination of tenets from all these religious teachings.

The Middle East–North Africa is the realm of the Islamic religion; Allah, the one God, is the supreme ruler, and people's lives are subject to God's will. There religion is a particularly strong

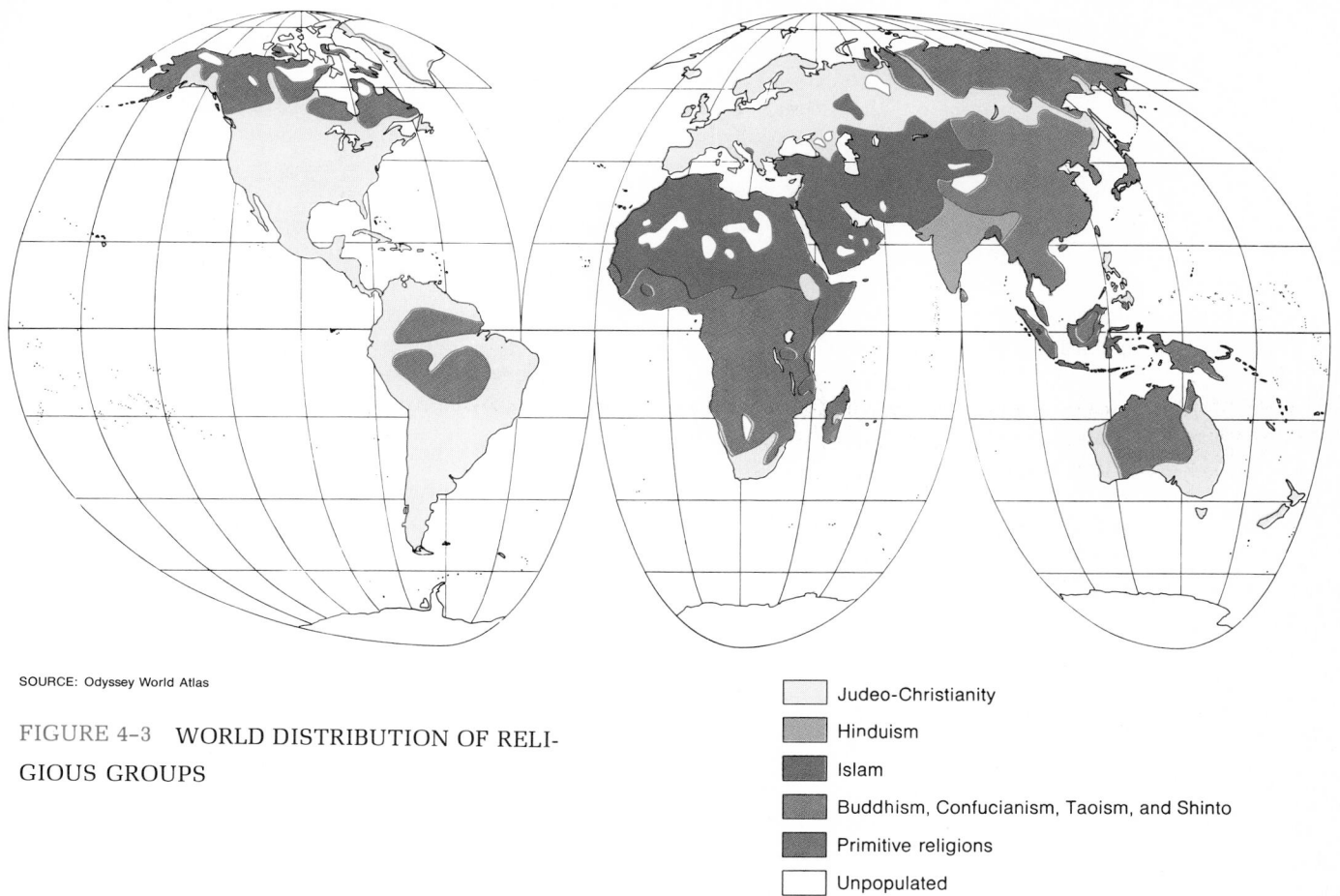

FIGURE 4-3   WORLD DISTRIBUTION OF RELIGIOUS GROUPS

Judeo-Christianity

Hinduism

Islam

Buddhism, Confucianism, Taoism, and Shinto

Primitive religions

Unpopulated

force in the daily life of the people. Europe, the center of the Judeo-Christian religious heritage, has still other emphases, such as the personal relationship between God and the individual, and human stewardship of God's resources on earth.

The importance of religion in shaping the practices and perspectives of a culture can scarcely be overstated. In today's world Hinduism in India is perhaps the best-known example of a religion with a very strong impact. This is graphically illustrated by the fact that cattle compete with the human population for foodstuffs, even though famine is an ever-present danger. India has the largest cattle population of any country in the world, yet religious practices forbid the killing of cattle for meat. In fact, people are generally reluctant to kill any animal. Monkeys and other pests destroy an important share of the harvests and complicate sanitation and health problems.

The Hindu religion teaches that all created things are one and that a person's soul may have been previously born in animal form or may be so born at another future time. This belief is fundamental to the Hindu view of human existence. Thus, obviously, control of the animal population is far more than a matter of proper governmental decree or education. Change is not going to take place without basic modifications in the Hindu concept of human life in the world. Moreover, the average Westerner's perception of the Indian cattle predicament stems from his or her particular view of humanity's place in the world and what seems right.

Hinduism has also provided scriptural sanction to the rigid system of social caste that is recognized as one of the great cultural obstacles to Indian economic progress. The caste system is a continuing force in the Indian village, despite its official abolition by the government. It is too deeply rooted in the ways of village life to be eliminated overnight.

The impact of Hinduism on Indian life is but a single illustration of the importance of religion. In the Moslem world there are taboos on eating pork and drinking liquor, but religious doctrine teaches that all running water is good to drink, even though modern scientific research has shown that such water may in fact be polluted.

The Christian religion has given distinctive character to human values as well. Like the Moslem and Jewish religions, but unlike other religious groups, it places particularly high value on the individual. It asserts that each human being is of special concern to God. Some have argued that this stress on the individual has been important to the development of democratic political tradition, even though democracy was practiced among the Greeks before the birth of Christ. Many dictators have also ruled in the name of Jesus Christ—and many battles have been fought under the sign of the cross.

Some people also believe that the stress on the individual and on what has come to be termed the Protestant ethic is essential to the free enterprise system of Western economies. It would probably be more accurate to say that individualism has become an important aspect of Western

culture as a whole and is no longer related to any religious tenet. Although the church has lost some of its influence as political and economic leader, it still plays an important role. The effect of Roman Catholic opposition to birth control and abortion legislation is a prime example of this.

As might be expected, contrasting religious views have presented some of the greatest difficulties to intercultural contacts. Aggressive religious groups have been driven by evangelistic zeal to convert the misguided to the "true" faith. This has been a particular characteristic of European Christians as they attempted to expand. One of the drives in the Spanish colonization of the Americas was to convert the Indians to Christianity. In the time of empire building during the nineteenth century, many people joined the Christian crusade to bring the gospel to the "heathen" of Africa and Asia. Missionary effort was an integral part of the colonial domination of Africa and Asia, and as a result Christianity has been identified as an aspect of European colonial exploitation.

In the modern interconnected world where people are more affluent and better educated and where other diversions compete with the church, religion has become a less important force in society. Many of the competing secular forms are, however, virtually religions in themselves. Communism in the Soviet Union could be viewed as a state religion. Lenin and Marx are revered as prophets or saints, and their teachings are as essential to Soviet life as the Bible is to practicing Christians and Jews.

Although the United States has not evolved an official state religion, in some homes the U.S. flag is as important as the crucifix, and a great many families have become more ritualistic about their television watching than their churchgoing.

In Western Europe and the United States a large proportion of the population no longer participates in any organized religious activity. Social critics have pointed out that a concern for material goods and scientific progress have replaced religion in establishing the value system of these societies. The "good life" is said to come from attaining material goods and position in society. In other words, belief in the modern interconnected system as the source of satisfaction has replaced traditional Judeo-Christian values. It is significant that recent protests against the modern industrial complex and the affluent society have fostered a return to various forms of religion including Oriental mystic beliefs and fundamental Christianity.

POLITICAL STATES: FUNDAMENTAL DECISION-MAKING UNITS

Political fragmentation is another result of the separation of people into separate culture groups. Just as distinctive religious values have emerged in different cultures, so have varying political beliefs and political systems. Thus different forms of tribal rule, feudal fiefs, monarchies, dictatorships, democracies, and a range of other forms of government have all appeared. No matter what the *form* of rule, the authority of the political regime has had a geographic spread. And in early days, like culture itself, that authority had a very limited reach.

Today there are more than 150 different sovereign political states in the world, most of them with populations the size of some individual states of the United States. Such tiny nations seem to be anachronisms in a world where emphasis is on large-scale power and where modern businesses function on a worldwide basis. Following the general pattern of the fragmentation of other cultural factors such as language and religion, the nations of today stem from a different era and represent a totally different scale of interaction.

The division of the world into many separate political nation-states is even more significant than the fragmentation of language and religion. Nations are political organizing units. They demand the primary loyalty of people living within their borders and are probably the most important decision-making bodies in the world today. A nation-state not only establishes the basic economic and social rules of life for people within its borders, it is also the primary agent for its citizens in dealing with the rest of the world. Political states negotiate peace and make war.

Perhaps the nation-states, with their origins in cultural diversity, best characterize the fragmented nature of the world today. Moreover, as decision-making entities they are perpetuating a parochial view of the world. If modern civilization had originally emerged with the present system of worldwide transportation and communication, there might well be world government now. But, like language, nation-states are deeply rooted in cultural patterns and represent a very important part of present-day cultural perspectives. Thus they are likely to remain with us for a long time.

TECHNOLOGY: THE BASIC DIFFERENCE BETWEEN THE TWO MODELS

The villager is limited to simple tools—hoes, wooden plows, and simple threshing floors—and human and animal energy sources. By contrast, American and European farmers employ modern tractors, self-propelled combines, and other machines using gasoline and electrical energy. Such contrasts in technology, as we noted in formulating the geographic equation, are often the most obvious and basic differences between cultures in the world today. That is why countries in the modern system are often called *industrial* or *technologically advanced,* whereas countries in the traditional, locally based system are called *nonindustrial* or *technologically less advanced.*

The outward thrust of Europe and its success in colonizing and dominating other cultures both politically and economically were, to a high degree, outgrowths of the technological advantage

TABLE 4-1 ESTIMATES OF AVERAGE DAILY CAPACITY OF
TRANSPORTATION MODE

	Mode	Estimated Weight Carried and Distance Moved in Average Day of Travel	Ton-Miles per Day
Muscle and Wind Power	1. Human carrier	60 lb for 20 miles	0.5
	2. Pack animal	300 lb for 20 miles	3
	3. Team and wagon	1.4 tons for 20 miles	28
	4. Sailing ship (19th-century American clipper ship)	2,500 tons for 200 miles	500,000
Fuel-Powered Transportation	5. Tractor-trailer truck	23 tons for 1,000 miles	23,000
	6. Air cargo, via 707	60 tons for 10,000 miles	600,000
	7. Train (75–100 cars	50 tons per car for 500 miles	2–2.5 million
	8. Cargo ship	30,000 tons for 400 miles	12 million
	9. Tanker ship	300,000 tons for 400 miles	120 million

held by the Europeans. Moreover, the struggle of non-European cultures against European colonialism has not been merely a campaign to retain political freedom. It has also been an effort to develop economic independence and acquire the kind of living standard that European culture has achieved through the development of modern technology.

Technological progress has been one of the most important threads in human cultural development. If people depend on only their own muscle power, they have little control over the natural environment. But they have been able to use their intellect to accomplish work far beyond their limited physical capacity. They have developed tools and machines that utilize other energy sources (Table 4-1).

Equally important in the development of technology is the organized structure it demands. More sophisticated machines have created large-scale production, and modern transportation allows for mass marketing. Factories call for subdividing work assignments and tremendous capital investments. Although craftspeople in their own shops can manage their business, handle its sales, and still take an active part in the production process, a factory requires not only a work supervisor but also an office full of support personnel—persons who keep the books, make the sales, and determine long-range company plans. Besides manufacturing, the modern economic system also depends on giant transportation systems, great financial organizations, and huge wholesale and retail complexes.

TRANSPORTATION TECHNOLOGY AND CAPACITIES PRIOR TO AND SINCE ABOUT 1850

The technology to harness fossil fuels as power for locomotion was not known until the late eighteenth century, and the use of fuel-powered transportation was of little significance until the middle of the nineteenth century. Muscles and wind were the only sources of power for moving goods and people before about 1850. Overland transportation capabilities were especially limited under these conditions. Only ships had capacities and speeds that were comparable to some of today's fuel-powered modes of transportation. Before 1850, where did a large urban area have to be located in order to develop and to be able to support its population? What might have made it possible for some urban development to occur away from seacoasts before the nineteenth century?

The development of fuel-powered transportation is one of the major prerequisites for the development of a modern economy. It would be impossible to move the huge quantities of mineral resources, food, forest products, and fuels needed to produce the material wealth of modern societies, or to distribute the products of modern industry, without efficient land transportation. Fuel-powered transportation technology, which has developed in little more than a century, has changed our capabilities to move goods and people over land by several orders of magnitude. Widespread modernization was not possible prior to the fuel-powered era, and the larger urban places were ports where ships could deliver the resources that had to be drawn from distant places.

Transportation is a major consumer of fossil-fuel energy. In 1970 transportation consumed one-quarter of the total energy used in the United States. Different modes of transportation have large differences in the efficiency with which they use fuel. For example, the following figures compare modes in terms of ton-miles per gallon of fuel consumed: large trucks, 50; air cargo, 8 to 10; 100-car train, 250; tanker, 930 (data from G. A. Lincoln, "Energy Conservation," *Science* 180, no. 4082 (April 1973): 157). The automobile is mainly used for moving people, not goods, but for the United States it accounts for nearly 60 percent of the energy used for all transportation, or 14 percent of the total energy. The high degree of dependence on the automobile in the United States and many other modern countries may be temporary. As fossil-fuel resources are depleted and their costs rise, the automobile may become more and more a luxury.

THE EVOLUTION OF TECHNOLOGY

The first tools were designed specifically to make human muscles more efficient. Shovels, knives, bows, baskets, and carrying poles were among the tools created to be powered and controlled by human muscle. With the discovery of other energy sources such as animals, water, wind, and fire, new tools were designed, including plows, water wheels, sails, and cooking devices. With more powerful and dependable sources of energy, people shifted from simple tools to machines that could do the work of the people who had formerly operated the hand tools. Wagons, grinding mills, cannon, spinning wheels, clocks, and hundreds of other machines were made with moving parts. Finally, with the discovery of **inanimate energy** sources such as coal, oil, natural gas, electricity, and atomic energy, a further technological revolution took place. Tools and machines can now be created that operate on almost any scale, at any speed, and with tremendous flexibility. Today computer-controlled machines produce automobile engines and load coal trains without any direct human energy or control.

Human capability for creating material wealth has been revolutionized by the use of tools. From small producers using just their own muscles, people have become the managers of an amazing array of machines that operate day and night, year after year, in their presence or absence. But the technology of tools and machines has left a further heritage to any civilization. It has bequeathed a complex of artifacts constructed with that equipment: buildings, transportation routes, clothing, books and phonograph records, and all the rest. As a result every generation has a large supply of manufactured goods with which to work. These represent **capital** that the present generation can utilize. When one considers the time and effort necessary to build a house, complete a road, develop a water or sewer system, the great value of this technological legacy becomes apparent.

Like other inheritances from the past, the capital accumulation of material, tools, and machines is also a liability. No generation can wipe out its past; it must build on what is there. Houses built in the era when coal was used to heat have to be converted to gas or oil and then air-conditioned. Roads built for horses and buggies must be adapted to automobile commuters. The cost of wiping out all the capital accumulation from the past seems too great. Nowhere is this problem more sharply apparent than in the great metropolitan centers of the world. Urban change calls for removing or updating outmoded parts of a city; but these things cannot be done cheaply or quickly. As a result, people live in inadequate housing and drive on inadequate streets. Moreover, who is to say that, if by some special effort all of a metropolitan area could be renewed today, the renewed area would fit the needs of tomorrow?

DEVELOPMENT OF ENERGY SOURCES

Since tools and machines depend on energy, the greatest human technological achievement has been the development and control of energy sources. Human power has been mobilized on a large scale throughout history. Big families have been used to increase production on the family farm. Slaves have been captured in battle or bought to do heavy physical labor. Masses have been organized in China and elsewhere to build roads, dams, and factories.

Other sources of energy have become far more important than the human one. More than 10,000 years ago the first domestication of animals began, and since then horses, water buffaloes, camels, and other animals have been put to work as beasts of burden. They have not only transported goods, they have also pulled plows and run machines such as the Indian threshing machine. The discovery of fire enabled people to derive energy from timber, grasses, crop wastes, animal manure, coal, oil, and other combustible materials. The force of running water, winds, steam and hot water from the earth, and even the tides has been used. Now we are beginning to harness the forces resulting from the release of nuclear energy and from direct radiation from the sun. Throughout most of human history, however, people had little energy at their command and were not always effective in producing enough, even for their own domestic needs.

Until some 200 years ago the energy available for human use was largely limited to that derived from humans, animals, trees and other plants, and from wind and water flows. Generation and use occurred at the same time. Very little energy could be stored. At best, wood could be cut and stored for consumption at some other time, and water dammed for later use.

The Energy Revolution:
Tapping the Earth's Resources

In the eighteenth century the Industrial Revolution in England opened a whole new storehouse

TABLE 4–2   WORLD ENERGY CONSUMPTION BY PERCENTAGE
OF SOURCE, 1973

Traditional Sources	Percent	Industrial Sources	Percent
Human power	1	Coal and lignite	29
Work animals	4	Petroleum	39
Fuel wood, manure, waste	6	Natural gas	19
		Waterpower, nuclear power, geothermal power	2
Total	11	Total	89

SOURCE: United Nations

of energy. This revolution followed the invention of new machines and the organization of the factory system, which replaced traditional handicraft. Most of all, it was an energy revolution that has transformed human life.

The Industrial Revolution started with harnessing the force of running water in the small streams on the slopes of the Pennine Chain in England. It has advanced through the invention of machines to harness the power released from combustion of coal, oil, and natural gas, and in the twentieth century the radioactive decay of uranium. Energy from the sun, absorbed by living plants, had been captured and stored in rocks as coal and oil or gas. The discovery of ways to utilize this energy in engines and power plants released a new wealth of power. Now in our automobiles we burn in an instant the gasoline refined from petroleum that accumulated in the earth over millions of years, and our power plants make electricity from coal that took equally long to form.

The Industrial Revolution meant that humans, for the first time, could tap the inanimate energy of coal, oil, and natural gas that had accumulated during eons of geological time. Life was trans-

formed for those on earth with the knowledge to use inanimate energy. Today the traditional energy sources available before the Industrial Revolution account for only 11 percent of total world energy consumption (Table 4–2). Eighty-nine percent of total energy used in the world in 1973 came from sources essentially unavailable to humanity 250 years ago. This energy bonanza has made possible the modern interconnected system as we know it today. But we have used up this inheritance of earth history so rapidly that the deposits of liquid petroleum are expected to be depleted in less than a century and coal will be gone in a few hundred years.

The Significance of Modern Energy Use

The geography of energy consumption over the earth is fundamental to an understanding of "industrial" and "nonindustrial" worlds (see Figure 4–4 and Table 4–3). Today's traditional society depends in large part on the same sources of energy that humans have used throughout recorded history. These sources are comparatively poor in their ability to produce material wealth. On the other hand, modern society has harnessed sources stored in the earth from past geological

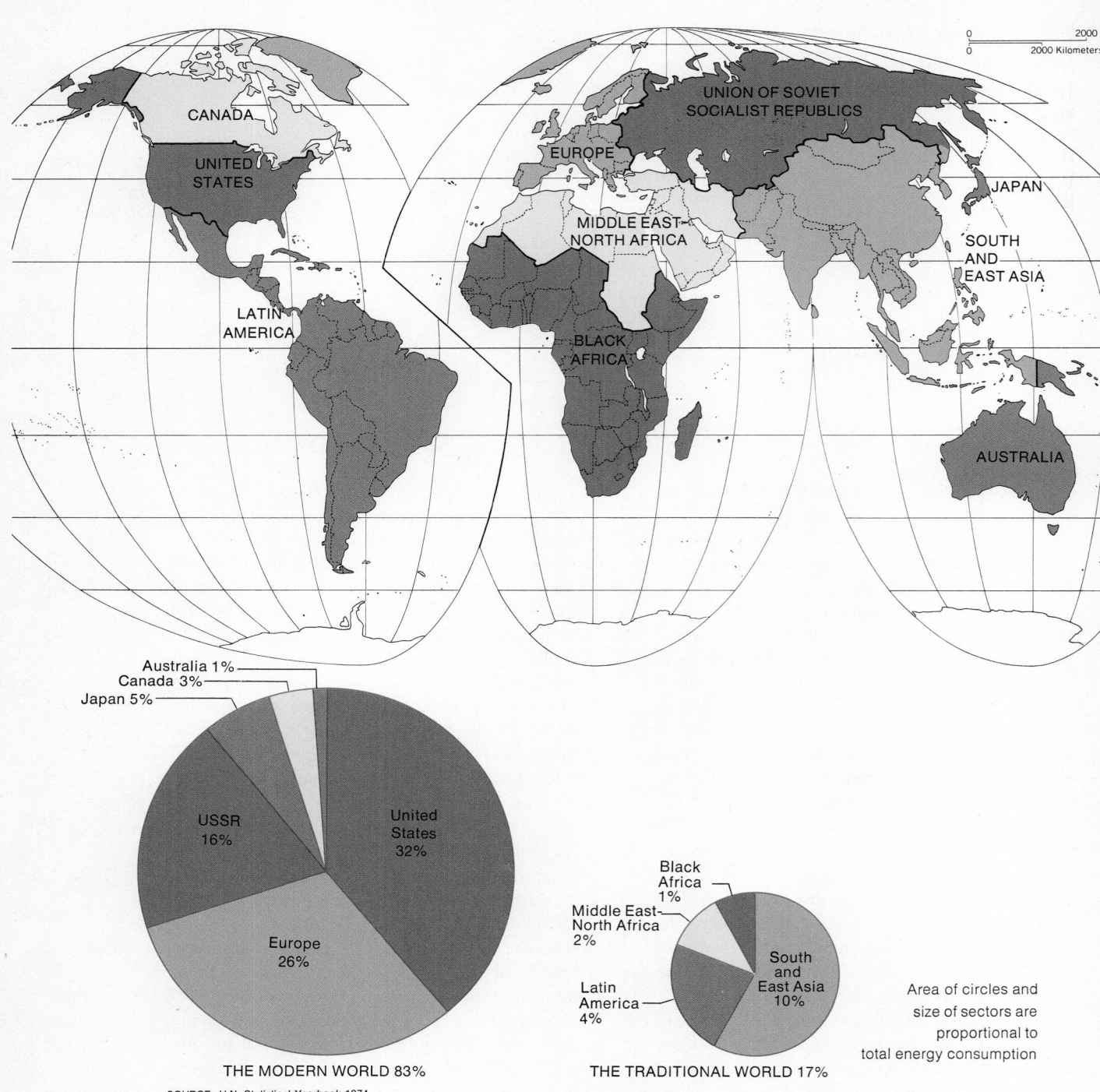

FIGURE 4–4   WORLD ENERGY CONSUMPTION

THE MODERN WORLD 83%

SOURCE: U.N. Statistical Yearbook 1974

THE TRADITIONAL WORLD 17%

Area of circles and size of sectors are proportional to total energy consumption

TABLE 4-3 TOTAL ENERGY CONSUMPTION, BY WORLD
REGIONS, 1973

Region	Total Consumption (Million Metric Tons of Coal Equivalent)	Percent of World Consumption
United States	2,516.44	32
Europe[a]	2,023.80	26
USSR	1,230.44	16
Japan	390.20	5
Canada	248.61	3
Australia	78.22	1
Subtotal	6,488.	83
Latin America	301.20	4
Black Africa	102.26	1
Middle East–North Africa	148.11	2
South and East Asia[b]	756.88	10
Subtotal	1,308.	17
Total	7,796.	100

SOURCE: *U.N. Statistical Yearbook, 1974.* Regional data have been recalculated to conform to the regional divisions used in this book.
[a]Excluding the USSR.
[b]Including Oceania, excluding Japan.

eras to develop a degree of material productivity unbelievable 200 years ago. A few industrialized nations consume most of the world's energy. Six countries—the United States, the Soviet Union, China, West Germany, the United Kingdom, and Japan—consumed two-thirds of the industrial sources of energy in the world in 1972.

The difference between the energy sources of the industrial and nonindustrial societies is more than just the relative quantity of energy available to the industrial world. Inanimate energy can be transported long distances and concentrated in a particular place to do prodigious jobs impossible in previous times. One thinks of the Egyptian pyramids built by human labor as an amazing feat calling for the assemblage of tens of thousands of workers. But the concentration of power in one nuclear explosion is far greater than the power that ever could have been assembled in the preindustrial world. In addition, modern energy can be stored until needed, used in measured quantities, and turned off and on. In contrast to human slaves or animal power, the "energy slaves" of inanimate power are completely at humanity's bidding.

In the traditional society the individual producer simply is not efficient enough to accumulate wealth. A family is able to provide its basic needs only with the expenditure of tremendous effort. In such a society wealth comes from being able to command a share of the productivity of many workers. This is why slaves have been important: they were the only supplement or substitute for one's own work. Before the Industrial Revolution only a tiny fraction of any group had wealth. As a rule, the wealthy were those who controlled large tracts of productive land and drew a share of the output in exchange for the privilege of using the land. Thus the wealthy in feudal times were the landlords. The affluent in the American South and in Latin America were the plantation owners whose land was worked by slaves. It is the same today where traditional society exists. The headmen of Ramkheri are the traditional landowners who have both affluence and power.

In modern society the vast quantities of energy slaves that power the technological equipment increase productivity per person to such a degree that it is possible not only for the factory owner to be wealthy but for the workers to be well paid. In fact, the workers' standard of living is greater than that of feudal landlords of the past and is certainly higher than that of the village headmen of India. The material affluence that reaches down to workers in today's society comes not only from the superior productivity of the workers themselves but from the economy's utilization of quantities of coal, oil, natural gas, and other energy slaves. The system is so productive that there is wealth unknown in the preindustrial world, wealth that may be widely spread among workers.

To illustrate the point, consider a worker on a modern automobile assembly line in the United States. The worker is paid five to seven dollars per hour even though the particular task may be only to attach door handles by using a pneumatic screwdriver. Such a person cannot be considered a highly skilled worker; a few hours' training is sufficient to learn to do the job efficiently. The worker is highly paid not because of personal skill, but rather because the assembly line on which he or she works can complete a 5,000-dol-

lar automobile every minute, thanks to a multitude of power-driven tools and machines. Since the power-driven equipment is constantly being redesigned to operate more and more efficiently, the assembly process has become more productive. The benefits of this greater productivity are increased profits for management and higher wages for the workers.

The assembly-line worker is so well paid that he or she can afford to buy the 5,000-dollar automobile and many other machine-made products. Industry benefits as well from the worker's being a consumer of the products of this automation. This ever-increasing utilization of cheap, controlled energy slaves supports the affluent industrial society.

In contrast to the assembly-line operator, consider the highly skilled artisan in India or Iran doing intricate metalwork for which several years' training was needed. The payment is only a few cents per day. Without machines, it takes days or weeks for the artisan's deft, trained hands to produce a single brass tray. Poorly paid, this worker can scarcely afford the basic essentials of life. The worker and most of the other people in the country are poor and can buy very little to stimulate economic growth.

Differences in technology have resulted in more than different levels of ease and affluence. Decision-makers, producers, and consumers all have very different outlooks. Recently, much has been said about the dehumanizing nature of assembly-line work. One worker may attach the same three bolts on car frames as the cars move by each minute, eight hours per day, forty hours per week. Such work, day after day, year after year, offers little personal satisfaction and causes psychological strain for these workers. Weekends for recreation may be essential therapy.

In contrast, the Indian artisan's life is centered on work. The artisan is as much artist as worker and, like the artist, has the satisfaction of making a beautiful finished product from an amorphous mass of material. The artisan works many more hours than the assembly-line worker but has little time, and less money, for recreation. But compared to the assembly-line worker, perhaps the artisan has less need to get away from work.

People in the two different human systems also face very different energy problems. In the industrial countries and in the urban outposts of the modern system throughout the world, the problem of increased oil costs has loomed great in recent years. Longer-term problems are depletion of oil and gas and the need to develop new energy sources such as nuclear power and solar power. In the traditional society an increasing shortage of fuel wood has resulted in a great increase in its price. At least half the timber cut in the world is still used for cooking and to heat dwellings. The majority of people in most poor countries still depend on firewood for fuel, and the growing population has meant that the number of fuel consumers is surpassing fuel production. A load of wood that cost six or seven rupees in Katmandu, Nepal, two years ago now sells for twenty. In Niger a worker's family now spends nearly one-fourth of its income on firewood; in Upper Volta the figure approaches 30 percent. People search the hillsides and garbage dumps for fuel. In rural villages in India and Pakistan people used to gather wood for free; now landlords are harvesting the timber to sell it in cities. Under such pressure trees are becoming increasingly scarce. People who used to find firewood within an hour's walk from their homes now sometimes must spend a day searching remote areas for it.

THE "POPULATION EXPLOSION" AND THE MALTHUSIAN PRINCIPLE

The growth of world population has accelerated rapidly over the past 200 years, and especially during the twentieth century. World population is now doubling approximately every thirty-eight years, and for some of the most rapidly growing nations the doubling time is less than twenty years. The present rate of growth, close to 2 percent per year, is larger than at any other time in world history; the long-term average rate is about 0.1 percent. About 74 million people are now added to the world's population each year (Table 4-4). South and East Asia alone (not including Japan) gains about 41 million people per year, more than the entire population of New York and California combined.

This rapid population growth, called the **population explosion,** is occurring in both traditional and modern societies. As we shall see, however, it has very different results in the two systems.

A long-established principle of social science, originally put forward by English economist Thomas Robert Malthus before the Industrial Revolution, has been applied to population growth. It is called the **Malthusian principle.** Malthus believed that population increased in **geometric progression:** 2, 4, 8, 16, 32, . . . , where each number is *multiplied* by a constant number (in this example, 2). Agricultural production, however, the primary support of the population, could increase only in a strict **arithmetic progression:** 1, 3, 5, 7, 9, . . . , where a constant number is *added* (in this example, 2). In the eighteenth century agricultural production could be increased only by clearing new land for additional crops. Ultimately, the population would overtake people's ability to clear more land in England or anywhere else. Competition for available food would then intensify. Famine and warfare would inevitably follow, which in turn would control population growth.

Malthus's predictions never came to pass in Europe. Today four times as many people live there as in Malthus's day, and most live at a much higher standard than was common in the eighteenth century.

What happened to produce results so different from Malthus's predictions? The Industrial Revolution changed his formula completely, at least

TABLE 4-4   POPULATION GROWTH, BY WORLD REGIONS, 1976

Region	Total Population (Millions)	Population Added per Year (Millions)	Annual Rate of Growth (Percent)	Years to Double Population
United States	215	1.7	0.8	87
Europe[a]	476	2.9	0.6	116
USSR	257	2.3	0.9	77
Japan	112	1.3	1.2	58
Canada	23	0.3	1.3	53
Australia	14	0.2	1.5	46
Subtotal	1,097	8.7		
Latin America	326	9.1	2.8	25
Black Africa	313	8.4	2.7	26
Middle East–North Africa	221	6.2	2.8	25
South and East Asia[b]	2,063	41.3	2.0	33
Subtotal	2,923	65.0		
World	4,020	73.7	1.8	38

SOURCE: Population Reference Bureau, Inc., *World Population Data Sheet, 1976*, Washington, D.C., 1976. Regional data have been recalculated to conform to the regional divisions used in this book.
[a] Excluding the USSR.
[b] Including Oceania, excluding Japan.

in the short run. Contrary to his equation, the total amount of production increased even more rapidly than did the population, thanks to the new technology. The combination of machinery, fertilizer, new seeds, better animal breeds, better harvesting methods, and insecticides expanded output rapidly. Moreover, transport ties to food-producing areas on different continents made other foods available to Europeans. The number of people who worked in factories, trade, and shipping grew larger as the modern interconnected system spread over the world, with Europeans supplying manufactured products and managing the commercial networks. Today Europe's population problem consists of congestion, not starvation. But the question still remains: Can technology continue to move at such a pace

that it will outdistance population growth and improve material existence in the modern world?

POPULATION GROWTH
IN THE MODERN WORLD

The data in Table 4-4 show that modern societies have much lower rates of natural population growth than do traditional ones. This can be explained by the differences in their birth rates compared to their death rates. **Birth rate** is the number of births in a country or any geographical area per unit of time divided by the average population during that time. The birth rate in the United States in 1973 has been calculated as 15.0 per 1,000 population. **Death rate** is computed similarly; the U.S. death rate in 1973 was 9.4. The rate of natural population growth is the difference between the birth rate and the death rate. The U.S. natural growth rate in 1973 was 5.6. These rates are usually expressed per 1,000 people. In comparison, the birth rate in Indonesia during the period 1965 to 1970 was 48.3, the death rate was 19.4, and the natural growth rate was 28.9 per 1,000 population.

Most modern areas have low birth rates and low death rates. This results in low rates of natural growth. Traditional areas, on the other hand, have higher rates of natural growth because they have relatively high birth rates but modest or low death rates. In general, death rates have declined for most areas of the world. The decline is due to widespread efforts to control infectious diseases, especially those that afflict children. Declines in birth rates, however, are particularly associated with modern systems and reflect attitudes toward family size and means of contraception. Some industrialized nations have very low population growth. The United Kingdom and West Germany, for example, add fewer than 200,000 persons per year to their respective populations. Although neither country produces enough food for its present population, each has an economy that creates wealth with which food and other basic necessities can be obtained elsewhere.

An increase in production seems to have overcome the Malthusian principle in the modern world. That does not mean, however, that population growth does not create problems in industrialized countries. The problems are those of congestion and consumption. Millions of people are crowded into urban centers, which have thousands of people per square mile. The movement of people to and from work in congested areas produces traffic problems such as the ones we saw in Washington.

The high level of consumption in the modern world may not appear to be a major problem today. The U.S. agricultural surplus, for example, has been so great that the idea of overpopulation in this country seems absurd. Yet that agricultural productivity is largely dependent on inanimate energy, which for the present is primarily oil. Should supplies of oil be drastically reduced, the United States might quickly shift from food surplus to food deficit and thus become overpopulated. The very dependence of modern productive technology on massive exploitation of the world's environments severely limits the proportion of the world's population that can live in affluence.

POPULATION GROWTH IN
THE TRADITIONAL WORLD

The population explosion has had quite different results in the traditional, locally based society, where the Malthusian principle is still at work. The past 200 years have brought a tremendous increase in population but little basic change in the system of resource use. Thus with each generation, the pressure of population on the resource base increases. In places like Ramkheri, population growth is almost outrunning the ability of the economy to produce; without imported food supplies, famine would be rampant.

Today most traditional areas face rapid growth of population, and there is little prospect that it will be brought under control in the very near future. Most of these nations, moreover, cannot afford to support adequately their present populations, let alone any additional people. Egypt, for example, must find food for about 1 million more people each year even though it does not produce sufficient food for the existing 38 million. India and China face similar problems, but the numbers are much larger.

Population in India has more than doubled since 1900, from 240 million to over 600 million. Yet traditional agriculture and the farm village have not changed substantially for most of the Indian people. It is not hard to imagine that rural living a hundred years ago might have been more comfortable than it is today. Each year the agricultural output has to feed more and more mouths.

Population densities today are much higher in India and China than in most parts of Latin America and Africa, which have some of the highest growth rates in the world. They too may face famine in the near future unless they check population or make a major change in the economic base. In recent years countries have engaged in programs of birth control education, distribution of birth control devices, sterilization programs, and abortion. Several states in India, which has the second largest population in the world, have passed laws to require sterilization of a couple with more than two or three children.

These countries need far more than a way of stabilizing population. They require some means of dramatically increasing production. One answer would be for the traditional model to reap the technological benefits of industrialization and the modern system, using modern energy sources and scientific knowledge. In Ramkheri and in other traditional societies, people utilize their environment intensively and carefully, but modern technology could significantly increase their production. Hybrid seed, fertilizers, steel-tipped plows, and insecticides would all improve the crop. Waste could be reduced if the harvest could be stored in rodent-proof granaries. The economy needs a drastic change, such as the one that occurred in Europe as a result of the Industrial Revolution. Production must expand more rapidly than population in the underdeveloped world if living standards are to be raised.

Any change that takes place within the tightly structured life of a village like Ramkheri inevitably leads to other changes. Undoubtedly, economic changes must be made if the people of the traditional world are to survive even at present levels. The basic question is not just whether changes can take place, but whether the society can change economically and survive culturally.

Any economic change threatens the basic structure of the culture. Still, economic change must be made to prevent famine.

Moreover, if the society becomes part of the modern industrial system, it will have other problems to face. Even in industrial countries, rural primary producers have had difficulty keeping their incomes comparable to those of city workers. With modernization come problems of obtaining capital, of taxes, and of competition from other producers elsewhere. In addition, if modernization comes, are the earth's resources adequate to support the whole population of the earth at higher consumption levels?

What happens in the traditional societies will have an important effect on the modern world. Though population growth is low in modern societies, densities are high, and to sustain their populations the industrialized countries reach out for food and resources to other areas of the world. Can the European countries and Japan expect continued access to food from other areas? The United States, too, faces uncertainties in the future. Food production currently supplies large exports in addition to meeting domestic demand, but the nation is increasingly dependent upon energy resources drawn from other areas of the world. Considering population growth and needs elsewhere in the world, to what extent is the present U.S. demand on world resources realistic or even acceptable? Keep these questions in mind as you look at world patterns of living standards, transportation, and decision making in the next chapter.

SELECTED REFERENCES

Barnett, Harold, and Chandler Morse. *Scarcity and Growth.* Resources for the Future, Johns Hopkins Press, Baltimore, 1963. Includes a classic argument counter to that of Malthus (Chapters 11 and 12). The idea is that resources are not lost, only changed in usefulness; hence scarcity is a function of the technology to create new resources.

Broek, J. O. M., and J. Webb. *A Geography of Mankind.* John Wiley, New York, 1973. An introductory human geography textbook that develops the concepts of culture and race, livelihood, settlement, and population change.

Cipolla, Carlo. *The Economic History of World Population,* 6th ed. Penguin, Baltimore, 1974. Historical perspective on the use of energy and world population. The importance of energy control in the development of the world of today is argued, and the ultimate problems of population growth are explored.

Cook, Earl. *Man, Energy, Society.* W. H. Freeman, San Francisco, 1976. An excellent treatment of the role and importance of energy in the evolution of human societies. Chapters 7 and 8 are most appropriate to the ideas developed in this chapter.

Hall, Edward T. *The Silent Language.* Doubleday, Garden City, N.Y., 1959 (paper ed., Fawcett, Greenwich, Conn., 1966). Explores how a culture establishes behavioral and perceptual patterns that are used by its individual members and are unrecognized by members of another culture.

Jones, Emrys. *Human Geography.* Praeger, New York, 1966. An English geographer's perspective on human geography.

Simoons, Fredrick J. *Eat Not This Flesh: Food Avoidances in the Old World.* University of Wisconsin Press, Madison, 1961. Reviews how food taboos are related to religious and other beliefs of culture.

Sopher, David E. *Geography of Religions.* Prentice-Hall, Englewood Cliffs, N.J., 1967. Shows how the many facets of religious beliefs and associated practices are integral to the ways humans have organized their lives and have significance as characteristics of places. Cultural traits other than religion are also discussed.

Spencer, J. E., and W. L. Thomas. *Introducing Cultural Geography.* John Wiley, New York, 1973. A basic text that emphasizes the humanization of the earth. Human cultural evolution, its regional expression, and the processes of technological development are major themes especially pertinent to the ideas discussed in this chapter. The book is well illustrated with world maps of cultural characteristics of past and present human populations.

Zelinsky, Wilbur. *A Prologue to Population Geography.* Prentice-Hall, Englewood Cliffs, N.J., 1966. Introduces both population geography and its cultural implications.

CHAPTER 5

A Global Perspective

This chapter provides an overview of people and
their living conditions. The focus is on broad
world patterns that have evolved from many
different solutions to the geographic equation.
The basic pattern of world population is presented, along with discussion of how the major
variables of the equation might have influenced
that pattern. Selected indexes of material well-
being and their variation over the world are discussed. Measures of world transportation and
communication are presented to represent the
real and potential spatial interactions between
peoples and places.

Sacks of rolled wheat awaiting distribution in rural
Haiti. (CARE photo)

Transfer of goods from ship to truck in Brooklyn, New
York. (Alex Webb/Magnum Photos, Inc.)

Left: Loading cargo in a German port. The sling, shown here, is rapidly being replaced by containers of uniform size. (Owen Franken/Stock, Boston)   *Above:* Cargo in weather-proof aluminum containers being loaded aboard a Boeing 747. This airplane can carry 305 passengers and 27,000 pounds (12.25 metric tons) of freight. (Courtesy of American Airlines)

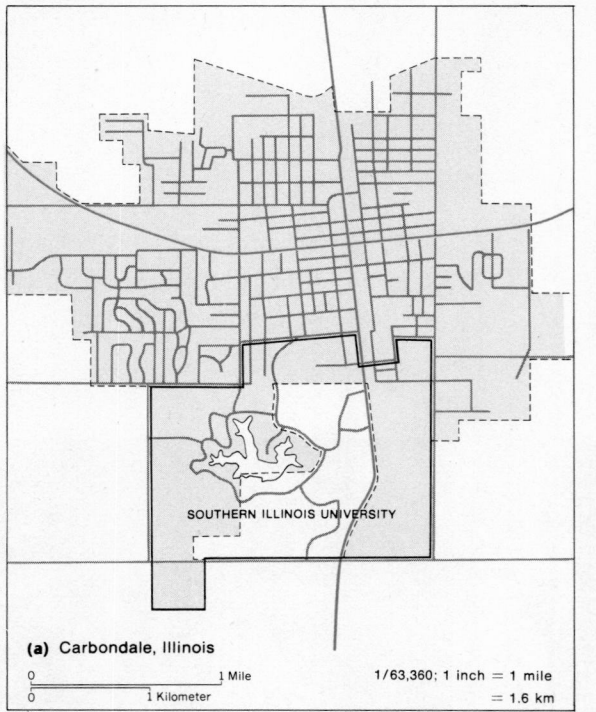

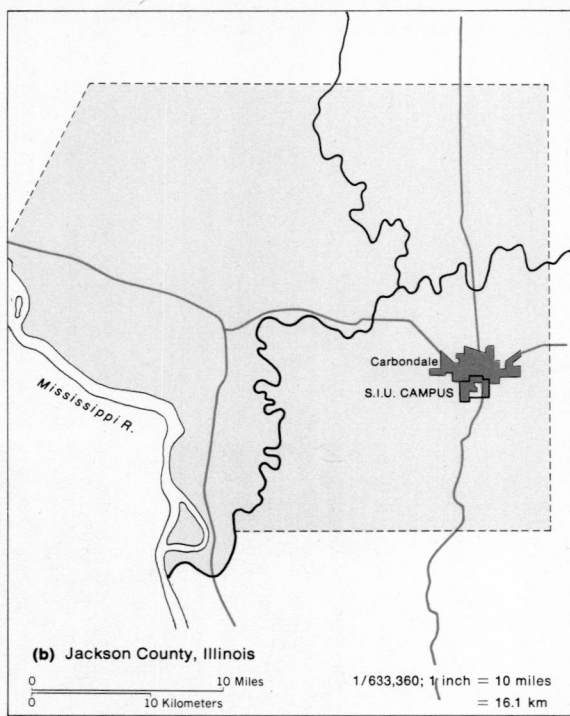

(a) Carbondale, Illinois

0   1 Mile   1/63,360; 1 inch = 1 mile
0   1 Kilometer   = 1.6 km

SOUTHERN ILLINOIS UNIVERSITY

(b) Jackson County, Illinois

0   10 Miles   1/633,360; 1 inch = 10 miles
0   10 Kilometers   = 16.1 km

Mississippi R.

Carbondale

S.I.U. CAMPUS

FIGURE 5–1   DIFFERENCES OF SCALE

What is your view of this world, of other people and places, or of how others live? What do you really know about other parts of the world? Most of us know about the world through our direct experience, which is generally quite limited, consisting of familiar people and nearby places. It is dominated by personal friends, who are part of a somewhat larger population with a similar culture; it contains known places of study, work, leisure, and family life; it consists of commonly held sets of mores, social behavior, and economic expectations. Those who have traveled extensively will have a broader direct experience. In addition, we are exposed through books and mass media to a great deal of information about people and places. From the sum of these expe-

riences we build our personal interpretation of the world.

For most of us the dominant experience is that of the "home" locality, and thus we may hold a biased and narrow view of the world. The information in this chapter supplements our local knowledge by characterizing people and places at continental and global scales. At these scales, information that is important to the description of local places—for example, the city of Carbondale shown in Figure 5–1—must be combined with information from many other local places. At the scale of the United States as a whole, the detailed characteristics of Carbondale cannot be retained and are not likely to be particularly significant. Likewise, the information presented

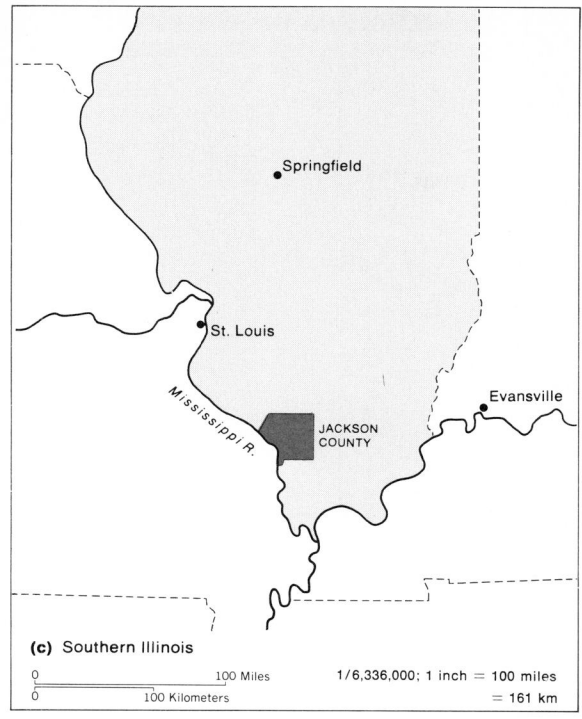

(c) Southern Illinois

0	100 Miles
0	100 Kilometers

1/6,336,000; 1 inch = 100 miles
= 161 km

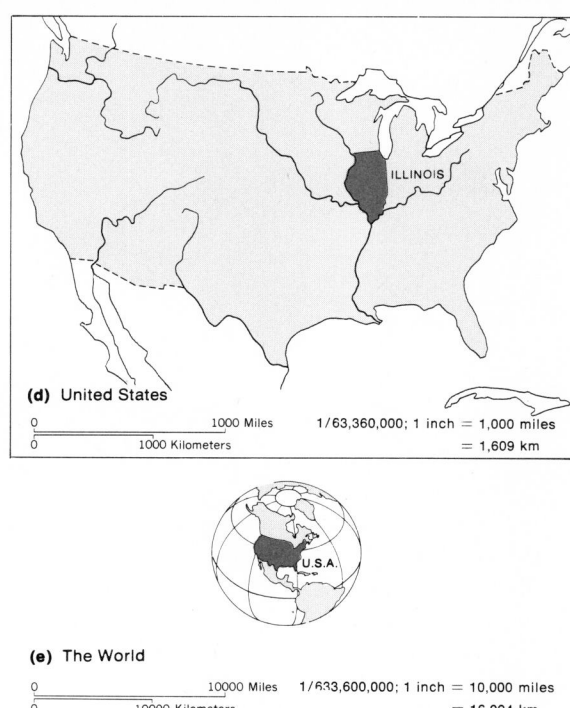

(d) United States

0	1000 Miles
0	1000 Kilometers

1/63,360,000; 1 inch = 1,000 miles
= 1,609 km

(e) The World

0	10000 Miles
0	10000 Kilometers

1/633,600,000; 1 inch = 10,000 miles
= 16,094 km

in this chapter for continental areas or the entire world reduces the diversity of local places into groupings and patterns so as to emphasize broad differences and contrasts. The process of making such generalizations involves statistical summarizations of data, the categorization or classification of descriptive facts into a few broad groupings, and the making of many judgments based on experience and purpose.

THE USE OF MAPS

In geography the generalized information from statistical analysis or classification is often put on maps to show its areal patterns. The patterns of world population characteristics described later in this chapter, for example, are the result of many statistical tabulations. In Chapter 3 the wide range of climatic diversity was generalized into a few descriptive categories through processes of statistical summation and classification. By reducing the great diversity of places into appropriate generalizations at world or continental scales, we can better visualize and comprehend the major differences over the world. These broad generalizations, in turn, provide perspectives from which to compare the circumstances of specific places.

Thinking geographically is very much a matter of fitting the characteristics of specific places into

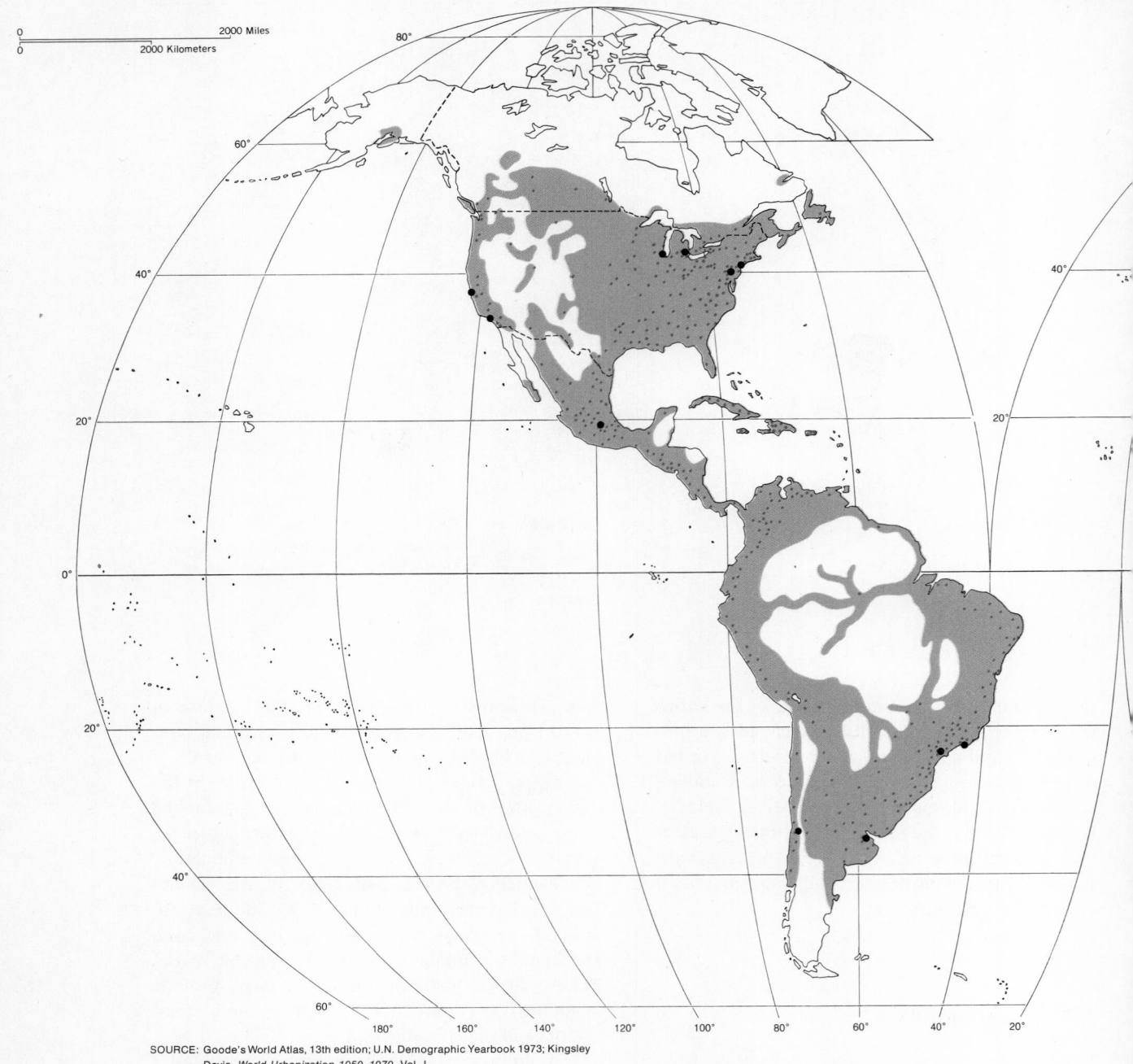

SOURCE: Goode's World Atlas, 13th edition; U.N. Demographic Yearbook 1973; Kingsley
Davis, *World Urbanization 1950–1970*, Vol. I

FIGURE 5-2 WORLD POPULATION DISTRIBUTION

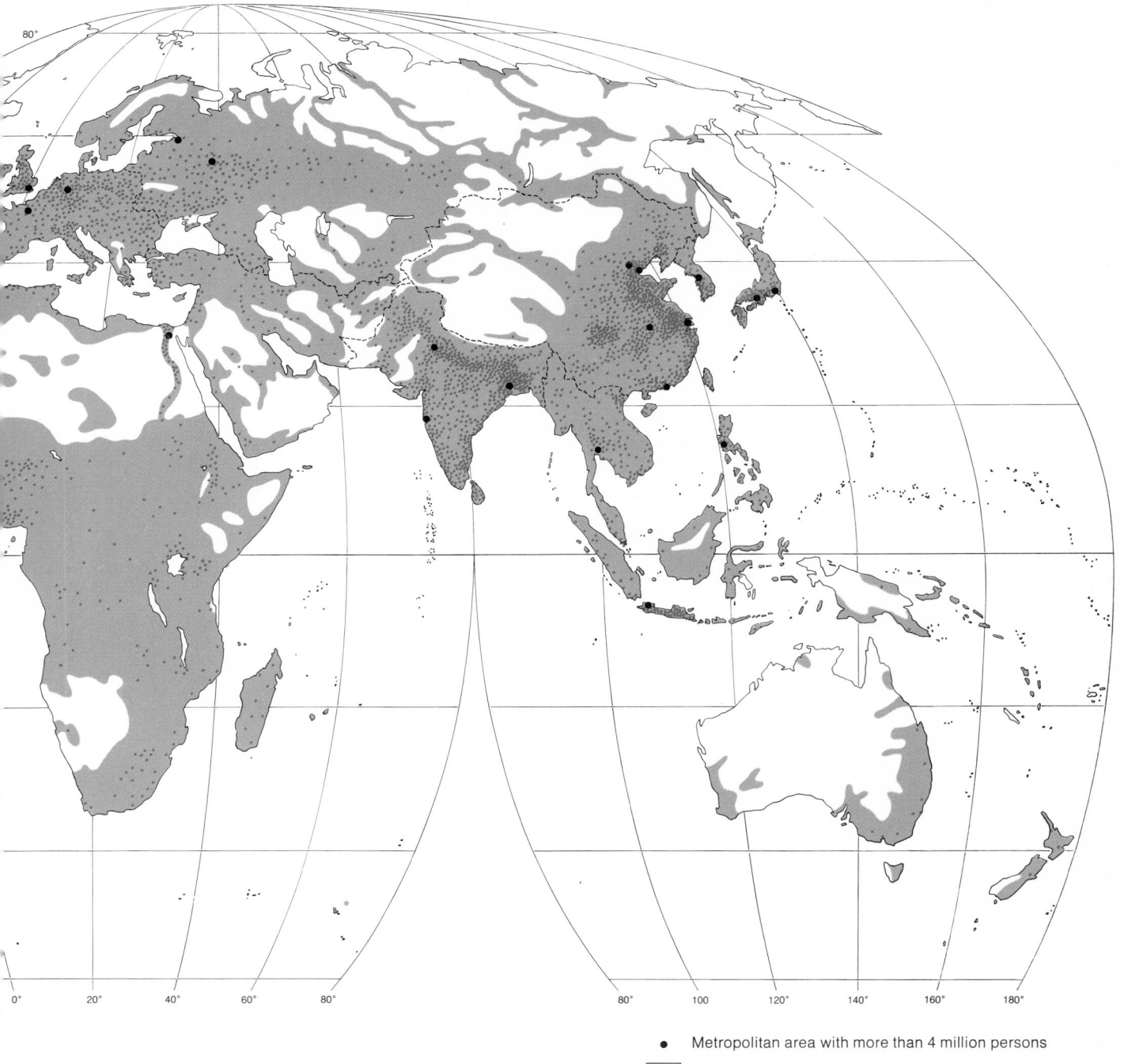

● Metropolitan area with more than 4 million persons

▨ More than 2 persons per square mile
(0.8 per square km)

broader patterns and relationships at regional, continental, and global scales. Being able to relate the particular to the general is very important for understanding the world we live in and our place in it.

As you consider the information and maps of this chapter remember that they are abstractions and summations of the diversity that characterizes specific places and people. For example, a comparison of the average values of gross national product or the average per capita calorie intake by country sacrifices the specifics of individual experiences in particular places, but such information helps us visualize major variations in the economic and organizational milieu in which people exist.

The global perspective for this chapter is built around a selection of maps, graphs, and tables. The topics for analysis are world patterns of population, selected physical resources, and economic conditions. Each map can be viewed as a piece of evidence about our contemporary world; together, they provide a context into which the specific regions discussed in the remaining chapters can be placed. Do not try to memorize the specific details of each map. Rather, concentrate on comparing and contrasting the patterns and distributions they show. You will often have to compare maps in different parts of the book in order to see relationships between different factors. Such comparisons, though sometimes inconvenient, are a normal part of geographical study. If your interests dictate, you may also find yourself comparing maps in this book with maps in various atlases and other sources in order to detect even more relationships.

As you compare and evaluate the maps try to categorize and group sections of the world according to common characteristics of population, physical environment, and economic well-being. This form of geographic generalizing is called **regionalization.** Regionalization simplifies the great variety of detail and diversity among places by grouping those with greatest similarities or greatest organizational cohesiveness together into the same areal category, or **region.** The purpose and problems involved in regionalization are analogous to those of classification in other sciences.

In this book we view the world as having essentially two kinds of places—traditional and modern. Though this simplifies the real world, it provides a useful framework for the study of places. In Part Two we look at countries or groups of countries that illustrate the characteristics of the modern world, and in Part Three we consider multicountry areas of mostly traditional ways of living. The countries or groupings of countries in each part illustrate real-world variety among these two distinctive forms of spatial organization. They fit the framework and serve the purpose set forth in Part One. Other books may use different regions, ones that are consistent with their particular framework and purpose.

Some regional divisions you may come across have an obvious and practical basis. The regions used by the United Nations for its statistical reports serve as an example. The reporting units are countries and they are grouped by continents, which is a very practical choice. Most summary information, however, is grouped along lines of political power blocs: the "centrally planned economies" (Communist countries), the "developed market economies" (such as the United States and Western European countries), and the "developing market economies" (including many countries of Africa, Asia, and Latin America, where much of the population lives at a subsistence level). Such a division, incidentally, has given rise to the idea of **Third World** countries, which are more or less those with developing market economies.

It should be apparent that no set of regional divisions or groupings can serve all purposes. Those of other books and reference sources are likely to be different because of different goals. The regions used in this book should be viewed as a means for learning, and not as the only possible divisions of our world.

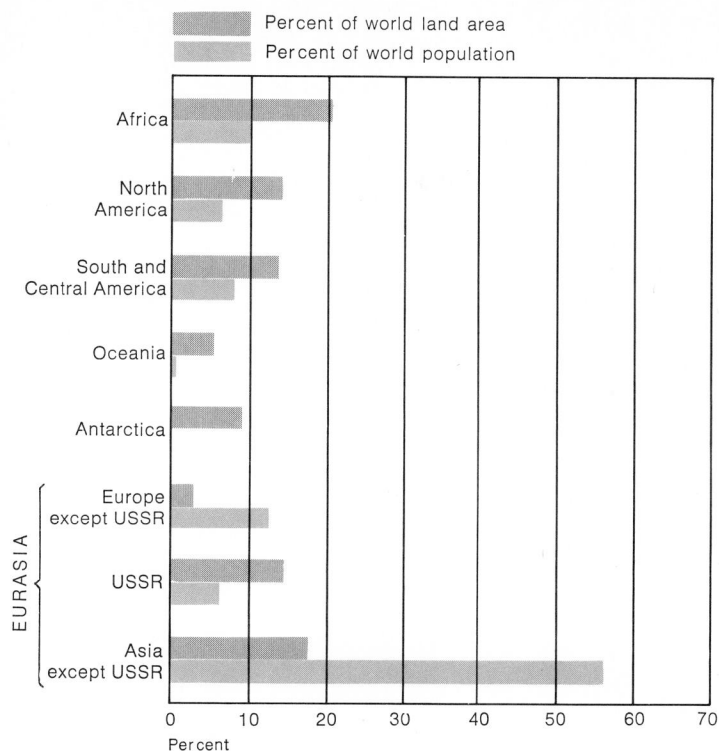

Percent of world land area
Percent of world population

SOURCE: World Population Data Sheet 1976

FIGURE 5-3   LAND AREA AND POPULATION
COMPARED BY CONTINENTS, 1976

WORLD POPULATION: A BASIC GEOGRAPHIC GENERALIZATION

The pattern of people on the earth's surface provides a good starting point for thinking about the world today. A striking geographical fact at the global scale is the unevenness of **population distribution,** as shown in Figure 5-2. This map suggests three or perhaps four major concentrations of population in the world. Eastern China and Japan constitute one such area; Pakistan, India, and Bangladesh comprise another; and the continent of Europe contains a third (a map showing the location of countries appears on pages 140–141). The eastern United States and adjacent parts of Canada might be considered as a fourth area of population concentration. Conversely, most of central and northern Asia, northern North America, interior South America, and North Africa, and most of Australia are very sparsely populated (Figure 5-2).

The striking unevenness of world population distribution is also graphically shown in Figure 5-3. Asia, excluding the Soviet Union, has more than half the world's population. Six countries of South and East Asia account for about 80

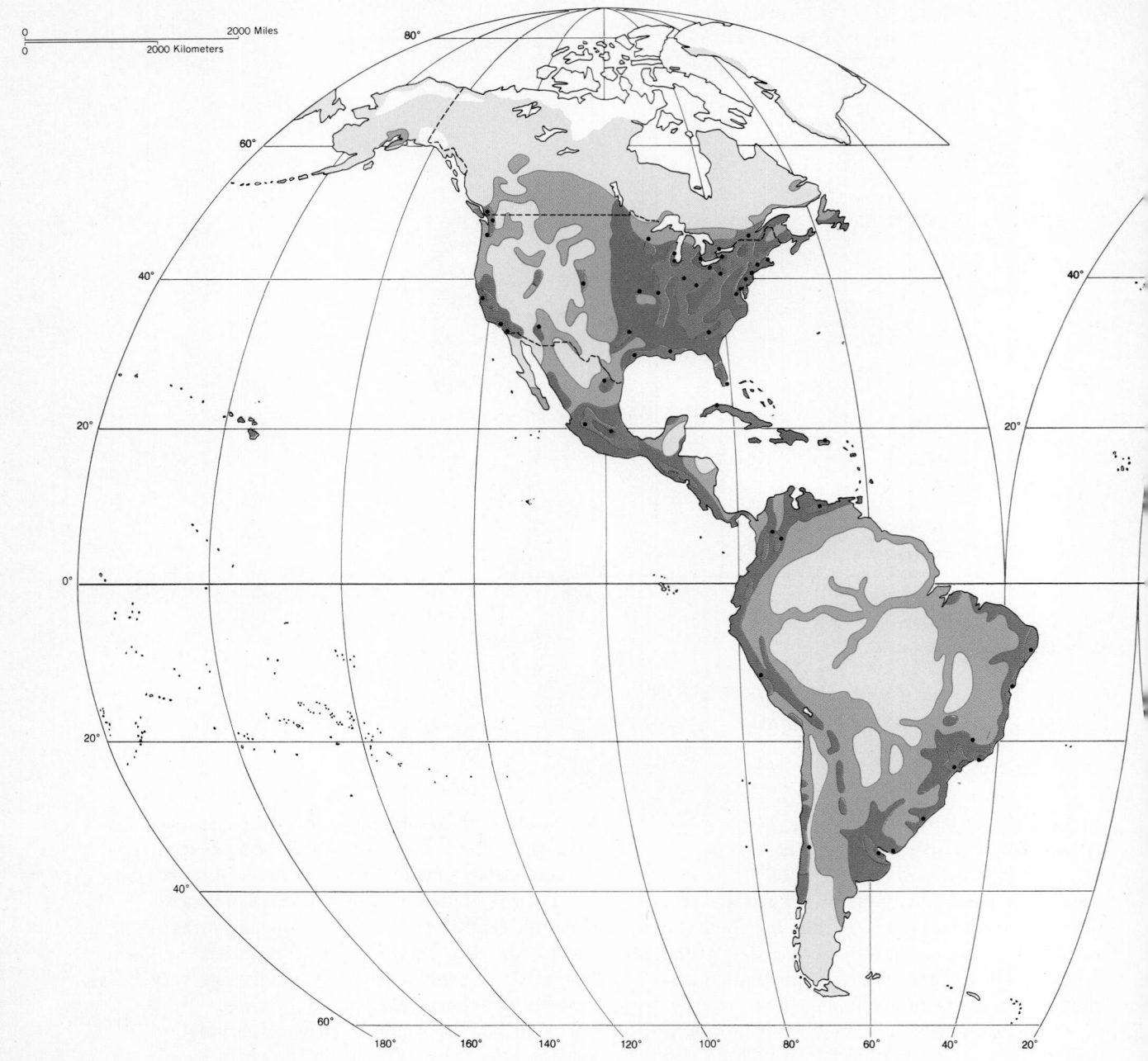

FIGURE 5–4 PATTERN OF WORLD POPULATION DENSITIES

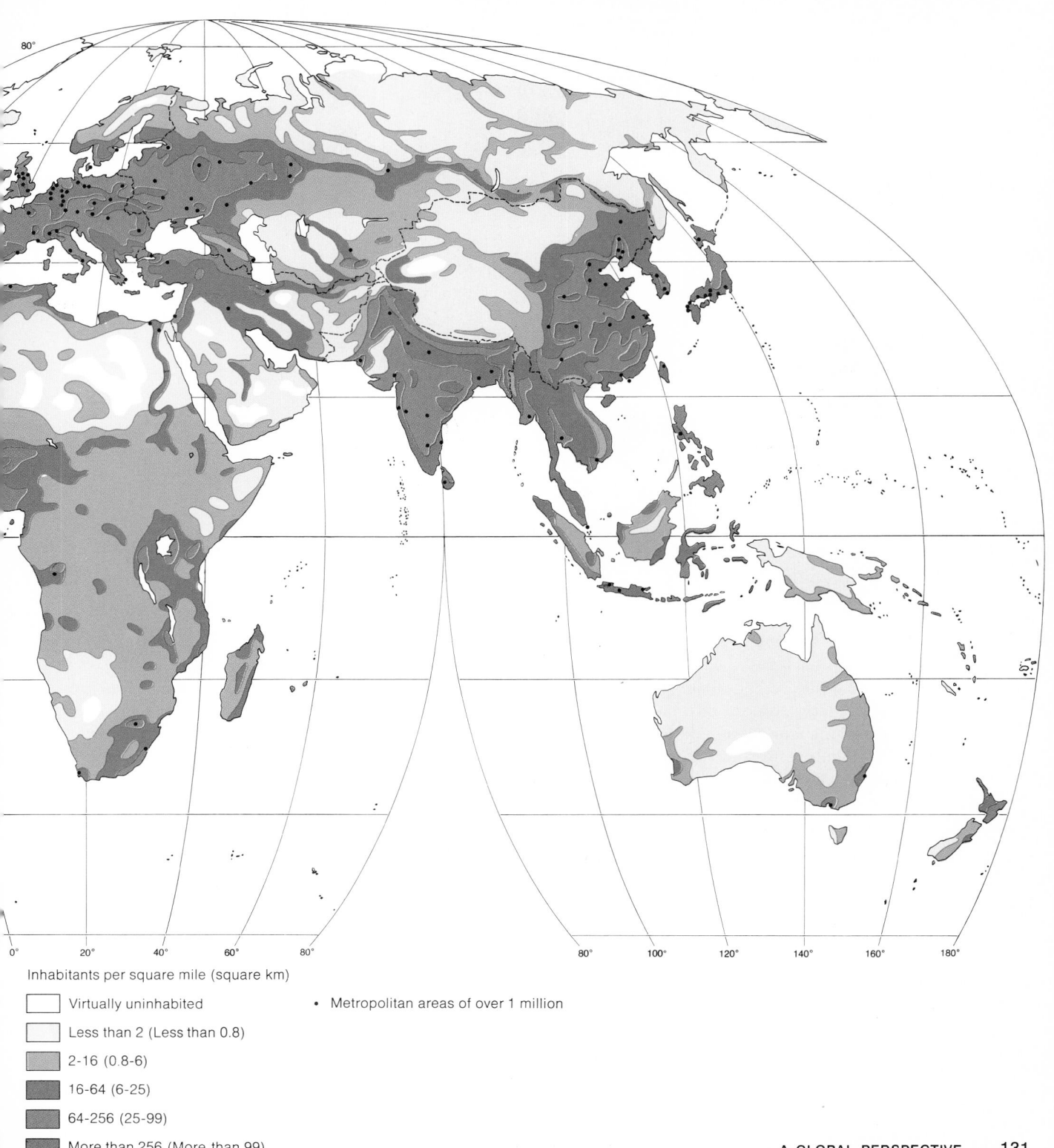

Inhabitants per square mile (square km)

☐ Virtually uninhabited

☐ Less than 2 (Less than 0.8)

▨ 2-16 (0.8-6)

▨ 16-64 (6-25)

▨ 64-256 (25-99)

▨ More than 256 (More than 99)

• Metropolitan areas of over 1 million

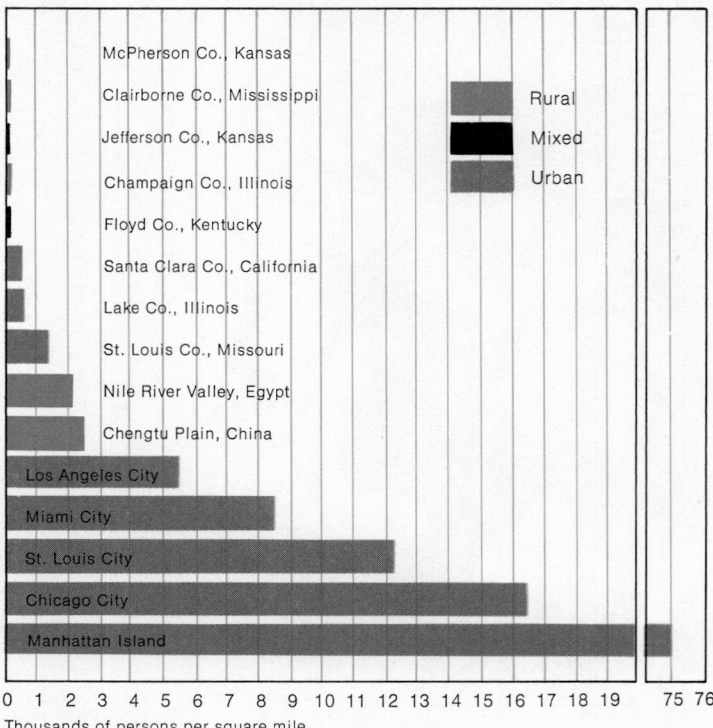

Rural
Mixed
Urban

McPherson Co., Kansas
Clairborne Co., Mississippi
Jefferson Co., Kansas
Champaign Co., Illinois
Floyd Co., Kentucky
Santa Clara Co., California
Lake Co., Illinois
St. Louis Co., Missouri
Nile River Valley, Egypt
Chengtu Plain, China
Los Angeles City
Miami City
St. Louis City
Chicago City
Manhattan Island

0 1 2 3 4 5 6 7 8 9 10 11 12 13 14 15 16 17 18 19   75 76
Thousands of persons per square mile

FIGURE 5–5   POPULATION DENSITIES OF SE-
LECTED LOCALITIES

percent of its population: China, India, Indonesia, Japan, Bangladesh, and Pakistan. Europe (including the European part of the Soviet Union) adds nearly 20 percent more. Thus three-quarters of the world's population is concentrated in less than one-fifth of the world's land area.

The concentration of people in these two limited areas of the earth raises several issues. Why are so many people there? Why have they crowded together rather than spread out more evenly? Is South and East Asia, with over half the world's people, of greater importance to our contemporary world than any other comparable area? These questions have no ready answers. We do discuss the first two in this chapter, and deal more extensively with the third in Chapter 16.

Another way to view the distribution of population is in terms of **population density,** which is expressed in persons per unit area, usually square kilometers or square miles (Figure 5–4). Geographers often use population density because it has a common base by which to compare places. The box on the meaning of densities and Figure 5–5 illustrate that point. Since a density measure states population in terms of the land area that the people occupy, it suggests a relationship between people and their land **resource base,** that portion of the earth environment upon which they depend for the support of life. Population density can be a useful measure, especially when comparing traditional, locally based societies where people depend mostly on the biological productivity of their local area. For modern

MEANING OF DENSITIES: IN RURAL TERMS AND IN URBAN TERMS

Let's examine the densities shown in Figure 5–4 in terms of the real crowding in local areas. For example, Figure 5–5 shows local densities, rural and urban, for a number of places. First consider rural areas.

1. Density of less than 2 persons per square mile (640 acres) would mean there would be less than one farm family per square mile (assuming a family to have 4 members). The farm would be at least 640 acres—like those in many parts of the U.S. Middle West.

2. Density of 10 to 20 people per square mile would allow about 3 farms of 200 acres each.

3. Density of 20 to 75 per square mile would allow 5 to 20 farms of 30 to 150 acres. Such farms approximate the size of farms in western Europe.

4. Densities of 100 to 250 would permit about 25 to 60 farms per square mile—like the larger Asian farms.

5. Densities of 500 persons would mean that 125 farms of 5 acres each would squeeze into one square mile. This is approximately the farm size in Ramkheri.

We get a different view of densities if we consider land from an urban point of view. In the city of Chicago, streets are measured off twelve to the mile in an east-west direction, and eight to the mile in a north-south direction. There are 96 blocks to a square mile.

1. Density of less than 2 persons per square mile would mean fewer than 2 people lived in 96 blocks.

2. Densities of up to 250 persons place $2\frac{1}{2}$ persons per block at the maximum density. Since this is smaller than urban family size, there would be less than one family per block.

3. In densities of up to 500 persons per square mile, over 5 persons at the maximum would live in a block.

These simple calculations suggest that from a rural or agricultural standpoint the densities on the map are indeed high. They approach those of Ramkheri. It takes good farmland and very productive agriculture to support such populations locally. Even so, the living standard is likely to be low. The lower densities can produce much more than a marginal living for farmers if they use machinery and inanimate energy. Low densities may also be a measure of poor terrain. In some places one square mile of land does not produce enough for subsistence. Ranches in the western United States can be 100,000 acres in size. Nomadic herders might traverse even larger tracts in their migration through the year.

The densities have different meaning when viewed from an urban perspective. The most recent census data show that the average density for all U.S. metropolitan areas is around 360 persons per square mile. For metropolitan Washington it is more than 1,200 per square mile, and for Jersey City, New Jersey, it reaches over 17,000 persons per square mile.

In the central city, residential densities can be even greater. Not only are dwellings close together, but in high-rise buildings, people can be stacked on top of one another. In Chicago large public housing units run fifteen stories high and as many as 1,200 persons live in each unit. But the people who live in those buildings go elsewhere to work or school. Conversely, very few people make their homes in the industrial, commercial, and institutional parts of the city, but such areas can be very crowded during working hours.

What is the significance of population density? Compare how it affects life in Washington, D.C., with how it affects life in Ramkheri.

areas, however, the meaning of population density must be considered more carefully, because here people do not depend so fully or directly on the land they occupy. A case in point is metropolitan Washington, described in Chapter 1.

The four areas of population concentration mentioned previously, plus significant areas in West Africa, Indonesia, Mexico, and along the eastern and western margins of South America, are crowded, with more than 64 persons per square mile (25 per square kilometer). At the other extreme, more than half the land area of the world has fewer than 2 persons per square mile. In fact, all continents except Europe have large areas in which population densities are low.

Why should there be such stark contrasts in population density? The answer, as we shall see, is to be found in the combined effects of such factors as physical environment, historical circumstances, and regional differences in rates of population growth. The population pattern of the world today is the result of the past blending of these factors.

THE ENVIRONMENT AS A RESOURCE BASE

The interdependence of population and the physical environment is important to both modern and traditional societies. All people, no matter what their technology, must employ resources drawn from the physical environment and must depend on the processes of the environment to maintain a suitable living habitat. Nevertheless, the relationships differ in nature in each kind of society. A close relationship between the pattern of biological potential of environments and population distribution is normal for traditional societies because of their dependence on local land resources for food, clothing, and shelter. In traditional societies people must live near the sources of food and of materials for clothing and shelter. In modern societies, by contrast, the technology of transportation and communication permits people to live in places away from the

TABLE 5-1 ESTIMATED PERCENTAGE OF COLD AND DRY AREAS BY CONTINENTS AND FOR THE WORLD

Continent	Cold[a]	Dry[b]
North America	45	9
South America	1	11
Africa	nearly 0	43
Eurasia	19	23
Australia and New Zealand	0	52
Antarctica	100	0
World	24	23

SOURCE: Data from E. H. Hammond.
[a] Fewer than ninety frost-free days per year.
[b] Includes both arid and semiarid environments.

locations of basic resources. As a result, modern societies can concentrate to a greater extent in large metropolitan areas.

THE CLIMATIC FACTOR

The clearest relationship between population and environment can be seen in the consistent association of low population densities with the cold and dry areas of the world. Table 5-1 provides estimates, by continents, of the extent of low-work environments, and Figure 3-7 in Chapter 3 shows their pattern. Approximately half the land areas of the world as a whole, except for South America, lie within low-work environments.

A few areas do not fit the pattern. For example, Pakistan, Egypt, and other parts of the Middle East are in dry environments, but they have large population concentrations. The dense settlement in such areas is a response to the presence of major rivers and adjacent lowlands, where intensive agriculture under irrigation can be practiced.

The map of population density (Figure 5-4) also shows a few major areas of sparse population in equatorial regions where there is high year-round rainfall (see Figure 3-7 for the pattern of high-work environments and Figure 3-4 for the world pattern of rainfall). The Amazon Basin in South America, the interior of the Congo Basin in Africa, and New Guinea are such areas. As was discussed in Chapter 3, conventional crop agriculture has not done well in these areas despite the high biological productivity of the natural vegetation. Nutrient-poor soil, the competition from the rapid growth of other plants, diseases of all kinds, and abundant insect life are all detrimental to conventional agricultural practices. Such conditions have hampered the growth of human populations in these areas.

By contrast, more than one-quarter of the world's population is concentrated in the tropical latitudes of South Asia. These dense populations, however, are in seasonal (wet and dry) environments for the most part, not year-round wet environments. Furthermore, they are associated with special land resources, such as nutrient-rich soils on the island of Java and the Deccan Plateau of India, or riverine lowlands where paddy rice cultivation dominates. Paddy rice cultivation is an especially appropriate type of agriculture for tropical environments because paddy flooding limits the growth of competing vegetation and algae in the water provide a source of nitrogen.

By the process of elimination we can now say that most of the world's population is concentrated in areas of seasonal environments. This is not to say, however, that all areas of seasonal environments are well populated. Population densities may vary widely, and areas of similar seasonal environment may have very different population densities. Eastern China, India, and Western Europe, for example, are much more densely settled than are areas with comparable climatic environments in North America, South America, or Australia. The conclusion we are led to at this point is that climatic extremes of too hot, too wet, too cold, or too dry normally are associated with sparse populations, but that elsewhere the relationship between population density and climatic environment is not consistent.

LAND SURFACE CONFIGURATION

We might expect people to concentrate on land surfaces that they can most easily cultivate for food and that provide the fewest barriers to movement of goods. It follows that people should concentrate on flatter land and should avoid mountains and other rugged and steeply sloping land, where there are erosion and other problems in managing the environment. Compare the map of population density (Figure 5-4) with the world elevation map (Figure 3-1). You will see that the humid lowlands of China, India, some parts of Southeast Asia, and Western Europe are densely populated. Low densities, however, prevail on the flat lands of South America, over much of the interior plains of North America, and in western Siberia. Conversely, highlands in parts of Latin America, China, and Japan are rather densely settled. In general, the relationship between population density and land surface configuration is not consistent.

POPULATION CONCENTRATIONS AS HISTORICAL DEVELOPMENTS

The unusual concentration of population in a few areas of the world is not explained adequately by environmental factors alone, but is better explained by the cumulative effects of historical circumstances. Table 5-2 shows that as long as 2,000 years ago China and India were major centers of population. The reasons for their early development are not entirely clear, but they are probably rooted in the early development of productive agriculture, the attainment of a high degree of internal social and economic stability and effective governmental authority, and the ability to defend occupied areas from outside invaders.

TABLE 5-2 ESTIMATES OF POPULATION ABOUT 2,000 YEARS AGO

World	250–300 million
Indian subcontinent	100–140 million
China and fringe areas	70–80 million
Roman Empire	50–60 million

SOURCE: G. T. Trewartha, *A Geography of Population: World Patterns*, John Wiley, New York, 1969.

Both China and India, as well as the Middle East, are recognized as centers of origin for domesticated plants and animals (tea, rice, jute, and pigs are important examples). The Chinese and Indians also were early to practice intensive cropping and to use irrigation extensively. All these circumstances could have made food supplies more certain and permitted populations to flourish.

The concentration of population in Western Europe is an outgrowth of the new ideas and explorations of the Renaissance period of the fourteenth through sixteenth centuries and the Industrial Revolution in the eighteenth and nineteenth centuries. These developments, along with the colonial bonds that were formed with other lands, gave Europeans greater access to material wealth and food supplies. After the Industrial Revolution Europe experienced rapid population growth and by the early nineteenth century had become a major concentration of world population (Table 5-3). Also, migrants from Europe formed the nucleus for population growth and technological advances on the more sparsely populated continents of North and South America and Australia (see Figure 5-6). Population concentration in eastern Anglo-America (Canada and the United States) can be seen historically as an extension of European population concentration. As is true for each of the other three major regions of population concentration, the contemporary population of Anglo-America is primarily the result of domestic growth rather than migration.

INDEXES OF MODERNIZATION

The distribution of world population provides us with important information about where people live, but it is less revealing about how they live.

TABLE 5-3 POPULATION GROWTH, 1650–2000 (*in millions*)

	1650	1750	1850	1900	1950	1970	2000[a]
North America	1	1	26	81	166	228	354
Latin America	12	11	33	63	162	283	638
Europe and USSR	100	140	266	401	559	705	873
Asia (except USSR)	330	479	749	1186	1302	2056	3458
Africa	100	95	95	120	198	344	768
Oceania	2	2	2	6	13	19	32
World	545	728	1171	1857	2400	3635	6123

SOURCE: W. S. Woytinsky and E. S. Woytinsky, *World Population and Production: Trends and Outlook.* © 1953 by the Twentieth Century Fund, New York, p. 34, Table 14; and Population Reference Bureau, Inc., *World Population Data Sheet, 1970.* Washington, D.C., 1970.
[a] U.N. medium growth projection.

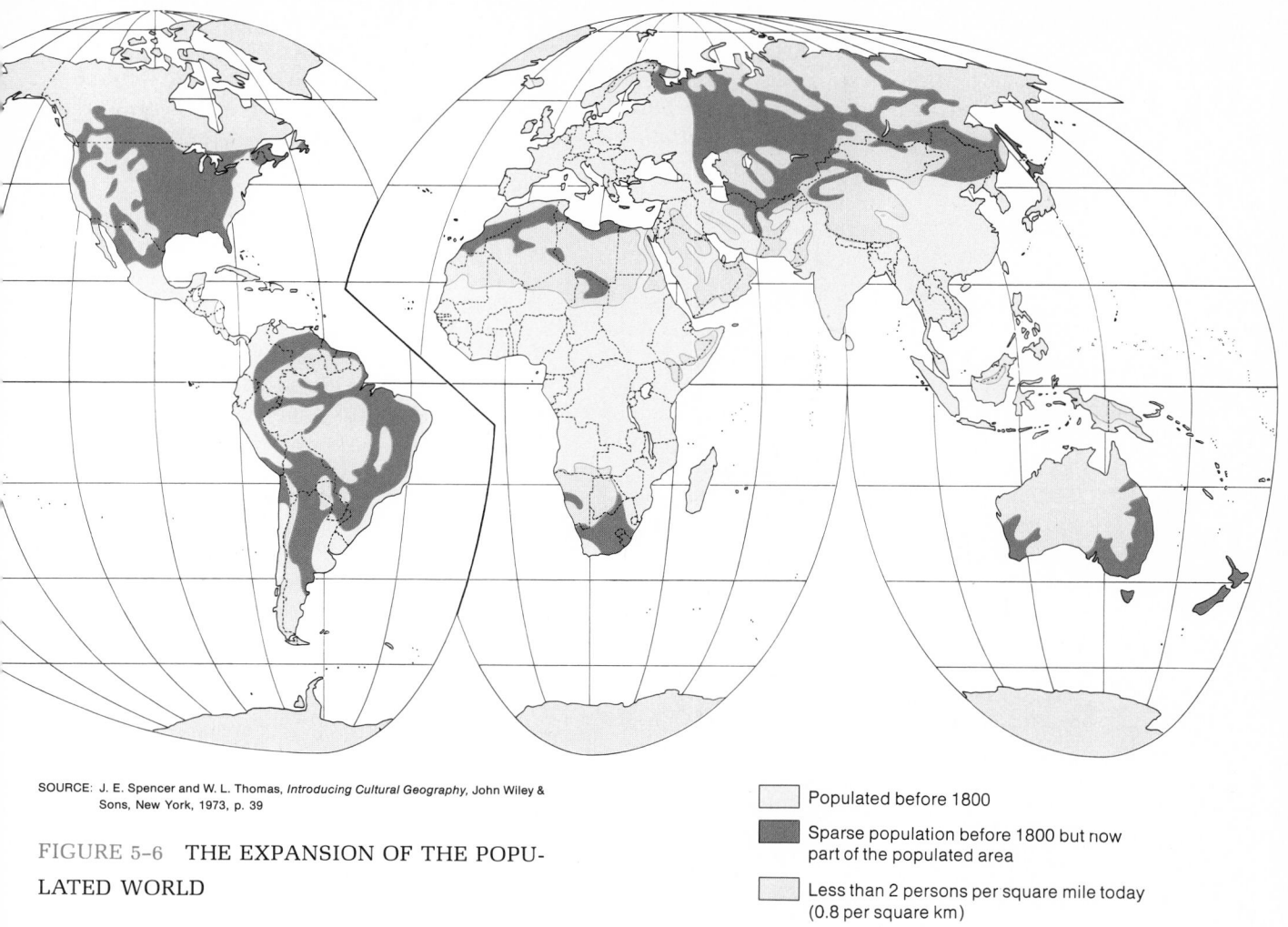

SOURCE: J. E. Spencer and W. L. Thomas, *Introducing Cultural Geography*, John Wiley & Sons, New York, 1973, p. 39

FIGURE 5–6 THE EXPANSION OF THE POPU-
LATED WORLD

☐ Populated before 1800

■ Sparse population before 1800 but now
  part of the populated area

☐ Less than 2 persons per square mile today
  (0.8 per square km)

The varied ways of life over the globe are generalized in this book, as a first step toward understanding, into a simple dichotomy between modern and traditional. To place this dichotomy in
geographical perspective, the following sections
review major differences in urban and agricultural settlement, in a few key measures of material well-being, and in the interconnectedness of

places. Urban and agricultural settlement patterns represent contrasting ways of living that
can serve as indexes of modernization. The key
measures of material well-being and their world
patterns that we shall review are dietary affluence, gross national product, and energy consumption. The patterns of transportation networks and communications provide measures of

the accessibility and connectivity of places and are crucial indexes of modernization. The chapter concludes with a brief discussion of interactions and transitions between modern and traditional societies and their spatial organizations.

COUNTRIES: UNITS OF DECISION
MAKING AND POLITICAL POWER

Our focus so far has been on continental regions of the world. We have spoken of Southeast Asia, the interior of South America, and Europe, for example. Countries, however, are the basic operational units of government today and are thus much more important in the functional world than continents or any other divisions.

National governments are decision-makers and controllers. They set the basic rules for individuals and organizations within their boundaries. They also represent the interests of their people in relations with other countries, control the movement of trade, and set the stage for war or peace.

Almost the entire land surface of the earth has been incorporated into political units. Only the icy continent of Antarctica is not under the political control of some government; yet even there, some countries have made political claims.

The political map (Figure 5-7) points up the compartmentalized nature of the human use of the earth. The contemporary world is made up of more than 150 separate political entities, each claiming legal power over the people who live within its boundaries. Each government expects the primary loyalty of its people, who see themselves as citizens of a particular political state, not of the world. Today, unlike thirty years ago, almost all political units are independent. The few dependent smaller places, mainly islands and port areas, do not show up easily on a world map.

The political map reveals great differences in the land areas of countries. The vast Soviet Union encloses 15 percent of the land surface of the earth. China, Canada, the United States,

Australia, and Brazil are also huge. At the other extreme, Singapore is a single metropolitan center that is politically independent. The few remaining feudal kingdoms, such as Liechtenstein, Monaco, and Andorra, are too small even to show up on a world map. The area of Monaco is less than one square mile.

How well does the size of a country correspond to its political power today? Some of the patterns already examined in this chapter might suggest an answer to this question. Imagine the political map (Figure 5-7) superimposed on the map of world population distribution (Figure 5-2). Some of the largest countries in the world are in areas of low population density; some of the smallest are in areas of high density. Because few countries have their populations spread evenly over their territories, the population distribution map shows countries in a new perspective. Huge countries such as the Soviet Union, China, Canada, and Brazil take on new shapes in our minds. Let us keep these "shapes" of countries in mind as we study the spatial pattern of the two systems—the modern interconnected system and the traditional, locally based system—throughout the world.

URBAN PATTERNS

Places where people live near each other and are mostly engaged in nonagricultural livelihoods are usually referred to as **urban.** In earlier times the most common activities in urban places were government, religion, trade, and commerce. More recently, industrial activities, service activities, and the business of management and decision making have become prominent urban functions.

Urban populations have existed for several thousand years, especially in the areas of early civilization such as China, India, and the Mediterranean region, and until recent times they were small. World population has grown rapidly over the past few centuries, but urban population has grown at much higher rates and has been

TABLE 5-4 POPULATION IN METROPOLITAN PLACES OVER 100,000 IN 1800

World Region	Number of Metropolitan Places	In Places over 1 Million	In Places 500,000– 1 Million	In Places 100,000– 500,000	Total Metropolitan Population
Asia	30	1.1	1.5	4.7	7.3
Europe	18		2.0	2.6	4.6
USSR	2			0.5	0.5
Latin America	1			0.1	0.1
Anglo-America	0				
Africa	1			0.3	0.3
Total	52				12.8

SOURCE: Tertius Chandler and Gerald Fox, *3000 Years of Urban Growth*, 2nd ed., Academic Press, New York, 1978.

especially associated with industrialized areas. Therefore the proportion of the population that is urban can be used as a rough measure of a nation's position on the continuum between traditional and modern economic and political development.

A better indicator of modernization is the porportion of population concentrated in metropolitan centers (large cities). **Metropolitan areas** are a new phenomenon in the world. Less than 200 years ago, only one city had as many as a million people, and, as the estimates in Table 5–4 show, only fifty-two cities exceeded 100,000 population. All but four of these were in the areas of early civilization and population concentrations of Europe and Asia.

Metropolitan areas like Washington, D.C., are not limited to the resource base in their immediate vicinity but rather draw on the resources of distant places. The development of more efficient means of overland transportation in the past 150 years was therefore essential to the development of metropolitan areas. Until the appearance of railroads and motor vehicles, land transportation was poor and only ships could move large volumes of materials and people. Historically, the largest cities were seaports because only in such

locations was there sufficient access by ships to the food and material resources needed to support large numbers of nonagriculturalists. Nevertheless, many of today's modern inland metropolitan centers, especially in Europe and Asia, had their beginnings as local political or trade centers.

Because large metropolitan centers are particularly identified with the development of the modern interconnected system and that system is unevenly distributed over the world, these areas follow a pattern rather different from that of world population in general. The greatest growth of metropolitan centers has taken place in industrialized areas where much of the activity of the modern interconnected system has focused. These centers are the hubs of the world's transportation and communication networks and dominate such activities as wholesale and retail trade, manufacturing, finance and business, medical services, and government.

The highest concentrations of population in large metropolitan centers are found in Anglo-America, Australia and New Zealand, Europe, and Japan (Table 5–5). Metropolitan development is also well established in parts of the Soviet Union, Latin America, and India and China.

FIGURE 5-7 POLITICAL MAP OF THE WORLD

NORWAY
SWEDEN
FINLAND
DEN.
E.
GER.
POLAND
W.
GER.
CZECH
AUS. HUNG.
RUM.
YUGO.
BULGARIA
ITALY
ALB.
GREECE
TURKEY
SIA
CYPRUS
LEB.
SYRIA
IRAQ
IRAN
ISRAEL
JORDAN
KUWAIT
LIBYA
EGYPT
SAUDI
ARABIA
QATAR
UNITED
ARAB
EMIRATES
OMAN
CHAD
YEMEN
YEMEN
DEM. REP.
SUDAN
AFARS AND ISSAS
ETHIOPIA
CENT.
AFRICAN
EMPIRE
SOMALIA
UGANDA
KENYA
RWANDA
ZAIRE
BURUNDI
TANZANIA
NGOLA
ZAMBIA
MALAWI
MOZAMBIQUE
RHODESIA
MALAGASY
REPUBLIC
BIA
RICA)
BOTSWANA
SWAZILAND
SOUTH
AFRICA
LESOTHO
TRANSKEI
(S. AFRICA)

SOVIET UNION
MONGOLIA
AFGHANISTAN
JAMMU
& KASHMIR
CHINA
N
KOREA
S
KOREA
JAPAN
PAKISTAN
NEPAL
SIKKIM
BHUTAN
INDIA
BANGLA
DESH
BURMA
LAOS
TAIWAN
THAI-
LAND
VIETNAM
KHMER
REP.
PHILIPPINES
SRI
LANKA
BRUNEI
MALAYSIA
SINGA-
PORE
INDONESIA
PAPUA
NEW GUINEA
AUSTRALIA
NEW
ZEALAND

20°   40°   60°   80°   80°   100°   120°   140°   160°   180°

TABLE 5-5   POPULATION IN METROPOLITAN PLACES OF MORE
THAN 1 MILLION PEOPLE, 1970, BY WORLD REGIONS

| | Metropolitan Places of Over 1 Million Population | | |
Regions	Number of places	Sum of population (millions)	% of total population
Asia	69	175	8
Europe	43	96	21
Anglo-America	36	87	38
Latin America	17	51	18
USSR	10	23	9
Africa	8	15	4
Oceania[a]	2	5	26
World	185	452	

SOURCE: Kingsley Davis, *World Urbanization 1950–1970, Vol. 1; Basic Data for Cities, Countries, and Regions,* Population Monograph Series, No. 4, University of California Press, Berkeley, 1969, Table 4; and *U.N. Demographic Yearbook.*
[a] Including Australia and New Zealand.

Africa is the continent least dominated by large metropolitan areas.

The small number of metropolitan areas in Africa reflects the dominance of traditional, locally centered life. Metropolitan development in Latin America and the Soviet Union raises interesting points about the urban character of these two areas, which we shall explore in more detail in later chapters. Very large metropolitan centers of Latin America, such as Buenos Aires, Rio de Janeiro, São Paulo, Mexico City, Caracas, and Lima, are major "islands" of modernization within their respective countries. Away from them, the traditional form of living generally dominates. By contrast, most of the population of the Soviet Union is tied into the country's industrial-urban economy, even though people are no more concentrated in large cities than they are in Latin America and not nearly so concentrated in large cities as they are in the United States and Western Europe.

AGRICULTURAL PATTERNS

One of the best indexes of traditional living is the proportion of total employment that is agricultural. Because most people in traditional societies must devote the majority of their labor to activities that produce food and the other essentials of life, the percentage of a country's work force employed in agriculture indicates the degree to which that country is within the traditional, locally based system (Figure 5–8). Note especially the areas that have more than 75 percent employed in agriculture; they are dominated by locally based, rural ways of living. Most countries of Africa, South and East Asia, and Central America are only slightly less dominated by the traditional system, having over 50 percent of their labor forces in agriculture. The industrialized countries of Western Europe, of Anglo-America, and Australia and New Zealand, on the other hand, have less than 15 percent of their

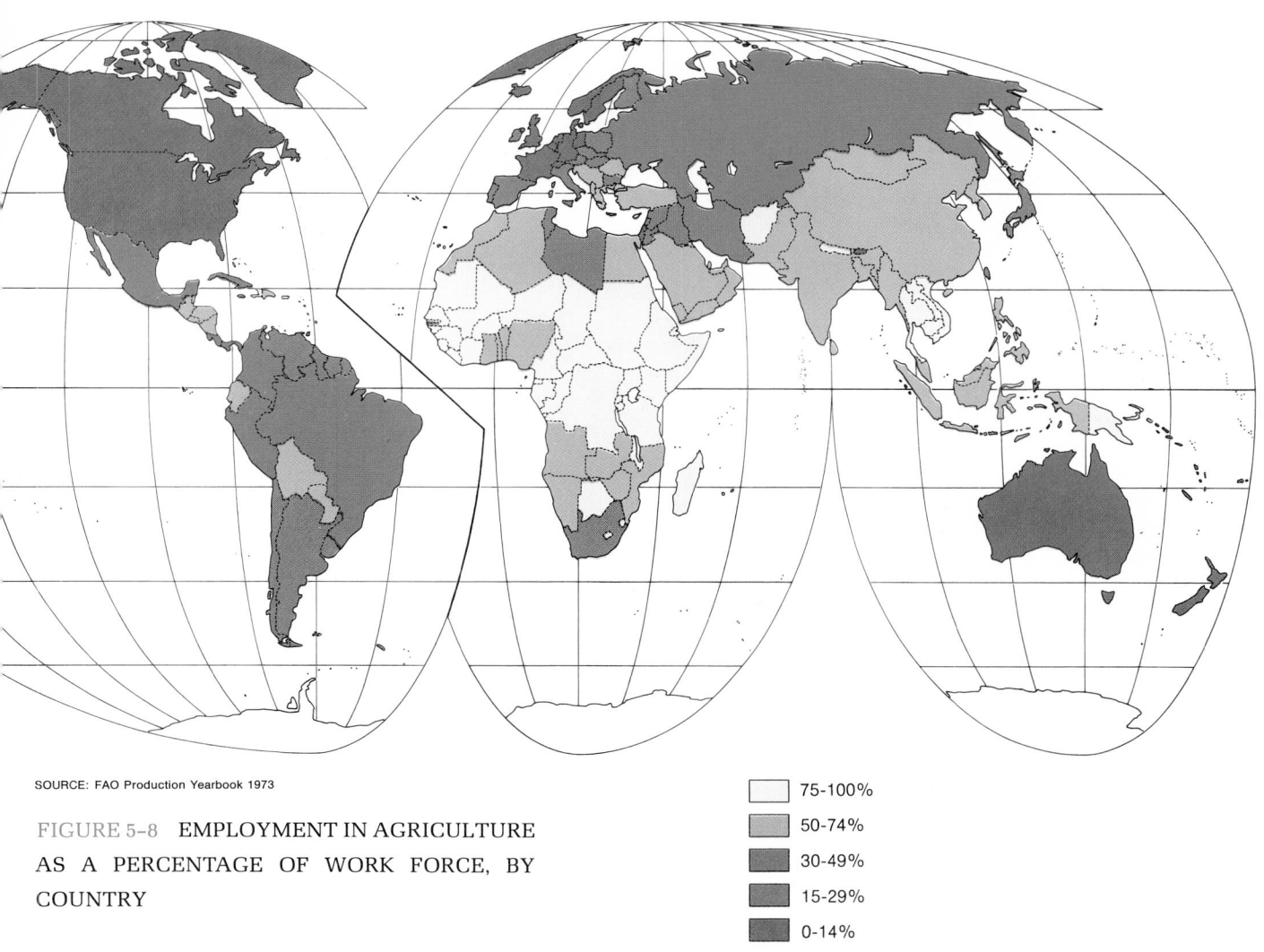

SOURCE: FAO Production Yearbook 1973

FIGURE 5–8   EMPLOYMENT IN AGRICULTURE
AS A PERCENTAGE OF WORK FORCE, BY
COUNTRY

	75-100%
	50-74%
	30-49%
	15-29%
	0-14%

labor forces in agriculture. The countries of
Eastern Europe and South America and the Soviet Union constitute a middle group, with one-
third to one-half of their employment devoted to
agriculture.

Comparison of Figures 5–8 and 5–9 shows that
agricultural employment is not tied to the availability of land for cultivation, as one might expect. The United States is an example of how
modern agricultural technology results in low
employment in agriculture. At present, employment in agriculture is less than 5 percent of the
U.S. labor force, even though cultivated land
comprises more than 20 percent of U.S. land

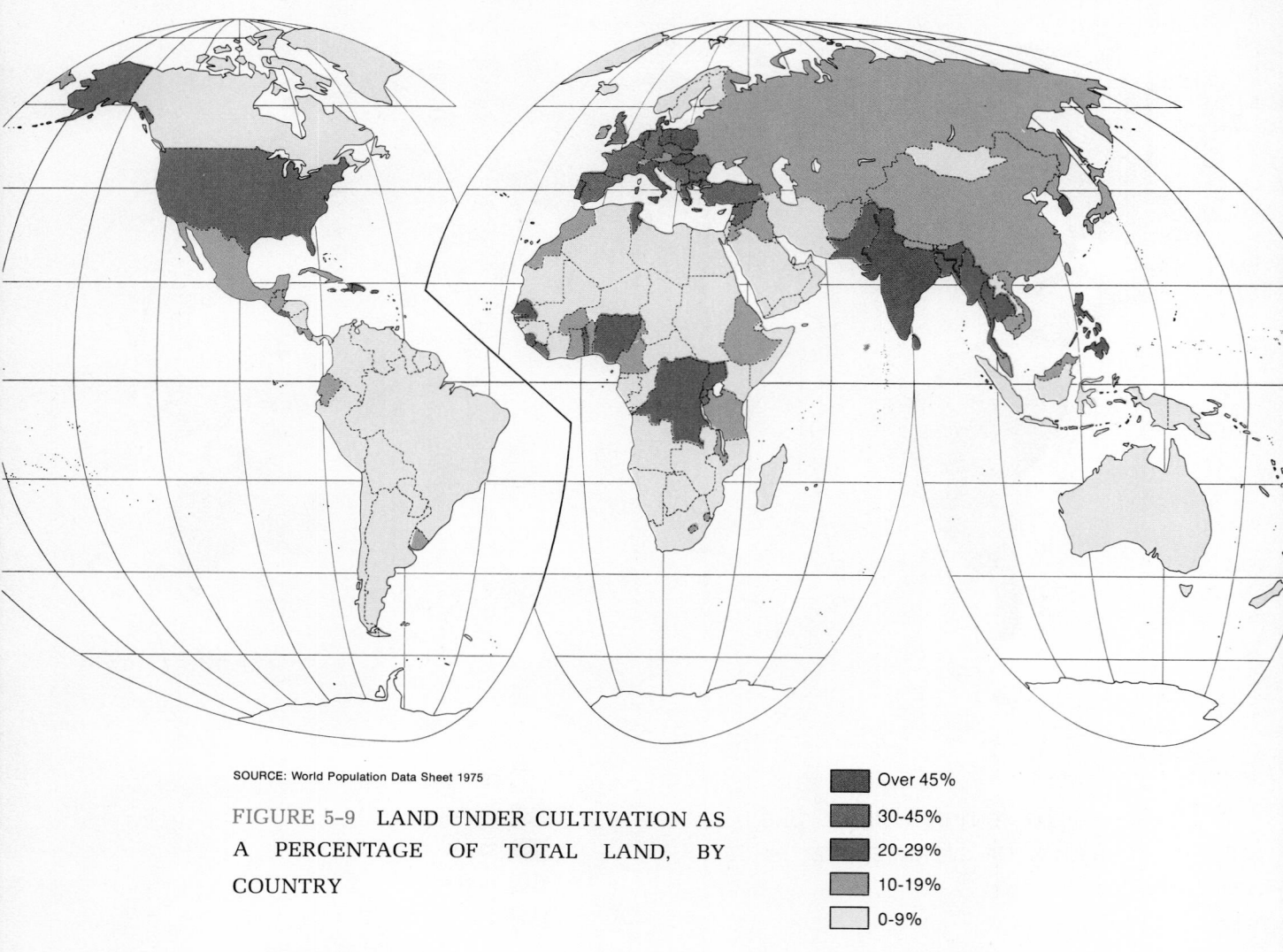

SOURCE: World Population Data Sheet 1975

FIGURE 5-9 LAND UNDER CULTIVATION AS A PERCENTAGE OF TOTAL LAND, BY COUNTRY

- Over 45%
- 30-45%
- 20-29%
- 10-19%
- 0-9%

area. At the other extreme are countries such as those of central Africa, where more than 75 percent of the labor force is employed in agriculture but less than 10 percent of the land is under cultivation. India and China illustrate another contrast. Over 70 percent of their labor forces are on the land, and the density of rural population is high. In the best agricultural areas of these countries, densities are similar to those of urban residential areas in the United States. The type of agriculture practiced in India and China is labor intensive, requiring a great deal of human effort. United States agriculture, on the other hand, requires only a low labor input but intensive use of machines and fuels for energy.

The pattern of agricultural settlement can also be quite different over the world. In places such as Ramkheri and elsewhere in South and East

Asia and in most of the rural parts of Europe, people live in villages surrounded by the fields, grazing land, and woodland on which they labor and from which they sustain life. In other places, such as rural Anglo-America, farming people have generally lived in isolated dwellings dispersed over the land.

WORLD DIFFERENCES IN
LEVELS OF LIVING

The several measures of standards of living to be discussed here are based on averages for nations and therefore may mask major disparities within countries. Nevertheless, such data provide clues to differences in material well-being between regions and countries.

Food: Calorie and Protein Intake

Material affluence is often thought of in terms of the production of goods people use, but food consumption is perhaps more fundamental. To eat, either food must be produced by the family, as in Ramkheri, or else something must be produced to exchange for food and some other person and place must produce the surplus food that is exchanged, as is the situation in Washington, D.C.

Calorie intake, as shown in Figure 5–10, is a rough measure of the energy available to the human body to carry on work and activity. The actual number of calories that individuals need varies substantially. People in the tropics require less energy than do people in cold climates, where extra energy is needed to sustain body temperature. Persons engaged in heavy work need more calories than people who do little manual labor. Individuals of small stature or body size need fewer calories than large people do. Growing children use more energy per unit of body weight than do adults.

In Figure 5–10, high per capita calorie intake exceeds 3,000 calories per day, and low intake is fewer than 2,500 calories per day. Many coun-

TABLE 5–6   PROTEIN CONSUMPTION PER CAPITA BY WORLD AREAS

Region	Daily Protein Consumption (Grams per Person)
North America	96
Europe	87
Oceania[a]	106
Israel, Japan, South Africa	77
USSR	92
Africa (except South Africa)	58
Latin America	68
Middle East (except Israel)	70
Far East (except Japan)	57
South Asia	48

SOURCE: *The Food and People Dilemma* by Georg Borgstrom (Scituate, Mass.: Duxbury Press, a division of Wadsworth Publishing Company, 1973), p. 126.
[a] Including Australia and New Zealand.

tries have per capita intakes of fewer than 2,100 calories, indicating that many of the people in these countries must have barely adequate or even deficient diets in terms of their food energy needs.

Although it is useful to compare food energy, the calorie is not an adequate measure of the quality of an individual's diet. A balance of essential nutrients—proteins, mineral elements, and vitamins—is needed to build and maintain body tissues. Protein intake, however, serves as a good measure of the nutrient quality in the human diet (Table 5–6). Sixty grams of protein per day is a reasonable minimum requirement for adults to maintain good health and vigor. By this standard, about half the world's population, primarily people in Africa and South and East Asia,

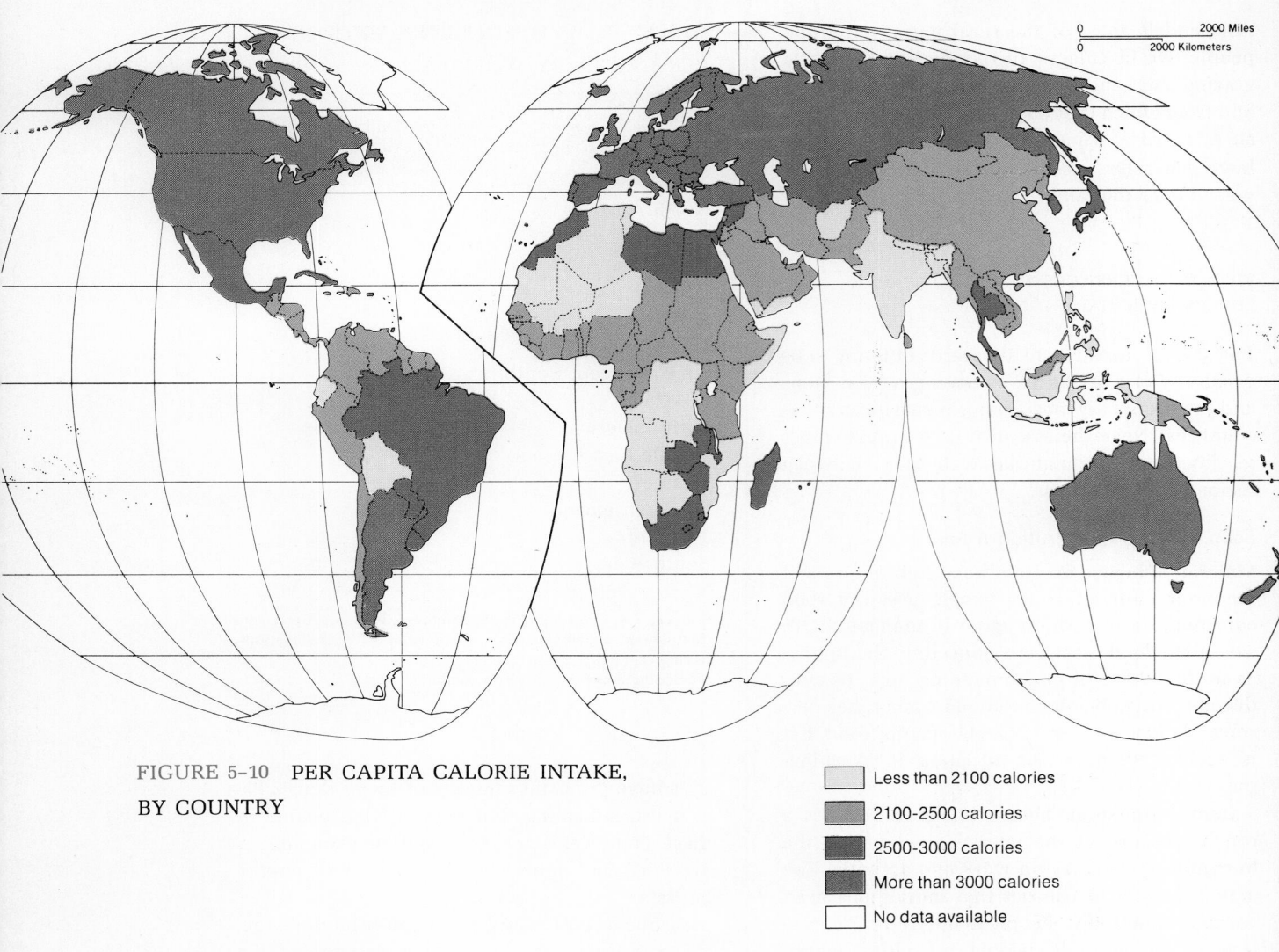

FIGURE 5–10   PER CAPITA CALORIE INTAKE,
BY COUNTRY

Less than 2100 calories

2100-2500 calories

2500-3000 calories

More than 3000 calories

No data available

have marginal or inadequate protein in their diets. By contrast, people in North America and Oceania far exceed the minimum, and might well be considered excessive in their protein consumption.

Figure 5–10 and Table 5–6 provide strong evidence that populations in several parts of the world live with chronic dietary deficiencies. In-

adequate energy as well as malnutrition from protein deficiencies are continuing problems in these areas. Famine and starvation are recurrent threats, especially for the very young, the aged, and the landless poor. On the other hand, some parts of the world have ample, perhaps even excessive food intake. These, for the most part, are the modern areas of the world.

Gross National Product:
A Measure of Overall Output

Economists commonly use **gross national product** (GNP) per capita as a measure of a country's productivity. GNP measures the total production of goods and services in a country during a given period of time, usually a year. It is difficult to obtain accurate figures on GNP in modern countries, but it is even more difficult to gather suitable data in areas where a large share of production is used locally by people to feed themselves or to trade with others in local markets. As a result, GNP estimates may not adequately account for subsistence-type production.

It is hard to imagine how people survive in places where per capita GNP is very low, stated as the equivalent of fifty to a hundred dollars per person per year. In terms of our exchange economy, this is just not sufficient to support life. However, much of the output of the subsistence producers is simply not counted; for countries where subsistence producers account for a large share of total production, the value of production is undoubtedly greater than the GNP indicates. Another complication in obtaining comparable GNP estimates is that output must be expressed in some common value, usually U.S. dollars. This value must be based on currency exchange rates, which can be misleading. Per capita GNP is thus at best only an approximate measuring stick for comparing productivity among countries, and the national averages do not reveal the striking inequalities that can exist among people within a nation. For example, in several of the oil-rich nations where average per capita wealth may approach or exceed that of the United States, that wealth is unevenly distributed. A small minority is very wealthy, while the vast majority of the people may be as poor as agriculturalists in most traditional societies. Still, GNP is one of the best measures we have, and since productivity varies so greatly among countries, it illustrates the extremes in the world. For our purposes, GNP is a useful index of the extent to which each country's population is a part of the modern system, since it is primarily a measure of modern economic activity.

The range between the most productive and the least productive countries is wide (Figure 5-11). In the United States, production of goods and services per person is more than 10 times that of many Latin American countries, more than 20 times that of many African and Asian countries, and more than 100 times that of some countries with very low per capita GNP.

When we study Figure 5-11 and the table in the Appendix, we find that Asia, with more than half the world's population, has low production of goods and services per capita, with the exception of Japan. Low productivity also characterizes most of Africa and Latin America. Most of the people who enjoy the benefits of modern productivity are in Europe, North America, the Soviet Union, Japan, Australia, New Zealand, and South Africa. Some of the oil-rich nations of the Middle East also show high per capita GNP, but as mentioned above, the averages do not fairly represent the economic well-being of the majority of the people in those countries. Collectively, the affluent populations of the world are a minority, no more than one-third the total world population.

Energy Consumption: A Measure of Production

Per capita energy consumption (Figure 5-12) is perhaps the most important indicator of the two contrasting worlds. The industrial world depends on inanimate energy to run its machines, and people in the modern interconnected system use enormous amounts of energy for comfort, convenience, and transportation. Most of this energy comes from coal, oil, and natural gas. The traditional world, in contrast, relies on people and animals for power and on biological materials, such as wood, for fuel.

Vast differences in energy consumption are evident from the map (Figure 5-12). Per capita energy consumption in the United States is about twice that of most Western European nations or Australia; almost 10 times that of Mexico; nearly

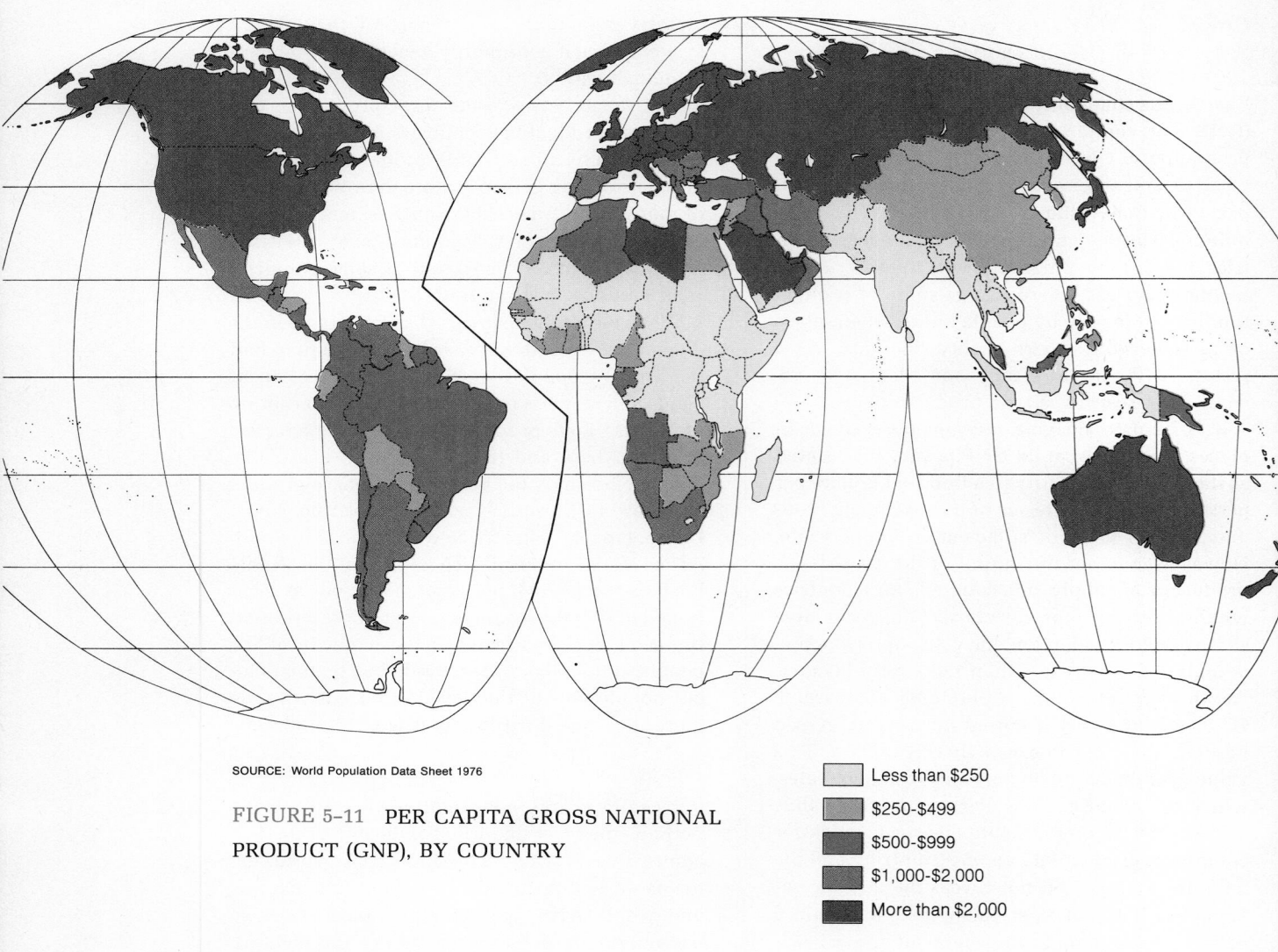

SOURCE: World Population Data Sheet 1976

FIGURE 5–11   PER CAPITA GROSS NATIONAL
PRODUCT (GNP), BY COUNTRY

Less than $250
$250–$499
$500–$999
$1,000–$2,000
More than $2,000

100 times that of India; and as much as 200 times that of China. With the exception of a few small countries, the areas or countries with energy consumption above the world average are industrialized: Anglo-America, Europe, the Soviet Union, Japan, Australia, New Zealand, and South Africa.

Even among the industrialized countries, per capita energy consumption varies by large amounts. What do these differences imply? Is the United States, with at least twice the per capita consumption of most Western European countries, that much better off in per capita goods, comforts, and general standard of living, or is it that much more wasteful in energy use? Consider Japan, which uses only one-third the

148   THE GEOGRAPHIC PROBLEM: FINDING A FOCUS FOR THE WORLD

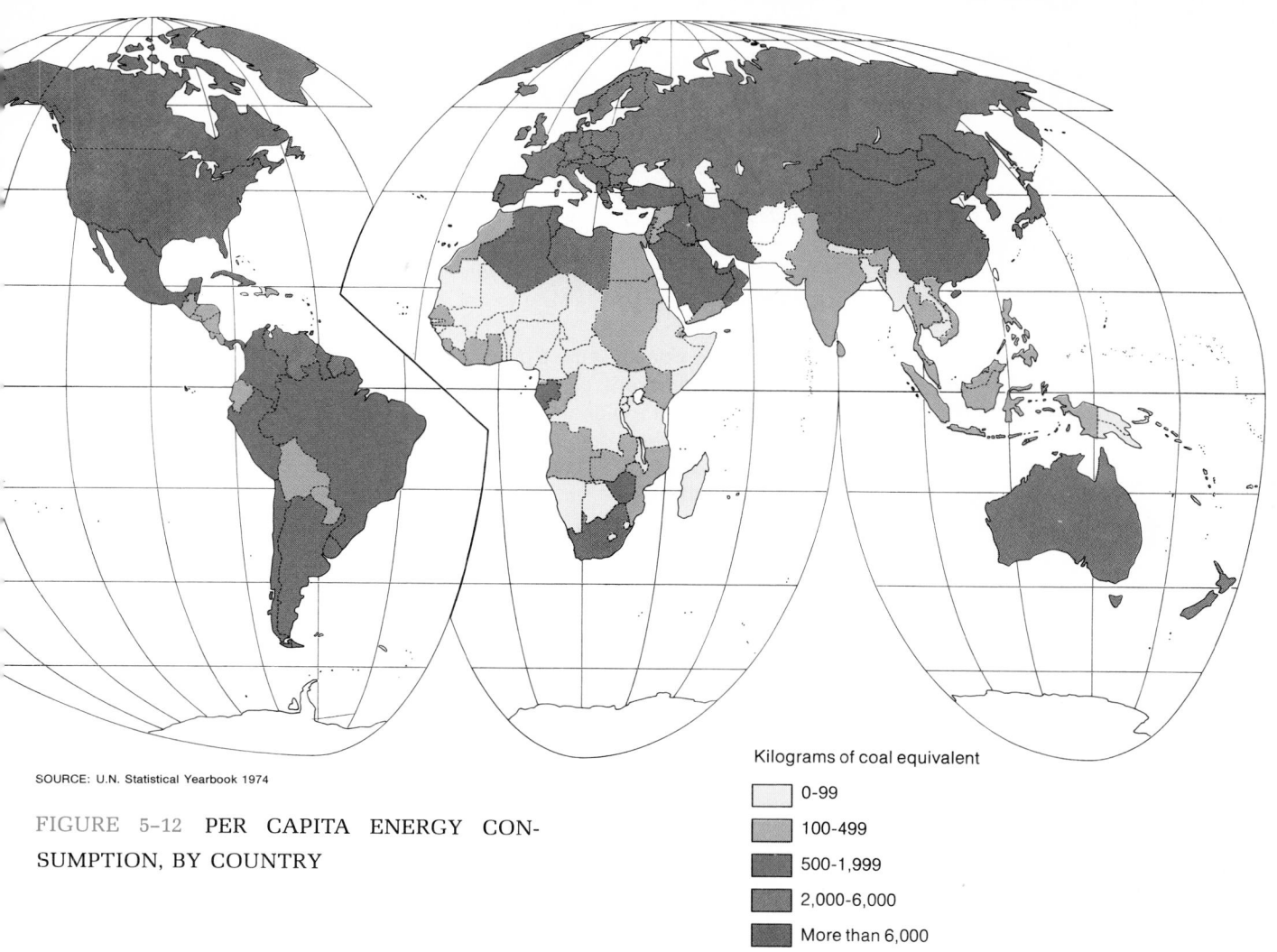

SOURCE: U.N. Statistical Yearbook 1974

FIGURE 5–12 PER CAPITA ENERGY CON-
SUMPTION, BY COUNTRY

Kilograms of coal equivalent

0-99

100-499

500-1,999

2,000-6,000

More than 6,000

energy per capita of the United States and 15 to 20 percent less than most countries of Western Europe.

Such differences in energy use among modern nations may in part reflect environmental differences. For example, countries with long, cold winters may require more energy for heating purposes, or countries with large land areas may need more energy for transportation. Nevertheless, between countries large discrepancies still remain that point to differences in efficiency or wastefulness. Even with its more northern location and high level of industrialization, Sweden uses only 60 percent as much energy per capita as the United States. Evidence such as this indicates that U.S. use of energy in particular may be

Ocean shipping is a very important indicator of international movement of bulk commodities. It is dominated today by the movement of petroleum, which shows up clearly in a comparison of loadings and unloadings. For example, the values for Iran and Saudi Arabia (oil producers) and Japan (a major user) reveal the importance of petroleum. The larger unloadings for Europe reflect its dependence, like Japan's, on imported oil and many other raw materials. In general, the more traditional areas of the world have larger tonnages loaded than unloaded, whereas the more modern areas have larger tonnages unloaded. Is it reasonable to expect these differences in view of demands for raw materials by modern industrial areas and their marketing of manufactured goods? Might these balances change if the figures were in value (for example, dollars) rather than tonnage?

Civil aviation data (Table 5–8) provide a good measure of the movement of people at regional and world scales by showing the number of passengers carried by airlines on both domestic and international flights. Note that the United States and Canada (Anglo-America) account for more than half the world's air passengers. Africa, Latin America, and Asia (except for Japan), on the other hand, account for less than 15 percent of the world's air passengers, even though they have more than half the world's population. Here again we see the contrast between the modern areas and the traditional ones.

The figures for international passengers show that about one-fourth of all air passengers travel across national boundaries. Anglo-America and Europe account for nearly three-fourths of all international air passengers. Can you explain the exceptionally high proportion of international air passengers for Europe?

inefficient and wasteful. Because energy is becoming more expensive and shortages of certain kinds of fuels may become more frequent, efficient use of energy will probably get more attention in the future than it has in the past.

Until a few years ago the trend in energy costs was down. Now it is moving upward. In addition, reserves of the most widely used source of energy—petroleum—are likely to be exhausted within the next fifty years, and as of now most of the proven reserves are concentrated outside the major consuming areas. If, because of rising costs or diminished access to supply areas, the modern nations have to use less energy per capita, how might this decrease affect the material wealth and lifestyles of their populations? Conversely, can poor nations with low per capita energy consumption expect to obtain a larger share of the world's limited energy resources? If not, how might this affect their expectations for industrialization and economic development? Keep these questions in mind as we study the different regions of the world.

WORLD CONNECTIVITY: TRANSPORTATION AND COMMUNICATION

The circumstances of human existence at every place in the world are strongly influenced by the **accessibility** of each place to the resources, ideas, and people of other places. In areas of traditional societies the only convenient access is to a limited, nearby territory. On the other hand, people of modern societies have partial access to most of the world and widely available access to the

TABLE 5-7 INTERNATIONAL SEA-
BORNE SHIPPING, 1972 (*Thousands
of Metric Tons*)

Regions	Goods Loaded	Goods Unloaded
Africa	392,699	98,485
Anglo-America	308,248	402,979
Latin America	381,001	204,008
Asia	1,134,961	780,058
Iran	(236,300)	(5,200)
Saudi Arabia	(247,504)	(3,000)
Japan	(53,198)	(517,396)
Europe	408,043	1,298,950
Oceania	131,275	37,807
USSR	109,331	29,971
World[a]	2,865,558	2,852,258

SOURCE: *U.N. Statistical Yearbook, 1974.*
[a]Petroleum products account for 58 percent of world shipping tonnage. The difference between the world totals for goods loaded and goods unloaded represents goods in transit.

TABLE 5-8 WORLD CIVIL AVIATION,
PASSENGERS CARRIED, 1973,
BY WORLD REGION

Region	Total Passengers (Millions)	International Passengers (Millions)
Africa	9.6	4.5
Anglo-America	217.5	21.7
Latin America	24.0	6.0
Asia[a]	46.4	11.8
India	(3.8)	(0.8)
Japan	(25.4)	(2.4)
Europe	94.5	50.8
United Kingdom	(19.3)	(12.6)
France	(12.0)	(6.3)
Oceania	11.4	1.6
USSR	84.2	1.4
World[a]	487.6	97.8

SOURCE: *U.N. Statistical Yearbook, 1974.*
[a]Does not include China.

resources, ideas, and people of other areas. One of the most important aspects of modern development has been the build-up of transportation and communication facilities into regional and world networks. This infrastructure of transportation and communication routes and facilities binds the modern world together; supplies it with its resources; makes possible the control and direction of economic, social, and political institutions; and permits rapid dissemination of new ideas and technology.

TRANSPORTATION:
ROADS, RAILROADS, AND AIR ROUTES

The extent of modern air and land transportation is shown in Figures 5–13 and 5–14. The modern world depends on these means of transportation, and on ocean shipping (Table 5–7), to provide the place-to-place **connectivity** essential to bring resources, people, and production facilities together and to tie places of production with the markets for their goods.

Roads, railroads, and air and sea routes provide the links by which goods and people move from one part of the world to another. Each mode of transportation functions as part of the total world network of interconnectedness. Individual routes connect with each other and in some places merge into major corridors of interregional flows. Obviously, these corridors are much more important than single routes, and some of the various routes are much busier than others. The wilderness and frontier areas, which lack enough flow to justify rail tracks or roads, are brought into the flow network by the airplane or helicopter. When all available modes of

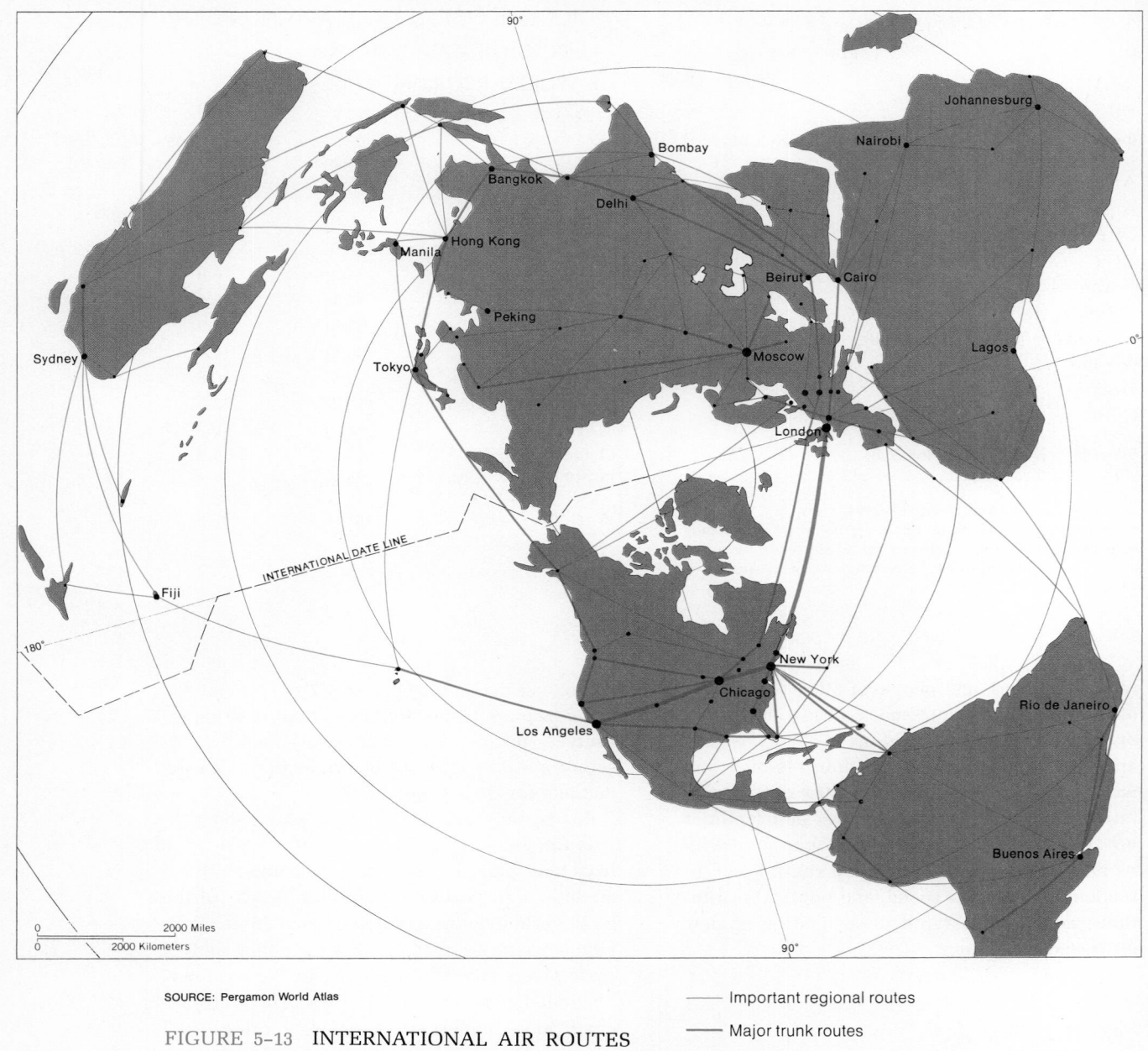

Sydney

Fiji

INTERNATIONAL DATE LINE

180°

Tokyo

Manila

Hong Kong

Bangkok

Peking

Delhi

Bombay

Beirut

Cairo

Moscow

London

Nairobi

Johannesburg

Lagos

0°

90°

Los Angeles

Chicago

New York

Rio de Janeiro

Buenos Aires

90°

0   2000 Miles
0   2000 Kilometers

SOURCE: Pergamon World Atlas

FIGURE 5–13 INTERNATIONAL AIR ROUTES

—— Important regional routes

—— Major trunk routes

—— Air corridors

INTERNATIONAL TRADE CONNECTIONS

The following generalizations about the order of trade connections between major world areas can be made.

1. First-order trade occurs between the industrialized areas of the world. These are the major producing areas in terms of manufacturing, and the major consuming areas for manufactured products and for the resources needed to support manufacturing. The importance of Europe in international trade connections results partly from the number of international boundaries within Europe. Why is trade between the United States and the Soviet Union so small?

2. Second-order trade occurs between the industrial and nonindustrial areas. Each major industrial area has its own particular links with the nonindustrial areas. Though Anglo-America, like Western Europe and Japan, has worldwide links, it also has especially strong regional links with Latin America. Western Europe is more strongly linked to Africa and the Middle East, and Japan with Oceania and the rest of the Far East. The Soviet Union, by contrast, is much more self-sufficient or depends on other Communist countries.

3. Third-order trade is the trade between nonindustrial countries. In general, the trade between these countries, even neighboring ones, is small compared to their trade with industrial countries.

transportation are considered, no part of the world is beyond the reach of the modern system of transportation.

The contrasts between industrial and nonindustrial parts of the world are obvious on a map such as that of world railroad densities (Figure 5–14). Yet even the United States, Europe, and the Soviet Union include areas that are outside the reach of the fixed network of regularly scheduled transportation services. In other modern countries such as Canada, South Africa, and Australia the transportation network is highly concentrated, leaving very large areas without fixed routes. On the other hand, some parts of the less developed world have well-developed transportation networks. Note the density of railroads in parts of Argentina and Brazil. India has a transportation network that appears more highly developed than the Soviet Union's.

In general, the best developed parts of the world's transportation network are found around the major metropolitan centers. Areas of greatest metropolitanization have the greatest concentrations of transportation facilities. Conversely, most areas of sparse population are outside the system, except for isolated locations with important resources needed by the modern world or places through which the routes linking populated areas have to pass.

COMMUNICATIONS:
A MEASURE OF CONTROL OR ISOLATION

Life today is more than just producing and obtaining food and material wealth. Events and changes in every part of the world can affect our lives, so it is important to be informed about what is going on in our own locality and in the larger world. Rapid dissemination of information is accomplished by modern means of communication, particularly the newspaper, radio, television, and telephone. These inventions use industrialized techniques in their production and

FIGURE 5–14   WORLD RAILROAD DENSITIES
AND MAIN TRANSCONTINENTAL ROUTES

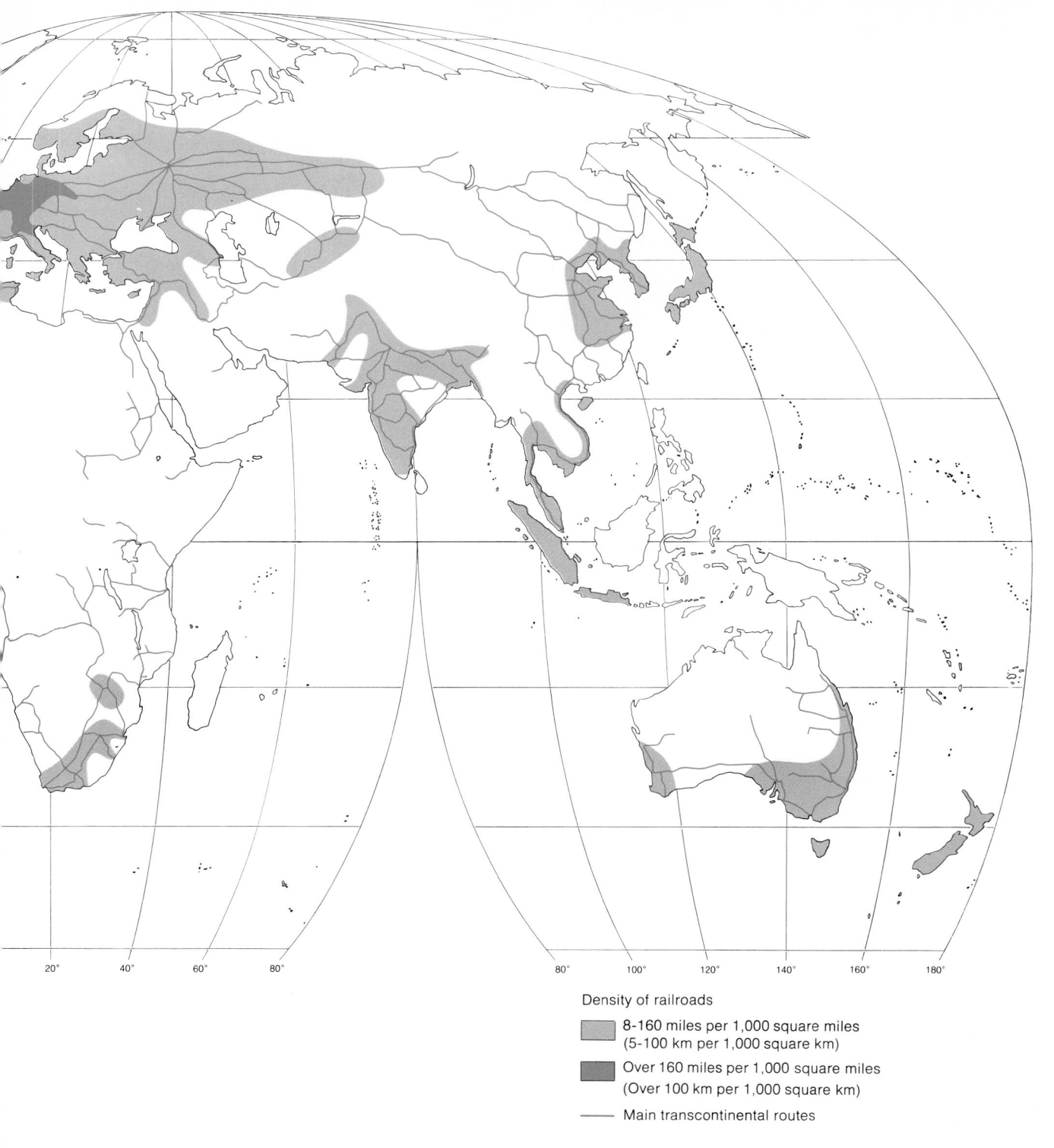

Density of railroads

■ 8-160 miles per 1,000 square miles
(5-100 km per 1,000 square km)

■ Over 160 miles per 1,000 square miles
(Over 100 km per 1,000 square km)

—— Main transcontinental routes

20°   40°   60°   80°   80°   100°   120°   140°   160°   180°

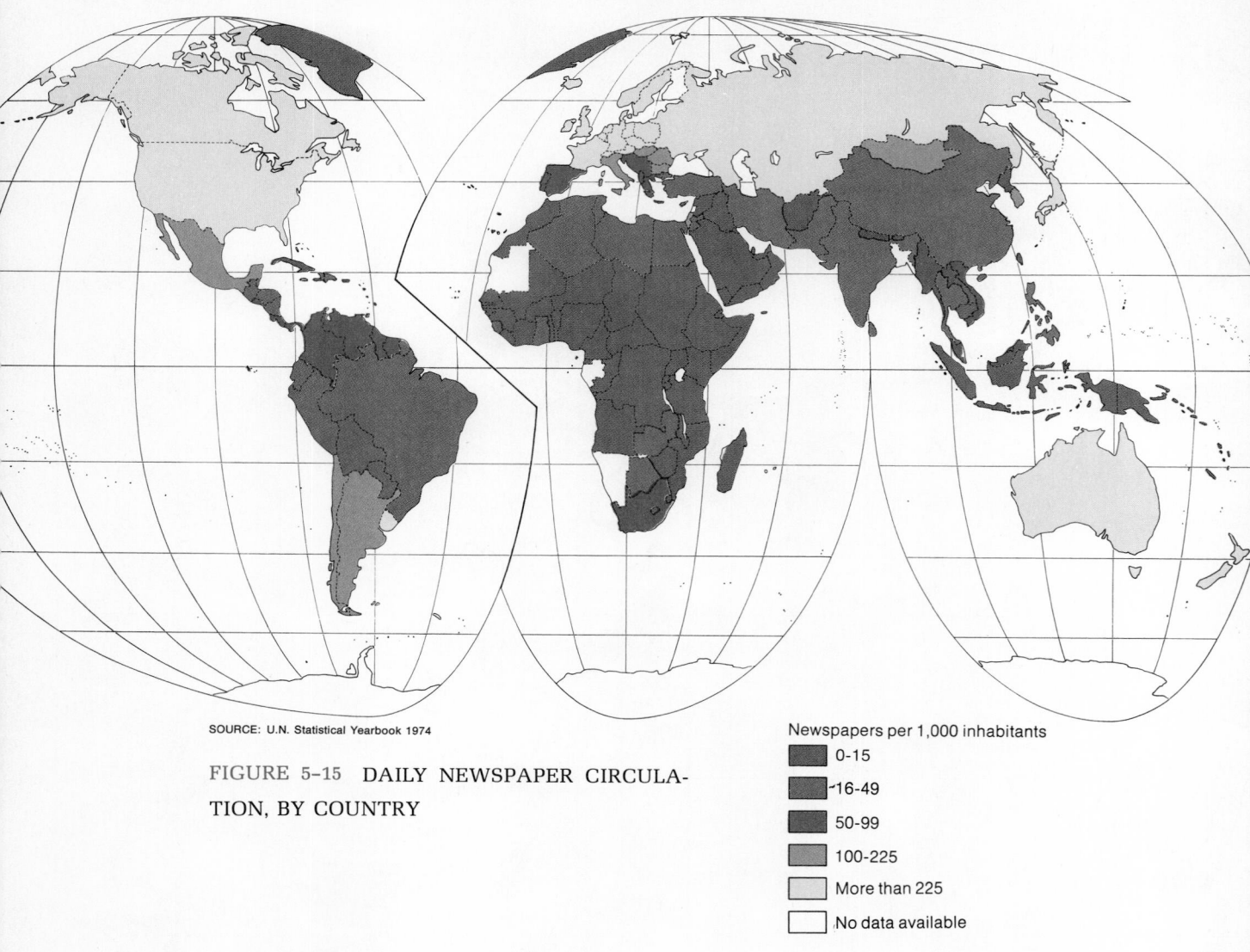

SOURCE: U.N. Statistical Yearbook 1974

FIGURE 5-15 DAILY NEWSPAPER CIRCULA-
TION, BY COUNTRY

Newspapers per 1,000 inhabitants
- 0-15
- 16-49
- 50-99
- 100-225
- More than 225
- No data available

operation, and we would therefore expect them to correlate closely with modern, industrialized areas. Figure 5–15 and Table 5–9 show that this is indeed so.

The smooth operation of the modern world depends on rapid and far-reaching communication links. Imagine trying to manage an interna-

tional corporation without a worldwide network of rapid communications. Most transportation systems, such as airlines, could not operate efficiently without instantaneous communications. Think how difficult it would be to manage our personal business without the telephone or the postal service.

TABLE 5-9 TELEPHONE COMMUNICATIONS

Region	Number of Telephones per 100 Inhabitants
Africa	1.1
South Africa	7.5
Anglo-America	64.6
Latin America	3.5
Asia	2.2
Israel	20.8
Japan	35.7
Europe	21.7
Oceania	30.6
USSR	5.7
World	8.6

SOURCE: *U.N. Statistical Yearbook, 1974.*

Modern techniques of communication are more than measures of connections; they offer a means to control and influence people's thoughts and actions. Mass media reach large numbers of people and can bring to their attention problems of national and international importance. Of course, they can also be corrupted to control information. In any case, communication is one means by which attitudes can be changed and action brought about. Conversely, lack of information is a measure of individual and cultural isolation.

BETWEEN TWO WORLDS

Obviously, all countries do not fall nicely into either the industrial group or the nonindustrial group. The world's countries, ranging from the most affluent to the least productive, can better be seen as existing on a continuum. The United States and Sweden, for example, are at the high end of that continuum; Burundi and Rwanda are at the low end; and other countries lie somewhere between.

There are no completely industrial or completely nonindustrial countries. Even in the United States not all people are fully within the modern system of living, nor do all people of Burundi lead a traditional existence. For example, the vast majority of the people in the United States live as part of the modern world, either as participants in urban industrial and commercial activity or as rural residents with connections to the urban centers. Even farmers have become specialists, producing products for sale in the national market and consuming the vast range of goods and services offered by the modern system. Nevertheless, islands of locally based life remain in the United States. Examples are found on some Indian reservations and among subsistence farmers in various parts of the country, such as the Amish and other similar groups. The poor in the ghettos of large cities constitute another important local group that is not effectively integrated into the modern system. These people are not participants because of discrimination, because of lack of usable skills, or because the cultural values and habits they hold do not mesh well with the modern world around them. As a result they are often by-passed and regarded as peripheral to the system. Unlike subsistence farmers, who can provide for their needs outside the modern system, ghetto dwellers are caught because they must depend on the modern system for survival.

On the other hand, most people in Burundi may live in agricultural villages even more primitive than those in India and must coax the adjacent land to produce enough to keep their families alive. Yet Burundi has a capital city with a population of about 50,000, modern buildings, automobiles and trucks, an airport, and telephones. Moreover, large numbers of the population work in government offices and in stores and shops. In many ways the lives of these city workers are more like the lives of people in the Washington metropolitan area than like those of the people in the rural parts of Burundi. Burundi is represented by a delegation in the United Nations and in Washington. More and more of its farmers are raising coffee, cotton, and tea for shipment to markets in Europe and the United States. In Burundi, however, the traditional system still predominates.

Individual people also are not clearly modern or traditional in their ways of living. Some people are in transition from traditional to modern systems of living, especially where the two exist in close proximity. Such situations are common in many of the developing countries, where urban places are mostly modern and the rural surroundings are still largely traditional. Consider the transition that must be made by rural Indians of Peru trying to become part of the urban system of Lima; or Bantu herders from East Africa going to work in the South African gold mines; or individuals from a nomadic group of central Asia going to work at an industrial job in a Soviet city. Many people in today's world must live in this transitional manner. They must break with the life they have known and try to succeed in a new way of living. For most of us, that transition was made for us by a previous generation.

The following chapters emphasize the dichotomy between modern and traditional, even though each of the individual nations or areas placed in each category has worked out its own particular version of the geographic equation. Keep in mind, however, that the dichotomy is a simplification of the real world. Many peoples exist *between* these two ways of living—between two worlds—and most places represent a complex mix of these two ways of life.

SELECTED REFERENCES

Encyclopaedia Britannica Atlas. Encyclopaedia Britannica, Chicago (revised frequently). An elaborate atlas containing exceptionally good world and regional maps. The atlas also contains a very useful collection of maps and tables giving world and regional information on population, economic, cultural, and environmental characteristics.

Espenshade, Edward, and Joel Morrison, eds. *Goode's World Atlas,* 14th ed. Rand McNally, Chicago, 1974. (Available in paperback and hardcover.) A well-balanced world atlas containing world and regional maps of selected measures of economic, environmental, and social characteristics, and an extensive place-name index.

Oxford Economic Atlas of the World, 4th ed. Oxford University Press, New York, 1972. Mostly thematic maps with some descriptive text. They cover information on the environment, society and politics, demography, disease, transportation, communication, and primary and secondary production. A useful appendix of economic and population data covers all countries of the world for which the United Nations has statistics.

Population Reference Bureau, Inc., 1754 N Street, N.W., Washington, D.C. 20036. Publishes *Population Bulletin, Intercom,* and *World Population Data Sheet,* and distributes *Worldwatch Papers.* These are timely and valuable sources of information on world, regional, and national populations, and related problems of food, resources, and health.

World Population Growth and Response, 1965–1975, A Decade of Global Action. Population Reference Bureau, Washington, D.C., April 1976. Reviews the major trends in population at world, regional, and country scales. The review of the world population situation is especially relevant background and is well supported with graphics.

Statistical Sources

Demographic Yearbook, United Nations; *Statistical Yearbook,* United Nations; *Production Yearbook,* United Nations Food and Agriculture Organization. These three yearbooks, revised annually, provide comprehensive statistical information on population, food and other agricultural production, and economic characteristics for most countries of the world. They are probably the best sources for information about all countries in the United Nations.

Central Intelligence Agency, *Handbook of Economic Statistics.* CIA, Economics Accounts Section, Office of Economic Research, Washington, D.C., published annually. A very useful summary of economic data for selected countries, regions, and the world. The first part is a graphic summary and the rest a wealth of statistical tables.

PART TWO

Industrialized Areas: Where the Modern Interconnected System Dominates

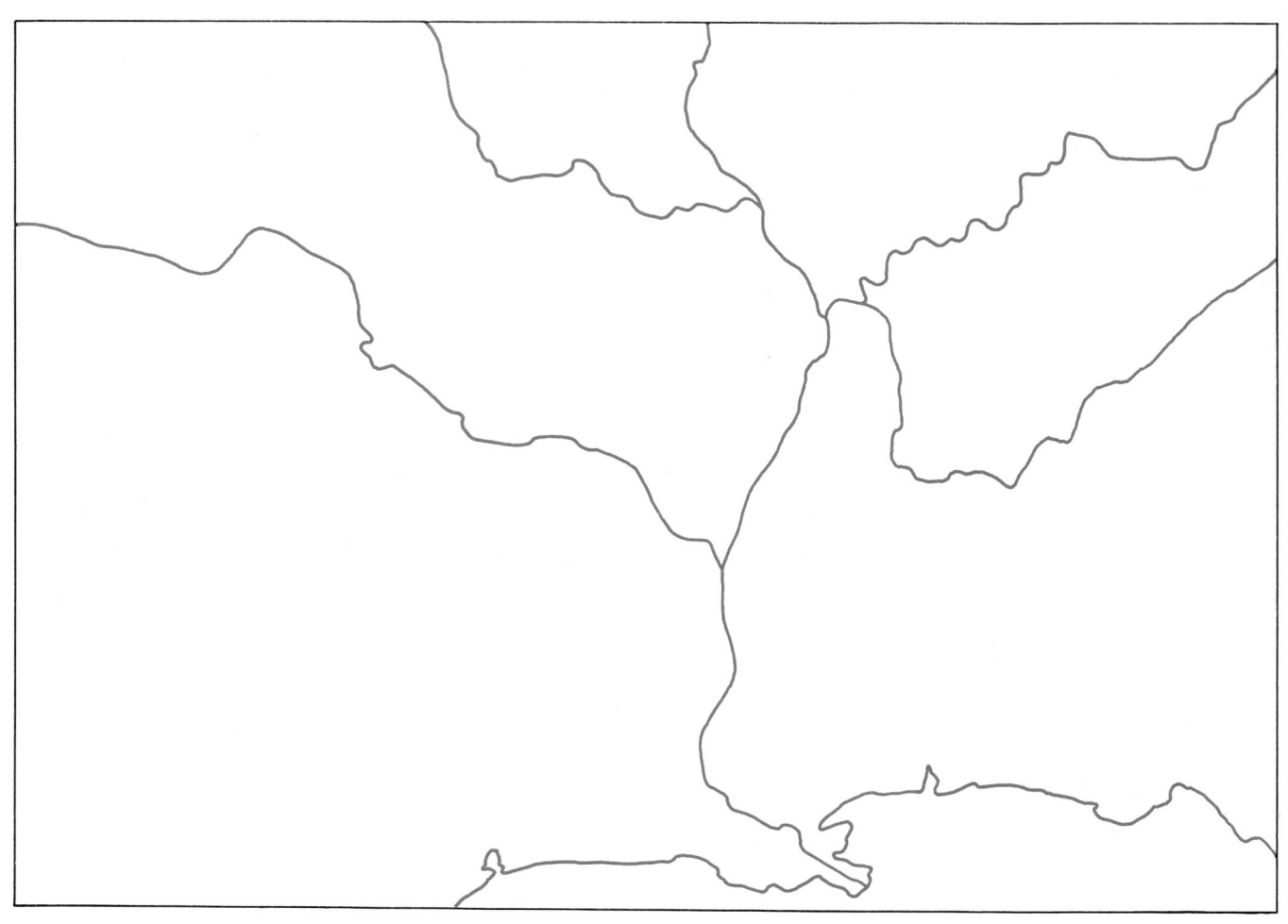

CHAPTER 6

The United States: Changing Evaluations of the Environment

Our first in-depth study of an example of the modern interconnected system begins with a consideration of the earth environment. The agricultural potential, fishing and mining, and other natural resources have been viewed differently in various periods of U.S. history. The variety of the environmental base has lent itself to development of a transcontinental system with industries organized on a national scale. Vestiges of the past, however, affect the functioning of the present-day system.

Left: Coal breaker. (Brown Brothers)   *Above:* West Virginia landscape. (Fred Maroon/Louis Mercier)

Above left: Coal miners in the early days of Appalachian mining. (Brown Brothers)
Above right: Lumber mill in the mining town of Putney, Kentucky, in the Appalachian hills. (Bill Strode)   *Below left:* Strip mining of hard coal in Pennsylvania. (Grant Heilman)

The modern interconnected system depends on three fundamental components in addition to advanced technological skills: (1) a large and varied environmental base from which to draw needed resources, (2) a large, affluent consumer population, and (3) a modern transportation and communication system to tie all producing and consuming areas together. Although one or two of these components were present in the Roman Empire and the British Empire, the United States was the first political entity in the world to exhibit all three. A single nation of more than 3.6 million square miles, it is exceeded in land area only by the Soviet Union, China, and Canada. Only China, India, and the Soviet Union have more people, and none of these countries has as many economically affluent persons able to buy consumer goods.

The United States has the major asset of large physical size. It is 2,800 miles across the conterminous United States and 1,600 miles from the southernmost point in Texas to the 49th parallel at the Canadian border (see Figure 6–1). This does not include Alaska, which is twice the size of Texas, or Hawaii. Despite its large size, the conterminous United States lies entirely within the middle latitudes. Hawaii is the only part of the country south of the Tropic of Cancer. In contrast, the Arctic Circle passes across northern Alaska, and most of the state is above 60° north latitude. Even the southern tip of the Alaskan panhandle reaches only as far south as 55° north latitude.

Such a large area offers tremendous variations in environmental conditions, which are ideally suited to the workings of the modern system. Few other countries have had such a varied environmental base for primary production. With modern transportation and communication, the resources of particular areas of the country can be readily drawn together in manufacturing centers, and the products of manufacturing (secondary production) can be readily distributed to consumers.

Today, more than 350 years after its first settlement by Europeans, the United States is a modern industrial country. In that time it has evolved into the modern interconnected system. Early settlers viewed only a small portion of the present area of the country, and they also viewed it in very different terms. Life in the United States has evolved from this particular heritage, a heritage of more than just ideals and values. The present United States of cities, farms, transport and communication networks, and regional specialization has emerged from decisions and investments made in other times in response to very different economic and cultural conditions.

THE ENVIRONMENTAL BASE: NATURAL RESOURCES OF CLIMATE, LAND, SEA, AND LOCATION

To examine the U.S. resource base, we start by constructing a model of environmental conditions. First, let us turn to agricultural potential. This means investigating the potential for plant growth in particular areas, ignoring for the moment the effect of technology.

PLANT GROWTH POTENTIAL

The most important factor in determining growth is evapotranspiration, which in turn depends largely on latitude. Potential evapotranspiration (PE) is greatest along the southern border of the country, for the closer a piece of terrain in the midlatitudes is to the equator, the greater the solar energy for plant growth. The farther from the equator, the less energy is available. Thus the tip of Florida, southern Texas, and southern Arizona appear to have the greatest potential evapotranspiration of any part of the United States (refer to Figure 3–3). The length of the frost-free season, also related to latitude (see Figure 6–2), is

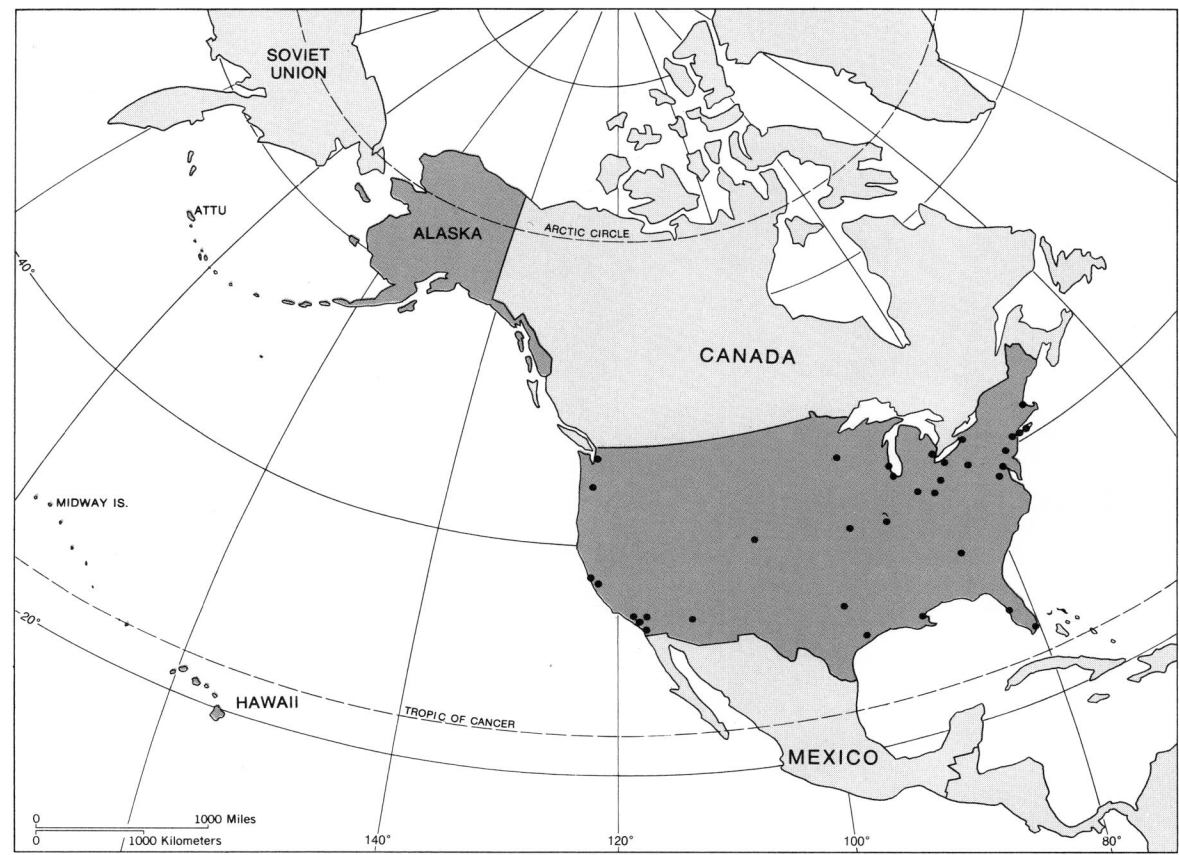

SOURCE: Statistical Abstract of the U.S. 1975

● U.S. Metropolitan areas over 1 million population

FIGURE 6-1 THE UNITED STATES AND ITS
CITIES OF OVER ONE MILLION POPULATION

another dimension of growth potential. It is in-
structive to compare the length of the growing
season in different parts of the country, for example, in Alabama and Iowa.

Together, potential evapotranspiration and
length of frost-free season provide a rough index
of the possibilities for plant growth. The patterns
in Figures 3–3 and 6–2 are similar. The most

noticeable difference is the shorter frost-free season in the Rocky Mountain region, where increased elevation has an effect similar to that of
higher latitude.

As we saw in Chapter 3, plant growth depends
not only on solar energy and absence of frost, but
also on an adequate supply of moisture in the
soil. Unfortunately, moisture from precipitation

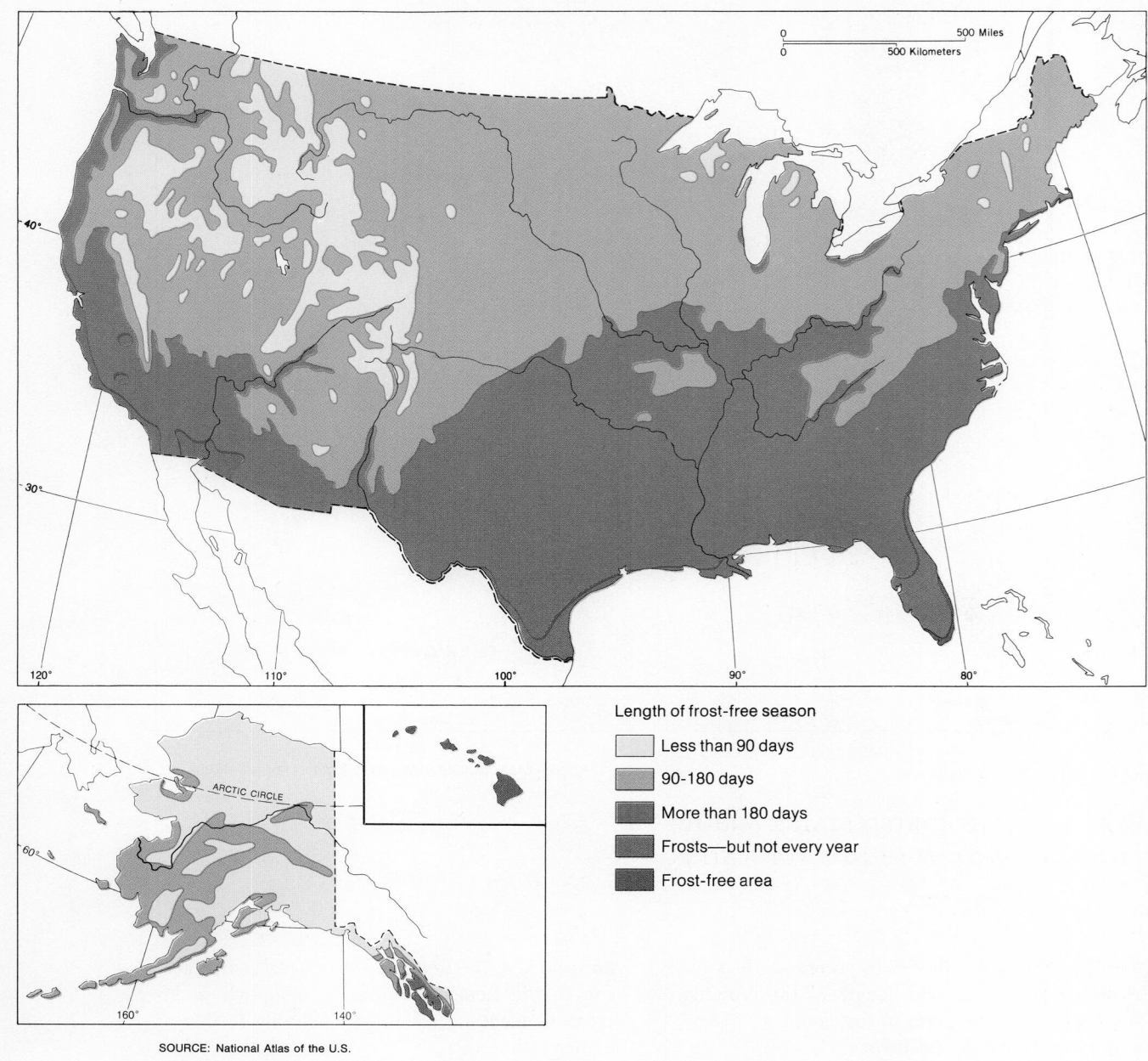

Length of frost-free season

☐ Less than 90 days

☐ 90-180 days

☐ More than 180 days

☐ Frosts—but not every year

☐ Frost-free area

ARCTIC CIRCLE

SOURCE: National Atlas of the U.S.

FIGURE 6-2 LENGTH OF FROST-FREE SEA-
SON IN THE UNITED STATES

is not necessarily available at the same time as the maximum sun energy; PE and precipitation do not always coincide. As a result, there are times of moisture surplus—commonly during the cool season when there is little evapotranspiration—and times of moisture deficiency—during the hot season when evapotranspiration is at a maximum. Parts of the country near the Pacific Coast and in the Northeast have particularly large surpluses of moisture, while most of the western half of the country has a deficit (see Figures 3–5 and 3–6). The southeastern part of the country has less surplus than farther north.

This problem of moisture availability markedly changes the evaluation of different regions. In the Southwest, for example, artificial water supplies are necessary for agriculture. By contrast, New England and most of the rest of the East have enough moisture for their relatively modest potential evapotranspiration. The moisture-holding capacity of soil, however, may vary considerably from place to place. Any place in a generally humid area may suffer dry years or weeks of inadequate moisture during the growing season, as happens often in New England.

Most of the United States is in a seasonal-work climatic environment (see Figure 3–7). The western mountain region, however, is dominated by a low-work environment because of dryness. The sequencing seasonal-work environment is found in the southeastern part of the country and in the Pacific coastal zone in Washington, Oregon, and northern California. The northeastern United States has a cold winter that produces a dormant season, and in southern California natural growth is handicapped by a summer dry season that requires irrigation. Alaska in the Far North has a small area of sequential work in its mountainous panhandle along the coast, but the rest of the state is handicapped by the long cold season that produces seasonal work in the south and a year-round low-work environment in the north. Hawaii has a high-work, tropical environment.

Plant growth potential is affected by the nature of the surface terrain, as we saw when comparing the maps of potential evapotranspiration and length of frost-free season. Temperature and moisture conditions vary markedly with differences in elevation. Except for the major mountain areas, most of the country is relatively flat rather than rugged (Figure 6–3), although there is some rough terrain in the Ohio Valley, across southern Missouri and northern Arkansas, and in the Great Lakes area. Moreover, areas of lowland along the Atlantic Coast from Cape Cod southward and along the Gulf Coast lie so low that they are poorly drained and are swampy. Poorly drained land has also been a problem in the lower Mississippi Valley.

The natural vegetation that European settlers found in the United States as they moved across the country provides a reasonably good measure of environmental growth possibilities. Forest vegetation dominated the areas with moisture surpluses in the eastern half of the country, but timber growth was much more rapid in the areas of high evapotranspiration in the South. Fine forests of giant trees—some of the largest in the world—were found in the areas with surprisingly mild winters and high moisture surpluses in the Pacific Northwest.

Along the western margins of the eastern forests was tall grass mixed with scattered woods. These grasses and woods, in turn, opened into almost treeless grasslands farther west, where moisture deficits were common. Trees appeared almost exclusively along stream valleys, which provided an adequate supply of soil moisture. Westward toward the Rocky Mountains the grasses became shorter and shorter as moisture deficits grew more severe. Then, in areas of the Southwest with extreme moisture deficiencies, the grasslands gave way to patches of drought-resistant plants, which form the incomplete cover of vegetation that characterizes all deserts.

Forests are found on most mountain slopes of the country because higher elevations experience lower temperatures and thus lower evapotran-

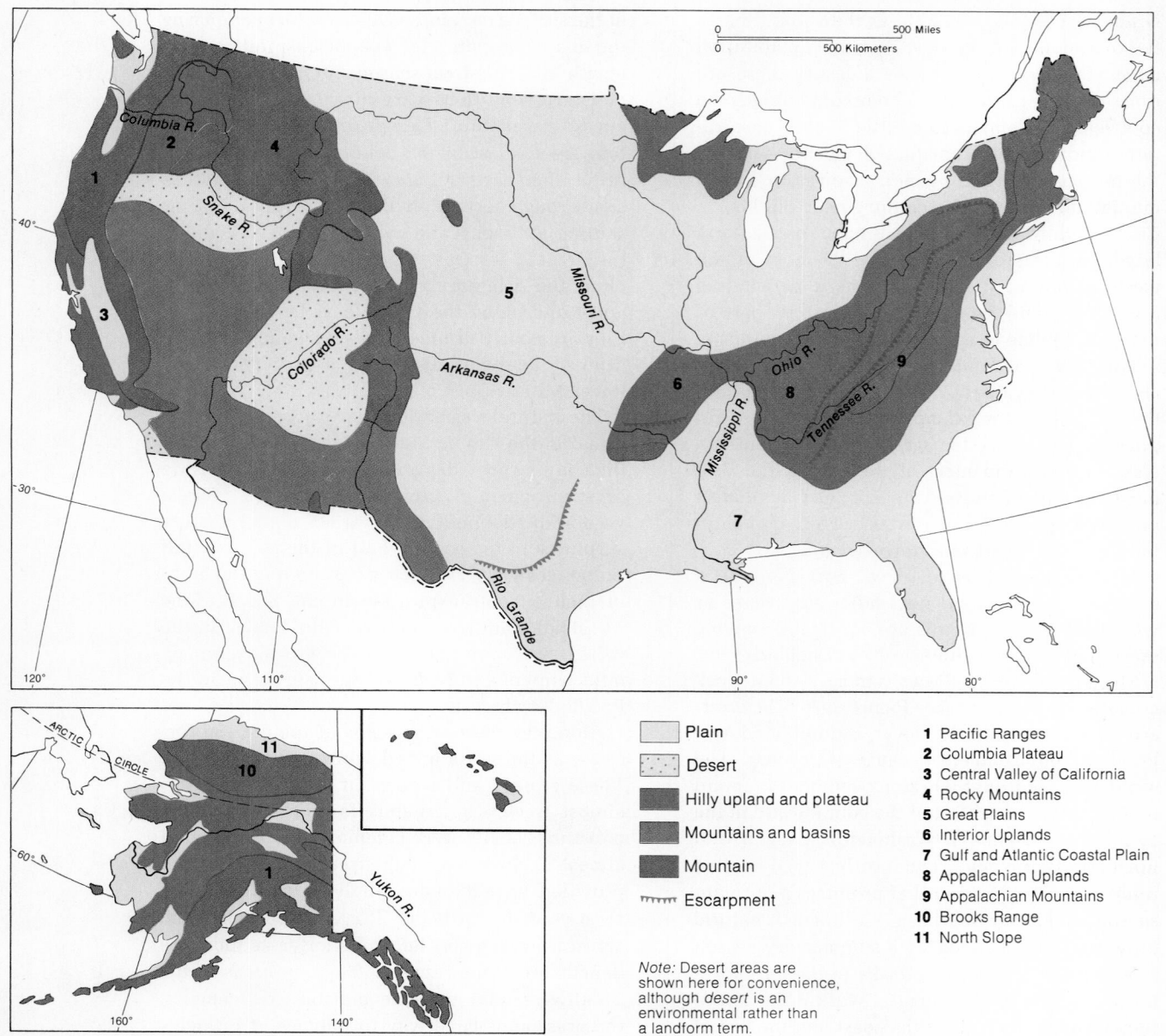

	Plain
	Desert
	Hilly upland and plateau
	Mountains and basins
	Mountain
TTTTTT	Escarpment

1 Pacific Ranges
2 Columbia Plateau
3 Central Valley of California
4 Rocky Mountains
5 Great Plains
6 Interior Uplands
7 Gulf and Atlantic Coastal Plain
8 Appalachian Uplands
9 Appalachian Mountains
10 Brooks Range
11 North Slope

Note: Desert areas are shown here for convenience, although *desert* is an environmental rather than a landform term.

SOURCE: Adapted from *A Geography of Man*, 3rd ed., by P. E. James, Copyright © 1966 by John Wiley & Sons, Inc. By permission of John Wiley & Sons, Inc.

FIGURE 6–3 LAND SURFACE FEATURES OF THE UNITED STATES

spiration. At the same time greater precipitation is usually found in the cooler mountain temperatures. Only on the highest slopes in the West, where temperatures are so cold that growth is stunted, are the mountains bare.

PRESENT EVALUATION OF GROWING AREAS

In today's modern interconnected system the location of agriculture, herding or ranching, and lumbering enterprises—all of which are primary activities—depends on variations in growth potential from place to place and on the relationship of the particular enterprises to markets. Decisions regarding location are based on far more than specific environmental tolerances and particular conditions for maximum growth of each commodity. Certain areas may be ideally suited to a variety of crops and livestock and also to timber production. Thus there is potential competition for land uses in some areas, and choices have to be made. The question is not where a crop will grow best, but where it can be most profitably produced in terms of growing conditions, dietary preferences, market locations, government farm policy, and other factors. Moreover, these evaluations must be made on a national scale rather than a regional or local one.

Today the major U.S. wheat-producing areas are not those where wheat grows best. Corn brings a better financial return than wheat on the best wheat lands; thus corn is grown on the best wheat land, and wheat is grown near the outer limits of its range, beyond the range of corn and most other crops. On the Great Plains, wheat can be grown on large mechanized farms on land that is fit to grow few other crops that are within American dietary patterns. Because wheat must be raised on marginal land, much effort has been made to create strains that will grow in such areas and to develop farming practices and equipment especially for wheat in such an environment. In parts of the South, where both to-

bacco and corn grow well, the farmer's chief concern is with tobacco, a crop narrower in range and with a more specialized use than corn, but one that brings a higher financial return per acre than corn—as long as tobacco smoking remains a part of the lifestyle of many Americans and as long as tobacco prices continue to receive heavy government support. In general, the narrower the range of the crop and the more specialized its use, the higher its priority as a crop to be planted near the location ideal for its growth.

Agricultural land use generally brings a higher return per acre than does ranching or lumbering land, and thus it usually has priority. As a result, ranching has largely been pushed to the drier margins, and forestry to the rougher, more poorly drained lands and lands with soils ill suited to agriculture.

Figure 6-4 shows the complexity of **primary land use** in the United States. The areas of grazing land are beyond the western margins of Middle Western and Southern agriculture, and beyond them the timber regions are found largely in the rougher areas of the country.

Notice how much of the country has been cleared and put into agricultural use. In the interior lowland almost all the land is in farm use, and virtually all of it is in crops. In a sense, the land from Columbus, Ohio, to Omaha, Nebraska, and beyond is one giant farm divided into many individually owned and worked units. In contrast, the traveler through most of the agricultural regions of the eastern seaboard and the South is struck by the large amount of farmland kept as woods, pasture, and unused fields. Many of these areas were farmed in the past but have been given over to pasture or reverted to forest as the developing national transport system brought about the growth of the Middle West as the nation's breadbasket.

Actually, although the U.S. population increased 55 percent between 1930 and 1970, the amount of cropland declined slightly, a measure of the agricultural revolution that occurred in

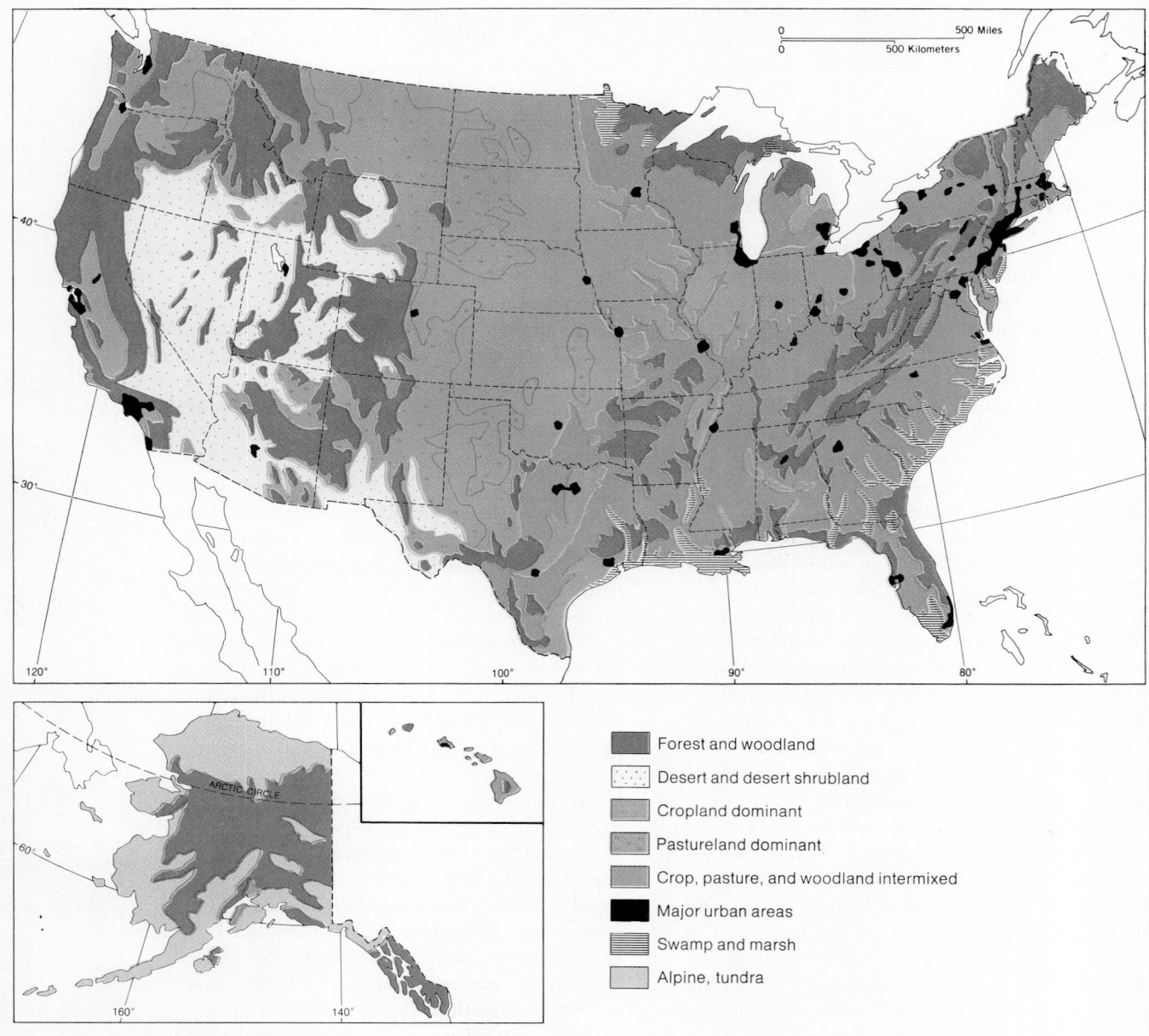

Legend:
- Forest and woodland
- Desert and desert shrubland
- Cropland dominant
- Pastureland dominant
- Crop, pasture, and woodland intermixed
- Major urban areas
- Swamp and marsh
- Alpine, tundra

SOURCE: National Atlas of the U.S.; F. J. Marschner, "Major Land Uses in the United States," U.S. Department of Agriculture, Bureau of Agriculture Economics

FIGURE 6-4   GENERAL LAND USES IN THE
UNITED STATES

that period. The development of hybrid seeds, insecticides, fertilizers, and complex machinery and the greater capitalization of agriculture greatly increased per acre yields, thus reducing the amount of land needed for production. The new, highly capitalized units invested money and labor in the best land rather than in the marginal land. The evaluation was national, not local or regional.

Undoubtedly, the important U.S. agricultural areas lie in the South and the Middle West. Notice that the middle four regions in Table 6–1—Middle West, Great Plains, Southeast, and South Central—account for 67 percent of the land in farm, 85 percent of the farms, 85 percent of the cropland, and 75 percent of agricultural sales. Here we find the prime stronghold of agricultural United States as well as the environmental base that produces most of the staple agricultural products for both domestic consumption and export.

In Table 6–2 the statistics substantiate what we already inferred from Figure 6–4 (the map of land use): the distribution of forests and productive forest land is very different from the distribution of agricultural land. Surprisingly, almost half the forests in the United States are in the Mountain and Pacific regions, not in the eastern region, which once was densely forested. However, the figures for the Pacific include Alaska with its vast forest tracts. The two western areas with all their forests account for less than 30 percent of commercial timberland, indicating that much of their timber is too remote to be exploited profitably. Nonetheless, these two areas still account for almost six dollars out of every ten dollars of lumber sales, indicating that the commercially operated forests are very productive. Species such as the giant Douglas fir and redwood trees of the Pacific Coast area yield more board-feet of lumber per tree and per acre than do other species.

The Southeast is the other great forest and forest products region. It has the largest share of commercial forest land of any region and accounts for almost one-fourth of all lumber production. The Southeast is also the most important region for pulp and paper production.

Like the agricultural producing areas, the major areas of timber production are far from the major center of population in northeastern United States, and we can expect considerable long-haul movement of timber.

FISHING AND MINERAL RESOURCES

The environmental factors used to select fishing and mining locations are different from those used to select agriculture locations. Fishing (limited in our discussion to saltwater fishing) depends, of course, on variations in the ocean environment: water temperature, salinity, depth, underwater terrain, and availability of plankton or other food. The waters off the United States have great variations, and fishing possibilities are strikingly different on either coast.

Not too distant from New England lie the Grand Banks, one of the most productive fishing grounds of the world. From New York south into the Gulf of Mexico, series of large shallow bays and estuaries are the habitats of seafood such as clams, oysters, and crabs. In the deeper waters of the Atlantic and the Gulf of Mexico, fishing centers on the catching of menhaden, small fish that are a source of oil and meal, which is used mainly as animal feed. In the Pacific the chief catch is large tuna and salmon. Nationally, fishing is the least important primary industry; its output is only 1 percent of the total value for agriculture.

Remember that mineral deposits are produced by forces in the earth's interior and by solar energy. Stored in rocks from prehistoric times, minerals may show little correlation with patterns of sun energy today. Moreover, various minerals usually occur under very different conditions. Areas with deposits of metallic minerals rarely have mineral fuels, and vice versa.

TABLE 6-1 REGIONALIZATION OF AGRICULTURE

(Percent of Totals)

Region	Number of Farms 1974	All Land in Farm 1974	Cropland 1974	Total Agricultural Sales 1974	Crop Sales 1974	Livestock Sales 1974	Regional Population 1974
New England[a]	1.0	1.0	0.5	1.2	0.8	1.7	5.7
Middle Atlantic[b]	4.8	2.0	2.8	3.5	2.2	5.2	17.6
Middle West[c]	32.9	17.8	36.9	33.4	33.5	33.1	24.8
Great Plains[d]	8.4	17.0	23.2	13.4	13.4	13.4	2.4
Southeast[e]	29.2	12.5	11.7	16.3	16.9	15.6	22.0
South Central[f]	14.5	19.1	13.0	12.3	12.0	12.8	9.7
Mountain[g]	4.3	24.0	7.7	7.8	6.3	9.6	4.5
Pacific[h]	5.0	7.0	4.3	12.0	14.7	8.5	13.2

SOURCE: Number of farms and land in farm: U.S. Department of Agriculture, Economics Research Service. Other data: U.S. Census Bureau, *Statistical Abstract of the United States, 1975,* Washington, D.C., 1975.

[a] Maine, New Hampshire, Vermont, Massachusetts, Rhode Island, and Connecticut.

[b] New York, New Jersey, and Pennsylvania.

[c] Ohio, Indiana, Illinois, Michigan, Wisconsin, Minnesota, Iowa, and Missouri.

[d] North Dakota, South Dakota, Nebraska, and Kansas.

[e] Delaware, Maryland, District of Columbia, Virginia, West Virginia, North Carolina, South Carolina, Georgia, Florida, Kentucky, Tennessee, Alabama, and Mississippi.

[f] Arkansas, Louisiana, Oklahoma, and Texas.

[g] Montana, Idaho, Wyoming, Colorado, New Mexico, Arizona, Utah, and Nevada.

[h] Washington, Oregon, California, Alaska, and Hawaii. For cropland data, Alaska is not included in total for the United States.

TABLE 6-2 REGIONALIZATION OF FORESTRY

(Percent of Totals)

Region	Land in Forest 1970	Commercial Forest Land 1970	Lumber Production 1973	Regional Population 1974
New England[a]	4.4	6.5	2.1	5.7
Middle Atlantic	5.0	6.9	2.2	17.6
Middle West	11.2	16.2	4.9[b]	24.8
Great Plains	0.6	0.8		2.4
Southeast	22.9	33.7	23.6	22.0
South Central	8.9	10.3	9.4	9.7
Mountain	18.2	12.1	11.9	4.5
Pacific	28.8	13.5	46.7	13.2

SOURCE: U.S. Census Bureau, *Statistical Abstract of the United States, 1975,* Washington, D.C., 1975.

[a] See Table 6-1 for states included in each region.

[b] Middle West and Great Plains combined.

THE REGIONALIZATION OF AGRICULTURE

See what you can infer from Table 6–1 about **regional specialization** in U.S. agriculture.

1. What are the major agricultural areas of the country? How important are they?

2. The states of the Middle Atlantic, Middle West, and Southeast regions have a larger share of the nation's farms than of its farmlands. On the other hand, the Great Plains, South Central, Mountain, and Pacific areas show the opposite relationship. What conclusions do you draw?

3. The Middle West and Great Plains areas have a higher proportion of cropland than land in farm. What inferences do you draw from those figures?

4. The Southeast, Pacific, and Middle Atlantic areas and New England have a higher percentage of total agricultural sales than of either cropland or land in farm, but in most other areas the opposite is true. What characteristics of agriculture might these variations indicate?

5. Some areas show relatively high proportions of crop sales as compared with livestock sales; in others the reverse is the case. What is the significance of these variations? Keeping in mind that livestock sales nationally are about one-third larger than crop sales, indicate the relative importance of livestock and crops in each region.

6. See if you can develop from the table a general classification of agriculture in each region.

7. Compare regional agricultural sales with share of population. What are the major surplus and deficit areas of agricultural production? What changes in crops and livestock would you predict?

In the United States, as elsewhere, the mineral fuels—coal, petroleum, and natural gas—are associated with sedimentary rocks. Although petroleum and natural gas are commonly found together, the major coalfields have a different distribution pattern. Some petroleum is produced in important midcontinent coalfields in Illinois, Indiana, and western Kentucky, but little comes from the most important coalfields of the western Appalachians. The most important oil fields are along the Gulf Coast and in Texas, Oklahoma, and Kansas. Major production also comes from California.

New fields of large proportions have recently been discovered on the remote North Slope of Alaska. In 1973 proven reserves in Alaska were almost as great as the remaining reserves of Texas, the greatest oil-producing state. The questions of how to get Alaskan oil to market from this distant, desolate place and how to exploit the resource without severely damaging the fragile arctic environment were largely resolved, following years of investigation and debate, by constructing a pipeline across Alaska from the Arctic Ocean to the Pacific Ocean at Valdez. Now, new questions are being raised about the most suitable route and method for transporting natural gas found in association with the petroleum.

The major producing coalfields have been close to the large population centers of the country in the Northeast and Middle West, but oil and gas production has come from the South and Great Plains, far from the country's major markets. The result has been the development of an elaborate, long-range system built around oil and gas pipelines and special coastal tankers and river barges to move these fuels to the chief areas of consumption. In 1972, 90 percent of U.S. coal was produced east of the Mississippi River, but new, high petroleum prices are encouraging rapid development of the large coal resources of the western United States.

Metallic minerals—iron, copper, lead, gold, and silver—form in igneous and metamorphic rock; they appear in sedimentary rock only after they have been eroded from the original deposit and redeposited as sediment. In the United States metamorphic and igneous rocks are less common than sedimentary rock and are more widely scattered. They are generally found in rougher, more remote areas, necessitating long-haul transportation. Except for iron ore, these minerals are transported in much smaller quantities than mineral fuels and carry higher per unit values. As a result, transportation is not as great an economic factor as it is with coal and oil.

More than three-fourths of the iron ore, the most important metallic mineral, comes from mines around Lake Superior. Until recently, iron ore was shipped directly without processing, since its iron content was substantial, one-third or more. Now most ore with high iron content has been mined, and it has become necessary to build large concentrating plants in the minefields to reduce shipping costs. Similarly, most of the specialty metallic minerals mined in the western United States are low grade, and smelters and refineries are required to process the ores and reduce the bulk that must be shipped.

Limestone, clays, sand and gravel, and salt, among the nonmetallic minerals, are widespread, and major production occurs near large population centers. Other nonmetallic minerals are more localized. Sulfur comes from along the Gulf of Mexico in Louisiana and Texas, phosphate rock for fertilizer from Florida, and specialty mineral salts from arid parts of California.

The regional locations of mineral production can be seen in Table 6–3. The populous area of the New England and Middle Atlantic states has only a little more than 5 percent (by value) of the national output of all minerals. The largest regional output, with 43 percent (by value) of the total, is in the South Central area (Arkansas, Louisiana, Oklahoma, and Texas). This reflects

TABLE 6-3 REGIONALIZATION OF MINERAL PRODUCTION, INCLUDING MINERAL FUELS, 1973 (*Percent of Totals*)

Region	Value of Mineral Production	Regional Population 1974
New England[a]	0.5	5.7
Middle Atlantic	5.1	17.6
Middle West	12.0	24.8
Great Plains	2.5	2.4
Southeast	14.8	22.0
South Central	43.1	9.7
Mountain	14.9	4.5
Pacific	7.1	13.2

SOURCE: U.S. Census Bureau, *Statistical Abstract of the United States, 1975,* Washington, D.C., 1975.
[a]See Table 6–1 for states included in each region.

the great value of oil and gas as compared with other minerals. The western Mountain region, with all its specialty metallic ores and such exotic minerals as uranium, is a distant second in the value of its mineral output.

RESOURCE UTILIZATION AND MARKETS

Tables 6–1, 6–2, and 6–3 show that there is a poor correlation between the concentrations of population in the country and the locations of its **primary producing regions.** The Middle West and the Southeast are both populous regions and major producing regions, particularly in terms of agriculture. However, the New England and Middle Atlantic regions, with almost a quarter of the population, have less than 6 percent of any of the kinds of primary production presented in the three tables. On the other hand, the Great Plains and the South Central and Mountain regions all have a higher percentage of total primary production than they have of total population. All this indicates that, in the modern interconnected system, food and raw materials must move on a large scale between the producing regions and the populous regions with low primary production.

The discrepancy between the places of production and the places of consumption necessitates a flow of food, timber, minerals, and other products toward the major centers of population, especially the Northeast. Just as raw materials and food flow from the underdeveloped countries into the United States in exchange for manufactured goods, within the United States goods flow similarly between the highly industrialized, metropolitan areas of the Northeast and Middle West and the primary producing areas in other regions.

ENVIRONMENTAL AMENITIES AS RESOURCES

Another aspect of the environmental base needs consideration. People have long been conscious of the pleasures the environment has to offer and of their effect on lifestyle. Many people value the wild beauty of nature, and the recreational possibilities of the environment are important to them. Unlike other primary resources, however, the amenities cannot be harvested and carried somewhere else; the market must come to the environment.

Until recent times, limitations of disposable income, private transportation, and short vacation times restricted most people to firsthand experience of only those natural environmental amenities of their own locality. Today, however, a large proportion of the population has sufficient disposable income, an automobile, and adequate vacation time or early retirement to give them the opportunity to directly experience all but the most remote wilderness areas of the country. Now recreational resources are being evaluated nationally in the same way that the environment has been judged in terms of agricultural and mining use (Figure 6–5).

Like many other activities, recreation in the United States has become a business, developed and managed on a national scale. Even before 1900 the federal government set aside outstanding natural attractions as national parks. In this century, conservationists have called for the expansion of preserves to include forests and seashores. Recreational and other aesthetic land use is still considered in the federal construction of dams for irrigation and flood control. States and local governments have also established parks and forest reserves.

The widespread development of businesses based on amenities has been undertaken by private interests. Those of greatest commercial success and ultimate effect on changing the residences of people have focused on pleasant climates rather than scenic vistas. Marshy coastal areas in Florida and dry wastes in California and Arizona, which had little value in terms of agricultural or other primary resources, have emerged as major population centers because of their subtropical climates. California's population almost doubled in the thirty years between 1940 and 1970, and the populations of Florida and Arizona increased two and one-half times during the same period. The Miami metropolitan population expanded more than three and one-half times.

The people who rush to these areas are not only the retired, vacationers, and workers needed to provide services. Industry is following, too. Manufacturing firms not closely dependent on suppliers or easy access to markets have been moving into areas of year-round "open" climates, which offer savings on heating and building construction, and are also attractive to personnel, particularly executives and researchers. Manufacturing employment in Phoenix, Arizona, and Miami, Florida, increased over nine times between 1947 and 1972. These cities are not manufacturing centers of the order of major northeastern cities, but the change has been significant, considering that these are areas where there was almost no manufacturing before. Phoenix, which had 7,000 manufacturing workers in 1947, had 72,000 twenty years later.

Bodies of water

 Areas of numerous small water bodies

Outline area of large water bodies

− Reservoirs with more than 810,000 acre ft. capacity

• Remaining usable water bodies with capacity greater than average annual inflow

Areas suitable for skiing

 Local relief 500 feet or greater and more than 64 inches of snowfall per year

 Local relief 500 feet or greater and less than 64 inches of snowfall per year

 Local relief less than 500 feet with less than 64 inches of snowfall per year

── Southern limit of 5 or more days per year with 1 inch or more snowfall

Warm-weather vacation areas

 Mild winters: average monthly winter temperature more than 50°F; minimum monthly winter temperature more than 32°F

 Mild summers: July average temperature is less than 70°F; July maximum temperature is less than 80°F; July average relative humidity is less than 60%

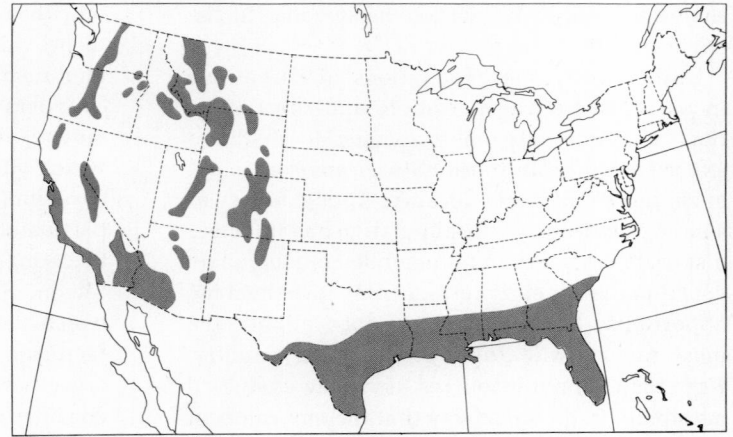

SOURCE: National Atlas of the U.S.

FIGURE 6–5 RECREATIONAL RESOURCES OF THE UNITED STATES

In some vacation centers the population during "the season" may be more than ten times its size during the rest of the year. Others are centers for weekend and holiday visitors or even for single-day transients. Thus recreation centers are national, regional, and local.

EVOLUTION OF THE MODERN SYSTEM: CHANGING VIEWS OF AN ENVIRONMENT

The current socioeconomic system in the United States has evolved from that begun in Virginia more than 360 years ago. During the intervening period a revolution in the technology available for developing and managing the earth's environment has occurred. Even more important, purposes have shifted. For more than 150 years the settled territory of the United States functioned as a series of colonies of first several European countries and later, one. Under those conditions, the system's primary purpose was to support the "mother country" both economically and politically. Independence brought a shift in the system's goals. The new nation sought to reorder its economy so that the primary benefits would be for itself rather than for an external power. In addition, it made major efforts to absorb new regions into the nation.

The evolutionary changes that have occurred in the United States in the past 300 years reflect the same problems that the countries of the underdeveloped world now face in shifting from colonial rule to independence and control over their own socioeconomic affairs.

THE COLONIAL SYSTEM: 1608–1776

Since the U.S. heritage has had such an influence on the present, we must examine its evolution. The colonial era lasted almost 200 years. Yet by the end of the Revolutionary War only scattered settlements were found west of the Appalachian Mountains, and most European settlers lived within 200 miles of the Atlantic Coast.

The colonial era preceded the Industrial Revolution in the United States and, largely, in Europe itself. Manufacturing was a **craft industry,** and most people were farmers. Ocean travel was in sailing ships scarcely larger than many of today's cabin cruisers. Overland wagons and carriages were standard means of transport, but most people traveled by foot or on horseback. Wood and stone were used for construction, and the emerging factories used waterpower, not steam.

The economic and political focus of colonial America was England. From the British point of view, the colonies were an extension of their own resource base, providing useful primary products. Each colony made its own contribution to the British system. Moreover, each colony tried to develop a twofold economic base. On the one hand, it sought to produce goods that it could sell to the mother country. These items had to be high in value and low in bulk—such as tobacco, dried fish, and animal furs—to be carried in the small ships of the time. On the other hand, each colony also needed its own primary support base. It required production of staple foods, building materials, and basic crafts. Little trade occurred between colonies. As in underdeveloped areas today, little was produced in one colony that another colony needed in quantity, and transportation and communication systems were designed primarily to connect the colonies with England.

These midlatitude U.S. colonies, however, were not the prize possessions of the period. Rather, the tropical islands of the Caribbean with their output of sugar and other products that Europe wanted were highly valued. The midlatitude colonies that belonged to England had to search for products that could be sold to the mother country.

In colonial days the great environmental variations along the Atlantic seaboard from Maine to Georgia were judged by each colony in terms of the products they offered for subsistence and for export. England measured the value of a colony by its contributions to the British economy.

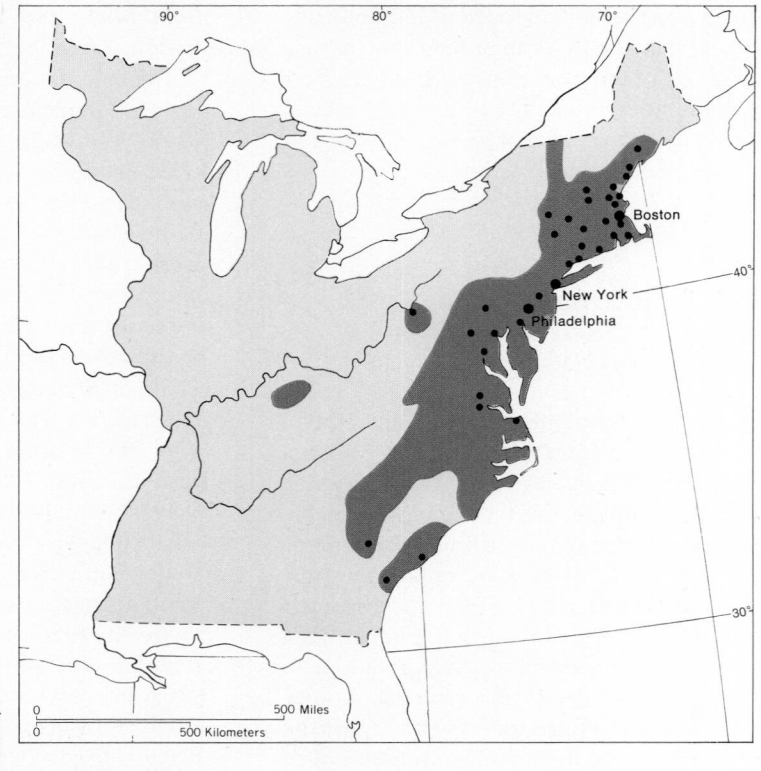

Large towns

Smaller settlements

Areas with at least 6 persons per square mile (2.3 per square km)

SOURCE: Redrawn, with permission, from the *Geographical Review*, Vol. 57, 1967.

FIGURE 6–6 SETTLEMENT OF THE UNITED STATES BY 1790

New England, with its rough terrain and short growing season, was in most respects more limited agriculturally than England and thus could produce nothing of agricultural interest to the mother country. It was, however, adjacent to the rich fishing grounds of the Grand Banks, and its forests contained excellent trees for construction. With an interest in the sea and with the materials at hand for building, New Englanders soon built their own ships, carried their own cargoes to Europe, and engaged in trade with Africa and the West Indies (Figure 6–7). In port cities New England businesses engaged in shipping and finance. They controlled a portion of the transatlantic trade, so in contrast to most colonies, New Englanders made profit not only

from producing raw materials, but also from shipping and financing them. With the rise of populations in the port cities, the agricultural villages and towns back from the coast supplied the basic food needs of the city dwellers.

The South's specialty products were very different, and so was its way of life. The growing season was longer, and an agricultural economy based on specialty crops developed. The primary crop was distinctly American: tobacco. Having become popular in Europe, it was an almost ideal export because the leaves weigh little and it does not deteriorate badly in shipment. In South Carolina and Georgia long-season crops included rice and indigo.

The nature of the southern coastline, with

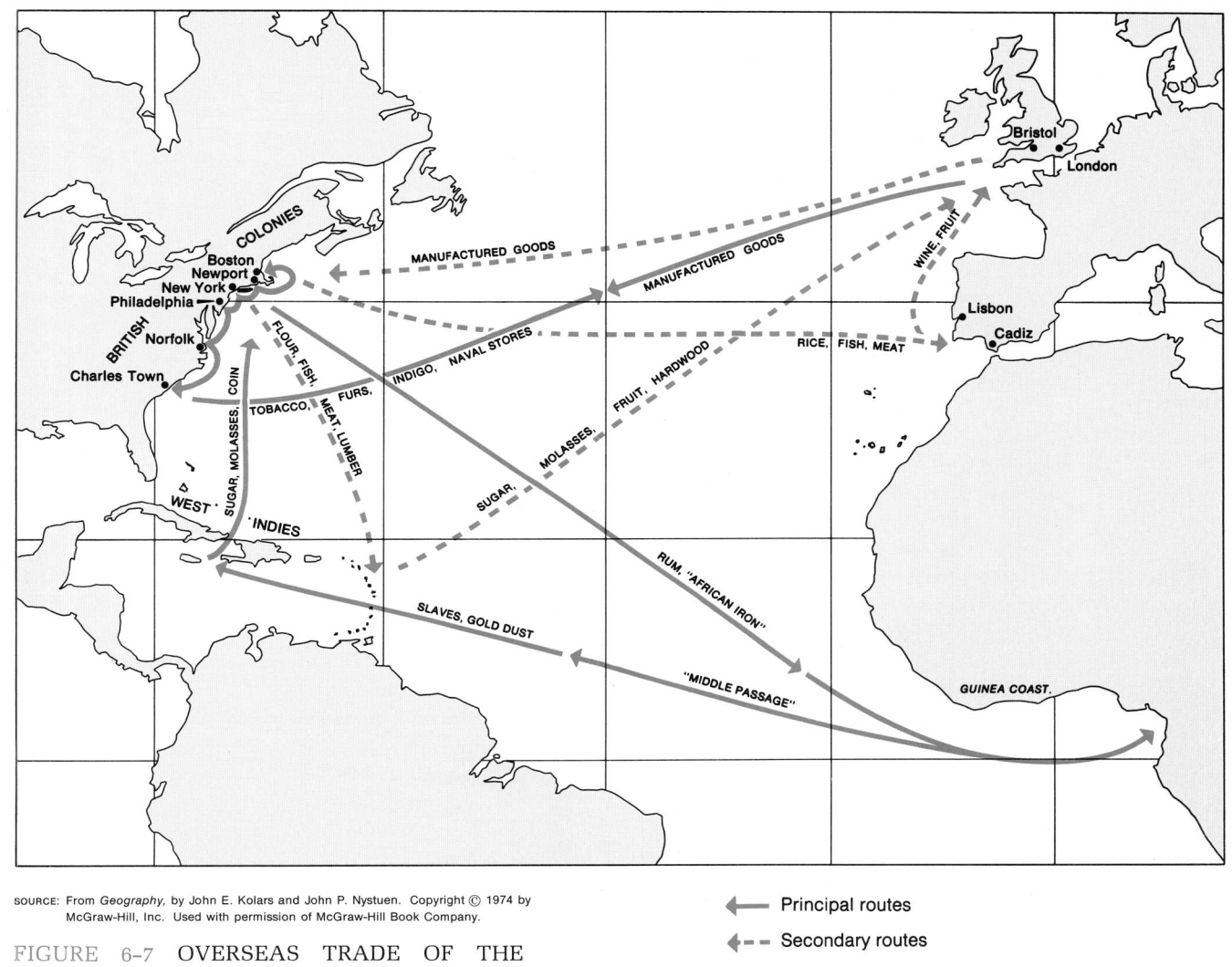

Principal routes

- - - - ◀ Secondary routes

FIGURE 6-7 OVERSEAS TRADE OF THE AMERICAN COLONIES

many small streams flowing into large estuaries and bays, was an aid to shipping. Plantations on tidewater could be reached directly by the small sailing ships, eliminating the need for an urban intermediary. With direct shipment from dockside and the production of its basic crops and livestock, the **tidewater plantation** was in large part a self-contained unit connected only to Europe. In turn, Europe supplied the plantation with needed manufactured goods. Unlike New Englanders, southern plantation owners did not own ships or handle the overseas trade. In fact, New England vessels often carried their goods to and from the southern plantations.

Colonies in the area of New York City and Philadelphia seemed to have few advantages. They did not grow exotic crops and had no particular nonagricultural resources. They could provide an adequate food base for themselves but had no crops in demand in Europe. Instead, flour and barreled pork from farms in the area were shipped to the West Indies to feed the slaves on sugar plantations. Local iron deposits were found, and iron smelting was developed in eastern Pennsylvania and nearby New Jersey.

During the colonial era between 80 and 90 percent of the population lived in rural areas or in villages supported by agriculture. City dwellers were a small minority, and there was no such thing as a metropolitan area. Each of the largest cities had fewer than 50,000 people. Each colony, functioning separately from the other colonies, had its own major center, usually the port through which it was connected with Europe. The size of the community largely depended on the importance of the trade moving through it. Each operated virtually independent of the others, and all trading communities were satellites of trade centers in London, the West Indies, and Africa.

Except for these port connections with Europe, each community was isolated because of transportation difficulties. Poor land routes meant dependence on the local area for the food base. They also meant that news traveled slowly. Moreover, each community had to provide the basic cultural life for its people. The church was the social as well as the spiritual center, and each town had its own days of celebration, fairs, and entertainment. Many towns developed their own weekly newspapers.

The South needed towns and villages less than other colonies did, since each plantation functioned not only as a separate economic entity with direct ties to Europe, but as a social and religious community as well. Like the haciendas that dominated rural settlements in Latin America, the plantation commonly had its own chapel and its own close social contacts. Workers seldom left the community. Thus the large-scale plantation agriculture of the South established a society very different from small-town life in New England.

THE FIRST ATTEMPTS AT A
NATIONAL SYSTEM: 1776–1840

With independence from England came the first attempts to reorient the system developed under colonial ties. Europe was still the cultural focus, the source of manufactured luxury items, and the chief market for goods. Breaking ties with England necessitated a reorganization of the separate colonies into a more cohesive and functional political and economic unit. To supplement the use of coastal waters, great effort was made to provide land routes from Georgia to Maine. And as settlers pushed out beyond the Appalachians, interregional trade took on a new dimension.

This period marked the beginnings of industrialization as well as national organization, but the effects of technological change did not really take hold until later. Wagons, barges, and sailing ships continued to be the dominant forms of transportation, and it still took four days to get from New York to Washington. A journey to Cincinnati from the East took more than two weeks. In rural America human power and animal power unaided by modern agricultural implements still did the farm work.

The new factories were powered by running water and thus represented only the first stage of industrialization. Swiftly flowing rivers channeled into millraces turned the water wheels to provide power.

Regional specialties of colonial times continued to develop in the post-Revolutionary period, but they were influenced by new technologies. The shipping and trading interests of New England flourished, and this region became the country's first manufacturing area. Some manufacturing had been done in the area during the Revolutionary War, and by 1820 there was a boom of new mill towns at key waterpower sites from Maine to Connecticut.

The hilly countryside with its humid climate had many small streams that could easily be used to produce waterpower. As in England, textiles developed as the leading new industry, first using wool and then turning to newly popular cotton. Locally raised sheep provided the raw material for wool, but cotton, which requires a long growing season, could not be grown in New England. Local iron deposits and charcoal from the forests provided the base for a small metalworking industry.

In the South, cotton became the base for an expanded plantation economy that moved back from the tidewater of southern Virginia into Alabama and Mississippi. Like tobacco, cotton was suited to large plantation units, and a great deal of labor was required to weed and harvest it. The importation of black slaves from Africa, often in ships operated by New Englanders, provided the necessary labor force and was the basis for the localization of black population in the rural South.

The plantation South did not organize itself to manufacture its chief crop into textiles, and the market for cotton remained mostly in England. Thus the shift of plantations away from the tidewater resulted in the growth of new southern seaports. As New England mills emerged, some interregional trade began.

The Appalachians presented an almost unbroken barrier that had to be crossed to get to the interior of the country. From Atlanta as far north as Maryland, the Blue Ridge Mountains offered virtually no easy breaks, and no major routes directly west developed from the coast to the interior. In the Middle Atlantic states the Hudson–Mohawk gateway led to Buffalo, and crossing points in the mountains were also found farther south.

The Great Valley route behind the Blue Ridge and the Potomac route were popular, too. The Great Valley route connected westward through the famous Cumberland Gap to the Ohio and Mississippi river system. One could float on a barge to the southwest to Knoxville and the Tennessee River, or to Nashville on the Cumberland

River and thence to the Ohio and Mississippi rivers. The Potomac route followed river breaks through the series of parallel ridges that blocked westward movement farther north and then worked its way into the valleys leading to the Ohio River.

Using these routes, the first settlers found not the rich plains of the Middle West but the hilly region that is the western extension of the Appalachians in Kentucky and Tennessee and along the Ohio Valley. Major settlements were in the rich Bluegrass Basin of Kentucky and in the Nashville Basin. It was not until the opening of the Erie Canal and the development of the National Road to Illinois in the 1820s and 1830s that settlers had direct access to the rich agricultural lands farther west (Figure 6–8). In the far South settlers moved around the Appalachians at their end point and on into the lands of the Gulf Coastal Plain in Alabama, Mississippi, and Louisiana.

During this period the problem of all settlers along the western frontier was less of producing goods than of getting them to market. Without water transport and good roads it was almost impossible to ship bulky farm produce back over the mountains to eastern settlements, although herds of cattle and hogs could be driven over the trails and corn whiskey could be transported in jugs. Thus the Ohio–Mississippi river system, which led away from the settled areas toward New Orleans, offered the best highway for goods moving out. Cargo could be floated downstream by raft or barge, and the raft as well as the cargo could be sold at New Orleans for shipment to Europe, the West Indies, or the coastal states.

The Appalachian barrier was bridged with difficulty. The National Road, with connections to Baltimore and Philadelphia, reached the Ohio River at Wheeling in 1818. More important, the Erie Canal to Buffalo was opened in 1825 and proved a great boon to the Middle Atlantic states. As a result of the National Road and the Erie Canal, the ports of Baltimore, Philadelphia, and particularly New York became gateways to the interior, and their function as national trade centers was established. Settlers on their way

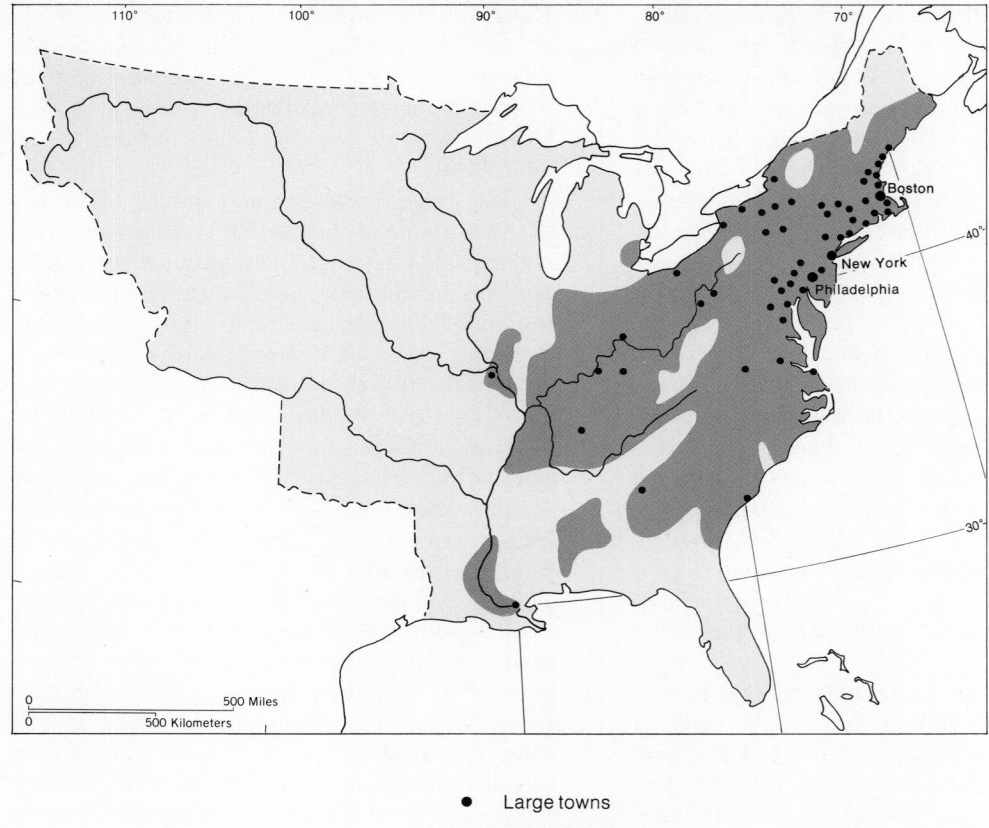

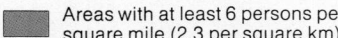

● Large towns

• Smaller settlements

▓ Areas with at least 6 persons per square mile (2.3 per square km)

SOURCE: Redrawn, with permission, from the *Geographical Review,* Vol. 57, 1967.

FIGURE 6–8 SETTLEMENT OF THE UNITED STATES BY 1830

from Europe to the frontier invariably came through these cities. Merchants and shippers prospered at these focal points of **interregional trade,** and from then on, the movement of immigrant settlers westward expanded the influence of these cities. Increasing trade, in turn, meant more jobs for more people in the seaboard cities. Moreover, iron manufacture in Pennsylvania and New Jersey and manufacturing in New England towns were stimulated by these new interregional contacts.

In the early nineteenth century transportation and communication were still very inefficient. Most regions remained semi-isolated, and life continued to center around the local church, the local government, and local customs. Each community needed its own food base and its own craftspeople. New ideas filtered slowly westward from the port cities and Europe. Immigrant groups settling frontier areas tended to cluster together, with ethnic cohesion lending distinct character to various settlements.

The lack of an easy flow of ideas on an inter-regional basis resulted in subcultures as well as regional economic specialization. This was most obvious in the contrast between the plantation South and the industrial and commercial North. Moreover, the agricultural Middle West evolved in still other, distinctive ways. Most decisions were in the hands of local entrepreneurs—shop-keepers, mill operators, and plantation owners. Only the ocean shippers in the large ports of the seaboard could function efficiently on an inter-regional basis.

Towns and cities themselves were tightly packed. Since workers had to walk to work, residences were crammed close together. Only the wealthy who owned carriages could live any distance from places of business and shopping. The compactness of the cities of that day led people to believe that little urban expansion would ever occur. Government officials in Washington returned a portion of the District of Columbia to Virginia—they did not believe the city would ever reach the limits of the District.

THE EMERGENCE OF THE MODERN INDUSTRIAL SYSTEM BY THE 1870s

By 1870 the continental boundaries of the United States as they exist today had been established. Except in the Southwest and the northern Great Plains, the government had taken Indian lands, and the break between the North and South, which resulted in the Civil War (1861–1865), had been resolved. But the job of transforming this huge territory into a functioning economic unit remained. The tools of technology that could pull the regions of the United States together were at hand. The railroad and the telegraph made interregional mass movement possible for the first time. By 1860 railroads connected the three areas of major settlement (Figure 6–9). In addition, packet steamships sailed the rivers and coastal waters, and large freighters carried traffic on the Great Lakes.

In the South the railroads were less a network and more a series of local routes connecting agri-cultural areas with port gateways. Notably absent were any trans-Appalachian lines between Atlanta and Roanoke, Virginia. In the North an interstate system had developed that crossed the Appalachians along four different routes connecting Boston, New York, Philadelphia, and Baltimore with St. Louis, Chicago, and other Middle Western cities. In the Middle West the railroad system crisscrossed the newly settled agricultural lands of what was quickly to develop into the Corn Belt. Almost from the start farmers found that they could easily market produce by rail to cities in the Northeast.

Interregional rail connections meant that regions no longer had to depend on local food production, and the production of staple foods in New England and the Middle Atlantic area began to decline. At the same time, specialized mining areas developed at coal, iron ore, and copper deposits. For the first time, large-scale industry could draw on resources from areas thousands of miles apart. An iron and steel complex developed that was based on iron ore from around Lake Superior and coal from the Appalachians.

Over the railroads from the Northeast came masses of emigrants from Europe. Large numbers of people from the East also settled farmlands of the Middle West (Figure 6–10). This increased the close cultural as well as economic interaction between the two regions. But following the Civil War the links between these two areas and the South were less close than ever.

By 1870, before large-scale settlement of the frontier, a transcontinental railroad connection had been made with the Pacific Coast, and a rush of settlers to the Great Plains and the Far West began. Settlers were often recruited in Europe by railroad companies and then carried across country to their new homes by train. These people found the railroad ready to ship their products from the start, but the western areas had just begun to contribute to the country's economy with some mineral production and limited live-stock shipments. The important ties were still between the Middle West and the Northeast.

At this time, too, industry completed its changeover from wood and waterpower to coal.

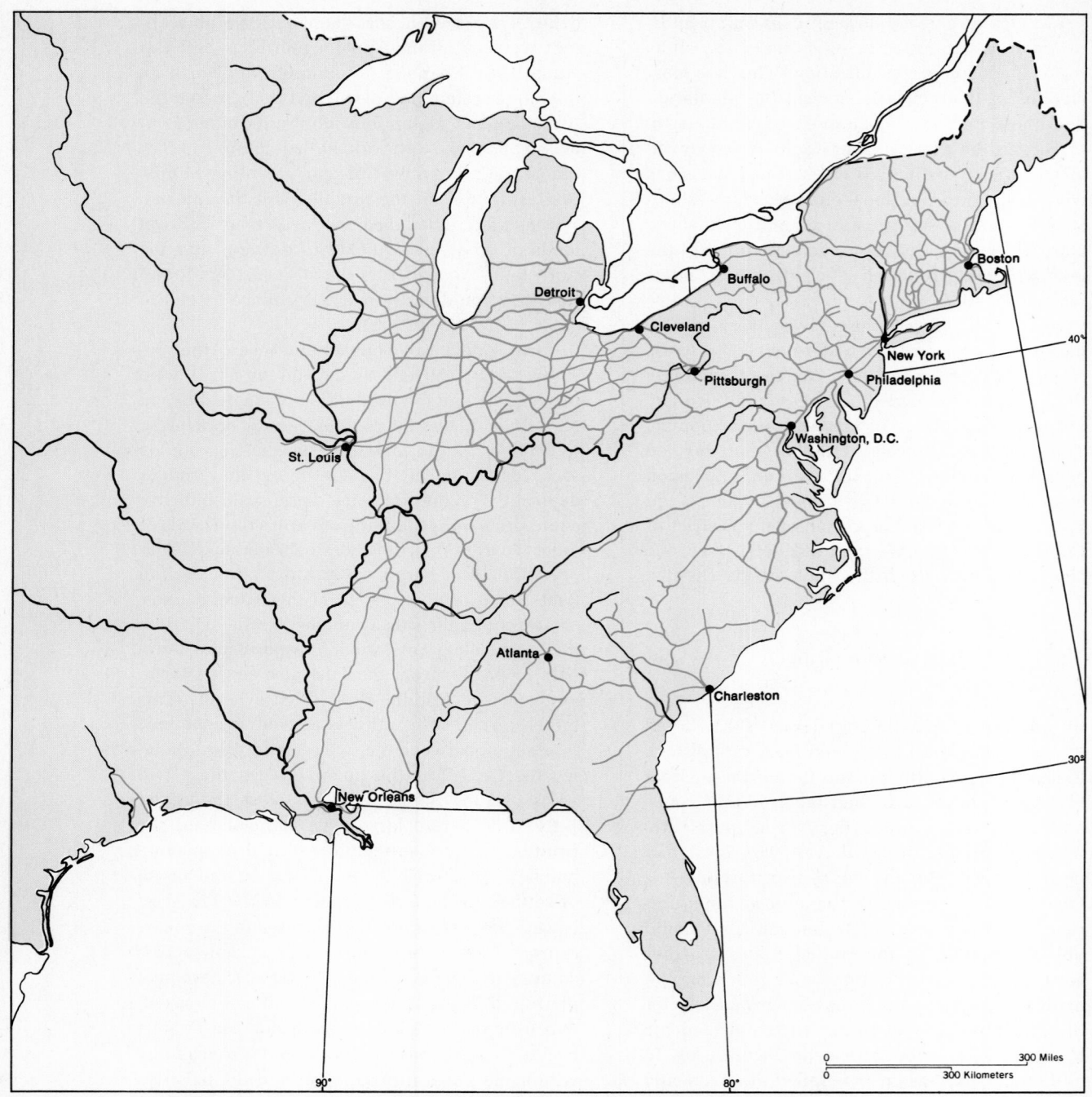

Detroit

Buffalo

Boston

Cleveland

New York

Pittsburgh

Philadelphia

Washington, D.C.

St. Louis

40°

Atlanta

Charleston

30°

New Orleans

90°

80°

0   300 Miles
0   300 Kilometers

SOURCE: *Atlas of the Historical Geography of the United States*, Carnegie Institution of
Washington and American Geographical Society, New York, 1932, Plate 139B

FIGURE 6-9   EXTENT OF RAILROAD CON-
NECTIONS IN THE UNITED STATES BY 1860

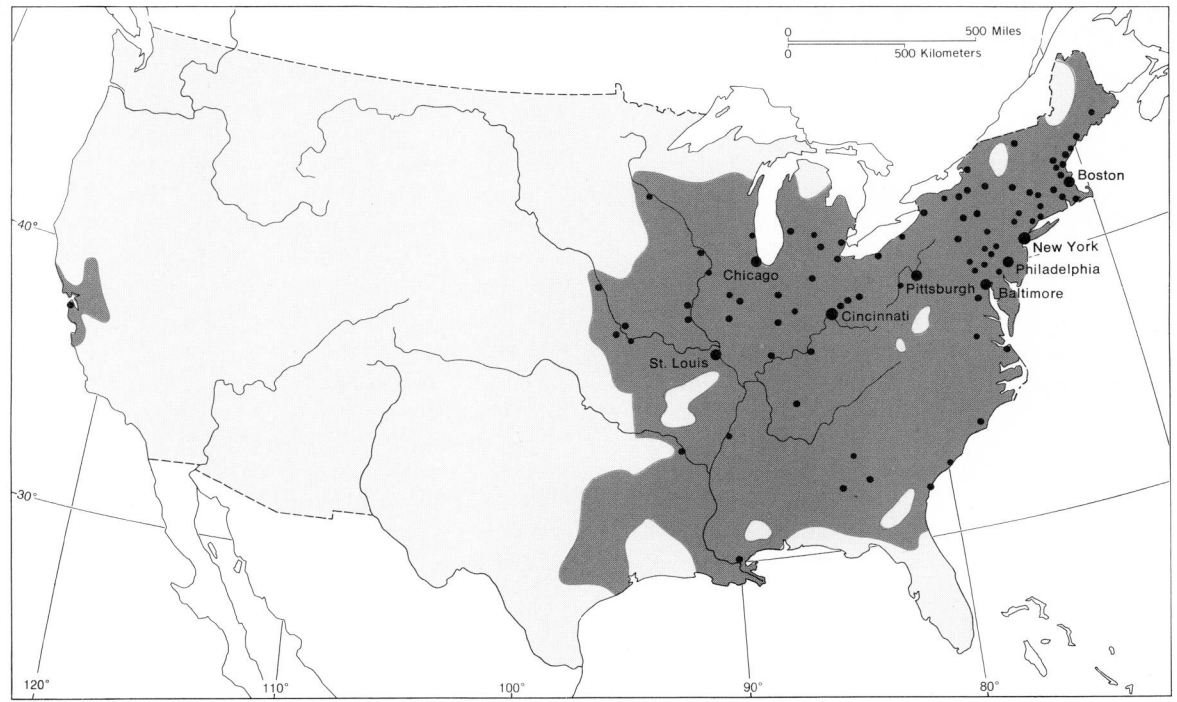

SOURCE: Redrawn, with permission, from the *Geographical Review*, Vol. 57, 1967.

FIGURE 6-10   SETTLEMENT OF THE UNITED STATES BY 1870

●   Large towns

•   Smaller settlements

� Areas with at least 6 persons per square mile (2.3 per square km)

This required a major reassessment of the country's resource base. New England, the traditional center of manufacturing, had no major deposits of coal and iron, the basic resources of the new industrialization. Although some deposits were found in the Middle Atlantic area along the eastern flank of the Appalachians, the major resource base emerged when the Great Lakes waterway linked the bituminous coal region of the western Appalachians with the iron region near Lake Superior. Thus the Middle West, developing as the great producer of agricultural staple foods, also became the most important center of metalworking.

Despite a lack of the raw materials needed for the new industrialization, New England's indus-

try continued to expand. Coal was brought in from the Appalachians by rail and ship. More important, New England capitalized on its entrepreneurial experience from colonial times, its skilled labor, and its access by rail to eastern seaboard and Middle Western markets and by ship to overseas markets.

The function of the large Middle Atlantic gateway ports as a "hinge" between the United States and Europe continued to develop rapidly. These centers were focal points in the growing interregional transportation system. The flow of people and goods meant an acceleration of business in such areas as trade, finance, and supporting activities. The financial and commercial interests in these gateway ports were in a position to manage interregional and international transactions, much as New England traders had been earlier. The businesses in these cities made the basic financial and management arrangements for economic developments in other parts of the country. Finally, with new interregional communication and with the increased power of the federal government in the late nineteenth century, Washington, D.C., began to emerge as an important decision-making complement to the Middle Atlantic business centers.

The South remained predominantly rural despite attempts to build its own industrial base to support its war effort. Cotton and tobacco, the traditional cash crops, were the mainstays, and the European market was of first importance. Railroads connected agricultural regions with ports. Disruption and destruction caused by the Civil War handicapped the region, but the plantation system survived the freeing of slaves by shifting to a sharecropping system, in which a former slave worked a small plot of ten or more acres of plantation land and paid rent in the form of a share of the crop. Such tenants, completely uneducated and inexperienced in business affairs, had little hope of gaining an adequate living standard and were usually in debt to the local supply stores.

The parts of the South that did not produce cash crops on plantations—particularly the Appalachian uplands—were left out of the emerging **interregional system.** Their populations continued to live primarily on a subsistence basis, just as they had in frontier days.

For the first time, small towns could be in constant touch with the outside world by telegraph and rail. Local newspapers could print national news, and trains could bring big-city papers. Local stores could be sure that trains would bring ready supplies of needed stock.

The larger urban centers, however, benefited most from the new transportation and communication systems. They were able to expand their economic and cultural influence at least as far as a train could travel in one night. Towns and villages within this distance served as collecting points for agricultural products and other primary resources, and were the retail outlets for large urban-centered wholesalers. The big urban places were not only hubs connecting the small towns and rural areas around them, but were also part of a network of interregional connections with large cities elsewhere. Because they were centers of transport and telegraph, big cities could develop regional businesses. Entrepreneurs and interregional management emerged in all the large cities of the eastern seaboard and the Middle West. Boston, New York, Philadelphia, Baltimore, Chicago, St. Louis, Cincinnati, and New Orleans each had over 100,000 people by 1860, at which time 15 percent of the population of the country lived in cities of 10,000 or more. In the largest cities horse-drawn streetcars provided the first municipal transport and allowed the cities to spread out, since workers could ride to work. Further expansion, however, was not anticipated.

GROWTH OF THE MODERN INDUSTRIAL SYSTEM ON A TRANSCONTINENTAL BASE: 1870–1920

From 1870 to 1920 the modern interconnected system expanded rapidly, and for the first time a national system operated from coast to coast. All major regions of the country were inhabited and connected into the system, and the interregional connections themselves were greatly improved. Railroads used steel rails, instead of iron, and large steam locomotives became the most efficient long-haul movers of both people and goods. With the invention of refrigerator cars, tank cars, and livestock cars, virtually all types of goods could be moved. Steam-powered, steel-hulled ships, larger and faster than earlier vessels, multiplied the volume that could be transported over water. The first oil pipelines were completed.

Inanimate energy was applied on a large scale to secondary production and began moving into agriculture and mining. Machines for planting and harvesting crops were invented, and by the end of this period tractors to pull the equipment

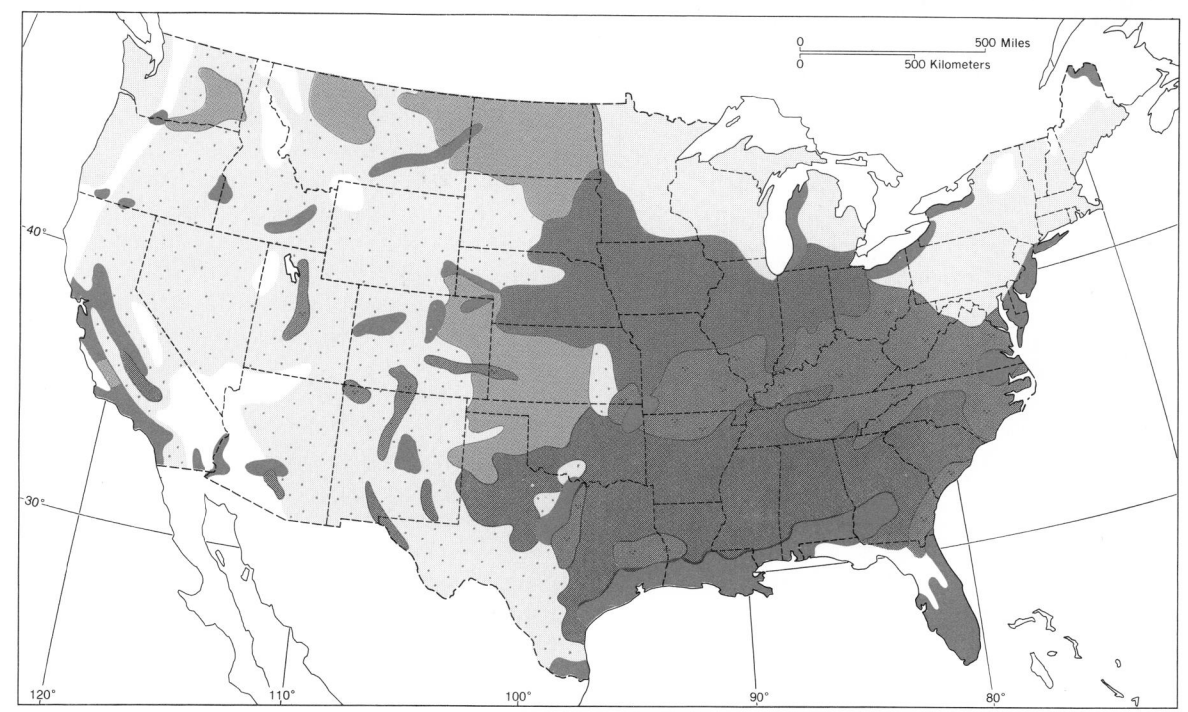

SOURCE: American Heritage Pictorial Atlas, pp. 270–271

FIGURE 6–11 REGIONAL AGRICULTURAL SPECIALTY PRODUCTS OF THE CONTERMINOUS UNITED STATES IN 1920

Dairy, cattle, hay

Corn and livestock

Wheat and small grains

Cotton

General farming, tobacco

Fruit, truck, special crops

Grazing livestock

Little or no agriculture

had been developed. Trucks slowly replaced horse-drawn wagons on the farms. In factories electricity provided easily controlled and flexible power for a great variety of operations, and large electricity-generating plants were able to supply whole cities and their factories with this form of power. Such organization enabled newly settled parts of the country to contribute specialties, while older producing areas became more specialized.

Regional Variations

By 1920 most of the present agricultural regions had been established (Figure 6–11). The Cotton Belt extended from North Carolina into western Texas. The Middle Western Corn Belt was clearly defined, as were the dairy regions in the northern lake states (Minnesota, Wisconsin, and Illinois), New York, and New England. Specialized tobacco-growing areas, each producing its own particular type of leaf, were found from northern Florida to Connecticut. Citrus fruits came from both Florida and California. The newly settled lands of the Great Plains still produced cattle; however, with machines to work the land, the eastern margins had already become established as the country's most important wheat-growing areas. Orchard fruit came from the Pacific Northwest and upstate New York, and grapes from New York and California. In the Mountain West newly irrigated areas using water from giant dams produced sugar beets, potatoes, specialty crops, and alfalfa.

The coalfields of southern Illinois and western Kentucky had also come into operation late in the nineteenth century, and the vast midcontinent and Texas oil fields were opened. The lead deposits of eastern Missouri proved the most productive in the world, and zinc was mined in a productive field along the border of Missouri, Kansas, and Oklahoma. The copper fields of western United States had been opened, along with a variety of other new mineral sources in the western mountains.

Most of all, this was the period when heavy industry developed: steel mills, metal foundries, railroad equipment plants, oil refineries, chemical works, and meat-packing plants. These industries generally occupied sites in the established metropolitan centers of the northeastern seaboard and the Middle West. A series of urban-centered districts of the new heavy industry sprang up near major cities from New York to St. Louis. These complexes sprawled over the landscape, belching noise, smoke, and dirt. Nearby were some of the most tightly packed workers' communities in the country, occupied mostly by new immigrant groups. Other features of the industrial districts were the miles of railroad tracks and switchyards needed to deliver raw materials and to ship heavy finished goods.

The Rise of National Industries

At this time national rather than regional control made its first appearance in industry. John D. Rockefeller gained virtual control of the U.S. oil industry until the government forced him to break up his empire. Andrew Carnegie laid the foundation for the United States Steel Corporation with plants in the Northeast, the Middle West, and the South. Multiplant corporations became the rule in the meat-packing, mining and smelting, metal-processing, and chemical industries. All this resulted from improved interregional transportation and communications and offered the advantage of industrywide control by a firm. The heavy investment costs could be financed only at the highest banking level, and management control had to be sophisticated. Operations on such a scale, beyond the capabilities of the local or regional entrepreneur, enabled goods to be produced at a lower cost per unit. Such savings resulting from large-scale operation are called **economies of scale.**

Cities became known not only for their specialized manufacturing operations, but also for the special financial and business services they offered in connection with those industries. Pittsburgh was the steel center; Chicago, the center of meat packing, railroads, and railroad equipment; Detroit, of automobiles; Akron, of rubber; Cincinnati, of soaps; and Toledo, of glass. The headquarters of companies in these centers managed networks of plants. Banks and financial institutions specialized in loans and money management for the city's particular industry.

Manufacturing outside the northeastern seaboard and the Middle West was largely tied to particular resources, such as oil, lumber, and specialty metals. A major migration of long-established northern industry into small Appalachian towns in the South took place—not everywhere, but just along either side of the mountains. Since the textile industry has no particular demands for massive power and special transportation, it began moving south from New England in search of large quantities of unskilled labor. The usual textile mill produces different weights of standard "gray cloth," which is later dyed in other plants nearer to markets and so requires only simple machines and semiskilled workers. With the growth of other industries in New England, wages began to rise. New England was one of the first areas to become unionized, and the unions called for better working conditions.

The South had no unions, few other industries to compete for labor, and communities eager to assist new factories with building subsidies and special tax relief. New regional networks of electrical power provided all the necessary electricity.

Other light industries dependent on large quantities of unskilled labor moved to the South, to mining communities, and even to farm villages. Shoe and clothing factories were established in small rural towns. Such firms commonly employed mainly female labor—wives who supplemented the incomes of their farmer or miner husbands.

A major part of the clothing industry remained firmly centered in the heart of New York City, where high fashion clothing, particularly women's wear, is still made today. In a sense the clothing industry has really been a handicraft rather than a factory industry. It has always been located in New York City, the largest, most fashion-conscious U.S. clothing market, where it can adapt quickly to the latest style trends. If a new style of garment sells well in New York shops and department stores, dozens of clothing manufacturers can produce copies in quantity almost overnight.

The Development of Metropolitan Areas

By the beginning of this century the largest cities were becoming metropolitan areas. New developments in urban transportation—the electric street railroad, the elevated railroad, and commuter railroads—allowed cities to sprawl. Outlying communities were supplied by urban public utility networks. The first suburbs grew up beyond city limits along commuter railroad routes. The wealthy and the upper middle class moved out along Philadelphia's Main Line or Chicago's North Shore.

In the cities streetcars could handle masses of people and allowed working-class people to live away from the neighborhoods in which they worked. By the end of this era, large apartment buildings close to streetcar routes were replacing the old inner-city row houses. Outlying shopping districts appeared at major intersections of streetcar lines, and shops sprang up along important streetcar routes, forming commercial streets.

Ethnic groups arrived in great numbers from Europe. Slums grew as the immigrants crowded into row houses abandoned by the wealthy and middle classes in their exodus to the suburbs.

Some aspects of city life improved. With water and sewer lines to individual homes, gas and electricity, and paved streets and sidewalks, cities had more conveniences than before. Public schools and new civic centers with concert halls and museums were built. On the other hand, the masses of low-income groups in tenements and row houses created problems of health, congestion, and education on a scale never before experienced in any U.S. urban community.

TECHNOLOGICAL CHANGES AND MASS
PRODUCTION: 1920 TO THE PRESENT

Since 1920 few new areas of the country have been added to the system. Rather, this period has been characterized by continual adjustment to accelerating technological changes within the industrial system: highway and air transportation; regional electrical power networks; and the automation of agriculture, mining, and many manufacturing processes. Most important, perhaps, has been the tremendous increase in the scale of operations. The population of the country more than doubled from 105 million in 1920 to over 215 million today. At the same time the living standard of most people has risen rapidly. Unions have gained higher wages and better working conditions for laborers, and the demand for sophisticated services has brought about better educated, better paid specialists. The result has been **mass consumption,** with many millions of people able to purchase goods and services for their own use and convenience.

The development of mass consumption has been facilitated by installment credit, which allows people to pay for a relatively expensive item over a period of time. More recently, installment

THE EXAMPLE OF APPALACHIA

As the socioeconomic system of the country has evolved, the whole basis on which a region or city has developed has changed. Consider the case of the southern Appalachian Mountains (Figure 6–12). Before the coming of railroads, people had to walk, ride horses, or use Conestoga wagons to cross the mountains. Some of the most important routes from the eastern seaboard to the Middle West were through the valleys of western Virginia, eastern Kentucky, and Tennessee. Along these routes, frontier settlements flourished, occupied by some of the most intelligent and adventuresome people of the time. Prospects for the future looked bright for these settlers with their access to important transport routes.

The opening of the Middle West and the coming of the railroads changed the pattern of movement between the east coast and the interior of the country. The major railroads were built from New York and Philadelphia through upstate New York and Pennsylvania. Few went west from Virginia or the Carolinas. With the coming of the railroads the early overland routes were abandoned, and the people in the valleys along these byways found themselves isolated from the rest of the country. As agriculture shifted more and more to commercial products, the farms of Appalachia proved too small and isolated to compete with new agricultural regions.

Later, when coal was discovered, a second wave of population entered different parts of the Appalachian region. Railroads were built to the coal-producing districts, and for a time the new coal towns boomed. Immigrants from Europe were attracted to work in the mines. But in recent years coal mining has become highly mechanized; strip-mining techniques and automated equipment in underground mines have reduced mine employment by as much as 90 percent. Miners have found no job alternatives in these small communities built specifically as mining towns. The nationwide shifts to metropolitan life and to centers of modern transportation and communication have widened the gap between the people in Appalachia and those in the rest of the country. Today this is one of the nation's most seriously depressed areas, but new interest in coal to supply energy for power plants promises to breathe some life into Appalachian mining areas. This will bring wealth to local coal operators and landowners, and it has increased mining employment. Mechanized mines do not, however, need the large numbers of workers required in the past, and so the bulk of the population remains trapped in poverty.

credit has been extended to virtually all types of consumer purchases through the use of bank credit cards. Mass consumption has caused a shift to **mass production**—not only in the factory, but also on the farm and in the mine, and even in schools. A market of millions of affluent people has ensured the successful operation of large-scale, highly specialized enterprises turning out vast quantities of a single product.

Both the size and geographic distribution of the market have stimulated the growth of mass production and consumption. The growth of metropolitan areas and the increasing concentration of population in them has brought not only economies of scale, but also savings resulting from the presence of a large number of consumers within a relatively small geographical area. Such savings resulting from increasing concentration, or "agglomeration," of the market, including reduced transportation costs, are known as **agglomeration economies.**

Mass production has caused a further redistribution of the productive units of the country. Large "factory farms" with corporate ownership have replaced many small family farms as the mainstay of agricultural production. Most steel is

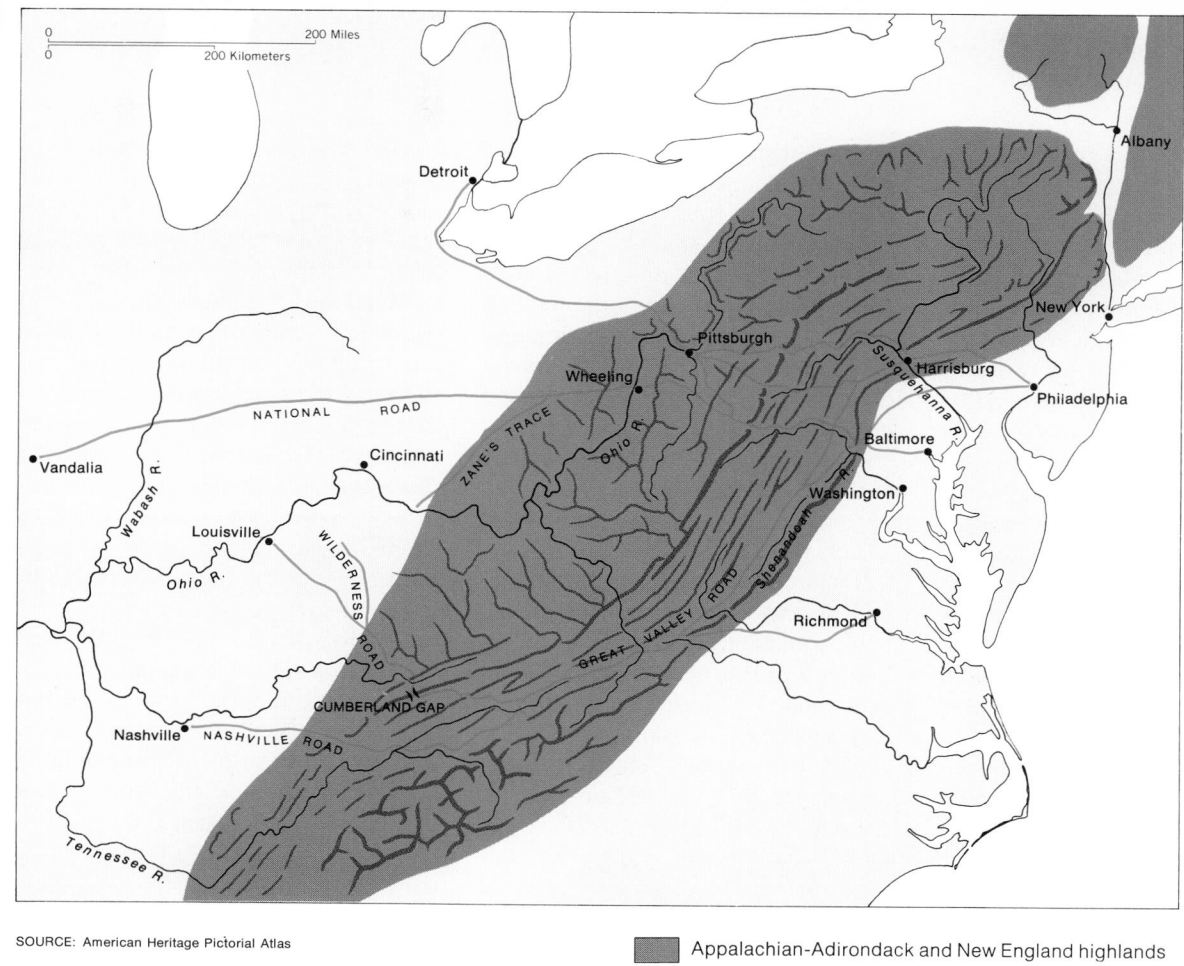

SOURCE: American Heritage Pictorial Atlas

FIGURE 6-12 SOUTHERN APPALACHIA IN THE PRERAILROAD ERA

▨ Appalachian-Adirondack and New England highlands

〰 Main ridges and escarpments

— Main trans-Appalachian routes

produced by a dozen major corporations, most of which have only one or two basic manufacturing plants. All Cadillac automobiles are produced in a single factory, for example.

Now national entrepreneurs evaluate the resources of particular parts of the country to find the area best suited to the production of a particular item for consumption by the whole country. People who specialize in industrial location study the cost of production at and distribution from different centers to find the location that offers "least cost" operations. Such studies do not "start from scratch," but build instead on existing patterns—markets, production sources, and transportation networks—and result in the choice of only a few localities.

In selecting the location that lends itself to the most economical operation, an industry considers,

THE EXAMPLE OF BUFFALO

We often speak of the city of Buffalo as if an important city at that site were predestined by environmental conditions. Located at the eastern end of Lake Erie, it has the advantages of being at the western end of the best natural route across the Appalachians in the Northeast, and of being just above Niagara Falls (Figure 6–13). Buffalo is the ideal **break-in-bulk point** for goods and people changing between lake transportation and cross-country carriers, either land or canal. It is the most advantageous place to transfer from one transportation mode to the other, because it allows maximum use of low-cost water transportation and minimum use of more expensive land transportation in moving goods from the Middle West to markets in the Northeast.

With the opening in 1825 of the Erie Canal from the Hudson River to Lake Erie, the rise of shipping on the Great Lakes, and the coming of the railroads, Buffalo rapidly grew to be a major center in the U.S. transport system. It became one of the most important grain-storage terminals, the site of a large steel mill, and a major oil terminal. Low-cost electric power generated at Niagara Falls attracted industry. The industrial-transportation center of Buffalo has become the twenty-fourth largest U.S. metropolitan area with a population of over 1.3 million people.

But suppose the U.S. transportation system were now being developed completely from scratch with no heritage of the past. Would Buffalo be needed? The St. Lawrence Seaway–Welland Ship Canal water route allows ocean-going ships and ore carriers to by-pass Buffalo and go directly on to Montréal on open water. Ocean cargoes that once came to the port of New York for shipment by rail or barge to Buffalo can now go directly to Chicago, Detroit, or Milwaukee.

Moreover, the New York Central Railroad, which developed the Hudson–Mohawk rail route, has become part of a more extensive railroad system that offers more direct service between the eastern seaboard and the Middle West across Pennsylvania. New limited-access roads across Pennsylvania also provide more direct connections east and west than the Hudson–Mohawk gateway, and even the electric power from Niagara Falls goes into a regional grid that can deliver electricity far from the power site. But Buffalo is too large to become a ghost town. Thanks to its origins in another era, Buffalo has remained an important U.S. city.

among other factors, the distance from the raw materials (resources), the distance from the intended market, and transportation costs. The nature of the manufacturing process has an important effect on transportation costs and therefore on location. A **resource-oriented** industry is one that processes bulky raw materials into a product that is lighter and therefore less costly to ship than are the raw materials. This kind of industry tends to locate near the source of the raw materials to avoid the cost of shipping them. For example, the steel industry is centered in Pittsburgh, Cleveland, and Chicago, near the iron ore deposits of Minnesota and the coalfields of the Middle Western states. A **market-oriented** industry is one that manufactures a product more costly to ship than are the raw materials. It is more advantageous for this type of industry to do its manufacturing in the vicinity of its market, to reduce the cost of shipping the manufactured product. The electronic equipment industry shows a market orientation; its principal centers are the suburbs of Boston and San Francisco.

The change to large-scale units of primary production has brought a shift in the use of the resource base on which agriculture rests. Large mechanized operations place a premium on level land where machines can operate easily and where water is available for irrigation. This development has put a premium on Middle Western

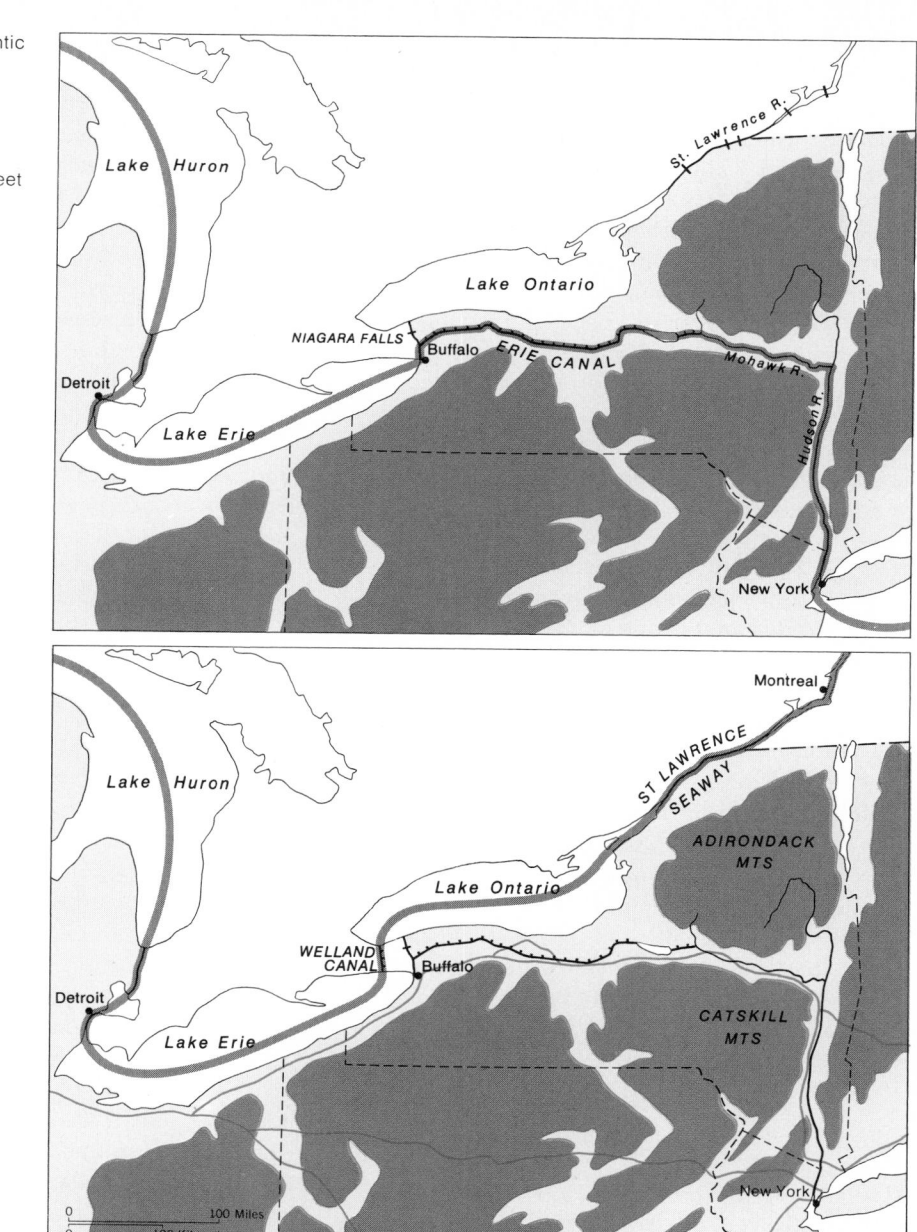

Legend

- Great Lakes and Atlantic shipping
- Interstate highways
- Falls and rapids
- Elevation over 1,000 feet (305m)

Top map labels: Lake Huron, Lake Ontario, St. Lawrence R., NIAGARA FALLS, Buffalo, ERIE CANAL, Mohawk R., Detroit, Lake Erie, Hudson R., New York

Bottom map labels: Lake Huron, Lake Ontario, Montreal, ST LAWRENCE SEAWAY, ADIRONDACK MTS, WELLAND CANAL, Buffalo, Detroit, Lake Erie, CATSKILL MTS, New York

0 — 100 Miles
0 — 100 Kilometers

FIGURE 6–13   BUFFALO BEFORE AND AFTER
THE ST. LAWRENCE SEAWAY

agricultural lands and on river lowlands such as the lower Mississippi Valley. In the West, floodplains and river terraces have been supplied with irrigation water, and the great, almost level Central Valley of California has become the most important oasis of all. Even the high plains of Texas have been intensively farmed by using well water for irrigation. Hill lands, meanwhile, have largely been abandoned, despite rich soil and potential productiveness, because they do not lend themselves to large-scale operations.

The most highly prized level lands for large-scale farming are those with adequate moisture and long growing seasons. Thus the irrigated lands of the Southwest, some of the most remote areas in terms of the major market for agricultural production, have been most successful in growing specialty crops because of their high solar energy, high potential evapotranspiration, and controlled application of water by irrigation. With fertilizers Florida, too, produces winter vegetables despite its lack of good soil. In contrast, much of the Great Plains presents a problem because rainfall does not provide adequate moisture and most land cannot be irrigated. Vast stretches of the central and western margins of that region are without adequate irrigation sources and so remain pasture.

Under present technology, soil fertility is the least important ingredient in the environment. It is now possible to add chemical nutriments to soil "by prescription" to produce lush growth, provided the land is flat and has adequate solar energy and moisture supply.

TODAY'S ECONOMIC SYSTEM AND THE BURDEN OF THE PAST

The distribution of economic activities in the United States today reflects many features that are remnants of the past when the country had different objectives. Yet these vestiges—places, people, ideas, and prejudices—are very much a part of economic interplay today. Perhaps the greatest problem in making the national economic system work is that no region can be wiped clean of these physical and psychological remnants of the past. Decisions made in a different cultural and technological environment have become established features of the system. Life today must take into account established railroad and road networks, cities, functioning industries, and farming areas.

This is a constant problem in urban areas. Eastern cities, built originally in the era of the sailing ship, coach, and wagon, have had to adjust to the steamship, train, car, bus, truck, and airplane. Each advance produced a structural change, either removing or modifying the old. Projects to build new freeways and to clear slums in the 1960s destroyed neighborhoods, uprooted families, and affected the lives of hundreds of thousands of people. Moreover, those affected were usually the people who did not want the change and did not benefit from it. Whenever a new freeway or a new apartment complex was built, it created new problems that required further adjustments. Today slum clearance and freeway construction in metropolitan areas have virtually ceased. The problems now are how to overcome increasing congestion on the streets and how to upgrade to modern standards those structures built fifty to one hundred years ago.

Ideas inherited from the past are often major deterrents to change, too. Some of the regulations enforced by today's labor unions date from earlier decades when conditions in many industries were quite different. The federal government's reluctance to subsidize failing railroads in the 1960s and 1970s may have reflected a widespread belief in the advisability of private ownership of all means of transportation, even though government-operated rail systems have been very successful in other countries. The present organization of governments in the United States embodies many accepted ideas from the past, such as the separation of city and suburb, and the continuation of separate township, county, and state governments. Most of these ideas are not likely to change very soon.

SELECTED REFERENCES

Billington, Ray Allen. *Westward Expansion: A History of the American Frontier.* Macmillan, New York, 1949. Treats the history of the dispersal of settlement across the continental United States in a geographical manner. It offers insight into the factors associated with the sequence and spatial pattern of U.S. settlement.

Borchert, John R. "American Metropolitan Evolution." *Geographical Review* 57, no. 3 (July 1967): 301–332. Analyzes the development of the pattern of metropolitan places from 1790 to the present. The relationships between transportation and technology, industrial development, and physical resources are discussed in terms of their effects on the periods of growth peculiar to groups of metropolitan places.

Brown, Ralph. *Historical Geography of the United States.* Harcourt Brace Jovanovich, New York, 1948. Still the classic text on the historical geography of the United States.

Comparative Metropolitan Analysis Project. *Contemporary Metropolitan America: Twenty Geographical Vignettes,* vol. 1. Ballinger, Cambridge, Mass., 1976. Covers the twenty metropolitan centers: Atlanta, Baltimore, Boston, Chicago, Cleveland, Dallas, Detroit, Hartford–Central Connecticut, Houston, Los Angeles, Miami, New Orleans, New York–New Jersey, Philadelphia, Pittsburgh, St. Louis, St. Paul–Minneapolis, San Francisco, Seattle, and Washington, D.C. Separate paperback editions are available for most of these city vignettes, each of which deals with the historical background and development, the contemporary functions, and major problems of a metropolitan center. As a group these twenty examples reveal much about the national and regional character of U.S. economic, social, political, and decision-making systems.

Gottmann, Jean. *Megalopolis: The Urbanized Northeastern Seaboard of the United States.* Twentieth Century Fund, New York, 1961; M.I.T. Press, Cambridge, 1964. Analyzes the development of the Boston-to-Washington urbanized area, focusing not only on how this particular region functions but also on how its development has been part and parcel of the development of the national territory. It provides evidence of the dominant role of this urbanized region in making decisions about the development of the rest of the nation.

Hart, John Fraser, ed. *Regions of the United States.* Harper & Row, New York, 1972 (paper). Also appears as *Annals of the Association of American Geographers,* 62 (June 1972). A collection of essays on U.S. regional geography. Articles by Meinig on "American Wests"; Durrenburger on "The Colorado Plateau"; Mather on "The American Great Plains"; Hart on "The Middle West"; and Borchert on "America's Changing Metropolitan Regions" are broad in scope and focus on the major changes within each region. Other articles have narrower topics for the regions they consider.

Perloff, Harvey, and Lowden Wingo, Jr. "National Resource Endowment and Regional Economic Growth," in *Natural Resources and Economic Growth,* J. L. Spengler, ed. Resources for the Future, Washington, D.C., 1961, pp. 191–212. Also appears as Chapter 11 in *Regional Development and Planning,* John Friedmann and William Alonso, eds. M.I.T. Press, Cambridge, Mass., 1964. An overview of the changing role of resource endowments and their locations in the economic development of the nation. It integrates a wealth of historical economic data into a rather brief interpretive statement.

Pred, Allan R. *The Spatial Dynamics of U.S. Urban-Industrial Growth, 1800–1914: Interpretive and Theoretical Essays.* M.I.T. Press, Cambridge, Mass., 1966. Contains an especially relevant analysis of the growth of major U.S. metropolitan centers from the impetus of industrial location and growth (Chapter 2).

Ward, David. *Cities and Immigrants: A Geography of Change in Nineteenth Century America.* Oxford University Press, New York, 1971. Looks at changes in U.S. cities in the last century in terms of people and cultures.

CHAPTER 7

The United States: The Interconnected System on a Grand Scale

This chapter is a study of the workings of the modern interconnected system within the United States today. Metropolitan areas act as nodes in a network of transportation and communications and as centers of economic activity. Major decisions about the functioning of the system are made at the headquarters of large, multifaceted corporations. Washington, D.C., is the decision-making center of the federal government. A closer examination of the United States reveals that some parts of the country are outside the modern system.

Left: Macy's department store in New York City. (Bruce Anspach/EPA Newsphoto) *Above:* The Chicago Commodity Exchange. (Roger Malloch/Magnum Photos, Inc.)

Above: Wholesale produce market serving the Boston area. (Arthur Grace/Stock, Boston) *Right:* Tour group considering a statue of the Hindu deity Vishnu in the Asian Art Museum, San Francisco. (San Francisco Convention and Visitors Bureau)

The modern interconnected system of the United States was a development of the industrial age. All other large countries in the world have developed out of cultures that have been evolving for several thousand years. Their long-established patterns of settlement, routes of transport, producing areas, and cultural structure predate development of the modern interconnected system. Change to new ways has had to come through remaking the physical plant of past generations and modifying cultural patterns to fit new conditions. Cities built in preindustrial times have had to adapt to the airplane age. Methods of farming have changed to take into account scientific developments. With all these changes has come pressure to alter organizational, social, and personal patterns to fit the modern interconnected system.

In contrast, only the U.S. eastern seaboard was settled by Europeans in preindustrial times. United States history has been less a matter of changeover from the traditional system to the modern interconnected system than of constant modification and improvement of the interconnected network in a rapidly changing technology. Basically, the change in the United States has occurred in the *scale* of operation—from local, to regional, to national, and finally to international interconnection.

Modern transportation allows rapid movement of masses of goods and people from virtually any location in the country, and modern communication provides instantaneous flow of information; thus it is possible for each place to make its particular contribution to the overall system. At the same time, centralized political and cultural forces can replace local cultural differences with a single, nationwide structure.

The modern interconnected system, operating over the huge territory of the United States, has had the opportunity to develop separate, distinctive distributions for production, consumption, and decision making. Thus primary production such as agriculture, mining, forestry, and fishing is closely tied to the variations in the environmental base, although it must take into account such things as accessibility to customers, governmental regulations, and the location of competitors.

Secondary production, which depends as much as primary production on availability of labor and accessibility to markets, shows very few ties to the environment and tends instead to concentrate in major urban nodes. In addition, the customers are also workers, so they locate where the jobs are, mainly in major metropolitan centers. The managers and decision makers, who now can manage operations spread out over the whole country or even the world from a single location, prefer the few largest metropolitan centers, for they have the best transportation and communication connections. These centers also have the advantage of specialized management communities with unique supporting services.

A DISTINCTIVE EMPLOYMENT STRUCTURE

Table 7–1 shows the gross employment pattern of the United States in 1974, divided into four types, two production (primary and secondary) and two nonproduction (supporting services and management and decision-making activities). The table reveals the relative importance of the major economic specializations of the United States. Note that about one-third of the working population is employed in the production of goods. Almost two-thirds of the workers are not employed in making anything: they neither extract earth resources in primary production nor manufacture goods. Rather, they are employed in servicing, helping, and managing, and they sell goods, provide personal and professional services, and work in government. What influence would we expect employment structure to have on population distribution?

TABLE 7-1 EMPLOYMENT STRUCTURE OF THE
UNITED STATES, 1974

	Employment (Thousands)		Percent	
Production				
Primary				
Agriculture	3,492		4.3	
Mining	672		0.8	
		4,164		5.1
Secondary				
Construction	3,985		4.9	
Manufacturing	20,016		24.5	
	24,001		29.4	
		28,165		34.5
Supporting Services				
Transportation and utilities	4,699		5.7	
Trade	17,011		20.8	
Other services	13,506		16.5	
		35,216		43.0
Management and Decision Making				
Finance, real estate	4,161		5.1	
Government	14,285		17.5	
	18,446			22.6
Total		81,827		

SOURCE: U.S. Bureau of Census, *Statistical Abstract of the United States, 1975,* Washington, D.C., 1975.

THE MAP OF POPULATION AS AN
INDEX OF ECONOMIC OPPORTUNITY

The map of population (Figure 7-1) shows the contrast between two different distributions of people in the United States. On the one hand, it presents the thin veneer of rural population spread widely over the country with variations in density, presumably providing a measure of dif-

fering resource utilization. On the other hand, it presents the concentration of population not simply in urban areas but in the network of a relatively few large metropolitan centers.

The two distributions shown in Figure 7-1 are those of **metropolitan population,** the population living in metropolitan areas, and **nonmetropolitan population,** the people who live outside the metropolitan areas. From the map itself it is

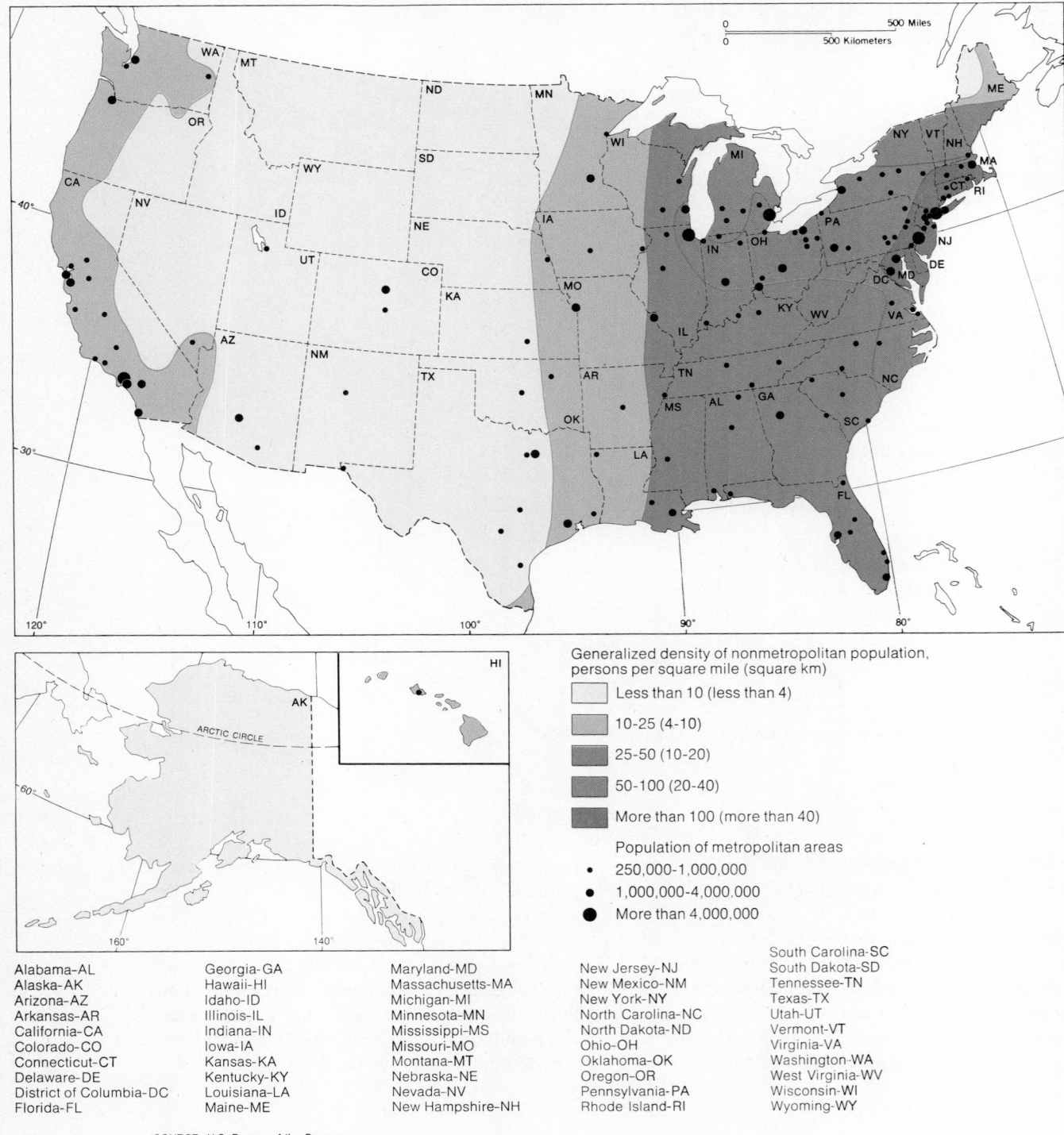

Generalized density of nonmetropolitan population,
persons per square mile (square km)

- Less than 10 (less than 4)
- 10-25 (4-10)
- 25-50 (10-20)
- 50-100 (20-40)
- More than 100 (more than 40)

Population of metropolitan areas
- • 250,000-1,000,000
- ● 1,000,000-4,000,000
- ⬤ More than 4,000,000

Alabama-AL
Alaska-AK
Arizona-AZ
Arkansas-AR
California-CA
Colorado-CO
Connecticut-CT
Delaware-DE
District of Columbia-DC
Florida-FL

Georgia-GA
Hawaii-HI
Idaho-ID
Illinois-IL
Indiana-IN
Iowa-IA
Kansas-KA
Kentucky-KY
Louisiana-LA
Maine-ME

Maryland-MD
Massachusetts-MA
Michigan-MI
Minnesota-MN
Mississippi-MS
Missouri-MO
Montana-MT
Nebraska-NE
Nevada-NV
New Hampshire-NH

New Jersey-NJ
New Mexico-NM
New York-NY
North Carolina-NC
North Dakota-ND
Ohio-OH
Oklahoma-OK
Oregon-OR
Pennsylvania-PA
Rhode Island-RI

South Carolina-SC
South Dakota-SD
Tennessee-TN
Texas-TX
Utah-UT
Vermont-VT
Virginia-VA
Washington-WA
West Virginia-WV
Wisconsin-WI
Wyoming-WY

SOURCE: U.S. Bureau of the Census

FIGURE 7-1 NONMETROPOLITAN POPULA-
TION DENSITY IN THE UNITED STATES

TABLE 7-2 REGIONAL DISTRIBUTION OF NONMETROPOLITAN
POPULATION OF THE UNITED STATES, 1950 AND 1973

	1950		1973		Change	
	Population (thousands)	Percent of national total	Population (thousands)	Percent of national total	Number (thousands)	Percent
Northeast[a]	8,794	13.4	7,774	13.5	−1,020	−11.6
Middle West[b]	16,787	25.5	14,574	25.4	−2,213	−13.2
South[c]	28,928	44.0	24,305	42.4	−4,623	−16.0
Great Plains– Mountain[d]	7,115	10.8	7,107	12.4	−8	−0.1
Pacific[e]	4,124	6.3	3,610	6.3	−514	−12.5
Total	65,748		57,370		−8,378	−12.7

SOURCE: U.S. Bureau of Census, *Statistical Abstract of the United States, 1958* and *1975,* Washington, D.C.
[a]Maine, New Hampshire, Vermont, Massachusetts, Rhode Island, Connecticut, New York, New Jersey, and Pennsylvania.
[b]Ohio, Indiana, Illinois, Michigan, Wisconsin, Minnesota, Iowa, and Missouri.
[c]Delaware, Maryland, District of Columbia, Virginia, West Virginia, North Carolina, South Carolina, Georgia, Florida, Kentucky, Tennessee, Alabama, Mississippi, Arkansas, Louisiana, Oklahoma, and Texas.
[d]North Dakota, South Dakota, Nebraska, Kansas, Montana, Idaho, Wyoming, Colorado, New Mexico, Arizona, Utah, and Nevada.
[e]Washington, Oregon, California, Alaska, and Hawaii.

difficult to determine the relative importance of the metropolitan and nonmetropolitan populations, but Table 7-1 leaves no question about which population is the greater. Only about 5 percent of the working population is now engaged in primary production activities, which are essentially nonmetropolitan. According to the 1970 census, 69 percent of the population lived in metropolitan areas. By 1973 the percentage of the population living in metropolitan areas had risen to almost 73 percent.

RURAL POPULATION AND
PRIMARY PRODUCTION

If we assume that knowing the nonmetropolitan population gives us the proportion of Americans dependent on primary use of land resources, ei-

ther using the resources directly or in supporting towns and cities, the distribution of nonmetropolitan population should provide a reasonably good measure of agricultural, mining, forest, and fishing regions on which the interconnected system depends.

Statistical information on U.S. metropolitan areas is usually based on definitions established by the U.S. Office of Management and Budget. A **Standard Metropolitan Statistical Area** (SMSA) is defined as an integrated economic and social unit with a large population nucleus. The criteria for drawing the exact boundaries of metropolitan areas change as the population distributions change.

About two-thirds of the nonmetropolitan population is located in two areas—the South and the Middle West (Table 7-2). None of the other three areas accounts for as much as 20 percent of

TABLE 7–3   REGIONAL DISTRIBUTION OF PRIMARY PRODUCTION, 1973

	Output (Millions of Board-Feet)	Value of Output ($ Millions)				Percent of National Total	Percent Non-metropolitan Population
	Lumbering	Fishing	Mining	Agriculture	Total		
Northeast[a]	1,667	158	2,070	4,097	6,442	5.0	13.5
Middle West	[b]	9	4,412	28,721	33,142	25.7	25.4
South	12,445	395	21,315	26,443	49,024	38.0	42.4
Great Plains–Mountain	6,478	0	6,389	19,364	26,206	20.3	12.4
Pacific	18,005	345	2,601	9,965	14,171	11.0	6.3
Total	38,595	907	36,787	88,590	128,985		

SOURCE: U.S. Bureau of Census, *Statistical Abstract of the United States, 1975*, Washington, D.C., 1975.
[a]See Table 7–2 for states included in each region.
[b]Included in Great Plains–Mountain.

the total. Moreover, the nonmetropolitan population of the country as a whole declined almost 13 percent over the twenty-three-year period covered by the table. During this period the total population of the country increased almost 40 percent, from 151 million to 209 million. The lack of growth of nonmetropolitan population surely does not fit the Malthusian idea of the geometric progression of population growth. Where are the population increases that should have occurred in the rural areas? Table 7–2 also shows that there was almost no redistribution of the nonmetropolitan population between 1950 and 1973. The percentages of the total national population remain virtually unchanged in all regions.

The close correlation between the proportions of nonmetropolitan population and primary production is evident in Table 7–3. Here is rural America. By contrast, the Northeast has a higher proportion of nonmetropolitan population than of production. Large numbers of this nonmetropolitan population probably are "exurbanites" who work in adjacent metropolitan areas. Further examination of Table 7–3 shows that agri-culture is by far the leading primary activity, more than twice as important as mining. Lumbering and fishing are much less important. The South is a major producer in all categories, whereas the Middle West is predominantly an agricultural area.

METROPOLITAN POPULATION AND THE BULK OF EMPLOYMENT OPPORTUNITIES

While nonmetropolitan population has remained almost static during recent decades, metropolitan areas have shown tremendous growth, increasing almost 80 percent between 1950 and 1973. This large growth shows a close correlation with the rise in employment in urban-centered industries. Figure 7–2 shows that total urban-centered employment more than doubled during the period from 1940 to 1974. Growth was important in all categories, but the greatest growth occurred in consumption services and management. Manufacturing, which accounted for over one-third of urban-centered employment in 1940, grew at a

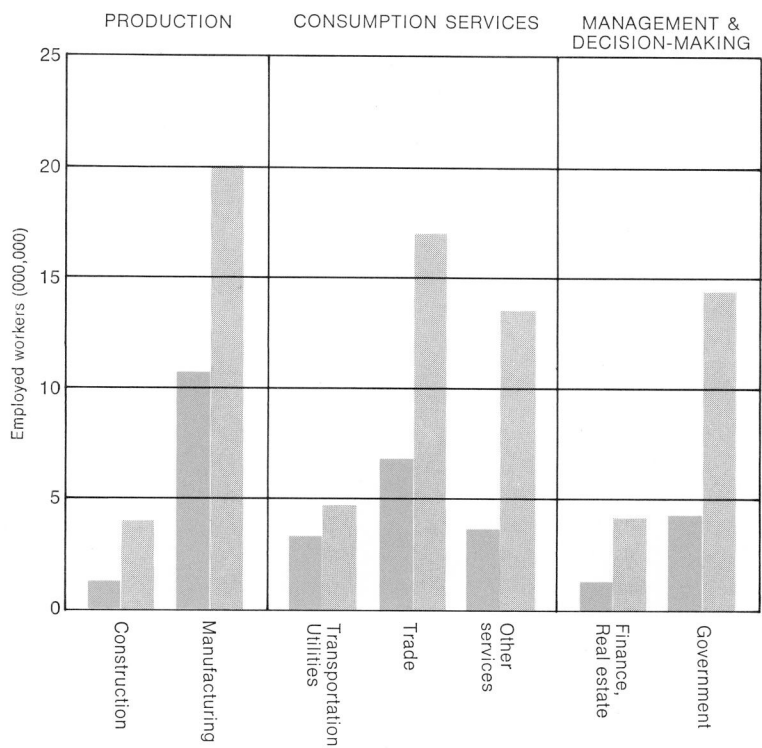

FIGURE 7-2 CHANGING STRUCTURE OF METROPOLITAN-CENTERED EMPLOYMENT IN THE UNITED STATES, 1940 TO 1974

lower rate than other activities and totaled only slightly more than a quarter of total urban-centered employment by 1974. It has been collectively outstripped by nonproductive metropolitan-centered activities, particularly government, trade, and other services. Together the three activities have accounted for more than seven out of every ten new jobs in metropolitan areas since 1940.

Metropolitan growth has occurred in all parts of the country (Figure 7-3). In contrast to the scarcely changed pattern of nonmetropolitan population, however, the distribution of metropolitan population has shifted considerably in recent decades. In 1950 almost two out of every three people in metropolitan areas were in the Northeast and the Middle West. By 1973 those two areas accounted for only slightly more than half the total metropolitan population, even though the metropolitan population of the two regions had grown by more than 25 million. The

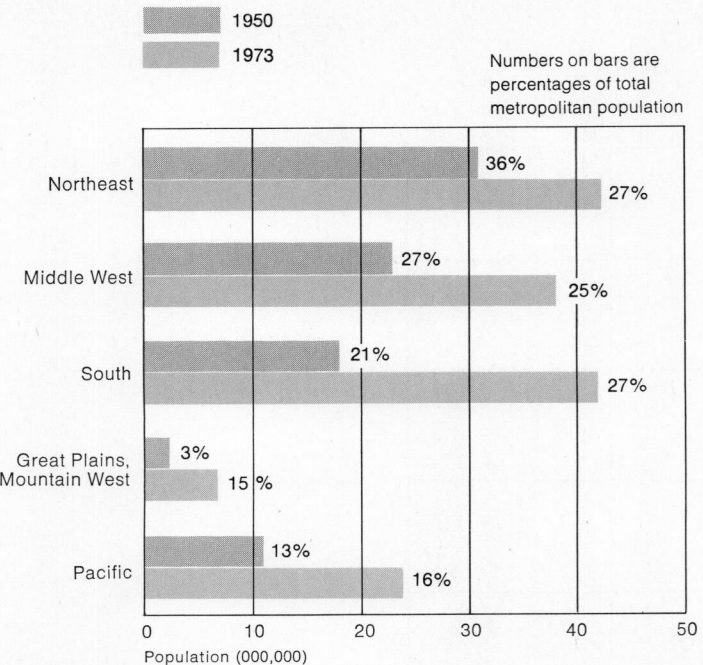

SOURCE: Statistical Abstract of the U.S.

FIGURE 7-3 CHANGING DISTRIBUTION OF METROPOLITAN POPULATION IN THE UNITED STATES, 1950 TO 1973

largest absolute growth occurred in the South, especially in Texas and Florida. The Pacific surpassed the Northeast in total numbers of people added and showed a much greater percentage growth (number of people added expressed as a percentage of 1950 metropolitan population). The Great Plains–Mountain area, with the smallest total numbers, had the highest percentage increase.

Since 1970 the shift of metropolitan population to the South and West has accelerated. According to 1975 U.S. Census Bureau estimates, the percentage increase in population for the United States as a whole between 1970 and 1975 was 4.8 percent. For the South and West it was 8.4 and 8.7 percent, respectively—Arizona had the highest increase, 25.3 percent, and Florida was second with 23.0 percent. Although these figures are for total population, not only metropolitan population, the shift they indicate probably results largely from the growth of major metropolitan centers, such as Atlanta, Houston, and Phoenix, in the South and West. In contrast, the growth rates in most northeastern states were below the national average; in fact, New York and Rhode Island experienced a decline in total population.

Coupled with the stable character of nonmetropolitan population across the country, metropolitan growth has caused sharp changes in the character of each region. In 1950, when the population of the country was almost balanced between metropolitan and nonmetropolitan (Table 7–4), only one section—the Middle West—was close to the national average. The Northeast and the Pacific areas were dominated by their metropolitan populations, and the South and the Great Plains–Mountain states remained predominantly nonmetropolitan. By 1973 only one region—the Great Plains–Mountain states—was evenly balanced between metropolitan and nonmetropolitan. On the other hand, the Northeast and the Pacific areas were overwhelmingly metropolitan; more than four out of five of their residents lived in metropolitan areas.

The regions of the country show important contrasts. Some areas, such as the Northeast and the Pacific, are largely metropolitan in population and, presumably, in functional base. The South and the Rocky Mountain regions have a near balance between metropolitan and nonmetropolitan populations. Only the Middle West shows the national ratio of approximately two metropolitan residents to every one nonmetropolitan resident.

Obviously, different parts of the country have different economic problems and aims, different political ideas, and even different cultural patterns. The large rural populations of the Middle West and South are greatly concerned with federal farm legislation such as acreage controls and subsidy payments. They follow development of hybrid seeds and feel threatened by environmentalists who object to the use of certain insecticides and fertilizers. They worry over the linking of cigarettes to lung cancer or the prospects of a large wheat crop in Australia. In turn, the nation's political parties recognize the importance of programs that will "carry the farm states" and the significance of the "farm bloc" in Congress.

On the other hand, the urban Northeast, Mid-

TABLE 7–4 PERCENT OF POPULATION OF EACH REGION IN METROPOLITAN AREAS, 1950 AND 1973

	1950	1973
Northeast[a]	77.8	84.3
Middle West	58.2	72.2
South	37.9	63.3
Great Plains–Mountain	26.1	50.3
Pacific	72.8	86.9
Whole of U.S.	56.5	72.7

SOURCE: U.S. Bureau of Census, *Statistical Abstract of the United States, 1958* and *1975*, Washington, D.C.
[a]See Table 7–2 for states included in each region.

dle West, and Pacific Coast fight for revenue sharing between the federal government and cities, and struggle with welfare questions and labor strikes. These areas react to increasing imports of steel and automobiles from Western Europe and Japan. Politicians speak of winning the "industrial states" and the "big-city vote."

THE NODAL NATURE OF METROPOLITAN CENTERS

In contrast to rural population, which is spread widely over the landscape, metropolitan populations are, by definition, **nodal.** Population is concentrated in a small area and, more important, economic and other human activities of surrounding areas are focused there. Except in the most highly developed metropolitan areas (Megalopolis from Boston to Washington along the northeastern seaboard, or the Chicago–Milwaukee or Los Angeles–San Diego corridors), each metropolitan area is separated from others by the nonmetropolitan world. Metropolitan areas stand out as separate islands of concentrated population in a sea of rural population that extends almost without interruption over the eastern two-thirds of the country and breaks into separate, isolated clusters only in the western Great Plains and most of the Mountain West. On the Pacific Coast the rural population again forms a more continuous distribution. As Figure 7–1 shows, the metropolitan islands are sometimes large, sometimes small and sometimes closely clustered, sometimes not; they have their own pattern of distribution.

Keep in mind the separateness of the metropolitan centers when considering their part in the functioning U.S. economy and culture. Such centers cannot operate in isolation. Each has a **hinterland** that it serves and from which it draws workers and economic support. Metropolitan areas also depend on connections with the primary producing regions for food and the basic resources used by manufacturers, but their most important connections are with one another. Modern business is a complex of interactions and transactions involving the producers, consumers, financiers, supporting business services, managers, and government support and regulatory agencies of the metropolitan nodes. But the nodes do not function as entities; rather, they form a functioning **network,** laced together by transportation and communication services.

TRANSPORT FLOWS AS INDEXES OF THE IMPORTANCE OF METROPOLITAN TIES

Examination of the transportation and communication system shows it is designed primarily to form links between the metropolitan nodes. Consider the interstate highway system shown in Figure 7–4. Although nonmetropolitan centers have many entry and exit points in the system, the pattern unmistakably focuses on major metropolitan areas.

The concentration of the air transportation centers is even more striking. Together the 100 places listed in Table 7–5 accounted for almost 92 percent of all domestic airline passenger traffic in 1972. Except for the three locations in Hawaii, all are metropolitan areas.

The airline figures, of course, give only movement of people. Some of these people are tourists; some are persons going to visit relatives or friends; some are military personnel; and others are students. But the predominance of male passengers and individual ticketholders suggests that by far the largest share of passengers traveling by air are moving from one metropolitan center to another on business. As a result, airline traffic, except where it represents an obvious major flow of tourists to resorts, gives us a crude indication of intermetropolitan business connection (almost the only one readily available in public statistics), and it shows that the first-order movement of business personnel is between the metropolitan nodes rather than back and forth between the metropolitan centers and the many nonmetropolitan communities.

Figure 7–5 highlights the dominance of the few largest metropolitan areas in the business interactions of the country. The fifty busiest passenger air routes between pairs of cities in the country account for almost one-fourth of all passenger movements by air. One can quickly see the importance of resorts such as Miami, Fort Lauderdale, West Palm Beach, and Tampa. The interisland movement in Hawaii might also be expected. But the major interactions are between the largest metropolitan areas. All the metropolitan centers on the map, except the resorts and island centers, have populations of more than 1 million persons. If we consider that more than half the passengers on flights between metropolitan areas are traveling for business, this map reveals the major intercity interactions of business. Notice that the interaction is not between the metropolitan areas with 1 million to 2.5 million people, such as Atlanta, Dallas, Seattle, Kan-

SOURCE: National Atlas of the U.S.

FIGURE 7–4 UNITED STATES INTERSTATE
HIGHWAY SYSTEM

sas City, Cleveland, Pittsburgh, or St. Louis. Rather, it is between those cities and the three largest centers—New York, Chicago, and Los Angeles. Of these three, New York is by far the most important hub. You will note the major traffic between New York and the other two cities and between Los Angeles and Chicago. Notice also that Washington has major connection to the three largest metropolitan areas and to Boston as well. New York's links to Boston are very important, too; Boston is in fact more significant as a node than either Philadelphia or Detroit, both of which are larger metropolitan areas.

Although the primary movement of people is between metropolitan centers, we might expect the first-order movement of goods to be between metropolitan nodes and the nonmetropolitan primary producing regions. What data we have, however, make this conclusion doubtful.

As with airline passengers, most of the goods from manufacturers remain within the network of metropolitan centers. This is really not surprising, since two out of every three people in the United States live in these centers, and more than three out of every four dollars of manufactured output come from them. These individuals and

TABLE 7-5   100 CENTERS ORIGINATING OR TERMINATING

DOMESTIC AIRLINE TRIPS, BY PASSENGER TRIPS, 1972

(In Thousands)

	Population	Passenger Trips		Population	Passenger Trips
New York[a]	16,480	24,835	Louisville	888	1,449
Chicago	7,729	14,904	Salt Lake City	744	1,429
Los Angeles[b]	8,932	11,530	Dayton	857	1,330
Washington, D.C.	2,999	9,046	Rochester	969	1,314
San Francisco	3,132	7,854	Charlotte	571	1,225
Boston	3,417	7,753	Syracuse	643	1,217
Miami	1,423	7,165	Nashville	716	1,209
Detroit	4,489	5,809	Norfolk	339	1,181
Atlanta	1,684	5,644	Jacksonville	636	1,200
Dallas–Fort Worth	2,446	5,469	Oklahoma City	736	1,123
Philadelphia–Camden	4,878	5,416	Omaha	569	1,101
Denver	1,320	4,602	Albuquerque	357	1,024
St. Louis	2,400	3,886	Raleigh	439	950
Pittsburgh	2,396	3,851	Birmingham	779	943
Cleveland	2,046	3,847	Albany	793	937
Houston	2,168	3,748	West Palm Beach	378	937
Minneapolis–St. Paul	1,996	3,741	Tulsa	560	935
Honolulu	660	3,262	Tucson	387	853
Seattle	1,400	3,236	Providence	783	850
Tampa	1,189	3,137	El Paso	374	798
Las Vegas	296	2,957	Greensboro	745	756
Kansas City	1,304	2,833	Des Moines	325	752
Phoenix	1,053	2,684	Sacramento	851	745
New Orleans	1,077	2,667	Kahului, Hawaii		708
Ft. Lauderdale	685	2,427	Reno	121	700
Baltimore	2,125	2,208	Lihue, Kanai HI[d]		683
Cincinnati	1,391	1,965	Richmond	553	654
San Diego	1,443	1,964	Columbia SC	336	637
Buffalo–Niagara	1,353	1,954	Little Rock	336	621
Portland	1,036	1,932	Witchita	376	613
Memphis	848	1,914	Spokane	302	595
Indianapolis	1,128	1,876	Knoxville	421	594
Orlando	506	1,835	Charlestown SC	342	580
Hartford–Springfield[c]	1,425	1,820	Hilo HI	26	560
Columbus	1,058	1,674	Austin	349	530
Milwaukee	1,423	1,544	San Bernadino	1,179	526
San Antonio	937	1,459	Jackson MS	267	520

TABLE 7–5 (*Cont.*)

	Population	Passenger Trips		Population	Passenger Trips
Shreveport LA	340	494	Salinas–Monterey	254	387
Grand Rapids	549	463	Madison WI	300	384
Harrisburg–York	761	463	Greenville–Spartanburg	497	373
Boise	112	445	Roanoke	181	372
Toledo	781	436	Moline IL	361	364
Mobile	386	429	San Jose	1,127	360
Colorado Springs	263	425	Savannah	208	360
Fresno	431	420	Newport News VA	339	355
Sarasota–Bradenton	217	416	Pensacola	252	346
Huntsville	286	399	Lubbock TX	179	345
Charleston WV	259	399	Tallahassee	109	341
Chattanooga	381	394	Evansville IN	389	336
Akron	682	389			

SOURCE: Civil Aviation Board, Bureau of Accounts and Statistics, *Handbook of Airline Statistics, 1973*, Washington, D.C., 1974.
[a]Population of New York–Northeastern New Jersey–Long Island.
[b]Population of Los Angeles–Anaheim and Oxnard Metros.
[c]Population of Hartford and Springfield Metros.
[d]Less than 25,000.

AIR TRAFFIC AND THE IMPORTANCE OF U.S. METROPOLITAN CENTERS

Assume that air passenger traffic is a useful measure of the importance of business centers. Turn to the data in Table 7–5.

1. Group the cities into five categories: first importance, second importance, third importance, fourth importance, and fifth importance.
   a. What was the basis for your grouping?
   b. What are the problems of grouping?

2. Map the categories on a map of the United States. What geographic pattern do you have for each category?

3. How many centers in the lowest category would it take to equal the traffic of New York? How does the traffic of Chicago and Los Angeles compare with the traffic of New York?

4. How does air traffic correlate with the population sizes of the centers?
   a. What important anomalies do you find?

   b. In your judgment why are they anomalies?
   c. Can you pick out the resort centers?

5. Describe the hierarchy as you see it.

6. Comparing Table 7–5 with Figure 7–5, describe the workings of the air traffic system of the United States.

7. What problems do you see in using air traffic as a measure of the importance of business centers?

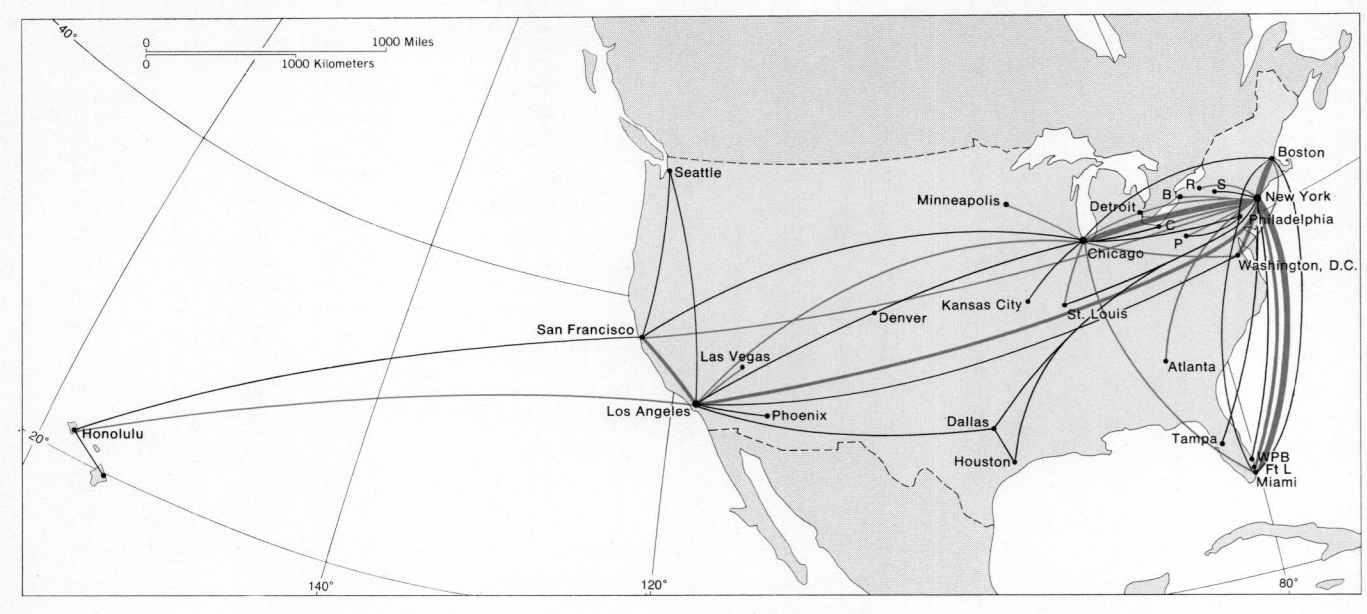

Percent of all passenger traffic

━━━ Over 1.4%

━━ 0.7–1.4%

── 0.4–0.7%

── 0.2–0.4%

B–Buffalo
C–Cleveland
Ft L–Fort Lauderdale
P–Pittsburgh
R–Rochester
S–Syracuse
WPB–West Palm Beach

FIGURE 7–5   FIFTY BUSIEST INTERCITY AIR
LINKS IN THE UNITED STATES, 1972

firms are the major producers and consumers of
the nation's goods. Not only do manufacturers in
metropolitan areas sell to consumers in metropolitan areas, but they also sell components and
subsystems to other manufacturers in the metropolitan areas.

The major importance of metropolitan centers
in manufacturing is apparent from Table 7–6.
Most of the major manufacturing centers are the
large metropolitan areas of the Northeast or the
Middle West.

METROPOLITAN CENTERS AS THE KEY
NODES IN THE FUNCTIONING SYSTEM

The pattern of traffic flow might be likened to the
body's circulatory system. A main trunk system
connects the major components of the system,
and an outreaching system ties to each tiny capillary throughout the body. However, the outlying parts do not need to be connected with one
another; the system is thus designed for flows
between the important centers and outlying areas.

TABLE 7-6 METROPOLITAN AREAS PRODUCING OVER $1 BILLION IN VALUE ADDED BY MANUFACTURING, 1972

	Population (Thousands)	Value Added ($ Millions)		Population (Thousands)	Value Added ($ Millions)
Over $5 billion					
New York Consolidated[a]	16,480	$28,383	Philadelphia	4,878	$9,239
Chicago Consolidated[b]	7,729	$19,734	Cleveland	2,046	$5,213
Los Angeles Extended[c]	8,932	$17,990	St. Louis	2,400	$5,190
Detroit	4,489	$11,792			
$4-$5 billion					
Boston	3,417	$4,845	Houston	2,142	$4,180
Rochester	969	$4,383	Pittsburgh	2,396	$4,157
			Dallas–Ft. Worth	2,446	$4,094
$3-$4 billion					
San Francisco–Oakland	3,132	$3,818	Cincinnati	1,391	$3,566
Minneapolis–St. Paul	1,996	$3,731	Baltimore	2,125	$3,469
Milwaukee	1,423	$3,697	Buffalo	1,353	$3,146
			Lousiville	888	$3,017
$2-$3 billion					
Kansas City	1,304	$2,915	Atlanta	1,684	$2,479
San Jose	1,127	$2,896	Seattle–Everett	1,400	$2,235
Greensboro–High Point	745	$2,598	Dayton	857	$2,177
Indianapolis	1,128	$2,527	Columbus	1,058	$2,000
$1-$2 billion					
Providence	783	$1,960	Syracuse	643	$1,205
Toledo	781	$1,879	San Diego	1,443	$1,185
Denver	1,320	$1,853	Wilmington	512	$1,162
Allentown–Easton–Bethlehem	680	$1,806	Canton	399	$1,158
Youngstown–Warren	544	$1,752	Nashville	716	$1,150
Portland	1,036	$1,569	Peoria	353	$1,138
Akron	682	$1,548	Richmond	553	$1,096
Flint	521	$1,514	Rockford	268	$1,088
Greenville–Spartanburg	497	$1,458	Washington	2,999	$1,068
Hartford	834	$1,390	Ft. Wayne	372	$1,063
Phoenix	1,053	$1,379	Beaumont	352	$1,034
Memphis	848	$1,352	Lorain–Elyria	261	$1,033
Albany–Schenectady–Troy	793	$1,322	Northeastern Pennsylvania	632	$1,028
Charlotte	571	$1,238			
Miami	1,423	$1,219			
Total of table	212,956 = 60.1%				

Source: 1972 Census of Manufacturers.

[a] Figures for New York include Long Island, Newark, Jersey City, and the rest of the New Jersey side of the Hudson River.
[b] Figures for Chicago include Gary-Hammond.
[c] Figures for Los Angeles include Anaheim-Santa Rosa-Garden City.

Eighteen of the thirty-three metropolitan areas with 1 million or more people in 1970 were located in the Middle West or the northeastern seaboard, including ten of the twelve largest. Giant metropolitan agglomerations in these parts of the country are closely spaced, almost merging with each other. Because of the concentration of nodes in the Middle West and the Northeast, these two areas together function as the **core region** in the U.S. interconnected system.

Along the eastern seaboard from southern Maine to Washington, D.C., live almost two out of every five Americans in a series of large metropolitan centers that are so close together that their suburban fringes coalesce. The resulting supermetropolitan complex has been called a "megalopolis." Five of the top twelve metropolitan centers and thirty-four smaller centers are located in this belt. This area is important not only for its size—the scale of functional interaction between these centers makes the region unique.

The regions outside the Northeast and Middle West can be called **peripheral regions.** The Pacific Coast area is the only other part of the country with a significant number of cities with over 1 million population; in the South, despite the recent growth of metropolitan population, only six metropolises have a million or more residents; and in the Mountain West only Denver has slightly more than 1 million people. For the most part the individual large metropolitan centers stand well apart from each other as regional capitals for wide areas. They are the focuses for regional transportation links and are nodal points that tie into the national system.

CENTERS OF ECONOMIC ACTIVITY

Metropolitan areas are more than places to live. Although the population of metropolitan areas is about 68 percent of the U.S. total, they account for 70 percent of retail sales and 80 percent of wholesale sales (see Table 7–7). Personal services, business and repair services, entertainment, hotels, and manufacturing are overwhelmingly metropolitan activities in the United States. Figures are not readily available, but we might suspect that metropolitan areas would also dominate in the concentration of the nation's educational, medical, and other specialized services.

Table 7–7 shows that the share of total national employment accounted for by the major metropolitan areas is greater than their share of population, and that their share of national income is greater than their share of employment. Thus they are more important as places to work and to make money than as places to live. In other words, these areas are even more significant as centers of the nation's economy than as centers of population concentration.

Metropolitan areas are the dominant consumers of goods, services, and ideas—the marketplaces where food and other products of primary production, resources, and manufactured goods, not only from the entire country, but also from much of the world, are bought and sold. They are the major consumers of books, television and radio, fashions, political ideas, advertising, and other forms of culture. More and more the advertising agencies, innovators, politicians, and activists have been turning their attention to metropolitan audiences rather than to rural, small-town America. The result is an increasingly metropolitan-oriented culture and economy.

Large metropolitan areas in the twentieth century have been thought of first of all as centers of manufacturing. The rise of great cities of the United States in the last century was based in large part on manufacturing. Not all major metropolitan centers, however, are primarily manufacturing places. Re-examine Table 1–3. Washington, New York, Boston, Baltimore, and San Francisco-Oakland all had less than one in four workers employed in manufacturing. The fifty-six metropolitan areas listed in Table 7–6 account for three-fifths of the value

	All Metropolitan Areas (SMSAs)	Metropolitan Areas of 1 Million or More		All Metropolitan Areas (SMSAs)	30 Largest Metropolitan Areas
	Percent of total for U.S.			Percent of total for U.S.	
Population	68	39	Headquarters 500 largest industrial corporations	93	75
Employment (all)	69	40	Headquarters 50 largest banks	100	88
Retail	70	40	Headquarters 50 largest insurance companies	96	56
Wholesale	80	50			
Services	75	44	Headquarters 50 largest retailing companies	100	92
Manufacturing	69	38			
Income	72	43	Headquarters 50 largest transportation companies	100	84
			Headquarters 50 largest public utilities	98	74
			Scientists on national register of scientific and technical personnel	77	47

SOURCES: U.S. Bureau of Census, *Current Population Reports*, Washington, D.C., 1975, Series p-23, No. 55; material for the 50 largest world industrials reprinted from *The Fortune Directory*, © 1976 Time Inc.; National Science Foundation, *American Science Manpower*, Washington, D.C., 1970.

added by manufacture in the country. Two facts stand out in Table 7–6. Twenty-seven centers (66 percent) are in the Northeast and Middle West. More important, of the top twenty-seven metropolitan areas, which together account for two-thirds of total value added by manufacturing, only Los Angeles, San Francisco, Houston, and Dallas–Fort Worth are located outside the Northeast or Middle West.

Manufacturing is still predominantly a function of the Northeast and, particularly, the Middle West, but important changes have been taking place. The Pacific Northwest, using the energy resources of the Columbia River, has developed into a specialty industrial region devoted to aircraft and power-using industries. Industry has also been moving southward. Houston and Dallas-Fort Worth have emerged as major manufacturing centers, as we have seen. New industry has also moved into the large metropolitan areas of the Southeast, from Richmond southward through Greensboro, Charlotte, and Greenville-Spartanburg to Atlanta, into Miami, and into Nashville and Memphis in western Tennessee. Meanwhile output has declined slightly in moderate-sized manufacturing centers such as Hart-

TABLE 7-8 LOCATION OF THE HEADQUARTERS OF THE 158 LARGEST
U.S. PUBLIC CORPORATIONS, LISTED BY KIND OF ECONOMIC ACTIVITY
(ALL LISTED BUSINESSES HAD ASSETS IN 1975 OF AT LEAST $2.7 BILLION)
(Millions of Dollars)

	Industrial Corporations	Banks	Insurance	Other Financial Corporations	Retail Firms	Transport Companies	Public Utilities	Total
New York City	10—$ 99,808	9—$216,628	5—$ 78,056	5—$28,767	1—$ 3,226	1—$ 3,418	4—$ 96,511	35—$ 526,014
Suburbs	4—$ 34,935		2—$ 42,696				2—$ 7,187	8—$ 82,104
	14—$134,743		7—$120,752				6—$113,698	43—$ 620,832
San Francisco–Oakland	1—$ 12,898	4—$ 92,658		1—$ 4,896		1—$ 3,724	1—$ 6,621	8—$ 120,797
Chicago	2—$ 13,364	4—$ 46,967		1—$ 6,540	2—$15,113	1—$ 2,884	1—$ 5,180	12—$ 90,008
Los Angeles	4—$ 17,884	3—$ 37,612		3—$16,940			1—$ 4,650	11—$ 77,086
Detroit	3—$ 41,952	4—$ 16,659					1—$ 3,650	8—$ 62,261
Hartford			3—$ 28,939	2—$26,704				5—$ 55,643
Pittsburgh	5—$ 31,747	2—$ 12,237						7—$ 43,984
Philadelphia	1—$ 4,384	4—$ 17,808	1—$ 2,909	1—$ 4,308		1—$ 4,392	1—$ 3,961	9—$ 37,762
Houston	2—$ 13,594	2—$ 9,101		1—$ 4,083				5—$ 26,778
Boston		1—$ 8,614	2—$ 17,410					3—$ 26,024
Minneapolis–St. Paul	1—$ 3,127	2—$ 14,560				1—$ 3,279		4—$ 20,966
Dallas–Fort Worth		2—$ 11,571					1—$ 3,248	3—$ 14,819
Cleveland	1—$ 4,220	2—$ 6,904				1—$ 2,817		4—$ 13,941
Milwaukee		1—$ 3,656	1—$ 7,918					2—$ 11,574
Buffalo		1—$ 11,104						1—$ 11,104
Atlanta		1—$ 3,083					1—$ 7,237	2—$ 10,320
23 other metros	11—$ 43,692	8—$ 27,767	3—$ 12,354	1—$ 3,679			5—$ 18,025	28—$ 105,517
3 nonmetros	2—$ 10,391						1—$ 3,203	3—$ 13,594
Total	47—$331,996	50—$436,529	17—$190,282	16—$95,917	3—$18,339	6—$20,474	19—$169,473	158—$1,363,010

SOURCE: Based on material appearing in *The Fortune Directory*; © 1975 Time Inc.
Bold numbers indicate the number of firms with headquarters in the listed cities.

THE LARGEST BUSINESSES IN THE UNITED STATES: LOCATION OF THEIR HEADQUARTERS

Table 7–8 shows the location of the headquarters of the 158 largest businesses in the United States. To be included in the table, a company had to have assets of $2.7 billion or more in 1975. In fact, the largest public utility (American Telephone and Telegraph) had assets of $80.2 billion, the largest bank (Bank of America) $66.8 billion, the largest life insurance company (Prudential) $39.3 billion, the largest industrial corporation (Exxon) $32.8 billion, the largest other financial institution (Aetna Life & Casualty) $15.5 billion, the largest retailer (Sears, Roebuck) $11.6 billion, and the largest transportation company (Penn Central) $4.4 billion. The assets of these 158 corporations amount to eight times the combined total revenues of all the state and local governments in the United States in 1972.

Although 50 banks, 47 industrial corporations, 17 life insurance companies, 16 other financial institutions, 19 public utilities, 6 transportation companies, and 3 retail firms are included, these businesses can be grouped into two categories on the basis of the character of their assets: (1) those whose assets are largely in physical plant and equipment and inventories (the industrial corporations, retail firms, transport companies, and public utilities), and (2) those whose assets are mostly in money holdings (banks, life insurance companies, and other financial institutions). Note that the companies with predominantly money holdings have more total assets than those with predominantly physical plant facilities. Remember that the assets shown are the total amounts controlled by the companies. They include factories, rights of way, power plants, and stores located throughout the country and even the world, or financial loans and other investments supporting building and production anywhere in the country or the world.

What conclusions can you draw about the location of the headquarters of the nation's largest businesses?

1. Much is said about the importance of New York City as a business center. What is your opinion after studying the table? Some large headquarters are moving out of New York City. Why? Do you think that is a wise move?

2. Group the metropolitan areas into classes: first order, second order, third order, fourth order, fifth order. Do you see a hierarchy? Are there national and regional centers? How does Washington, D.C., fit into the system of headquarters centers?

3. The table includes both companies with predominantly physical assets used in production or sales and those with assets largely in money holdings. Are there specialized financial centers and production centers among the metropolitan areas, or are most metropolitan areas diversified? Do your five classes above fit the twofold breakdown?

4. Describe the location of the headquarters of the country's largest businesses.

ford and Bridgeport in the Northeast and Indianapolis, Dayton, and Youngstown in the Middle West.

Today the largest metropolitan areas are more important as places from which to manage the economy than as production places. The metropolitan areas listed in Table 7–8 are the headquarters locations for the nation's largest business firms—industrial corporations, banks, insurance companies, retailing companies, transportation companies, and public utilities. In all categories, over 90 percent of the headquarters of the largest firms are located in metropolitan areas, making them the management and decision-making centers for the country.

The largest headquarters are concentrated in just a few of the largest metropolitan areas in the country. The 16 centers listed in Table 7–8 are only 8 percent of the 223 metropolitan areas in the United States, but they control over 90 percent of the assets of the nation's largest public corporations. Of them, only Hartford has fewer than 1 million people. New York City alone is headquarters for 35 corporations with almost 39 percent of total assets; and the Greater New York area, including northeastern New Jersey, Long Island, and Stamford, Connecticut, contains 43 businesses with more than 45 percent of the assets of the country's largest businesses. There are, however, more than 15 metropolitan areas whose populations exceed 1 million but that have not been chosen as headquarters for any of the 158 largest U.S. businesses. Among them are Washington, D.C., Baltimore, and St. Louis.

As we saw in Chapter 2, the most distinctive feature of the modern interconnected system is specialization. Specialization results in Florida citrus growers concentrating on juice oranges, Texas citrus growers on grapefruit, and southern California citrus growers on oranges for eating. One California district produces much of the country's lettuce during the winter months; another does so during the summer. In the same way manufacturers locate plants over the country in response to particular locational factors: availability of cheap power in one area, cheap labor in another, access to perishable food products in a third. The nearness of a supplier or a particular customer and the availability of financial support are also factors. This is also the case with business management. Auto management is in Detroit, steel in Pittsburgh, and petroleum in Houston. New York is pre-eminent as the major financial center and headquarters for the greatest variety of corporations. Other financial centers, particularly Chicago, Los Angeles and San Francisco, are important management centers, too.

PRODUCING AND MARKETING NETWORKS

In the United States, with its great range of environmental and consumer possibilities, many decisions about where to produce and where to sell must be made. Who makes those decisions?

Each individual and each family makes daily decisions that may be influenced by ideas from local schools, neighborhood discussions, or advertising on television or in magazines. These decisions are made within the framework of the individual's local world, as we saw in the discussions of families in the Washington area. One's lifestyle, although influenced by outside forces, is formed within the framework of a daily world that rarely extends beyond one's own life space.

Similarly, businesspeople in small towns know how to operate only within their own local areas. They know their customers, who all come from within a radius of 25 miles, and they deal with suppliers in cities less than 100 miles away. Their other contacts are with local bankers and businesspeople. Bankers in small cities may work within a larger scene. They know a portion of their state or even several states. Their loans may be to farmers and businesspeople who live no more than a hundred miles away, and they have contact with banks in large regional metropolises 200 miles away.

On the other hand, the country's major decision making takes place in a few great metropolitan centers. These centers are the nodes of transportation and communication networks that allow managers to be in constant touch with plants and markets throughout the country. In the industrial, financial, and trading world today, some of the giant corporations restrict themselves to operating within a specialized niche, such as clothing or motels. In so doing, they organize their operations on a national basis.

In recent years giant multifaceted corporations (conglomerates) have come to dominate a wide spectrum of businesses that may or may not show any relation to one another. The headquarters of

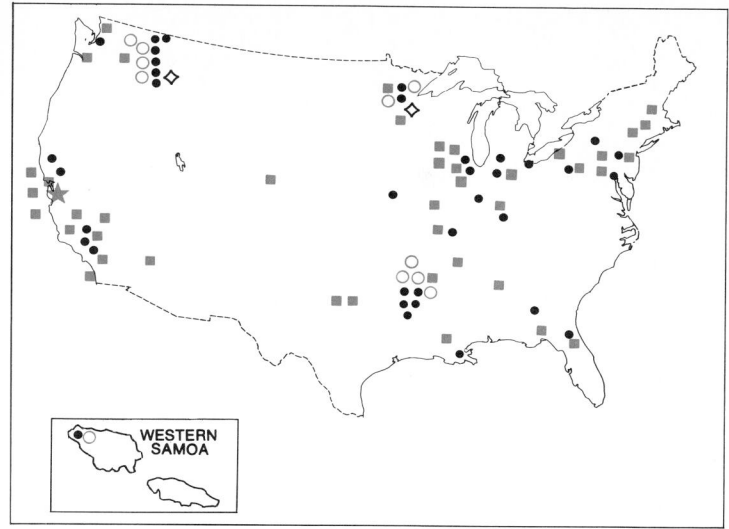

○ Timber reserves

◇ Research centers

● Manufacturing facilities

■ Sales offices

★ Executive offices

SOURCE: Courtesy of Potlatch Corporation, San Francisco

FIGURE 7–6 DISTRIBUTION OF FACILITIES OF POTLATCH CORPORATION

such a corporation manage huge complexes of factories and production facilities scattered over the country or over the world. At the same time they control a marketing network that may have a very different distributional pattern.

THE EXAMPLE OF POTLATCH CORPORATION

Potlatch Corporation, with headquarters in San Francisco, was ranked 342 of the 500 largest companies in the United States in terms of sales in 1976. It deals in wood products and factory-built structures, in paperboard and packaging, in business and printing papers, tissues, and paper plates. Figure 7–6 shows how Potlatch has chosen to locate its subsidiaries across the United States. Note that they fall into several distinct categories, each with its own distribution pattern. Potlatch owns and manages timberlands, manufactures a variety of wood and paper products (each with particular locational needs), and then sells the products to customers throughout the country.

For the timber it needs as raw material, Potlatch has gone to three of the major timberlands of the country: the northern Rockies, the South, and the upper Middle West. Because logs are difficult to ship, the company's heavy manufacturing operations—lumber, plywood, veneer, and papermaking—are located in lumber towns adjacent to timberlands. But distinctions are made according to the particular character of the timber resource in a given area. The company makes veneer and plywood in Idaho, where logs are large and of high quality; wood is converted into pulp and the pulp into paper in Minnesota and Idaho. Paper is also made in northern California, where there is no pulp mill, but this operation specializes in stationery, which does not depend entirely on wood pulp, and Potlatch purchases pulp for this factory from noncompany sources.

Whereas lumber and paper plants are located in the forests, the manufacture of paper specialties and the fabrication of lumber are generally located close to markets. It is cheaper to ship the paper and lumber to the plant than to move the manufactured product to the market. Paper specialties are made in California, milk cartons in California and Pennsylvania, paper plates in Missouri and California, and facial tissues in Pennsylvania. Most of these plants are in metropolitan centers or strategically located towns. Cartons and boxes, the most difficult items to ship, are manufactured in a number of plants in large metropolitan areas, located so that together they serve the major regional markets. The one wood-fabricating operation, making relocatable buildings, produces structures that are also difficult to ship. To get coverage in major markets, three plants have been built, one near San Francisco, one in Indiana, and one near Pittsburgh.

The whole Potlatch operation, with over 10,000 employees, is supervised from headquarters in San Francisco. The final decisions on the purchase of new timberland, construction of new plants, closing of old ones, development of new products, and strategies for selling are made there. Notice, however, that Potlatch operations show little orientation to San Francisco, or even

to California. It is a national firm producing for a national market. Its raw material sources and manufacturing plants are located to serve that market in accordance with present business conditions and the operations of a variety of competing firms. Its thirty-three major sales offices are spread across the country in the large metropolitan centers where the largest retailers and industrial customers are found. Recently, Potlatch has also gone overseas for additional sources of timber. It has timberlands and lumber mills in Western Samoa, but has not yet opened up overseas sales offices.

GENERAL MOTORS AND SEARS AS THE ULTIMATE EXAMPLES

Potlatch is far from being the largest industrial corporation in the United States, yet its operations are representative of how a corporation develops a network of resources, manufacturing plants, and sales units where management plans to fit them into the interconnected system. Some firms do not worry about immediate control of natural resources. Some buy directly from other manufacturers; some sell to retailers and wholesalers, some only to other manufacturers. Most successful large corporations have tended to gain control over their sources of supply; this is called **vertical integration.**

The operations of two major corporations are outlined for study in the boxes on pages 219–221. Notice how General Motors, the second largest industrial corporation in the country, has developed or purchased specialized divisions, each with its own part to play in the company's operations.

Imagine the task of managing one of these operations. General Motors has 108 plants in the United States alone, each a major operation in itself with the number of workers ranging from several hundred to several thousand. Although it produces a large share of its own component parts, General Motors buys products from thousands of suppliers. In selling its products it must

AN INVENTORY OF GENERAL MOTORS CORPORATION

General Motors Corporation: Second largest industrial corporation in the United States in 1975
Headquarters: Detroit, Michigan
1975 sales: $35,724,911,000
Employees: 681,000

1. Automobile Assembly
Buick Division: 1 plant in Michigan
Cadillac Division: 1 plant in Michigan
Chevrolet Division:

a. Automobile assembly: plants in Georgia, Maryland, Michigan (2), Massachusetts, Wisconsin, Missouri (2), California, Ohio, New York

b. Carburetors: 1 plant in Michigan

c. Boxes for automobile export: 1 plant in New Jersey

d. Axles: plants in New York, Michigan (2)

e. Transmissions: plants in Ohio (2), Indiana, Michigan

f. Sheet-metal stamping: plants in Ohio, Michigan, Indiana

g. Forgings: plants in Michigan, New York

h. Engines: plants in Michigan (3), New York

i. Springs and bumpers: 1 plant in Michigan

j. Engine castings: 1 plant in New York

k. Gray iron foundries: plants in Michigan, New York

l. Miscellaneous automobile parts: 1 plant in Michigan

m. Coke plant for foundries: 1 plant in Illinois

Oldsmobile Division: 1 plant in Michigan
Pontiac Division: 1 plant in Michigan
Fisher Body Division:

a. Automobile bodies: plants in Georgia, Maryland, Michigan (5), Massachusetts, Wisconsin, Missouri (2), California, Ohio, New York, Illinois

b. Truck bodies: plants in Michigan, Ohio

c. Automobile stamping, trim: plants in Ohio (2), Michigan (5), Indiana, Pennsylvania, Illinois

GM Assembly: automobile assembly plants in Texas, Georgia, California (2), Missouri, New Jersey, Delaware (serves Buick, Oldsmobile, and Pontiac)

GMC Trucks and Coaches: 1 plant in Michigan

Canada: 7 plants

2. Automobile Distribution:

a. About 11,700 dealers

b. 50 warehouses

3. Automobile Components:
AC Spark Plug: 1 plant in Michigan
Central Foundry: plants in Indiana, Illinois, Ohio, Arkansas, Michigan
Delco Moraine (brake components): 1 plant in Ohio
Delco Products (shock absorbers, etc.): plants in Ohio, New York
Delco Radio: 1 plant in Indiana
Delco-Remy (ignitions, batteries): plants in California, Indiana, New Jersey, Kansas
Guide Lamp (lights): 1 plant in Indiana

Harrison Radiator: 2 plants in New York
Hydramatic (transmissions): 1 plant in Michigan
Inland Manufacturing (steering wheels, brake linings): 1 plant in Ohio
New Departure–Hyatt Bearing (bearings): plants in Connecticut (2), New Jersey (2), Ohio
Packard Electric (wiring, switches): 1 plant in Ohio
Rochester Products (carburetors): 1 plant in New York

4. Defense:
A.C. Electronics: 1 plant in Wisconsin
Allison: plants in Indiana, Ohio

5. Engines:
Detroit Diesel Engine: 1 plant in Michigan
Diesel Equipment (fuel injectors): 1 plant in Michigan
Electro-Motive (locomotives, large diesel engines): 2 plants in Illinois
Earth-moving Equipment: 2 plants in Ohio

6. Household Appliances:
Frigidaire: 1 plant in Ohio

7. Research and Training:
GM Defense Laboratories: California
GM Technical Center: Michigan
GM Training Centers: 30 over the country

8. Finance
General Motors Acceptance, Motors Insurance, Motors Holding: offices over the country

General Motors is first of all an automobile manufacturing company. Making and selling an automobile call for a variety of operations: designing and engineering the car, making the components that go into it, assembling the car, and then marketing it. Each of these operations is distinct and could be done independently—either by a separate company or in a separate location.

Questions

1. Is there a distinctive geographic distribution to each of these separate operations as far as General Motors is concerned? Where are the design installations? Is the distribution of components plants similar to that of the assembly operations?

2. Buick, Cadillac, Oldsmobile, Pontiac, and GMC each has only one assigned assembly plant, but Chevrolet has eleven. Why? How do the divisions other than Chevrolet get national distribution of cars?

3. Where are the parts suppliers for assembly plants in Georgia and California?

4. What other businesses is General Motors in besides the automobile industry? Does the distribution of the nonautomobile plants show any pattern different from the others?

5. Does the distribution of General Motors plants show any close relationship to the location of the company's headquarters in Detroit? General Motors has just completed in New York City a new office skyscraper in midtown Manhattan. Why do you suppose it has done this?

AN INVENTORY OF SEARS, ROEBUCK AND COMPANY

Sears, Roebuck and Company: Largest retailer in the United States in 1975
Headquarters: Chicago, Illinois
1975 sales: $13,639,887,000
Employees: 377,000

1. The company operates 851 stores in the United States and Puerto Rico:
a. 319 complete department stores in large metropolitan centers
b. 375 medium-sized department stores in smaller markets
c. 157 appliance, sports equipment, and auto supply stores in outlying shopping areas and smaller communities
d. Location of Sears Stores:

Northeast	
Connecticut	12
Delaware	3
District of Columbia	3
Maine	8
Maryland	12
Massachusetts	21
New Hampshire	4
New Jersey	21
New York	49
Pennsylvania	49
Rhode Island	4
Vermont	2
West Virginia	6
	193

Pacific	
Alaska	1
California	72
Hawaii	5
Oregon	7
Washington	19
	104

Middle West	
Illinois	52
Indiana	39
Iowa	16
Michigan	28
Minnesota	14
Missouri	15
Ohio	50
Wisconsin	30
	244

Southeast	
Alabama	17
Florida	34
Georgia	18
Kentucky	11
Mississippi	12
North Carolina	19
South Carolina	8
Tennessee	16
Virginia	14
	149

Great Plains–Mountain

Arizona	12
Colorado	7
Idaho	4
Kansas	10
Montana	4
Nebraska	7
Nevada	2
New Mexico	4
North Dakota	4
South Dakota	4
Utah	6
Wyoming	1
	65

South Central

Arkansas	6
Louisiana	12
Oklahoma	8
Texas	53
	79

Latin America

Brazil	7
Colombia	5
Mexico	26
Venezuela	13
Peru	2
Panama	5
Costa Rica	1
San Salvador	2
Puerto Rico	7
Nicaragua	1
Honduras	2
	71

Europe

Spain	2
Belgium	10
	12

2. In addition, it has 2,886 catalogue offices in its stores and in small-town locations.

3. It operates thirteen catalogue-order plants to handle distribution of catalogue merchandise.

4. It owns several subsidiaries manufacturing products for its outlets:

a. A plumbing-fixture company with plants in New Jersey, Texas, Pennsylvania (2), California, Wisconsin, Georgia, Arizona; sales offices in Boston; Atlanta; Sherman Oaks, California; Chicago; Camden, New Jersey; Houston.

b. A chemicals and furniture firm:
Chemical coatings: plants in California (2), Illinois (2), Ohio, North Carolina, Texas, New York
Chemical products: plants in Wisconsin, Illinois, California
Engineering specialties: 1 plant in Illinois
Furniture: plants in Florida, Oregon (3), Pennsylvania, Illinois, Georgia New Hampshire, New Jersey, Texas (2), Washington, Mississippi (5), Arkansas (4), Iowa
Other: plants in New Jersey, California, Texas, New York, Illinois

c. A home furnishings firm with plants in Missouri (2), New York (2), Pennsylvania (3), Tennessee (2), Massachusetts, California (2), Illinois (2)

d. A company that makes cooking ranges, lawn mowers, and accessories with plants in Illinois (2), Tennessee, Ohio, Maryland (3)

5. It has a research center in the Chicago metropolitan area.

Questions

1. How do Sears stores adapt to selling to both metropolitan and non-metropolitan markets?

2. How does the distribution of Sears stores correlate with the distribution of population in the country? Where are the major concentrations of stores?

3. Why would Sears go into manufacturing? Why into particular types of manufacturing?

4. How does the distribution of factories compare with the distribution of stores?

5. Does the pattern of Sears operations show any relation to the fact that Sears headquarters are in Chicago?

6. Where would you place the thirteen warehouses or distribution centers to serve the Sears national market?

deal with more than 11,000 auto dealers, with appliance distributors, railroads, and industrial firms, and must enter into major contracts with the government.

Sears supplies 851 stores in the United States and 2,896 catalogue outlets. Even though it manufactures some of its own products, it must deal with a large number of suppliers. It must also bill a large number of customers!

THE UNIQUE ADVANTAGES OF NEW YORK CITY

Undoubtedly, the headquarters of these corporations must be in a center of communications. They must be able to communicate with each unit in the network, with suppliers and customers, with banks and financial institutions, and with supporting business services such as advertising agencies, shippers, and lawyers. Headquarters must be placed so that the companies know what their competitors are doing.

For them the metropolis, both as the center of a business community with all its services and as a focal point for worldwide transportation and communications networks, is the key place from which to manage their complex operations. The larger the metropolitan center, the larger its supporting business community is likely to be and the better its transportation and communication services. Office buildings in the largest metropolitan areas have grown spectacularly in size since World War II, and the growth continues despite increasing congestion on city streets and pollution of city air. Indeed, it is likely to continue until there is a breakdown in transportation and communication. Workers and executives in New York City complain about the increasing problems of commuting, the rise of crime in the streets, and rising taxes. Yet skyscrapers continue to increase in number and size.

In the late 1960s, however, the first real threats to business in New York began to appear. The telephone system in Manhattan became overloaded, the installation of new phones was delayed, and the message-switching system broke down. The airports serving the city became overloaded, and it was necessary to ration the number of commercial flights during certain hours. Even so, at rush hours twenty or more planes were waiting to take off.

In the 1970s financial uncertainties began to hamper the operations of the New York City government. The large banks on which the city had depended for huge loans on a regular basis refused to extend further credit until the city improved its financial status. A Municipal Assistance Corporation was formed by New York state to oversee the city's finances, and federal loans were granted to help the city balance its budget. Despite state and federal aid—in fact, as a condition of it—the New York City government was required to make extensive cuts in services in such areas as transportation, education, hospitals, and police.

All this has caused some companies to move their headquarters from New York City. Up to now most have moved from the skyscrapers of downtown or midtown Manhattan to suburban locations in Westchester County, Long Island, New Jersey, or southwestern Connecticut. A few have completely abandoned the New York metropolitan area for smaller centers such as Houston and Phoenix that have attractive climates. It is notable that the move in such cases has been out of the East Coast megalopolis altogether.

The key management question for the future is whether the dominance of New York City, encouraged in recent years by the mergers of industrial corporations into ever larger management units, will continue. Theoretically, improvements in transportation and communication that link metropolitan centers make it increasingly possible to manage a financial-industrial complex from any major metropolitan center in the network. With modern telephone connections and jet aircraft, the firm based in Phoenix is within instantaneous reach of the entire New York business community. The only thing missing is the direct personal contact that many businesspeople claim is important.

New York City has also generally been recognized as a special place to live as well as to work. As the theatrical, musical, and artistic center of the United States, it has had a special attraction for the wealthy and ambitious of the country. As the largest U.S. city and its primary connecting link with the rest of the world, New York has a varied and cosmopolitan atmosphere. But now these urban attractions are competing with the attractions of outdoor living with its recreational possibilities. Moreover, with increased congestion, pollution, crime, and other problems, the quality of life in New York has deteriorated.

A corporation's decision to locate where it chooses is considered an unchallenged right, even though thousands of new workers may overload a small city's infrastructure, making excessive demands on its housing, roads, and schools. These urban problems have been seen in the United States as part of a different "sector" of the economy—the public as contrasted to the private sector. Usually, the **private sector** makes a decision based on its own economic self-interest, then leaves the burden of providing back-up facilities to the **public sector.** It is said that many of the real costs of production, both supporting facilities and costs in quality of life, are not being met by the private sector in the United States. Rising taxes and deterioration of public services stand as evidence to support this position.

Up to now there has been no attempt to control the ever-growing centralization of economic activities in New York City, despite the problems of overextension of supporting services, the growing congestion on city streets and in airports, and pollution. In contrast, the government of the United Kingdom has been conscious of the problems of overcentralization in London. It has placed limitations on the amount of new industrial and management expansion that can take place in the capital. New relocation in London can take place only after the government has studied its impact and issued a certificate of necessity.

THE MANAGEMENT NETWORK

The big centers such as New York, Chicago, and Los Angeles represent the highest level of management in the United States. Decisions are made and management functions at all levels—the metropolitan center, the city, the small town, the village, the farm unit, and the individual family. The hierarchy of decisions is not simple; rather it is a complex of links, transfers, and overlaps. The classic model in regional economics shows a "nesting" of villages within the trade area of a town, of towns within the larger trade area of a small city, of cities within the larger territory of a metropolitan area. The term **nesting** in this context refers to the fact that the market area of the village lies totally within the market areas of the town, of the small city, and of the large metropolitan area; the market area of the town is part of the market areas of the small city and of the metropolitan area; and the market area of the small city is within the market area of the metropolitan center.

The village, town, small city, and metropolitan area comprise a **hierarchy of places.** The term *hierarchy* may suggest that each level interacts only with the levels immediately above and below it, but that is not the case in this kind of hierarchy. In a **nested hierarchy,** movement of goods, people, and ideas takes place simultaneously at all levels. The farm family deals with the village for some things, the town for others, the city for still others, and the metropolis for some. City dwellers know that the same kind of multiplicity exists between the neighborhood, the large shopping center, and "downtown." Either way, the system works through a series of nodes that serve both as points of transaction and as points of decision making and management. The whole operation, of course, depends on transportation and communication. The better the transportation and communication, the larger the scale on which the system can work.

Banking provides a good illustration of the flexibility and overlapping of the system. There are more than 13,500 federally insured commercial banks in the country, ranging in size from huge international banks—such as the Bank of America in San Francisco, with more than $22 billion in deposits—to small-town banks. At the top, the fourteen largest banks in the country hold about 25 percent of total deposits. The bottom quarter, the smallest banks, holds less than 2 percent of deposits.

The commercial banks provide service to all financial customers in the country ranging from the largest corporation to the smallest child. Customers include corporations, the federal government, local governments, and pension accounts as well as individuals. Money flows from one bank to another as needed. For example, a farmer in the Corn Belt may need $55,000 of operating capital a year, and in the surrounding community there may be dozens of farmers with similar requirements. This demand for loans exceeds the resources of the small-town local bank. If the loans are approved, however, the money can usually be obtained by the bank through larger correspondent banks in regional centers.

At the other end of the scale, a giant corporation may need a multimillion-dollar loan for expansion purposes. This may be beyond the capacity of even the largest banks for any single loan. If several banks join together, a "package" of money can be put together to provide the necessary funds.

Banks are classified as international, regional, or local according to the basic range of their operations. Only twelve to fourteen U.S. banks are recognized as operating regularly on an international scale. That is only 0.1 percent of all commercial banks. Over half these international banks are in New York City, and all but one other are in Chicago, Los Angeles, and San Francisco.

About 500 banks—less than 5 percent of all banks—operate on a national scale (not necessarily those termed "national bank," meaning that they have a national rather than a state charter). They deal with other banks and national corporations. On the next level are 1,500 regional banks, mostly in metropolitan areas and the larger regional cities. Their range is regional rather than national. Finally there are the local banks—totaling over 90 percent of all banks—which operate only in a small town or a neighborhood of a city. A small town or village may have only one small local bank. A large metropolitan area such as New York has local banks in neighborhoods, regional banks, national banks, and international banks, each serving a particular function.

Money and credit must flow throughout the system and even be available for international transactions. U.S. international banks have offices in foreign capitals—such as London, Paris, Rome, and Tokyo—and foreign international banks have offices in U.S. financial centers such as New York, San Francisco, and Chicago.

ORGANIZATIONS AND COMMUNITIES AS INTERMEDIARIES

The bank serves as a marketplace or an **intermediary** between persons and organizations wishing to save money and earn interest on deposits and those wanting to borrow money. The bank makes a profit by charging for its services as a catalyst or conduit. This is also true of other organizations in the system. All depend on a two-way interchange: an inflow of goods and ideas on the one side and an outflow of services or finished products on the other. This is the way a manufacturer operates. A retailer or wholesaler is likewise an intermediary. So is the life insurance company, which collects on policies and invests money. Government takes in taxes and distributes services.

Communities also serve as intermediaries, providing a service to some and receiving services from others. Each organization and individual, then, is dealing on two levels. It collects on one and distributes on another, operating both on a local level and on a broader one. The grain elevator in the local community collects produce grown by many farmers in the area and ships it to a warehouse or user; it receives fertilizer or feed from a company and distributes that product to local farmers. Like the international bank, only the largest center simultaneously operates on all levels—international, national, regional, and local.

THE NETWORK OF CULTURAL AND POLITICAL IDEAS AND DECISIONS

Government activity and cultural influences have impact beyond the effects of the movement of goods, services, and money. Ideas move through the system from the individual rural family to the largest metropolitan area. At the center are people in New York and Washington whose social and economic contacts are largely international. Others function mostly at a national level. The sphere of most people is restricted to their own neighborhoods. People functioning in these three different "worlds" can all live within the same metropolitan area.

Individual values are shaped by local schools, churches, and social groups, by regional influences such as state universities and church dioceses and conferences, and by national forces such as television commercials, church denominations, book publishers, and fashion houses. In the communications field a local newspaper may be read along with a regional big-city newspaper or the *New York Times*. Local radio and television stations intersperse local news and advertising with programs from the national broadcasting networks. Before the era of instantaneous communication, national influence was felt only after days or weeks of delay; local influences were primary; communities were distinctive.

Now, with network programs conceived in New York, Washington, and Los Angeles, national forces are reflected immediately in any community. Popular topics spread from center to center overnight. However, each community, each family, and each individual has its own filter for accepting or rejecting to one degree or another these stimuli. Moreover, each community presents its own feedback, its own tiny input, to the system. In turn, each community prospers in accordance with how it fits economically, politically, and culturally into the national system on the basis of values established at the national level.

Freedom of choice varies sharply within the hierarchical system. The wealthy have maximum choice regarding sources of supply. The middle class has more limited choice, and the poor have the least choice of all. Consider the purchase of furniture or food. The wealthy individual furnishing a house calls in an interior decorator with access to the full range of furnishings from local shops, department stores, and specialty shops throughout the system. Goods may be purchased in the community where the buyer resides, from a New York importer, or from an antique shop in the part of the country where a "period piece" might most easily be found. Middle-class people are likely to make the selection of furnishings themselves, so their choices are limited to goods in the stores in their community. The family of modest means without a car has to choose furniture from the neighborhood dealer.

The range in food buying is much the same. The wealthy party-giver may telephone to have fresh Maine lobster flown in; the middle-class shopper may drive to the supermarket in a city twenty miles away; the low-income buyer walks to the neighborhood grocery. The choice of recreation follows a similar pattern: the wealthy can fly to any part of the world; the middle-class family buys a camper and goes off to see the United States; the low-income family goes to the local park.

With the impact of national cultural and economic influences has come the rise of national government. As we saw in our earlier study, Washington, D.C., has grown from a rather minor metropolitan area and become second only to Los Angeles in population growth. It now ranks seventh among the largest metropolitan areas of the country and has reached this position despite its unimportance in manufacturing, wholesaling, and banking. But it is the decision-making and management center of the federal government.

PARTS OF THE COUNTRY OUTSIDE THE MODERN SYSTEM

So far we have presented the modern interconnected system of the United States as reaching out over the entire resource base to its many diverse regions. If we examine the system more closely, however, we find that this is not completely true. As local, regional, and national industries have made decisions about the best places to locate, they have centralized activities in a few key locations and have discarded the places with less potential. We saw this in the early nineteenth century when New England discovered that agricultural products could be delivered to its cities efficiently from the fine farmlands of the Middle West and therefore began abandoning its own marginal local farming.

The problem is that **marginal areas**—areas that are relatively unsuccessful in the modern interconnected system—do not fit the system any longer. Nor does it look as if their situation will improve in the future. There are large regions, such as most of Appalachia, that are no longer being considered by local or regional decision makers. They have been measured against national—and even international—alternatives and found unattractive. Efforts at regional economic development have been going on for years. Farm assistance programs and special government projects try to provide incentives for development. Governmental organizations such as the Tennessee Valley Authority and more recently the Appalachian Regional Commission have also made efforts, yet the most depressed areas of the country remain so. Agricultural possibilities are poor; most manufacturing and service activities will not locate in small-town, rural environments; and mining and manufacturing are becoming more automated all the time and therefore not providing additional jobs. Moreover, the people of these areas are among the most poorly educated and hence least prepared to take their places in the modern system.

The depressed rural areas are predominantly agricultural, since agriculture is the most important primary activity. However, one can also find lumbering and mining areas where resource depletion or automation has left populations without an economic base. The farm areas include those of Appalachia, which were settled early and then by-passed as the modern interconnected system developed. Areas of the South, where machines and large-scale farming have displaced intensive sharecropping of cotton, also fit the pattern.

It may be surprising that such large portions of the United States are depressed, and Americans were shocked several years ago by disclosures that large numbers of their rural families were not only poor but undernourished. The presence of such depressed areas in a modern interconnected system that has the technical capacity to service all areas shows how selective the system has become. It uses only the resources and regions that it finds useful. In the process it abandons other areas that supported viable economies when the system operated at local or regional levels. Moreover, in the process of abandoning regions, the system abandoned the people who lived there.

One of the major problems of the United States today is what to do with the people in these largely abandoned areas. Government programs such as those for Indian reservations and Appalachia call for investment in new roads and new water and sewage facilities, loans for building factories, and subsidies to manufacturers who locate in such areas. Unfortunately, small towns and villages in remote areas populated by unskilled workers have little to offer the metropolitan-oriented business community.

Some people have advocated increased encouragement of migration from depressed areas. They suggest abandoning the regions. Is this a viable alternative? Certainly, it is not a politically attractive answer to local politicians and to congressional representatives from such regions. Many people living within the depressed regions of the country love their home territory. They do not want to move, and the challenge of finding a meaningful lifestyle for them remains.

People who do migrate from depressed areas of rural United States have almost nothing to offer the system operating in metropolitan areas. In their rural schools they have not been taught the skills needed in specialized business office jobs or the technical skills required in modern manufacturing. Like peasant European immigrants in the past, they are qualified only for unskilled jobs. Today, unlike the situation fifty to seventy-five years ago, expanding industries no longer need masses of unskilled workers in packing plants, steel mills, and clothing shops.

Figure 7–7 shows the geographic distribution of the poor of the country by percentage of each state population. Nationally, 12 percent of all families were determined to be below the poverty level, according to an index developed by the Social Security Administration. This index is based on family size and composition, sex and age of the family head, and farm or nonfarm residence. Note that in four states in the Deep South—Arkansas, Louisiana, Mississippi, and Alabama—poor families totaled more than 20 percent of all families. All the other states of the South from Texas and Oklahoma to Virginia and West Virginia were above the national average, as were North Dakota and South Dakota. All these areas are largely rural, and predominantly agricultural. In the South they include remnants

of the old labor-intensive plantation areas where technological change and a shift away from cotton production have left much of the farm population unemployed. There too are the small farms of the hill country of the Appalachians. From these states came much of the labor force for the industrial growth of northern and western metropolitan areas during the 1950s and 1960s. People were as much pushed from the region by a lack of employment opportunities as they were pulled by the factory jobs elsewhere. But the out-migration was not sufficient to overcome the population surplus of the region.

In absolute terms, however, the largest numbers of poor families are in California and the populous states of the Northeast and Middle West—New York, Pennsylvania, Ohio, and Illinois. Here much of the poverty is in the cities, much of it among the people, both black and white, who emigrated from the poor rural areas of the South and among Spanish-speaking emigrants from Puerto Rico, Mexico, and other parts of Latin America. These people have developed a subsistence lifestyle, existing on far less than is necessary for an adequate life in a metropolitan economy. Through little fault of their own, they have not become part of the modern interconnected system any more than have Indians on reservations or Appalachian farmers. They are concentrated in ghetto areas of cities instead of being spread over the rural areas of many states. The New York metropolitan area has more families on public assistance than have all the southern states east of the Mississippi.

These "inner cities" are usually in old residential areas that have been allowed to deteriorate. Houses have been divided into multiple units with several families per floor, and population densities are extremely high. Buildings have inadequate plumbing and waste disposal. In these big-city ghettos masses of poverty-stricken people occupy block after block and mile after mile, isolated from the rest of the city except for their television sets. They have little opportunity to break away from the mold. They have no place

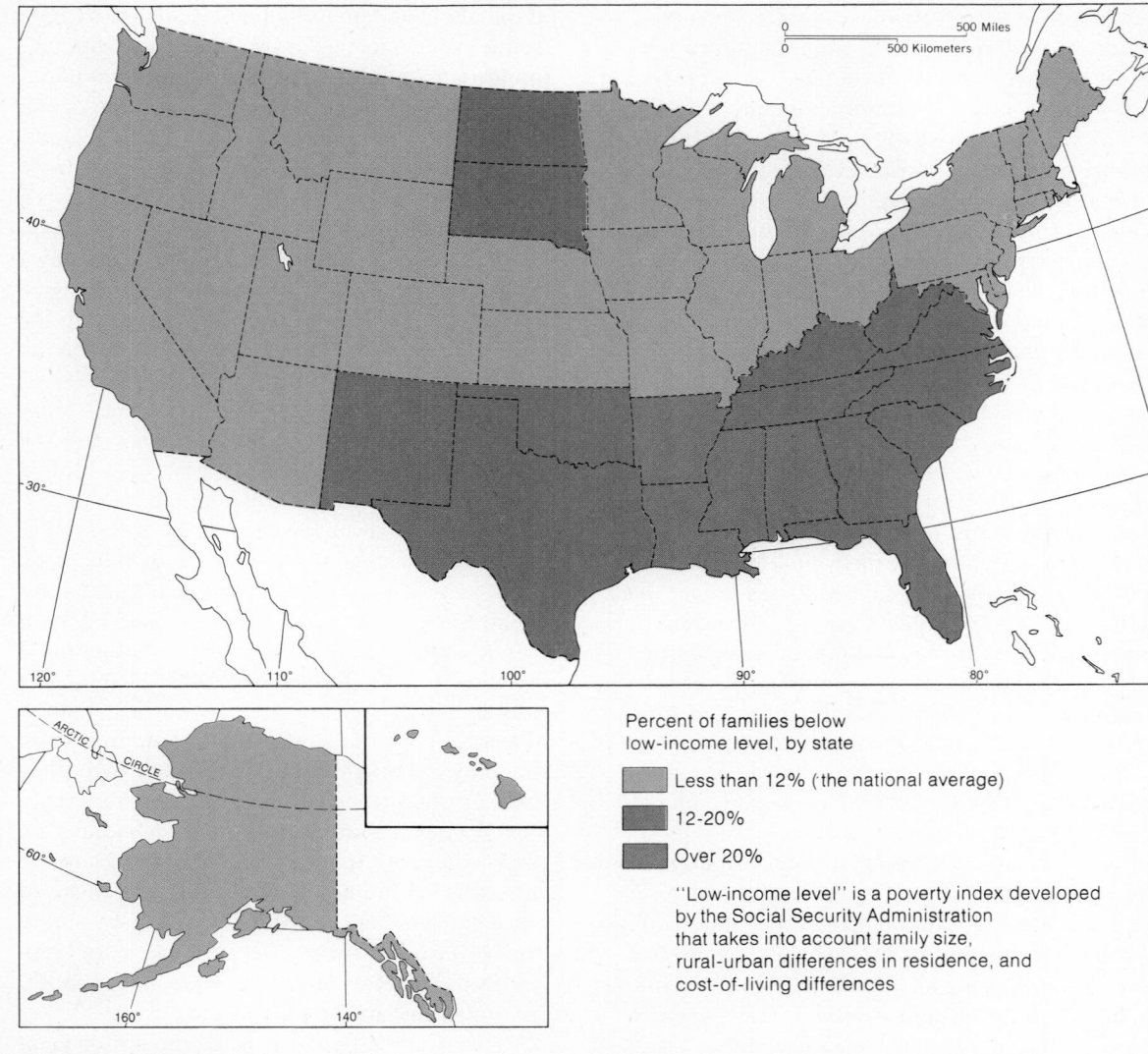

Percent of families below
low-income level, by state

Less than 12% (the national average)

12-20%

Over 20%

"Low-income level" is a poverty index developed
by the Social Security Administration
that takes into account family size,
rural-urban differences in residence, and
cost-of-living differences

SOURCE: Statistical Abstract of the U.S. 1975

FIGURE 7–7   FAMILIES BELOW LOW-INCOME
LEVEL IN THE UNITED STATES

to go, no way to get out of the world in which they are confined.

AN INTERNATIONAL SYSTEM WITH HEADQUARTERS IN THE UNITED STATES

We have been examining the United States primarily as a self-contained system. However, the United States has important links to the outside world, links that have gained in importance as it has increased its affluence and technological leadership in the world. Traditionally, the U.S. system has reached out into tropical areas of the world for specialty products and has supplemented its mineral resources with production from other countries. With the increasing affluence of its people, it has become a market for sophisticated manufactured goods such as luxury automobiles, complex optical goods, and high-fashion clothing and furnishings from other countries.

Much of this international trade has been controlled by the large corporations that once operated on a national scale within the United States. Now, with modern transportation and communication throughout the world, they can manage systems of global dimensions. Originally, they reached beyond U.S. borders for raw materials and exotic agricultural specialties that could not be produced here. However, they are now looking increasingly to other parts of the world for markets for U.S. goods. U.S. exports totaled $98 billion in 1974, with machinery accounting for 24 percent of this figure. Agricultural exports are also important; cereal grains, including wheat, made up 11 percent of U.S. exports in 1974.

In recent years large U.S. corporations have begun to locate manufacturing facilities in other countries—in industrial countries to tap markets there, and in underdeveloped countries to take advantage of cheap labor. Major U.S. companies now have facilities producing automobile components, electronic equipment, and clothing in foreign countries primarily for markets in the United States itself. Foreign trade now accounts for more than 10 percent of U.S. gross national product, and much of this trade is in goods controlled by large corporations with production facilities and markets overseas as well as in the United States. But these corporations still maintain their management headquarters in New York or other large U.S. metropolitan centers. Foreign multinational corporations are in turn expanding their marketing and even manufacturing in the United States.

The outreach of the modern interconnected system toward other parts of the world, not only from the United States but also from other modern countries, is the subject of Chapter 12. There we shall see interactions similar to those that occur in the U.S. system but on an international scale.

THE MODERN SYSTEM: THE PROBLEMS OF SUCCESS

The modern interconnected system operates on its most massive scale within the United States. The result has been affluence and the replacement of much physical labor by machines, which in turn has produced declining employment in agriculture, mining, and manufacturing. However, the results vary in different parts of the country and on different levels of the society. The largest metropolitan centers have grown spectacularly both in numbers and in importance while most rural areas have declined.

In most regions of the United States, particular districts and communities have had healthy, often booming, economic growth while others have lost their economic base. The South is a case in point. Traditionally, it has been an area of agriculture and timber products, based, since the Civil War, on small farms and small, portable sawmills working a few acres. Today certain parts of the agricultural South are booming with lands devoted to specialty crops and winter vegetables. Some of the most highly mechanized and most efficiently managed farms in the country are found there. These pockets of prosperity include processing plants, fertilizer dealers, banks, stores, and attractive residences sprawling out into the countryside. But over much of the South, the old small-scale agriculture struggles to support a declining rural population. Congressional investigations in recent years have discovered many cases of malnutrition in these places.

The same contrasts are found in the cities and towns of the South: Atlanta and Miami are among the most modern and rapidly growing metropolitan centers of the country, whereas such established centers as Savannah (Georgia), Lake Charles (Louisiana), and Gadsden (Alabama) have suffered. The South has become increasingly fragmented between these areas of increasing poverty and centers of great economic viability.

These contrasts are the result of a shift in the economic, political, and cultural life of the United States to a national scale. Before the twentieth century the economic units were predominantly individual family units—farm families, shopkeepers and their families, and lawyers and their families. Decisions were made within a local or regional framework. With the coming of radio, television, and the telephone, both individuals and businesses have been put in contact with happenings throughout the entire interconnected system. Everyone is regularly pressured by advertising to buy, and everyone is exposed to a steady stream of mass-produced culture. The scale of today's national economic system, made possible by modern technology and communication, calls for a different basic organizing unit. Economic efficiency is achieved by developing units that can function throughout the nation as a whole.

The individual is virtually lost in this system.

Large corporate and governmental units make decisions that affect regions, communities, and individuals alike. Government decisions on agricultural subsidy payments, tax write-offs, irrigation dams, and power networks affect the lives of millions of people. The decision to place a new factory in a town may breathe new life into an area that has been declining. On the other hand, the decision to close a mine, or even to automate one, may mean the end of economic support to the community where it is located.

The growth of metropolitan areas has been an important part of the shift to a national scale. Metropolitan areas are nodes in the functioning of the system, and they provide large markets for mass-produced consumer goods within a concentrated geographical area. Yet the quality of life in many metropolitan areas has declined as congestion and pollution have increased. Waste disposal has become a serious problem. Some large cities have encountered financial difficulties, with resulting cutbacks in the services they provide to their populations. Despite these problems, the metropolitan areas of the United States remain the greatest centers of affluence and technological expertise in the world. The overriding problem is how to channel these resources efficiently toward solutions to the problems that exist in the modern interconnected system.

SELECTED REFERENCES

Berry, Brian J. L. *A Geography of Market Centers and Retail Distribution.* Prentice-Hall, Englewood Cliffs, N. J., 1967. A basic survey of retail trade and cities as central places, with case studies and results of methodological studies.

Brzezinski, Zbigniew. *Between Two Ages: America's Role in the Technotronic Era.* Viking Press, New York, 1970. A popular book on the order of *Future Shock.* It provides another dimension of the question of how the United States functions in the world.

Comparative Metropolitan Analysis Project. *Contemporary Metropolitan America: Twenty Geographical Vignettes,* vol. 1. Ballinger, Cambridge, 1976. See Selected References for Chapter 6.

Estall, Robert. *A Modern Geography of the United States.* Penguin, Baltimore, 1972 (paper). Uses a systematic approach to examine the modern U.S. geography. The book probes in some depth the regional patterns and variations on each topic. Three chapters review population, land use, and employment structure; eight chapters examine the major productive systems of agriculture, mining, and manufacturing; and two chapters focus on urban changes and the geographical role of the federal government. The charts and maps provide excellent supporting information for this chapter.

Galbraith, John Kenneth. *The New Industrial State.* Signet, New York, 1967. A look at the industrial system that dominates the United States by the well-known economist.

Gottmann, Jean. *Megalopolis: The Urbanized Northeastern Seaboard of the United States.* Twentieth Century Fund, New York, 1961; M.I.T. Press, Cambridge, 1964. A thoughtful study of the large metropolitan centers of the Eastern Seaboard that dominate the country's decision-making process. See the chapters on the main street of the nation, the continent's economic hinge, and the white-collar revolution, and the conclusion.

Higbee, Edward C. *Farms and Farmers in an Urban Age.* Twentieth Century Fund, New York, 1963. An agricultural geographer looks at the revolution in agriculture created by the rise of urban centers.

Paterson, J. H. *North America: A Geography of Canada and the United States,* 5th ed. Oxford University Press, New York, 1975. A basic regional geographic approach in which the regional divisions are based on physiographic properties for the most part. Several early chapters examine population, agriculture, industry, transportation, and foreign trade. Chapter 8 sets the regional framework, and subsequent chapters explore each region, emphasizing its special geographical characteristics and role.

Starkey, Otis P., J. Lewis Robinson, and Crane S. Miller. *The Anglo-American Realm,* 2nd ed. McGraw-Hill, New York, 1975. A study of the United States and Canada using regional divisions that relate to their functional unity or particular roles in Anglo-America. This text has fewer regional divisions and somewhat less detail than the other texts cited.

U.S. Bureau of the Census, *Statistical Abstracts of the United States.* Washington, D.C., annual. A comprehensive summary of census and other information compiled from government agencies. Information is presented primarily in tables of data.

U.S. Department of Interior, Geological Survey. *The National Atlas of the United States of America.* Washington, D.C., 1970. A comprehensive atlas of environmental, historical, economic, administrative, and population characteristics of the United States.

White, C. Langdon, Edwin J. Foscue, and Tom L. McKnight. *Regional Geography of Anglo-America,* 4th ed. Prentice-Hall, Englewood Cliffs, N.J., 1974. A basic regional approach to the geography of the United States and Canada. Most of the regional divisions are based on physical criteria except for French Canada and Megalopolis, which are identified by their distinctive cultural or urban qualities. Chapters 7, 19, and 20 concern regions largely in Canada.

CHAPTER 8

Europe: Where the Modern Interconnected System Began

Europe was the first part of the world to develop the modern interconnected system. Present-day Europe, however, still shows the imprint of many centuries of life in the locally based, traditional system. Some of these vestiges of the past make it difficult for Europe to compete with the United States and the Soviet Union. Recently, to improve its position in the modern world, Europe has moved toward functioning as an economic unit, though the twenty-six countries retain their political and cultural distinctiveness.

Left: Dairy plant situated on a canal in the Netherlands. (Standard Oil Co., N.J.) *Above:* West German intercity passenger trains. (Werner Müller/Peter Arnold)

Above left: Immigrants from Southern Europe at work in Munich, West Germany. (Gilles Peress/Magnum Photos, Inc.) *Above right:* Oil drilling and production platform being assembled near Stavanger, Norway, before being towed to Beryl Field, in the British sector of the North Sea. (United Press International) *Below left:* Partial view of the town of San Gimignano in northern Italy. (Fritz Henle/Photo Researchers, Inc.)

Europe is not just another example of the modern interconnected system. It is the part of the world that was first dependent on trade with other continents and where industrialization began.[1] Sea exploration in the Atlantic Ocean by Spaniards and Portuguese, begun in the fifteenth century, brought trade connections with all other continents over the next 200 years. With these early explorations began the European political and cultural domination over the Americas, Africa, most of Asia, and Australia that has begun to fade only in this century. Industrialization—based on the harnessing of inanimate energy—was a European development that provided the basis for even greater economic and political dominance over most of the world in the nineteenth and early twentieth centuries.

Today Europeans are more affluent than ever. Only two of the twenty-six countries in Europe—Yugoslavia and Albania—have a per capita gross national product below the world average of $1,360, and Europe in general, with a per capita GNP of $3,680, is the most affluent region of the world after Anglo-America ($6,580) and Australia–New Zealand ($3,800). Sixteen countries have per capita GNP more than twice the world average, and six of those—Switzerland, Sweden, West Germany, Denmark, Luxembourg, and Iceland—average more than four times the world average. In fact, calculations by the World Bank indicate that Switzerland and Sweden have higher living standards than the United States. Travelers from the United States feel at home in most European cities, with their fashionable shops and department stores, streets congested with automobiles, and residents dressed in the latest styles.

European countries remain among the most important industrial nations in the world. Euro-pean-made automobiles, cameras, optical goods, fashions, and fabrics are highly prized by Americans. European financial organizations, among the most successful in the world, have worldwide investments. European capitalists even control such giant industrial organizations operating in the United States as the Shell Oil Company (owned jointly by Dutch and U.S. interests) and Lever Brothers (British).

Yet the largest and most economically advanced countries of Europe—the United Kingdom, France, West Germany, and Italy—are no longer among the great powers of the world, either economically or politically. They have lost their empires and since World War II have been far surpassed in political power by both the United States and the Soviet Union.

THE ENVIRONMENTAL BASE

Although the modern interconnected system dominates most of Europe, the continent's basic geography is very different from that of the United States. In the first place, Europe is very small and crowded in comparison with the United States. Its population of 476 million people in 1976 was more than twice that of the United States and slightly more than the combined populations of the United States and the Soviet Union. Yet this large population is crowded into a land area only half the size of the United States. As a result, European population density is more than four times that of the United States. In 1976 this density was more than 274 persons per square mile, compared to fewer than 60 in the United States and fewer than 30 in the Soviet Union. Pressure on the environmental base of Europe to produce primary products would seem to be much greater than that on the environmental base of the United States or the Soviet Union; Europe's interconnected system therefore draws more resources from other continental areas.

[1] In this chapter when we speak generally of Europe, it is to be understood that the Soviet Union is excluded. Our purposes are better served by one discussion of the Soviet Union as an entity in Chapter 9.

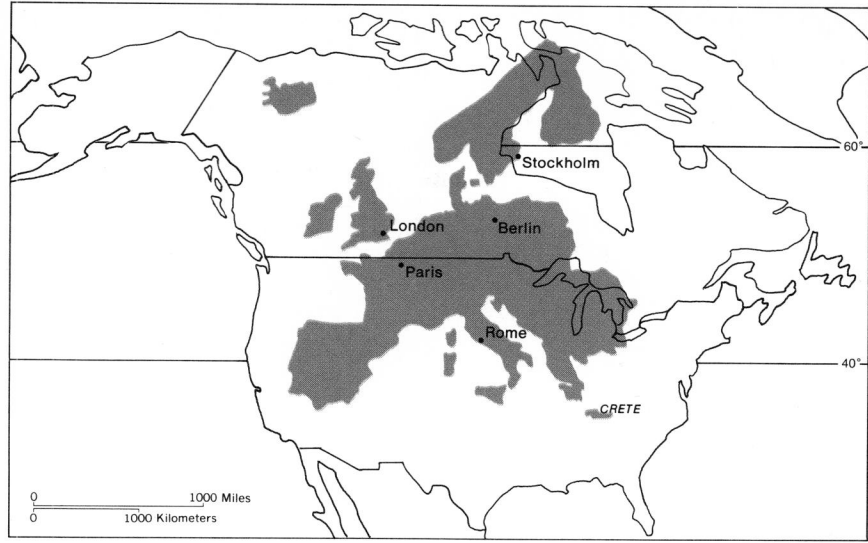

FIGURE 8–1 EUROPE (EXCLUDING THE SO-
VIET UNION) AND ANGLO-AMERICA COM-
PARED IN AREA AND LATITUDE

MODEST ENVIRONMENTAL ENDOWMENTS

Although Europe, like the United States, is a midlatitude area, its environmental base, in terms of both climate and terrain, is very different. It also has a dissimilar land configuration resulting in growth potential different from that of the United States.

High-Latitude Location

Several important facets of Europe's environmental base can be seen in Figure 8–1. Notice that while Europe is still mostly in the midlatitudes, it is located in higher latitudes than the United States. The southernmost part of Europe, the Greek island of Crete in the Mediterranean Sea, is 35° north of the equator, the same latitude as North Carolina. Rome has the same latitude as Chicago. Moreover, a large share of Europe lies north of the 49th parallel in the latitude of Canada, not of the conterminous United States.

A Peninsula of Peninsulas and Islands

As Figure 8–2 shows, Europe appears as a large **peninsula** extending out from the Eurasian landmass with seas on three sides. Only the east is landlocked. The European "peninsula" comprises a series of secondary peninsulas—Norway and Sweden, Denmark, northwestern France, Spain and Portugal, Italy, and Greece—and numerous islands, including the British Isles. On a smaller scale these and other portions of Europe encompass many more local peninsulas, **promontories,** and islands. As a result the sea seems to lap around and into Europe, and most places are no more than 400 miles (640 km) from the sea, whereas parts of the U.S. interior are more than 1,000 miles (1,600 km) from salt water.

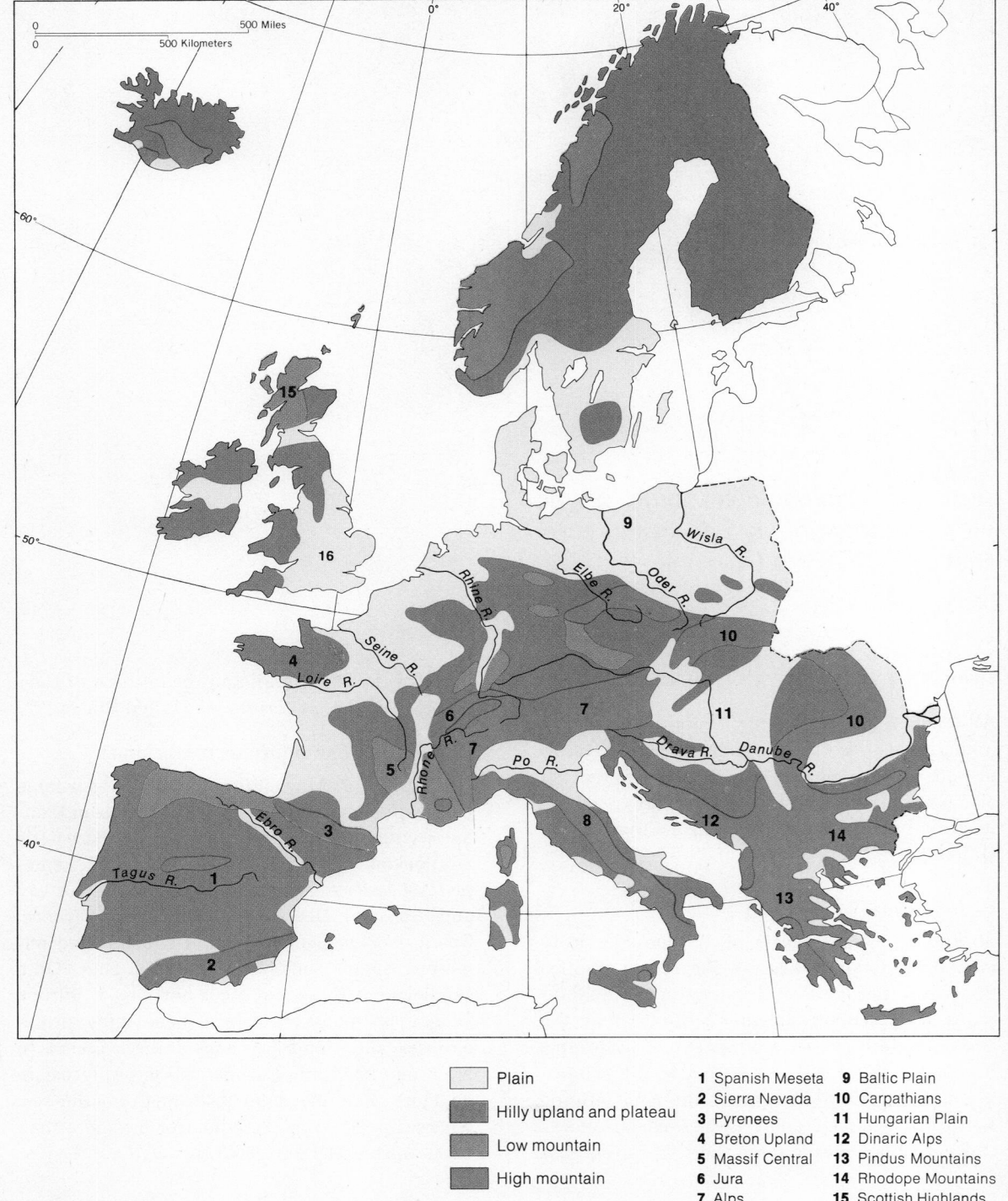

	Plain
	Hilly upland and plateau
	Low mountain
	High mountain

1	Spanish Meseta	**9**	Baltic Plain
2	Sierra Nevada	**10**	Carpathians
3	Pyrenees	**11**	Hungarian Plain
4	Breton Upland	**12**	Dinaric Alps
5	Massif Central	**13**	Pindus Mountains
6	Jura	**14**	Rhodope Mountains
7	Alps	**15**	Scottish Highlands
8	Apennines	**16**	English Lowlands

SOURCE: Adapted from *A Geography of Man,* 3rd ed., by P. E. James, Copyright © 1966 by John Wiley & Sons, Inc. Reprinted by permission of John Wiley & Sons, Inc.

FIGURE 8-2 LAND SURFACE FEATURES OF EUROPE

A brief survey of the land surface features of the "continental" portion of the European landmass will aid in our understanding of Europe's environmental base. The Iberian Peninsula (Spain and Portugal) is dominated by uplands and plateaus, except for small areas of lowland along the coast of Portugal and in southwestern Spain. The peninsula is bounded on the northeast by the Pyrenees, a mountain chain extending in an east-west direction along the border between Spain and France.

The Breton upland of northwestern France is a promontory extending into the Atlantic Ocean. Except for this region, most of the land within 150 miles (240 km) of the western and northern coast of the European continent, from France to Poland, is low-lying plain. The broad plain extending inland from the North Sea and the Baltic Sea is known as the Baltic Plain. A few areas in the Netherlands and West Germany are below sea level.

The central part of the European landmass contains several upland regions, among them the Bohemian **massive,** the Jura along the border between France and Switzerland, and the Massif Central of south-central France. The more southerly region of Europe is dominated by mountain ranges, the highest being the Alps extending in an east-west direction through southeastern France, northern Italy, Switzerland, and Austria. Reaching southward through the Italian peninsula are the Apennines, a less rugged range, while along the eastern coast of the Adriatic Sea run the Dinaric Alps. The southern extension of the latter range into the Balkan peninsula is called the Pindus Mountains. East of the Pindus Mountains lies the Rhodope massive.

The Carpathians, a J-shaped mountain range in eastern Europe, partially encloses the fertile Hungarian plain, and the enclosure is completed by the Alps and Dinaric Alps. Two other large plains areas, the Po Basin of northern Italy and the Rumanian plain, also contribute significantly to commercial livestock and crop farming in Europe.

Europe's landscape is thus characterized by great variety within surprisingly short distances.

In almost any part of the continent there are valleys and hills, usually within sight and always within a few hours' drive. Farming practices and settlement patterns have had to adjust to these variations.

MARITIME CLIMATE

As we have seen, access to the sea is comparatively easy from most places in Europe. Moreover, the winds from the seas have a major influence on the climate and the plant growth potential of the continent. In these latitudes, as in the United States, air flows predominantly from west to east, exposing Europe to full influences of winds from the Atlantic. Since the major mountains and highlands of Europe follow an east-west axis except in the north (Norway and Sweden), no terrain obstructs the air from the sea as it crosses the continent. This penetration of maritime air across the relatively small European landmass results in the general availability of moisture throughout the continent. Europe does not have any great tracts of desert, although Southern Europe, like California, receives moisture only during the winter half of the year. Another influence on Europe's climate is the North Atlantic Drift, a warm ocean current that flows in a northeasterly direction toward and along the coast of Europe. The North Atlantic Drift warms the air above it, thus contributing to the mildness of the European climate.

The influence of air from the sea "air conditions" Europe, giving it a **maritime climate.** Since water heats and cools more slowly than land, air coming off the water is comparatively warm in winter and cool in summer. This fact is particularly important in winter because of Europe's high latitude. Instead of winter lows of $-20°F$ ($-29°C$) as in much of the populated area of Canada, temperatures in most of Europe rarely fall below $0°F$ ($-18°C$). Average January temperatures in London are higher than those in New York City and almost as high as those in Washington and U.S. cities farther south. Farther inland from the Atlantic coast, winters are more severe, but they do not approach those of the northern Great Plains, except in the highlands of northern Scandinavia. Cities in Poland have January averages roughly like those of Chicago and Detroit. In summer the cooling effect of the sea means both lower temperatures and more cloudiness than in most of the United States. Over virtually all of Europe in July average temperatures are less than 70°F (21°C), several degrees lower than those in most of the United States.

The result is that Western Europe as far north as southern Norway has a seasonal-work environment like that of the southeastern United States. However, under these climatic conditions growing potential is generally less than that of the United States. Most of Europe, except for the Alps and areas far to the north, has more than ninety days of summer growth, so the length of the growing season is much the same as that in the United States. But the cooler summers and high latitudes result in considerably less potential energy for growth (see Figure 3–3).

The presence of the Mediterranean Sea influences the climate south of the major east-west mountain ranges. This large, deep, and almost landlocked body of water tends to retain heat; it warms up gradually in the spring and cools slowly in the autumn. The result is mild winters. Southern Spain and Greece are essentially frost-free. Small parts of Spain, Portugal, Italy, and Greece have only one or two cold spells each winter. The regions around the Mediterranean are also characterized by dry summers with little cloud cover and, therefore, abundant sun energy. The potential evapotranspiration (PE) of these regions is much the same as that of the southern United States north of Florida and of southern California, but it drops off sharply inland. The climate of the lands around the Mediterranean is often labeled a **Mediterranean climate.**

Most of the British Isles, northern France, Belgium and the Netherlands, the northern and central parts of West and East Germany, Denmark, Poland, and southern Sweden have annual potential evapotranspiration figures similar to those

of Canada north of the agricultural frontier. Even in the more interior areas of Hungary, Rumania, and Bulgaria, potential evapotranspiration is comparable only to that in New England and the northern Great Lakes area of the United States.

The winds from the west off the Atlantic supply continual moisture over most of the continent. Water deficits are not present during the growing season in Eastern Europe (see Figure 3–6); they are very small in Western and Northern Europe and generally less than deficits in the eastern United States. Thus actual crop growth in most of Europe can reach the full PE for each district. Unlike the situation in the United States, the relatively low sun energy, not moisture, presents the major limitation to crop growth.

Only Southern Europe is faced with serious moisture deficiencies. But this is the region of Europe with the longest growing season, and people have worked for thousands of years to overcome the problem. During the winter wet season there is little water surplus except in the mountains (see Figure 3–6). During the hot summer, the time of greatest growth potential, there is almost no precipitation, and the deficit becomes great. Since the time of the ancient Greeks and Romans, Southern Europeans have developed reservoirs to hold winter rainfall in the highlands for use in summer irrigation, and they have chosen crops and livestock that can adjust to the moisture-poor environment. Wheat is grown during the wetter winter season, and corn and barley are summer crops that do well in hot, sunny conditions. Olive trees and grapevines with deep root systems are staple crops. Sheep and goats that can scavenge on dry pasture are as important as cattle. Moreover, farmers use special moisture-holding methods, called **dry farming,** to allow the soil to carry over as much moisture from the rainy season as possible. The basis of dry farming is cultivation of bare fields to condition the soil to absorb and hold moisture.

In summary, we might speak of the environment of Europe as "manageable," but not outstanding in its potential for agricultural and timber production. While water management is a problem only in the Mediterranean area, energy potential for growth in that region is most like that of the southern United States. Over most of Europe moisture shortages do not hamper plant growth, but sufficient sun energy is simply not available for high evapotranspiration in summer. The relatively high crop yields in Western Europe reflect both the adequate moisture for full plant growth and the extremely intensive, enlightened care given the land by a hard-working, knowledgeable labor force. Farmers have long rotated crops and used fertilizers to maintain soil. The farming system utilizes livestock not only to provide meat in the diet but also to provide fertilizer for crops.

USABLE LAND SURFACE AND SOIL

Much of the European terrain is mountainous, with no vast lowlands comparable to those in the U.S. interior. The plain extending from western France along the southern margin of the North and Baltic seas is generally less than 200 miles (320 km) wide. In Eastern Europe the plain is even narrower, having been broken by the same kind of glacial deposits that produced the rolling landscape and blocked drainage ways in southern and central Michigan and Wisconsin. Thus, like the climatic environment, Europe's terrain is best described as "manageable."

Most soils in Europe are generally adequate—with modest amounts of plant nutrients—but they are not the best. In the north, particularly in Finland, much of the original soil cover was removed by the action of continental glaciers long before human settlement, leaving barren, rocky land similar to the Canadian Shield. Much of the plain that lies south of the North and Baltic seas is covered with glacial deposits providing a varied soil cover: sandy in some areas, rocky in others, and poorly drained in other places. Moist regions in Northern and Western Europe were covered with forests before people began clearing the land. In the midlatitudes forest soils are leached of nutrients by water moving through them and are inferior to those found in the drier grasslands like those of the United States. Around the Mediterranean Sea the combination of rugged terrain and semiarid conditions along with a sparse vegetative cover has resulted in soil of varying quality. In those areas where soils have been worked for thousands of years on more rugged slopes, severe erosion has occurred.

A MANAGEABLE ENVIRONMENT

In Europe, people have found a usable environment rather than a bountiful one. Europeans have experimented so persistently and successfully, especially in the past three centuries, that they have learned to manage the European resource base with relative ease. Except in the south, there is adequate moisture during the growing season and a low probability of drought, flood, tornado, and hail. In 1976, however, Western Europe was hit by an unusually severe drought, the worst in centuries.

Furthermore, human management has generally increased the productivity of the European environment. Through thousands of years of land use, the population increase has imposed ever greater pressure on it to produce. Along with this has come a growing technology for agriculture and forestry. As a result, Europeans have expended much effort to make their land more productive. They have applied to production a greater scientific knowledge of land use, crops, and livestock than anyone except the Japanese. To a high degree the agricultural land of Europe today is the product of human effort, the result not only of clearing but also of modifying both the surface and the soil.

Europeans have also led the world in forest management. Most of the forests of central and northern Europe are the result of replanting after the original timber had been cleared. Forests have been restocked with the most valuable species, and much of the timberland is expected to yield an annual harvest, as any other crop does.

The combination of workable environment and human effort has resulted in a fuller utilization of the land in Europe than on any other continent. Almost one-third of Europe's surface area is cultivated, whereas on no other continent is the proportion more than 18 percent. Only slightly more than 20 percent of Europe is considered unsuitable for the production of food and fiber. Elsewhere, however, the proportion is at least 26 percent, and frequently it is over 33 percent. This is as much a result of **intensive farming** as it is an indication of the quality of the environment.

Europe's land surface has presented few real barriers to human movement. The presence of a rather complex system of small rivers, particularly in humid central and Northern Europe, has provided people with both waterways to the sea and land routes along the valleys through rough country. Since early days, however, Europeans have overcome the most rugged terrain barriers. Even the Alps have been crossed by established routes since the days of the Roman Empire.

Moreover, the environment has provided essential resources for building and industry from early times. Timber from European forests and rock have been basic building materials since prehistoric days. Waterpower from small streams has been employed since the beginnings of industrialization, and windpower has been used on land and sea even longer. Although no longer considered a major center of world mineral production—particularly not of oil and nonferrous minerals—Europe has adequate and accessible deposits of coal and iron. These provided the basis for the spread of the Industrial Revolution in the nineteenth century. Coal is produced in quantity in Poland, the United Kingdom, and West Germany; and iron in Sweden and France. Although Rumanian oil fields have been producing for some time, significant deposits of oil and natural gas are now being developed in the northern part of Europe, mostly by placing drilling platforms in the North Sea.

THE HUMAN COMPONENT

A major contrast between Europe and the United States is seen in Europe's political and cultural fragmentation. Europe is divided into twenty-six sovereign states and a few tiny city-states (Figure 8–3).

POLITICAL AND CULTURAL FRAGMENTATION

Each of these twenty-six European countries represents a separate functioning unit not only in political terms but also, to a high degree, in economic and cultural terms. The citizens of each country represent a different group that has been conditioned through training at home and in school to think of itself as loyal to the country, its government, and its heritage, rather than to Europe as a whole. Instead of a single version of the modern interconnected system operating on a single "gameboard," there are twenty-six different "games" on country-sized "gameboards." Each country has its own solutions to its own geographic problems.

Unlike the United States, Europe did not develop as a single functioning entity. Its twenty-six separate countries stand today as evidence of that fact. Indeed, it was in Europe that the concept of the national political state, the nation-state, functioning in terms of its own self-interests, emerged. Each European country has established itself as a political, economic, and cultural entity distinct from its neighbors. In consequence, it has spent extra effort managing its particular resources. Throughout much of European history certain international economic and cultural forces, such as the Hanseatic League and the Roman Catholic church, have been important. The European Economic Community represents a modern-day version of such intra-European forces, but it is still in conflict with deeper-rooted nationalistic sentiments of European peoples.

FIGURE 8-3 POLITICAL MAP OF EUROPE

London National capitals

• Metropolitan areas over 1 million

Political Diversity and Different Economic Systems

The importance of the political sovereignty of individual European countries must not be underestimated. Each government has authority over the people and institutions within its territory. Each establishes its currency and sets down the rules and laws under which institutions of all sorts operate. Different European governments have very different ways of looking at economic and cultural goals and also have different political aims.

The Eastern European countries, under the dominance of the Soviet Union since World War II, have established one-party political regimes and have followed the Soviet economic system under which essentially all land and structures of the business and manufacturing community are controlled by the central government. The original group of Soviet-dominated countries included Poland, East Germany, Czechoslovakia, Hungary, Rumania, Bulgaria, Yugoslavia, and Albania. Yugoslavia, under the leadership of Marshal Tito, resisted Soviet domination and has become more Western-oriented than the other Communist countries. Albania, the country most remote from the Soviet Union and the most inaccessible because of its rugged terrain, has developed ideological ties with China.

In other parts of Europe a variety of political and economic forms can be found. In some countries—Norway, Sweden, Denmark, the Netherlands, and the United Kingdom—kings and queens still reign as vestiges of ancient traditions, but most have only ceremonial power. In others, particularly the Eastern European countries, dictators and one-party governments are in control, granting varying degrees of freedom to both individuals and business enterprises.

The United Kingdom (consisting of England, Scotland, Northern Ireland, and Wales) has only two major political parties, but in Italy and France party structure is more fragmented. Because no one party can gain a majority in the Italian government, Italy has had more than one government change each year since World War II. Socialistic economic systems have developed in some European countries since 1900. In the United Kingdom such a system allows individual ownership of rural properties but calls for strict control of land around urban centers and for government ownership of economic services deemed essential to the public good. Thus the British government operates not only the postal service and radio but also coal mining, the steel industry, electricity production, the railroads, and the airlines. At the same time, shops, large businesses, most industrial enterprises, and farms are privately owned but operate under strict government regulations.

In Europe the management question has been how to develop each of the particular parts of the continent within the limits imposed by twenty-six political entities. The resources and people of the continent are not organized into a single functioning system despite the recent changes that have pulled Eastern European countries into a Soviet-dominated bloc and have brought a degree of economic unity to the countries of the European Economic Community (West Germany, France, Italy, the United Kingdom, the Netherlands, Belgium, Luxembourg, Ireland, and Denmark).

The situation is complicated by the fact that European countries are comparable in size to individual states within the United States. France, the largest European country in area, is four-fifths the size of Texas, and tiny Luxembourg is about the size of the smallest U.S. state, Rhode Island (Table 8–1). In densely populated Europe most countries have larger populations than U.S. states of comparable area. Even so, each of the four most populous countries has only slightly more than one-fourth the population of the United States. Over half the countries of Europe have populations smaller than Ohio's.

Diversity: A Handicap to the Modern System

Whether measured by land area or population, European countries are diminutive when compared with the United States or the Soviet Union. As a result, they operate on a much

TABLE 8-1 SIZE AND POPULATION COMPARISON OF EUROPEAN COUNTRIES WITH STATES IN THE UNITED STATES

Size: Square Miles			Population, 1976: Millions of People		
Larger than any U.S. state but Alaska and Texas			*Larger than any U.S. state*		
	Alaska	586,000	West Germany	62.1	
	Texas	267,000	United Kingdom	56.1	
France	210,000		Italy	56.3	
Spain	195,000		France	53.1	
Sweden	174,000		Spain	36.0	
			Poland	34.4	
The size of the states in the West					
	California	159,000	*The size of the more populous U.S. states*		
Finland	130,000		Yugoslavia	21.5	
Norway	125,000		Rumania	21.5	
Poland	121,000			California	21.2
Italy	116,000			New York	18.1
Yugoslavia	99,000				
West Germany	96,000		East Germany	16.9	
United Kingdom	94,000		Czechoslovakia	14.9	
Rumania	92,000		Netherlands	13.8	
	Washington	68,000		Illinois	11.1
The size of states in the Middle West and South			*The size of medium-population U.S. states*		
	Illinois	56,000		Ohio	10.7
Greece	51,000		Hungary	10.6	
Czechoslovakia	49,000		Belgium	9.8	
Bulgaria	43,000		Greece	9.0	
East Germany	42,000		Bulgaria	8.8	
Iceland	40,000		Portugal	8.5	
Hungary	36,000		Sweden	8.2	
Portugal	35,000		Austria	7.5	
Austria	32,000		Switzerland	6.5	
	South Carolina	31,000		Massachusetts	5.7
The size of states in the Northeast			*The size of small U.S. states*		
Ireland	27,000				
	West Virginia	24,000	Denmark	5.1	North Carolina 5.1
Denmark	17,000		Finland	4.7	
Switzerland	16,000		Norway	4.0	
Netherlands	14,000		Ireland	3.1	
Belgium	12,000		Albania	2.5	
Albania	11,000		Luxembourg	0.4	Alaska 0.4
Luxembourg	1,000	Rhode Island 1,000	Iceland	0.2	

SOURCES: Size: Edward Espenshade and Joel Morrison, eds., *Goode's World Atlas*, 14th ed., Rand McNally, Chicago, 1974. Population of European countries: Population Reference Bureau, Inc., *World Population Data Sheet, 1976*, Washington, D.C., 1976. Population of U.S. states: *Encyclopaedia Britannica, Book of the Year, 1976*, Chicago, 1976.

smaller scale than does either of the great powers. From our consideration of mass production and mass consumption in Chapter 6, European countries would seem to be at an important disadvantage in economic terms. The modern interconnected system in the United States is based on mass production and mass consumption for a market of 215 million Americans. Europe as a whole would offer a market over twice as large. But since it is divided into twenty-six separate political entities, each with the power to control the goods moving into and out of its territory, the "real" Europe can be viewed as twenty-six different markets, many of which have just a few million people. The economy of large-scale operation, so important in U.S. mass production, has until recently been unattainable in most European countries.

Even with the European Economic Community, which in recent years has broken down important national barriers, the lack of European economic and cultural integration can be seen by the importance attached to traditional local and national interests. Fifty or more different languages are spoken in Europe; twenty different types of currency are used; the people are governed by twenty-six different sets of laws and business codes. We can get some idea of the situation by imagining each of the U.S. states east of the Mississippi River functioning as a sovereign power with its own money, language, laws, businesses, and military forces. Obviously, none of these states could operate entirely in isolation from the rest. Goods, ideas, and people would move across political borders, particularly between adjacent states. But each state would be chiefly concerned with developing its own economy and raising the living standards of its own people. Thus in competing with strong neighbors, it might well encourage trade with areas overseas rather than expand interstate trade to its fullest.

European countries do have economic, social, and political dealings with one another, but they do so as separate entities with freedom to en-

courage or cut off such contacts, to ingratiate or antagonize. The interrelationships of European countries are similar in character to those each country holds with separate political units in other parts of the world.

Some European countries, particularly those that have been politically and economically powerful, have found their basic ties outside the continent. In so doing they have been encouraged by both negative factors—such as economic and sometimes military rivalry within Europe—and the positive factor of ready access to the seas of the world, the cheapest and most flexible trade routes. When closer economic ties were being planned between Western European nations after World War II, the United Kingdom stood aloof, believing that its ties to Commonwealth countries such as Canada, Australia, New Zealand, and India were more important. The Commonwealth still maintains some language, currency, trade, and legal ties among its members. France also continues trade preferences and cultural ties with its former colonies in Africa and other parts of the world.

This does not mean that important movements of people, ideas, and goods between parts of Europe do not go on. Indeed, intra-European trade is greater than trade with other continents. But there is not the free movement that characterizes the U.S. system. The boundaries of European countries, even within the European Economic Community, represent obstacles, real or potential, to free movement. The smaller countries, less viable as individual entities, often feed products into the economies of larger countries. Denmark, for instance, provides pork products and eggs for the United Kingdom and other Western European countries, and Ireland raises cattle for British markets.

Neither the environmental nor the human resources of the continent have ever been assessed in terms of a unified Europe. Europe simply has not been developed as a single, modern interconnected system in the way that the United States and the Soviet Union have. Instead, its resources

COMPARISONS OF SELECTED EUROPEAN COUNTRIES WITH U.S. REGIONS

In geography it is useful to compare unfamiliar places with familiar ones to have a base for examining other areas. However, geographers recognize that each place is a unique mix of location, environment, culture, technology, and spatial interaction. Keeping in mind both the importance of place comparisons and the ever-present liabilities, examine the data in Table 8–2 and see what conclusions you can draw about the selected European countries.

1. France is a leading industrial country in Europe. Thus we have compared it with the most industrial part of the United States—an area roughly comparable in size and population. However, France is also the most important agricultural country of Europe, whereas the northeastern United States is only of minor importance to this country agriculturally. What conclusions do you draw from the comparisons?

2. The Netherlands is one of the most densely populated and highly developed parts of Europe. So, too, is southern New England, an area roughly comparable in size. Elaborate on this comparison.

3. Sweden is the most industrial area in northern Europe, but its northern location hampers agricultural development. Wisconsin and Minnesota together have much the same population on a smaller land area. But the economies of the two different areas seem quite different. Why would this be so? Would you say each is successful?

4. Yugoslavia is developing from a traditional peasant base and is just beginning to industrialize, while Tennessee and Alabama are moving ahead industrially from their traditional agricultural base. The two areas are about the same size, but notice the much larger population in

Yugoslavia. How would Yugoslavia perform as a region within the United States?

5. Czechoslovakia, one of the most advanced of the Eastern European countries, is roughly comparable in size to Pennsylvania. Like Pennsylvania, it has a mix of industry and agriculture. How do they compare?

6. You may wish now to make comparisons among the five different European countries. Which is the most developed? the most industrial? the most productive agriculturally?

7. As you have been making comparisons, have you kept in mind the geographic components of each place—growing potential, mineral resources, culture, stage of technology, form of government, degree of self-sufficiency, and so on? Briefly summarize these geographic components for each of the five European countries in Table 8–2.

---

have been measured and developed in a compartmentalized fashion that reflects its political separation (Table 8–2).

Traditional Society as a Basis for Diversity

Why has Europe, faced with competition from the much larger interconnected systems of the United States and the Soviet Union, not organized itself into a single political and economic unit? With more population than the two rival powers combined, a skilled scientific community, and a highly advanced technology, a united Europe would indeed be a formidable rival to the great powers. The answer may lie in the cultural diversity of the continent. So why is there such cultural diversity in Europe, which is so much smaller than the United States?

Through the thousands of years of Europe's development, limited technology for transportation and communication determined that most

TABLE 8-2 COMPARISONS OF SELECTED EUROPEAN COUNTRIES WITH REGIONS OF THE UNITED STATES

	France	New England–Middle Atlantic States–West Virginia
Land area (thousands of square miles)	210.0	216.0
Population, 1976 (millions)	53.1	51.3
Value added by manufacture, 1972 ($ billions)	66.5	95.5
Gross national product, 1973 ($ billions)	255.1	Personal income 277.0
Grain production, 1973 (millions of metric tons)	41.9	25.4
Cattle population, 1973 (millions of head)	21.9	2.6
Electrical energy production, 1973 (billions of kilowatt-hours)	174.1	390.6

	Netherlands	Connecticut–Massachusetts–Rhode Island
Land area (thousands of square miles)	14.0	14.5
Population, 1976 (millions)	13.8	9.8
Value added by manufacture, 1972 ($ billions)	8.7	17.0
Gross national product, 1973 ($ billions)	59.7	Personal income 53.3
Grain production, 1973 (millions of metric tons)	0.9	1.3
Cattle population, 1973 (millions of head)	4.7	1.2
Electrical energy production, 1973 (billions of kilowatt-hours)	78.1	55.3

	Sweden	Wisconsin–Minnesota
Land area (thousands of square miles)	174.0	140.0
Population, 1976 (millions)	8.2	8.4
Value added by manufacture, 1972 ($ billions)	10.3	15.0
Gross national product, 1973 ($ billions)	50.1	Personal income 41.7

TABLE 8-2   (Cont.)

	Sweden	Wisconsin–Minnesota
Grain production, 1973 (millions of metric tons)	4.7	49.0
Cattle population, 1973 (millions of head)	1.8	3.7
Electrical energy production, 1973 (billions of kilowatt-hours)	78.1	55.3

	Yugoslavia	Tennessee–Alabama
Land area (thousands of square miles)	99.0	94.0
Population, 1976 (millions)	21.5	7.6
Value added by manufacture, 1972 ($ billions)	3.7	12.7
Gross national product, 1973 ($ billions)	14.4	Personal income 30.0
Grain production, 1973 (millions of metric tons)	13.4	4.5
Cattle population, 1973 (millions of head)	5.3	2.5
Electrical energy production, 1973 (billions of kilowatt-hours)	35.1	119.6

	Czechoslovakia	Pennsylvania
Land area (thousands of square miles)	50.0	45.0
Population, 1976 (millions)	14.9	11.9
Value added by manufacture, 1972 ($ billions)	21.1	23.5
Gross national product, 1973 ($ billions)	41.8	Personal income 59.4
Grain production, 1973 (millions of metric tons)	9.2	12.8
Cattle population, 1973 (millions of head)	4.4	0.8
Electrical energy production, 1973 (billions of kilowatt-hours)	53.5	102.5

SOURCES: *Encyclopaedia Britannica, Book of the Year, 1975,* Chicago, 1975; U.S. Bureau of the Census, *Statistical Abstract of the United States, 1975,* Washington, D.C., 1975; Agricultural Statistics 1974; *U.N. Statistical Yearbook, 1973;* Population Reference Bureau, Inc., *World Population Data Sheet, 1976,* Washington D.C., 1976.

activities follow the traditional, locally based pattern for managing the environment. Even though Europe was crisscrossed with trade routes dating back to antiquity, each group developed its own ties to the resource base, its own values, and its own language. These groups were aware of their neighbors and of groups in distant corners of the continent. Contacts, however, were infrequent; they often resulted from military campaigns.

These cohesive group forces still are significant in the European way of life. Unique to each political state are its interrelationships, memories of traditional friendships and hatreds, and carefully developed traditions and skills. No unifying force—political, military, economic, or spiritual—has been able to overcome the deep-rooted localism. Many have tried—the Roman Empire, Holy Roman Empire, Catholicism, Napoleon, and Hitler—but each has failed. Some groups feel that the present political states are too large, not too small. The Scots and Welsh call for separation from the rest of the United Kingdom, and the Bretons from France. Tyroleans and Serbians also think of themselves as separate peoples, not as Austrians or Yugoslavs. It remains to be seen if the strong economic pressures for international European cooperation can succeed where other forces have failed.

Physical variations in European topography, geographic barriers, and sheer distance have tended to separate the continent into different regions. The east-west direction of the Alps and the seas that separate Northern Europe from the rest of the continent have made north-south contacts more difficult than east-west ones. Early migration into Europe apparently came from the east and soon diverged into distinctive streams—one into Southern Europe south of the Alps, another across Eastern and Western Europe north of the Alps, and a third, less well defined, into Northern Europe beyond the Baltic and North seas. Westward movement within Southern Europe was made difficult by the rugged, broken terrain, and migrations here were commonly by sea. North of the Alps, however, over-land movement was comparatively easy on the east-west plain along the northern seas and on the uplands along the northern margin of the Alps.

Along each of these east-west paths, different peoples occupied particular territories. In Southern Europe the Greeks and Romans could be distinguished from east to west, and the Franks, Germanic peoples, and Slavs each occupied certain lands north of the Alps. Farther north the Scandinavians formed another distinctive cluster of cultures. These **subcultures** can be seen in the physical appearance of the people of Europe and in the distribution of languages and religions. Roman Catholicism is centered in the Mediterranean lands from Italy westward, Protestantism has its stronghold in the northern areas, and the domain of the Eastern Orthodox church extends northward from Greece.

Five hundred years ago Europe functioned largely in local terms. Under the feudal system monarchs ruled through arrangements with local landlords, and for most people life was centered within the local fiefdom, which provided all the necessities of life. Only the nobles and their entourage traveled much beyond the fief or had access to foreign goods. The nation-state, a European invention, did not really emerge until the fifteenth century and the beginning of industrialization. Italy and Germany, two of the largest countries of Europe, were not consolidated until the nineteenth century.

Thus the cultural and political diversity of present-day Europe, so very different from the situation in the United States, is largely a product of roots extending far back in time. In this regard it is important to remember that Europe's pattern is typical of world diversity. Asia and Africa, the other parts of the world where culture groups trace their beginnings back to earliest time, show a similar cultural diversity. Large-scale cultural homogeneity, such as one finds in the United States, is the exception in today's world. Moreover, this unity stems from the fact that the United States was settled in modern times over a relatively short period.

TABLE 8-3   DISTRIBUTION OF EUROPEAN POPULATION, 1976

Division	Percent of Land Area	Population (In Millions)	Percent of Population	Population per Square Mile (Square Kilometer)
Western Europe	28	217.5	46	419 (194)
Southern Europe	27	133.8	28	260 (101)
Eastern Europe	20	107.0	22	277 (107)
Northern Europe	25	17.2	4	37 (14)

SOURCES: Population Reference Bureau, Inc., *World Population Data Sheet, 1976,* Washington, D.C., 1976. For percents of land area: U.N. Food and Agriculture Organization, *Production Yearbook,* 1976.

DISTRIBUTION OF POPULATION

The population of Europe, like the population of the United States, is unevenly distributed. The continent can be divided into southern, eastern, western, and northern areas, which in large measure reflect different earth environments and distinct cultural variations. Each of these four divisions constitutes roughly one-quarter of the total land area of the continent (Table 8–3). But almost half the population is in Western Europe. Notice that the population there is just slightly larger than that of the United States. Two other divisions—Southern and Eastern Europe—each have close to their expected quarter of the total population. Northern Europe is comparatively sparsely populated, with only 4 percent of the total.

Figure 8–4 shows that there is considerable variation in population density within each of the four large divisions of Europe. Moreover, one can see bands of dense population that extend almost all the way across Europe. The most densely populated areas have densities similar to those in the eastern megalopolis and in other U.S. metropolitan areas. These areas cross Europe in both east-west and northwest-southeast axes that merge along the English Channel between England and France and Belgium. One axis extends from the Scottish Lowlands, across the Channel, up the Rhine, and then south of the Alps into northern Italy. The other axis reaches from southern England eastward through Germany and on into the Soviet Union. The most sparsely populated area is northern Scandinavia, but much of the rest of Europe has densities comparable to most nonmetropolitan areas in the eastern United States.

In Europe, as in the United States, the distribution of population has two components: (1) the population that has spread over the landscape and is engaged in agriculture and herding in rural, small-town settings; (2) the population that is involved in manufacturing, services, and management, and has settled overwhelmingly in large metropolitan centers.

The Rural, Agricultural Component

In Europe, where settlements have developed for centuries, the traditional, locally based society was deeply rooted long before the rise of trade. Although hundreds of years ago Europe had trade routes and trading centers such as twelfth-century Venice, by tradition most Europeans have been part of an agriculturally based, peasant society. Without modern energy sources and machinery, great human effort was needed to turn the mediocre European agricultural base

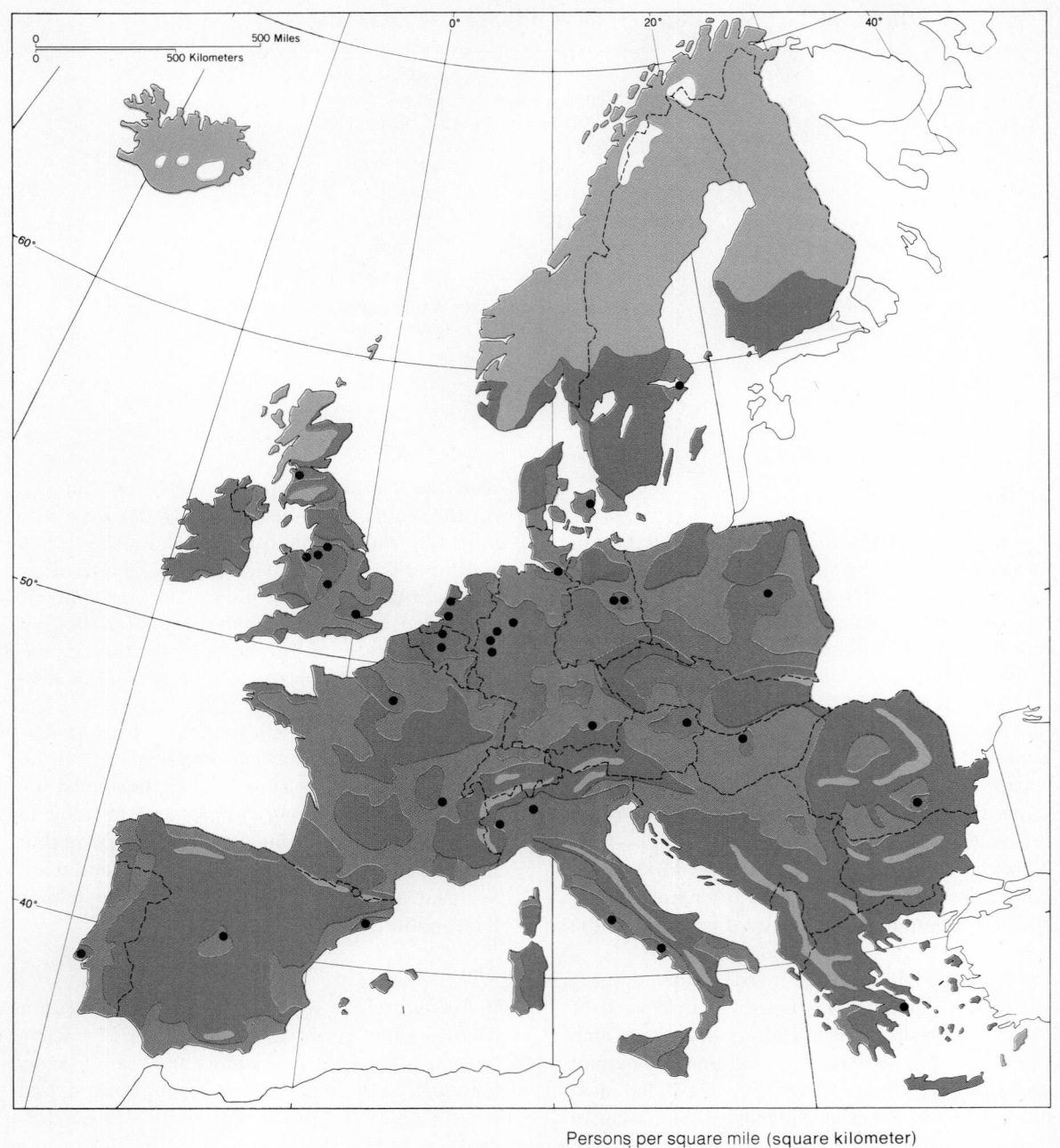

Persons per square mile (square kilometer)

▫ Virtually uninhabited

■ Under 32 (Under 12)

■ 32-128 (12-49)

■ 128-256 (49-99)

■ Over 256 (Over 99)

● Metropolitan areas with more than 1,000,000 population

FIGURE 8–4 POPULATION DENSITY OF EUROPE

into production. As in Asia, a tremendous amount of human energy was expended in intensive agriculture. The need for so much manual labor resulted in small working units, large families, and high rural and village densities.

In many ways life then was like life in Ramkheri today. Most people lived in agricultural villages and walked each day to the surrounding fields. Work was done by hand and with the aid of oxen or horses. Life was organized around the agricultural cycle, with planting and harvest being busy times requiring the efforts of all able-bodied members of families, even all village workers. There were religious ceremonies associated with planting and great religious celebrations at harvest. A large share of the crop was turned over to the landlord by the peasant farmers and the remainder was used by the farm family. The tiny surplus was used to pay for the services of blacksmiths, coopers, and other artisans and to pay religious tithes.

The locally oriented, traditional European culture we see today is based on the **feudal system.** Peasant farmers, first serving as serfs under local feudal landlords, then working as tenants or freeholders on their own small farms, were concerned primarily with providing basic food and other family needs from their land and then selling their surpluses in local market centers. Sometimes their goods moved on to other markets or cities, but the peasant farmers and their families, like Indian farmers today, lived their lives within the local environment. Their community, their local officials, and their priests or ministers set the cultural temper of their lives.

This peasant tradition of agriculture and small-town living established the basic pattern of population over Europe. Higher densities meant better, more intensively used agricultural areas, and lower densities implied areas where rough terrain, a short growing season, dryness, or some combination of unfavorable circumstances reduced the possibilities for agriculture.

In most of Europe today, farmers not only own their own land, but play an important part in the modern interconnected system. As in the United States, their first objective is to produce regional specialties that can be sold to supply the huge metropolitan population of their country or other parts of Europe. They use farm machinery, haul their goods to market by truck over paved rural roads, and keep up with market prices in the newspapers or on radio. A large percentage of the farm homes are served by electricity.

But the small size of European farms compared to those in the United States reflects peasant beginnings and a greater prevalence of family ownership, along with a need to use the land more intensively. The farming system, while based on cash sales, still shows the mixture of crops and livestock traditional in the old peasant system. Like the Pennsylvania Dutch in the United States, the European farm families still raise a mixture of the crops that will grow in their particular region instead of concentrating on one crop. Some crops in the mixture are important for building soil fertility in **crop rotation** systems. Livestock not only provide cash sales of meat, milk, and poultry products, but contribute valuable animal manure for use on the fields as fertilizer.

Crop and livestock mixtures vary sharply from one part of Europe to another. Wheat is most important in northern and western France, along the Rumanian-Bulgarian border, and as a winter crop in Mediterranean lands. Corn is largely confined to Hungary and Rumania in Eastern Europe and to Italy and the eastern Mediterranean. Rye is the chief grain of Germany, Poland, and Czechoslovakia; in Czechoslovakia it is commonly grown in association with potatoes. Barley and oats are more widely spread as secondary grain crops. Cattle are important in most farming areas, but hogs are largely confined to the rye-potato areas and sheep are most prominent in the dry Mediterranean area, where pastures are sparse, and in the western and northern United Kingdom, where it is too cloudy and moist for most crops to mature but where grass grows well.

The **specialty crops** in the farming system show distinctive regional patterns: olives and citrus fruits in the Mediterranean; tobacco there

TABLE 8-4 THE REGIONAL DISTRIBUTION OF AGRICULTURE IN EUROPE, 1974

Land Use	Northern Europe		Western Europe		Southern Europe		Eastern Europe	
	Million acres	*Percent*	*Million acres*	*Percent*	*Million acres*	*Percent*	*Million acres*	*Percent*
Arable land (cropland and cultivated land)	19.0	5	99.6	28	122.1	34	113.2	32
Pasture land	16.8	8	91.2	42	72.2	33	36.1	17
Woodland	123.6	35	73.1	21	89.0	25	68.4	19
Crop Output	*Millions of metric tons*	*Percent*	*Millions of metric tons*	*Percent*	*Millions of metric tons*	*Percent*	*Millions of metric tons*	*Percent*
Cereals	11.9	5	95.6	41	51.7	22	75.9	32
Potatoes and sugar beets	7.3	3	97.8	40	32.3	13	109.3	44
Livestock	*Million head*	*Percent*	*Million head*	*Percent*	*Million head*	*Percent*	*Million head*	*Percent*
Cattle	10.4	8	62.0	49	21.2	17	32.2	26
Horses	0.2	3	1.2	18	1.7	26	3.4	52
Pigs	5.0	3	65.1	42	26.8	17	58.0	37
	Thousands	*Percent*	*Thousands*	*Percent*	*Thousands*	*Percent*	*Thousands*	*Percent*
Tractors	545.2	8	4,009.7	60	1,291.9	19	858.5	13
	Millions	*Percent*	*Millions*	*Percent*	*Millions*	*Percent*	*Millions*	*Percent*
Farm workers (1970)	1.4	3	7.9	18	18.0	40	17.4	39
Total population (1976)	20.2	4	214.4	39	133.8	24	187.0	34

SOURCE: U.N. Food and Agriculture Organization, *Production Yearbook*, 1974.

and in southeastern parts of Eastern Europe; sugar beets in northeastern France, Belgium, the Netherlands, northwestern Germany, and the Po Valley of northern Italy. Grapes are grown in much of the southern half of Europe as far north as central France, the Rhine Valley of Germany, and Hungary and Rumania. This crop, more than any other perhaps, illustrates the importance not only of local growing conditions, but also, in the case of wines, of local winemaking techniques and local taste in wine that extend back for many centuries.

The most important agricultural area through most of European history was Southern Europe. Table 8-4 shows that it is still the most agricultural area of Europe in terms of land use. It has a

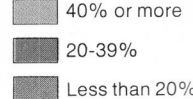

40% or more

20-39%

Less than 20%

SOURCE: FAO Production Yearbook 1973

FIGURE 8–5 PERCENTAGES OF ECONOMI-
CALLY ACTIVE EUROPEAN POPULATION EM-
PLOYED IN AGRICULTURE, BY COUNTRY

greater area of productive land than any other European region. In productive land Western Europe is close behind, and Eastern Europe ranks with the other two areas in total cultivated acreage. Northern Europe is far less important than the others: It has less than half the productive farmland of the two ranking areas and only 4 percent of cultivated land. It is significant only

for its pasture and woodland, as might be expected in a cool region with a short growing season, rough terrain, and considerable thin, rocky, or poorly drained soil.

Farm population does not show a perfect positive correlation with productive farmland. Notice in Table 8–4 and Figure 8–5 that both Eastern and Southern Europe have larger proportions of

EVALUATING EUROPE'S AGRICULTURAL REGIONS

1. Notice that Southern Europe, Western Europe, and Eastern Europe each has about one-third the continent's farmland. However, the agricultural output is quite different. Western Europe produces two-fifths the crop output while Southern Europe accounts for only slightly more than one-fifth. Only Eastern Europe has a crop output close to its share of farmland. Western Europe's importance in livestock production—especially the major meat animals, cattle and hogs—is even more marked.

2. At the same time, Southern and Eastern Europe each has more than one-third the farm workers of Europe. Western Europe, the most productive region in terms of output, has less than one-fourth all farm workers.

3. In terms of horses, used primarily as work animals, Southern Europe has almost two-fifths the European total. Eastern Europe has almost as many, but Western Europe has only slightly more than one-fifth.

Questions

1. How do you interpret these results?
2. Compare your findings with the results of your comparison of the South, the Middle West, and the Great Plains in the United States.

3. Do these data provide a measure of variations in the modern interconnected system of Europe?

farm workers than of total productive agricultural land and arable land. Combined, these two areas have almost three out of four of all farm workers in Europe, yet they total only about half of Europe's population. Western Europe, with 48 percent of total population, has only 23 percent of farm workers. Thus the distribution of farm population of Europe is quite different from that of total population, just as it is in the United States. In most European countries, however, the rural component is much more important than in the United States.

Further study of Table 8–4 reveals that neither Southern nor Eastern Europe is the most productive farmland of the continent. Western Europe is the leading agricultural area in total output of major crops including cereals. Some inferences about agricultural methods can also be drawn from the table. The higher agricultural productivity of Western Europe is accomplished with far fewer farm workers and fewer horses. The low farm population indicates that Western Europeans have moved to city jobs, leaving a modern, productive agricultural system. The small population of horses supports this view, as we can assume that horses have been replaced by machines.

In Southern and Eastern Europe, even now, a high dependence on traditional farm methods continues with considerable reliance on human and animal labor. This is true even in Eastern Europe, where after World War II the properties of large landowners were confiscated to form collective farms after the Soviet model and where former peasants are now part of work teams on large-scale farms. In Greece, Spain, Portugal, and Ireland traditional peasant agriculture still dominates.

The Urban Component

In agricultural Europe as early as the eras of the ancient Greeks and Romans, there were already long-range interconnections across Europe and by sea. Cities, though small by present standards, emerged as the major centers in this developing network. Nobles wanted luxuries from other lands, the Church needed to communicate its decrees, and rulers wished to keep up to date on political happenings. These cities served as the intermediaries between the long-distance, inter-

TABLE 8-5 LARGE METROPOLITAN CENTERS OF EUROPE
(*Population in Thousands*)

	1800[a]	1900	1974		1800	1900	1974
Western Europe				*Southern Europe*			
London	959	4,537	10,675	Milan	170	539	3,625
Paris	547	2,774	9,000	Madrid	160	540	3,580
Essen-Dortmund		252	5,425	Barcelona	115	533	3,165
Berlin	172	1,889	3,900[b]	Rome	153	423	3,025
Manchester		544	2,880	Athens			2,540
Birmingham		522	2,635	Naples	350	621	1,935
Hamburg	130	706	2,300	Lisbon	180	356	1,640
Brussels		599	2,000	Turin		330	1,625
Glasgow		762	1,970	Genoa	100	378	
Vienna	247	1,675	1,940				
Amsterdam	201	511	1,810				
Munich		500	1,765				
Cologne		373	1,715				
Stuttgart			1,650	*Eastern Europe*			
Frankfurt		289	1,615	Budapest		732	2,325
Liverpool		685	1,600	Katowice			2,070
Leeds			1,550	Warsaw	100	638	1,775
Copenhagen	101	401	1,460	Bucharest		276	1,575
Newcastle-upon-Tyne			1,360	Prague		202	1,175
Mannheim			1,265	East Berlin	—	—	1,085
Rotterdam		319	1,125				
Lyon	110		1,100				
Dusseldorf	—	—	1,080				
Antwerp			1,050	*Northern Europe*			
Marseille	111	459	1,015	Stockholm		301	1,352

SOURCES: 1800 and 1900: W. S. Woytinsky and C. S. Woytinsky, *World Population and Production: Trends and Outlook.* © 1953 by the Twentieth Century Fund, New York, p. 120, Table 55. 1974: Edward Espenshade and Joel Morrison, eds., *Goode's World Atlas,* 14th ed., Rand McNally, Chicago, 1974.
[a] Only cities over 100,000 are listed for 1800 and 1900, and cities over 1 million for 1974.
[b] Includes East Berlin. The population of West Berlin alone is 2,084,000.

regional connections, on the one hand, and the rural towns and villages of their own region, on the other.

Since the people, goods, and ideas moving in interregional trade were largely limited to those involving the wealthy nobility and religious orders, and since the local area served by any center was small, few large cities developed. Even as late as 1800, only seventeen cities in all of Europe had 100,000 people, and they accounted for only 3 percent of Europe's population (Table 8-5). Of these, ten were political capitals, five were ports, and the other two, Milan and Lyon, were ancient trade centers. The locations of these cities of less than 200 years ago show their ties to the traditional centers of cultural leadership. Only four countries had more than one large city: France had three, and Spain two; Italy, which had not

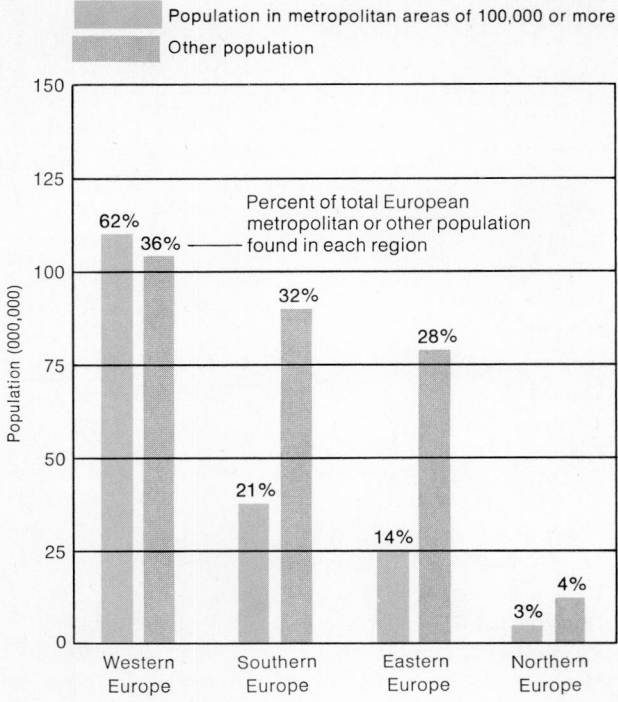

Population in metropolitan areas of 100,000 or more

Other population

Percent of total European metropolitan or other population found in each region

SOURCE: U.N. Demographic Yearbook

FIGURE 8-6 METROPOLITAN AND NONMET-ROPOLITAN POPULATIONS OF REGIONS OF EUROPE, 1970

yet been united as a single country, had four; and Germany, also still divided into local provinces, had two.

Since the change in European life that resulted from the Industrial Revolution, metropolitan centers have risen to dominate much of the continent. Now more than eighty metropolitan centers in Europe have 200,000 people, and they total more than 35 percent of Europe's population. Forty metropolitan areas in Europe have at least 1 million people, and greater London has more than 12 million.

Rural Europe has not disappeared, but metropolitan Europe has risen with a considerably different population distribution. As Figure 8-6 shows, nearly two-thirds (62 percent) of the population in large metropolitan areas is found in Western Europe.

Just as the Northeast and Middle West contain the largest metropolitan populations of the United States and constitute the core region of the country, so Western Europe is the core region of Europe. This is not surprising, since here industrialization had its beginning and the countries located here are still the industrial leaders.

INDUSTRIALIZATION AND ITS CONSEQUENCES

The Industrial Revolution completely changed human capacity for application of energy, and hence for increasing productivity. First people developed water-powered factories; then they learned to utilize fossil fuels for factories and transportation. There followed the railroad, the steam-powered ship, electricity, mass-produced steel, the radio, the internal-combustion engine, the use of petroleum, and myriad other engineering developments. At the same time a system of banking, business management, and trading functions was developing to take advantage of technological change. The result was a revolution in the life of Western Europe, and the first modern models of the long-range, interconnected system emerged in England, France, Germany, Belgium, the Netherlands, and Switzerland.

INDUSTRIAL METROPOLITAN CENTERS

The pattern of industrialization in Western Europe grew out of the established commercial system; the political capitals of the preindustrial era (London, Paris, Brussels) were the logical bases for the new industrial-business world—the centers of decision making and management, both governmental and private. Moreover, they were major markets in themselves and focal points for

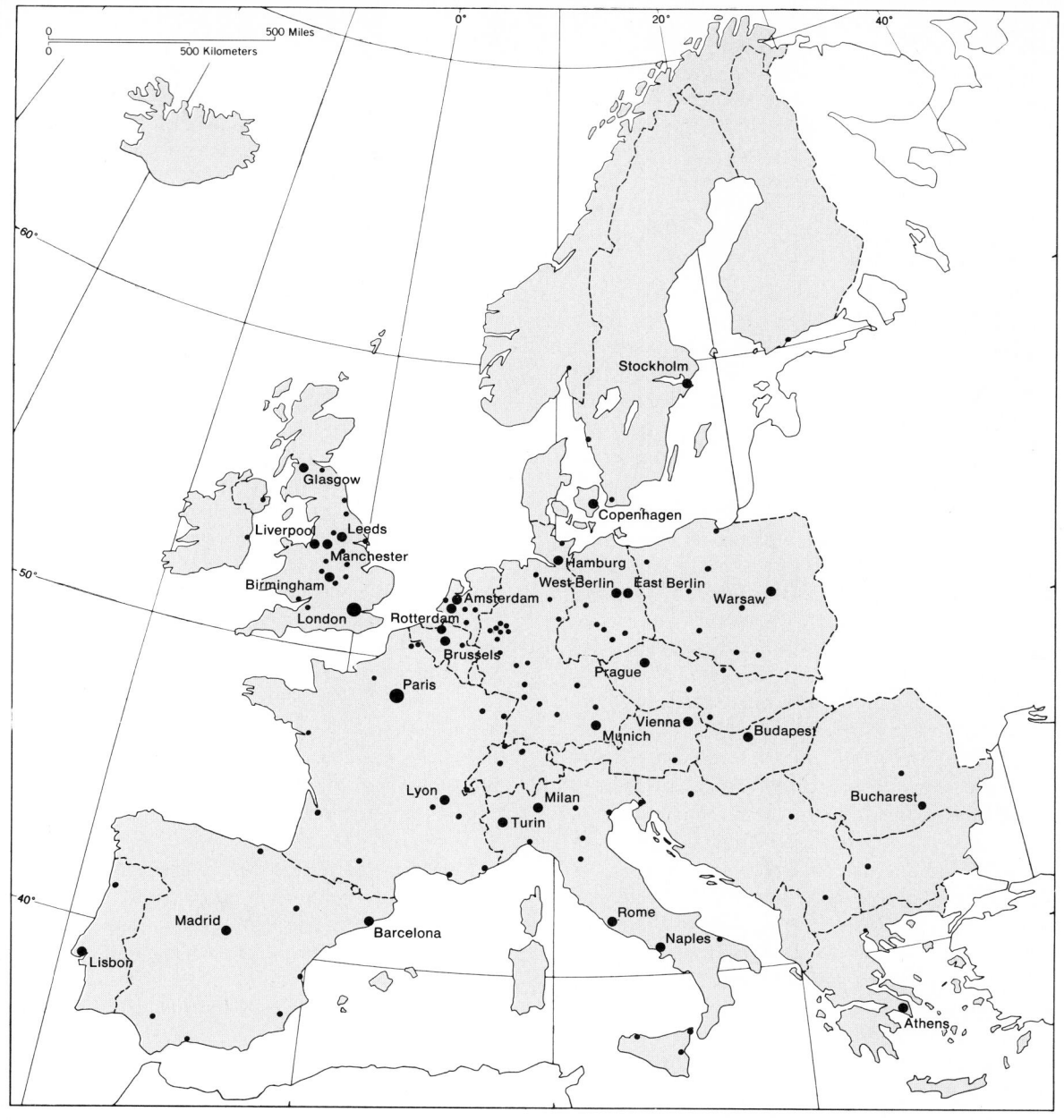

500 Miles
500 Kilometers

Stockholm

Glasgow

Copenhagen

Liverpool Leeds
Manchester   Hamburg
                 West Berlin   East Berlin
Birmingham   Warsaw
London   Amsterdam
Rotterdam
Brussels
                       Prague
Paris
                              Vienna   Budapest
                    Munich
Lyon
        Milan   Bucharest
      Turin

Madrid   Rome
Barcelona   Naples

Lisbon

                                                 Athens

• 250,000 to 1,000,000 population

● 1,000,000 to 4,000,000 population

● More than 4,000,000 population

FIGURE 8–7 EUROPEAN CITIES WITH 250,000
OR MORE PEOPLE

transportation and communication. The same was true of the ports of the earlier period (Amsterdam, Antwerp, Hamburg, and London), which had not only physical facilities for shipping but also management and financial expertise.

The development of modern industry, commerce, and management changed these old cities into very different places. They have grown many times larger in size and have also become centers of a much bigger and more complex system. As a result of this growth into large urban agglomerations, they have provided business with the same economies of agglomeration that we observed in the large U.S. metropolitan centers (Chapter 6). But there were great difficulties in adapting the new, larger urban center to the needs of the more complex modern system. If cities in the eastern United States have had problems fitting in railroads, highways, and new public utilities, imagine the difficulties of European cities dating back to medieval times. Here buildings built 500 or more years ago now house industrial and commercial activities or serve as homes for city families. Streets built for use by pedestrians and horsecarts must carry automobile, bus, and truck traffic.

The old network of capitals, trade centers, and ports did not, however, prove sufficient to meet the needs of the modern large-scale industrial system. New industry used great quantities of power and minerals; therefore industrial cities developed at waterpower sites, in the coal and iron fields, or at transport sites nearby. Today the United Kingdom has six metropolitan centers of more than 1 million people; West Germany has three; Italy has four; and even tiny Belgium and the Netherlands have two (see Figure 8-7).

Traditional trading centers and ports are among the large Western European metropolitan areas today. Cologne, Frankfurt am Main, and Mannheim were trading centers in the older system. Hamburg, Liverpool, Glasgow, Newcastle-upon-Tyne, Rotterdam, and Antwerp were preindustrial ports. But the importance of new industrial sites can be seen in the Ruhr metro-politan agglomeration (made up of the converging cities of Essen, Dortmund, and Duisburg in West Germany), which is the third largest metropolis in Europe, and in Manchester and Birmingham, which rank among the seven largest metropolitan areas. Like Detroit or Pittsburgh, they are highly specialized industrial nodes where modern industry linked to initial heavy industries has contributed to expansion. More important, they have developed as management and financial centers serving the interests of their particular manufacturing operations.

The importance of early resource sites, developed when the rules of the economy game were very different, can be seen in the location of these centers. Manchester and Leeds-Bradford, at the sites of waterpower, were centers of the early textile industry that opened the industrial era. The Ruhr complex and Manchester grew up in coalfields, and Birmingham was adjacent to iron deposits as well as to coal.

Western Europe has emerged as a largely metropolitan-centered society, whereas the other regions of the continent are still largely rural (Figure 8-8). The United Kingdom has by far the largest percentage of its population (71 percent) in metropolitan centers, but most other Western European countries have over one-fourth their populations in big cities. Greece is the only European country outside of Western Europe with such a large metropolitan component. In southeastern Europe, Rumania, Bulgaria, Yugoslavia, and Albania still have less than one-tenth of their populations in major metropolitan centers.

The concentration of employment in Western Europe in urban-type activities such as manufacturing, construction, trade, transportation, and services can be seen in Figure 8-9. Most Western European countries have over three-fourths of their employment in urban-type activities. Only Sweden and Norway in Northern Europe and Italy in Southern Europe show similar concentrations. Eastern and Southern European countries in most cases have less than half their working populations in urban-type employment. Western Europe, other than Ireland, is today pri-

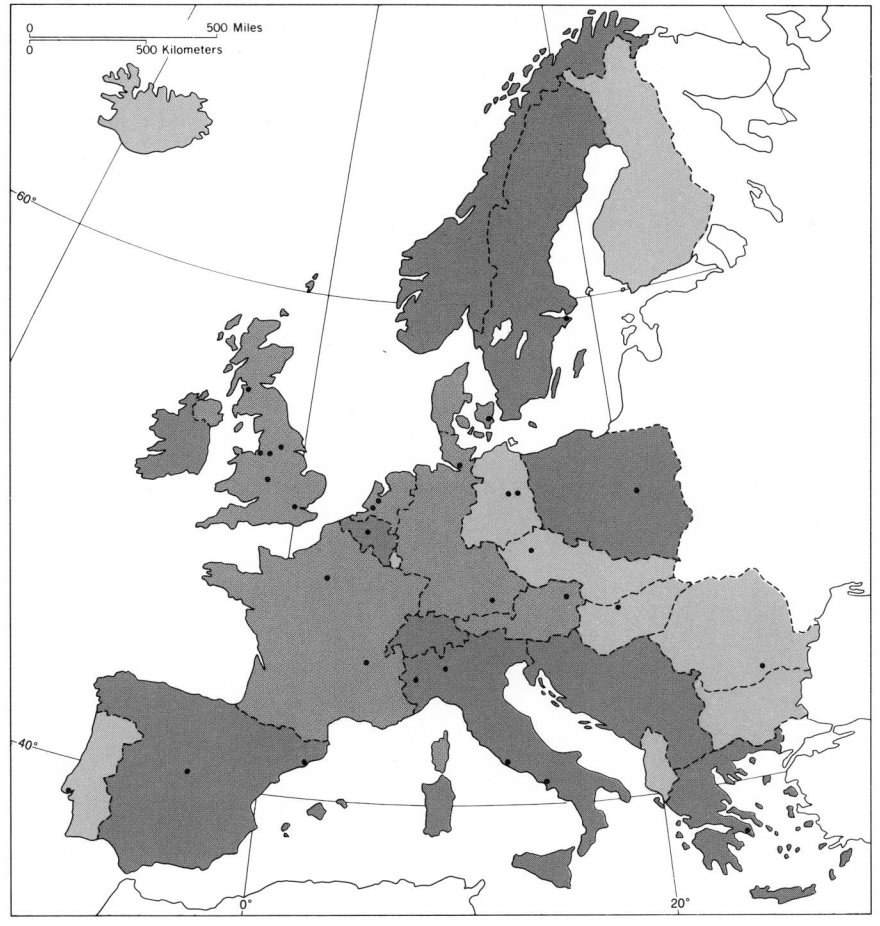

Less than 25%

25-35%

Over 35%

• Cities over 1 million

SOURCE: Kingsley Davis, *World Urbanization 1950–1970*, Vol. I

FIGURE 8–8 PERCENTAGES OF EUROPEAN POPULATION LIVING IN METROPOLITAN AREAS OVER 100,000, BY COUNTRY

marily an area of industry and business centered in large metropolitan areas. The same is true of Sweden, Norway, and Italy. In these countries, as in the United States, much of the population is engaged in the myriad specialized kinds of labor required by the modern system.

Figures 8–10 and 8–11 show per capita GNP and consumption of industrial energy. The contrast between affluent, energy-consuming Western and Northern Europe, on the one hand, and Eastern and Southern Europe, on the other, is striking. East Germany and Czechoslovakia fit

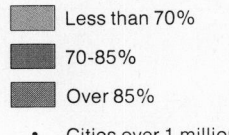

Less than 70%

70-85%

Over 85%

• Cities over 1 million

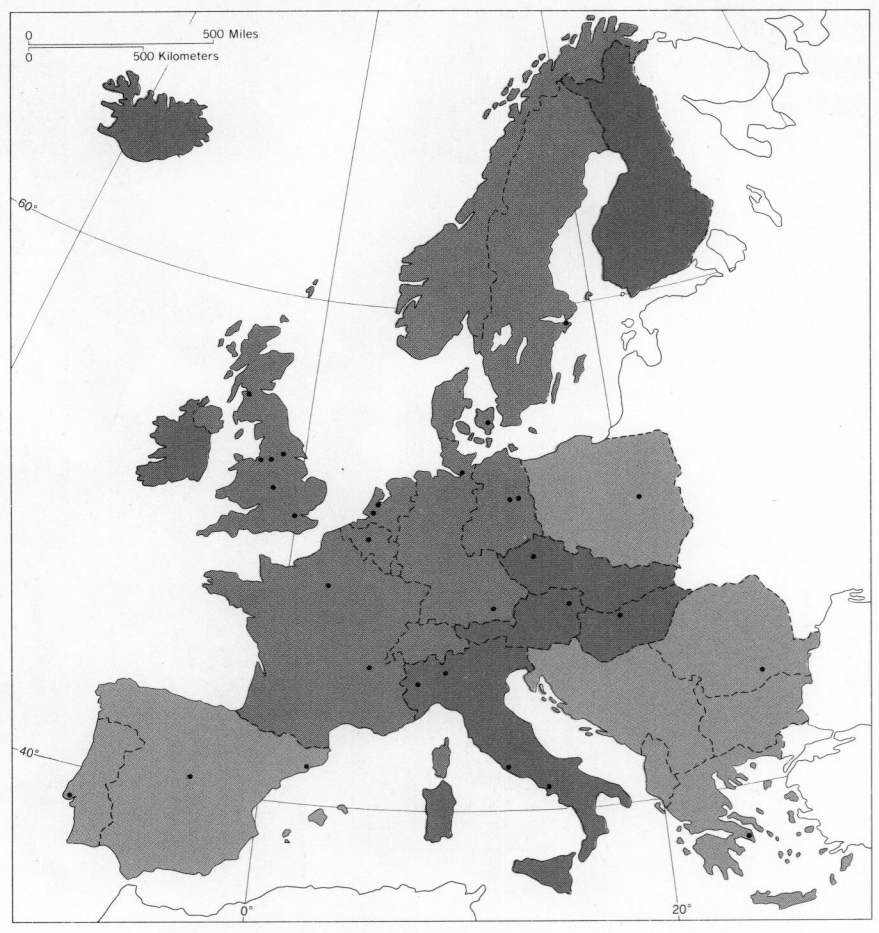

SOURCE: FAO Production Yearbook 1973

FIGURE 8–9 PERCENTAGES OF ECONOMI-
CALLY ACTIVE EUROPEAN POPULATION EM-
PLOYED IN NONAGRICULTURAL (URBAN-
CENTERED) ACTIVITIES

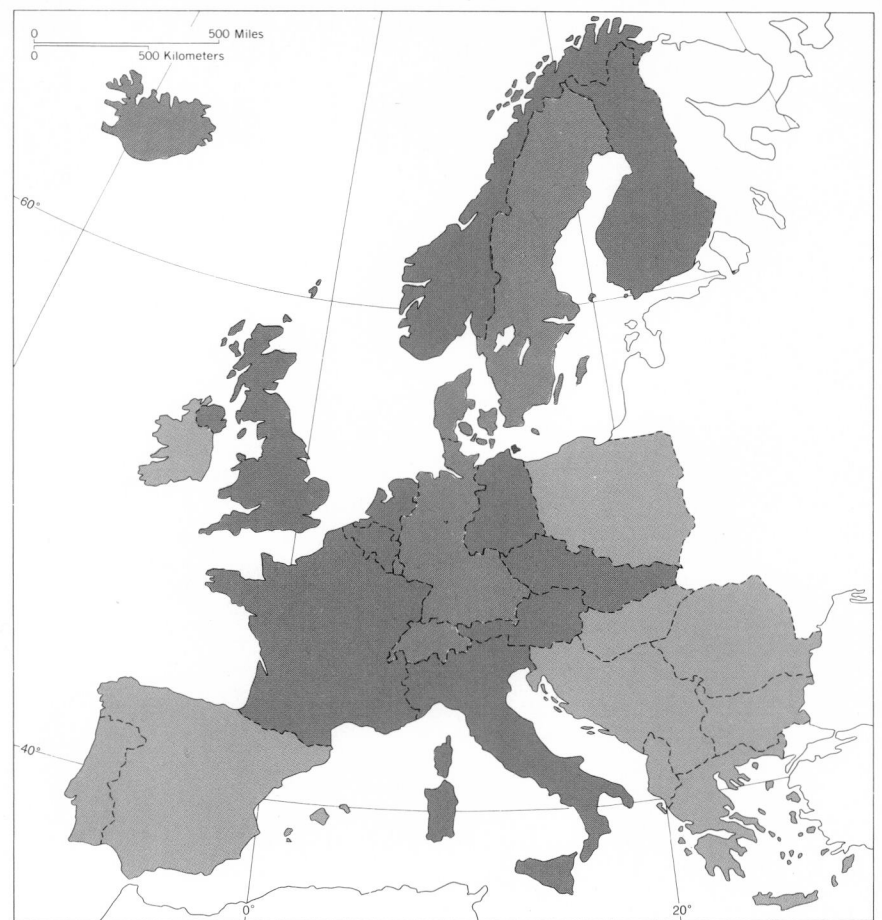

Less than $2,720 per person

$2,720-$5,440

More than $5,440

500 Miles

500 Kilometers

60°

40°

0°

20°

SOURCE: World Population Data Sheet 1976

FIGURE 8-10   PER CAPITA GROSS NATIONAL
PRODUCT, BY COUNTRY, FOR EUROPE IN
1971–1972

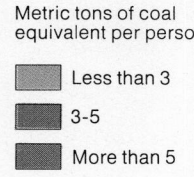

Metric tons of coal
equivalent per person

Less than 3

3-5

More than 5

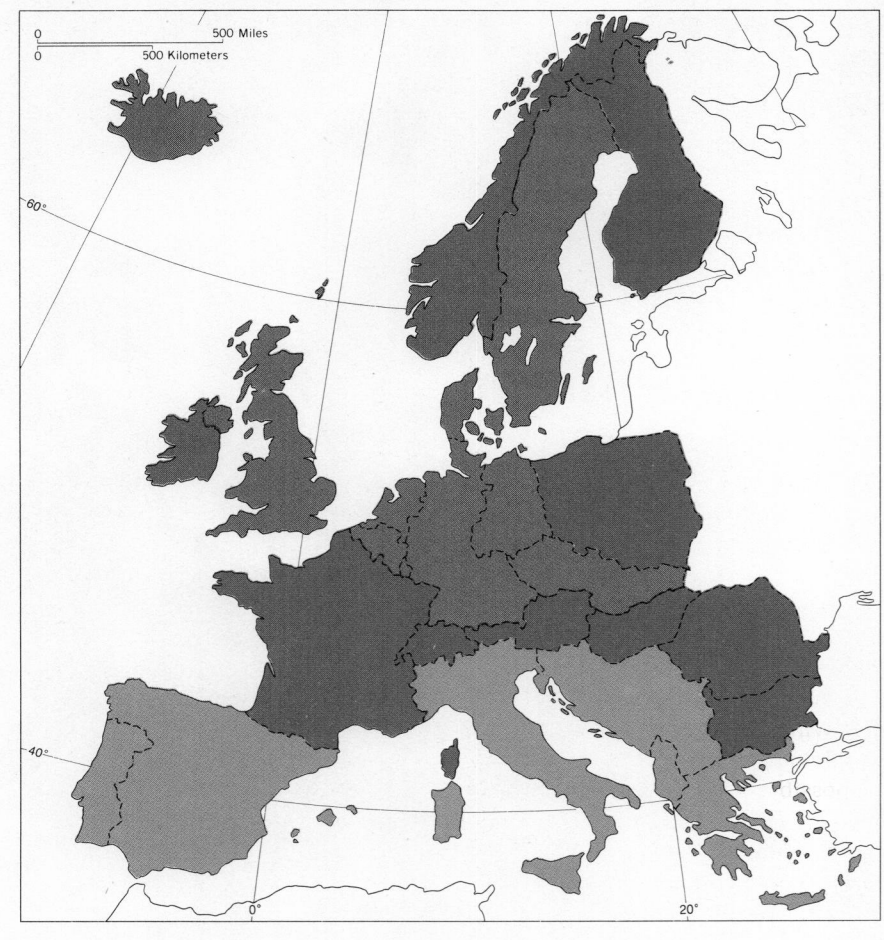

SOURCE: U.N. Statistical Yearbook 1974

FIGURE 8–11 PER CAPITA INDUSTRIAL EN-
ERGY CONSUMPTION FOR EUROPE

the affluent, energy-consuming mold as well, but most of the countries of Southern and Eastern Europe show much less intensive industrialization. Ireland is the only country in Western Europe that is not highly industrialized. Note that we are not thinking in world terms now: all countries in Europe except, perhaps, Albania would rank above the world average. The point is that most of those in Southern and Eastern

Europe are much less developed than their European neighbors to the west and north.

THE IMPORTANCE OF POLITICAL CAPITALS

A look at a map showing major metropolitan centers (Figure 8–7) also points up a difference between the more and less industrialized countries of Europe. Notice that in the still largely agricultural countries, virtually the only major metropolitan centers are the political capitals. Excepting Italy, in Eastern and Southern Europe, only the industrial city complex of Katowice-Zabrze-Bytom in Poland and Barcelona in Spain are noncapitals that have reached the level of 1 million people. In the urban, industrial countries, such as the United Kingdom and West Germany, however, there are a half-dozen or more such cities besides the national capitals. Even there, except in West Germany and the Netherlands, the capital is also the largest metropolitan area in the country.

One would expect capital cities in even largely agricultural countries to be major metropolitan centers. Government, after all, is a metropolitan function. A capital must be at the focal point of transportation and communication if it is to carry out its role of decision maker for the country. It is the country's point of contact with the rest of the world and with all places within its boundaries. It is also likely to be the place in the country that will have international airline connections and domestic air and overland transport links to all possible parts of its territory. It will be the heart of most communications with the rest of the world and the central node of its own mail and other national communication links. A country may have only a few road and railroad connections across its boundaries, but its internal network of roads and railroads will probably focus on the capital. As a result it is often necessary to travel via the capital even when going from one smaller city to another.

The capital of an agriculture-oriented country is usually more than a political center; it is often the cultural center that tends to dominate smaller regional fragments of the country; and, as the largest center, it is the most important marketplace for goods produced within the country or outside. Further, it is usually the intellectual hub and focus of the most advanced technical thought in the country. As the largest labor source and marketplace, it is generally the logical point at which to carry on whatever manufacturing might

occur other than special-purpose manufacturing tied to particular resources.

Such an exceedingly important role for the capital has been traditional in European history since the rise of the national state, and we shall see it emerging now in most of the developing countries of the world. In only a few countries, such as West Germany, the Netherlands, and Switzerland, is the capital not the major city.

Europe has a more intricate pattern of connectivity than does the United States. In the United States major decision making takes place in a few of the largest metropolitan areas, especially New York, Chicago, and Los Angeles, with Washington serving as a specialized decision-making center for the federal government. Europe viewed as a whole has a larger number of nodal centers, most of them capital cities. The reason for the greater complexity of spatial interaction within Europe is the fact that Europe has not been integrated into a single interconnected system. A few of the very largest cities, such as Paris and London, function in some ways as nodal centers for all of Europe, but most large metropolitan areas function primarily as nodes for interaction within their own countries and with adjacent countries.

DEVELOPMENT OF EXTRA-EUROPEAN TRADE

Industrialization made it possible for individual countries of Europe to organize an interconnected system on a worldwide resource base, with the capital city as the decision-making center of the system. England did this more extensively than any other European country. London was the focal point of the vast network of sea lanes that reached to Canada, Australia and New Zealand, India, and to colonies in Africa, Latin America, the Middle East, and the Pacific. Ports such as Liverpool, Glasgow, and Manchester (by means of a canal) were busy links in the system. Yet London was without doubt the focal point because the British Empire was a vast political system, and basic management decisions were made there. The empire was also the core of a great economic system through which England tapped the resources of regions throughout the world. In the same way France developed its own empire with colonies in Africa, Latin America, and Asia centered on Paris. By World War I Germany, Italy, the Netherlands, and Belgium had smaller worldwide empires.

The British Example

After its industrialization the United Kingdom increasingly became an area of urban-centered activities. Like the United States, less than 10 percent of the population of the United Kingdom is engaged in primary production. But this is less the result of mechanization of primary production than a shift away from dependence on the resources of its own country to dependence on the rest of the world. In a sense, the United Kingdom functions in much the same way as the megalopolis of the northeastern seaboard of the United States. It concentrates on business, trade, and industry and draws its basic food, raw materials, and specialty manufactures from other areas. In the case of the United Kingdom, the basic producing areas lie outside the country and to a large degree outside Europe. The British long depended on the resources of their empire and on those of other areas with which they developed ties. Although the empire is gone and the British Commonwealth, which replaced it, is a weak political organization, some aspects of the economic system developed during empire days still remain.

Just as Megalopolis in the eastern United States depends on the Middle West and Great Plains for grain and meat, so the United Kingdom utilizes the midlatitude farming regions of Canada, Australia, New Zealand, Argentina, and Uruguay. As Megalopolis gets cotton, citrus fruits, wines, and tobacco from the South and California, the United Kingdom draws on production from subtropical areas throughout the world, including those of the United States. Megalopolis is tied to oil fields in Texas and the Gulf Coast; the United Kingdom is tied to those of the Middle East,

though it is developing its own in the North Sea. Just as industrial Megalopolis ships goods throughout the United States, British motor vehicles, machinery, chemicals, and textiles are shipped overseas to the specialized primary producing areas that do not have adequate manufacturing to meet their own needs.

Since empire days the whole system of worldwide economic outreach has been managed from the United Kingdom, essentially from London. The British not only built the ships but owned and operated them, financed the operation through their banks and insurance companies, controlled trade through their great commercial organizations, and provided supporting legal, accounting, and other services. London, as the management and decision-making center, was the business capital of the international, interconnected system until the rise of the large international corporations headquartered in New York.

Today the empire is gone, as are government revenues from former colonies. Moreover, British manufacturing at home has lagged behind U.S. mass-production operations and, because of the late entry of the United Kingdom into the European Economic Community, behind those of other countries in Western Europe as well. London, however, is still the headquarters of large multinational corporations and an important center of banking and finance.

The Importance of Extra-European Trade

The model of worldwide dependence on resources and markets overseas also fits the other industrialized countries of Europe, although on a smaller scale. France, Germany, Italy, Belgium, and the Netherlands had overseas empires and trading ties. Today, despite the loss of political empire, these worldwide relations continue.

Some measure of the importance to European countries of extracontinental trade can be seen in Table 8-6 in per capita data for some European countries. As a point of reference, U.S. overseas trade (total foreign trade except with Canada)

TABLE 8-6 EXTRA-EUROPEAN TRADE
ON A PER CAPITA BASIS, 1973

Country	Per Capita Value of Imports and Exports
Netherlands	$888
Belgium	878
Switzerland	832
Norway	717
United Kingdom	635
West Germany	627
Sweden	627
Denmark	588
United States (Foreign trade except with Canada)	510
Finland	473
France	433
Italy	350
Austria	257
Portugal	205
Spain	190
Greece	184
Yugoslavia	126
Ireland	96

SOURCE: U.N. Statistical Office, *World Trade Annual 1973*, Walker, New York, 1975.

totaled $510 per person in 1973. From this perspective, most major European countries depend on overseas resources and markets more than the United States does. Within Europe the range is very great, however. Western and Northern European countries except Ireland, Austria, France, and Finland depend on overseas trade more than the United States does. The small countries of Belgium, Switzerland, and the Netherlands find overseas trade most important, with per capita figures more than half again the U.S. figure. Overseas trade is also more important to West Germany and the United Kingdom, two of

Europe's most populous and industrial countries, than to the United States.

Table 8-7 shows the gross pattern of overseas trade for European countries outside the Eastern European bloc. The great range in size of total overseas trade is striking. Notice that the United Kingdom and West Germany have by far the largest value of trade outside of Europe. Their trade is much greater than France's or Italy's, even though both France and Italy are large industrial countries of similar populations. Regardless of the size of the trade, the pattern is remarkably similar. The leading trading partner for all countries except France, Italy, Finland, Greece, and Portugal is not the underdeveloped areas supplying raw material; rather it is industrialized Anglo-America. What must this mean in terms of the kinds of goods moving in trade? (Remember that Table 8-6 is based on dollar value, not tonnage.)

Notice also that trade with the Middle East and North Africa is usually the greatest of any connection with nonindustrialized areas. This indicates the importance of petroleum shipments from that part of the world to European countries, which have few productive oil and natural gas deposits within their boundaries.

The isolation of the large Communist countries, China and the Soviet Union, is evident. China, with the largest population of any country, is insignificant in the total trade of all non-Communist European countries. The Soviet Union, which in area and population is the largest industrialized country in the world and a close neighbor to European countries, accounts for less than 10 percent of the trade of any of the countries in Table 8-7 except for Finland and Austria. Europe's trade with the United States and Canada is many times greater than that with the Soviet Union. Notice, also, that Europe's trade with industrial Japan is generally greater than with the Soviet Union.

Study the **balance of trade** in Table 8-6, the ratio of exports to imports. Virtually all European countries have a **trade deficit** in overseas

TABLE 8-7 OVERSEAS TRADE CONNECTIONS OF SELECTED EUROPEAN COUNTRIES, BY VALUE, 1973

Country	Africa South of the Sahara (Percent)	South and East Asia (Percent)	Middle East–North Africa (Percent)	Oceania (Percent)	Latin America (Percent)
United Kingdom	13	11	16	9	9
West Germany	10	9	18	3	10
France	16	5	28	4	9
Italy	11	4	28	3	11
Netherlands	11	8	26	2	17
Belgium	16	7	21	3	10
Switzerland	9	7	10	2	9
Spain	9	3	24	2	19
Sweden	9	8	12	4	16
Denmark	10	9	16	2	11
Norway	16	4	11	2	13
Finland	4	4	10	1	10
Austria	10	7	22	3	9
Portugal	35	3	10	2	11
Greece	7	2	31	7	6
Ireland	6	5	10	4	9

SOURCE: U.N. Statistical Office, *World Trade Annual 1974*, Walker, New York, 1974.

OVERSEAS TRADE OF SELECTED EUROPEAN COUNTRIES

1. Notice the importance of the four largest countries in Europe's overseas trade. What share of total trade shown in the table is accounted for by the United Kingdom, West Germany, France, and Italy? These countries together with the Netherlands, Belgium, Luxembourg, Denmark, and Ireland make up the European Economic Community (EEC). How important is the EEC in the overseas trade of Europe outside the Communist bloc?

2. Compare the importance of the trade with industrialized regions (Anglo-America, Japan, and the Soviet Union) with that of the developing world (Black Africa, South and East Asia, North Africa–Middle East, and Latin America). What conclusions do you draw?

3. What does the table indicate about the cost of oil to the industrialized countries of Europe and the profits gained by the oil-producing countries of the Middle East–North Africa?

4. South and East Asia have over half the population of the world. How does Europe's trade with this area compare with that with other developing regions?

5. China and the Soviet Union are among the world's largest countries. Moreover, the Soviet Union is a major industrial power and China has been making rapid progress. How do you explain the small amount of trade they have with Northern, Western, and Southern Europe?

China (Percent)	Japan (Percent)	Anglo-America (Percent)	USSR (Percent)	Imports ($ Billions)	Exports ($ Billions)	Total Trade ($ Billions)
0.7	6	30	2	20.9	14.6	35.5
1	6	34	5	19.3	19.6	38.9
1	4	23	4	12.7	10.4	23.1
1	3	26	4	10.4	8.8	19.2
1	4	27	2	8.2	3.8	12.0
1	5	32	4	5.0	3.9	8.9
1	8	20	2	5.2	3.2	8.4
0.4	5	36	2	4.6	2.0	6.6
2	8	33	6	2.3	2.8	5.1
1	10	36	4	1.7	1.3	3.0
0.2	19	31	2	1.7	1.1	2.8
1	7	19	43	1.2	1.0	2.2
2	9	27	12	1.0	.9	1.9
1	9	28	—	1.1	.6	1.7
1	16	26	5	1.3	.4	1.7
3	5	40	1	0.7	0.4	1.1

trade: their imports exceed their exports in value. Only West Germany and Sweden show a larger value of exports than imports, and all the largest trading countries buy significantly more than they sell overseas. This seems to contradict the usual view that advanced countries such as the industrialized European nations have a trading advantage in buying low-value raw materials and foods, and in selling high-value manufactures. What reasons might there be for the unfavorable balance? How is this imbalance resolved?

Data on the trade of individual Eastern European countries are not readily available. However, U.N. figures for the whole Eastern European bloc indicate that less than 10 percent of its 1973 exports went to the world outside Europe and the Soviet Union. More than any other part of Europe, Eastern Europe is tied to its neighbors in Europe rather than to the world.

INTRA-EUROPEAN TIES

We have made the point that the twenty-six European countries function politically, culturally, and economically as separate entities in contrast to the fifty U.S. states, which operate as a single functioning whole. This is obvious in the political and also, to a high degree, cultural realms. But perhaps we have overdrawn the economic situation by stressing overseas ties rather than intra-European economic links. The economic situation is somewhat similar, as we have seen in Table 8–7, yet overseas trade presents only part of

TABLE 8-8 FOREIGN TRADE OF EUROPEAN COUNTRIES, 1973

	Intra-European		Extra-European		
	Total ($ billion)	Percent	Total ($ billion)	Percent	Total ($ billion)
West Germany	83.1	68	38.9	32	121.9
France	49.5	68	23.2	32	72.7
United Kingdom	33.9	49	35.5	51	69.4
Italy	30.8	62	19.2	38	50.0
Netherlands	35.9	75	11.9	25	47.8
Belgium-Luxembourg	35.4	80	8.9	20	44.3
Sweden	17.7	78	5.1	22	22.8
Switzerland	15.7	75	5.4	25	21.1
Spain	8.1	55	6.6	45	14.7
Denmark	10.5	78	3.0	22	13.5
Austria	10.4	84	2.0	16	12.4
Norway	8.1	74	2.8	26	10.9
Finland	5.7	72	2.2	28	7.9
Yugoslavia	5.2	66	2.6	34	7.8
Portugal	3.2	64	1.7	36	4.9
Greece	3.3	67	1.6	33	4.9
Ireland	4.0	80	0.9	20	4.9

SOURCE: U.N. Statistical Office, *World Trade Annual 1973*, Walker, New York, 1975.

the economic picture and, taken by itself, over-emphasizes the independent functioning of European countries. We must now adjust our assessment of Europe's economic functioning by considering trade between European countries.

INTRA-EUROPEAN TRADE

Intra-European trade is more valuable for most European countries than is overseas trade (Table 8-8); the United Kingdom is the single exception. In terms of relative importance, the small, advanced countries of Western and Northern Europe depend most on intra-European trade. What are the ratios of intra-European to extra-European trade for small industrial countries?

All countries of Europe have extra-European connections, but the large European industrial countries that had nineteenth-century empires have the greatest volume of intercontinental ties today. The smaller countries mostly feed goods into the interregional trade of Europe and draw other products in return. This is particularly true of Austria, Switzerland, Norway, Sweden, Finland, and Ireland. Like the outlying regions of the United States, they function as **complementary regions** to the major industrial countries: they contribute to the functioning of the major industrial countries in addition to being a market for those countries. In particular, note the importance of the Netherlands and Belgium at the mouth of the Rhine. They provide major ports for West Germany, Switzerland, and Austria.

Trade with	Western Europe (Percent)	Northern Europe (Percent)	Southern Europe (Percent)	Eastern Europe (Percent)	Value of Total Trade in Europe ($ Billion)
EEC countries					
West Germany	67	8	20	5	83.0
France	72	4	21	3	49.5
Netherlands	82	4	9	2	36.8
Belgium and Luxembourg	87	4	8	1	35.4
United Kingdom	66	17	14	3	33.9
Italy	82	3	10	5	30.8
Denmark	57	31	8	3	10.5
Ireland	90	4	5	1	4.0
Other Western European countries					
Switzerland	73	6	17	3	15.7
Austria	70	6	15	9	10.4
Northern European countries					
Sweden	70	19	8	3	17.7
Norway	63	26	7	3	8.1
Finland	62	28	7	3	5.7
Southern European countries					
Spain	77	6	13	3	7.9
Yugoslavia	50	4	22	24	5.2
Greece	68	4	20	8	3.3
Portugal	73	11	15	1	3.2

SOURCE: U.N. Statistical Office, *World Trade Annual 1973,* Walker, New York, 1975.

Their intra-European trade is greater than that of either the United Kingdom or Italy.

Even though Europe has not been as consciously developed as an economic entity as the United States has been, regional trade has been built into the interchange of local specialty production, as Table 8–9 clearly indicates. Northern Europe, particularly Sweden and Finland, provides timber products, especially paper and pulp. The Mediterranean, with its warm, sunny climate, provides agricultural specialties, citrus fruits, wine, and tobacco, mostly for the metropolitan core in Western Europe.

Minerals, too, move from one part of Europe to another. Coal and iron ore have long moved across national boundaries between France, West Germany, Belgium, and the Netherlands. Iron ore moves into the metropolitan core from Sweden by way of a Norwegian port and from northern Spain; coal comes from Poland.

Table 8–9 gives a picture of the internal trade of Europe except within the Communist bloc in Eastern Europe. Notice the dominance of the Western European countries in intra-European trade. Not only do countries in Western Europe—West Germany, the United Kingdom,

France, the Netherlands, Belgium, and Luxembourg—dominate the trade in total value, but Western Europe is the focus of at least 50 percent of the intra-European trade of each country shown in the table. In the case of Ireland, Belgium, the Netherlands, Italy, and Spain, more than three out of every four dollars of intra-European trade is centered in Western Europe.

The countries of the other parts of Europe are "outsiders." Like the peripheral regions of the United States, their significant trade flows are not with one another. The important ties are really each outlying area and the countries in the inner trade flow, between each peripheral region and the core region. The most important international trade flows within Europe are between the major countries of Western Europe. Next in importance are the links between Western Europe and other regions of Europe. Of little importance are the ties between countries on Europe's periphery, although Table 8–9 does show some interaction between countries within Northern Europe and Southern Europe.

Northern Europe, with only 4 percent of total European population, shows up reasonably well in intra-European trade. But Southern Europe, with 28 percent of the population, does not get its share, and Eastern Europe is least involved in trade with the rest of Europe. The Soviet dominance over Eastern Europe since World War II has been economic as well as political. The countries are bound, along with the Soviet Union, into COMECON (Council for Mutual Economic Assistance), an economic and trading organization dominated by the Soviet Union. This has reduced the traditional trade between the countries of Eastern Europe and other parts of Europe, particularly West Germany and Italy.

The Soviet Union is the leading foreign trading partner of all Eastern European countries, but its share of total exports varies greatly from country to country. Over half of Bulgarian trade is with the Soviet Union, but only about a quarter of Rumania's. Most of the rest of the trade of Eastern European countries is among themselves. Several Eastern European countries, however, now have increased trade with Western European countries, particularly West Germany. Rumania and Czechoslovakia both have important trade links with other parts of Europe, but Poland is the most important trading partner of other European countries. Polish coal and food are exported to West Germany, Belgium, the Netherlands, and the United Kingdom. Nevertheless, as Table 8–9 shows, for most Western European countries, trade with Eastern Europe is less than trade with any other region. Only Yugoslavia, Austria, and Greece have significant proportions of their trade with Eastern Europe.

West Germany, the most populous and industrialized country on the continent, appears most important within the core region. But perhaps the most surprising fact is the importance of the Netherlands-Belgium-Luxembourg cluster. These three small countries seem to be almost as significant in the pattern of trade flows as France and the United Kingdom.

THE EUROPEAN PATTERN OF
SPATIAL INTERACTION

We can now begin to see that both the economic and political facets of the pattern of spatial interaction in Europe are quite different from those of the United States. Three distinct types of interaction are all occurring simultaneously.

1. The first order of interaction is within each of the individual countries of the continent. As political entities they make a great effort to educate their populations on the importance of being French, German, Swedish, etc. The national government reinforces the local cultural heritage and determines the degree of interaction with other countries. Most European countries have also worked to integrate their countries economically. The capital, usually the largest city, is both the major market for production from outlying parts of the country and the center of the industrial-business economy. The first-order importance of national integrity can be seen in each nation's transportation and communication networks, which provide a channel linking the capital with other parts of the country.

2. The large Western European countries that formerly had empires still depend on overseas connections more heavily than the United States does. Their most important overseas ties are not, however, with underdeveloped nations as sources of raw materials and tropical products. Trade with those nations continues, but now the United States is the center of economic action both as supplier of advanced manufactures and as a market for European products. Furthermore, the migration of Europeans to the United States has created social and cultural ties, and because both areas are affluent they share an interest in similar activities.

3. With improved transportation and communication and with greater affluence, total intra-European interaction is more significant than overseas ties. As we have seen, European trade centers on the major countries of Western Europe, particularly West Germany, France, Belgium, and the Netherlands. Only in Eastern Europe is this focus broken, by the influence of the Soviet Union. However, as the Communist governments of Eastern Europe mature, they tend to trade more with the West. As in the United States, first-order interaction, both of products and business communication, occurs between the major industrial metropolitan areas of Western Europe. Second, movement takes place between the centers at the core and the less developed producers of specialized food and raw materials in Eastern and Southern Europe. In recent years there has also been greater movement of workers shifting jobs; large numbers of Italians and Yugoslavs, for instance, have moved to France and Germany. More affluent Europeans are also traveling to other European countries for recreation. Eastern Italian and Yugoslav coastal resorts cater to Germans; the coast of Spain and the island of Majorca attract the British; and Switzerland and the Riviera are the mecca for a full range of Europeans. The Black Sea resorts in Rumania and Bulgaria are gaining in popularity.

The fact that both intra-European and overseas trade are focused on the countries of Western Europe suggests an analogy with the economy of the United States. Just as the Northeast and the Middle West constitute the major U.S. center of manufacturing and business, so does Western Europe within Europe. Table 8–10 shows the overwhelming importance of manufacturing in Western Europe—in West Germany, France, and the United Kingdom, in particular, but also in such small countries as Belgium and Austria. Only Italy in Southern Europe and Sweden in Northern Europe are of major significance in manufacturing. Western Europe is second only to Anglo-America as a manufacturing region, as indicated by total **value added by manufacture,** the difference between what a manufacturing firm receives for a finished product and what it pays for the materials used in the manufacturing process. Moreover, per capita manufacturing production in West Germany and Belgium is greater than in the United States.

In Western Europe manufacturing is centered in the largest metropolitan areas and in surrounding cities. Major centers include the Ruhr cities of West Germany, known for their heavy industry; London, a diversified manufacturing center, and the more specialized manufacturing cities of Birmingham, Manchester, Liverpool, and Glasgow in the United Kingdom; Paris, known for its high-value luxury products; and Brussels-Antwerp and Rotterdam-Amsterdam, both centers of diversified manufacturing. In a lesser way major manufacturing has spread into northern Italy at Milan, Turin, and Genoa, which produce a variety of advanced manufactured products.

Perhaps more important, Western Europe is a major management center, not only for national businesses, but also for multinational corporations. From Table 8–11 you can see that twenty of the world's largest industrial companies have headquarters in Western Europe. Among these are three of the ten largest companies in the world: Royal Dutch Shell, British Petroleum, and Unilever. All have major subsidiaries operating in the United States. Notice from the table that most of the large European multinational corporations have headquarters in West Germany or the United Kingdom, but that France, the Netherlands, Italy, and Switzerland all are represented

TABLE 8-10   MEASURES OF MANUFACTURING, EUROPEAN
COUNTRIES, JAPAN, AND THE UNITED STATES, 1972

	Population (Thousands)	Employment (Thousands)	Value Added by Manufacture[a] ($ Millions)	Per Capita
West Germany	61,674	8,018	105,496	1,709
France	51,700	5,996	67,258	1,301
United Kingdom	55,788	7,685	52,667	944
Italy	51,177	3,322	24,937	487
Sweden	8,122	879	11,503	1,420
Spain	34,494	2,096	9,233	268
Belgium (1971)	4,963	1,236	8,810	1,762
Yugoslavia	20,772	1,458	4,492	216
Denmark	4,992	415	4,351	870
Austria (1971)	7,456	720	4,531	604
Finland	4,634	491	3,463	753
Norway	3,933	360	3,455	886
Portugal	8,590	561	1,682	196
Greece (1971)	8,851	269	1,447	163
Ireland (1971)	2,971	196	1,085	362
Japan (1971)	104,661	11,541	86,010	827
United States (1971)	206,200	17,358	313,300	1,519

SOURCE: *U.N. Statistical Yearbook*, 1974.
[a]Data not available for the Netherlands or Switzerland.

among the headquarters. London stands out among the large metropolitan areas as the headquarters for several large corporations.

The concentration of major world banks in Europe is even more striking. Europe has twenty-three of the fifty largest commercial banks in the world, as compared to only ten in the United States. These include six West German banks; four each in the United Kingdom, France, and Italy; three in the Netherlands; and two in Switzerland. Again, London stands out as the most important center.

MOVES TOWARD ECONOMIC
INTEGRATION WITHIN EUROPE

Political and economic changes need to be added to the story of functioning Europe today. Each year the amount of trade between Eastern European countries and the rest of Europe increases. The amount varies from country to country; it is especially important, however, between West Germany and Austria and their Eastern European neighbors: Poland, East Germany, and Czechoslovakia.

TABLE 8-11 THE FIFTY LARGEST INDUSTRIAL COMPANIES IN THE WORLD, 1976

Company	Headquarters	Products	Sales ($ Millions)
Exxon	New York	Petroleum	44,865
General Motors	Detroit	Automobiles	35,725
Royal Dutch/Shell	London/The Hague	Petroleum	32,105
Texaco	New York	Petroleum	24,507
Ford Motor	Dearborn, Michigan	Automobiles	24,009
Mobil Oil	New York	Petroleum	20,620
National Iranian Oil	Teheran	Petroleum	18,855
British Petroleum	London	Petroleum	17,286
Standard Oil/California	San Francisco	Petroleum	16,822
Unilever	London	Food, toiletries	15,016
International Business Machines	Armonk, New York	Computers	14,437
Gulf Oil	Pittsburgh	Petroleum	14,268
General Electric	Fairfield, Connecticut	Electrical equipment	13,399
Chrysler	Highland Park, Michigan	Automobiles	11,699
International Telephone & Telegraph	New York	Conglomerate	11,368
Philips Gloeilampen- fabrieken	Eindhoven, Neth.	Electronics	10,746
Standard Oil/Indiana	Chicago	Petroleum	9,955
Cie Française des Pétroles	Paris	Petroleum	9,146
Nippon Steel	Tokyo	Iron, steel	8,797
August Thyssen-Hutte	Duisburg, Germany	Iron, steel	8,765
Hoechst	Frankfurt	Chemicals	8,462
ENI	Rome	Petroleum	8,334
Daimler-Benz	Stuttgart	Automobiles	8,194
U.S. Steel	Pittsburgh	Iron, steel	8,167
BASF	Ludwigshafen, Germany	Chemicals, petroleum	8,152
Shell Oil	Houston	Petroleum	8,143
Renault	Boulogne-Billancourt, France	Automobiles	7,831
Siemens	Munich	Electronics	7,760
Volkswagenwerk	Wolfsburg, Germany	Automobiles	7,681
Atlantic Richfield	Los Angeles	Petroleum	7,308
Continental Oil	Stamford, Connecticut	Petroleum	7,254
Bayer	Leverkusen, Germany	Chemicals	7,223
E. I. du Pont de Nemours	Wilmington, Delaware	Chemicals	7,222
Toyota Motor	Toyota City, Japan	Automobiles	7,194
ELF-Aquitaine	Paris	Petroleum	7,165

(continued)

TABLE 8-11   (Cont.)

Company	Headquarters	Products	Sales ($ Millions)
Nestlé	Vevey, Switzerland	Food products	7,080
Imperial Chemicals Ind.	London	Chemicals	6,884
Petrobrás	Rio de Janeiro	Petroleum	6,626
Western Electric	New York	Electronics	6,590
British-American Tobacco	London	Tobacco, paper products	6,146
Proctor & Gamble	Cincinnati	Food, toiletries	6,082
Hitachi	Tokyo	Electrical equipment	5,916
Westinghouse Electric	Pittsburgh	Electrical equipment	5,863
Mitsubishi Heavy Ind.	Tokyo	Machinery, automobiles	5,694
Union Carbide	New York	Chemicals	5,665
Tenneco	Houston	Conglomerate	5,600
Nissan Motor	Tokyo	Automobiles	4,480
Goodyear Tire & Rubber	Akron, Ohio	Tires	4,452
Montedison	Milan	Chemicals	4,418
British Steel	London	Iron, steel	4,340

Totals	Companies	Sales ($ Millions)
United States	23	315,080
Europe	20	194,734
Japan	5	33,081
Other	2	25,481
		568,317

SOURCE: Reprinted from The Fortune Directory; © 1976 Time Inc.

The important flow of trade between West Germany, France, Belgium, the Netherlands, and Italy is partially explained by a remarkable economic enterprise. These countries, together with tiny Luxembourg, were the original members of the European Economic Community.

The European Economic Community

Initiated in 1958, the European Economic Community (EEC), which is often referred to as the Common Market, comprised six nations. It gradually eliminated tariff barriers between these countries and established a single set of duties for certain types of goods entering any of them.

Essentially, the countries remained separate political units, each with its own rules governing the conduct of business and its own currency, but the barriers to the movement of goods from one member country to the others were substantially lessened. Instead of six separate national markets for goods, products were able to flow freely throughout the six countries.

All the charter members were recognized as among the leading examples of industrialization and the modern interconnected system; they included three of the largest countries in Europe: West Germany, France, and Italy. Together the original six EEC countries had an affluent population comparable to that of the United States. The European Economic Community overcame what had been a major handicap to the development of modern mass production: the small size of European national markets.

The EEC developed from an earlier agreement of the original six nations to eliminate barriers to trade in coal and steel, the basic ingredients of modern industrialization. This Coal and Steel Community called for the rationalization of resource production: the closing of inefficient mines in one country and the expansion of efficient ones in another; and the development of steel capacity wherever it seemed best to do so, regardless of country. It also called for the free movement of workers to wherever they were needed. This freedom of movement has now been generally expanded to include all industries and is resulting in international movements of workers in addition to the rural-urban movement common during industrialization.

The United Kingdom and the EEC

Although its leaders were among those pushing for international economic cooperation in Europe after World War II, the United Kingdom turned down the opportunity to join the European Economic Community in its early days. Among other things, it felt it was a world power whose Commonwealth ties were of first importance.

Recall that the United Kingdom is the only European country that depends more on overseas trade than on intra-European transactions. However, the United Kingdom found it increasingly difficult to compete with the new economic scale of the EEC. To strengthen its position, the United Kingdom joined with six other European countries—Norway, Sweden, Denmark, Austria, Switzerland, and Portugal—in a more limited trade oganization aimed primarily at increasing trade among its members by reducing tariffs between them. All but Portugal were advanced industrial-trading countries.

An initial goal for all members of the British-sponsored European Free Trade Association was membership in the EEC, and in 1972 the United Kingdom, Norway, Denmark, and Ireland were invited to join. The United Kingdom, Denmark, and Ireland accepted the invitation, but Norway turned it down. The United Kingdom's joining was significant because it was the most important extra-European trading and business country of the continent. As a result of the new members, the population of the EEC nations surpassed that of the United States and became comparable to that of the Soviet Union. Table 8-12 shows that as an economic entity the EEC compares favorably with the two great powers.

Thus most of industrialized Europe outside the Communist bloc has become a single economic market. All this movement toward a single functioning economy would seem to portend some form of political unity in Europe. Framers of the original Coal and Steel Community had this in mind, and the European Economic Community has a single institutional framework that includes a council of ministers, a parliament, and a court of justice. But national boundaries remain important in Europe, both politically and culturally.

The organizations that have taken greatest advantage of the EEC have not been companies within the EEC countries themselves, but large international corporations, most of them American. At first thought this might seem surprising; however, the long experience of U.S. corporations in centralized decision making and decentralized production makes their involvement understandable. By 1967, $14 billion of U.S. investment had been made in plants and equipment in EEC countries and accounted for one-third of all such overseas investment since 1958. By 1973 the value of U.S. private investment in EEC countries was over $31 billion; this was 29 percent of total U.S. private international investment. All types of major corporations have been involved: automobile manufacturers, food processors, chemical companies, oil companies, and even hotel and motel chains. Particularly important have been industries involving the most advanced technology such as electronic equipment, semiconductors, computers, and integrated circuits. By the mid-1970s, however, rising labor costs, the increasing cost of procuring raw materials, and the inflated currencies of West Ger-

TABLE 8-12 COMPARISONS WITHIN THE EUROPEAN ECONOMIC
COMMUNITY AND WITH THE UNITED STATES, THE SOVIET UNION,
AND JAPAN, 1973

	Population	Arable Land	Wheat	Barley	Oats	Corn	Potatoes	Cattle	Pigs	Horses	Sheep	Tractors
	Millions	Million acres	Millions of metric tons					Million head				Millions
West Germany	62.1	20.0	7.1	6.6	3.0	0.6	13.7	13.9	20.0	0.3	0.9	1.4
France	53.1	45.9	17.8	10.8	2.2	10.6	7.4	22.5	13.4	0.4	10.2	1.5
Italy	56.3	30.2	8.9	0.5	0.4	4.9	2.9	8.7	8.0	0.3	7.8	0.7
United Kingdom	56.1	17.8	5.0	9.0	1.1	—	6.6	14.4	9.0	0.1	27.9	0.5
Netherlands	13.8	2.0	0.7	0.4	0.1	—	5.8	4.7	6.4	0.1	0.7	0.2
Belgium– Luxembourg	10.2	2.2	1.0	0.8	0.3	—	1.5	3.0	4.3	0.1	0.1	0.1
Ireland	3.1	2.5	0.2	0.9	0.2	—	1.3	7.0	1.1	0.1	4.2	0.1
Denmark	5.1	5.7	0.5	5.5	0.5	—	0.7	2.8	8.4	0.05	0.1	0.2
EEC	259.8	100.3	41.2	27.9	7.8	16.1	39.9	77.0	70.6	1.4	51.9	4.7
United States	215.3	472.2	46.6	9.2	9.6	143.3	13.6	121.5	61.0	8.0	17.7	4.4
USSR	257.0	622.9	109.7	55.0	17.5	13.4	121.5	104.0	59.2	8.5	139.1	2.2
Japan	112.3	13.1	0.2	0.2	0.04	0.02	3.3	3.6	7.5	0.08	0.02	0.3

	Electricity	Coal and Lignite Production	Crude Petroleum	Steel	Aluminum	Sulfuric Acid	Cement	Cotton Yarn	Automobiles	Merchant Vessels	Telephones	Radios
	Millions of kilowatt-hours	Millions of metric tons							Millions	Millions of gross registered tons	Millions in use	
West Germany	298,995	221.7	6.6	49.5	0.9	5.1	41.0	0.2	3.6	2.0	17.8	20.6
France	174,080	29.1	1.3	25.3	0.5	4.4	30.7	0.3	3.2	1.1	11.3	17.0
Italy	145,518	1.3	1.0	21.0	0.4	3.0	36.3	0.1	1.8	0.8	12.6	12.4
United Kingdom	282,128	132.2	0.1	26.6	0.5	3.9	20.0	0.1	1.7	1.0	19.1	37.5
Netherlands	52,628	1.7	1.5	5.6	0.2	1.5	4.0	6.04	0.1	0.9	4.3	3.8
Belgium-Luxembourg	43,232	8.8	—	20.8	—	2.6	7.4	0.07	—	0.2	2.6	3.8
Ireland	7,348	0.1	—	0.1	—	—	1.7	—	—	0.03	0.4	0.6
Denmark	18,004	—	0.1	0.4	—	—	2.9	—	—	0.9	2.0	1.6
EEC	1,021,933	394.9	9.6	149.3	2.5	20.5	144.0	0.8	10.4	6.9	70.1	97.3
United States	1,958,741	542.9	454.2	136.8	5.1	28.5	77.5	1.4	9.7	0.9	138.3	386.6
USSR	914,653	614.7	429.0	131.5	1.4	14.8	109.5	1.5	0.9	18.2	14.3	110.3
Japan	470,082	22.5	0.7	119.3	1.6	7.1	78.1	0.5	4.5	15.7	38.7	70.8

SOURCE: *U.N. Statistical Yearbook, 1974.* Population data from Population Reference Bureau, Inc., *World Population Data Sheet, 1976,* Washington, D.C., 1976. Data for France, Italy, the United Kingdom, Belgium and Luxembourg, and Denmark are for 1972; data for Ireland are for 1969.

PRODUCTIVE CAPACITY OF THE EUROPEAN ECONOMIC COMMUNITY

Table 8–12 not only offers the opportunity to compare the productive output of the European Economic Community with that of the other major industrial powers—the United States, the Soviet Union, and Japan—in agriculture, industry, and energy production. It also allows you to make comparisons between the countries of the European Economic Community.

Which is the most important agricultural country? Which is the most important industrial country? Or can such simple distinctions be made? Are there signs of possible trade movement in terms of one country providing an agricultural base for others, or shipping energy or manufactured goods to others? What are the agricultural specialties of particular countries? the industrial specialties? What countries seem to have the greatest domestic fuel deficiencies?

many and France had reduced the attractiveness of the EEC to the multinational corporations. Instead of U.S.-based companies entering the EEC countries, European companies were establishing production units in the United States.

Difficulties Within the EEC

Despite good intentions and the obvious advantages of economic unity, the EEC has found it difficult to overcome the traditional self-interest of its member countries. The individual governments view the proposals and prospects of the EEC from their own perspectives as sovereign states. Each finds it hard to overcome old rivalries, and to disregard opportunities for self-advantage in economic affairs and for gaining political advantage over the other members. Rivalry between France and West Germany for a position of power has characterized the EEC since its early days, and the presence of the United Kingdom in recent years has complicated the situation.

It has been particularly difficult to establish an agricultural policy that is fair to both the farmers of the different countries and to the consumers. Nearly three-quarters of the EEC's 1976 budget was allocated to support for agriculture. The program was designed to provide price supports to farmers and to keep out cheap foreign imports. The result has been that prices of basic food products within the EEC countries remain high. In addition, a program was instituted to compensate farmers in countries whose currencies had increased in value. These farmers' products were more expensive as a result of the increase in the value of the currency, and therefore were not competitive with farmers' products in some other countries. The program did not work satisfactorily, however. In one particular case, France, to protect its wine growers, prohibited the entry of inexpensive Italian wine.

Currency exchange rates have been a problem, too. The members agreed to a "snake" of jointly floating currencies. Under this arrangement, the currency of one country would not be sharply revalued up or down in relation to others; instead, the adjustments would be moderate. By 1976, however, the British pound and the Italian lira had fallen so drastically in relation to the German mark that the currency rates were no longer within the range specified by the "snake."

In 1974 the EEC adopted an energy policy aimed at reducing dependence on imported oil by 1985. This was to be accomplished by increasing the use of coal and lignite (a low-grade, brownish black coal) and of nuclear energy, and by tapping natural gas deposits under the North Sea. It remains to be seen whether this policy can be carried forward in a unified manner. When the oil embargo was imposed by the Middle East oil-producing countries in 1973–1974, each EEC member moved swiftly to make its own peace with the oil suppliers. The Netherlands, which was excluded from Arab oil sources because of alleged aid to Israel, was left by the other EEC members to solve its own oil problems.

The European Economic Community has begun to act on social as well as economic issues. A forty-hour workweek and a four-week vacation are to be adopted by the end of 1978. A program to facilitate employment of people under twenty-five and to retrain workers affected by partial or total closing of firms in the iron, steel, and coal industries is under way. Some studies have called for an EEC passport. Other proposed measures are improvements in family allowances and educational facilities for migrant workers.

The EEC has set up a Regional Development Fund to improve living standards in depressed areas such as southern Italy, Brittany in France, and northern England. However, the EEC has no overall regional development policy; the fund allocates money to individual governments to implement their own policies. In general, the EEC has made less progress toward European integration than had been hoped. It must be remembered, though, that the EEC is less than twenty years old. In comparison with the individual sovereign states of Europe, which have developed control of their territories over the course of centuries, it is a very young organization.

MOVEMENT OF WORKERS WITHIN EUROPE

In Europe, with its strong local loyalties and its many international borders, large movements of population from one region to another would seem unlikely. In fact, it is estimated that more than 15 million people have migrated from one European country to another since World War II. In addition, people have migrated from India, Pakistan, the West Indies, Turkey, and North African nations.

The movement began after World War II when the rapidly growing West German economy absorbed 3.5 million refugees from East Germany. At the same time Italy was plagued by unemployment and poverty in its crowded and poor rural areas in the south. Some of these southerners moved to the industrial cities of Milan and Turin in the north, but by the late 1950s they were also welcomed by the growing industries of Switzerland, West Germany, France, and even the United Kingdom. In the United Kingdom nearly a million workers came from Ireland and large numbers from India, Pakistan, and the West Indies.

The EEC further facilitated the movement of workers. According to its charter, workers as well as resources and finished products could move freely between member countries. As a result the movement of Italians to other member countries increased. But the rapid growth of the economies of Western Europe and Sweden required more workers than Italy alone could supply. Soon workers began to come from Spain, Portugal, Yugoslavia, and Greece; later, Turks and Algerians were recruited.

At first the movement seemed to benefit all. Workers facing unemployment and low rural incomes in their home countries in the less developed parts of Europe and the developing world found jobs with good pay in the rapidly industrializing Western European countries. Rapidly growing sectors of the job market in large metropolitan areas—building construction, transportation, assembly lines, and garbage collection—got the workers they needed. Money made by the workers in their new city jobs was sent home to help support people in low-income rural areas of home countries. This money also helped to ease the trade deficits of the poorer countries and stimulated their economies.

The movement has turned out to be greater than anyone expected. No one thought the workers would move permanently to the industrial cities; instead, it was believed that they would come alone, work two or three years, build a "nest egg," then return to the home village, buy land, build a good house, and return to the rural life. But many have lived in the city for ten or more years, and often their families have come to join them.

Today Greeks make Swedish crystal, and workers from as many as forty different countries assemble cars in Sweden, France, West Germany, and the United Kingdom. Almost half the workers in Belgian coal mines are foreigners. Non-French workers account for nearly half of France's steel production. Most of the workers clearing Paris slums are Portuguese; Arabs run West Germany's railroad system, and Indians and Pakistanis run London's buses and subways.

Industrial cities had not planned on this. Housing is inadequate; workers crowd into single rooms in the poorest, most run-down parts of cities. Children speaking foreign languages crowd city schools in West Germany and France. The problem is concentrated in the few largest cities. In West Germany, for example, Munich, Frankfurt, and Cologne are known as foreign-worker cities. The situation has been complicated in recent years by economic recession, as in 1974–1976. Factories and other businesses laid people off, and for the first time, a significant number of people were unemployed in several Western European countries, where near-total employment has been the norm. Now native workers called for preferential treatment over foreigners in the job market and for deportation of the immigrants.

The migration of workers has had a profound impact on the traditional European culture system. Unlike the United States, European countries have not thought of themselves as "melting pots." People have been proud of their national ways. They saw the incoming workers as "guest workers," as useful people who did the jobs that increasingly well-off nationals were not interested in doing—collecting garbage, cleaning streets, working as waiters and waitresses. But the guests now occupy large areas in the industrial cities. Like the blacks and poor whites from the southern United States who occupy the vast ghettos of U.S. cities, foreigners speaking different languages live in the poorest areas of European cities. They follow their own customs, but they also expect to be given rights equal to those of native-born citizens. The result is that the industrial European countries are experiencing problems similar to those of the United States. The countries from which the refugees come have also found migration a problem. They are exporting a valuable resource: the young people of their countries, the most mobile and skilled members of their work force.

THE GEOGRAPHIC EQUATION OF EUROPE

Europe, the first part of the world to develop the modern interconnected system some 200 years ago, still shows the imprint of hundreds of years of life in the locally based, traditional system. Europe's modern cities retain the narrow streets and property boundaries that date back to pre-railroad and preautomobile days. The farms, the human institutions, and the rest of the culture also have vestiges of the past. These reminders of the past do much to give European countries their charm, but some are obstacles to these countries' ability to compete with the United States and the Soviet Union. The United Kingdom, France, and Germany, the first great examples of the modern interconnected system, and among the most industrialized and most sophisticated business centers in the world, have been overshadowed by the new superpowers, the United States and the Soviet Union. The superpowers have been able to organize vast territories containing not only rich agricultural and industrial resources but also populations four times those of the largest European countries, populations that serve both as workers in the system and as consumers of goods and services.

Faced with that competition, Europeans have been changing their economy. The expanded European Economic Community is a bold venture that involves the most important industrial countries of Europe and provides a strong resource base, market, and work force even larger than those of the United States or the Soviet Union. But within the EEC are nine separate countries whose people see themselves as distinct from their neighbors. These people speak six different major languages plus numerous dialects. They see themselves first of all as members of their culture groups; then, like the Scots or the Bretons, secondarily as citizens of their particular country; and only peripherally as members of the European Economic Community. At present the EEC's future is still in question because of the strong allegiance of people and governments to the traditional political-cultural system. The regions of Western Europe, Southern Europe, and Northern Europe, though not Eastern Europe, are beginning to function economically like the United States, with specialized production in particular regions and interregional trade. But Europe has not begun to function politically or culturally as a unit.

SELECTED REFERENCES

Beck, Robert H., et al. *The Changing Structure of Europe: Economic, Social and Political Trends.* University of Minnesota Press, Minneapolis, 1970. An exploration of different aspects of economic and political trends toward integration that have been especially strong in the post–World War II development of Western Europe.

Gottmann, Jean. *A Geography of Europe,* 4th ed. Holt, Rinehart and Winston, New York, 1969. A regional geography with provocative and perceptive interpretations of Europe and its individual countries.

Hoffmann, George, ed. *A Geography of Europe.* Ronald Press, New York, 1975. A basic regional geography of Europe.

Jordan, Terry G. *The European Culture Area.* Harper & Row, New York, 1973. A systematic geography of Europe, including the western part of the Soviet Union. The emphasis is historical-cultural in each of the topical sections. Chapters 2, 3, 7, 8, 10, 11, and 12 are of greatest relevance to the context of this book.

Nystrom, J. Warren, and George W. Hoffman. *The Common Market,* 2nd ed. New Searchlight Series, D. Van Nostrand, New York, 1976 (paper). An overview of the economic integration of Western Europe with brief profiles of recent economic development in each member country.

Reynolds, Robert L. *Europe Emerges.* University of Wisconsin Press, Madison, 1967. A historian's presentation of the antecedents to the Industrial Revolution.

Scargill, D. I., general editor, Problem Regions of Europe Series. Oxford University Press, New York. A series of short (less than fifty pages) monographs that focus on specific problem regions of Europe and are authored by European geographers, most of whom are British. The general focus is on the important economic and social problems that each region must confront as part of its future change and development. Three of the problem regions are recommended as illustrative of the changing geography of Europe as discussed in this chapter. More than fifteen other problem regions are covered in this series.

Clout, Hugh. *The Massif Central.* 1973. Sets the problems of this traditionally rural area in terms of agricultural development and strengthening the larger urban centers.

Hellen, J. A. *North Rhine–Westphalia.* 1974. Exemplifies problems of urban and industrial obsolescence, resource decline, and maintenance of agricultural activities in a major industrial area of Europe.

Thompson, Ian B. *The Paris Basin.* 1973. Focuses on the problems arising from Paris's dominance in the functioning of France and from its major population and economic change.

Servan-Schreiber, J. J. *The American Challenge.* Atheneum, New York, 1967. The invasion of large U.S. corporations into the European Economic Community countries seen as a challenge to Europe's future by a French journalist-politician.

Statistical Office of European Communities. *Basic Statistics of the Community,* biannual. A summary of basic statistics for the European community of nations and other selected modern nations of the world.

CHAPTER 9

The Soviet Union: A Highly Centralized Interconnected System

Like the United States, the Soviet Union has a large and diverse resource base and relatively sparse population. For political reasons, however, the Soviet Union must depend largely on its own resources. A highly centralized government functions somewhat as a giant corporation with Moscow as headquarters. Two kinds of nodes, administrative and industrial, characterize the Soviet system; they are concentrated in an area of denser population called the core triangle.

Left: Recreation among the senior members of a collective farm in the Turkmenian Soviet Socialist Republic. (Tass from Sovfoto)   *Above:* The Trans-Siberian railway near Sverdlovsk. (APN/Photo Trends)

Above: Oil tanks at Surgut, center of an oil-producing region on the Ob River in Western Siberia. (E. Ettinger/Photo Trends)   *Below:* The Kremlin in Moscow, the heart of the centralized Soviet system of government. (Inge Morath/Magnum Photos, Inc.)

The Soviet Union, our third example of the modern, long-range, interconnected system, seems to have all the essential ingredients for success. A single, highly centralized government has complete control over the economic, cultural, and political development of the world's largest country—a country about the size of all North America plus the islands of the West Indies. Such a huge area is likely to have the natural resources necessary for the functioning of a modern interconnected system.

Nevertheless, the Soviet Union has not yet achieved the material prosperity of the United States and most European countries. Although it has more people than the United States, its gross national product is only about one-third as large. Its per capita GNP is less than half that of West Germany or France and is even lower than Italy's. As a result, the Soviet Union's position as the second great political and economic power in the world rests more on the size of the country's economy than on the intensiveness of its industrial development. If its per capita GNP were increased to the level of West Germany's, its total GNP would be almost twice the present level.

SELF-CONTAINED VERSION OF THE INTERCONNECTED SYSTEM

The Soviet Union more closely resembles the United States model of the interconnected system than it does the models of European countries. Its environmental management problem is much the same as that of the United States: how to develop the resources of a continental-sized territory for the benefit of a population relatively large by world standards but relatively sparse in terms of density. In fact, the population density of the Soviet Union is almost the same as that of the area of the United States and Canada combined: about 33 persons per square mile (about 13 per square kilometer).

More than any other major industrial country of the world, the Soviet Union depends on the resources of its own territories. The major capitalist countries have not looked with favor on the Soviet communist political and economic experiment, because the experiment has been avowedly anticapitalist and has threatened an onset of world revolution against private capitalism. Moreover, the Soviet Union has tried to carry out its method of governing and managing the country's economy in isolation from the contamination seen in capitalism. For these reasons Soviet international trade totals less than one-third that of the United States. As Table 9–1 shows, in 1973 almost half of Soviet trade was with other countries within the Communist bloc in Eastern Europe.

THE EMPIRE OF THE CZARS

The Soviet Union's huge size is difficult to comprehend. It covers 15 percent of the land area of the earth and is the only large country that is an important part of two continents: Europe and Asia. It stretches almost halfway around the earth from 20° east longitude to 170° west longitude. A person taking a train from Leningrad on the Baltic coast in the west to Vladivostok on the Pacific Ocean in the east, a trip across the longest dimension of the country, would be traveling the same distance as from San Francisco straight to London, England. The north-south dimension is not as great as the north-south dimension of North America, but at its maximum the distance is equivalent to a trip from Memphis northward to the northernmost island off the coast of mainland Canada (Figure 9-1).

Although the Soviet Union has reoccupied land lost in World War II, the vast expanse of national territory is not the creation of an aggressive outreach by the Soviet Union. Rather, it is the inheritance of the empire of the Russian czars, who, at the time that the Western European countries

TABLE 9-1 INTERNATIONAL TRADE, 1973

	Total Trade ($ Millions)	Per Capita ($)	Trade with USSR ($ Millions)	Per Capita ($)
European countries outside the Communist bloc[a]	534,030	1,477	8,420	23
United States	138,879	660	1,375	7
Japan	75,244	694	1,325	12
USSR	42,570	170		
Trade with other European Communist bloc countries	20,590	82		

Source: *U.N. Statistical Yearbook, 1974.*
[a]The Communist bloc consists of Albania, Bulgaria, Czechoslovakia, East Germany, Hungary, Poland, and Rumania.

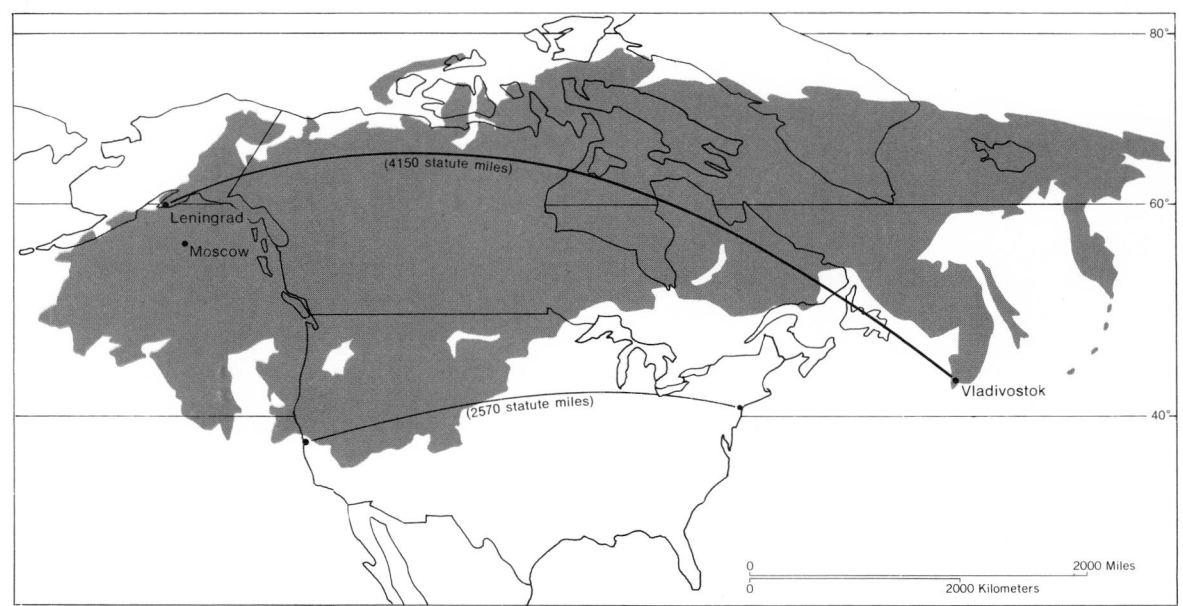

FIGURE 9-1 THE SOVIET UNION AND ANGLO-AMERICA COMPARED IN AREA AND LATITUDE

were developing their sea empires, pushed a land frontier and the European culture eastward from the Russian **culture hearth,** in the area of Moscow and Leningrad, to the Pacific Ocean and beyond to Alaska and the coast of northern California.

Like the sea empires of the Western European countries, the empires of the czars brought peoples of very different cultural backgrounds under control. The czars, with their home base near the Baltic in the northwest, were Great Russians, a Slavic people. Slavic people are found over much of Eastern Europe from Poland, Czechoslovakia, and Yugoslavia eastward; this is the Slavic culture realm. Slavs are distinguished by a particular language family. All Russian Slavs are descendants of an early group that occupied land near the present city of Kiev in the Ukraine, but, through time, they diffused to the north, west, and south.

The Great Russians, the northern group, later pushed south and southwestward to occupy lands of two other Slavic peoples, the Byelorussians, who lived east of Poland, and the Ukrainians, who lived north of the Black Sea. In the same way they subjugated non-Slavic Europeans along the Baltic Sea and on the Rumanian and Finnish borders.

Movement to the southeast and east brought non-Europeans into the czars' domain: the peoples of the Caucasus Mountains and Central Asia, who in language, religion, and economic practices were Middle Eastern. Mostly Moslem rather than Christian, they were nomadic herders and farmers experienced in irrigation agriculture. To the east, other Turkic peoples and various scattered Mongol tribes had spread over the flatlands of southwestern Siberia. This ethnic variety can still be found in the population of the Soviet Union (Table 9–2).

The dominance of the Slavic peoples, particularly the Great Russians, persists even today. In conversation we often hear the term "Russians" applied to the people of the Soviet Union as a whole, but for greater accuracy we shall here use the term "Soviets."

TABLE 9–2 ETHNIC COMPOSITION OF THE SOVIET UNION, 1970

Ethnic Group	Percent	
Slavs	74	
Russians		53
Ukrainians		17
Byelorussians		4
Other Europeans	7	
Jews		1
Baltic peoples		2
Moldavians		1
Armenians		2
Iranians	1	
Turkic peoples	13	
Caucasian peoples	2	
Uralians	2	
Mongols and other small groups	less than 0.5	

SOURCE: *Translations on the USSR: Political and Sociological Affairs,* No. 143, U.S. Joint Publications Research Service #53061, July 1971; nondepository entry 11311–29.

SOVIET REORIENTATION OF MANAGEMENT

World War I and the revolution and counterrevolution that followed it brought the end of the czarist regime and the beginning of the world's first government based on the writings of Karl Marx. The new Soviet government promised a very different alternative to capitalism as an example to the world, and assumed that Communism, once proved successful in the Soviet Union, would sweep the world. The new Soviet government inherited territory larger than that of the United States and Canada, and immediately began to organize it politically and economically in a very different way. Thus we have the opportunity to see how a very different system organizes space on the scale of the United States.

The czarist empire had been run along the lines of a traditional early European empire, strictly for the benefit of the ruling power—in this case, the Slavs of the Moscow area and the European northwest. The resources of outlying regions were the property of the Crown and thus were exploited for shipment to the center of power in the area of Moscow and Leningrad (then St. Petersburg). Moreover, the non-Slavic peoples were subjugated, and their ways were at best tolerated. The Russian language and customs were the instruments of political and economic power in all parts of the empire, with Russian officials in charge and Russian culture superimposed on native groups. As in most European empires, the native peoples were expected to learn the official language—Russian—if they hoped to rise in the economic system.

Political Organization

When they seized power, the Communists were faced with much the same problem that confronts many underdeveloped countries throughout the world today: how to build a functioning modern state with a population divided among a great many minority cultures. Empires traditionally subjugate minorities, particularly those geographically isolated in different parts of the country, but the doctrine of the Communist party called for the equality of peoples and an international communist movement.

The Soviets replaced the empire with a political organization whose name promised change. In 1923 the Russian Empire became the Union of Soviet Socialist Republics. The word "socialist" suggests a system in which the political power and the means of production and distribution rest with the people. In practice, the Communist party is overwhelmingly dominant in all aspects of the Soviet system. "Soviets" are councils whose members are chosen by elections from candidates approved by the Communist party. An extensive network of such councils is found at all

levels of political organization, from village, district, city, and region to the national level. Representatives from these soviets attend the Congress of Soviets held every two years; they also maintain law and order at the local level and carry out policies of the national government. The Supreme Soviet of the USSR is a two-house legislative body consisting of the Soviet of the Union and the Soviet of the Nationalities.

The ostensibly democratic constitution of the Soviet Union allows member republics to secede and maintain independent control over agriculture, law, education, and public health. Although the Ukrainian and the Byelorussian republics are represented separately in the United Nations, the law of the central government prevails over the law of a union republic, and most republican governments have very little autonomy in fact.

The "union republics" were designated along ethnic lines. Since ethnic groups had been sharply localized in particular parts of the country, it was comparatively easy to draw the boundaries of the republics with regard to populations of the major groups (Figure 9–2). This scheme has been modified, but today there are fifteen Soviet Socialist Republics, seven named after European peoples, three after people living in the Caucasus Mountains between the Black and Caspian seas (the Caucasian Republics), and five after peoples of central Asia (Soviet Central Asia). Note that the Russian Republic occupies 76 percent of the land area and contains 54 percent of the population of the entire Soviet Union. It stretches from the Baltic to the Pacific and from the eastern shore of the Black Sea to the Arctic Ocean. It therefore encompasses many of the smaller ethnic groups. Although it is named for the Russians who dominate its population, the presence of other ethnic groups is recognized by calling it the Russian Soviet *Federated* Socialist Republic (RSFSR). Within the RSFSR are a number of Autonomous Republics, Autonomous Regions, and Areas, each with its own Supreme Soviet and Council of Ministers. These administrative units are ostensibly self-governing but in

FIGURE 9-2   POLITICAL MAP OF THE SOVIET
UNION

Moscow National capital

Kiev Capitals of Soviet Socialist Republics

• Metropolitan areas over 1 million

1: Lithuanian S.S.R.   2: Latvian S.S.R.   3: Estonian S.S.R.
4: Moldavian S.S.R.   5: Georgian S.S.R.   6: Armenian S.S.R.
7: Azerbaidzhanian S.S.R.

TABLE 9-3   MEMBER REPUBLICS OF THE SOVIET UNION: THEIR
SIZE, POPULATION, AND LEADING ETHNIC GROUPS, 1970

	Land Area[a] (Percent of National Total)	Population (Percent of National Total)	Percent of Population That Is		
			Namesake ethnic group	Russian	Slavic
Russian SFSR	76.7[b]	54	83	83	86
European republics					
Ukraine SSR	2.7	20	75	19	96
Byelorussia SSR	0.9	4	81	10	94
Estonia SSR	0.2	1	68	25	28
Latvia SSR	0.3	1	57	30	39
Lithuania SSR	0.3	1	80	9	19
Moldavia SSR	0.2	1	65	12	28
Caucasian republics					
Georgia SSR	0.4	2	67	9	10
Armenia SSR	0.1	1	89	3	3
Azerbaijan SSR	0.4	2	74	10	10
Central Asian republics					
Kazakhstan SSR	12.4	5	32	43	52
Turkmen SSR	2.1	1	66	15	16
Tadzhik SSR	0.6	1	56	12	13
Uzbek SSR	1.8	5	65	13	14
Kirghiz SSR	0.9	1	44	29	34

[a] Data from tables in each regional chapter of J. P. Cole and F. C. German, *A Geography of the U.S.S.R.*, 2nd ed., Butterworths, London, 1961.
[b] Includes Western and Eastern Siberia and the Far East.

fact are an integral part of the RSFSR; they are represented in the Soviet of the Nationalities.

Notice in Table 9-3 how minor most of the member republics are in relation to the total population of the country. Eight of the fifteen republics have only 1 percent of the population each. Only the RSFSR and the Ukraine each have over 5 percent. From the table we can see how local ethnic groups dominate each member republic. Only in the Kazakhstan and Kirghiz republics of Central Asia, where new agricultural settlements along the northern border have been introduced in formerly sparsely settled areas by largely Slavic settlers, does the ethnic group for which the territory is named constitute a minority of the population. But in almost all republics other than the RSFSR, Slavs, and particularly Russians, constitute a significant minority group within the population. Although some of these Slavs settled in the territories under the czars, most of them have migrated since the Revolution. This, too, reflects the problem of the development of a land with minorities having different traditions and a different technology, the problem of **cultural pluralism.** The modern state of the Soviet Union has been built on a complete lifestyle involving the economics and social structure that evolved within one particular ethnic group, the Slavs.

Each ethnic group has had to be taught the Soviet lifestyle, and its economy has had to be developed to fit the modern interconnected system controlled from Moscow, center of the Great Russian homeland. For this, cadres of trained leaders were needed. Since trained leadership was scarce among ethnic groups that had been subjugated by the czarist empire, Slavs were sent to the member republics to assume positions of leadership. Today young non-Slavic people are trained and educated in schools in the republics and in Moscow, but Slavs are still in key political and economic positions in member republics and Slavs dominate the powerful Communist party. Moreover, the small populations of the ethnic minorities mean there are no masses of people to draw on for development schemes such as the agricultural settlement of the "virgin lands" in northern Kazakhstan in the 1950s or new mining-industrial complexes in non-Slavic areas. In these programs the Slavs dominate, thus adding to the non-native minority of the republic.

Soviet policy has encouraged the ethnic groups to develop their local cultures. Groups have been urged to continue national costumes at festivals and holidays, and they have developed written languages where there were none and written ethnic histories. All this has been done within the strict limitations of the doctrines of national government and the Communist party. Thus a local language has become a device for communicating communist ideas to minority peoples, and their local history is written to emphasize ethnic bonds to communist doctrine. Ethnic identity is confined within the narrow constraints of Soviet perspective of the proper values of life.

At the same time, however, strong centralized political and economic power has been maintained in Moscow, and Russian is the language of trade in all parts of the country.

Centralized Economic Control

Among modern industrial countries the distinctive feature of the Soviet Union is **centralization,** the virtually complete control that the central government maintains over the national economy. As soon as they were able to consolidate power after the Revolution, the Communists rejected the private enterprise and capitalist systems, and the idea that supply and demand should control the production and sale of goods and services. Instead, the Communist government took charge of economic production and concentrated almost all decision making within the central government. From that point on it set the economic course for the country and established growth priorities. In the Soviet Union the government operates as landlord, banker, traffic manager, and director of the board for the whole country.

The national interest, as determined by the government, has essentially replaced individual and corporate interests as a basis for decisions. The economy functions according to a national plan that determines what is to be produced and in what quantities. Then the state allocates resources to enable the various aspects of the plan to be carried out. Central planners work under priorities approved by the central government to establish production quotas for all types of goods and services. These national totals are then broken down into subquotas for each region and each producing unit—factory, mine, and farm. Each unit calculates what supplies and materials will be needed to fulfill its obligations and sends the report to the central planning agency, which then allocates the necessary materials as approved.

Prices, too, are fixed centrally in terms of what is considered best for the state, rather than in terms of the cost of production. Thus consumer items in short supply, such as automobiles, may be deliberately priced out of the reach of most people, while buses, considered more important to the country's economy, may carry a low price when sold to governmental transport agencies.

In such a controlled economy it would seem that the goal would be to have each producing unit meet its quota exactly, and thus to have all the economic parts mesh. Surprisingly, however, producing units and individual workers are given incentives not only to meet quotas but also to exceed them. Units and individuals who do not meet their quotas are scorned and even punished, and those who surpass their quotas receive special awards and recognition. Presumably, if all units were to surpass their quotas, the plan would operate on a level higher than can reasonably be expected, which would benefit the state. In most modern industries, however, products are dependent on the assembly of a variety of different parts. If, for example, those who make engines for buses surpass their quotas, but those who make the tires do not, the total number of buses will remain limited. This is not true of food quotas or quotas on small items, such as shirts, that are made in one factory.

Emphasis has usually been on quantity rather than quality of output. Annual production quotas are part of larger, long-range plans, usually covering five-year intervals. These plans establish priorities and determine what should be produced, and even where new factories and other units needed to produce the desired output should be built.

In recent years the extreme centralization of control has loosened somewhat. Agencies in the union republics are now given more say about capital investments and work forces in their areas. Instead of sheer physical output of goods being emphasized, factories are being asked to operate in terms of "profits," or the value of production measured against the capital invested and the amount of sales. In this way the output of goods is somewhat responsive to demand.

However, the slight decentralization and shift from physical quotas do not represent in any way a turning away from the centralized, state-controlled economy. In examining the environmental base and its development by the Soviets, we must consider them in terms of centralized planning by the state in contrast to the multifaceted economies of the United States and Europe with their many corporate and individual entrepreneurs. The Soviet central government is the landlord, entrepreneur, and decision maker for virtually the entire economy—primary and secondary production and marketing—of this huge country.

Soviet production priorities are somewhat different from those in the United States. The production priority is to produce the things that are needed to enhance the strength of the Soviet Union as a modern industrial power. The Soviets have been pushing hard to catch up with the United States and European countries. Because the Soviets have been industrializing for less than fifty years, their new economic programs give highest priority to developing the infrastructure needed by a modern industrial state. High priority is also given to defense expenditures, while consumer demands have a lower priority.

On the other hand, the Soviets have made a conscious effort to eliminate as much as possible the sharp economic class distinction that characterized czarist Russia, when a small upper class lived in wealth while the masses of the people, either as peasants or city workers, had a very low living standard. Now goods and services in the consumer sector are broadly distributed, and their goal is to have no one live in poverty. The Soviet housing program, for example, aims at mass production of living quarters, not only for middle-class workers but for everyone. When measured against middle-income housing in the United States, Soviet apartments seem cramped, poorly built, and without style. But the Soviets seem to have virtually eliminated slums. Medical and educational services, though less sophisticated than in the West, provide basic services to all people.

Because Soviet per capita productivity is still below that of the United States and Western European countries, the amount of consumer goods also remains well below Western standards. Nevertheless, the workers of the Soviet Union have a much higher living standard than did their forebears under the czars.

Government decisions concerning the development of the Soviet Union have political as well as economic bases. The Far East, close to Japan and China but far from the core of the Soviet economy, has been developed as a largely self-sufficient base although it is half a world away from Moscow. The Soviets feel they must be strategically represented in the Far East, but their transport system cannot supply this faraway region.

Western Siberia was first developed in the late 1930s as a sanctuary from a feared invasion by Germany. This decision proved wise when most of the European part of the country was lost to the German armies in 1941 and 1942. The country's prime agricultural base west of the Volga River, the iron and coal resources of the Ukraine, and major industrial centers were lost. But grain fields of Siberia, its new iron and coal supply, and the recently built factories east of the Volga helped provide the base for an offensive that eventually drove out the Germans.

At first glance, one would expect that the central planners of the Soviet Union could organize production very differently from the way it is organized in the United States. The planners would analyze the potentials of different parts of the country and then allocate capital and other resources to complete the development. In fact, the constraints in Soviet decision making are almost the same as those of any large corporation making locational decisions. First, they realize that it is impossible to duplicate, in new locations, the physical plant they inherited from the czars. It is cheaper to expand operations in existing locations. Whether the existing facilities are in the best locations or not, they form the base upon which new investment decisions must be made. Second, although political and other noneconomic factors can be considered in setting planning priorities, only so much capital is available to the state, and it must be invested wisely. New developments out in the wilderness must be built from scratch, which calls for investment not only in the physical plant but also in transportation links with established sources of supply and markets. Whole towns for workers and their families must be created for workers who must come from other areas. As in the United States, it is more expensive to develop remote resources, rich as they may be, than to use those in or near already established centers.

The vast size of the Soviet Union presents particular problems to developmental planning. With the Arctic Ocean and north-flowing rivers frozen most of the year, and with outlets to the Black and Baltic seas opening only into roundabout international waterways that depend on passage through narrows controlled by other countries, the major burden of tying the huge land area of the country together rests with overland transportation or air service.

Because of this, the Soviet land transport system has greater pressure on it than the systems of other countries have on them. In order to lessen this burden, the Soviets have had to modify the model of the modern interconnected system that calls for each region to specialize in the things that it can produce best, while, in return, depending on other regions for its basic needs. In the Soviet Union each region is encouraged to produce the specialties the country needs but, at the same time, is asked to supply as many of its own needs from its own local resources as possible. Thus regional specialization exists, as in the United States, but it is modified by an ideal of **regional self-sufficiency.** Each region is to have its own agricultural food base and its own essential industrial economy so as to reduce the need for transportation hauls from other parts of the country. Thus the irrigated areas of southern Central Asia are the best areas in the country for growing cotton, but farms there are also expected to produce grains, livestock, and other food needs of the local population. The region ships nonferrous metals to industrial districts elsewhere in the country, but a local steel industry provides for basic regional needs.

The situation somewhat resembles the economic development in California, an area of the United States remote from the productive Northeast and Middle West. Faced with the high prices of goods shipped from the East, California has developed a remarkably well-rounded economy but still ships its specialties to the rest of the country. But all of Central Asia has fewer people than California, and the Far East has less than a fifth as many. Thus it is questionable whether Soviet Central Asia has sufficient markets to support such major investments as steel mills, oil refineries, and chemical plants.

The Soviet experiment has not been totally carved from a wilderness; rather it has been superimposed on a previously established base. New agricultural, mining, and urban areas have been developed, but the most important economic regions today are, in large part, those inherited from the czars. The major agricultural areas remain the Ukraine and the lands south of Moscow in the RSFSR. The most productive iron and steel complex is still that established at the turn of the century by transport links between the coal of the Don River basin and iron ore near the Dnieper River in southern European Russia and the Ukraine.

Agricultural Reorganization

Agricultural production under the czars was just emerging from a feudal state before the Revolution. Most of the productive land had been held in large estates of individuals favored at court. The peasants who worked the land lived in farm villages and went daily to till small plots of soil, using human and animal power and such equipment as scythes and sickles. In the period before and after the Revolution, the peasants gained title to the land, but like sharecroppers in the southern United States after the Civil War, they continued to work tiny plots in traditional ways.

The Soviets saw the need to modernize the agricultural system that existed after the Revolution. Human power from the farms was needed in the factories that were the first priority of the Soviet crash program to join the ranks of the modern industrial powers. Moreover, because foreign loans from capitalist countries were unavailable to them as a result of their communist experiment, they needed to draw as much capital for industrialization from agriculture and other primary production as possible. Finally, they had to maintain agricultural output in order to feed their own population.

Collective farms The Communist regime rejected the private ownership of land, whether by farm family or corporate organization, and instituted collective farms. These farms were not as revolutionary as they seemed to people in the United States. Unlike U.S. farmers, Russia's rural population had traditionally settled in villages where fifty to several hundred farmers lived close together, each with a barn, a garden, and a few animals on a lot. Like farmers in Europe and Asia, they went out every day to work farmlands around the village. Under the czarist landlords the village was considered the working unit. Villagers shared a common pasture, and families did not work the same allotted land each year. Even when the large estates were broken up, little attachment remained to particular family fields. The focus of ownership was on the home plot in the village.

Not surprisingly, Soviet planners chose the farm village as the basic unit in their new agricultural plan. Under the reorganization, known as **collectivization,** the Soviet government became the new landowner, and the village, with its farmlands, became a single functioning agricultural unit that worked the land around the village as a team. The work force of the village became the work team of the collective farm. The villagers were to elect their own leaders who, in turn, would manage the farm according to quotas assigned by the central planners and would assign individuals to the particular tasks needed to produce crops.

Like the landlord of the past, the government made the decisions about what to grow and received a share of the crop as "rent" or taxes. Each farm was given production targets for particular crops, and, in addition, basic quotas of farm products had to be turned over to the government at very low fixed prices, whether the harvest was large or small. The government, like the landlord, was assured of its share, and the farmers on the collective assumed the risk of farming. If the farmers worked efficiently and growing conditions were favorable, the net output to be shared in proportion to the hours worked was large. If for any reason the crop was bad, so was the return to collective workers.

Collectivization meant the loss to farm families of all private agricultural property. As might be expected, the collectivization drive carried out in the early 1930s was met by considerable peasant opposition. Riots occurred, and a large proportion of the livestock, farm wagons, and other equipment was destroyed.

Faced with such opposition, the Soviet government compromised by allowing individual families to keep title to their individual small plots in the village; such title represented the norm of private ownership under the czars. Garden crops, livestock, and poultry raised on the home plot were not part of collective farm production and could be used to feed the family or sold through open markets.

Private plots remain an important part of the Soviet agricultural system. Although they total less than 4 percent of all sown area, they account for almost half the country's production of potatoes and vegetables, over half the dairy cows, and nearly a third of the hogs. This productivity is in part the nature of small-plot production: gardens carefully tended are among the most intensive forms of agriculture in the world, and many types of livestock production fit well on small units. However, the incentive to produce on "your own land" is much greater than the incentive to work

as part of a team managed by someone else, particularly since that work has no fixed reward. Individual farmers can feel that they are in control of production on their plots, and they profit from their own work investment.

Collectivization was designed not only to give the government direct control over the distribution of a large share of agricultural production through its quotas, but also to release farm laborers for work in newly rising factories and in other jobs in the modern system. Thus mechanization of farming was essential.

The big collective farm units were well suited to large fields worked by tractors and mechanized planting and harvesting equipment, but the machinery itself had to be made available. This involved first the task of purchasing manufacturing equipment abroad, and then building production lines to produce the tractors and other machines in the Soviet Union. It also required a system of allocation among the collectives of the country.

In the early days, mechanization was brought to the farms. The government established machine tractor stations, each designed to serve a large number of collectives in a particular area. Equipment was outfitted with lights so it could be used in late-shift work, and the stations had trained operators and mechanics to keep operations moving as efficiently as possible. Collectives contracted with the stations for tractors for plowing and combines for harvesting. In this way the scarce equipment could be kept in almost continuous use. In the 1950s, when more agricultural equipment was available, the stations were abolished, and the equipment was distributed among the collectives.

Since the mid-1950s the number of collective farms has been reduced drastically. There were more than 254,000 collectives in 1950, but the number had been reduced to 32,100 by 1972. This reduction was accomplished by combining collectives centered in individual villages into larger units consisting of the lands of several villages. The average size of the collective has risen from about 1,000 acres to over 15,000 acres. The original hope was that the villages could be combined into "agricultural towns" of 5,000 or more people with more urban services, but this plan has not been fully successful.

State farms Collectives were the solution to the problem of mechanizing the traditional agriculture of the villages, but a different unit was developed for lands that had never been settled before or those the government fully took over from estates. This was the "state farm," a unit completely controlled and managed by the government; workers were paid wages just as factory workers were. Not only was this more in keeping with the Soviet philosophy, but it was particularly well suited to new agricultural units on the dry margins where the risk from year to year is high, making collective farm operation almost impossible. State farms are designed also as highly specialized undertakings in which the maximum amount of governmental management and technical skills can be applied. Since a large number of these farms are located in the drier areas where per acre yields are low, they are larger than collectives. State farms tend to be concentrated in the newer developments and collectives in the better established agricultural areas. In 1972 there were 15,744 state farms in the Soviet Union.

POPULATION DISTRIBUTION

THE EUROPEAN MAJORITY

The distribution of population in the Soviet Union largely reflects the European heritage. Over three-fourths of the people live in European Russia and, as Tables 9–4 and 9–5 show, almost half live in the historic lands of the Great Russians, Byelorussians, and Ukrainians. In fact, as Figure 9–3 indicates, most of the population lies within a great triangular zone with its base along the boundary of Eastern Europe from the Baltic Sea to the Black Sea and stretching eastward to the city of Irkutsk near the western shore of Lake Baikal in Siberia. This is the land of the European peoples of Russia, their homelands in the west and their zone of eastward migration after the empire had been established. It is the core region of the Soviet Union, and we shall refer to it as the "core triangle." If the area between the Black and Caspian seas north of the Caucasus Mountains and the "virgin lands" area of north Kazakhstan are included, this zone of primarily European settlement includes 76 percent of the population of the Soviet Union.

In contrast, the areas of the traditional non-European peoples who were encompassed by the czarist empire are of minor importance. As Table 9–5 shows, south of the core triangle the Caucasus area accounts for only 5 percent of total population; Central Asia, for 14 percent; and the vast area of eastern Siberia and the Far East, for about 5 percent.

The contrast between the land area of primarily European settlement and the predominantly non-European areas is striking. European areas account for only slightly more than one-third the land area of the country, but four out of every five people live there.

VARIATIONS IN DENSITY AND THE LOCALIZATION OF POPULATION

Population densities are greatest in the Slavic homelands. Not only do these areas contain almost half the population of the country, but each, except the extreme northwest, has a density of more than one hundred persons per square mile (about forty per square kilometer). The only other area in the country with such a density is the Caucasus, a non-European area. It is important to keep in mind that these densities are much less than half those of Western Europe. They are more like those in the U.S. Middle West.

TABLE 9-4 POPULATION DISTRIBUTION OF THE SOVIET UNION, 1970

Area	Total Population (Thousands)	Number of Metros of 100,000+	Population in Metros of 100,000+ (Thousands)	Nonmetro Population (Thousands)	Percent of Population of Area in Metros
European Settlement					
Slavic homelands					
Northwest	12,160	10	5,679	6,481	47
Center	43,998	36	16,198	27,800	37
Byelorussian SSR	9,003	9	2,216	6,787	25
Ukrainian-Moldavian SSRs	50,708	44	13,768	36,940	27
Subtotal	115,869	99	37,861	78,008	33
Other European homelands					
Baltic	7,583	7	2,302	5,281	30
More recent European settlement					
Volga	18,377	16	6,604	11,773	36
Urals	15,184	16	5,728	9,456	37
Western Siberia	12,110	14	4,991	7,119	41
North Caucasus	14,285	15	3,632	10,653	15
Subtotal	59,956	61	20,955	39,001	35
Total	183,408	167	61,118	122,290	33
Non-European Settlement					
Caucasus	12,292	10	3,843	8,449	31
Central Asia	32,804	27	7,096	25,708	22
Eastern Siberia	7,464	8	2,402	5,062	32
Far East	5,760	9	1,718	4,062	30
Total	58,340	54	15,059	43,281	26
Grand total	241,748	221	76,177	165,571	32

SOURCE: *Translations on the USSR: Political and Sociological Affairs,* No. 143, U.S. Joint Publications Research Service #53061, July 1971; nondepository entry 11311–29.

Most of the areas of more recent European settlement in the Soviet Union have less than half the population densities of the predominantly Slavic lands. Only the North Caucasus region, with eighty-five people per square mile (about thirty-five per square kilometer), approaches the density of the older, established centers.

Except for the Caucasus, the predominantly non-European areas are the most sparsely settled

of all. This is particularly true of the vast regions of the north and east, which include over half the total land area of the country. There densities are around five persons per square mile or less (less than two per square kilometer), like those in Montana.

In area, the heartland of European Russia occupies a territory about equal to that part of the United States east of the Mississippi River and

TABLE 9-5 DISTRIBUTION OF METROPOLITAN AND
NON-METROPOLITAN POPULATION IN THE SOVIET UNION, 1970

Area	Percent of Total USSR Nonmetropolitan Population	Percent of Total USSR Metropolitan Population (Metros of 100,000+)	Percent of Total USSR Population
European Settlement			
Slavic homelands			
Northwest	4	7	5
Center	17	21	18
Byelorussian SSR	4	3	4
Ukrainian-Moldavian SSRs	22	18	21
Subtotal	47	49	48
Other European homelands			
Baltic	3	3	3
More recent European settlement			
Volga	7	9	8
Urals	6	8	6
Western Siberia	4	7	5
North Caucasus	6	5	6
Subtotal	23	29	25
Total	73	81	76
Non-European Settlement			
Caucasus	5	5	5
Central Asia	16	9	14
Eastern Siberia	3	3	3
Far East	2	2	2
Total	26	19	24

SOURCE: Derived from Table 9-4.

has a population slightly less. The full land area encompassed by the core triangle is almost nine-tenths that of the United States.

In the United States and Europe we found that the distribution of total population was the sum of two distinctive population patterns: the non-metropolitan population largely tied to the country's primary resource production and the population of large metropolitan centers mainly engaged in urban-centered activities. In both the United States and Europe the populations had very separate patterns; some areas were dominantly metropolitan, others largely nonmetropolitan.

Although the same situation exists to a limited extent in the Soviet Union, the variations in the two kinds of populations are much less marked (Tables 9–4 and 9–5), perhaps because agricul-

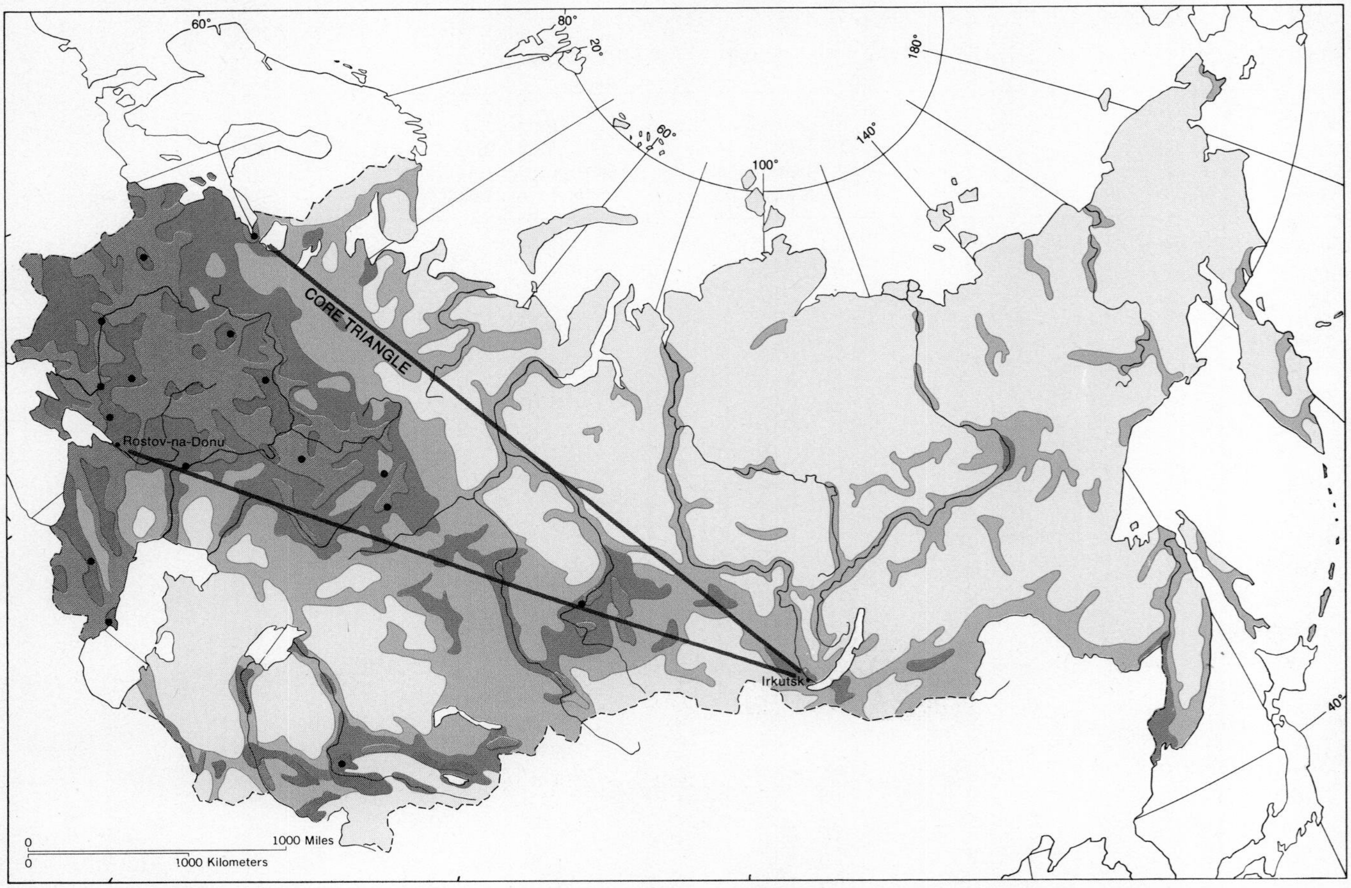

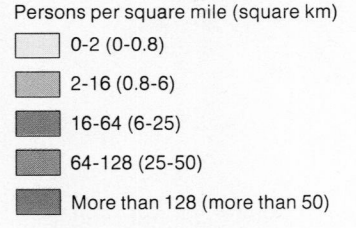

60°
80°
20°
180°
60°
140°
100°
CORE TRIANGLE
Rostov-na-Donu
Irkutsk
40°
0
1000 Miles
0
1000 Kilometers

SOURCE: Atlas Naradov Mira 1964

Persons per square mile (square km)

- 0-2 (0-0.8)
- 2-16 (0.8-6)
- 16-64 (6-25)
- 64-128 (25-50)
- More than 128 (more than 50)
- • Cities over 1 million

FIGURE 9-3 POPULATION DENSITY OF THE SOVIET UNION

TABLE 9-6   VARIATIONS IN METROPOLITANISM BY AREAS

IN THE SOVIET UNION, 1970

More Highly Metropolitan: 35 Percent and Above		Average Metropolitan: 27–35 Percent		Low Metropolitan: Below 27 Percent	
Northwest	47	Eastern Siberia	32	Byelorussia	25
Western Siberia	41	Caucasus	31	Central Asia	22
Center	37	Baltic	30	North Caucasus	15
Urals	37	Far East	30		
Volga	36	Ukraine–Moldavia	27		

SOURCE: Derived from Table 9–5.

tural employment is much higher in the Soviet Union. The last Soviet census (1970) reported that 35 percent of the labor force was employed in agriculture. Western European countries have up to 25 percent of their labor force in agriculture; in the United States the figure is less than 5 percent. At the same time, the populations of cities of 100,000 or more in the Soviet Union account for only 32 percent of the total population. Even considering differences in definition of *metropolitan* and *city,* the Soviet Union is undoubtedly far less metropolitan in its population distribution than is the United States or Western Europe.

METROPOLITAN POPULATION

Regional Variations

The national percentage of people in metropolitan centers is 32 percent, and only a few parts of the Soviet Union vary greatly from that average. The northwest and western Siberia are more highly metropolitan, and Byelorussia and the North Caucasus are less metropolitan. Still, a pattern of metropolitan importance can be seen by grouping the regions by the relative impor-

tance of their population that is metropolitan.

It can be noted in Table 9–6 that the more metropolitan areas form an east–west extension across the zone of heavy population from the old Russian centers of the northwest and center through the Volga to the Urals and western Siberia. These five regions of higher than average metropolitan population account for a higher share of metropolitan population than of total population. At the other extreme, the more rural areas lie to the south of the metropolitan zone except for lightly populated areas of the north and again form a zone from Byelorussia and the Ukraine to the southern boundaries of the country. The areas close to the national average are generally both peripheral and low in population.

Although the percentage of population in metropolitan areas in the Ukraine-Moldavian region is below the national average, this region still has the second largest metropolitan population in the country. In fact, as we saw from Table 9–5, the Center and Ukraine-Moldavia together have 39 percent of the metropolitan population.

The Soviet Union has far fewer large metropolitan centers than does the United States, even though it has a larger total population. Thirty-five U.S. metropolitan areas have a million or more people; only sixteen metropolitan areas in the Soviet Union do. Moscow, with 10 million people, is the largest metropolitan area and the major center of political, economic, and cultural decision making.

Cities Today and Czarist Urbanism

To outsiders unfamiliar with the Soviet Union, it would seem that the planners could simply create an urban development wherever it is needed. But establishing a new city, like any major investment, requires a tremendous amount of capital. Not only must the structure and industrial-commercial base of an entire local community be built, but essential transportation and communication links to the outside world must also be provided. Although the population of the Soviet Union has doubled since the Revolution, and urban centers have become more important in the functioning of the country, few large new cities have been established without a pre-existing base. In fact, only six of the seventy-two cities with populations of 250,000 or more have been created since 1920. Four resulted from further expansion of new mineral and heavy industrial developments in the Ukraine and the Urals. Only two were new enterprises in relatively remote areas that have emerged since the Revolution—Karaganda, site of a new coal complex in northern Kazakhstan, and Dushanbe, the capital of the Tadzikh republic (created from the former regions of Bokhara and Turkestan and admitted as a union republic in 1929).

Communities have been created either at new resource locations or at particularly strategic sites. However, only eight new cities having populations of over 100,000 have appeared outside the core triangle since 1920 (see Figure 9–4). Of these, only three—Norilsk in northern Siberia, Komsomolsk in the Far East, and Karaganda in arid Kazakhstan—represent developments in areas that are remote from existing cities. These cities in Siberia show that if development is important enough, cities in the wilderness can be established. Yet new cities in out-of-the-way

places are the exception, not the rule. They are not well tied into the Soviet transport system; they have no rail connections with the rest of the country and must be served at high cost by airplane and road.

The Hierarchy of Large Metropolitan Places

The hierarchy of major cities in the Soviet Union is simple. Political, economic, and cultural decision making are all combined under state control, and thus each of the most important cities functions as both an economic and a political administrative center. Of the thirty-one cities in the Soviet Union in 1967 with 500,000 or more people, only Krivoy Rog was not the center of its own administrative unit (Figure 9–5). The only large cities that are not administrative centers are those in mining and industrial districts where urban developments are close to one another. The most important of these are the areas of coal mining and metalworking in the eastern Ukraine and western Siberia and the iron mining and industrial areas of the Urals.

The hierarchy follows economic lines more closely than political ones. Six of the twelve cities of over 1 million population are old established industrial-trading centers in the RSFSR, and four are in the Ukraine. The two remaining are capitals of non-European republics, Baku and Tashkent. Baku is the center of a major oil field, and Tashkent is the leading industrial-commercial center of Central Asia. Only one of the fifteen capitals—Dushanbe—has fewer than 250,000 people.

NONMETROPOLITAN POPULATION

If the nonmetropolitan population can be taken as a measure of the people most closely tied to primary production and hence the resource base of the Soviet Union, the resources must be either unevenly distributed or unevenly developed. Almost half the nonmetropolitan population resides in the traditional homelands of the Slavs, which

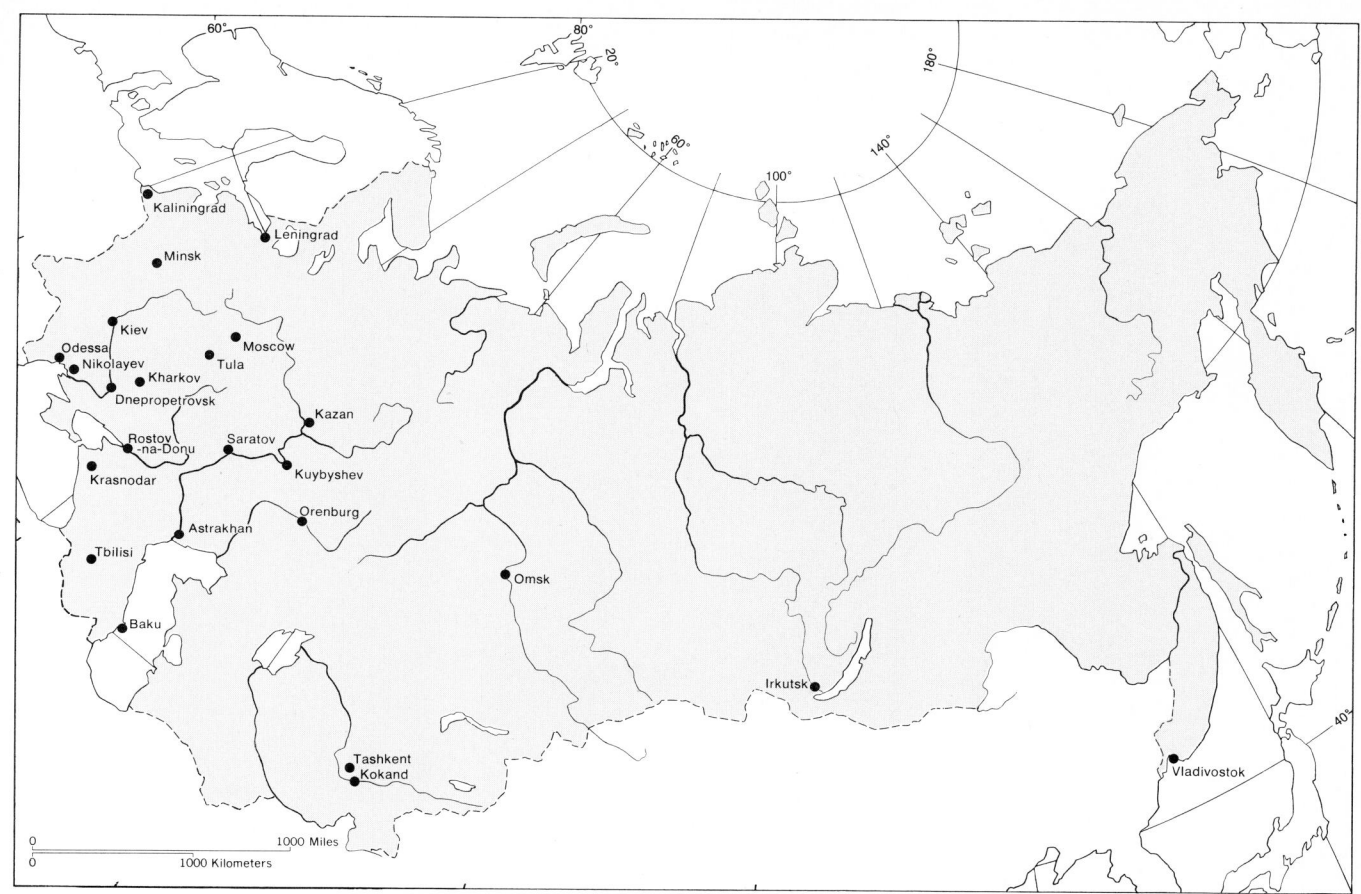

SOURCE: C. Harris, *Cities of the Soviet Union,* Association of American Geographers
Monograph Series, 1970, Table 27, pp. 256–267

FIGURE 9-4   CITIES WITH MORE THAN 100,000
PEOPLE IN THE SOVIET UNION IN 1920

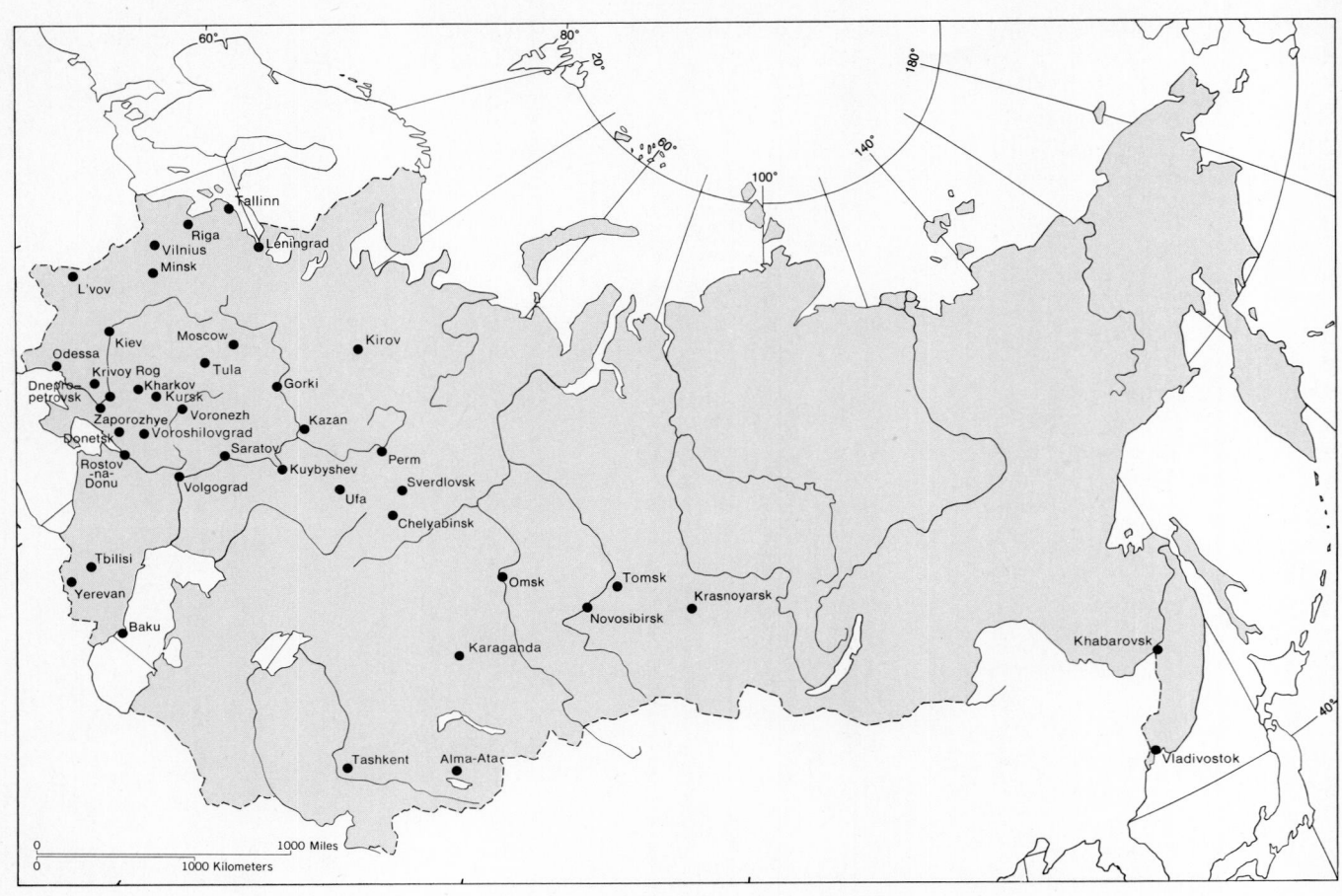

SOURCE: C. Harris, *Cities of the Soviet Union;* Administrative Map of the USSR, Moscow, 1973

FIGURE 9–5   CITIES WITH MORE THAN 500,000
PEOPLE IN THE SOVIET UNION IN 1974

is only 9 percent of the country's land area. On the other hand, the north, eastern Siberia, and the Far East with 52 percent of the land area have only 7 percent of the population. Almost four-fifths of the nonmetropolitan population of the Soviet Union is concentrated in the core triangle.

PRIMARY PRODUCTION AND THE ENVIRONMENTAL BASE

Since the emptiness of an area can usually be taken as a prime indicator of its lack of agricultural possibilities, it would seem that most of the Soviet Union has poor agricultural resources. Table 9-7 shows the lands sown to crops and confirms that most areas are little used for crop agriculture. Only 4 percent of the country's cropland is in the north and east, which total 52 percent of the land area. Only 10 percent of the sown area is in the south, which totals 19 percent of the land area.

These uncultivated areas do not result from a lack of interest in agricultural potential. Since the Revolution the Soviet Union has been struggling to increase agricultural production and has experimented with agricultural colonies in all areas where scientists and agricultural specialists see possibilities for farming. The possibilities are very much limited by environmental deficiencies, particularly poor drainage, short growing seasons, and dryness.

TERRAIN AND DRAINAGE

The major mountain systems in the Soviet Union form more or less a rim on its southern periphery. Eastern Siberia is also mountainous and hilly (see Figure 3-1). Elsewhere, with the exception of the Ural Mountains, which are more like hills than mountains, European Russia and western Siberia have mostly plains terrain.

TABLE 9-7 DISTRIBUTION OF SOWN LAND IN THE SOVIET UNION, 1967

	Percent of Total	
	Sown area	Land area
The North and East		
North	2	13
Eastern Siberia	1	25
Far East	1	14
	4	52
The South		
Caucasus	1	1
Central Asia (including Kazakhstan)	9	18
	10	19
Settled European areas		
Northwest	1	1
Center	18	4
Byelorussian SSR	3	1
Ukraine-Moldavia	20	3
Baltic	3	1
Volga	10	2
North Caucasus	8	2
	63	14
Urals-Western Siberia		
Urals	9	3
Western Siberia	11	5
Krasnoyarsk-Irkutsk	3	7
	23	15

SOURCE: *Narodnoe Schziaistvo SSSR*, 1967.

Because of the mountain rim on the south, most large rivers of the Soviet Union drain northward into the Arctic Ocean. Only the Volga flows southward, into the Caspian Sea. The large, north-flowing rivers of Siberia, some comparable in size to the Mississippi system, create special **drainage** problems. Not only is the land flat, so that drainage would ordinarily be slow, but the rivers also flow from areas with longer growing seasons into those with shorter seasons. In the spring, snow melts in the upper (southern) reaches, while the mouths and downstream parts of the rivers are still ice-bound. Spring flooding and poor drainage are especially severe under these circumstances.

SHORT GROWING SEASON OF
THE NORTH AND EAST

More serious than the drainage problem are the long, cold winters and the low potential evapotranspiration (PE) over vast stretches of the northern and eastern Soviet Union. Along the northern coast and over much of central and eastern Siberia, the frost-free season is less than ninety days, which precludes most cereal crops. In general, these lands have an annual potential evapotranspiration of less than half that of the U.S. Middle West (see Figure 3–3), and nowhere in the Soviet Union is annual potential evapotranspiration greater than that found along the U.S.–Canada border. Even though the summer months have almost continuous daylight, the short duration of sufficiently high potential evapotranspiration means that only grasses and root crops such as beets can grow to maturity.

The combination of the short, cool summer and the long, severely cold winter results in permanently frozen subsoil called **permafrost.** The Soviet Union has more area in permafrost than does any other country. Permafrost is almost continuous in Siberia north of 65° latitude, and patches are found as far south as the Chinese border. The frozen subsoil impedes soil drainage in summer and stunts tree roots. It makes mining, road and railroad construction, and any type of excavating, even plowing, difficult and expensive. The Soviet Union has done more experimentation on how to carry out agriculture, mining, and construction in permafrost areas than any other country, yet the problem remains a major one.

In the far north, even where there is no permafrost, vegetation is sparse and plant decay very slow. Thus soils are low in organic matter. Low-growing shrubs, mosses, lichens, flowering herbs, and grasses are the dominant vegetation in this tundra environment along the Arctic coast. All but the most southern areas of eastern Siberia have severely limited agricultural potential under today's technology. The extensive coniferous forests of the north, the **taiga,** form the largest existing forest area of any country in the world, but they are exploited in only a few places. Only the northern forests west of the Urals, relatively close to population centers, are significantly developed. Elsewhere the expense required to build transportation to and service centers in these remote forests has left most of them undeveloped.

THE DRYNESS OF CENTRAL ASIA

South of the core triangle, to the east of the Caspian Sea, the Soviets face another climatic problem for agriculture: dryness. This interior desert is walled off from the south by mountains, and is a long distance from its principal moisture source, the Atlantic Ocean to the west, from which moisture is carried by winds. Winters in Central Asia are cold, but summers are sunny and hot. The area has the greatest potential evapotranspiration in the Soviet Union, comparable to that in the northern Great Plains of the United States. Moisture supply, however, not temperature, sunlight, or length of frost-free season, is the critical missing factor.

Moisture deficits in Central Asia range up to 40 in. (100 cm). These deficits are comparable to those in the western United States (see Figure 3–6), and they make agriculture impossible without irrigation. Where streams or underground water sources can be tapped, irrigated agriculture flourishes, and these are among the most productive areas of the Soviet Union. Only 5 percent of the land area of Central Asia has been irrigated,

REGIONAL LOCATION OF MINERAL PRODUCTION

1. Much is said of the vast mineral resources in northern and eastern Siberia. Yet production of basic industrial minerals in the Soviet Union occurs still predominantly within the core triangle. Why does this apparent illogicality exist in a planned economy? Is the situation the same in North America?

2. In the United States the major iron-producing regions are far from the coal-producing regions, and oil and gas production is located in still other regions. Is this true in the Soviet Union? In the United States the Great Lakes have been a key factor in economically linking the coal and iron deposits. Has the Soviet Union any similar transportation link?

3. In the United States pipelines deliver oil and gas from fields in Texas and the rest of the south central region to the Middle West and Northeast. What would you expect the Soviet pipeline network to look like?

---

however, despite heavy Soviet investment in facilities. Most of the irrigated areas lie along the southern margin, where streams coming from the moist mountains are tapped in individual oases. Such irrigation dates back hundreds of years, but modern dams and distribution facilities have increased productivity.

Cotton is the most important of many specialized irrigated crops. The Soviet Union is now the world's leading cotton-producing country, and nine-tenths of its production comes from Central Asia. These irrigated areas, remote from the country's major markets, also specialize in other high-value crops such as orchard products, as do similar parts of California. Central Asia is also an important livestock region. Livestock herding is an outgrowth of the early development of nomadic herding along the desert fringe and in the mountains by native tribes. Cereals and fodder crops on irrigated lands supplement natural pastures.

INDUSTRIAL RESOURCES OF NORTHERN SIBERIA AND CENTRAL ASIA

The empty lands of the North account for half the country's land area, and probably include much of the potential mineral wealth. Major deposits of coal, petroleum, and iron are known to exist in northern Siberia, but most remain virtually untouched (Figure 9–6), while other, often less rich, deposits close to population centers are utilized.

As Table 9–8 shows, the North and eastern Siberia each account for less than 10 percent of the output of each of the major industrial minerals—iron ore, coal, crude oil, and natural gas. (The North consists of the most northerly sections of the Northwest, the Urals, and eastern Siberia in Figure 9–6.) Thus far, coal from the Pechora Basin in northern European Russia and the oil and gas in the northern part of western Siberia are the only major resource developments in the north. The overwhelming majority of the output of all basic industrial minerals still comes from the core triangle, particularly the European part of the RSFSR and the Ukraine.

Most of the mineral developments in the North are either on its southern fringe close to the main populous regions of the country or at remote sites where particularly rare, high-value minerals such as gold are found. Most famous are the nickel mining area at Norilsk in the far north of Siberia near the lower Yenesei River, and the gold fields of the upper Aldan River in eastern Siberia. These mining sites can usually be served by air,

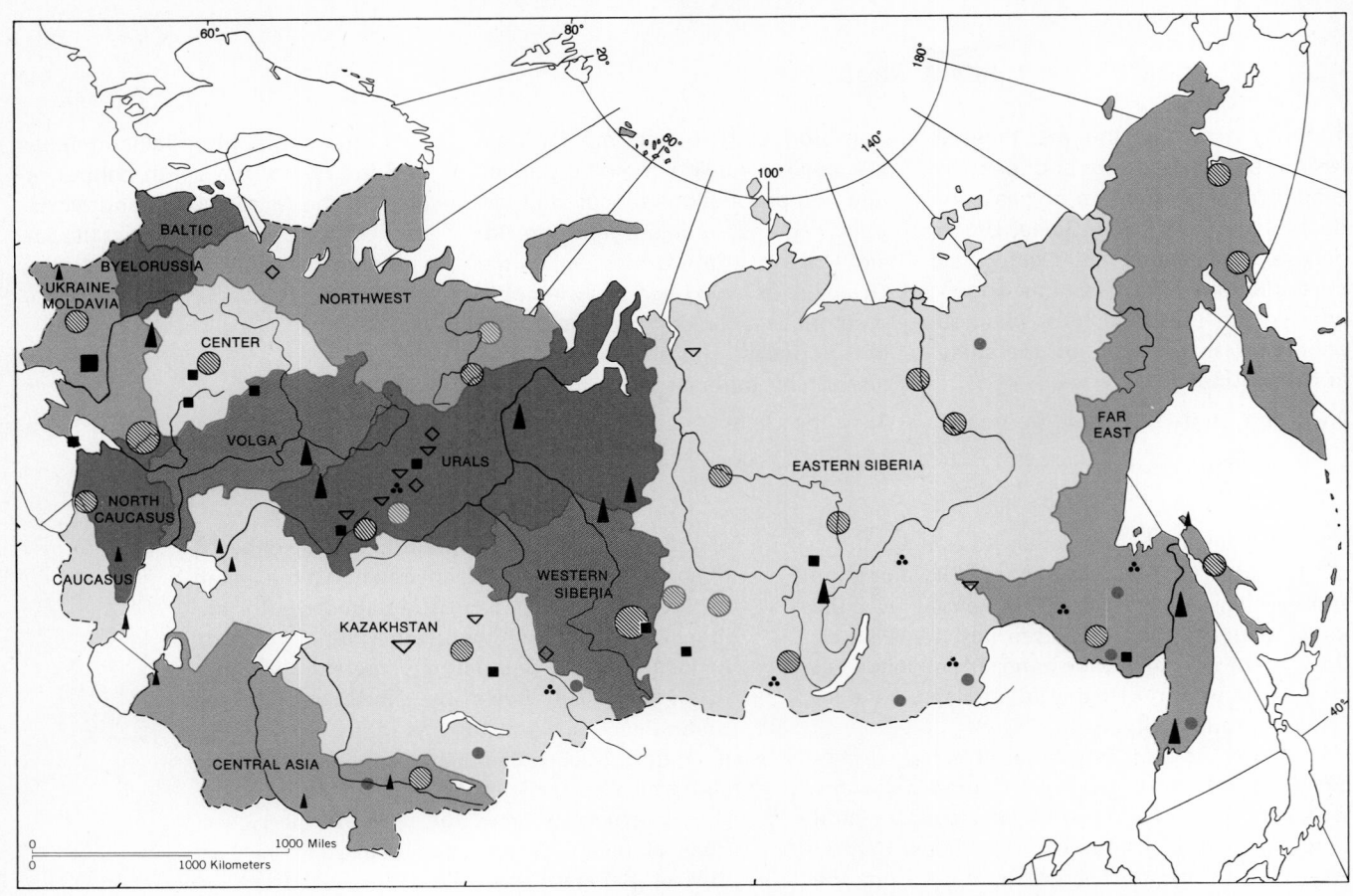

SOURCE: Wessermann Schulatlas, Grosse Ausgabe, 1973; Jack Petersen-Studienatlas, Copenhagen, 1975

FIGURE 9–6 ECONOMIC REGIONS AND IMPORTANT MINERAL DEPOSITS IN THE SOVIET UNION

Legend:
- ◎ Coal
- ◎ Lignite
- ▲ Petroleum
- ■ Iron ore
- ▽ Copper
- ◇ Bauxite
- ● Lead and zinc
- ⚬ Gold

TABLE 9-8 REGIONAL LOCATION OF THE PRODUCTION OF MAJOR
INDUSTRIAL MINERALS, SOVIET UNION, 1972 *(Million Metric*
Tons, Except Natural Gas in Billion Cubic Meters)

	Iron Ore		Coal		Crude Oil		Natural Gas	
	Quantity	Percent	Quantity	Percent	Quantity	Percent	Quantity	Percent
Areas of European Settlement								
European RSFSR	30	14	79	12	221	55	59	27
Ukraine	120	58	211	32	15	4	67	30
Rest of Europe	—	—	—	—	6	2	—	—
Urals	25	12	42	6	29	7	4	2
Western Siberia	14	7	152	23	63	16	11	5
Subtotals	189	91	484	73	334	84	141	64
Other Areas								
North	—	—	23	4	6	2	13	6
Eastern Siberia	—	—	63	10	2	0.5	—	—
Kazakhstan	18	9	75	11	18	4	4	2
Central Asia	—	—	9	1	19	5	55	24
Caucasus	1	0.5	2	0.3	17	4	7	3
Subtotals	19	9	172	26	62	16	79	35
Totals	208		655		400		221	

SOURCE: Theodore Shabad in *Soviet Geography,* American Geographical Society, May 1973, pp. 332–339.

and the tremendous investments for roads and railroads through the wilderness can be avoided. All of eastern Siberia and the Far East has virtually no railroads other than the Trans-Siberian, which follows the Chinese border closely. The dual-track Trans-Siberian Railroad, connecting Moscow with Vladivostok on the Pacific coast, passes through Omsk, Novosibirsk, and Irkutsk. Its economic and military importance to the Soviet Union is reflected in an excellent program of operation and maintenance.

The North also has the major waterpower resources of the Soviet Union, but they are tapped only along the southern margin close to the core triangle. The largest hydroelectric stations are at the eastern apex of the triangle, where the Soviets have developed major power-using industries, particularly aluminum, atomic power, and chemical plants.

Development of mineral deposits in dry Central Asia has been somewhat greater. In the more remote southern margins emphasis has been on high-value, strategic minerals—copper, gold, tin, and mineral salts—which are not readily available in the more accessible areas of the core triangle. Elsewhere, in keeping with the Soviet program of regional self-sufficiency, minerals needed in the local economy of the region, such as coal and oil, are produced. Coal, oil, and even iron ore are produced in Kazakhstan, but much of the production is in northern Kazakhstan near the new agricultural settlements on the southern fringe of the core triangle.

The Caucasus has been more closely tied to the core of the country than have other non-European areas because of its proximity and accessibility via the Caspian Sea–Volga River water route. It contains a small but important subtropical and humid (high-work) climate where citrus fruits and tea are grown. The oil fields at Baku were the first to be discovered in the Soviet Union, and for many years this was the only important oil-producing area of the country. Today, however, the Caucasus area accounts for less than 5 percent of oil and gas production.

THE REGIONAL PATTERN

Over half the Soviet Union functions essentially as **supplemental regions,** or "colonial regions," in the Soviet economy. Like the former colonies of European countries, their principal function is to support the core region by supplying needed raw materials. Unlike the complementary regions of the United States and Europe, the supplemental regions of the Soviet Union are not important markets for the products of the core region. The more important outlying areas—ports in the European north, coal and oil fields there, oases and mining areas of Central Asia and the Caucasus, and strategic settlements along the southern Pacific coast near China and Japan—are linked by railroad to the core triangle. But as Figure 9–7 shows, the flows are generally much smaller than those within the populated regions. The smaller, even more isolated developments in the Asian north are linked primarily by air transportation. Pipelines similar to that in Alaska or those in the Middle East have been built to carry the oil and gas to the core triangle. Efforts are also made to use the Arctic and Pacific oceans and the Siberian rivers, even though they are usable only for a few weeks during summer. One of the great

problems of the Soviet Union has been the development of a transport network to make the vast overland connections needed to tie the huge expanse together. Even with central planning, the task is overwhelming.

THE PRODUCTIVE CORE TRIANGLE

The majority of Soviet population and most economic development lie within the core triangle of European settlement. The economic importance of this region can be seen from the flow of traffic on the railroads (Figure 9–7), which are the basic means of freight transportation in the Soviet Union and carry about 80 percent of all goods moved. From the core triangle comes the major agricultural and industrial output of the country.

As Tables 9–9 and 9–10 show, the core triangle produces the basic food needs of the country. It accounts for 86 percent of the sown cropland, and its share of total food output is even greater. Emphasis is on grain, sugar beets, potatoes, hay, silage, and sunflowers. In the drive toward national self-sufficiency, sugar beets have replaced dependence on tropical cane sugar, and sunflowers replace soybeans and peanuts in providing essential vegetable oil. Except for sheep, which are prominent in Central Asia and the Caucasus, the basic livestock production of the country comes from this area, too.

Comparison with the Eastern United States

Even within the core triangle, the environment for plant growth is not so favorable as is that of the United States. In the Northwest, where there is adequate moisture for full crop production, summers are cool, as they are in Europe. Consequently, crops such as wheat cannot mature properly during the growing season. Along the southeastern margin where growth potential, in terms of sun energy, is more than 50 percent greater than in the Northwest, substantial moisture deficits normally occur during the growing

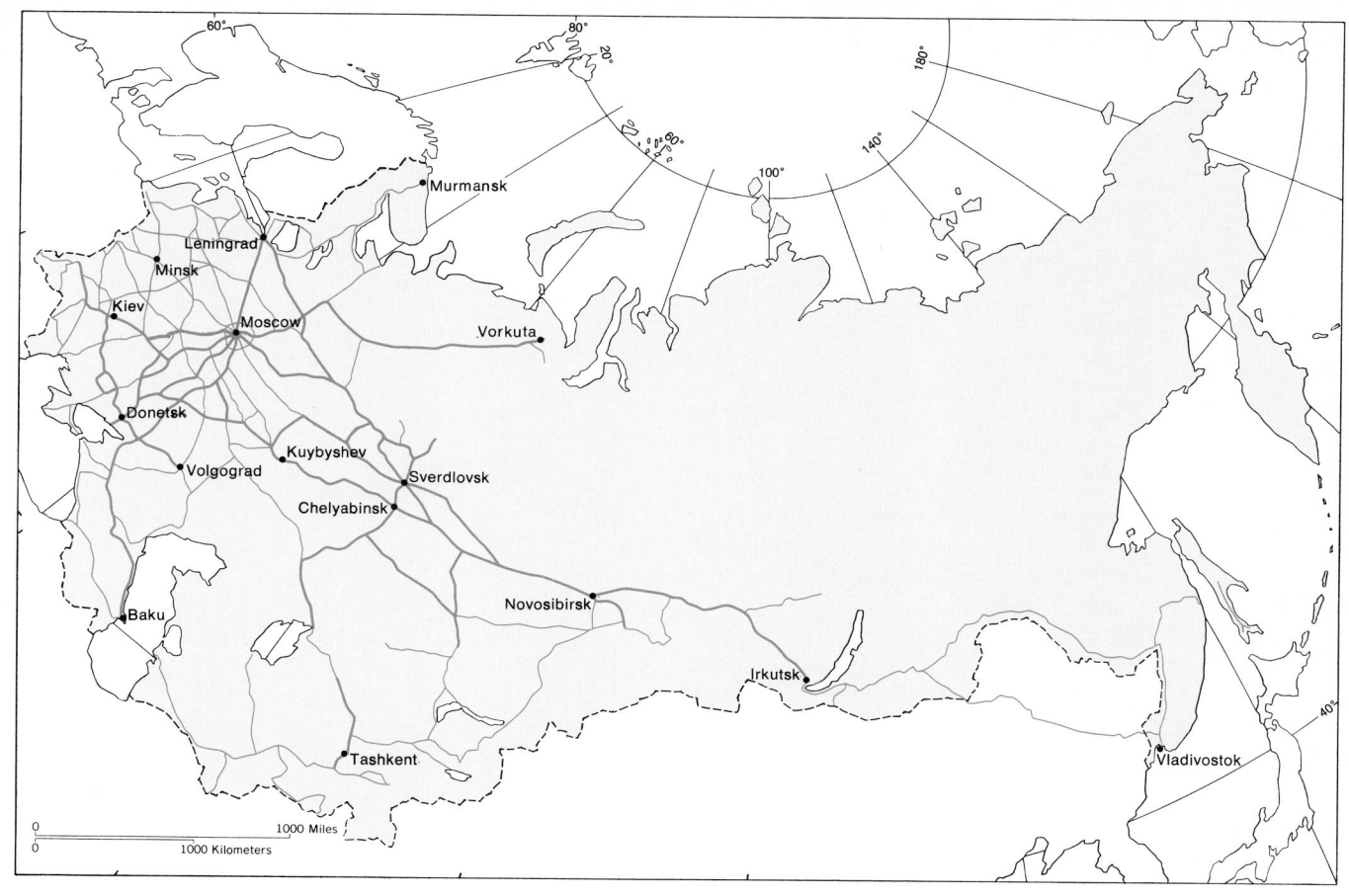

FIGURE 9-7 RAIL CONNECTIVITY IN THE
SOVIET UNION

——— Lines carrying heaviest freight traffic

——— Lines of lesser freight traffic

season (see Figure 3–6). Thus where it is wet enough to grow crops, solar energy is insufficient, particularly for cereals; and where it is warm enough for good midlatitude crop growth, drought is a major hazard, as in the western margins of the Great Plains of the United States. Most years have moisture shortages, and in some years drought ruins the entire crop.

The Soviet "Middle West" lies south of Moscow and west of the Volga, extending southward to include the Ukraine. In this "black earth" country, where rich prairie soil covers largely flat areas, over three-quarters of the land is in agricultural use, and 70 to 80 percent of it is in crops (Table 9–9). Yields here are the highest per acre in the country outside of irrigated regions. Traditionally, this has been the major wheat region of the country. In the early part of this century it

TABLE 9-9   SOWN AREA IN THE CORE TRIANGLE
AND KAZAKHSTAN

	Total Area	Sown Area	Percent of Total of Sown Area in Triangle	Sown Area as Percent of Total Area
	Thousands of square miles			
Old established centers				
Northwest	82.0	7.7	1	9
Baltic	67.0	17.0	3	25
Byelorussia	80.0	21.1	3	26
Center	370.0	121.5	18	33
Ukraine–Moldavia	245.1	134.8	19	55
North Caucasus	137.1	68.1	10	50
Total	981.2	370.2	54	
Recent areas of settlement				
Volga	191.0	71.4	10	37
Urals	291.0	63.9	9	22
Western Siberia	409.0	76.1	11	19
Kazakhstan	1,064.0	110.6	16	10
Total	1,955.0	322.0	46	

SOURCE: J. P. Cole and F. C. German, A Geography of the U.S.S.R., 2nd ed., Butterworths, London, 1961, pp. 88–89.

produced a surplus of wheat for export and still produces almost a third of the country's wheat (Table 9–10), although no longer a surplus for export. Most of the wheat is planted in the fall and harvested in the spring (winter wheat), which is comparable to wheat growing in the southern half of the Great Plains in the United States. In recent decades land has been shifted to corn, sugar beets, sunflowers, potatoes, and other higher-return crops, and an attempt has been made to establish a "corn belt" type of grain and livestock economy. This area now includes over one-third of all the cattle and almost half of all the hogs in the country.

As a result of these changes, wheat in the Soviet Union has been shifted to lands east of the Volga on the dry margins of the core triangle, largely in northern Kazakhstan. Spring wheat is grown in this area, since winters produce little snow cover and soil freezing would kill fall-planted wheat. These lands, which were mostly uncultivated early in this century, now account for almost half the Soviet wheat crop. Production east of the Volga in 1964 was more than three times that in 1940. As in the United States, production is possible only by using tractor-driven plows and mechanized combines that can cover hundreds of acres of land.

The areas north and west of Moscow are much less important because of the more limited physical environment. Like the New England region in the United States, however, they are close to markets and have increasingly specialized in market-oriented products such as dairy products

Area	Percent of Production in Population Triangle and Kazakhstan				
	Grain	Potatoes	Sugar beets	Whole milk	Meat
Old established centers					
Northwest	1	4	—	5	3
Baltic	2	7	1	8	7
Byelorussia	1	14	1	6	6
Center	13	29	20	22	17
Ukraine–Moldavia	22	24	60	25	27
North Caucasus	11	2	10	6	10
Total	50	80	92	72	70
Recent areas of settlement					
Volga	16	8	4	9	10
Urals	9	5	—	8	7
Western Siberia	9	4	1	7	5
Kazakhstan	16	2	2	4	8
Total	50	19	7	28	30
Percent of total production of the USSR within triangle	94	96	97	91	88

SOURCE: Gregory, *Russian Land, Soviet People*, Pegasus, Indianpolis, 1968, Appendix table 3, pp. 887–890.

and vegetables. Nevertheless, much of the land remains in forest, and large parts of the farmland are planted for hay crops.

Variations in Agricultural Use and Productivity

Within the core triangle there is considerable variation in the use of agricultural land. Table 9–9 shows that the area of sown crops is almost equally balanced between the established areas west of the Urals and the areas to the east in which the Soviets have encouraged settlement. But notice that the areas along the western margin of the triangle—the Northwest, Baltic, and Byelorussia—together account for only 7 percent of sown cropland. Each of the other areas con-

tributes at least 9 percent, and the Ukraine and Moldavia, the Center, and Kazakhstan each approaches 20 percent.

Table 9–9 shows that the proportion of each region used for sown crops varies greatly. It is as low as 9 percent in the Northwest and 10 percent in Kazakhstan in the southeast, around 20 percent in the Urals and western Siberia, and about 25 percent in the Baltic and Byelorussian areas. In the Center area it reaches one-third, and in the North Caucasus and Ukraine-Moldavia it peaks at 50 percent or more. This description gives some measure of the evaluation by the Soviets of the agricultural possibilities of the core triangle and some idea of the greater difficulties that are apparent in the northwestern, western, and southeastern margins.

The importance of the central areas in the triangle—the Center and the Ukraine—can be seen by looking at output of major agricultural products rather than at area sown to crops. In the outlying regions where it is too cool or dry for maximum crop growth, one would expect to find smaller crop areas and lower yields—thus less production. Table 9–10 shows regional proportions of production of the three major types of crops plus meat and whole milk in the core triangle. The triangle accounts for virtually all production of these agricultural products. Notice how much more important the areas west of the Urals are in total output of these major food products than are areas to the east. The western portion accounts for 70 percent of the triangle's meat, 70 percent of its milk, 80 percent of its potatoes, and 92 percent of its sugar beets. Only in grain production do the more recently settled eastern areas hold an equal share.

Only when the combined output of these five major agricultural products from the Center and Ukraine-Moldavian regions alone is considered can the importance of the central portion of the triangle in the agriculture of the Soviet Union be appreciated. These two regions account for 80 percent of the sugar beets, 53 percent of the potatoes, 47 percent of the whole milk, 44 percent of the meat, and 35 percent of the grain of the triangle. This is the Soviet "Middle West," by far its major food-producing area.

On the other hand, the importance of the Soviet drive to increase agricultural production east of the Volga, particularly in western Siberia and northern Kazakhstan, can be seen. The development of the "virgin lands" into a major grain-producing region was started in 1954. New state farms were established and railroads were built to serve the area. The amount of cultivated land in the Soviet Union was increased by almost 100 million acres. The program to turn these formerly unproductive lands into grain "factories" has resulted in the production there of almost half the country's grain needs. In fact, in 1964

they produced 60 percent of the total wheat crop of the country. The crop choice on these drier lands is limited, but they contributed an amount equal to almost 20 percent of the potato crop of the triangle and around 30 percent of its whole milk and meat output—certainly, significant figures in a country striving for increased agricultural output.

The areas along the western margin of the triangle, while generally less important, are still significant producers of some agricultural products. Over 20 percent of the potatoes and almost 15 percent of milk and meat products come from those regions. These cool, damp areas also contribute to agricultural output.

A COMPARISON WITH THE UNITED STATES

We have frequently compared regions of the Soviet Union with regions of the United States. Let us now attempt an overall comparison of the Soviet Union with the United States. The Central area, including Leningrad and Moscow, might be considered the national focus, comparable to the Northeast in the United States. The Ukraine, with its emphasis on steel and agricultural production, is the second key area, analogous to the Middle West of the United States. The area within the core triangle east of the Volga and extending into western Siberia is in many ways like the Great Plains–South Central region of the United States. The emphasis here is on agriculture in a high-risk environment and industry based on natural resources, particularly oil and natural gas.

Central Asia and the Caucasus, outside the core triangle, function much as does the U.S. West, with irrigation agriculture and specialty mineral production. But the Soviet Union has no special "oasis area" comparable to California. The Soviet North and the Siberian East are still largely wild, like Alaska. There is scattered development of key natural resources such as oil and gas, minerals such as gold and nickel are mined, and timber is produced on the southern

TABLE 9–11   INDEXES OF INDUSTRIAL REGIONALIZATION:
STEEL PRODUCTION AND ELECTRIC POWER
PRODUCTION, 1972

	Steel Production		Electric Power Production	
	Millions of tons	Percent	Billions of Kilowatt-hours	Percent
Areas of European Settlement				
European RSFSR	20	16	283	33
Ukraine	49	38	158	18
Rest of Europe	6	5	57	7
Urals	39	30	98	11
Western Siberia	8	6	118	14
Totals	122	95	714	83
Other Areas				
The North	—	—	—	—
Eastern Siberia	—	—	38	4
Kazakhstan	4	3	41	5
Central Asia	0.4	—	33	4
Caucasus	2	1	30	4
Totals	6.4	4	142	17

SOURCE: Theodore Shabad in *Soviet Geography,* American Geographical Society, May 1973.

fringe of the core triangle in Europe. The Soviet Union, however, does not have a sizable area that is comparable to the southeastern United States.

MANUFACTURING, THE
FIRST PRIORITY

Industrial output is of first importance to the Soviet Union. Agricultural output contributes less than one-fourth the total domestic production of the Soviet Union. Manufacturing and mining total twice as much, or slightly more than half of all production.

Manufacturing is highly concentrated within the core triangle, as is agriculture. Some 82 percent of primary fuel consumption and 83 percent of workers involved in manufacturing are found in the core triangle, not including Kazakhstan. The triangle produced 95 percent of the country's steel and 83 percent of its electric power in 1972 (Table 9–11). In contrast to agriculture, manufacturing is a much more localized activity. Not only does it tend to concentrate in focal points, usually urban and metropolitan places, but certain regions develop as major centers of manufacturing. Recall the localization of manufacturing in the United States. The same condition holds even in a state-controlled economy such as that of the Soviet Union despite the ideal of regional self-sufficiency, with a minimal industrial base in all outlying regions.

THE "BIG THREE" REGIONS
AND THEIR ORIGINS

Industry in all three of the leading industrial districts dates back to the czarist attempts at industrialization. Manufacturing was highly centralized in the Russian Empire. It was in the Center and neighboring northwest, where St. Petersburg, the capital, was located, that artisans producing the manufactured needs of both the court and the government were concentrated. Following the Western European model, the first factories were based on waterpower, and the production of textiles was most important. At first production was based on local supplies of wool and flax, but as in Western Europe, cotton manufacture soon followed. As the most advanced part of the country and its most important market, this area pioneered new manufacturing and engineering undertakings. Moreover, it was the hub of transport links from all major parts of the empire, and so it had access to necessary resources. The development of railroads in the nineteenth century tended to centralize this manufacturing even more.

The Central area, like the northeastern seaboard of the United States, did not contain much of the basic resources of modern industrialization—coal, iron, and petroleum. The first real iron industry developed in the Ural Mountains on the basis of local deposits of iron ore and charcoal. That area lacked the necessary large-scale coal deposits required to shift over to the mass production of steel in the late nineteenth century.

The Ukraine was the first major center of large-scale, modern steel production. Like the Middle West of the United States, this agricultural area had the deposits of coking coal and iron required as a base for a modern iron and steel complex. Coal was mined in the southeast corner of the Ukraine known as the Donets Basin. Iron was found some 200 miles to the west. By the time of the Revolution the Ukraine was the dominant steel-producing area of the country, contributing two-thirds of the country's

output. Most of the development was undertaken with Western European capital and technical knowledge.

INDUSTRIAL REGIONS UNDER THE SOVIETS

The Soviets have been on a crash program of industrial expansion ever since the Revolution. They consider the development of manufacturing to be the foundation for the political and economic power of their state. Expansion in the established centers inherited from the czars has been encouraged in addition to the establishment of new areas with access to resources and with good market orientation (Figure 9-8).

It is not surprising that in this centralized state the largest industrial complex exists in the area known as the Center. This is where the market is concentrated, and it is the focal point of the entire transport system of the country. Here is concentrated the intellectual, scientific, management, and engineering talent and the largest labor pool. The only major obstacle, as noted earlier, is the relative sparsity of the resources needed for modern industry. This can be overcome by importing the needs from other areas. The Center meets some of its energy requirements from local deposits of peat and lignite, but it draws most of its energy from coal, oil, and gas brought in by rail and pipeline. Like industry on the northeastern seaboard of the United States, industrial production in the Center emphasizes sophisticated engineering equipment and consumer goods, rather than basic steel and chemical production, which requires large-scale resources.

Today the Ukraine remains the Soviet Union's leading steel producer. It produced almost 40 percent of national output in 1972. Supporting this complex are resources of manganese, salt, and natural gas and large-scale hydroelectric power on the Dnieper River. Also, the region's productive agricultural base serves both as a supplier of raw materials and as a market for farm machinery, fertilizer, and other farm chem-

REGIONAL LOCATION OF INDUSTRIAL PRODUCTION

1. Steel is the basic heavy industry of any industrial country. What are the major regions of Soviet steel production? How does their distribution compare with the distribution of population?

2. Electric power is a measure of total industrial output. By this measure, what are the key industrial regions? How do they compare with the regions of steel production?

3. The European part of the RSFSR is the center of population; it is also the location of Moscow and the country's original industrial development. Is it therefore similar to the northeastern seaboard in the United States? This U.S. area is resource poor and not a major center of steel production. Is this also true of the most important industrial region of the Soviet Union?

---

icals. The result is an emphasis on metalworking activities, closely linked to the Center by the country's busiest transport routes. In these respects the Ukraine is similar to the Middle West of the United States.

Despite its czarist industrial origins, the Ural region is a Soviet creation. Ural development was seen in strategic terms. An industrial complex in this location is far from the European borders and well located with regard to developments farther east in Siberia. An almost completely new modern industrial base was established, despite the presence of a basic infrastructure of rail lines, cities, and mines. Industrial development in the Urals is based on the utilization of large deposits of mineral resources, particularly iron and nonferrous metal ores. It is the country's second major steel base and produces almost a third of the steel.

On the site of a major iron deposit in the southern Urals the Soviets built a steel industry and a new city, Magnitogorsk ("Iron Mountain"), in the 1930s. Coal for coking was brought from coal deposits in the Kuznetsk Basin near the eastern apex of the core triangle, and iron ore was hauled on the return trip. In this way the Soviets developed two iron and steel centers. The operation was nevertheless expensive be-

cause both iron ore and coal had to be hauled more than a thousand miles. Today the Urals industry has found a closer coking coal source in northern Kazakhstan, and iron mines have been developed in the Kuznetsk Basin in Siberia.

The other smaller industrial districts of the core triangle also consist mainly of heavy industry tied to iron, coal, and oil deposits. They include the Kuznetsk steel center and other communities in western Siberia, cities along the middle Volga River, and cities near hydroelectric power at the eastern apex of the triangle. The Siberian regions suffer from their remoteness from the major centers of population. Despite rich resources, the steel industry in western Siberia is only one-fifth as important as in the Urals, and the Siberian hydroelectric power complex is a specialized development comparable to the specialized industrial district of the Pacific Northwest of the United States.

The lower Volga cities could be called the Soviet "Gulf Coast." They have always had ready access to Caucasus oil, and now the most important oil fields in the country have been developed east of the Volga along the western flank of the Urals. Thus these cities specialize in oil refining and petrochemicals as well as in producing the equipment needed in oil production. As with the

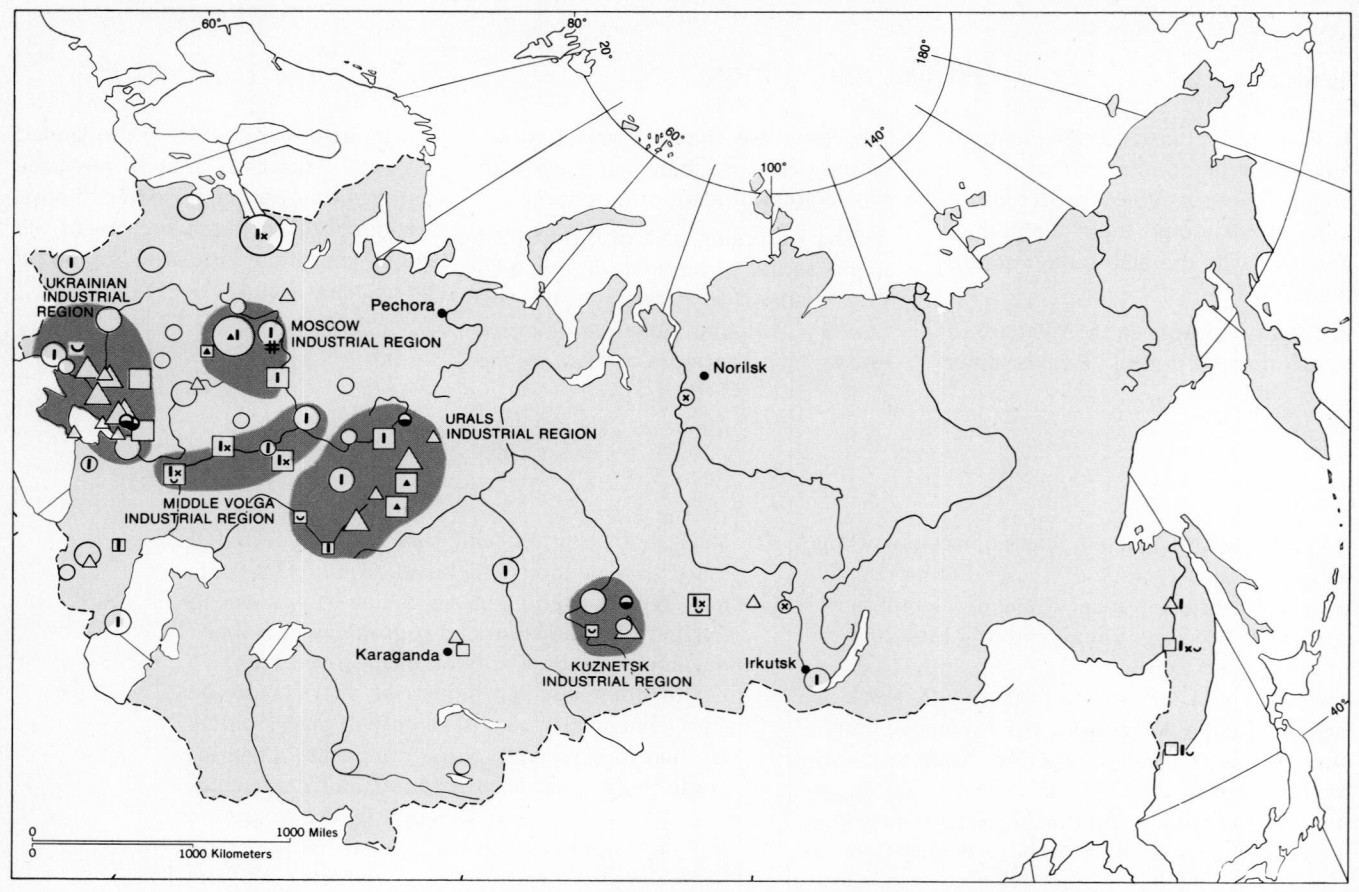

FIGURE 9-8 MAJOR INDUSTRIAL REGIONS
OF THE SOVIET UNION

Major industrial region

Major types of industrial centers
Classified by most significant type of manufacturing

△ Iron and steel

☐ Machine building

○ Diversified industries

Other important industries

❙ Petroleum refining

⊖ Chemicals

✕ Wood industries

⌗ Textiles

⌣ Food processing

Texas ports, access to cheap water transport is important—in this case, via the Volga River and its connections with other waterways.

THE FUNCTIONING INTERCONNECTED SYSTEM TODAY

The nonmetropolitan population of the core triangle of the Soviet Union has a relatively sparse distribution of people living in collective farm villages and on state farms. In the better agricultural areas, such as the Ukraine, rural population densities are high, whereas in the Northwest and on the fringes they are lower. Important agricultural outpost areas lie outside the core triangle in the Caucasus and in Central Asia.

Superimposed on the widespread agricultural population is a series of metropolitan and urban nodes of two sorts. There are administrative centers like Moscow and the capitals of the other fourteen member republics; and industrial centers, largely tied to the development of particular nonagricultural resources. Some of these are in traditional, early industrial locations that had their beginnings before the development of coal and steel manufacture. Some were developed in czarist times, and others are the result of recent expansion under the Soviets. Most cities are clustered in the interconnected industrial regions such as the Center around Moscow.

INTERNAL OPERATION

The metropolitan centers function much as they do in the United States and Europe, as marketplaces, distributing centers, and administrative and cultural centers for rural areas and smaller communities around them. The most important function of the national transportation system is to provide connecting links between the metropolitan centers. Its secondary function is to link the metropolitan centers to the communities and rural areas around each of them.

A hierarchy of places exists in the Soviet Union, reflected in their size. The highly centralized control of the Soviet system tends to make the hierarchy both simpler and more structured than in the United States or Europe. The lack of separate entrepreneurs and corporations reduces competition between places for national and regional markets or in the development of particular resources. The government determines the resource development plan, the production centers, and the distribution network. Through its system of production goals, allocation quotas, and consumer controls, it pushes for conformity to its master plan. In recent years there has been pressure to build incentives into the system by allowing plant managers more freedom in terms of procurement and production. In some respects, the Soviet producing system can be equated to the operations of a gigantic corporation with the goal of achieving the central government's political, economic, and planning goals.

Moscow, the seat of the planning bureaucracy of the highly centralized Soviet system, is by far the largest metropolitan center, the dominant national city, and the hub of the national transportation and communications systems. On a level below Moscow is Leningrad, only 300 miles (480 km) away. As Moscow's closest port, it continues to be an extension of the political, cultural, and economic control center of Moscow. It performs a national function, although it is also the regional center of the Northwest.

At the lower levels of the hierarchy, two different components determine a city's importance: administrative connections to the member republics, and economic links to the various producing centers. Although all major metropolitan areas carry out both of those functions, in the **administrative centers** manufacturing employment is generally less than 50 percent of the work force. In most **industrial centers,** on the other hand, at least 50 percent of employment is in

TABLE 9-12 HIERARCHY OF CITIES OVER 500,000 IN THE SOVIET UNION, BY POPULATION, 1974

Moscow	7,528,000
Leningrad	4,243,000

Administrative Centers[a]		Industrial Centers	
Kiev	1,887,000	Kharkov	1,330,000
Tashkent	1,552,000	Gorki	1,260,000
Baku	1,359,000	Novosibirsk	1,243,000
Minsk	1,095,000	Kuybyshev	1,140,000
Tbilisi	984,000	Sverdlovsk	1,122,000
Kazan	931,000	Odessa	981,000
Ufa	871,000	Chelyabinsk	947,000
Yerevan (Erevan)	870,000	Dnepropetrovsk	941,000
Alma-Ata	813,000	Omsk	935,000
Riga	776,000	Donetsk	934,000
		Perm	920,000
		Volgograd	885,000
		Rostov-na-Donu	867,000
		Saratov	820,000
		Voronezh	729,000
		Zaporozhe	729,000
		Krasnoyarsk	728,000
		Krivoy Rog	620,000
		Lvov	605,000
		Karaganda	559,000
		Yaroslavl	558,000

SOURCE: *U.N. Demographic Yearbook, 1974.*
[a] Capitals of union republics and autonomous republics.

manufacturing. The size of a metropolitan center in either line of the hierarchy is a measure of its relative importance (Table 9–12).

Note the rapid drop in the size of city populations near the top of Table 9–12. Moscow is by far the largest metropolitan center, and Leningrad has several million fewer people. The next largest cities, in both the administrative and in- dustrial hierarchies, are clustered in the range between 1 and 2 million people. In this pattern of city sizes, where population drops sharply near the top of the hierarchy, the largest city is called the **primate city.** Compare this pattern with that of the populations of New York City, Los Angeles, Chicago, Philadelphia, and Detroit (see Table 7–5), where city populations decline more gradually. What accounts for the difference in the patterns of city sizes in the Soviet Union and the United States? The centralization of the Soviet political and economic structure goes far toward explaining Moscow's dominance. Another factor is that the process of urbanization has not developed as much in the Soviet Union as it has in the United States. The network of urban centers in the Soviet Union is therefore simpler; it has not been so shaped by complex interactions and flows as has that of the United States.

The Soviet Union has fewer administrative centers than industrial ones. Among cities of over 500,000 population, ten are administrative centers and twenty-one industrial. The administrative centers, as capitals of union republics and autonomous republics, are widely dispersed, whereas the industrial centers are more concentrated. Almost all industrial centers are within the core triangle, eight of them in the Ukraine alone.

The larger member republics and economic regions usually have a larger number of minor metropolitan areas. In the Ukraine are twenty-nine metropolitan centers of over 100,000 people, four of them in the coal mining area and two adjacent to a major industrial center on the Dnieper. In the Center are thirty-one centers over 100,000, eight of them within the immediate Moscow district. Larger outlying places, such as administrative centers in eastern Siberia and mining centers outside the populated zones, usually have fewer than 100,000 people and may be tied into the national system only by airline and radio communication; it all depends on their importance.

A substantial portion of the Soviet Union is virtually outside the present-day functioning system. This is true of the vast expanses of the empty northern and eastern areas of Siberia, of the Central Asian desert outside economic islands—such as the new coal-mining center of Karaganda in northern Kazakhstan—and of the high mountains of the south and east. These areas are the home of peoples similar in level of development to the Middle Eastern Bedouins.

The Soviet system sees all people as tied to the state and its goals, and the state tries to bring them into the political and economic system as much as possible. Efforts have been made to indoctrinate even the most isolated peoples in the principles of the Soviet political-economic system and to provide them with the country's basic social and health services. Mountain peoples and arctic tribes now have their own collective farms. Similarly, efforts are made to encompass all city populations in the modern interconnected system.

People are expected to work in the system and, in turn, the system provides for their basic needs. As might be expected, the government-controlled system restricts freedom of movement from rural areas to cities; workers need working permits. The result is less movement of unskilled rural people than in the United States or Europe. At the same time, schools and training institutes in rural areas provide schooling designed to fit young people not only for rural work situations but also for urban ones. Moreover, major efforts have been made to raise standards of living and employment in rural areas. Doctors, engineers, and other technicians are commonly assigned to work in outlying cities and on collective farms.

LIMITED EXTERNAL CONTACTS

Because of traditional ideological differences with the capitalist world, the Soviet Union is a much more self-contained functioning unit than are other modern nations. It is not completely self-sufficient; it has, for example, had to import large quantities of grain from the West, especially the United States and Canada. According to the terms of a five-year agreement signed in 1976, the Soviet Union will purchase 6 to 8 million tons of U.S. grain per year. However, as noted in Table 9–1, the value of Soviet international trade is less than one-third that of the United States. Moreover, two-thirds of that trade takes place within the Soviet bloc, mostly with Eastern Europe. These areas provide needed resources such as coal and food, but also manufactured goods, especially machinery and consumer goods. The modern industrialized areas of East Germany, Poland, Hungary, and Czechoslovakia contribute importantly to the Soviet economy.

The United States has viewed trade by the Soviets with industrial countries as having strategic importance, because this trade has usually involved machinery and other items of advanced technology. For many years after World War II, the United States had an embargo on trade with the Soviet Union and exerted pressure on Western European countries to hold down their trade with the Soviets.

The Soviet Union, like other industrial countries, has used programs of foreign aid to gain the support of underdeveloped countries. Soviet assistance has gone to India, Egypt, Indonesia, North Vietnam, and Cuba. Since Cuba's break with the United States and with most Latin American countries in 1959, the Soviet Union has provided basic military and technical aid to that country and has become the primary source of foreign trade for the Castro regime. In the same way, the Soviet Union sent both capital goods and technicians to assist China when the Chinese Communist regime gained power after 1948. This aid stopped only when the Chinese broke with the Soviets in the mid-1960s and asked that assistance be discontinued. The Soviet Union has also provided India with grain and built a steel mill there. All these moves have had political as well as economic motives.

Since World War II the Soviet government has also undertaken programs of technical and military assistance to underdeveloped countries in Africa, Asia, and Latin America. It built the new Aswan High Dam in Egypt after the United States refused to do so, and in addition it provided Egypt with military equipment. It strongly supported the North Vietnamese military takeover of South Vietnam, and it has supplied Syria, Libya, and Algeria in the Middle East–North Africa and several of the new Black African countries.

There is only a limited private sector in the Soviet Union, and all foreign dealings are government controlled. Other industrial countries and their private trading enterprises fear this, because the Soviet government can enter any trading market and undersell corporations or dump underpriced goods on international exchanges, thereby disrupting existing patterns. Thus Soviet trade with underdeveloped countries is looked upon with great suspicion not only by capitalist governments but by corporate management as well. At present this trade is not very significant, even though the Soviet Union needs products from the underdeveloped world. These areas account for only a small fraction of Soviet trade.

The Soviet Union has also been pushing to open diplomatic and trade channels with Western Europe. Soviet production still has gaps that can be filled by trade with those countries, particularly with West Germany and France. European and U.S. corporations have obtained licenses to trade with the Soviet Union and to produce their products there. Fiat automobiles are being built there, and manufacturers in Europe and the United States have supplied essential production equipment to a giant new truck and engine complex along the Kama River. Even Pepsi Cola has been bottled in the Soviet Union.

Thus the Soviet Union seems to be moving toward greater participation in the modern interconnected system on the international level. In so doing, it seems to be following the U.S. model—trading overseas where there is an advantage in obtaining resources or selling products, and where political pressures dictate. And it seems likely that the Soviet Union, like the United States, will continue overwhelmingly to depend on its own resource base and market in expanding its modern interconnected system.

SELECTED REFERENCES

Central Intelligence Agency. *USSR Agriculture Atlas.* Washington, D.C., 1974. A nonclassified document that provides an excellent treatment of the agricultural environment, technology, organization, and output of this important facet of the Soviet economy.

Lydolph, Paul E. *Geography of the USSR,* 2nd ed., 1971; 3rd ed., 1977. John Wiley, New York. Two widely different editions. The second edition presents overall geographic patterns in the Soviet Union; the third gives detailed information on the physical, cultural, and economic components of regions of the country.

Mathieson, R. S. *The Soviet Union: An Economic Geography.* Barnes & Noble, 1975. A basic geography by an Australian geographer that touches briefly on the environmental base, human resources, major industries, and Soviet economic regions.

Smith, Hedrick. *The Russians.* Quadrangle/New York Times, New York, 1976. A perceptive journalist's look at the Russian people and the world in which they live.

Soviet Geography: Review and Translation. American Geographical Society, New York. A journal providing insight into the way Soviet geographers view their own country. A section of news notes gives current information on economic developments within the Soviet Union.

The Centre National d'Art et de Culture Georges Pompidou ("Beaubourg"), an architecturally innovative museum of art, opened in Paris in 1977. (Marc Riboud/Magnum Photos, Inc.)

Right: Grain produced in the U.S. Middle West is temporarily stored in elevators, like these in Indiana, until it is transported to its market. (Grant Heilman)   *Below:* Communications satellite, developed by the European Space Research Organization, being prepared for space simulation tests at the Toulouse, France, space center. (Alain Nogues/Sygma)

The infrastructure of transportation and communication networks is the basis of the modern interconnected system. Without such complex networks connecting various places in the world, the industrialized nations would have difficulty tapping resources and markets throughout the world. Thus technology, the third variable in the geographic equation, has greatly expanded the possibilities for spatial interaction, which is the fourth variable in the equation. In the modern world there is constant movement of people, goods, and information within towns and cities, across countries, and between producing areas, markets, and population centers throughout the world. Two main levels of interaction can be distinguished. The first is that of countries or regional units; at this level, we speak of the Soviet air transport system, the Japanese rail system, or the European satellite communications system. Each of these systems, however, interconnects with similar systems in other parts of the world to form a worldwide system operating at a higher level of complexity. This is the modern interconnected system on the global scale, where national boundaries and limitations are of less significance.

Newly opened airport at Izhevsk, located between the Volga River and the Urals in the Soviet Union. (Tass from Sovfoto)

Japanese express passenger trains travel between Tokyo and Osaka at a speed of approximately 120 miles per hour (almost 200 kilometers per hour). Mount Fuji is in the background. (Japan National Tourist Organization)

Agricultural produce of Israel awaiting shipment to its market. (Erich Lessing/Magnum Photos, Inc.)

Large metropolitan centers function as nodes in the modern interconnected system; here is where the interconnections between the national or regional systems and the global system occur. Cities are of major importance for the production and distribution of manufactured goods. They are even more important as places where ideas originate and are developed. Twentieth-century culture, including ideas about literature, the arts, and entertainment, has diffused from one metropolitan center to another. Technological innovations have also spread through-out the world by way of metropolitan nodes. The large populations of metropolitan areas allow for the development of the many specialized kinds of labor required by the modern interconnected system. Management of the modern system and its interconnected subsystems is heavily concentrated in metropolitan areas. Here, in the headquarters of large corporations, financial institutions, and government agencies, decisions are made that affect the day-to-day lives of people throughout the world.

Left: New York City, in a view centered on Park Avenue, where the headquarters of many large corporations are located. (Hiroji Kubota/Magnum Photos, Inc.)   *Above:* The Ginza area of Tokyo. (David Lomax/Photo Trends)

Above: Aerial view of the main plants of the BASF Group, a multinational corporation, in Ludwigshafen, West Germany. (German Information Center)   Below: Construction of the Concorde supersonic aircraft in England. (Malcolm Gilson/Black Star)

Finished coils of galvanized sheet steel being removed from storage area. (Bethlehem Steel Corporation)

The primary producing regions are as vital to the modern interconnected system as are the large metropolitan centers. Not only do they supply food to nearby metropolitan centers, but they also provide specialty crops and surplus production for export to other regions or countries. The application of scientific research and modern technology has increased per acre yields through the development of new hybrid seeds and the application of synthetic fertilizers and insecticides. New machinery is constantly being designed to plow the land, plant seeds, spread fertilizer and pesticides, and harvest crops. Specialty crops such as tea require specially designed machines; the small fields characteristic of Japanese farms necessitate machinery quite different from that used in the U.S. Middle West. The distribution system for agricultural products, including livestock and meat, is comparable to that for manufactured goods.

Above: Mechanical plow specifically designed for the wet rice fields of Japan. (Paolo Koch, Rapho/Photo Researchers, Inc.)   *Below:* Wheat combines in operation in the Middle West of the United States. (John Deere Company)

Wholesale meat market in Rungis, a suburb of Paris. (Alain Nogues/Sygma)

A typical large North American farm, with some crops planted in broad strips. (Soil Conservation Service/USDA)

Tea being harvested in the Georgian Soviet Socialist Republic. (Pictorial Parade)

CHAPTER 10

Japan: An Oriental Culture Adapted to the Modern Interconnected System

Like Europe, Japan has a large population crowded into a small area with modest environmental resources. In adapting to the modern interconnected system, Japan has shifted its relationship with the earth environment, tapping raw materials and food resources in other parts of the world and selling the products of its industry and business in worldwide markets. A tradition of centralized control and a rigid class system have influenced the nature of Japan's industrial system centered in Tokyo, Kyoto, and Osaka.

Crowded street in Tokyo. (United Nations)

Rock garden of Ryoanji Temple, a Zen Buddhist temple in Kyoto. (Consulate General of Japan, New York)

Above: International business meeting. (Sony Corporation) *Right:* Intensive land use is evident in this view on Kyushu Island. (George Gerster, Rapho/Photo Researchers)

If this book had been written a hundred years ago, Japan would be described in much the same terms used for India today. At that time Japan, long a traditional Oriental culture isolated by royal decree from the outside world, was in the initial stages of industrial development. The Imperial Court had great wealth and pomp, but 80 percent of the population were subsistence farmers who lived much as Indian farmers do now. The Japanese exerted great effort in growing crops on tiny farm plots which formed the resource base for each family. The living standards of most of the people were little better than those of present-day Indian villagers, and any disruption of crop production, by flood or fire, brought famine. With 34 million people, Japan was overpopulated.

Only forty years ago Japan was one of the leading economic and military powers in the world. Japanese ships carried textiles, silk, tea, and toys and novelties throughout the world, even to the industrialized countries. Japanese industry built airplanes, tanks, warships, and other military supplies for the armies that had overrun much of China and held Korea and Manchuria. With 72 million people, Japan was not overpopulated.

If this book had been written in the early 1950s, a few years after the Japanese defeat in World War II and during the U.S. occupation, it might have followed the popular view of the time that Japan was one of the problem areas of the world where population had outrun the means of production. In those days Japan's population badly outdistanced the ability of the land to produce. Its industry, destroyed during the war, was unable to compete with modern technology elsewhere, and Japanese goods were not welcome in many areas of the world. At that time Japan's economy was supported in part by economic grants from the United States. With 80 million people, it was again overpopulated.

Now Japan is again one of the most successful of the industrialized countries of the world, the one that showed the greatest increase in output during the 1960s. Its gross national product, greater than that of any individual European country, stands third in the world after the United States and the Soviet Union. With 112 million people, Japan is not overpopulated. Table 10–1 tells the story.

THE WORLD'S THIRD LARGEST ECONOMY

Japan is without question one of the great examples of the modern interconnected system. It is the third largest steel-producing nation and a leader in textiles and clothing; its electronic equipment, optical goods, and complex machinery—radios, televisions, cameras, sewing machines, calculators, and automobiles—are marketed in the United States, Europe, and elsewhere. In fact, Japan has a **trade surplus**: its exports exceed its imports. It has twelve of the world's fifty largest commercial banks in terms of deposits and twelve of the fifty largest corporations in terms of assets.

As Table 10–1 shows, the results of Japan's rapid growth in production during the 1960s are reflected in the increase in its per capita GNP. In 1958 it was about one-fourth that of the major industrial countries, but by 1968 it had increased almost fourfold and was not far from that of any of the leading European countries. We can interpret this as the shift of a majority of the Japanese people from a subsistence economy to a monetary economy in just one short decade. By 1976 Japan's per capita GNP was higher than the per capita GNPs of Italy, the United Kingdom, or the Soviet Union.

In 1974 Japan's economy experienced a downturn. The major factor was the large increase in oil prices imposed by the oil-producing countries. Because Japanese industry depends entirely on imported oil, the price increase meant

TABLE 10-1 TOTAL AND PER CAPITA GROSS NATIONAL PRODUCTS OF LEADING INDUSTRIAL COUNTRIES

	Total GNP ($ Billions) 1974	Per Capita GNP ($)		
		1958	*1968*	*1974*
United States	1,403	2115	3580	6640
Soviet Union	577	1966: 860		2300
Japan	422	285	1110	3880
West Germany	365	840	1670	5890
France	272	1005	1925	5190
United Kingdom	189	1015	1445	3360
Italy	152	480	1150	2770
Canada	135	1505	2250	6080

SOURCES: Figures for 1958 and 1968 per capita GNP: "National Income," *Encyclopaedia Britannica, Book of the Year, 1970*, Chicago, 1970, Table 3, p. 407. Figures for 1974 per capita GNP: Population Reference Bureau, Inc., *Population Data Sheet, 1976*, Washington, D.C., 1976. Figures for total GNP: computed from 1974 population estimates in Edward Espenshade and Joel Morrison, eds., *Goode's World Atlas*, 14th ed., Rand McNally, Chicago, 1974.

an increase in production cost that had to be passed on not only to buyers in Japan but, more important, to buyers of Japanese goods overseas. The major question during these and the following years was whether the overseas customers would continue to buy Japanese goods. If they did not, Japan's living standard would suffer. Such changes in Japan's fortunes illustrate how much Japan's success depends on conditions elsewhere in the world. Economic depression, trade restrictions, and wars in the countries with which Japan has links are likely to have an impact on Japan's economy.

Japan is the one non-Western country that has accomplished an industrial revolution. It can now compete as an equal with the United States, the Soviet Union, and the European Economic Community. Japan is an example of what other non-European underdeveloped countries might do with similar efficiency and skillful organiza-

tion. But to evaluate Japan's success as a model for others, we must first examine the country's resource base and note the circumstances that made the changes possible.

JAPAN'S GEOGRAPHIC BASE

Like most European countries, Japan has a large population crowded into a small land area with relatively modest environmental resources. From cultural roots very different from those of Europe, Japan has developed a very different sort of structure: first, the Japanese over the centuries developed a distinctive version of the traditional system; and, more recently, in adapting to the modern interconnected system, they have greatly modified the model of industrialization developed in Europe and the United States.

Japan's rise has been accomplished despite a population density higher than that of any of the major industrial powers and a relatively poor environmental base. Although Japan ranks third in GNP behind the vast countries of the United States and the Soviet Union, it is more comparable in area to the small European countries. Japan is two-thirds the size of France and 10 percent smaller than California. At the same time, Japan is one of the most populous countries in the world. Whereas California has over 20 million people, Japan has 112 million. It has almost twice the population of any of the European countries and ranks sixth in the world in population. Still, Japanese population densities are less than those of some smaller countries like the Netherlands and Belgium.

If, however, we consider only the amount of arable land, Japan is probably the most densely populated country in the world. Less than one acre in five is in crops, which means there are seven persons for every acre of cropland. Even in the Netherlands there are only five persons per arable acre.

If the population were dependent on the resources of its own country, as it was a hundred years ago, Japan would be greatly overpopulated. But today Japan has one of the most successful and rapidly growing economies in the world. It has achieved this by completely shifting its relations with the earth environment. Today the long-range, interconnected system taps raw material and food resources in many parts of the world, and, in exchange, the products of Japanese industry and business flow into worldwide markets. Japan has created wealth on a meager environmental base.

RUGGED MOUNTAINTOP ISLANDS

Japan is essentially a series of mountains that rise above the Pacific Ocean off the coast of Asia, forming a series of rugged islands. As Figure 10–1 shows, there are four large islands, each separated from the next by no more than 10 miles (16 km): Hokkaido, in the north; Honshu, the largest and main island of settlement; Shikoku, south of Honshu; and Kyushu, in the southwest. Honshu accounts for over half the land area, Hokkaido over one-fifth, Kyushu over one-tenth, and Shikoku only one-twentieth. In the northwest the tip of Hokkaido is about 150 miles (240 km) from the coast of the Soviet Union, and in the southwest Japan is only 125 miles (200 km) from the Korean peninsula.

A MIDLATITUDE, AIR-CONDITIONED CLIMATE

Like other highly industrialized countries, Japan lies in the middle latitudes. Southern Kyushu is in the latitude of the coast of Georgia, and northern Hokkaido is about as far north as eastern Maine in the United States (Figure 10–2). With such a great latitudinal range, Japan has important climatic variations. But the islands are very narrow, so the full air-conditioning effect of the sea is felt. No place in Japan is more than 75 miles (120 km) from the sea. Most of Japan has a maritime climate like that of Western Europe.

The result of the sea effect is that although the growing season is long in the lowlands—as much as 300 days in the extreme south and 150 days in northern Hokkaido—the relatively cool summers reduce the potential evapotranspiration (PE). The PE is no higher than in southern France or in the Great Lakes region of the United States. But with the long growing season, it is possible to grow rice in Hokkaido, oranges in the southern half of Honshu, and tobacco and tea over most of Honshu. In small portions of the extreme south, it is even possible to grow two crops of rice. Winter grain crops are grown in rice fields as far north as central Honshu (Figure 10–3).

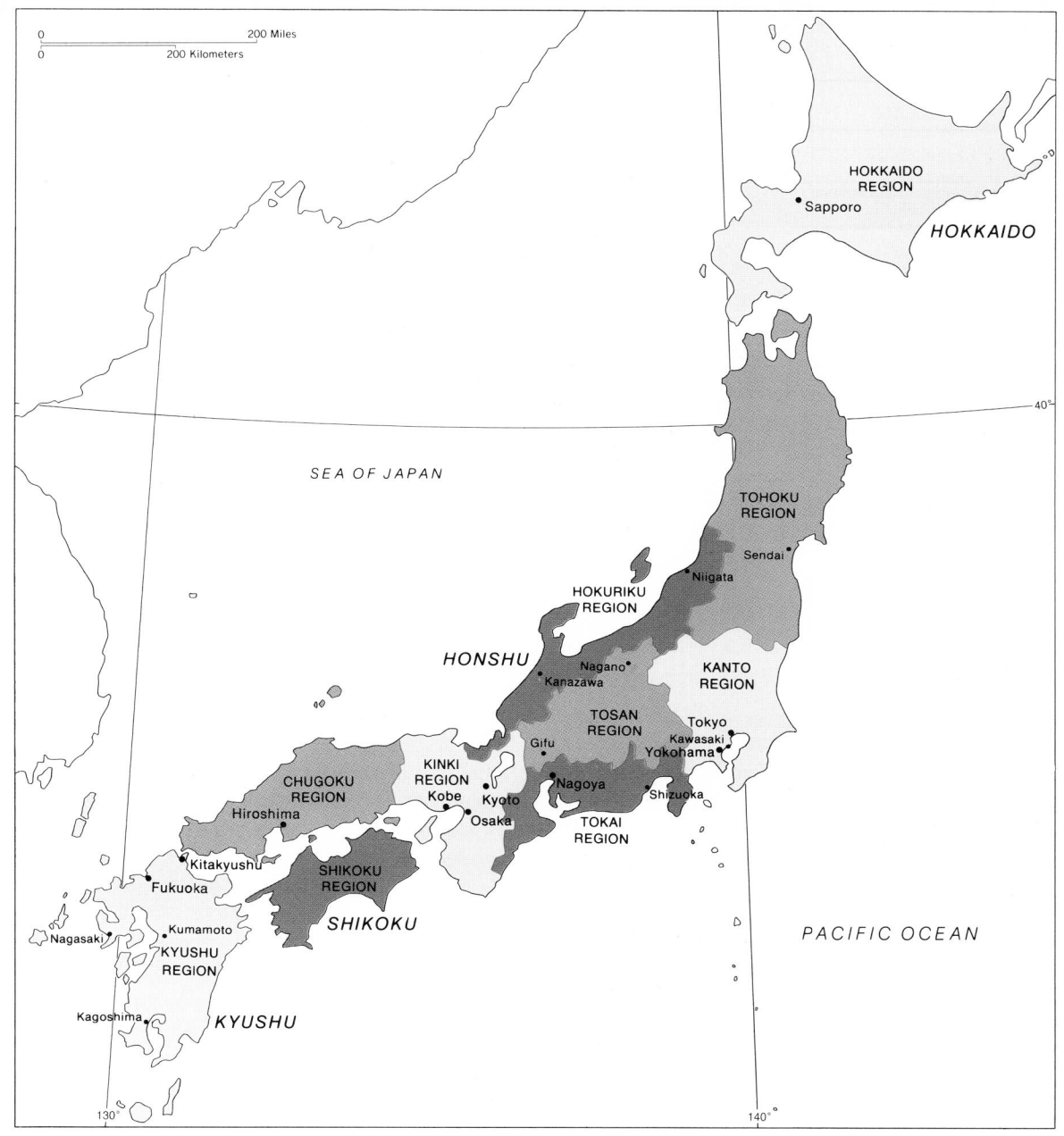

FIGURE 10-1 REGIONAL MAP OF JAPAN

- Cities over 1,000,000
- Other selected cities

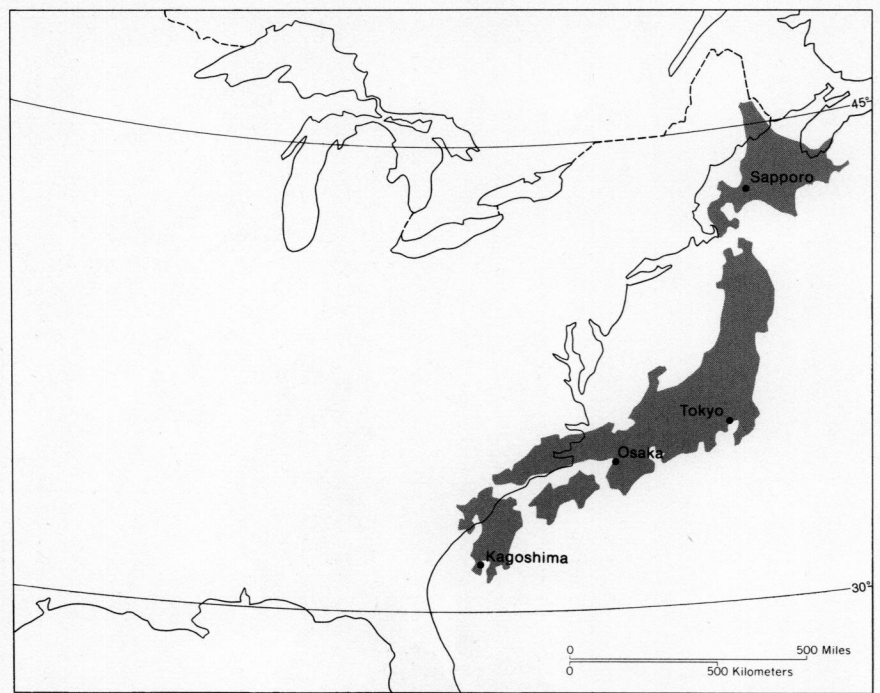

FIGURE 10–2 JAPAN AND EASTERN UNITED
STATES COMPARED IN LATITUDE AND AREA

With its maritime climate Japan has no short-age of moisture even in summer. Rainfall is over 40 in. (100 cm) a year in all areas except eastern Hokkaido and over 100 in. (250 cm) along the eastern side of the mountains in central Honshu (Figure 10–4). Much of that precipitation comes as snow. Japan is in the belt of **typhoons**—tropi-cal storms identical to hurricanes—in the spring and fall. Typhoons often do considerable damage with their high winds and torrential downpours.

LIMITATIONS OF TERRAIN

It is the terrain rather than the climate that most limits Japanese agriculture and settlement. Mountains in central Honshu reach heights over 10,000 ft (3,050 m), and only one-fifth of the total land area of Japan is lowland—small coastal areas and interior basins. The highlands are gen-erally too rugged for agriculture. More than 60 percent of the island is considered productive forest land, but large sections are scrubby and useful only for fuel wood. About 10 percent of Japanese timberland is in such rugged and inac-cessible mountain terrain that it is unused.

A TRADITION OF SELF-CONTAINMENT

Japan's environment has presented problems to be overcome, but its culture has been a major

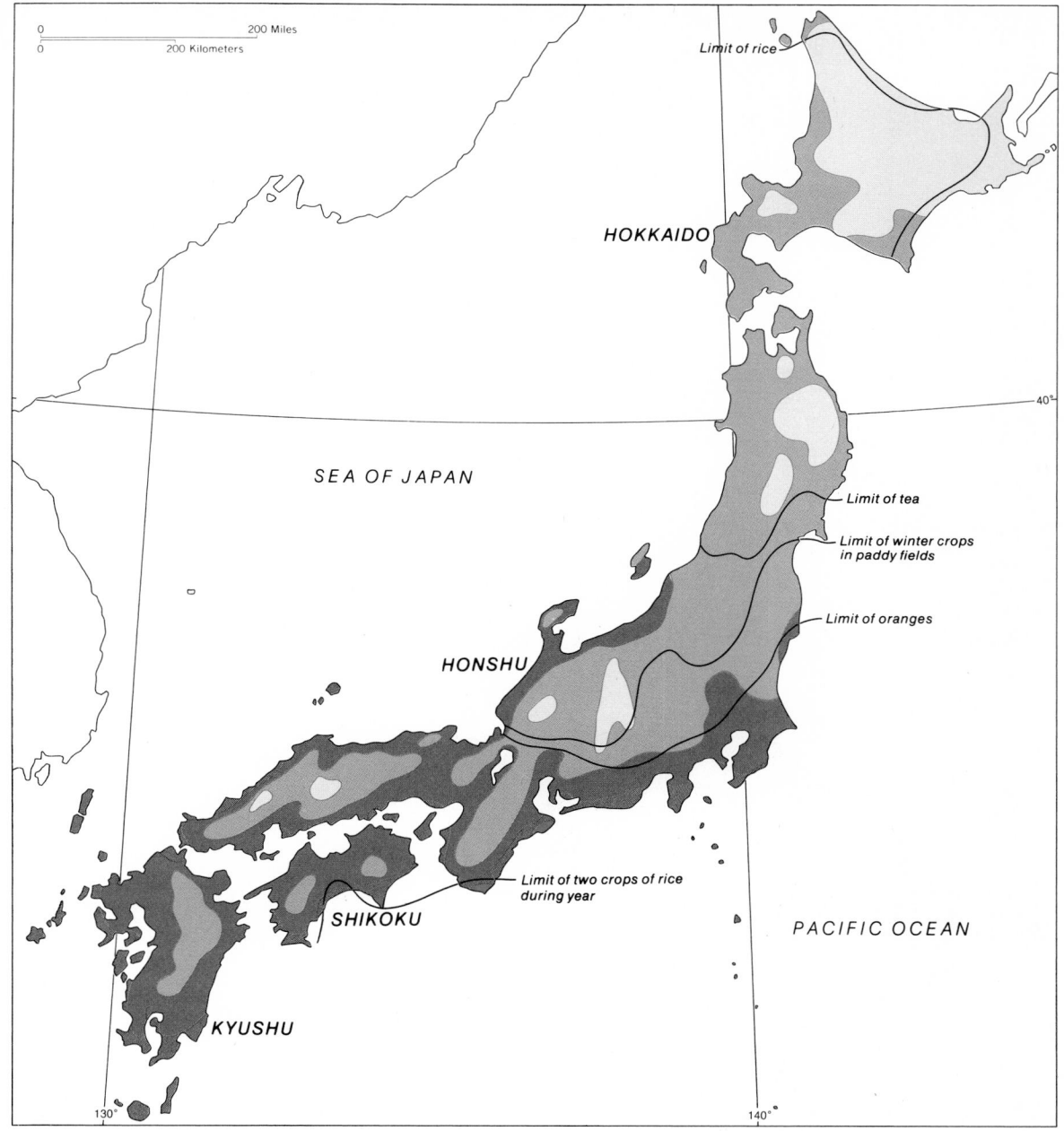

FIGURE 10–3   FROST-FREE SEASON IN JAPAN

Number of days between average dates of last spring frost and first autumn frost

☐ Under 150 days

▨ 150-200 days

▨ 200-250 days

▨ 250-300 days

▨ Over 300 days

337

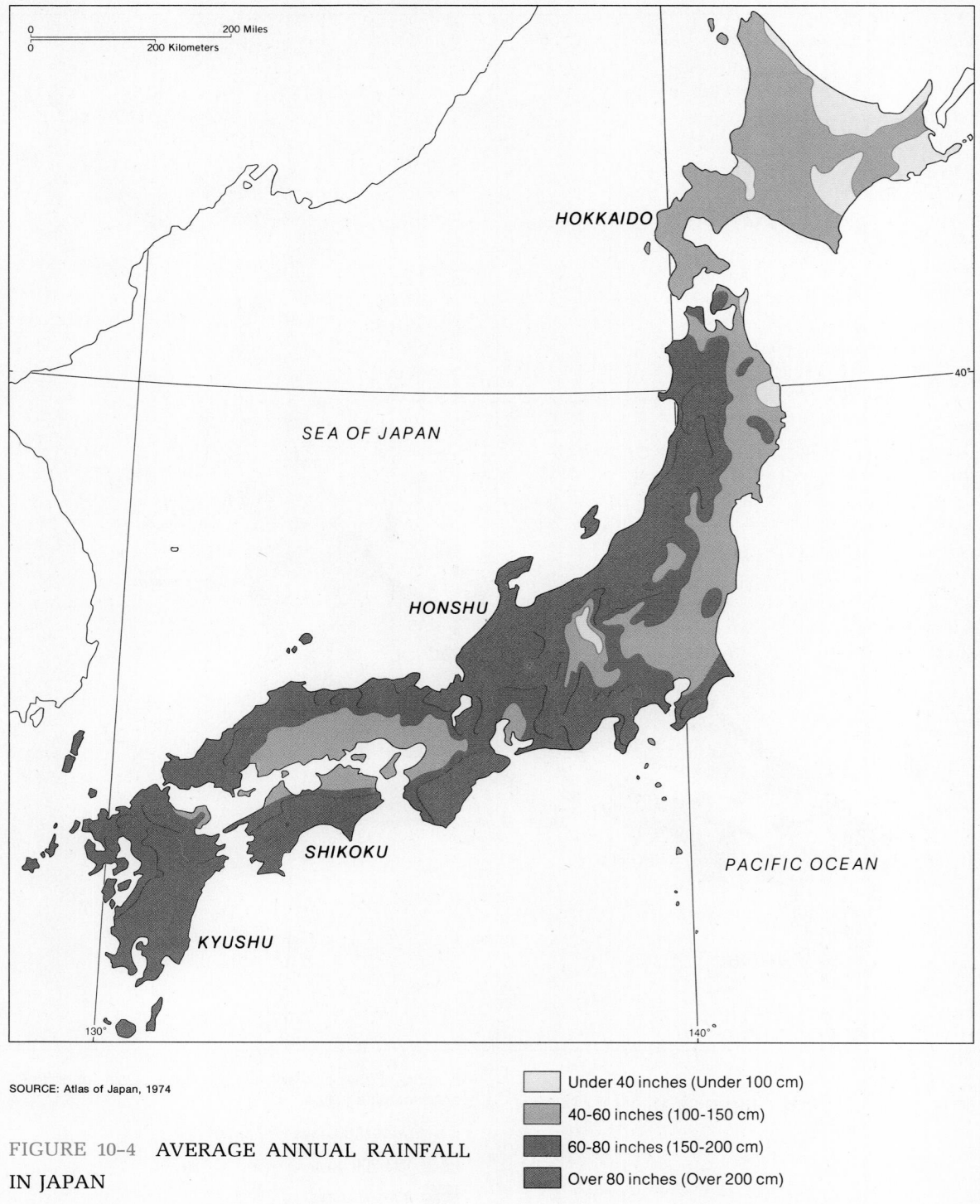

200 Miles

200 Kilometers

HOKKAIDO

40°

SEA OF JAPAN

HONSHU

SHIKOKU

PACIFIC OCEAN

KYUSHU

130°

140°

SOURCE: Atlas of Japan, 1974

Under 40 inches (Under 100 cm)

40-60 inches (100-150 cm)

60-80 inches (150-200 cm)

Over 80 inches (Over 200 cm)

FIGURE 10-4   AVERAGE ANNUAL RAINFALL
IN JAPAN

asset. Since the early seventeenth century Japan has functioned with an essentially centralized government. Moreover, since that time Japan has been notable for its **cultural homogeneity.** Until late in the nineteenth century Japanese were prohibited from leaving, and foreigners were not allowed to move freely through the country or to settle. Since then there has been almost no immigration from other areas. Except for the period of military occupation after World War II, Japan has functioned as an independent political state for hundreds of years. As a result, Japanese development is the story of a single people under a central government developing its own future.

If one were looking for the Asian country least likely to industrialize in the middle of the nineteenth century, Japan would have been a leading candidate. It fitted the traditional mold of a largely subsistence, agricultural people. Moreover, until the latter half of the nineteenth century, the Japanese government tried to keep the country sealed off from the outside world. Trade was allowed only through the single port of Nagasaki in the southwest and was limited to a few Dutch and Chinese merchants who were restricted to the limits of the port.

THE ISOLATED FEUDAL KINGDOM OF THE
EARLY NINETEENTH CENTURY

The cultural-economic system was essentially self-contained, and the people of Japan lived off the resources of their own small country. The Japanese government was theoretically an absolute monarchy, but the **shogun,** or military minister of the emperor, was the real head of state. He lived in a palace in Tokyo, and the emperor, whose role was limited to religious ceremonies, resided in splendid isolation in Kyoto. Although the shogun controlled vast lands, most of the country was divided into almost 300 feudal realms, each under the control of a hereditary family. Like European feudal lords, these local rulers taxed their people, had their own military force, and were responsible for the maintenance of roads and public works in their domains. Their rule depended on the shogun's authority. To guarantee this authority, the local rulers were expected to live half the time in Tokyo, and even when they returned to their own localities, their families remained in Tokyo as hostages.

An Agricultural Base and Local Orientation

In the nineteenth century the economic base of the system was agricultural. About 85 percent of the population were farmers who, except for the payment of a portion of their crop as taxes, lived on a subsistence basis. Traditional agriculture followed the Oriental model of farming tiny plots to produce the basic needs of the family. The country had almost no livestock except for beasts of burden and the horses used by the nobility for travel. The tiny plots were more gardens than farms. They were carefully tended by all working members of the family, and although animals were used in plowing, most of the cultivation and harvesting was done by hand with simple implements.

The chief crop was rice, grown in virtually all parts of the country. Vegetables were usually the supporting crops, and grains other than rice were planted in the paddies during the winter season. Little meat was eaten, but near the coast fish provided important protein for the diets of most of the population.

Population Distribution and the Land

Because the population was predominantly agricultural, a map of population density provides a measure of the varying quality of farmland. Where growing conditions were good, population was dense. Where conditions were poor, population density was low. Figure 10–5 shows the local population densities for 1903. Except for the densities in Tokyo and Osaka, the variations

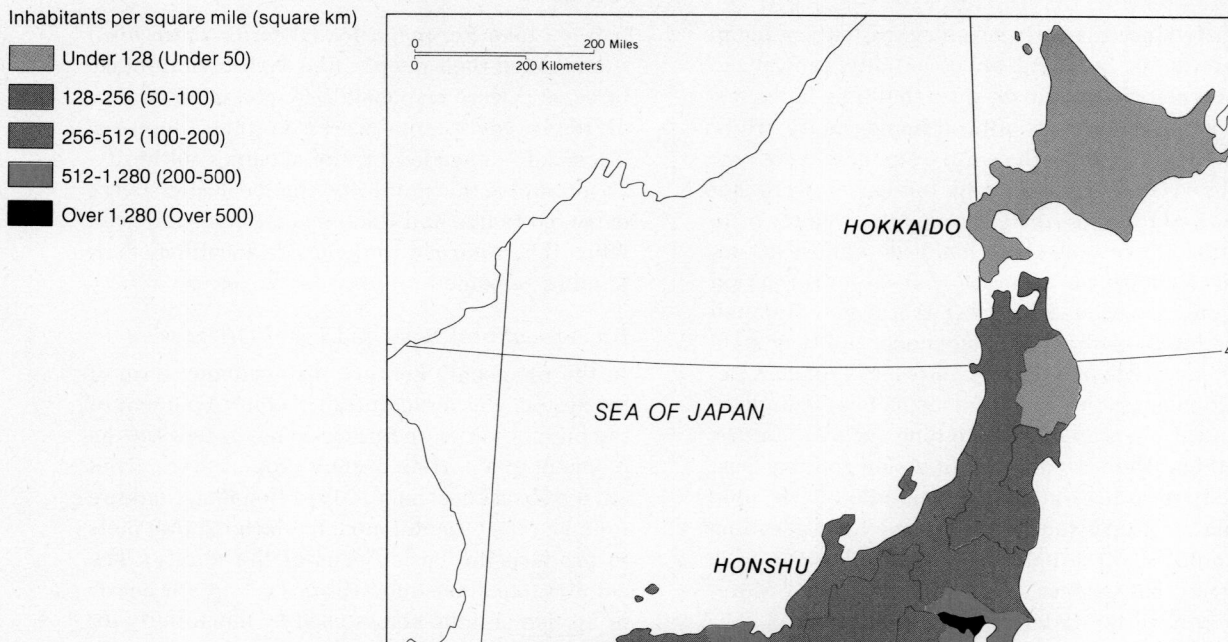

Inhabitants per square mile (square km)

- Under 128 (Under 50)
- 128-256 (50-100)
- 256-512 (100-200)
- 512-1,280 (200-500)
- Over 1,280 (Over 500)

SOURCE: *Japan Data Book,* Japan Unit, *Geography in an Urban Age,* High School
Geography Project (Washington, D.C.: Association of American Geographers,
1966), p. 120. Reprinted by permission of the Association of American Geographers.

FIGURE   10–5 POPULATION   DENSITY   OF
JAPAN IN 1903

in population density can be attributed to agriculture.

What conclusions about variations in distribution can be drawn from the map? The differences in densities between the north and south are clear, and the greater agricultural productivity of the south is obvious. This shows the importance of winter crops, which increase production capacity. With Japan's predominantly rural population concentrated in the south, it is not surprising that Japanese farms were only a few acres in size.

To understand the pressure of people on the land, it is helpful to compare Japan with California. At the end of the nineteenth century there were 34 million people in Japan—70 percent more than in California today, even though Japan is slightly smaller in land area. Only 5 percent of California's population, in contrast to 85 percent of the population of preindustrial Japan, depend directly on agriculture for a living.

A METROPOLITAN MINORITY AND A NATIONWIDE SYSTEM

The small urban population presented a different pattern. Most urban dwellers lived in one of three large metropolitan centers: Tokyo, the residence of the shogun; Kyoto, the residence of the emperor; and Osaka, the port for inland Kyoto. In 1800 Tokyo already was probably one of the three largest cities in the world.

A National Rather Than Local Base

Japan's population functioned on a national rather than a local base, although an important part of the food came from local areas. The big cities were connected by roads to all the feudal centers of the country, and over these roads moved not only the feudal lords on their annual processions to Tokyo but also food, porcelain, and other luxury products of Japan's local artisans. One traveler described the road between Tokyo and Kyoto as crowded with as many travelers as the streets of any European capital. In addition, coastal ships carried passengers and goods to Tokyo and Osaka.

The interconnections and metropolitan centers were primarily political and cultural, rather than economic. Tokyo and Kyoto were important places because they were the seats of national components of governmental and cultural authority for the whole country. Movements between these centers and the local feudal centers largely involved political control and management.

Economic movement occurred as well. The urban population had to be fed and otherwise supplied. Moreover, urban centers housed the wealthiest people in the country, who demanded the best work of artisans, the most exotic foods, and the best services. As the centers grew, they fostered increasing trade even though, under the Japanese system of social classes (based on the value of one's services), merchants were the lowest of the four leading groups. Still Japan remained predominantly agricultural, with fewer than fifty cities (Figure 10–6).

The importance of political ties can be seen by the fact that even as late as 1890 only four of the urban centers designated as cities were not centers of local government. Under the traditional system, Tokyo, Kyoto, and Osaka, the national centers, were linked to the local feudal centers throughout the country. In turn, each was the center of a locally based agricultural system. The trade flows were those needed to support the wealthy, their warriors, their servants, specialized artisans, and the growing merchant class.

Cultural Classes and Rules of Behavior

The basic sociocultural patterns established in the nineteenth century are important because of their implications for the later rapid modernization and the functioning of Japan today. The

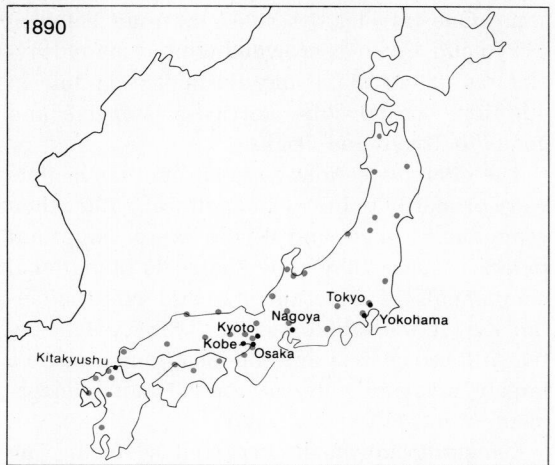

1890

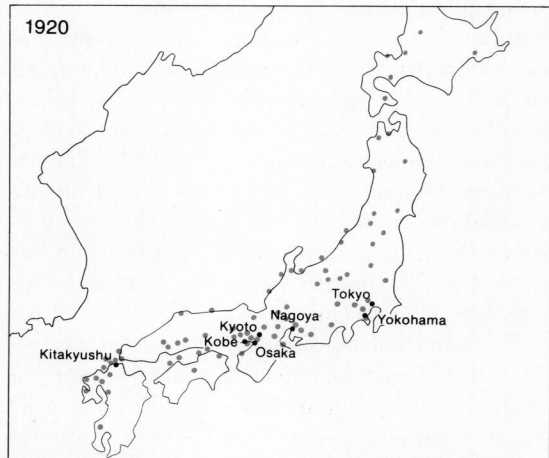

1920

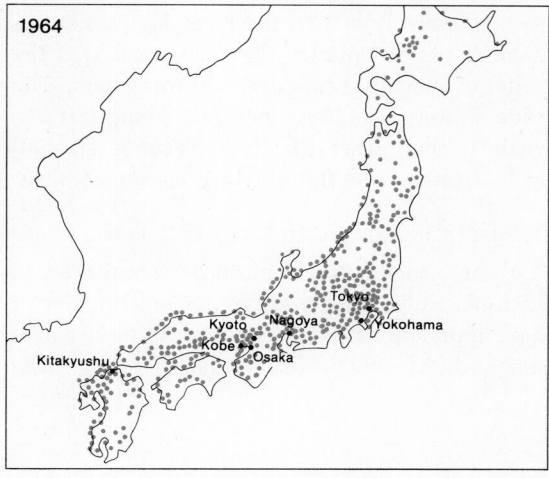

1964

population was divided into sharply distinguished social classes based on economic function, similar to the Indian caste system. Highest were the warriors, who provided military support for the feudal system. They were landless urban dwellers. Next was the mass of the peasant farmers, who provided the food, followed by the artisans, who provided necessary goods to the two higher groups. Lowest of the important groups was the merchant class, which, in a society with little trade, seemed least needed. Below all these, as in India, were those dishonored by the disagreeableness of their jobs: the ragpickers, tanners, and the like.

Rules of behavior were based on a hierarchy of loyalty bonds between these classes and also within the family. Wife was inferior to husband, son to father, vassal to lord. Not only did the inferior offer loyalty to the superior, but the superior had paternalistic obligations to subordinates. The whole system ascended to the power of the shogun and his loyalty to the emperor as head of the family state.

Under such a system little individuality was permitted, and personal freedom was further circumscribed by a dictatorial legal system. Groups rather than individuals were held responsible for crimes; whole families would be punished for an individual's activities. But even though the personal freedom usually associated with artistic societies was lacking, considerable attention was paid to education and culture. By the middle of the nineteenth century, probably 40 percent of the males were literate, and the large urban places contained centers of higher education, art, and theater.

SOURCE: *Japan Data Book,* Japan Unit, *Geography in an Urban Age,* High School Geography Project (Washington, D.C.: Association of American Geographers, 1966), p. 128. Reprinted by permission of the Association of American Geographers.

FIGURE 10–6 URBAN AREAS OFFICIALLY DESIGNATED AS CITIES IN JAPAN: 1890, 1920, 1964

JAPAN TURNS TO THE OUTSIDE WORLD: 1868–1945

In 1868 the whole political structure of Japan, which for more than 200 years had been attempting to develop a stable world of its own, was overthrown. This event is known as the Meiji Restoration. The new government was dominated by young leaders who believed that Japan needed a unified national government to equal the West in military and economic strength. Japanese leaders had awakened to the power of the industrial world of the Europeans. They were trying to maintain Japanese autonomy at a time when India and most of Southeast Asia had become colonies and China was locked into trading treaties that gave Europeans virtual economic control over resource development.

THE THRUST TOWARD MODERNIZATION

The result was a change almost as great as what Russia experienced under the Soviets. Centralized political and economic power was combined in the hands of a national leadership determined both to compete economically with the industrialized countries and to maintain Japan's political and cultural individuality. The policy of the new Japan was to leave agriculture alone and concentrate on building modern industry, to an extent even greater than in the Soviet Union. But the strategy of tiny, resource-poor Japan was very different from that of the giant Soviet Union. Instead of economic isolationism, Japan followed the path of colonial England, emphasizing trade with the outside world, even with the industrialized countries themselves.

In the early days of the new policy, Japan had virtually no steam engines or machines and no knowledge of modern industrial organization. Thus its early exports were traditional products of farmers and native artisans: silk, tea, porcelain, curios, copper products, and pearls. Japanese businesspeople traveled all over, studying modern systems of law, government administration, banking, and manufacturing.

From the beginning tea and silk formed the basis for trade with the industrialized world. Both were traditional agricultural products of China as well as Japan, but the new Japanese regime, by concentrating on the improvement of grading and processing, soon became the world's leading exporter. Silk accounted for between a quarter and a half of the value of all Japanese exports until the depression of the 1930s. Before the coming of nylon in the late 1930s, silk was the luxury fabric of the world, and Japan was the world's leading producer of raw silk. This high-value product paid many of the bills for early Japanese investment in industry.

TEXTILES AND THE BEGINNINGS OF INDUSTRIALIZATION

As in other developing countries, the Japanese textile industry first surpassed the shipment of raw silk as the leading export. The Japanese not only manufactured silk into finished products but also shifted to cotton textiles and clothing, importing the cotton and other raw materials. Farm girls were employed in the mills of Osaka for periods of two or three years, receiving food, lodging, and enough wages to provide them with a dowry when they left to marry. Cheaper Japanese textiles and clothing drove Europeans from selling in the Asian market. Osaka and its outport, Kobe, became the leading ports for textiles. At the same time Japan began manufacturing cheap pottery and glassware, matches, low-cost novelties, and paper products for export, but until the 1930s textiles accounted for over half the employment in Japanese manufacturing.

The comparatively low wages of Japanese workers, the tradition of hard work under unquestioned authority, and the education of trained technicians were the keys to Japanese success. Equally important was the Japanese effort to maintain organizational control over external trade. Like New Englanders in colonial America, the Japanese not only produced needed products but gained increasing control over the financing, selling, and shipping of their products.

New profit also came from overseas shipment and sales.

Decision making in Japan was more decentralized than in the state-controlled Soviet Union. But the group making decisions was the small hereditary group that had been the nobility of the old closed system. Its members were both the political and economic leaders of the country. Thus, whether decisions were made by the government or by private corporations, and whether production was state-owned or family-owned, the same people were involved in the decisions and reaped the benefits. The new industries gradually developed into giant industrial-commercial-financial organizations largely controlled by the leading families of the new regime.

This leadership did not remain content with specialization in light, high-labor-input industries when Japan's advantages over other industrial countries seemed clear. Fifteen thousand college students were sent to industrial countries to learn the technology of modern heavy industry. By 1930 Japan was producing steel, cement, and chemicals, and before long its productive capacity was in the full range of the heavy equipment needed by any modern military power.

Japanese farmers benefited somewhat from all this. Their children found jobs in cities, they sold more of their crops for cash, and they could buy some fertilizer and essential machine-made tools. But they were accessories to the system, not the essential components. Their living standards remained low, and they still were first of all subsistence peasants and not important markets for Japanese industrial goods. According to Japanese traditions, however, the farmers remained, accepting their place both in the society and in the economy.

CHANGES IN THE WAKE
OF INDUSTRIALIZATION

Links to other parts of Japan were soon improved by building modern roads and railroads, but the essential ties shifted to overseas markets. As industry evolved, the fledgling interconnected system that was then centered on Tokyo, Kyoto, and Osaka expanded, although Kyoto became less important when the emperor moved to Tokyo.

As these new urban-centered activities developed, other cities became part of the interconnected system, but overseas connections were overwhelmingly through the ports on Tokyo Bay and through Osaka-Kobe. The new industrial centers, by and large, were the established urban places of the past. Almost without exception, they arose along the southern coast from Tokyo in the east to the island of Kyushu in the west. Thus the country's traditional population centers became even more populous under industrialization. New agricultural, forest, and mining developments took place in Honshu's sparsely populated north and on Hokkaido, but these rugged regions got little of the new industry. An "industrial belt" developed on southern Honshu and northern Kyushu, while the rest of the country continued to focus on primary production. The result was an extreme concentration of population and economic activity in the core region of southern Honshu and northern Kyushu. This concentration continues today.

The resource base of Japan, like that of European industrial countries, expanded to include resources from throughout the world. Sixty percent of Japanese imports in the 1930s were raw materials, and another 20 percent were semifinished raw materials. Japan imported iron ore, petroleum, nickel, salt, and many other essentials. Continuous expansion of Japan's empire from the end of the nineteenth century through World War II added to the available resource base. Although the Japanese population had doubled since 1868, food accounted for less than 10 percent of imports during the 1930s. The efficient traditional farming system and worldwide fishing grounds continued to supply the country's food needs.

TABLE 10-2   POPULATION CHANGE IN JAPAN BY REGION, 1950-1974

Regions[a]	1950 Population (Thousands)	1950 Percent	1974 Population (Thousands)	1974 Percent	Change Total (Thousands)	Change Percent
Hokkaido	4,296	5.2	5,279	4.8	983	23
Honshu						
Tohoku	9,022	10.8	9,111	8.3	89	1
Hokuriku	5,179	6.3	5,244	4.8	74	1
Chugoku	6,797	8.2	7,267	6.6	471	7
Tosan	4,417	5.4	4,622	4.2	205	5
Kanto	18,242	22.0	32,098	29.2	13,856	76
Tokai	7,323	8.9	10,731	9.8	3,408	46
Kinki	11,607	13.9	18,550	16.9	6,943	60
Shikoku	4,220	5.1	3,972	3.6	−248	−6
Kyushu	12,097	14.5	13,179	12.0	1,082	9
National Totals	83,200		110,053		26,853	32

SOURCES: *Japan Data Book,* Japan Unit, *Geography in an Urban Age,* High School Geography Project, Association of American Geographers, Washington, D.C., 1966, data reprinted by permission of the Association of American Geographers; miscellaneous publication, National Science Foundation, 1966, pp. 74-77; *Japanese Statistical Yearbook, 1975,* Bureau of Statistics, Office of the Prime Minister, Tokyo, 1975.
[a]See Figure 10-1.

MODERN JAPAN SINCE 1945

Although Japan made the initial transition from a traditional, closed society to a modern industrial country in the seventy years or so prior to World War II, the emergence of the country as an affluent industrial power has been a post-World War II phenomenon. From a country whose cities and industry were badly destroyed, whose people were in a state of shock, whose markets had been closed off by the war, and whose territory was occupied by a foreign power, Japan has emerged as the most rapidly expanding, in economic output, of all industrial powers.

POPULATION SHIFTS

Even by 1960 the dramatic change in Japanese life could be seen in population shifts. A dramatic shift of people from rural areas to the largest metropolitan centers took place. As in the United States, the change resulted in regional shifts as well. Although the Japanese population increased by more than 25 million between 1950 and 1974, most regions of the country experienced relatively little growth, and one region even declined. As Table 10-2 shows, three of the ten regions had growth above the national average of 32 percent, and all three were the regions of the country's largest metropolitan centers. Together they form the country's most populous area, with over half the total population.

The only other region of rapid growth is the island of Hokkaido in the north, Japan's frontier area and the least densely populated part of the country. In most of the rest of Japan, the rural areas and small cities showed little net population growth. Obviously, as in other industrial

TABLE 10–3 POPULATION OF THE LARGEST
METROPOLITAN AREAS OF JAPAN, 1950–1974

| | 1950 | | 1974 | | Growth | |
Area	Population (Thousands)	Percent of national total	Population (Thousands)	Percent of total	Thousands	Percent
Tokyo						
Tokyo	6,228	7.5	11,519	10.5	5,291	85
Kanagawa	2,488	3.0	6,225	5.7	3,737	150
Saitama	2,146	2.6	4,655	4.2	2,509	117
Chiba	2,139	2.6	3,991	3.6	1,852	87
Subtotal	13,001	15.7	26,390	24.0	13,389	103
Osaka						
Osaka	3,857	4.6	8,160	7.4	4,303	112
Hyogo	3,310	4.0	4,932	4.5	1,622	49
Subtotal	7,167	8.6	13,092	12.9	5,925	83
Kyoto	1,833	2.2	2,374	2.2	541	30
Nagoya	3,391	4.1	5,847	5.3	2,456	72
Kitakyushu						
Fukuoka	3,530	4.2	4,173	3.8	643	18
Yamaguchi	1,541	1.9	1,532	1.4	−9	0
Subtotal	5,071	6.1	5,705	5.2	634	13
Total	30,463	36.7	53,408	48.5	22,945	75

SOURCES: *Japan Data Book,* Japan Unit, *Geography in an Urban Age,* High School Geography Project, Association of American Geographers, New York, 1966, data reprinted by permission of the Association of American Geographers; miscellaneous publication, National Science Foundation, 1966, pp. 74–76; and *Japanese Statistical Yearbook, 1975,* Bureau of Statistics, Office of the Prime Minister, Tokyo, 1975.

countries, people were migrating in large numbers from such areas to the largest metropolitan areas (Table 10–3).

Metropolitan Japan

The metropolitan population of Japan is sharply localized in one section of the country. Japan has essentially developed its own megalopolis along the southern coast of Honshu from Tokyo to Kobe. The three largest metropolitan areas of the country, Tokyo, Osaka, and Nagoya, are located in this 366-mile (585-km) stretch. Four out of every ten Japanese live here, and these areas accounted for over 80 percent of the population increase in Japan between 1950 and 1974.

The only other metropolitan areas with over 1 million people are part of an urban extension of the megalopolis along the southern shore of

TABLE 10-4 EMPLOYMENT PATTERNS OF METROPOLITAN AND
NONMETROPOLITAN REGIONS OF JAPAN, 1970

Regions	Workers Employed (Thousands)			
	Primary production	Secondary production	Tertiary activities	Total[a]
Metropolitan				
Kanto (Tokyo-Yokohama)	1,755	5,600	7,351	14,706
Tokai (Nagoya)	781	2,183	2,259	5,223
Kinki (Osaka-Kobe-Kyoto)	776	3,539	4,227	8,542
Kyushu (Kitakyushu)	1,695	1,286	2,726	5,707
	5,007	12,608	16,563	34,178
Metropolitan regions' share of national total	49%	72%	65%	66%
Nonmetropolitan				
Hokkaido	569	575	1,314	2,458
Tohoku	1,755	965	1,846	4,566
Hokuriko	971	889	1,143	2,823
Tosan	664	873	923	2,460
Chugoku	919	1,165	1,628	3,712
Shikoku	601	536	873	2,010
	5,299	5,003	7,727	18,029
Nonmetropolitan regions' share of national total	51%	28%	35%	34%
National total	10,306	17,611	24,290	52,207

SOURCE: *Japanese Statistical Yearbook, 1975,* Bureau of Statistics, Office of the Prime Minister, Tokyo, 1975, pp. 60–61.
[a]Totals are more than sum of parts because of rounding.

Honshu, northern Shikoku, and western Kyushu, which also includes seven other cities of over 250,000. They are Yamaguchi at the western tip of Honshu and Fukuoka-Kitakyushu in northern Kyushu. Kitakyushu is, in fact, an agglomeration of urban centers on both southern Honshu and northern Kyushu, connected by an underwater tunnel. Only four cities in Japan of over 250,000 population are not located in the great metro-politan-industrial belt from Tokyo to Kyushu Island.

The Two Different Population Components

The contrast between the metropolitan and non-metropolitan regions of Japan shows up in employment patterns (Table 10-4, Figure 10-7). The nonmetropolitan areas, which account for only 34 percent of total employment, have 51 percent

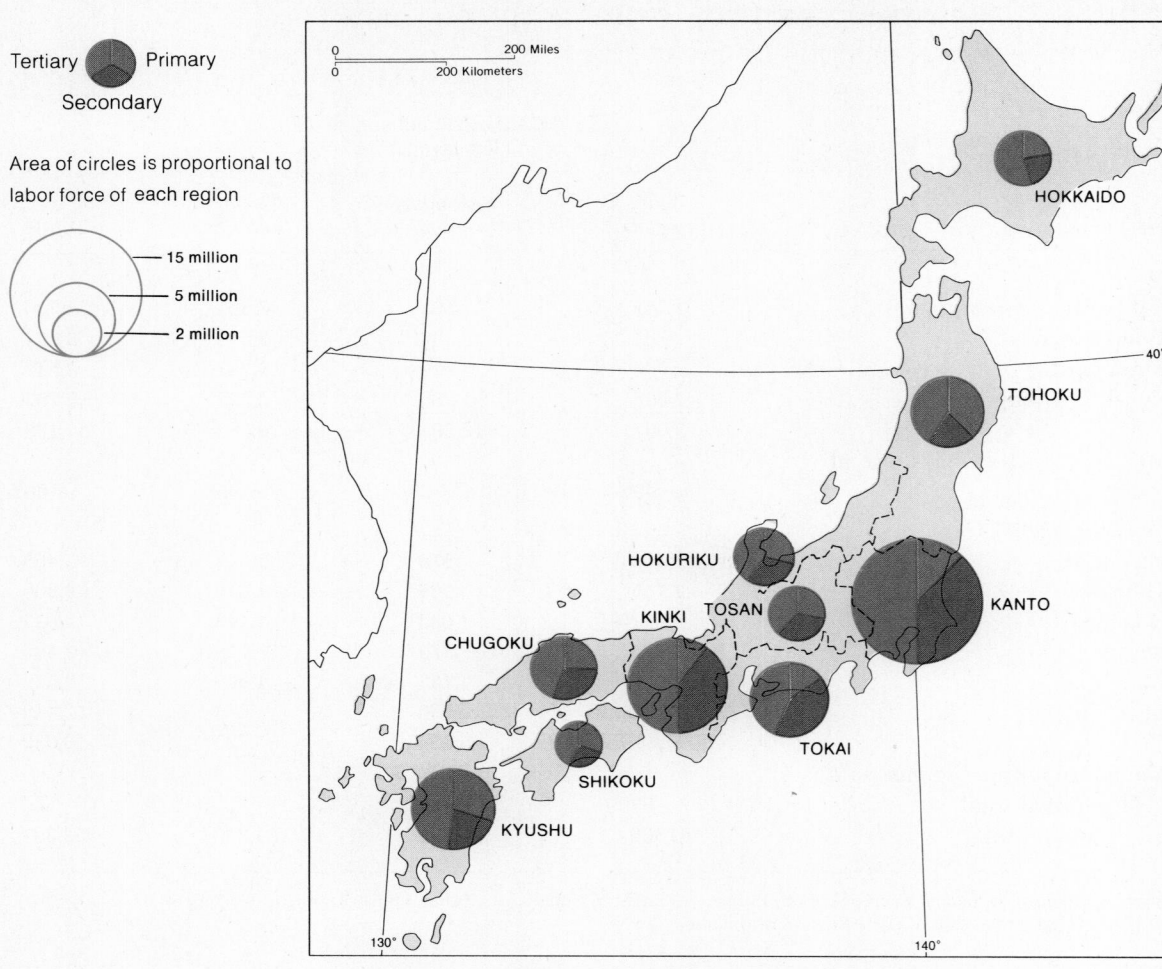

Tertiary   Primary

Secondary

Area of circles is proportional to labor force of each region

— 15 million

— 5 million

— 2 million

200 Miles

200 Kilometers

HOKKAIDO

40°

TOHOKU

HOKURIKU

TOSAN

KANTO

KINKI

CHUGOKU

TOKAI

SHIKOKU

KYUSHU

130°   140°

SOURCE: Japanese Statistical Yearbook 1975

FIGURE 10–7   EMPLOYED LABOR FORCE AND EMPLOYMENT STRUCTURE IN JAPAN, BY REGION, 1970

of their production in primary activities, over-whelmingly in agriculture. The metropolitan regions have an even larger share in traditional urban-centered occupations of secondary production (manufacturing and construction) and **tertiary activities** (services). Primary production is no longer the leading employment category in all nonmetropolitan areas. Now Japan employs more people in tertiary services than in secondary production. This is also true of the United States, but not of the Soviet Union. Service industries are the most important employment activity in Japan, in nonmetropolitan as well as metropolitan areas.

Japan's nonmetropolitan population, like that in other major industrial countries, is spread widely over the country (Table 10–4), and the metropolitan population is concentrated in a few large nodes. As a single functioning municipal entity, Tokyo is the largest city in the world and one of the most densely populated. Whereas New York City has about 126 square feet (11.90 square meters) per inhabitant, Tokyo has only 7 square feet (0.65 square meters). Moreover, the Osaka-Kobe metropolitan area is larger than Los Angeles, Chicago, or Paris.

"JAPAN, INC.": TRADITIONAL CONTROL OF A MODERN INDUSTRIALIZED SYSTEM

Japan's economic decision making is almost as centralized as the Soviet Union's, the government functioning as a close partner and decision maker with industry. Even after the democratization of political elections, the traditional close liaison between government and business in Japan seems to have re-emerged, with a few prominent Japanese families again in leadership positions.

Business analysts in the United States speak of "Japan, Incorporated" in explaining Japanese success in capturing world markets. They draw this analogy to a giant U.S. corporation such as General Motors to describe both the close alliance between the Japanese government and Japan's large industrial corporations, and the intimate cooperation that often exists among firms.

The key to economic operations is the Supreme Trade Council, chaired by the Japanese head of state, the premier, and made up of top business and government leaders who decide how firms will divide up world markets and set annual goals for every major product and country. The Japanese Supreme Trade Council is comparable to the corporate board and headquarters staff that do the planning and decision making for General Motors. Various commercial banks, industrial firms, and corporations operate like the semiautonomous divisions of GM. Just as GM's Chevrolet, Pontiac, Buick, Oldsmobile, and Cadillac divisions essentially compete with one another but share know-how, facilities, and overall management of the corporation, so do the giant Japanese companies.

To boost exports, the government backs corporations with whatever financial assistance they may need. Groups of exporters meet regularly to establish prices and lay plans for operating in overseas markets against foreign competitors. A government-owned company operates on a global basis to promote products and provide companies with export intelligence. The government foreign-aid program is another wedge into the economies of developing countries. This program provides long-term credit for development using Japanese goods and money for direct Japanese investment in foreign countries.

The unique international trading firms are also important. These three companies operate in countries with which Japan trades, and serve as "the eyes and ears of industry." Each has about a thousand employees and a hundred offices over the world, and serves a thousand different Japanese companies. These firms can carry out almost any kind of overseas trading arrangement, financing, or investment. They report market

TABLE 10-5   HOLDINGS OF MAJOR FAMILY-NAME FIRMS IN JAPAN,
1969   ($ *Millions*)

Mitsui Family		Mitsubishi Family		Sumitomo Family	
Mining/manufacturing		Mining/manufacturing		Mining/manufacturing	
Tuatsu chemicals (68)[a]	265	Heavy industries (1)[a]	2,612	Metal industries (11)[a]	984
Shipbuilding and		Electric industries (12)	857	Chemicals (30)	576
engineering (46)	478	Oil (32)	325	Electric industries (38)	269
Mining and smelting (58)	253	Chemical industries (33)	510	Metal mining (69)	200
Mining (90)	212	Rayon (43)	282	Cement (109)	210
Petrochemical		Metal mining (48)	271	Machinery (129)	141
industries (118)	220	Petrochemical (88)	325	Light metal (138)	113
Total	1,428	Mining (144)	143	Coal mining (196)	121
		Paper mills (151)	129	Bakelite (258)	58
		Cement	110	Total	2,672
		Steel manufacturing (190)	106		
		Plastics (247)	60		
		Edogawa chemicals (357)	63		
		Kakoki (393)	32		
		Total	5,825		
Banking		Banking		Banking	
Bank (7)	3,969[b]	Bank (3)	5,920[b]	Bank (2)	6,041[b]
Trust and banking (3)	727	Trust and banking (1)	844		
Total	4,696	Total	6,764		
Insurance		Insurance	0	Insurance	
Mutual life insurance (7)	528			Marine and fire (8)	149
				Mutual life (3)	1,210
				Total	1,359
Warehouse (2)	32	Warehouse (1)	36	Warehouse (3)	25
Shipping (2)	338				
Real estate (1)	264	Real estate (2)	376		
Trading firms (2)	2,588	Trading firms (1)	2,357	Trading firms (6)	889
Construction (11)	115			Construction (29)	61
Grand total	9,989		15,358		11,047

SOURCE: *The President Directory,* The President Magazine, Time, Inc., New York, 1969.
[a] Figures in parentheses indicate ranking among leading Japanese companies in that category. Most are rated in terms of sales or revenues rather than holdings.
[b] Deposits, not holdings. Holdings would be greater since they would include deposits plus physical holdings.

trends, industrial developments, customer habits, and investment opportunities from all over the world.

The great family-owned combines of the past, which function across the whole range of industrial, financial, and trading operations, also play an important role. Table 10–5 shows the major companies and business firms in Japan, each bearing one of three famous family names. Note the range of activities in which they operate. In terms of holdings, all are primarily involved in financing rather than in manufacturing, but notice the importance of the trading firms within each family grouping. The Mitsubishi holdings are greater than those of General Motors, although less than half those of American Telephone and Telegraph. These family holdings are reminiscent of those of famous U.S. families—the Fords, Mellons, Du Ponts, and Rockefellers—but represent a larger share of their country's total business holdings.

Although the family holdings in preindustrial Japan were in the rural areas, today they are in the largest cities. Land reform carried out under post–World War II U.S. military occupation eliminated the control of the largest landlords over farmland, but the re-emergence of Japanese business and industry after the occupation once more found the traditional families in charge. Just as the management function of large U.S. corporations is centered in New York City and Chicago, so the Japanese family enterprises are headquartered in Tokyo and Osaka. The Mitsui and Mitsubishi holdings are based in Tokyo, but the Sumitomo empire is headquartered in Osaka. These family firms control over 80 percent of the holdings of the one hundred leading industrial corporations in Japan. The Tokyo firms alone hold almost two-thirds. Twelve of the fifteen largest banks in Japan are based in these two areas, as are most of the largest trading companies. Family enterprises in Japan are as clustered together in the largest metropolitan centers as are the big business corporations of any modern industrial country.

MANUFACTURING FOR A WORLD MARKET

The close interaction of government and business is primarily designed for the world market rather than for Japan's domestic market. It assures that goods manufactured in Japanese factories travel overseas in Japanese ships and that the financial and trading arrangements are made by Japanese firms.

Japanese manufacturing has been geared to foreign markets. Textile exports to the United States have been so great that textile mills in the American South have been lobbying for quota limitations or protective tariffs. Just as textile manufacturing arose in the South because it offered low-cost production compared to high-wage New England, so less expensive Japanese labor threatens the South today.

The Japanese factories are new and very efficient. Modern plants without windows offer precise lighting and temperature controls. Machinery is highly automated and includes a water-jet loom developed in Japan that is three times as fast as conventional looms and much quieter. Moreover, the textile industry has been organized by the industrial combines on a much larger scale than have most locally owned mills in the United States.

Japanese firms have invested heavily in research on synthetic fibers and processing. To ensure profitability and hence protect this investment, the firms must keep their labor costs relatively low. As labor costs have increased at home, the Japanese, in order to maintain competitive prices in world markets, have moved production to sources of cheaper, unskilled labor in Taiwan, South Korea, and Thailand, where wages

Japan, like other modern industrial countries, has a core region and outlying peripheral regions. From our examination of the workings of the Japanese economy today, you should be able to identify the Japanese core. Examine the map of railway freight traffic (Figure 10–8) to test your generalizations. Remember that Japan has a small area, so air service within the country is not important, and vehicle ownership, though rapidly increasing, is still low. Thus trains form the major means of inter-city freight movement. The map of railroad freight traffic, then, is a much more accurate measure of movement within the system than it would be in the United States.

How would you describe the pattern of railroad freight traffic? Does it fit in with other evidence of the distribution of the modern system? From it can you identify producing and consuming areas?

are only ten to fifteen cents per hour, in contrast to about forty-five cents in Japan. A Korean-made shirt may sell for five dollars retail in the United States when a comparable American-made shirt costs nine dollars. Of course, the Japanese commercial intelligence system knows Americans' style preferences, and Japanese manufacturers make clothing to such specifications.

Today textiles account for only 11 percent of Japan's exports. Industry concentrates on the most advanced products of modern technology: motor vehicles, ships, electronic equipment, electrical appliances, machinery, component parts, and steel products. In 1952 Japanese decision makers decided to shift the economic base to heavy industry. The Machinery Industry Promotion Law was passed; banks made low-interest loans to companies in these industries.

Japan is now third in the world in steel production, with 117 million metric tons in 1974— more than nine times the production of 1955. As with textiles, most steel production comes from the most modern and efficient mills using some of the world's largest blast furnaces, modern oxygen converters, and the latest finishing facilities. These enable Japan to compete even though it must import more than 80 percent of its iron ore and 60 percent of its coking coal. Japanese steel production is the base for the country's new diverse metalworking-machinery industry, but basic steel products are among the country's leading exports. Agreements have been necessary to control Japanese steel imports to the United States, but Japanese combines and trading companies have been creating markets for Japanese steel by helping to develop steel-using industries in developing countries such as Ethiopia, Brazil, and Venezuela.

Japan has also become the world's leading shipbuilding country. It specializes in the new, giant oil tankers of 300,000 tons or more that revolutionized the shipment of oil after the closing of the Suez Canal in 1967 during the Arab-Israeli war. Japanese manufacturing has gained its greatest reputation in advanced electronic equipment and automobiles. Japan specialized early in electronic components—transistors, printed circuits, and semiconductors—for U.S. manufacturers. Then it moved into making radios and televisions, first marketed under U.S. labels and now under Japanese brands such as Sony and Panasonic. Again, Japan capitalized

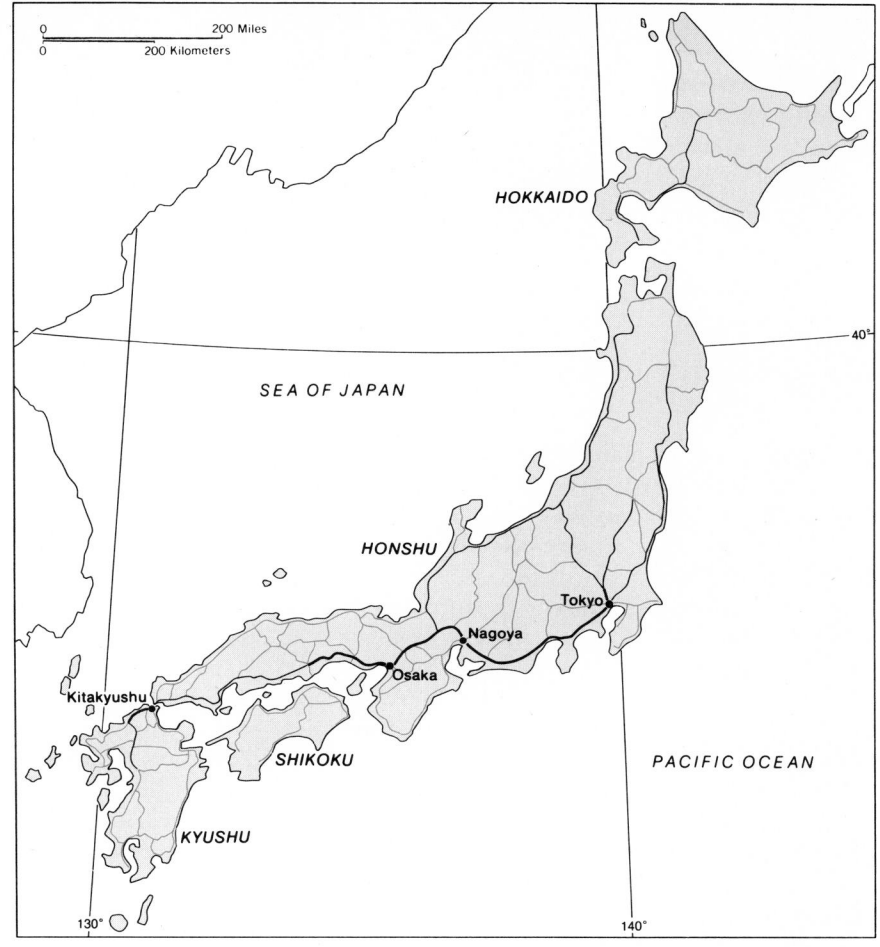

Average tons per day

Less than 10,000

—— 10,000-30,000

━━ Over 30,000

SOURCE: *Japan Data Book,* Japan Unit, *Geography in an Urban Age,* High School
Geography Project (Washington, D.C.: Association of American Geographers,
1966), p. 133. Reprinted by permission of the Association of American Geographers.

FIGURE 10–8   RAILWAY FREIGHT TRAFFIC IN
JAPAN

both on the technology that could be borrowed from other countries and on innovations by its own researchers. Motor vehicle production progressed from bicycles to motorcycles and motorbikes to automobiles. Automobile production was almost 4 million in 1974, compared to less than 3 million in West Germany and about 7 million in the United States.

THE INCREASINGLY IMPORTANT DOMESTIC CONSUMER

Until recent years Japanese businesses concentrated on overseas markets, but growing industrial and business activity within the country has begun to create an important **consumer market** at home. Prior to World War II the urban population had remained secondary to the rural family, but today the Japanese population is predominantly metropolitan. Wages have also risen. In 1950 the basic diet of rice and other essentials took 57 percent of family income. Under these circumstances few Japanese had any hope of owning a refrigerator, vacuum cleaner, or washing machine, much less an automobile. With higher incomes the share of income spent on food has dropped to about one-third, and the new city workers, though poorly paid by U.S. standards, now enjoy an affluence that has sparked domestic demand not only for clothing and food but also for televisions, radios, household appliances, motorcycles, and even automobiles. More than three-fourths of Japanese automobiles are now sold in Japan, and the leading export industries actually produce from 60 to 90 percent of their output for the home market.

Such a market would seem ready for imports of consumer goods from the other industrial countries of the world, and the United States and European countries have been trying to enter it. But Japanese decision makers, who have so efficiently worked their way into the markets of other countries, have built a protective barrier against foreign competition with Japan. Automobile tariffs have been very high. Until recently, imports of automobile engines were limited by quota to 1,000 per year. Moreover, foreign investments in Japan have been sharply limited.

MANUFACTURING OPERATIONS

To an important degree Japan's whole industrial-business system operates on the traditional basis of loyalty arrangements dating back through the centuries. It is said that the Japanese have been brought up with the concept that their own needs are secondary to serving others. Workers believe that their job is to help make more and better products, and that this will lead to the prosperity of the company and greater prestige for the country. Workers hold great loyalty to their employer and operate in a disciplined manner. Employment with one company is most likely to last the worker's full career. In turn, the company takes the traditional paternalistic posture of the feudal lord. Workers are often given family benefits, so that a worker with children is paid more than an unmarried worker. Firms provide medical services and hospitalization plans, day care nurseries, and sometimes even sports facilities and employees' clubs. It is also traditional to pay bonuses twice yearly, usually the equivalent of two to three months' pay. The side benefits are generally greater than those that U.S. workers receive, a factor that somewhat narrows the apparently wide wage gap between the two countries.

Although industry is dominated by the huge industrial-business combines, Japan retains a large number of traditional **cottage industries.** These are the Japanese sweatshops where work is from dawn to dusk with perhaps two days off per week. Wages are extremely low, and workers are mostly ill-educated people from rural areas. A firm may employ only a few workers outside the owner's family, or it may have up to fifty employees. These are the "subcontractors" who

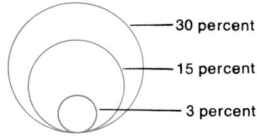

Circle size is proportional to
percent of national total
in each region

30 percent

15 percent

3 percent

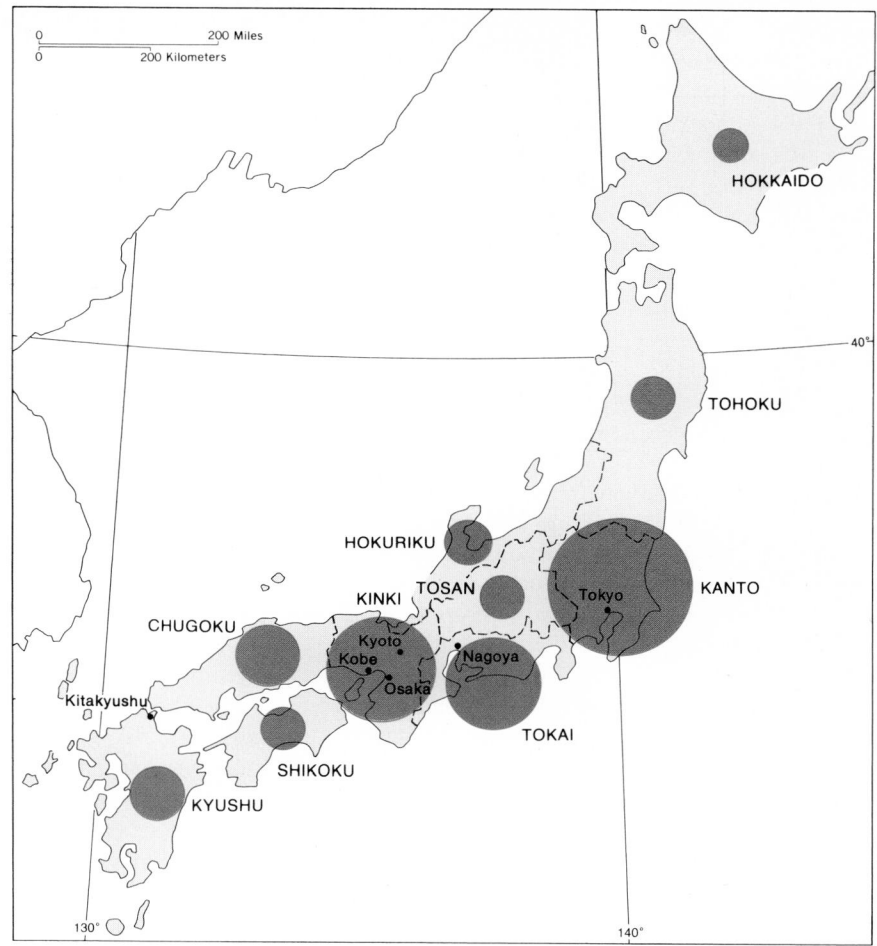

SOURCE: Japanese Statistical Yearbook 1975

FIGURE 10–9 MANUFACTURING DISTRIBU-
TION IN JAPAN BY PERCENTAGE OF VALUE
OF SHIPMENTS, BY REGION, 1971

make component parts—such as bicycle wheels
or electronic plugs—used on the assembly-line of
some large combine.

Figure 10–9 shows the tremendous amount of
manufacturing output concentrated in the large
metropolitan centers. Roughly two-thirds of the
country's manufacturing production comes from
the five largest metropolitan centers—Tokyo,
Osaka, Kobe, Nagoya, and Kitakyushu—and less
than a quarter of its output from regions outside
the southern core. This high concentration is not
surprising in a country that has been oriented to

external trade. Even now, two-thirds of all Japan's foreign trade moves through the ports on Tokyo Bay, Osaka-Kobe, and Nagoya. It is logical for most industries to locate near the ports to save on transportation costs. This applies both to industries that are usually considered market oriented and to those considered resource oriented. The resources and most of the market for Japanese industries are in foreign countries, which are accessible only through the ports.

THE AGRICULTURAL REVOLUTION

Amid the great postwar industrial expansion, a quiet revolution in agriculture has been going on. Basically, the farms remain small, family-owned enterprises, and the garden plots are still handled with utmost care. But now modern scientific knowledge is being applied to agriculture with the same intensity as to industry. The result is that the productivity of Japanese agriculture increased about two and one-half times between 1890 and 1965, and it has increased even more since then. Although some additional land was put into production during this period, the increase in output has come essentially from improved irrigation systems, better fertilizer, better insecticides, and plant breeding, all of which increase productivity per acre. Japan probably has the highest per acre agricultural production of any country in the world.

Only 15 percent of Japan's total land area is cultivated, and that is divided among more than 5 million farm households. Thus most farms are small; many are less than two acres, although larger farms are common in Hokkaido. These small holdings must be farmed intensively. The infertile soil requires careful management and application of fertilizer. The problems are compounded by the danger of soil erosion in many parts of Japan, where excessive rainfall may wash soil down the steep slopes and into the sea.

There has been a major shift in the products of agriculture as well as in productivity. Rice production has increased sharply as the result of a government program of price supports and the introduction of new "miracle" strains. Productivity increased by 50 percent in the late 1960s. By 1968 rice production was half again the total for the early 1950s. In fact, for the first time in Japanese history, rice was in surplus. Although more than 1 million tons were made available as livestock feed, still an eight-month supply was stored in government granaries. Faced with the cost of its rice program, the Japanese government began to pay farmers for not growing or harvesting rice. But since rice is the traditional crop on 90 percent of Japanese farms and the government price subsidy makes it one of the highest-value crops in Japan, the farmers continued production.

The new affluence of the Japanese worker has brought an important change in other crops of Japan's farming system. In recent years other grain crops have decreased rapidly, even though it is necessary for Japan to increase imports of feed grains. With city customers changing their diets toward more meat, dairy products, fruits, and vegetables, the output of these products has risen sharply. The production of oranges, grapes, apples, milk, and eggs, and the number of cattle and swine slaughtered, have increased substantially in recent years.

Modern roads and trucks enable farmers to organize in much the same way that the commercial farmers have organized in the United States and Europe. They are planting the crops that do well in their particular area and can be marketed in the cities. The move toward mechanization and electrification has also been rapid. Almost all Japanese farms use powered garden cultivators and power sprayers, and both farms and homes have electrical connections. On larger farms full-sized tractors are in operation.

Time and cultural change have eliminated the traditional surplus of farm labor; migration to the cities has left a shortage of farm workers. By tradition the eldest son became the manager of the farm, and the younger boys moved to other jobs. Today, however, most young people prefer city life, as in the United States, and the average age of the farming population increases each year.

Fish and other seafood have provided the basic protein in the Japanese diet for centuries. The Japanese have expanded and modernized their fish production. The country has become the leading fishing country in the world, and it is also the owner of one of the largest, most modern fishing fleets. Since the oceans are international waters, Japanese fishers work almost all the major fishing areas of the world from the coasts of California and South America to the whaling grounds off Antarctica.

A CULTURAL REVOLUTION

The rapid development of the modern interconnected system in Japan has produced a cultural revolution as well. Like people in other industrial nations, the Japanese have increased materialism and have greater material expectations. But a more fundamental change than that has taken place.

Before modernization Japan was a homogeneous culture largely isolated from outside influences. With world trade has come the influx of outside ideas and values, particularly those from the United States and Europe. Many Western ways have taken root in Japan: films, modern music, dress, and even baseball. The very different philosophical and moral values of the West are also present in Japan.

The result has been cultural conflict in Japan probably on a greater scale than in any other industrialized country, even the Soviet Union. This **cultural dualism** has both sociological and geographic dimensions. As might be expected, the new ways take hold particularly among the young. Thus the generation gap within Japan is one between the traditional Oriental values of parents and the strongly Westernized young people. At the same time, Japanese industrialists whose families set the standards in traditional Japan find that they must accommodate to the modern industrial world. They, more than almost any other Japanese, live in two worlds, in both of which they take leadership roles. The contrasting values of old and new are evident in any modern country between the people of the city and those who remain on the land. Thus modern Japan has risen in the cities, whereas traditional Oriental practices have remained entrenched on the farmsteads and in the rural villages. But even there, electricity and radios and televisions are bringing the new world into rural homes.

Undoubtedly, the outward appearance of Japanese culture will shift more and more toward what is becoming a worldwide metropolitan-centered culture. Just as the Japanese have developed their own particular version of the modern interconnected system, so we might expect that the culture of the "new Japan" will continue to bear the particular stamp of the Japanese heritage that has been so distinctive until now.

THE GEOGRAPHIC EQUATION OF JAPAN

In many ways Japan's rapid rise as a center of the modern interconnected system is surprising. In terms of the geographic equation, Japan has neither of the key components for development of a modern system: a large, varied resource base or a culture that has experienced the Industrial Revolution of the eighteenth and nineteenth centuries. Japan, however, has a different and very important cultural characteristic: a tightly knit people responsive to elightened, aggressive leadership. When Japanese leaders decided that they should embark on a program of industrial development, the population was readily mobilized. Moreover, despite the general shortage of industrial resources, the advanced Japanese culture had products that the rest of the world wanted. These products formed a base for the trade necessary to begin industrialization. Japan also had the advantage of technology that had been developed in Europe and the United States.

Like the eastern seaboard of the United States or the countries of Western Europe, Japan has built its economy not on its own resources, but on primary products obtained from other areas. Japanese factories have depended on the agricultural and mineral raw materials of other parts of the world. Also like the U.S. eastern seaboard and Western Europe, Japan has done more than manufacture products. It has maintained control over the management and business support activities necessary for industrialization—namely, financing, shipping, and marketing. Japan has therefore profited not only from manufacturing, but also from the shipping and trading of its products. Another similarity between Japan and Western Europe and the eastern seaboard of the United States is that the major market for Japanese business is not Japan alone, but the affluent consumers of the rest of the world.

The Japanese model of the modern interconnected system stands in sharp contrast to the United States as a whole and to the Soviet Union. These are both huge areas with great agricultural and mineral resources. However, the economic demands of even these countries are outstripping their own resource bases, and they must increasingly trade with the world for essential materials, as industrialized Japan has always done.

The Japanese model of the modern interconnected system, with its dependence on external sources of materials and external markets, is likely to set the pattern for other such systems in the world. These systems have the benefit of operating on a global scale, drawing on the resources and markets of the whole world. As we have seen in Japan, however, they have become increasingly dependent on international relations. They can be affected by trade sanctions, quotas, world economic depression, and armed hostilities in other parts of the world.

SELECTED REFERENCES

Bureau of Statistics, Office of the Prime Minister. *Japan Statistical Yearbook.* Tokyo, annual. A great wealth of statistical information for the nation as a whole and by prefecture (the basic civil subdivisions of the country).

Dempster, Prue. *Japan Advances: A Geographical Study,* 2nd ed. Methuen, London, 1969. A basic text with a topical approach to the character of Japan. About half the book consists of long chapters on agriculture and industrialization. Short chapters treat population, transportation, cities, and trade.

Gibney, Frank. *Japan: The Fragile Superpower.* W. W. Norton, New York, 1975. An interpretation of the Japanese people and their values, attitudes, and institutions by an American who lived in Japan both during the occupation and also more recently. Chapters 1, 3, 8, 9, and 14 deal most directly with the functioning of Japan's modernized economic system.

Hall, Robert B., Jr. *Japan: Industrial Power of Asia,* 2nd ed. New Searchlight Series, D. Van Nostrand, New York, 1976 (paper). A brief but balanced overview of Japan today. The urban-industrial character of Japan is emphasized, but chapters on agriculture and social conditions complement this focus.

Kornhauser, David. *Urban Japan: Its Foundations and Growth.* Longman, New York, 1976 (paper). A well-documented treatment of the development of Japan's urban-industrial land-scape. The review of the background and contemporary circumstances of modern Japan is excellent.

Noh, Toshio, and Douglas H. Gordon, eds. *Modern Japan: Land and Man.* Teikoku-Shoin, Tokyo, 1974. A brief but well-illustrated regional geography of Japan. In this excellent overview the text is effectively supplemented by many graphs, maps, and tables. Three-quarters of the book offers a regional survey of eight administrative districts of Japan. The final quarter presents an overview of recent changes in industry, agriculture, forestry, transportation, and trade for the entire nation. A final chapter deals with major land-use problems confronting Japan.

Trewartha, Glenn T. *Japan: A Geography.* University of Wisconsin Press, Madison, 1965. A comprehensive traditional geographic study dealing with the people and environment of Japan.

CHAPTER 11

Canada: Independent but Functioning as Part of a Larger Whole

Canada, with rich mineral resources but limited agricultural potential, functions as part of larger interconnected systems centered in Europe, Japan, and the United States. A small population sparsely distributed along the border with the United States has required heavy investment in transportation and communication links. Nevertheless, a lack of unity, both cultural and economic, persists. The cultural influence of the United States and the impact of foreign investment are of concern to Canadians, despite their affluence and success in the modern world.

Left: Baie Saint-Paul on the north shore of the St. Lawrence River in Québec. (Martin Litton/Photo Researchers, Inc.) *Above:* Large asbestos crusher being put into service by the Canadian Johns-Manville Company, Ltd. (Canada Wide)

The 1976 Olympic Games in Montreal. (Lafont/Sygma)

Left: The sport of curling, played on ice rinks. (Harrison Forman) *Above:* Containers in storage area of the container terminal in Vancouver, British Columbia. (Paolo Koch/Photo Researchers, Inc.)

Daily, Canadians and Americans by the thousands cross the border between their two countries. Some even live in one country and work in the other. Canadians spend winter vacations in Florida; Americans summer in Canada. But most of those who visit in the other country see little apparent difference in the lifestyles of the two countries. Roads and automobiles are the same; the people dress and look much alike; house types and the layout of cities and towns are similar. People compare rapidly growing Toronto to Houston, and Vancouver to Seattle. Highway signs advertise the same products—Kellogg's breakfast cereals; Kodak film; Ford, General Motors, and Chrysler automobiles. Canadian wheat farmers have the same huge farms and the same modern farm machines as U.S. wheat farmers. Except for the French Canadians, people speak dialects of English that are similar to the dialects of the northern United States.

At first glance, then, we would expect the modern interconnected system of Canada to be a carbon copy of that of the United States. After all, both are among the half-dozen largest countries in land area in the world; both were settled about the same time by similar European immigrant groups; both have benefited from the results of the Industrial Revolution and stand among the most advanced industrial countries in the world; both are pulled together by a modern system of transport and communication. Despite these similarities, however, one of the important variables in the geographic equation is different, and that variable has produced a very different version of the modern interconnected system. Canada has only about one-tenth the population of the United States—23 million people, only slightly more than in California, the most populous state in the United States.

Canada's small population does not provide the massive market that characterizes the United States. Its resource base is rich in minerals, and land resources for agriculture are substantial in relation to its population, even though Canada is much less well endowed agriculturally than the United States. Canada represents a particular form of the modern interconnected system: a country that is a peripheral, though important, part of a larger interconnected system. Other areas of European settlement—Australia, South Africa, and New Zealand—also exemplify this type of modern industrialized system. Canada will be our example, but consider how the others function within this class of countries.

THE CANADIAN ENVIRONMENT

The smaller population of Canada as compared with the United States in large part reflects the greater environmental problems its people have faced. Like the Soviet Union and the Scandinavian countries, Canada has a large part of its territory within the arctic and subarctic regions, where potential evapotranspiration (PE) falls below levels necessary to support commercial agriculture. The most favorable agricultural areas lie in southern Ontario and along the St. Lawrence lowland. The Canadian part of the Great Plains is a major crop area, but it suffers from frequent moisture deficits during the growing season, as do the Great Plains of the United States. Western Canada is mountainous, with agriculture restricted to a few major valleys; the bulk of the population is in the lower Fraser River valley, especially in and around Vancouver.

LOW GROWING POTENTIAL

Since the north-south dimensions of Canada are greater than those from east to west, we might expect to find a great range in potential evapotranspiration. The latitudinal distance from the southernmost point of Canada at Lake Erie to the northern tip of the northernmost island is the same as the latitudinal distance from the same southern point in Canada to Quito, Equador, in South America. But for most of that distance, in the arctic and subarctic, the frost-free season is less than ninety days and annual PE (see Figure 3–3) is less than 50 cm (20 in.). These limitations virtually preclude commercial agricultural production. Even the high PE of southern Canada is only 68 cm (27 in.), as compared to more than 150 cm (59 in.) in southern Florida. The highest PE in Canada is found in a few tiny valleys in the western mountains.

The best growing areas of Canada are similar to those in New England, Michigan, Minnesota, or North Dakota, and Canada has nothing comparable to the Middle West or the Central Valley of California. The boundary line between regions with potential for productive commercial agriculture and those with only marginal commercial possibilities runs roughly from the eastern tip of Lake Superior eastward to Arvida northeast of Québec City. In the west the line extends farther north, about halfway north in Saskatchewan and Alberta, some 300 miles (480 km) north of the United States border. Unfortunately, the western area suffers from frequent dryness, which makes agriculture highly risky.

A BARREN SHIELD INSTEAD OF THE "MIDDLE WEST"

The same succession of landforms appears in Canada as in the United States as we move from the Pacific Coast eastward. Two roughly parallel mountain ranges, the Coast Mountains and the Rocky Mountains, run in a north-south direction and are separated by plateaus. To the east of the mountains are the Great Plains and the central lowlands, which together are called the **prairies.** Canada's lowlands, however, are much more limited in extent than are those of the United States. Instead of the large expanse of agricultural land known as the Middle West, much of the central region of Canada is characterized by a vast expanse of very old metamorphic rocks, the **Canadian Shield.** This rock mass forms a gigan-

tic arc centered on Hudson Bay and reaches south to the northern side of the St. Lawrence Valley to the east of Montréal. The shield also extends into the United States in Minnesota, Wisconsin, and the Adirondack region of New York state.

Glaciation has removed soils from the Shield and left much of the surface as bare rock or gravelly surface materials; and much of the rest of the Shield is covered with lakes and bogs with poor drainage. The Shield area's potential for crop production is very low; as a result, most of the area has been left as forest. This is one of the largest remaining forests in the world.

Construction of transportation routes across the Shield, with its lakes and bogs, is difficult. North of Lake Superior the Shield environment not only blocks agriculture and settlement but is a barrier to east-west land travel in Canada. In the past, settlers going west by-passed it by sailing through the Great Lakes and then proceeding through Minnesota into the prairies. Today two railroads and a two-lane highway offer the only east-west crossing connecting Ontario with the Prairie Provinces, Manitoba, Saskatchewan, and Alberta (Figure 11–1).

PERMAFROST AND NORTHERN FORESTS

Most of northern Canada has sufficiently long and cold winters to keep all but the top few centimeters of soil frozen year round. Permafrost prevails under these conditions where soils are deep enough. During the summer the surface soil thaws and the subsoil remains frozen. Waterlogged soils and wet surface conditions prevail, making surface materials unstable. These are poor foundation environments for roads and buildings. In winter, heat from buildings that are not insulated from the permafrost may cause melting beneath them. When this happens, the ground surface becomes unstable and the building may shift and sink. Plant growth is also restricted by permafrost because of a shallow root

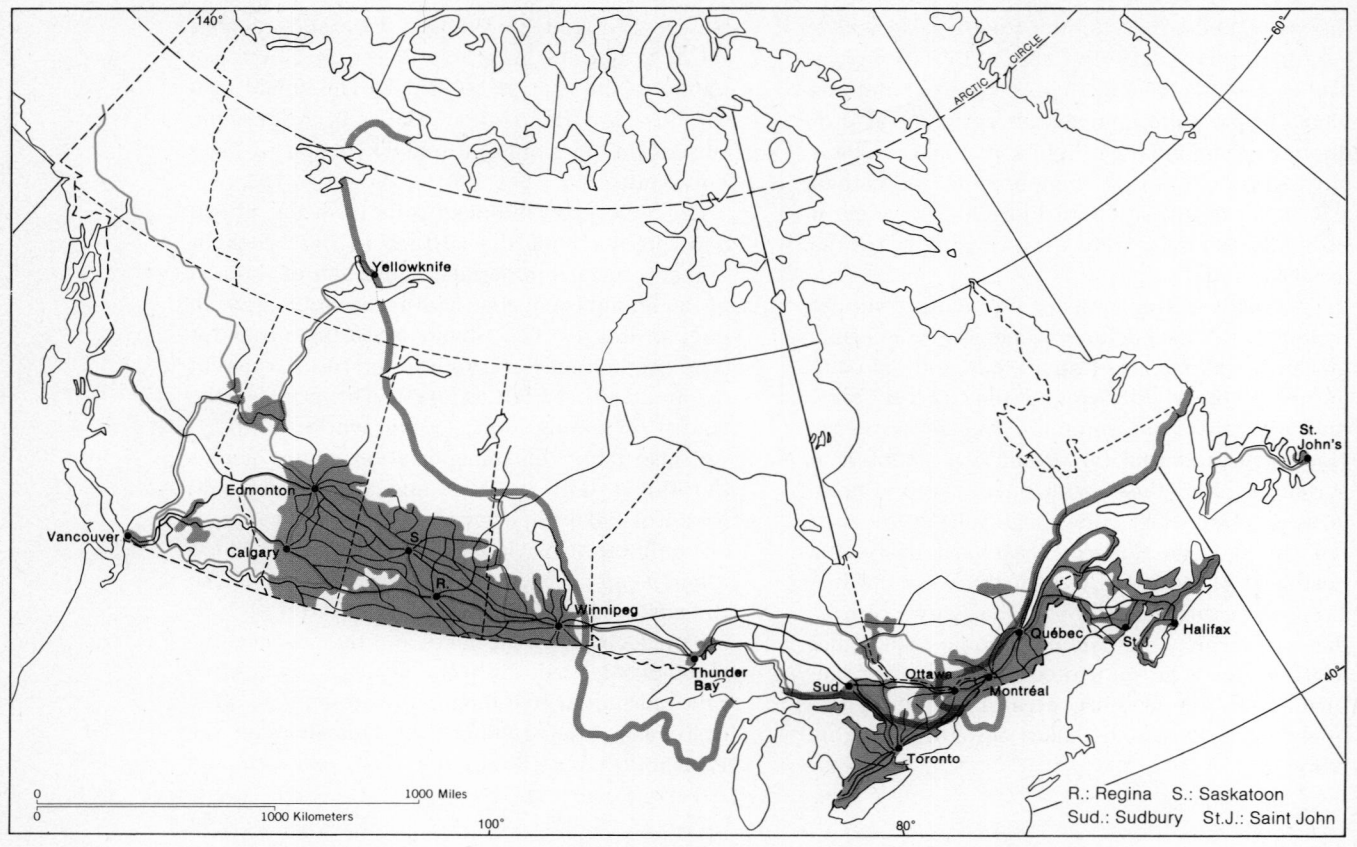

SOURCE: National Atlas of Canada; Oxford World Atlas

Roads closer than 4 miles apart

Major highways

Railroad lines

Boundary of the Canadian Shield

R.: Regina   S.: Saskatoon
Sud.: Sudbury   St.J.: Saint John

FIGURE 11-1 ROADS AND RAILROADS OF CANADA

zone and poor soil drainage. The extreme north of Canada is in fact a cold desert for most of the year. What vegetation there is consists of mosses, lichens, and scattered grasses and shrubs.

As in the Soviet North, most of Canada outside the agricultural areas in the south is forested. Also like the Soviet Union, Canada finds it unec-

onomical to use all but the most accessible areas of the forest, which are mostly on the southern fringe. Needleleaf forests predominate, especially white and black spruce, firs, and pines, with tamarack in wet sites. Throughout most of the forested areas of Canada low potential evapotranspiration permits only slow tree growth, making commercial timber harvesting more like mining than tree farming—removal of a resource rather than management of it.

Though the Canadian northland has little agricultural potential, it has developed as an important source of minerals and hydroelectric power and will probably be more important in the future. The rocks of the metamorphic Shield are rich in metallic minerals—iron, lead and zinc, silver and gold, and nickel. Good sites for hydroelectric power development occur where large streams drop from the Shield to adjacent lowlands; Canada is, indeed, a major hydroelectric power producer. Important petroleum resources are found in the sedimentary rocks underlying the Prairie Provinces, particularly Alberta. Canada's rich mineral resources are one of its important economic assets today, but a major problem is how to get access to the remote locations of many of these resources. Most mineral resource development, as a consequence, remains concentrated along Canada's southern margin.

THE CANADIAN HERITAGE

When Europeans first reached Canada, the native Indians and Eskimos lived in locally based societies that depended largely on hunting, fishing, and gathering. These native peoples have been pushed aside from the better lands in the south and now are left mainly in the north and on the fringe beyond the agricultural-industrial settlements and population of European ancestry. Today the Eskimo population is estimated at less than 20,000 and the Indian population at about 250,000. These native peoples make up only about 1 percent of the Canadian population.

The first explorers of Canadian territory were probably Vikings traveling by way of Iceland and Greenland, but by the fourteenth century Europeans were fishing the Grand Banks south of Newfoundland. By the sixteenth century English explorers had discovered the Northwest Passage into Hudson Bay, and the French had discovered the route up the St. Lawrence into the Great Lakes. The great interest was in fur trapping and trading with the Indians, not agriculture; and by the seventeenth century a busy trade in beaver and other furs had developed along each route. Agricultural settlements were not necessary at first and most settlements were trading posts, forts, and missions. Late in the seventeenth century, however, farm communities were established by the French in Nova Scotia and in the lower St. Lawrence valley. The region settled by the French was called New France.

NEW FRANCE AND THE BRITISH CANADIAN COLONIES

By the 1760s, when New France was ceded to England, there were 70,000 French settlers in Nova Scotia and the lower St. Lawrence. The French settlers of Nova Scotia, known as Acadians, fled, and many of the present-day residents of New Brunswick and of northeastern Maine are their descendants. By 1774, however, the British affirmed the right of the French in Québec to maintain their language and their own culture, including language, laws, and religion. This population and these rights established more than 200 years ago are the basis of French Canada today.

The French and later the British had easier access to the interior of what is now the United States from their settlements in Canada than did colonists east of the Appalachians. The interior was easily reached by the St. Lawrence–Great Lakes route, and the French, in fact, established settlements along the Mississippi as far south as New Orleans. The primary purpose of these settlements, however, was to obtain furs, not agricultural expansion. The early access to this interior area was the basis for French and British claims on the interior, though these were eventually lost to U.S. westward expansion.

BUILDING LINKS BETWEEN SEPARATE COLONIES

British Canada, like the earlier U.S. colonies, was a series of separate entities. There were six colonies: Nova Scotia, New Brunswick, Prince Edward Island, Newfoundland, Upper Canada (now Ontario), and Lower Canada (now Québec). In addition, the Hudson's Bay Company controlled the Canadian north and west, but settlements there were limited to forts and trading posts. Lumbering, fishing, and agriculture were the main activities in Nova Scotia and New Brunswick. Prince Edward Island and Lower Canada were mostly agricultural; Newfoundland was largely a fishing region; and Upper Canada depended mostly on agriculture, lumbering, and the fur trade.

By the nineteenth century all were aided by a growing trade with England, and a treaty with the United States allowed Canadian goods to enter that growing market. New Brunswick and Nova Scotia sent grains and basic foodstuffs to New England before the U.S. Middle West began supplying that area. Lumber from the two Canadas and New Brunswick was much in demand in the growing Northeast and Middle West of the United States. A series of canals around the rapids of the St. Lawrence and the Welland Canal between Lake Ontario and Lake Erie opened up navigation on the St. Lawrence–Great Lakes route and enabled timber to reach U.S. markets.

Much of the new lumbering was part of the process of clearing the farmlands of southern Ontario for agricultural use. The new farmland became one of the important wheat-growing areas of the new continent, with shipments by way of Montréal and the St. Lawrence. Wheat production in the French settlements along the St. Lawrence ended because of the depletion of the poor northern forest soil, but it was replaced by crops that remain today—potatoes and hay—and by dairying.

Each of the separate colonies had its own center. The ports of Saint John and Halifax were most important along the Atlantic, and Montréal emerged as the key port of the St. Lawrence route. Halifax, with its shipping and financial interests, was the Boston of the eastern colonies. Clipper ships built there sailed the world and Nova Scotian businesspeople financed the trade and owned the ships.

PROBLEMS OF DEVELOPING SUCH A LARGE LAND

The early colonies of Canada remained as separate centers of development well into the nineteenth century. Because of the rugged interior of New Brunswick and the lack of farmland around the St. Lawrence estuary, a great barrier of unsettled land separated the Atlantic colonies from the two Canadas along the St. Lawrence route. Two of the colonies were islands, Prince Edward Island and Newfoundland (Figure 11–2). Much of the new immigration was along the St. Lawrence route, and the two Canadas grew while there was little continuing immigration into the Atlantic colonies. The first railroad was from Portland, Maine—not Halifax—to Montréal. The pressure of population on the farmlands of Lower Canada caused French Canadians to migrate not to farms in other Canadian provinces but to the textile mills of New England.

As of 1850 the Canadian colonies had almost 2.5 million people, as compared with 23 million in the United States. Montréal, the port linking Upper and Lower Canada with the world, had 90,000 residents, compared with almost 700,000 in New York. Toronto was about half the size of Montréal, and Halifax and Saint John had more than 25,000 each. Virtually no settlement, other than forts and trading posts, took place west of Lake Superior. Several important factors called for the unity of the Canadian colonies: the increasing expenditures required of each colony in building canals and railroads were becoming a burden; the colonies had lost preferential trade treatment in Europe; Upper and Lower Canada

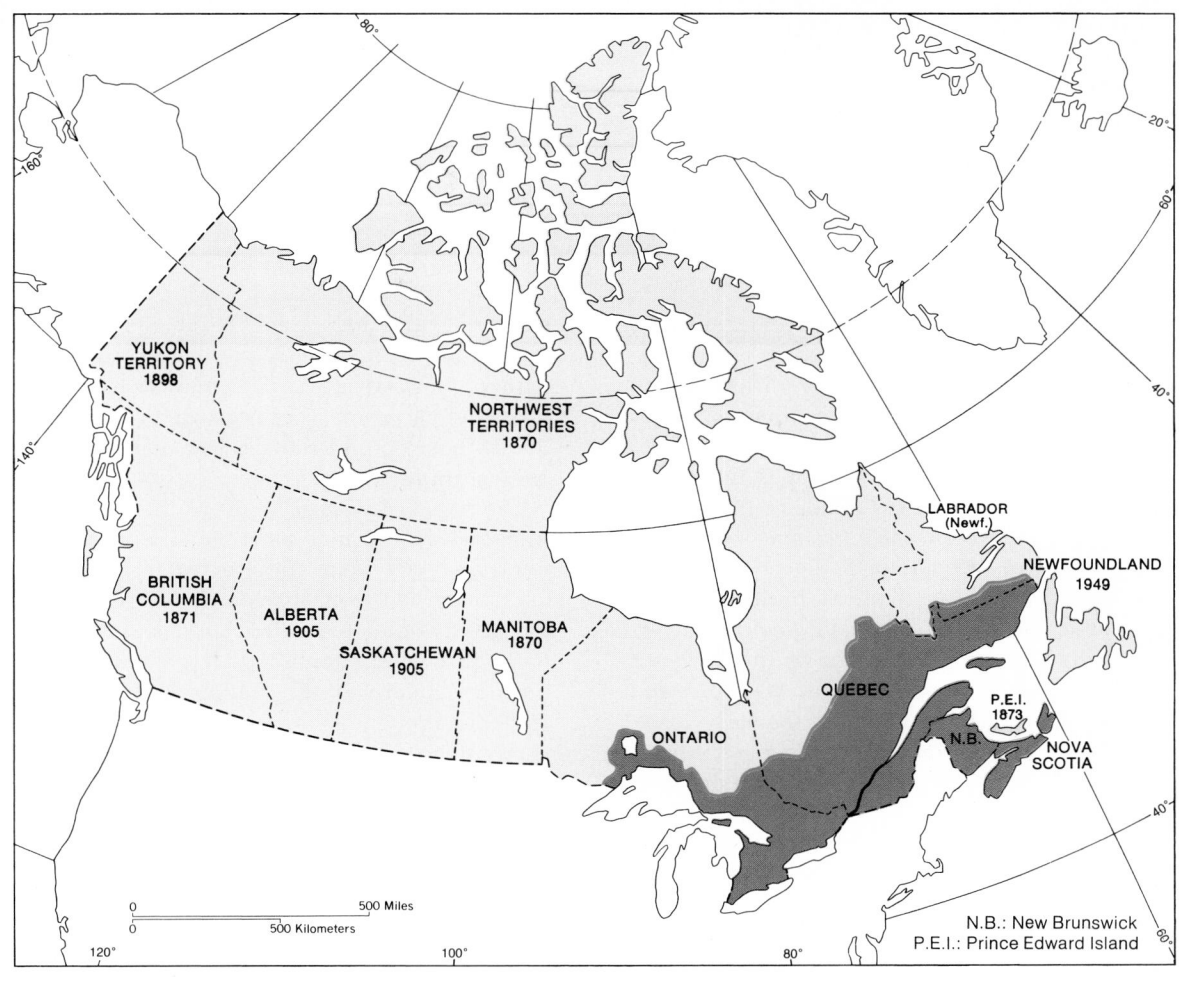

SOURCE: National Atlas of Canada

FIGURE 11-2   GROWTH OF THE CANADIAN
NATION

■ Province of Canada (formed by union of
Upper and Lower Canada, 1840)

■ Provinces added by British North America
Act, 1867

Dates are admission of provinces to the
Canadian Confederation

N.B.: New Brunswick
P.E.I.: Prince Edward Island

CANADA: INDEPENDENT BUT FUNCTIONING AS PART OF A LARGER WHOLE   367

were conflicting over governmental representation, and England was anxious to reduce expenditures on the Canadian colonies.

CONFEDERATION AND THE PROBLEMS
OF BUILDING A SINGLE NATION

Confederation of the colonies, but not full independence from England, came in 1867 (Figure 11–2). The colonies were united under a single government, but Canada retained its ceremonial and economic ties to England as a Dominion of the British Empire under Queen Victoria. Newfoundland refused confederation and Prince Edward Island hesitated six years before joining. Confederation meant that Canada could now try to play essentially the same game as the United States—that of attempting to develop an integrated system among different parts of a vast land area. Soon the Canadian government received the western lands owned by the Hudson's Bay Company. The provinces of Manitoba and British Columbia were added in the early 1870s, but Alberta and Saskatchewan remained territories until 1905. Newfoundland joined in 1949. The northern lands continue under control of the Canadian national government even now, although Indian and Eskimo groups have petitioned for autonomous regions.

Confederation changed the political and economic rules of the game, but the integration of Canada into a single functioning unit was not accomplished as easily as in the United States. Canada in 1867 had a population of fewer than 4 million people, about the same as the United States after independence. But the Canadian population was spread over an area many times larger than the thirteen original American states. Thus a much larger investment in transportation and communication was needed than in the U.S. colonies. Moreover, by the mid-nineteenth century the growing industrial technology had tied certain parts of Canada more with nearby areas in the United States and Europe than with other parts of Canada. The Maritime Provinces (New Brunswick, Nova Scotia, Prince

Edward Island) were linked to New England; Montréal had a rail link with Portland, Maine; and Ontario and the new settlements in the prairies were tied with the United States. The new province of Manitoba was reached by rail from St. Paul in the United States, and British Columbia had no rail connections to the rest of Canada. Thus we see the beginning of a series of north-south links between regions of Canada and of the United States; these links, often stronger than the east-west ties within Canada, remain important today. Parts of Canada were also connected with regions outside North America: Nova Scotia's apple growers were organized to serve the British market, and the farmers of the prairies grew grain for Europe.

The first task was to improve interprovincial transport and communication and to develop interprovincial links. Building of a transcontinental railroad from the Maritime Provinces to British Columbia was a tremendous undertaking for such a small population. It was built in sections as money became available, and not until 1885 did the Canadian Pacific Railway, a privately financed enterprise, complete the undertaking. With a railroad from sea to sea, the confederation seemed much more a real entity than ever before.

Government policy called for immigration programs to settle the western grasslands with farmers and for protective tariffs to encourage the establishment of modern industry. From confederation until 1900 more than 1 million immigrants had entered Canada, but the population was still only 5.4 million, compared to more than 76 million in the United States.

Even with transportation links between them, the Canadian provinces were not able to achieve the kind of integration that had occurred in the United States. Not only was there the problem of the smaller population and thus a smaller market, but there was less diversity in production. Except for Ontario, the environments of the original provincial members were much the same. All produced basic food staples and wood from the forests. Moreover, distances from each to points within the United States were less than from province to province. The Maritime Provinces were closer to New England than to settlements in Québec and Ontario, and Boston was the most accessible large center to them. In the same way, Ontario and Québec had easy access to New York State and to Michigan. Busy trade had developed and settlers from the United States had settled in those two provinces.

Then as now, Canada feared that its economy and culture would be overwhelmed by the larger United States. With most of Canada's population close to the U.S. border, the rapidly growing industry of the U.S. Northeast and Middle West could easily supply the Canadian people. Faced with this threat, the Canadian government imposed tariffs on U.S. goods to provide protection to new Canadian industry.

THE DIMENSIONS OF THE "REAL CANADA"

Almost all Canada's population lives in the south within 200 miles (320 km) of the U.S. border (see Figure 5–2). Edmonton (Alberta) and Prince George (British Columbia) are 300 miles (480 km) from the border, and the Peace River settlement in Alberta is 500 miles (800 km). The only other northern population islands are tiny settlements around mining developments, trading posts, and Indian villages. In 1971 the northern territories that account for almost 40 percent of the total land area had only 55,000 people—about 0.3 percent of the total population.

Functionally, then, Canada has very different dimensions than appear on the map. Canadian population forms a narrow band along the northern border of the United States. Even this narrow band has gaps in western New Brunswick and eastern Québec, in the mountains of British Columbia, and particularly north of the Great Lakes from Sault Sainte Marie to Thunder Bay. Nearly three out of four Canadians live east of Sault Sainte Marie, with Ontario accounting for 36 percent of the total population, Québec 28 percent, and the Maritime Provinces 10 percent (Table 11–1). The majority of Canadians still live along the St. Lawrence–Great Lakes corridor,

with the greatest number between Montréal and Windsor. The Prairie Provinces have only 16 percent of the population, and 10 percent of Canadians live in British Columbia.

As in the United States, Canada's early population was dominantly rural and small-town. Most were farmers, lumber workers, and trappers and the people in towns and villages supporting those primary activities. But today more than half the Canadian population lives in twenty-two large metropolitan areas each with over 100,000 population (Table 11–2). Included are the capitals of all provinces except Prince Edward Island, but twelve of the large metropolitan areas and two-thirds of the metropolitan population are within the St. Lawrence–Great Lakes corridor. This is obviously the core of the country, even more the central area than is the Northeast and Middle West in the United States. Montréal and Toronto, the two largest metropolitan areas in the country, each with almost 3 million people, are in this corridor.

Migration and population growth have been occurring away from the marginal rural areas and the successful agricultural areas for many years. The Maritime Provinces have had a net out-migration since confederation, and for some decades the Prairie Provinces have had an out-migration. Ontario and British Columbia are the chief centers of immigration. But, as in the United States, the big movement has been to the large cities. The metropolitan areas, which today have over half the population, had only a quarter of a much smaller population in 1921. The incoming population is both from rural areas and from foreign immigration. Close to one-quarter of Canada's population today represents immigrants from Southern and Eastern Europe, the West Indies, and India and Pakistan. Most of these new Canadians live in metropolitan Canada.

THE QUESTION OF CULTURAL
AND ECONOMIC UNITY

Today it is questionable whether the widely spread east-west ribbon of population across

TABLE 11-1 CANADIAN POPULATION BY POLITICAL
UNIT, 1973 ESTIMATES

	Population (Thousands)	Percent of Canadian Total
Maritime Provinces		
Newfoundland	541	2.4
Prince Edward Island	115	0.5
Nova Scotia	805	3.6
New Brunswick	652	2.9
Total	2,113	9.5
Canadian Core		
Québec	6,081	27.5
Ontario	7,939	35.9
Total	14,020	63.5
Prairie Provinces		
Manitoba	998	4.5
Saskatchewan	908	4.1
Alberta	1,683	7.6
Total	3,589	16.2
British Columbia	2,315	10.5
Northern Territories		
Yukon	20	0.09
Northwest	38	0.17
Total	58	0.3
Canada total	22,095	

SOURCE: Ministry of Industry, Trade, and Commerce, *Canada Year Book, 1974.*

Canada has much unity. Because they are so widely spread, Canadians in different parts of the country find people across the border in the United States closer and more accessible than Canadians in other regions. British Columbians have more contacts with people in Seattle and western Washington than with Canadians in the Prairie Provinces or the East. Manitobans probably visit Minneapolis as often as Ontario or Québec. People from the Maritimes travel frequently into New England, and large numbers of them have settled there. The French of Québec are still a group apart from other Canadians; in fact, there is strong sentiment in Québec for the establishment of a separate French Canadian country.

A further complicating force since World War II has been the immigration of new ethnic groups. In the period immediately after World War II most of the migrants were from Southern and Eastern Europe, and they formed Italian, Greek, and Polish neighborhoods in large cities such as Toronto and Montréal. More recently the immigration has shifted to West Indian and Asian people—mostly Pakistanis and Indians. These groups, too, have formed their own neighbor-

TABLE 11-2 POPULATION OF
CANADIAN METROPOLITAN CENTERS
OF MORE THAN 100,000, 1973
(*Thousands*)

Montréal	2,775
Toronto	2,692
Vancouver	1,116
Ottawa-Hull	619
Winnipeg	560
Edmonton	518
Hamilton	513
Québec City	493
Calgary	431
St. Catherines–Niagara	308
London	293
Windsor	264
Kitchener	235
Halifax	222
Victoria	203
Sudbury	155
Regina	147
Chicoutimi–Jonquière	136
St. John's	133
Saskatoon	128
Thunder Bay	113
Saint John	110

SOURCE: Ministry of Industry, Trade, and Commerce, *Canada Year Book, 1974.*

hoods in the large cities and have been even less assimilated into the Canadian population than the Southern and Eastern Europeans.

CANADA'S CULTURAL IDENTITY

Canadians take pride in their cultural achievements. A flourishing literature in both English and French, world-renowned academic institutions, and a broad range of festivals and expositions, including the Stratford Shakespeare Festi-val in Ontario—these are some of the things Canadians point out to indicate their rich and active cultural life. They also indicate the dynamic, cosmopolitan cities Toronto and Montréal, with their many links to metropolitan centers throughout the world; a larger variety of imported goods is available for purchase in these cities than in most large cities in the United States. The visitor to Canada often senses a fierce nationalism among its residents, especially young Canadians.

Nevertheless, with most Canadians living within 200 miles (320 km) of the U.S. border, the cultural shadow of the United States looms large over Canada's everyday life. American radio and television stations close to the Canadian border beam situation comedies based on life in the United States, music produced in the United States, and commercials pushing U.S. products. American magazines and books, with their domestic market of over 200 million people, are much more economical to produce than are Canadian magazines that must sell in a market just one-tenth as large. Thus big-budget U.S. magazines and mass-circulation best sellers are found on newsstands and in bookstores in Canadian cities alongside Canadian publications. Even Canadian radio stations feature music produced in the United States and sung by U.S. artists. With such constant exposure, it is hard for Canadians to avoid being engulfed by values created in New York, Hollywood, and the other culture-making centers of the United States.

French Canadians

The diversity of Canadians complicates the cultural situation further. The most important aspect of this diversity is the large French Canadian population. Their roots in Canada go back to New France before the English settled North America, and from the first coming of the British they have held tenaciously to their French traditions. Moreover, the majority have remained in the old settlements of New France along the lower St. Lawrence River while at the same time relatively few non-French settlers have come to this area (Figure 11–3). French Canadians account for about 80 percent of the population of Québec. There are also important French Canadian population centers in New Brunswick, where 34 percent of the population consider French their mother tongue, and in eastern Ontario along the border with Québec between the Ottawa and the St. Lawrence rivers. French Canadians are also found in scattered clusters in the Prairie Provinces. In all, French Canadians account for 27 percent of the Canadian population, though in most parts of Canada they are a minority.

Montréal, Canada's most important port and largest city, is part of French Canada. As a major business center of Canada and part of the international business community, it is a very cosmopolitan center. In addition, it has always had a large English-speaking population, with English-speaking persons in control of major businesses.

French Canada has tended to be rural and traditionally oriented, but in recent years a new group of young leaders, trained in modern engineering, management, and law, has surfaced. They have pushed for economic development in Québec and an increasing role for French Canadians in provincial leadership. Due to their efforts, technical and secondary education have been reformed to give better training to French Canadians and to preserve French culture. New laws call for all education in Québec to be in French and for French to be used as the primary language of business and culture. At the national level, French is recognized as coequal to English and government documents are printed in both languages. Still, the movement for separation of Québec from the Confederation and the establishment of an independent French Canadian state has grown. The 1976 provincial elections in Québec were won by a political party whose avowed goal is the eventual separation of Québec from the rest of Canada.

Non-Europeans as Citizens Apart

Recent large-scale immigration of non-Europeans to Canada presents a new dimension to the question of Canadian culture. West Indians, Indians, and Pakistanis still represent less than 2 percent of the total population, but their numbers are continuing to grow. These non-Europeans are now important minorities in Toronto and other large cities. They tend to be segregated into their own neighborhoods and thus to retain their traditional lifestyles. As with American blacks, their different skin color makes them immediately recognizable as minority members and subjects of possible prejudice.

A SUCCESSFUL BUT DISJOINTED ECONOMY

In its economy as in its culture, the country is not a unit. Canada can produce fewer of the basic range of foods and raw materials than the United States can, and it is costly to move products from one region to another over the long distances. Moreover, the specialized producing regions of Canada can produce far more than is needed domestically. For example, in a favorable growing year the Prairie Provinces can produce enough wheat for 90 million people, not just the 23 million of Canada. Western oil fields and mineral production from the Shield have been organized to produce far beyond Canada's needs. The same is true of the timber and fishing industries. Canadian manufacturing, too, depends not only on Canadian markets, but also on external markets.

Development by Outside Money for Outside Markets

The lack of complete integration of the Canadian economy also results from the interest of foreign investors in developing Canadian resources to supply industrial development or consumer markets outside Canada. Moreover, Canadians have needed to obtain foreign exchange to purchase goods from other countries. The relatively small Canadian population has had to depend, in large part, on technology developed in the United States and Europe. At the same time, it has not been able to generate the necessary capital for development of its vast resource base itself. Thus the expansion of Canada's mineral and timber

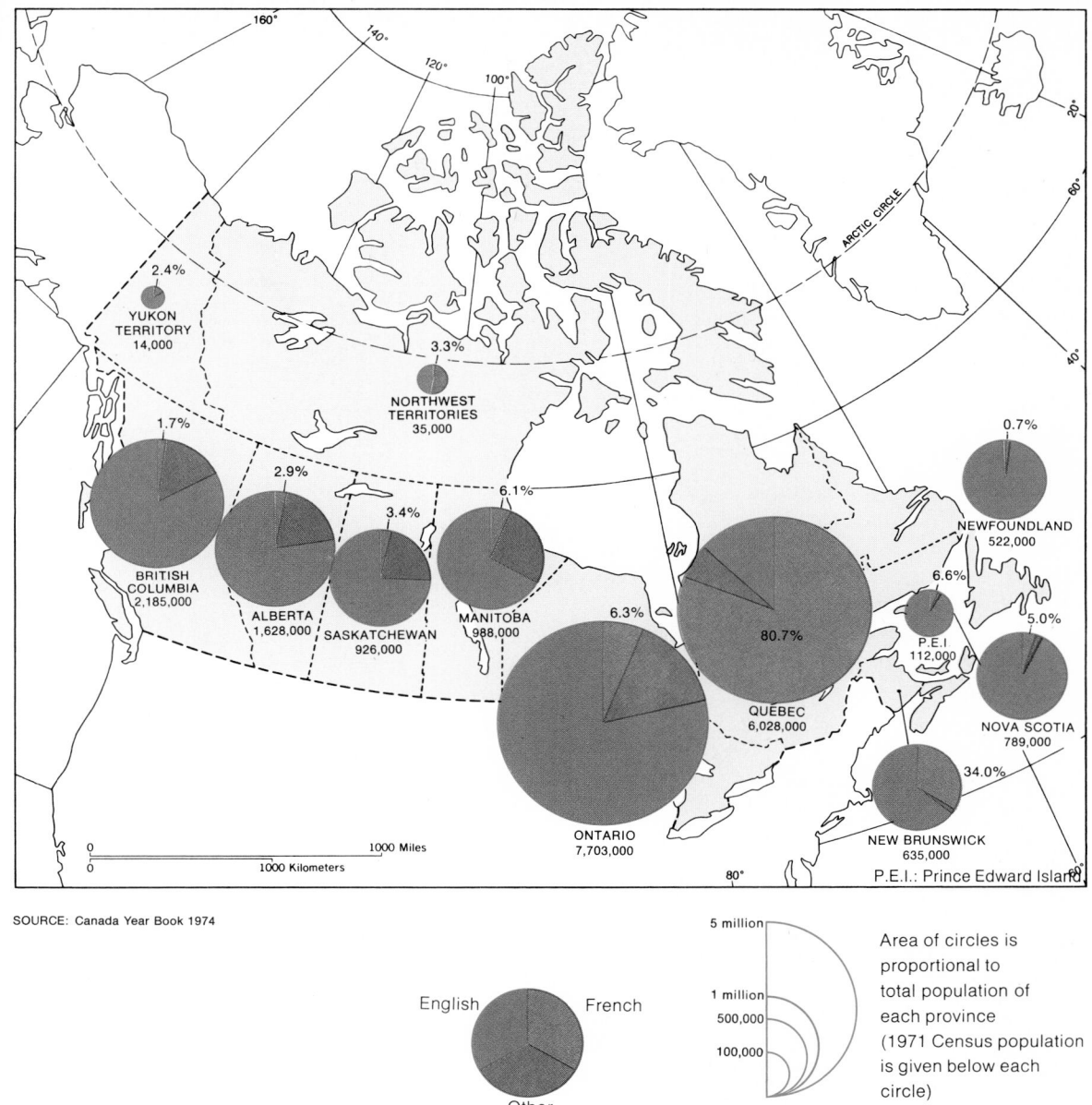

SOURCE: Canada Year Book 1974

English   French

Other

5 million

1 million
500,000

100,000

Area of circles is
proportional to
total population of
each province
(1971 Census population
is given below each
circle)

Map labels:

YUKON TERRITORY 14,000 — 2.4%

NORTHWEST TERRITORIES 35,000 — 3.3%

BRITISH COLUMBIA 2,185,000 — 1.7%

ALBERTA 1,628,000 — 2.9%

SASKATCHEWAN 926,000 — 3.4%

MANITOBA 988,000 — 6.1%

ONTARIO 7,703,000 — 6.3%

QUÉBEC 6,028,000 — 80.7%

NEWFOUNDLAND 522,000 — 0.7%

P.E.I 112,000 — 6.6%

NOVA SCOTIA 789,000 — 5.0%

NEW BRUNSWICK 635,000 — 34.0%

P.E.I.: Prince Edward Island

ARCTIC CIRCLE

0   1000 Miles
0   1000 Kilometers

FIGURE 11-3 PERCENTAGE OF CANADIAN
POPULATION WITH ENGLISH, FRENCH, AND
OTHER LANGUAGES AS MOTHER TONGUE,
BY PROVINCE, 1971

TABLE 11-3   VALUE OF CANADIAN EXPORTS AND IMPORTS,
BY TYPE OF COMMODITY, 1973   ($ *Millions*)

	Destination		
Exports	All countries	United States	United Kingdom
Edible commodities			
Fish and shellfish	434	282	50
Wheat and wheat flour	1,268	0.5	139
Other cereals	365	70	9
Raw materials			
Seeds	389	18	22
Iron ores and concentrates	462	304	51
Copper (ores, concentrates, scrap)	554	43	11
Nickel (ores, concentrates, scrap)	442	100	125
Other metals (ores, concentrates, scrap)	503	133	64
Crude petroleum	1,482	1,482	—
Natural gas	351	351	—
Other nonmetallic minerals	572	154	25
Fabricated materials			
Lumber	1,598	1,285	100
Wood pulp	1,059	617	63
Newsprint paper	1,287	1,068	77
Other wood products	516	288	110
Chemicals (including fertilizers)	527	368	54
Petroleum and coal products	312	293	0.5
Iron and steel	479	359	21
Aluminum	374	203	34
Copper and alloys	521	216	148
Nickel and alloys	377	224	43
Other nonferrous metals	331	245	55
End products			
Machinery	844	658	17
Passenger automobiles	2,373	2,316	—
Trucks and other motor vehicles	873	814	0.2
Motor vehicle engines, parts, accessories	2,092	1,994	3
Aircraft, engines, parts	414	332	13
Telecommunications equipment	301	177	26

TABLE 11-3   *(cont.)*

	Origin		
Imports	All countries	United States	United Kingdom
Edible commodities			
Fruits and vegetables	570	387	5
Raw materials			
Crude petroleum	941	—	—
Fabricated materials			
Textiles	493	225	41
Chemicals (including fertilizers)	349	245	27
Plastics	324	285	7
Iron and steel	653	355	45
End products			
Machinery	2,762	2,216	185
Passenger automobiles	1,775	1,438	32
Trucks and other motor vehicles	801	707	6
Motor vehicle engines and parts	3,487	3,350	37
Aircraft, engines, and parts	511	461	36
Other transportation equipment	397	247	25
Telecommunications equipment	812	505	34
Other equipment and tools	1,992	1,605	87
Apparel and footwear	458	70	30
Other personal and household goods	563	230	73
Printed matter	317	259	19

SOURCE: Ministry of Industry, Trade, and Commerce, *Canada Year Book, 1974.*
NOTE: Only commodities for which the value of exports or imports exceeded $300 million in 1973 are included.

resources—as well as much of its consumer goods industry, particularly automobiles and petroleum products—has been accomplished by foreign investors, especially U.S.-based companies.

Forest products   The impact of foreign investment shows sharply in the forest products industry (Table 11-3). The Canadian forests have been seen by foreign investors as a great source of both timber and pulp and paper. Big multinational corporations dominate the industry. Canada exports 60 percent of the lumber produced from its forests and 40 percent of its wood pulp and paper. The industry is concentrated along the lower St. Lawrence and the shores of Lake Superior in the east and in coastal British Columbia in the west (Figure 11-4). Timber is accessible not only to important centers of Canadian population, but also by water to the U.S. Middle West via the Great Lakes, to the U.S. Northeast via the lower St. Lawrence, and to the U.S. Pacific Coast and to Japan via the Pacific Ocean. Ships hauling paper can often make direct deliveries from the mills to the receiving docks of the newspapers. Gigantic integrated pulp and lumber mills have been developed since World War II in British Columbia.

Minerals The search for minerals also has been dominated by foreign investors, but this search has taken them into the largely empty areas north of the population belt of Canada. New discoveries are farther and farther from major settlements (Figure 11–4). Petroleum and natural gas, the most important minerals, have mostly been developed by the large U.S. and other multinational companies in the northern parts of the Prairie Provinces and eastern British Columbia, and more and more exploration now is taking place in the arctic. Copper is widely mined by foreign companies but particularly in mining areas of Ontario, Québec, and Manitoba north of agricultural and metropolitan settlement or in the mountains of British Columbia. Major nickel output comes from Sudbury in the Shield north of Lake Huron, and iron ore comes both from mines north of Lake Superior and from deposits developed since World War II along the Québec-Labrador boundary 350 miles north of the St. Lawrence valley. Zinc, which used to come from much the same areas of Ontario and Québec as copper, is now being produced in the Northwest Territories. Uranium, too, is produced well north of major Canadian settlement.

Northern production has occurred mostly along railroads built for other purposes. For instance, important producing areas have been developed along the railroad to Churchill on Hudson Bay, originally built as an alternative export route for grain from the prairies, and along the Canadian National Railway line across the Shield from Québec City to Winnipeg. To develop the Québec-Labrador iron deposits in the 1950s, it was necessary to build a new railroad to a port on the St. Lawrence that provided access to U.S. steel mills on the Great Lakes and to European mills as well. Similarly, a pipeline system built since World War II delivers oil and gas from the Prairie Provinces not only to major Canadian population centers but also to markets in the U.S. Middle West. Mining camps beyond the railroads to the north are serviced by air.

Farming and fishing Agriculture and fishing are still largely in the hands of Canadians themselves, but they, too, greatly depend on exports. Seventy percent of the value of Canada's fish catch is exported, and Canadian agricultural exports account for 11 percent of all exports. Canada is one of the six largest agricultural exporting countries in the world and as an exporter of wheat is second only to the United States. Wheat production for foreign markets is the basis for farming in the Canadian prairies.

Canada's Concern over Foreign Investment

Foreign investment is necessarily an important factor when 20 million people attempt to develop a country as vast as Canada. In 1972 foreign trade accounted for 38 percent of Canada's total earnings. Foreign investment in Canada and Canada's foreign trade are dominated by the United States; U.S. investors account for over 80 percent of all foreign investment in Canada, and over 70 percent of foreign trade is with the United States. Much of Canadian resource development has been undertaken by U.S. investors to provide raw materials for U.S. industry. Iron mines were developed in northern Labrador (the mainland part of Newfoundland) in the 1950s to supply the steel mills around the Great Lakes when it appeared that the deposits in the United States around Lake Superior were exhausted of high-grade ores. American oil companies dominated the control of most of the oil fields of the Prairie Provinces after World War II, and as late as 1970 almost half the oil production of Canada was going to the United States. Mines in Canada account for about half the nickel production of the non-Communist world, and about 20 percent of that output goes to the United States. Canadian asbestos production, too, has been developed largely to serve the U.S. market.

Such domination of investment and trade by a single foreign country is naturally of concern to Canadians, and in recent years they have passed laws to control foreign investment, particularly from the United States. The Foreign Investment Review Act provides for review of the extent of Canadian ownership of companies located in

SOURCE: National Atlas of Canada

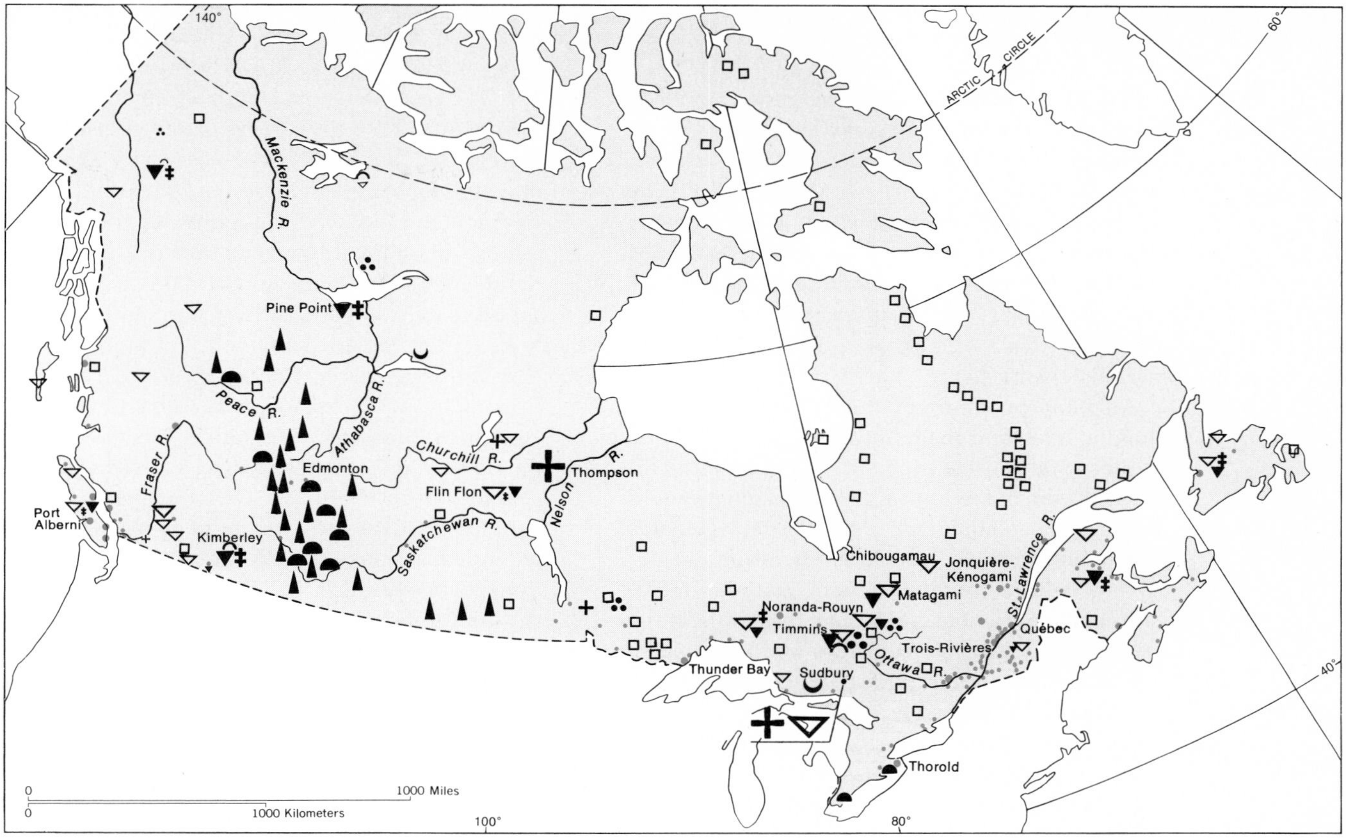

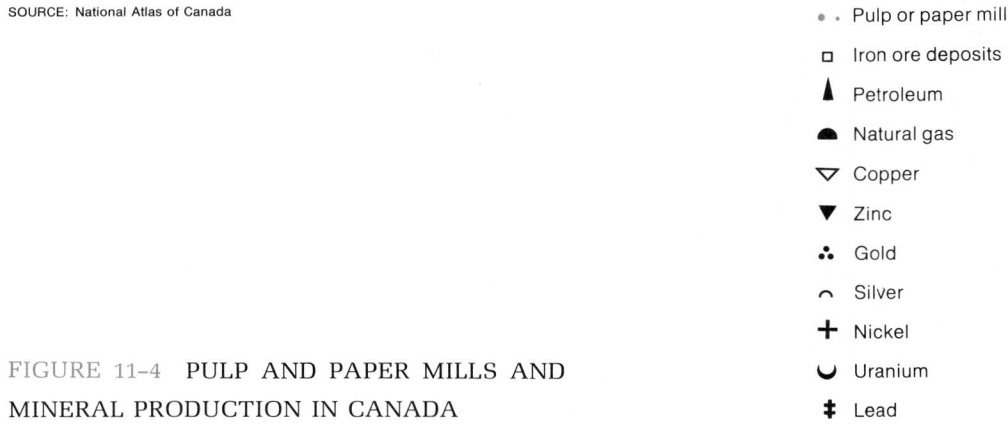

- Pulp or paper mill
- Iron ore deposits
- Petroleum
- Natural gas
- Copper
- Zinc
- Gold
- Silver
- Nickel
- Uranium
- Lead

FIGURE 11-4 PULP AND PAPER MILLS AND MINERAL PRODUCTION IN CANADA

Canada in order to help increase the Canadian-owned share of certain markets, particularly books, films, and magazines.

The United States is also the major source of Canadian imports, and a very large share of Canadian imports from the United States consists of industrial goods (Table 11–3). The Manufacturing Belt of the United States lies just across the border in the U.S. Northeast and Middle West, and it is not surprising that a wide range of the products of that industry move to nearby Canada. Included are tractors, aircraft, office machines, and various chemicals.

An important part of the industrial products moving back and forth between Canada and the United States is equipment, supplies, and finished products used by large U.S. companies with operations on either side of the border. The most important of these industries operating an integrated industrial complex with installations on either side of the border is the automobile industry. Detroit, center of U.S. auto operations, is on the Canadian border, and it appears logical that the parts and assembly plants that spread out from Detroit into Michigan and other nearby states would be found in southern Ontario as well. For a time, however, shipment of automobiles across the border was restricted, and it was necessary for U.S. auto firms to set up separate assembly plants for the Canadian market on Canadian soil. But in recent years these restrictions have been eliminated. The Canadian operations of the large automakers have now been integrated into each company's U.S. operations. Autos from Canadian assembly lines are sold in the United States, and others from U.S. production go to Canada.

It is not just the U.S. automobile manufacturers for whom 23 million affluent Canadians are an important market for their products. This is true for most major U.S. industries. Canada has been encouraging the development of its own industrial base since confederation, however, and has imposed import restrictions on many consumer goods produced in the United States. U.S. companies have responded by establishing branch plants, or **subsidiaries,** in Canada organized to do business in Canada. Three of the four largest Canadian companies are subsidiaries of U.S. firms or were organized by U.S. capital, and most of the largest U.S. consumer goods manufacturers have Canadian subsidiaries. Thus decisions affecting Canadian regions—whether to develop a new mining source in Labrador, to exploit a new timber stand in British Columbia, or to expand a factory or close it down in Ontario—are commonly made by large corporations with headquarters in New York, Chicago, or San Francisco.

As a step toward lessening the dominance of the United States in Canada's foreign relations, Canada in 1976 signed an agreement with the European Economic Community designed to expand trade and other exchanges between Canada and European countries. Canada is greatly concerned by outside investment in its resources and its industries because of the political implications. The question for Canada is: How much should foreign interests control economic development within Canada? On the one hand, the Canadian government has the power to exclude foreign investors; on the other, 23 million people cannot themselves generate enough capital or the range of technological expertise necessary to function without outside assistance.

THE CANADIAN SYSTEM IN RELATION
TO THE UNITED STATES

In relation to the United States, the modern interconnected system of Canada functions in some ways as an individual large state, such as California, does within the United States as a whole. California takes advantage of its own particular resources—tremendous agricultural potential and, in southern California, a year-round growing season, great forests in the north, and specialty minerals—to produce materials for shipment to the rest of the country. California has also developed specialized manufacturing industries—the movie industry, aerospace, patio and pool products—which produce goods for marketing elsewhere throughout the country. At the same time California, the most populous state, is a key market for products manufactured in other regions of the United States.

There are, however, important differences between Canada and California:

1. Canada is an independent country with its own government, with its own ties to other countries throughout the world, and with its own rules for the economic activities carried on within its boundaries.

2. Canada encompasses an area larger than the whole of the United States. Both the population and the producing districts are spread out far from one another. Lumber areas in British Columbia are hundreds of miles from wheat-producing regions in the prairies, which are in turn hundreds of miles from Ontario's industrial district, which is hundreds of miles from the lower St. Lawrence River, which is still hundreds of miles from Nova Scotia and the other Maritime Provinces. The products of these different producing regions go through different ports to different market areas. Each is tied to the modern interconnected system through different channels and routes. At the same time these various parts of Canada require different inputs of food, raw materials, and manufactured goods. The prairies produce food and fuel, but must import timber and manufactured goods. Québec is a major manufacturing area and has important timber production, but it is short of important food products produced in other parts of Canada and the world.

Each Canadian producing region is also in a particular position with regard to producing regions in its major supplier and consumer, the United States. Often, a producing region in the United States is better located to supply a Canadian region, through surplus production, than is a similar producing region within Canada. Or particular U.S. markets are closer for a Canadian producer than are most other Canadian regions. Thus from the early days of Canadian history complementary economies have developed across the Canadian–U.S. border at many places. This appears in U.S.–Canadian trade: coal from Appalachia and the Middle West is more accessible to the industry of Ontario and the St. Lawrence valley via the Great Lakes than is coal from Alberta and British Columbia. Oil from the Prairies is shipped not only to Ontario but also to the Middle West, while Eastern Canada uses some oil from refineries in Portland, Maine.

THE NORTH: STILL ONLY MARGINALLY PART OF THE SYSTEM

The bulk of Canada north of the major population concentrations still contributes only marginally to the functioning system. For most Canadians it is a world entirely foreign to their lives, only glimpsed occasionally on newscasts or documentary films. Only two Canadians per thousand live in the North, many of them Indians and Eskimos. Most of the northerners are government employees, military personnel, or employees of the mining operations.

The important part of the North is its southern fringe, where railroads and better highways reach. There minerals and timber can be produced for the outside world and tourists can come in summer for recreation. There, too, the water of the swift-flowing streams dropping off the Shield has been tapped for waterpower. Two-thirds of Canada's electricity is produced from water sources, much of it in large power grids based on a network of dams around the Shield operated by provincial electricity boards. High-tension electric wires carry power hundreds of miles to the main marketing network in the population corridor.

The North, like Alaska, has very limited agricultural possibilities, but is known to be rich in mineral resources. The problem is that it is at present too expensive to produce most of these resources. The cost of operations in the North, with its severely cold, dark, long winters, is great, and the cost of building transport routes over the Canadian Shield and areas of permafrost is even greater. In the Far North development is beginning—mining around Dawson in the Yukon and Yellowknife in the Northwest Territories—and these cities now have road-metal highways connecting them with the south. But most development in the North remains at the trading post, and the Indian and Eskimo village is only partly tied to the modern world.

Canadians have attempted to bring modern ways to the isolated communities of the North. Only in recent years, however, with the advent of the airplane and the radio, have the possibilities of cultural and economic integration really existed. Now, with the aid of its own system of space communications satellites, Canada is beaming television programs to communities in the Far North that may not even be tied into the national road system. People in Dawson and Yellowknife, and smaller communities, too, are viewing professional hockey and baseball games from Toronto, Montréal, and Vancouver and seeing national politicians make speeches. They are also viewing dramas, comedies, and documentary programs that give them a new perspective of life in the rest of Canada. It remains to be seen what the impact of this new cultural force will be.

THE WORKINGS OF THE CANADIAN SYSTEM

Many of the products of the Canadian economy—wheat, timber, minerals, and manufactured products—are bound for different destinations outside the country; capital for much of the development comes from other countries; and major decisions affecting Canadian business may be made in corporation headquarters in New York, Chicago, or London. Still, the Canadian producing systems are managed from the two largest business centers of the country, Montréal and Toronto. Money from abroad is funneled through banks in these cities, and profits from Canadian sales end up in these banks. The offices of organizational managers of mining companies, timber firms, and industrial corporations, and the headquarters of wholesalers and retailers, shippers and financiers, are housed in skyscrapers in these cities. Shares of business

firms are traded on national stock exchanges and agricultural products are handled in commodity exchanges. As the Canadian economy has grown, so have the office towers of Toronto and Montréal.

These two cities are closely linked to the larger U.S. system. Major air links tie both Montréal and Toronto to New York and Chicago, and to London and, increasingly, to Tokyo as well. Within Canada, the major air links are, first of all, between Toronto and Montréal, just as in the United States the major interaction is between New York and Chicago (Figure 11–5). Secondary ties are from Montréal and Toronto to the other large cities that serve as regional centers—Vancouver in British Columbia; Edmonton, Calgary, and Winnipeg in the Prairies; Saint John, Halifax, and St. John's in the Maritimes—and to the large industrial cities of Ontario and Québec such as Windsor, London, and Québec City; and, of course, to Ottawa, the Canadian capital. In the regional centers are subheadquarters of large firms operating in Canada as well as the headquarters of business and cultural enterprises operating only within those regions. It is not surprising that many of these large regional centers are also the capitals of Canadian provinces, for under confederation the provinces remain very powerful within the federal system.

THE PRODUCING NETWORK

The network of large Canadian metropolitan areas centered in Montréal and Toronto controls the system of production that spreads all over the country, producing goods for Canadians, the U.S. market, and consumers in other parts of the world. It also manages the marketing system that delivers products to people and businesses in all parts of Canada whether those goods themselves are Canadian made or imported.

Toronto and Montréal are, themselves, also the largest manufacturing centers in the country. Together they produce about one-third of the value of all goods manufactured in Canada. Another 25 percent of manufacturing comes from

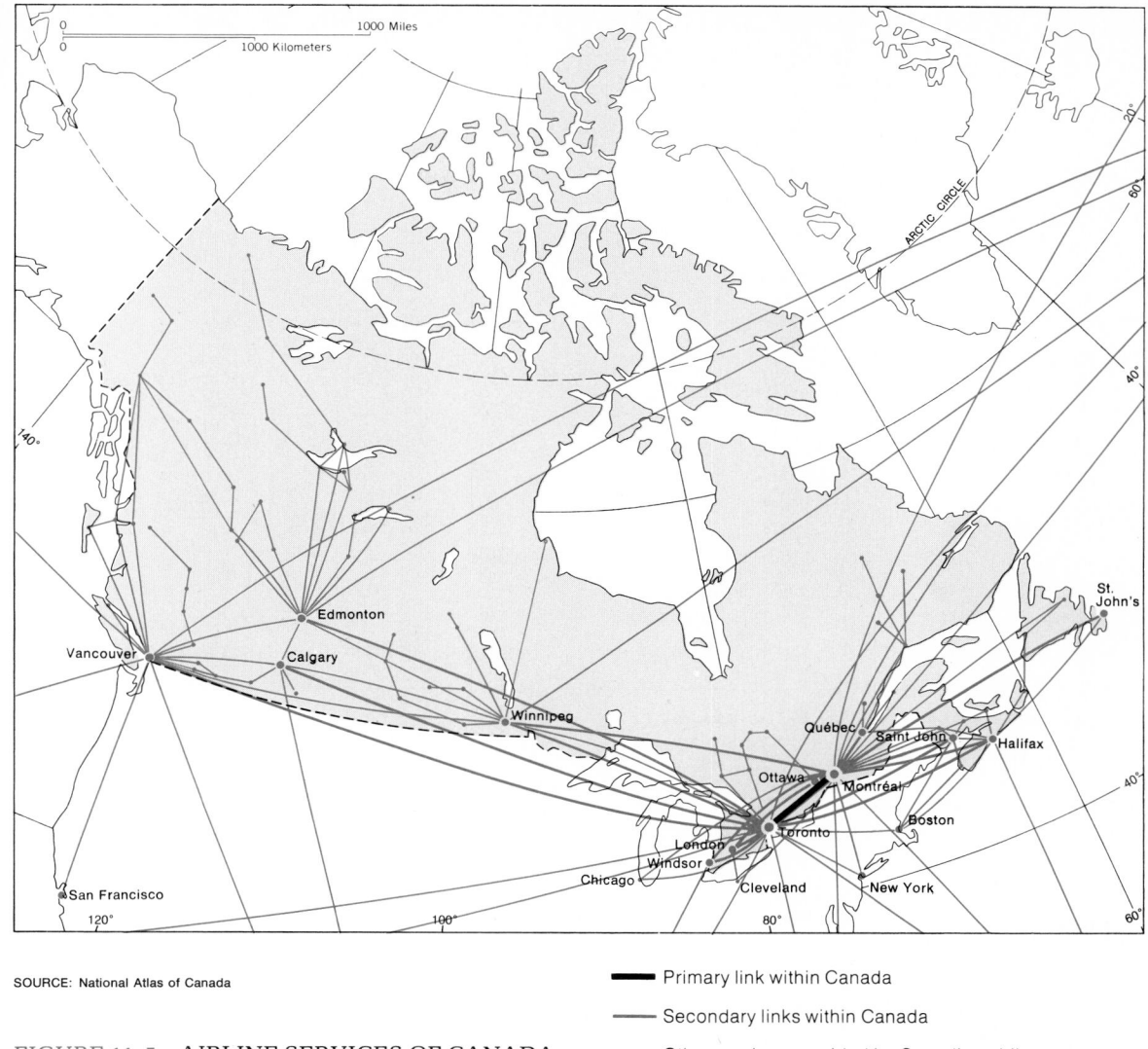

FIGURE 11-5   AIRLINE SERVICES OF CANADA

■ Primary link within Canada

── Secondary links within Canada

── Other services provided by Canadian airlines

the other large metropolitan areas with populations above 100,000. Moreover, almost 80 percent of all of the value of Canadian manufacturing is accounted for by the two provinces of Ontario and Québec (Table 11–4). These two provinces and, particularly, their large cities are the centers of Canada's most advanced metal and chemical industries. They not only export manufactures to the United States and other countries, but supply many of the manufactured goods for the whole of Canada.

Manufacturing is proportionally far less important to Canada than to the United States. Value added by manufacture in Canada is less than twice the value of primary production, whereas in the United States value added by

TABLE 11-4   VALUE OF CANADIAN PRODUCTION, BY PROVINCE

Province or Territory	Area (Thousands of Square Miles)	Agriculture Cash Receipts, 1973	Minerals, 1972	Forest Products,[a] 1971	Furs, 1972–73	Fishery Products, 1972	Value Added by Manufacture, 1971
				$ millions			
Newfoundland (including Labrador)	156.1	—	290.6	2.9	0.6	100.6	135.5
Prince Edward Island	2.2	72.0	—	—[b]	0.2	20.0	22.4
Nova Scotia	21.4	94.7	57.5	20.1	2.0	142.1	311.8
New Brunswick	28.4	95.6	119.9	52.1	0.5	86.4	296.8
Québec	594.9	962.4	782.6	205.1	4.9	25.9	6,406.2
Ontario	412.6	1,947.6	1,534.8	115.8	14.0	16.2	12,537.2
Manitoba	251.0	611.6	323.3	5.8	4.7 ⎫		558.9
Saskatchewan	251.7	1,435.5	409.6	13.4	4.3 ⎬ 15.4		217.9
Alberta	255.3	1,193.2	1,978.6	49.1	8.4 ⎭		785.3
British Columbia	366.3	316.6	678.0	930.6	4.7 ⎫		1,912.6
Yukon Territory	207.1	—	106.8	—[b]	0.3 ⎭ 159.2		1.1
Northwest Territories	1,304.9	—	120.3	—[b]	1.4	—[d]	2.2
Total	3,851.8	6,729.2	6,403.2	1,395.5	46.7[c]	545.6[e]	23,187.9

SOURCE: Ministry of Industry, Trade, and Commerce, *Canada Year Book, 1974.*
[a] All shipments of the sawmill and planing mill industry, including lumber.
[b] Confidential.
[c] Total includes furs not allocated to a province or territory, mainly seals.
[d] Included in combined total for Manitoba, Saskatchewan, and Alberta.
[e] Total differs from sum of provincial totals because shipments between provinces on the Atlantic coast have been subtracted.

manufacture is over three times the value of primary production. The proportion of the Canadian work force engaged in primary production is twice that in the United States. Moreover, within primary production mineral production is worth almost half again the value of agricultural output, a proportion quite the reverse of the case in the United States.

The outlying primary producing regions, then, are more important than in the United States, and mineral output is particularly important. As we have seen, most of the mineral output comes from the fringe of population outside the Toronto–Montréal corridor. Mining is done mostly in small, single-industry towns and villages. The output is shipped long distance to markets in Canadian industrial cities, to shipping points such as Montréal or Vancouver, or to specialized ports such as the iron ore port of Sept-Îles-Pointe Noire.

Canada is one of the great agricultural exporting countries in the world. To place this total output in perspective, however, keep in mind that total agricultural sales for all of Canada, domestic as well as export market sales, are less than those of California. Wheat is the most important agricultural crop and most of it is exported, but both cattle production and dairy products provide greater total return to Canadian farmers. The cool moist summers along the St. Lawrence valley, and also in southern Ontario, are well suited to dairying, and most production is to supply Canada's own needs. Beef cattle, too, are produced in southern Ontario, which has a

AUSTRALIA: A SELF-STUDY PROJECT

The geographical analysis of Canada as a national system and as part of the modern interconnected world can serve as a model for investigating the geography of Australia. These countries are similar in size and both have a large area of low-work environments. They were both also British colonies and were settled mostly by immigrants from England and Western Europe. In large measure they have evolved from a similar heritage and developed on resource bases that have much in common. Though the population of Australia is considerably smaller than that of Canada, it is also similarly distributed along the margins of the country. The economies of both countries are geared in a similar degree to primary production of mines and agriculture, although both have well-developed industrial bases and their populations are markedly concentrated in large metropolitan centers. Australia's location with respect to Europe and the United States, however, is much different from Canada's.

Selected basic information is provided below and in Figure 11–6. These facts and figures, in addition to information drawn from the world maps of environments, population, transportation, and communication in Chapters 3 and 5, provide a minimum essential body of evidence for an overview of Australia as part of the modern world. The questions at the end of this learning box may be useful in directing your thinking and lines of inquiry. Several references are suggested in case further information is desired.

Study Questions on the Geography of Australia

1. To what degree is Australia urbanized and industrialized compared to other modern nations, particularly Canada? What pieces of evidence support your characterization of Australia?

2. Explain the apparent contradiction between, on the one hand, the great importance of raw materials and primary products in the exports of Australia and, on the other, its highly urban and industrialized character.

3. Are the nations that are the major recipients of exports from Australia ones that consume and require large quantities of raw materials? Explain.

4. It is often assumed, as is true for Canada, that trade is greatest with nearby areas and least with more distant areas because of transportation costs. How many of Australia's leading trading partners (both for exports and for imports) are as close to Australia as the United States is to Europe? What do the locations of nations with which Australia trades suggest about trading patterns of industrialized nations?

5. In what ways does the relatively remote location of Australia have positive or negative effects on its development as an industrial nation?

6. How might the relatively small population of Australia affect its industrial development? Note that about 52 percent of the value of imports to Australia consists of capital equipment, transportation equipment, and finished consumer goods. Why is it reasonable that these manufactured products should dominate the imports of an industrialized nation? Might market size have something to do with the kinds of imports?

7. Is the proportion of the labor force employed in agriculture similar to or different from the proportion of GNP generated from farms? Does this suggest a relatively efficient or inefficient use of labor in agriculture? From this evidence would you expect agriculture in Australia to be mostly large mechanized enterprises or numerous small enterprises that heavily depend on family and hired labor? Explain.

8. Australia, like Canada, is a large land area with a relatively small population. How do the populations of its larger cities compare with the sizes of large cities in Europe, Japan, and North America? Does Australia appear to be more or less metropolitanized than Canada? Try to explain the high degree of metropolitanization of Australia's population and economic activity.

9. Describe similarities that you can see in the distribution of population in Australia and in Canada and also in their respective environmental limitations for settlement. What natural resource do you think might be most limiting to future growth and expansion of Australia's population and settlement? Is this same resource limitation true for Canada? If not, what is Canada's most limiting environmental resource?

Population:	Total Urban		10,915,435	85.6%
	Cities of 100,000 or more			65 %
	Rural		1,840,203	14.4%
	Total (1971)		12,755,638	
	Total (1973)		13,269,000	

Land Area:		7,686,000 km²	2,968,000 mi²
	Canada	9,976,000 km²	3,852,000 mi²
	United States	9,192,000 km²	3,549,000 mi²
	Cultivated land (1973)	43,361,000 hectares	
	Cropped	14,256,000 hectares	
	Sown pasture	26,130,000 hectares	
	Fallow	2,975,000 hectares	

Gross National Product (1973):	Farm	3,084,000,000	($ Aust.)[a]
	Nonfarm	37,899,000,000	($ Aust.)
	Total	40,983,000,000	($ Aust.)

Labor Force (1971): Total employed 5,240,428
Percent Employed by Economic Sector

Agriculture, forestry, fishing	7.4%
Mining	1.5
Manufacturing	23.2
Construction	7.9
Wholesale and retail trade	18.7
Transportation and communication	7.1
Finance, insurance, and business services	6.9
Public administration and defense	5.4
Community services	10.8
Entertainment, hotels, and personal service	5.1
Other	6.0
	100.0%

Agriculture (1973):	Value of production	3,084,000,000	($ Aust.)
	Value of crops	1,598,000,000	($ Aust.)
	Value of livestock and livestock products	1,486,000,000	($ Aust.)

Major crops by value: wheat, sugar cane, and fruits
Major states producing crops by percent of total value:

Queensland	28%
New South Wales	27%
Victoria	17%

Major livestock and livestock product by value: wool and cattle

Minerals (1973): Value of production 1,998,565,000 ($ Aust.)

	Metallic	995,366,000
	Coal	424,869,000
	Petroleum	311,903,000
	Construction	120,484,000
	Nonmetallic	95,943,000

Trade (1973): Value of exports: 6,213,704,000 ($ Aust.)
                       Value of imports: 4,120,727,000 ($ Aust.)

Percent of Trade by Major Commodity Sector

Exports		Imports	
Agriculture and fisheries	56.2%	Raw and other materials	
Forestry	0.2	of production	42.1%
Mining	21.1	Capital equipment and	
Subtotal for primary		transportation equipment	27.5
products	77.3%	Finished consumer products	24.5
Manufacturing	20.3	Fuels and lubricants	1.7
Refined oil	0.7	Other	4.3
Other	1.7	Total	100.0%
Total	100.0%		

Percent of Trade by Country of Origin or Destination

Exports		Imports	
Japan	31.1%	United States	20.9%
United States	12.2	United Kingdom	18.6
United Kingdom	9.7	Japan	17.9
New Zealand	5.2	Germany (F.R.)	7.0
Germany (F.R.)	3.3	Canada	3.3
France	3.0	New Zealand	3.2
Canada	2.7	Italy	2.1
Papua, N.G.	2.2	Sweden	2.0
Italy	2.2	Hong Kong	1.9
Singapore	2.1	France	1.8
Subtotal	73.7%	Subtotal	78.7%

Value of Major Commodities Exported in Percent of Total Value of Exports

Wool	18%
Frozen beef	11
Iron ore	7
Machinery and transportation	
equipment	6
Coal	5
Wheat	5
Drugs and chemicals	4
Sugar	4
Subtotal	60%

Source: Australia Bureau of Statistics, *Official Year Book of Australia, 1974,* Canberra, 1975.
[a]Australian dollars = $1.02 U.S.

10. What part of Australia is the economic and decision-making core area? Of what relative importance to the economic life of Australia is the northern fringe of settlement compared to that along the northeast coast, or to that of the far southwest? What is the evidence or basis for your conclusions?

Selected References

Blainey, Geoffrey. *The Tyranny of Distance.* Sun Books, Melbourne, 1966. More a history than a geography. The subtitle conveys the book's thesis: "How Distance Shaped Australia's History." The book provides a useful commentary on how the remoteness of Australia's location has shaped its development.

Commonwealth Scientific and Industrial Research Organization. *The Australian Environment,* 3rd ed. Cambridge University Press, London and New York, 1960. An excellent reference on the basic physical geography and agriculture of Australia. Fold-out, colored maps offer a great deal of useful information even if the text is dated. The book may be difficult to obtain if the local library does not have a copy.

Cumberland, Kenneth B. *Southwest Pacific.* Praeger, New York, 1968. Includes a specific discussion of Australia in Chapter 3. Chapters 1 and 2 provide an overview of the physical setting and the population and settlement of the southwest Pacific, including Australia. The chapter on Australia is well illustrated with maps of the basic population, resources, and economic patterns.

McKnight, Thomas L. *Australia's Corner of the World: A Geographical Summation.* Foundations of World Regional Geography Series, Pren-tice-Hall, Englewood Cliffs, N.J., 1970. Reviews the development of Australia as a continental nation and describes its contemporary geography as a modern industrial-urban nation. This book offers an inexpensive and relatively recent source on Australia that is at about the same level of generalization as this text.

MacInnes, Colin, and the Editors of Life. *Australia and New Zealand.* Life World Library, Time, Inc., 1964. Somewhat dated, but conveys a rich impression of the geography of Australia, including its historical development and cultural diversity.

Official Year Book of Australia, 1974. Australia Bureau of Statistics, Canberra, 1975. A basic source for current information and official statistics about Australia. A new edition is prepared annually.

grain-livestock farming system similar to that of the U.S. Middle West. There is also livestock ranching, similar to that in the western United States, in the Prairie Provinces and eastern British Columbia.

Canadian grain production consists mostly of wheat and barley grown in the Prairie Provinces, not corn and soybeans as in the United States, although corn and soybeans are produced in southern Ontario. The Canadian prairies, like the United States Great Plains, are "grain factories." There, farms of hundreds of acres use the same mechanized farming methods as in the United States to make efficient use of the relatively low actual evapotranspiration (AE) of the plains. Wheat is about four times more valuable than barley, and both are major exports for Canada.

The problem in Canada has been less how to produce the grain on the dry margins of agriculture than how to get the crop from producing regions either to the main Canadian market in the Toronto–Montréal corridor, located more than 500 miles away, or to the more important world market served by ports such as Montréal, Vancouver, and Churchill. Export has been made possible by modern rail transport and special low rail rates on bulk-handled grain, but moving the grain to market is still a tremendous task. Much of it must be stored along the way. For this

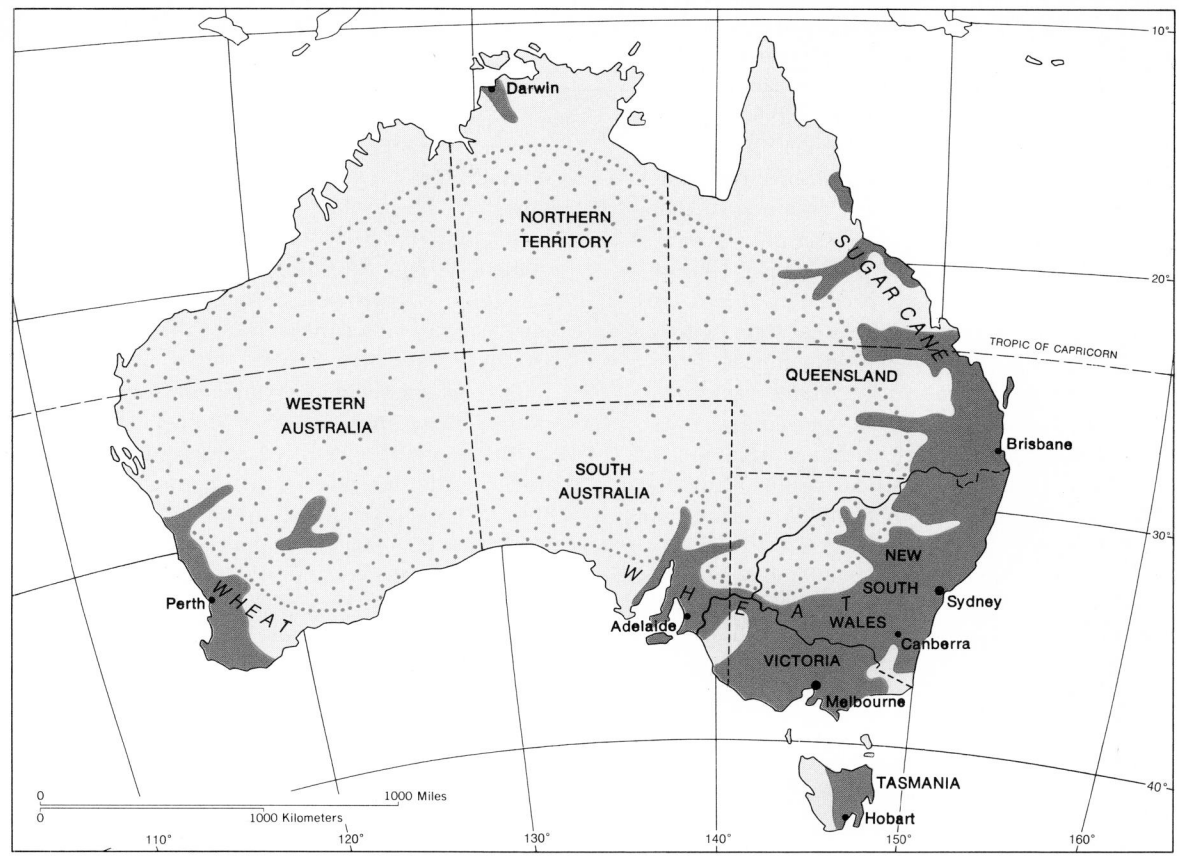

FIGURE 11-6   GENERAL MAP OF AUSTRALIA

Legend:
- Mostly arid or semiarid
- Population density greater than 2 persons per square mile (0.8 per square km)
- ● Cities over 1 million
- ● Other cities

purpose, almost every farm town in the prairies has grain elevators and Winnipeg, Thunder Bay, Montréal, and Vancouver have huge blocks of elevators. Much of the grain moves by rail to Thunder Bay at the head of the Great Lakes, where it is loaded on lake ships and carried to ocean-shipping elevators in Montréal and other ports along the lower St. Lawrence. But today, the largest amount moves westward through Vancouver, reflecting the importance of markets in Japan, China, India, and elsewhere in Asia.

CANADA: TOO LARGE AND TOO SMALL, BUT VERY SUCCESSFUL

Both as a cultural and as an economic entity, Canada is at the same time too small and too large. Its physical dimensions are too large for its population to adequately integrate economically. The North is still largely a frontier connected only by air and by gravel roads to the rest of the country. Wheat from the prairies, timber from the Maritimes and British Columbia, and minerals from the Shield move, not so much to Canadian population centers as to separate markets in the United States, Europe, and Japan. Thus different parts of Canada are connected not with one another into an economic system with an internal focus such as we saw in the United States, but are linked with systems centered in the United States or other industrial areas.

At the same time, even within populated Canada, local social and political values often conflict with the interests of Canada as a national economic and political unit. This conflict is sharpest in Québec, the second most populous province with more than one Canadian in four. But it is also seen in the sharp differences between the farm-centered governments in the Prairie Provinces and urban-centered Ontario, or in the distinctive values of the Maritimes as opposed to British Columbia.

With a population of 23 million, Canada simply does not have enough people or enough Canadian capital to develop this vast country. But by playing an important role in larger systems based in the United States and elsewhere, Canada has been able to prosper. Outside investment has contributed to building Canadian railroads and pipelines, factories and business enterprises. As part of larger systems, Canadian mines produce far beyond the size of the national market; Canadian wheat farmers are organized on a production scale made possible by world markets; and Canadian auto plants are part of the North American industry that has as its customers the 238 million people of the United States and Canada.

Yet the Canadian government and many Canadians see disadvantages in being part of larger systems, particularly of being so closely tied to the U.S. economy, which is so overwhelmingly large in Canadian terms. To a large degree U.S. dollars invested in Canada have produced the rapid postwar growth of the Canadian economy. American money has expanded mines and opened new deposits, pushed Canadian oil production, and financed industrial growth. Often, however, the price of such growth, with its welcome jobs and personal wealth, has been U.S. control of businesses and thus a major say in the Canadian economy. Canadians, like any sovereign people, value their independence and control over their own destiny.

The problem of national identity is compounded by the fact that the economic and cultural core of the United States—the Northeast and Middle West—lies just across the border from the main centers of Canadian life in Ontario and Québec. Just as Canadian resources and manufacturing production have been drawn into the economic network of the U.S. core, Canadian cultural ways have been affected by the mass media of the United States just across the border. Canadians cannot escape U.S. television and radio.

The Canadian dilemma is the necessity of functioning as part of a larger economic whole while at the same time retaining a distinctive identity and control of future development. In addition, Canadians are aware that prosperity has resulted from being tied to larger systems. Thus they find themselves torn between a wish to assert stronger control over their own destiny and a desire to maintain their position as one of the most prosperous peoples of the world.

SELECTED REFERENCES

Canada Year Book (annual). Ministry of Industry, Trade, and Commerce. A comprehensive summary of statistical data for Canada. Includes graphs, charts, and extensive descriptive textual material with each set of topical data.

Canadian Association of Geographers. *Studies in Canadian Geography.* University of Toronto Press, Toronto, 1972. Six-volume series for the 22nd International Geographical Congress. Each volume is a separate publication dealing with a major political division of Canada through several essays by different authors. The general purpose is to provide new perspectives on the regional entities that comprise Canada. The six volumes are:

Louis Gentilcore, ed., *Ontario*
Fernand Grenier, ed., *Québec*
Alan G. Macpherson, ed., *The Atlantic Provinces*
J. Lewis Robinson, ed., *British Columbia*
P. J. Smith, ed., *The Prairie Provinces*
William C. Wonders, ed., *The North*

Hamelin, Louis-Edmond. *Canada: A Geographical Perspective,* translated by M. C. Storrie and C. I. Jackson. Wiley, Toronto, 1973. An interpretation of the geography of Canada by a French Canadian. He addresses Canada's extensive cold climates, the development of the regional characteristics of Canada, population growth and cultural patterns, the economic structure of Canada, and major urban centers—their functions and linkages.

National Atlas of Canada, 4th ed., rev. Department of Energy, Mines and Resources and Macmillan, Toronto, 1974. A comprehensive atlas of environmental measures, population, economic indicators, agriculture, mining, manufacturing, and trade.

Paterson, J. H. *North America: A Geography of Canada and the United States,* 5th ed. Oxford University Press, New York, 1975. A systematic discussion of population, economic development, agriculture, industry, transportation, and foreign trade in Chapters 2 through 7, with appropriate attention to Canada on these topics. Chapters 11, 12, and 22 deal with regions that are mostly in Canada, and Chapters 13, 17, and 21 focus on regions that extend into Canada.

Starkey, Otis P., J. Lewis Robinson, and Crane S. Miller. *The Anglo-American Realm,* 2nd ed. McGraw-Hill, New York, 1975. Includes a discussion of population and settlement in Canada (Chapter 3). In addition, Chapter 9 focuses entirely on eastern Canada, particularly its major cities and principal industrial and agricultural regions. Chapter 10 uses an approach for western Canada similar to that for eastern Canada, and gives a broad overview of developments in northern Canada. These two chapters offer a good short geography of Canada prepared by a Canadian.

Warkentin, John, ed. *Canada: A Geographical Interpretation.* Methuen, Toronto, 1967. Prepared under the auspices of the Canadian Association of Geographers. A volume of essays about the geographical patterns of Canada on the occasion of its centennial of confederation. Twelve essays deal with topical aspects of the geography of Canada, and seven chapters are regionally focused essays. This is a valuable source for information and insight about the geography of Canada.

White, C. Langdon, Edwin J. Foscue, and Tom L. McKnight. *Regional Geography of Anglo-America,* 4th ed. Prentice-Hall, Englewood Cliffs, N.J., 1974. Covers regions of Anglo-America, including parts of Canada, in Chapters 6, 12, 13, 14, and 18. Chapters 7 on French Canada, 19 on the boreal forest, and 20 on the tundra are about regions that are almost exclusively Canadian.

CHAPTER 12

The Outreach of the Modern Interconnected System

In this chapter we shift again to a global perspective. The major links in world transportation and communication are those between the industrial centers of various nations and continents. Other important links are between the centers of the modern system and the locations of primary resources, especially minerals, tropical products, and food staples. The world oil industry exemplifies the outreach of the modern system for resources and the dominance of large corporations. The result in many underdeveloped producing countries is a dual economy—the worlds of Ramkheri and Washington, D.C., existing side by side.

Houston, Texas, the headquarters of several large oil companies. (Stewart/Photography for Industry)

Oil pipeline in Saudi Arabia. (Robert Motlar/Photography for Industry)

Above: Workers on an oil-drilling rig in Nigeria. (FAO photo) *Right:* Supertanker under construction in Japan. (George Gerster, Rapho/Photo Researchers, Inc.)

In each example of the modern interconnected system we have examined, a similar spatial pattern has appeared. A highly developed core region comprises several large metropolitan centers that form both the headquarters of political, economic, and usually cultural decision making and the major centers of manufacturing, trade, and other services. One needs only to recall the northeastern seaboard and Middle West of the United States, the Montréal–Toronto corridor in Canada, the concentration of metropolitan-industrial development in Western Europe, the core triangle in the Soviet Union, and the Tokyo–Kobe corridor in Japan.

In each case, the areas outside the central core play peripheral roles, serving mostly as producers of specialized primary production items or of certain types of manufactured goods tied to required resources such as minerals, perishable farm products, power, and labor. Each of the outlying areas also provides a secondary market for products emanating from the metropolitan decision-making core. The major link for each of the outlying regions is with the core, and the outlying regions have little interaction with one another. These regions are more often competitors in the production of goods for the core than complementary traders.

When we shift from the national and continental scale to a global view, we find the same phenomena on a much larger scale. World trade, is, first of all, between the different industrialized centers and then between those centers and the rest of the world. The amount of movement of either goods or people between one outlying area and another is negligible. Latin American countries generally do not trade extensively with one another or with African or Asian countries. Instead, they trade with European countries, the United States, Japan, Canada, and the Soviet Union. The same is true for Africa, Asia, and Australia, the parts of the world outside the industrial core regions. The model seems similar to the colonial United States when the primary ties of each colony were with the mother country.

The contacts between industrial core regions and peripheral areas in the nonindustrialized parts of the world are between separate cultural groups, just as they are in the Soviet Union or Europe. Each country in the underdeveloped world is a separate culture or a variety of distinctive cultures. Each has grown up largely in isolation and, like the feudal kingdoms of the Middle Ages, values its cultural and political independence.

THE SEARCH FOR PRIMARY RESOURCES

Before turning to the countries of the underdeveloped world, let us examine the outward reach of the modern interconnected system. Even before Columbus, the primary motive in long-range connection over the world was economic. There have been political and cultural reasons, of course, but the desire for gold, handicrafts, exotic agricultural products, and minerals has increased. Moreover, although people have migrated from the industrial countries to Latin America, Oceania, South Africa, and other places, the desire to obtain products through exploitation of the native economies has been much more prevalent. The search has been primarily for three sorts of products: minerals, tropical products that cannot be produced in the midlatitude industrial areas, and staple foods when there is insufficient production at home.

PLANTATIONS AND MINES REPLACE TRADING POSTS

In the period of European exploration in the fifteenth to eighteenth centuries, European traders were content to depend on the economic organization of the underdeveloped peoples to provide the basic output of desired goods. The ships used in trade could not carry large cargoes, and the demand at home was limited by the scarcity of money. As a result the inefficiencies of native production and the lack of quality control were of little consequence. In those days Europeans were truly "traders." They set up trading posts and forts to barter with indigenous peoples for furs, slaves, and regional specialties but did not become involved with production itself.

However, the demands of the expanding European marketplace became greater than the capacity of a largely subsistence native economy to supply, and Europeans established their own plantations and mines. Colonial territories were established in tropical lands such as islands in the Caribbean and the East Indies where key plantation crops included sugar cane, rice, and indigo.

Essentially, the purpose of the **plantation system** was to get more efficient production from areas that had the potential to produce a specialty crop. Native organization, developed primarily to supply the basic food needs of the local population, was replaced by a system of commercial one-crop production designed to provide large-scale output for shipment to markets overseas. This changeover was accomplished by substituting outsiders for native leaders and establishing market control. Thus local resources were turned from the traditional, locally based, direct support of local people to a specialized output for the overseas market. Wherever possible, not only local resources but also local labor forces were commandeered. Where it was not possible, as in tropical America, the additional labor of African slaves was imported.

In early plantation ventures European organization and management were most important. The Europeans planned the land clearing and crop planting, organized the labor force, established the land transport and port facilities, and handled shipping and financing. The native population, like the land, became a resource utilized by the new organizers. Although basic living patterns of the natives had been broken by Euro-

pean intervention, the plantation organizers gave low priority to the reordering of native lives. All they really wanted was the native input of labor. They disregarded the society they disrupted, leaving the native individual, family, and community to adjust to the political and cultural implications of the new economic system.

Even in colonial times crops were introduced to areas where they had never grown before. Sugar cane was brought from Spain by Columbus, and by 1600 plantations had been established on most Caribbean islands and the northeastern coast of Brazil. The sugar cane industry was probably the largest single industry in the world at the time. In the same way cotton was brought not only to the United States but also to India, Brazil, and other tropical countries. Rubber, first obtained in quantity from wild trees in the Amazon Valley, became the leading plantation crop in Southeast Asia. Coffee was brought from Arabia to Latin America and tea from China to Southeast Asia.

Following the pattern of agriculture, trading for gold, silver, and precious stones gave way to mining operations that were developed and managed by Europeans. Here, too, cheap native labor was utilized wherever possible, but primary in mining development was the organization of production under efficient European management.

SHIFT TO MASS-PRODUCTION SOURCES
AND MASS-PRODUCTION OPERATIONS

Industrialization, with its increasing technology and affluence in the modern industrial countries, greatly expanded the interest of these countries in the agricultural and mineral output from the underdeveloped world. Modern ships and railroads have enormously increased the capacity to move products and to make large-scale enterprises in remote areas economically feasible. Today, however, mass production in industrial areas and mass consumption of specialty foods have shifted business interests to areas that offer the best prospects for quality and quantity production. Just as the small local iron deposits used by iron workers in Europe and Anglo-America before industrialization were abandoned for the massive iron ore fields such as those of Lake Superior when the steel industry shifted to mass production, so overseas production has been localized in the areas that promise the greatest profit. Plantation areas have been slowly sorted out, with greater and greater emphasis on those that can be shifted to mass production. Mineral production has become concentrated in the parts of the world where there are massive deposits of easily accessible, high-quality resources.

LOSS OF COLONIAL CONTROL

One other major aspect of the developmental process has changed since colonial days. In the days of widespread European holdings in Latin America, Africa, and Asia, business organizations in the mother countries found it relatively easy to gain permission from their own governments to prospect and develop resources in foreign possessions. The whole philosophy of colonialism called for drawing production from the outlying empires.

Today, however, the colonies are gone, and in their place stand new countries very jealous of their sovereignty and very conscious of the importance of environmental resources. Like any landowner, these independent countries have control over the land under their domain. For this reason companies in modern industrial countries must not only obtain permission to prospect for minerals or to experiment with crops, they must also negotiate production agreements with the host governments. Although the corporation that invests in developing the mine or plantation and organizes the transportation usually retains basic management control, the rules under which it undertakes development are no longer simple.

The underdeveloped country is the landlord and at any time can foreclose on the leaseholder. Thus U.S. oil companies were shocked in the 1930s when Mexico nationalized its oil fields, told the oil companies to get out, and expropriated not only the resource rights but also the companies' wells, equipment, and shipping facilities. In recent years such actions have become common among oil-producing countries and in all sorts of resource developments. Sometimes national takeovers precipitate international incidents and involve long lawsuits to determine proper compensation to the companies for the facilities. But the rights of countries to resources found within their boundaries are generally accepted by development companies.

THE WORLD OIL INDUSTRY:
A CASE STUDY

The story of oil, today's most important energy source, illustrates the complexities of the outreach of the interconnected system into the rest of the world and of the relationships between the corporations, their governments, and the countries in which the resource is found.

A TWENTIETH-CENTURY
INDUSTRIAL RESOURCE

The use of oil and gas is the result of the second stage of the Industrial Revolution and is largely a phenomenon of the twentieth century. The widespread use of the internal combustion engine and oil and gas burners in modern countries is a major cause of the demand for oil and gas. The use of petroleum, at least, depends on a complex refining process that breaks crude oil down into a wide variety of derivatives ranging from asphalt or paraffin to jet fuel. As resources petroleum and gas are new, though they were formed millions of years ago. People have perceived a use for them only recently, and they are resources only to those who have industrial machinery to refine and utilize them.

MAJOR PRODUCING AND CONSUMING AREAS AND THE CHANGING GEOGRAPHY OF PRODUCTION

Petroleum and natural gas are critical resources today to parts of the world that have modern technology. Although one can find jeeps and kerosene lamps and outboard motors in almost every remote corner of the world, the overwhelming concentration of energy-using machinery occurs in the established industrial centers we have investigated—Anglo-America, Europe, the Soviet Union, and Japan. These areas, which account for less than 30 percent of the world's population, consume more than 80 percent of the petroleum and over 95 percent of the natural gas used in the world. Eight countries have daily consumptions of oil of over 1 million barrels: the United States, the Soviet Union, Japan, West Germany, Britain, France, Italy, and Canada, in descending order (see Figures 12–1 and 4–4).

Petroleum and natural gas, you will recall, are found in sedimentary rocks, and sedimentary deposits are not found everywhere in the world (Figure 12–2 and Table 12–1). In 1975 only one major industrialized area—the Soviet bloc, consisting of the Soviet Union and most of Eastern Europe—was producing a larger share of the world's petroleum than it was consuming (Table 12–2). Anglo-America, a major producing region with 21 percent of total world production, was also the leading consumer, with 32 percent of total consumption. Europe, consuming over 24 percent, was a major deficit area since it produced only 1 percent of its needs. Today all the industrial areas except the Soviet bloc depend on major energy sources in the underdeveloped parts of the world.

Almost 60 percent of oil and natural gas output in the world comes from underdeveloped coun-

tries. The Middle East–North Africa alone accounts for close to 40 percent of world production, but Latin America, particularly Venezuela, has also been a major producer since World War II. Latin American production seems to have peaked, but in 1975 it still contributed about 8 percent of total world output. Since 1970, Africa has moved ahead of Latin America, largely as a result of large-scale development in Nigeria; Africa produced 9 percent of all world oil and gas in 1975. None of these areas is a major consumer; all of them have oil surpluses that are exported.

The industrial countries' dependence on oil and gas production from underdeveloped countries is increasing rapidly. As Table 12–2 shows, between 1970 and 1975 production in Anglo-America decreased 11 percent, while consumption increased 6 percent. The United States, long the world's leading producer and an important exporter, has become increasingly dependent on supplies from the underdeveloped world. In Europe production remained negligible—some increase is expected when the North Sea fields start producing—but demand increased almost 8 percent between 1970 and 1975.

THE DOMINANCE OF LARGE INTERNATIONAL CORPORATIONS

In the first great period of oil exploration, from 1890 to 1950, virtually all effort in the underdeveloped world was carried out by a handful of giant oil companies—British Petroleum, Shell (jointly owned by British and Dutch interests), and five large United States corporations: Standard Oil of New Jersey, Mobil, Texaco, Gulf, and Standard Oil of California. Even in the late 1950s these companies controlled 90 percent of the oil concessions outside the United States and the Soviet Union.

Oil production in the Soviet Union has been a state-controlled monopoly, whereas in the United States it has been possible for companies large

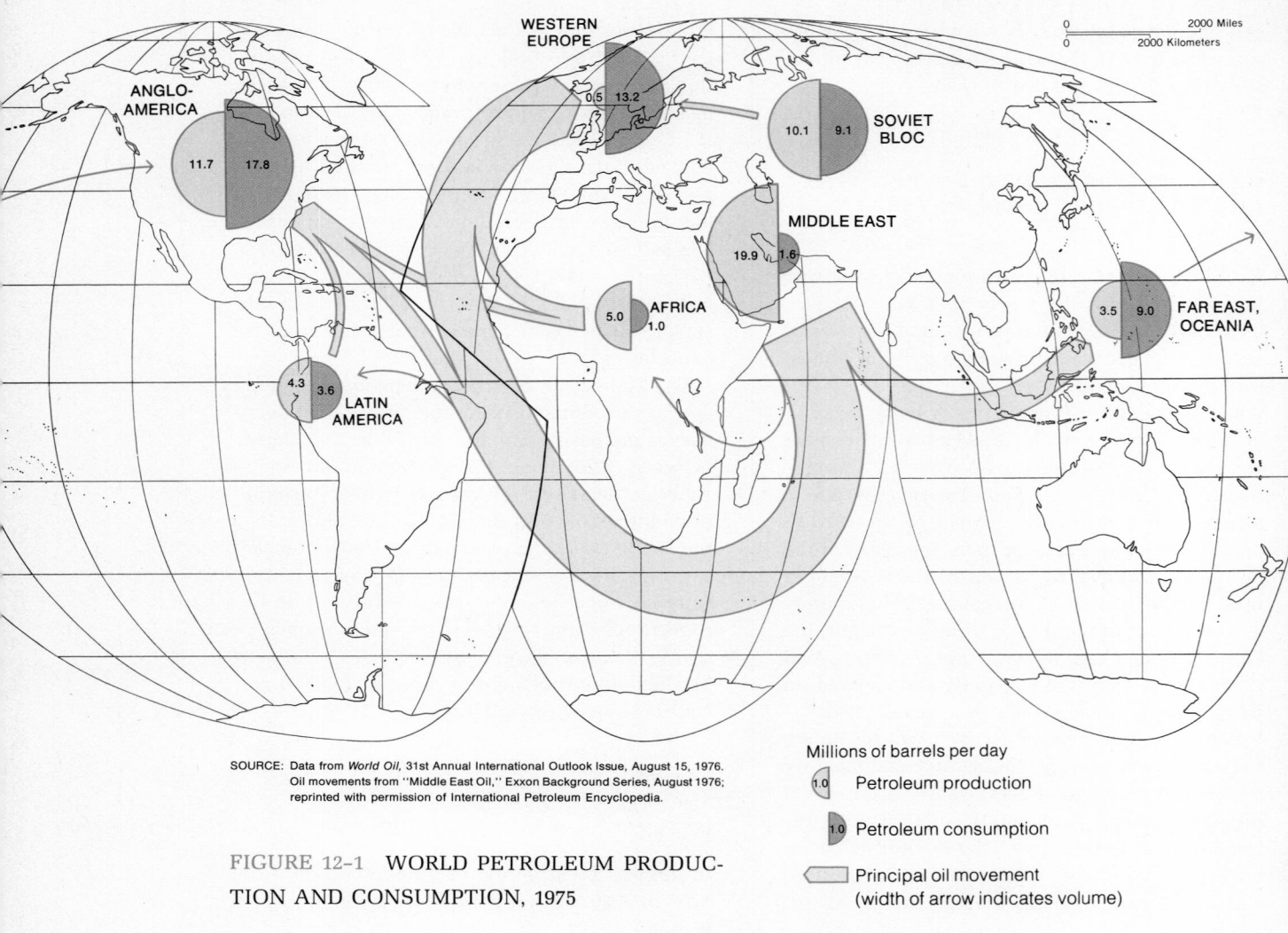

WESTERN
EUROPE
0.5 13.2

SOVIET
BLOC
10.1 9.1

ANGLO-
AMERICA
11.7 17.8

MIDDLE EAST
19.9 1.6

AFRICA
5.0
1.0

FAR EAST,
OCEANIA
3.5 9.0

LATIN
AMERICA
4.3 3.6

0   2000 Miles
0   2000 Kilometers

SOURCE: Data from *World Oil*, 31st Annual International Outlook Issue, August 15, 1976.
Oil movements from "Middle East Oil," Exxon Background Series, August 1976;
reprinted with permission of International Petroleum Encyclopedia.

Millions of barrels per day

1.0 Petroleum production

1.0 Petroleum consumption

Principal oil movement
(width of arrow indicates volume)

FIGURE 12-1   WORLD PETROLEUM PRODUC-
TION AND CONSUMPTION, 1975

and small to contract with landowners for drilling rights on particular pieces of property. Thus
oil drilling in the United States has been carried
out by large corporations and individual "wild-
catters." Since oil drillers need not be refiners,
the small operator has been able to compete with
the large.

Overseas, however, a company needs the expertise to deal with the government of a country
for a **concession** to drill for oil. Since in most
underdeveloped countries the government controls most of the landholdings, only a single drilling concession is commonly offered in any country or region. The company must prove that it

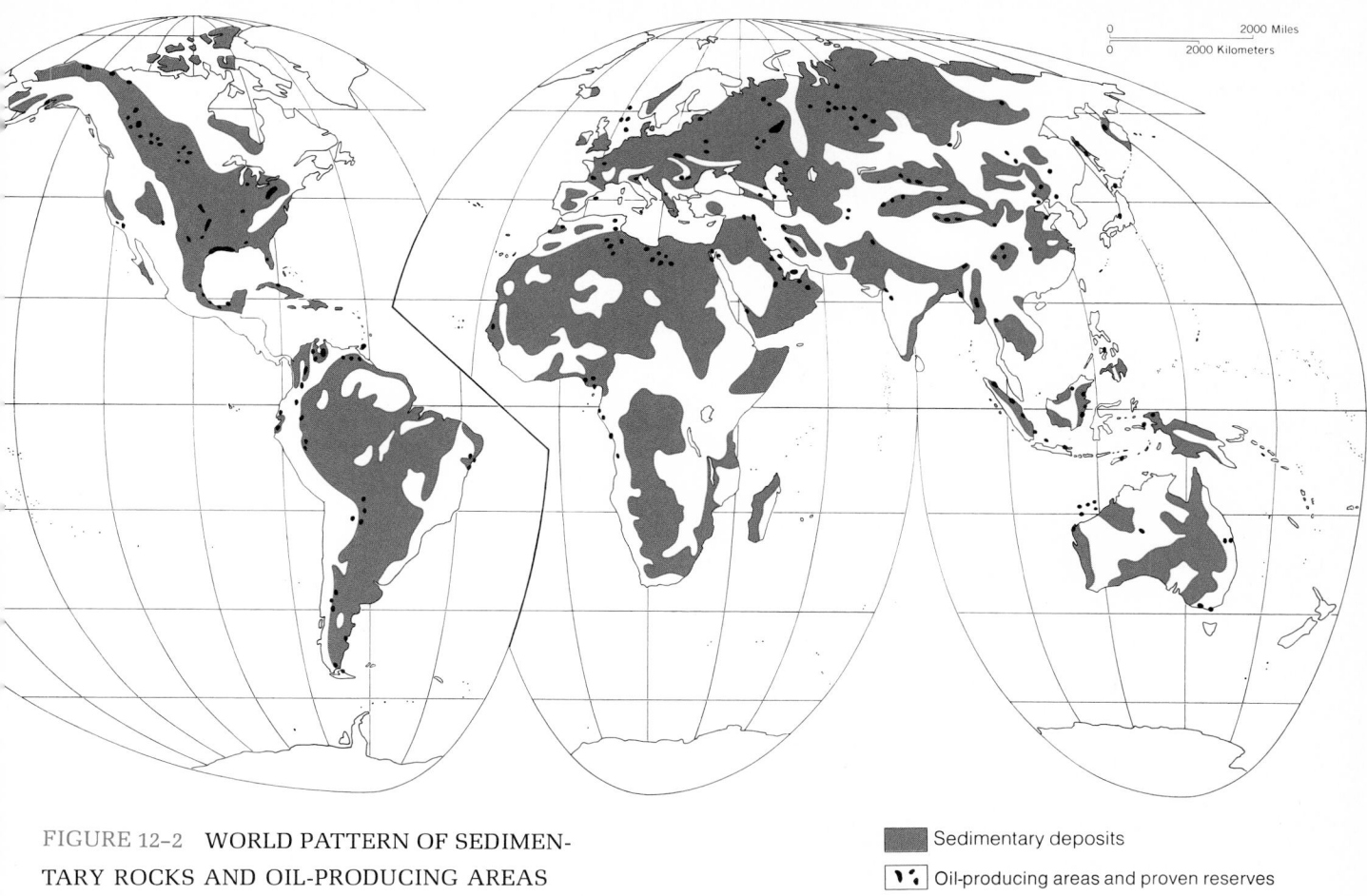

FIGURE 12–2 WORLD PATTERN OF SEDIMEN-
TARY ROCKS AND OIL-PRODUCING AREAS

███ Sedimentary deposits

▢ Oil-producing areas and proven reserves

has the financial capacity to mount the tremendous development costs, which often include not
only the cost of drilling but also of building
roads, pipelines, ports, and storage facilities. The
giant multinational oil companies own their own
tankers and refineries, although much of their oil
is transported by shipping companies. They have
also worked in close association to control the
entry into the field of competitors who might
threaten international marketing networks.

THE POLITICS OF OVERSEAS PRODUCTION

In the United States oil development and production depend on agreements made with landowners, providing them with a basic royalty for
each barrel of oil produced. In the Middle East
and Venezuela the agreements have been made
with the governments, and production in their
major oil pools has been a bonanza for the large
oil corporations. These oil fields are among the

TABLE 12–1   OIL AND NATURAL GAS RESERVES, 1975

Region	Crude Oil (Millions of Barrels)	Natural Gas (Billions of Cubic Feet)
Anglo-America	42,766	297,099
Latin America	25,557	68,420
Europe outside Soviet bloc	18,225	139,981
Middle East–North Africa	378,257	770,068
Africa South of the Sahara	15,767	54,827
Far East, Oceania	36,386	109,188
Soviet bloc	62,019	730,595
World total	578,979	2,170,175

SOURCE: *World Oil,* August 15, 1976, p. 44.

richest in the world, and since there is no competition in the concession area, wells can be scientifically spaced for most efficient production. As a result production costs in Kuwait have been only about six cents per barrel, and elsewhere in the Middle East they are no more than twelve cents per barrel, for oil that now sells for more than ten dollars per barrel.

Once the exploration and development stages have been completed, the major cost to producers has been the **royalty payments** to the local government. Originally, these were modest, consisting mostly of local taxes. But the governments have become increasingly aware of the value of their oil resources and of the importance of oil revenues to their countries. During World War II the first agreements that were reached set the royalty/profit split between government and oil company at fifty-fifty after taxes. In the 1960s such agreements were adjusted to seventy-thirty in favor of the government. Nevertheless, total producing costs in most overseas areas are still far lower than those in the United States, and in the 1950s imported oil flooded into the United States. As a result import quotas to protect domestic oil producers quickly followed.

In recent years, as the rulers of oil-producing countries and their advisers have recognized the increasing importance of their oil to the modern world, they have become dissatisfied with continued outside control of the profitable oil industry. Since 1970 virtually all the major producers have taken over the oil-producing facilities in their countries, although Mexico did so as early as the 1930s. In most countries this has meant the takeover of wells, pipelines, and port facilities only, since the multinational oil companies located most of their refineries near the market rather than the oil fields. However, Iran has taken over the world's largest refinery at its port of Abadan. In most cases the oil companies have not been forced out of the countries completely. Rather, they are now given management contracts to operate the oil fields for the government owners; but all decisions on how much to produce and at what price are made by the government. Of course, such decisions depend on world markets, and the oil companies still have important indirect influence.

THE COMPLEXITIES OF THE WORLD MARKET

The rise of the major multinational oil companies with headquarters in the United States, the United Kingdom, and the Netherlands in part has

TABLE 12-2 WORLD OIL AND GAS DEMAND AND SUPPLY, 1970–1975

(Thousands of Barrels per Day)

	Consumption				Production				Excess or Deficit			
	1970		1975		1970		1975		1970		1975	
Industrialized areas												
Anglo-America	16,711	36%	17,784	32%	13,242	27%	11,705	21%	−3,469	21%	−6,079	34%
Europe outside Soviet bloc	12,241	26%	13,205	24%	547	1%	480	1%	−11,694	95%	−12,725	96%
Soviet bloc	7,142	15%	9,156	17%	7,450	16%	10,137	18%	+308	4%	+981	11%
Total	36,094	77%	40,145	73%	21,239	44%	22,321	40%	−14,855	41%	−17,823	44%
Traditional areas												
Latin America	2,440	5%	3,592	7%	4,838	10%	4,298	8%	+2,398	98%	+706	20%
Middle East–North Africa	799	2%	1,645	3%	18,815	39%	19,894	36%	+18,016	225%	+18,246	1,109%
Africa South of the Sahara	1,183	3%	1,043	2%	1,457	3%	4,959	9%	+274	23%	+3,916	375%
Far East, Oceania	6,369	14%	9,042	16%	1,398	3%	3,513	6%	−4,971	78%	−5,529	61%
Total	10,791	24%	15,322	28%	26,508	55%	32,664	59%	+15,717	146%	+17,342	113%

SOURCE: *World Oil,* August 15, 1976.

resulted from the large investment required by oil operations. It also indicates the scale of operations needed in the uncertain market of world production and sales. Production and marketing involve not a single product but a great variety of items. Crude oil itself is not a single chemical substance; it contains hundreds of hydrocarbon compounds, plus other minor elements. Some crude oil is thick and asphaltic; some is almost gaseous; some is high in sulfur. Also, the finished refinery products are many and diverse, ranging from asphalt, waxes, and heavy fuel oil to liquefied petroleum gasoline. The composition of products used varies greatly from market to market. In industrialized countries like the United States, gasoline accounts for half the demand for petroleum products, but in nonindustrialized ones like India, the demand for kerosene (for lamps) is greater than for gasoline.

The discovery of new oil fields in countries that have not been significant producers has complicated the situation further. Countries seeming to have a near monopoly on world oil exports have found themselves faced with competition from new producing areas, some of which have larger and more productive fields. At the turn of the century, Russia, the United States, the Dutch East Indies, Rumania, and Mexico were the major exporters. After World War I Venezuela and the first Middle Eastern fields began to dominate the world market. From World War II until 1960 other major fields were discovered in the Middle East, which then became the principal source of world oil exports. Recently, major production has developed in the Sahara of North Africa, and now new developments are occurring in Indonesia, Nigeria, and China. With the major deposits on the land areas of the earth apparently already discovered, new exploration has recently turned to the fringes of the oceans. Exploration is now going on along the coasts of the United States, Australia, and Africa, and major new fields in the North Sea are the first important oil discovery in Western Europe.

With the control of most oil production in their hands, the large multinational oil companies have been able to respond quickly to both new markets and new production sources. All have multiple sources of supply, a series of refineries throughout the world, and marketing facilities in many countries. The management problem has been to keep this worldwide system functioning properly by shifting crude oil output from field to field in the way best to meet demands. Flow from a new field, for example, may call for a rethinking of plans to expand production in existing fields.

THE IMPACT OF OIL ON THE UNDER-DEVELOPED OIL-PRODUCING COUNTRIES

Although fluctuating arrangements may be very satisfactory for the multinational oil company, they are another matter to the producing countries. Libya, Iran, Saudi Arabia, Kuwait, Iraq, and Venezuela count on oil revenues for more than 90 percent of their **foreign exchange.** In their quest for internal economic development, these countries are continually seeking more oil income to provide for needed foreign purchases.

Recognizing their precarious position in the oil supply situation, most of the major oil-exporting countries banded together in the 1960s to form an organization to deal with prices and ratios of return, the Organization of Petroleum Exporting Countries (OPEC). Equally important, in recent years they have opened up their lands to additional oil companies. The Japanese, Italians, and a variety of smaller U.S. oil companies are now in international production operations. When Libya opened its promising Sahara fields after 1955, thirteen different companies applied for concessions.

The development of oil resources in a particular country often led to a conflict of interest. Whereas the company was concerned with fitting the output of that particular producing unit into its global operations and adjusting output to particular market conditions, the host country saw oil production in its area as its only source of foreign exchange with which to buy the products of the industrial world. To the underdeveloped country with only partial modernization, oil was an indirect asset; when it was sold to the industrial countries, the oil-producing country could buy more modernization for itself. But the amount of production from its wells was determined by the multinational company. The government could only exert political pressure to change decisions. It could threaten nationalization of the industry, but oil exports depended on refineries and ships controlled by the oil companies.

In the early 1950s when Iran tried to nationalize its oil production, the large multinational corporations simply boycotted Iranian oil while expanding production in other countries. As a result, production in the Iranian fields slowly came to a halt. Even though Iran took over one of the largest oil refineries in the world along with the producing facilities in the fields, and even though it could hire trained technicians to produce and refine the oil, Iran lacked the means to ship it. Within a few years Iran was ready to negotiate with a **consortium** of the major oil companies. Mexico has been able to nationalize oil largely because its growing economy has provided an internal market.

In the 1970s the situation is very different. Saudi Arabia, Kuwait, Iran, and other Third World countries control both the majority of oil production outside the United States and the Soviet Union and the greatest share of known reserves still in the ground. All industrial countries other than the Soviet Union depend, at least in part, on imports of oil from elsewhere in the world. Even the United States, formerly the world's leading producer, now finds its demand for petroleum products causing it to import as much as 40 percent of the oil it consumes.

This is the reason for the power of OPEC. The major Third World exporters, united in a common front, have cut off most of the alternative sources for the large multinational oil companies and the new oil companies that have entered the business since 1960. If they want oil for their worldwide markets, the companies must deal with the OPEC countries on OPEC's terms. Thus OPEC has been able to increase oil prices four-fold between 1973 and 1976 and individual countries within OPEC have been able to push forward plans for the takeover of their own oil production from the large companies.

Even so, the great multinational oil companies still dominate manufacture, shipping, and marketing, and so continue to play a major decision-making role in the distribution and pricing of oil. But in this, too, OPEC countries are pushing for change. With their new surpluses of cash from oil, they are beginning to develop their own tanker fleets and are building their own refineries. There have been reports of OPEC countries seeking to buy chains of gasoline stations in the United States. If the oil industry of the OPEC countries is successful, it will become vertically integrated like the large multinational oil companies—that is, they will control all facets of the oil industry from source to end use, or from oil field to service station. The OPEC organization will be a new force in competition with the multinational corporations in world markets.

Oil production in underdeveloped countries did not mean new energy sources for national development, and not until recently has it had an appreciable effect on the common people. In all cases it has meant jobs. Workers have been needed in construction and on the oil rigs, to drive trucks, and to work in refineries and in stores and shops associated with the new oil operations. Even in industrialized countries, however, the oil industry is not a large employer. Far fewer workers are needed in the oil fields than have been traditionally required in coal and iron mining. Refineries are among the most automated of modern industries, and pipelines deliver quantities of oil with far fewer workers than a railroad would require. Even at the ports, tankers can be loaded by attaching a pipe from the storage facilities. Thus most people have continued their traditional ways until the government has used oil money to bring changes such as irrigation water, roads, hospitals, and retirement and disability payments.

THE SITUATION OF UNDERDEVELOPED PRODUCING COUNTRIES IN THE INTERCONNECTED SYSTEM

The story of the oil industry shows the outreach of the modern interconnected system, with a happy outcome for most OPEC countries. Many underdeveloped countries, however, that depend on overseas markets in the great industrial countries have not been so fortunate. Their precarious position can be illustrated by examining another example of a critical mineral resource: copper. Although it is far less important than petroleum, copper is essential to the modern world and is the foreign-exchange base of several underdeveloped countries.

Copper's two key properties are its abilities to conduct electricity and heat and to resist corrosion, which make it most important in the electrical and communications industries and the heating and plumbing industry. It is also vital for heat transfer in engines.

The world's leading copper-producing countries are the large industrial countries—the United States and the Soviet Union. Together they produce over a quarter of the world's output (Table 12–3). Europe has little copper production, and U.S. consumption is so great that the United States must depend on imports as well as domestic production.

Underdeveloped countries are also major producers of copper and it is of great importance in their export trade. Chile, Peru, Zambia, and Zaire account for about 60 percent of international copper trade. Zambia's exports in 1974 were 98 percent copper; Chile's were 86 percent; Zaire's, 62 percent; and Peru's, 23 percent. Thousands of workers in each country were employed in mines.

Like petroleum, copper has been developed by U.S. and European interests. Fewer than a dozen companies have controlled most of the world's supply. But in recent years the leading copper-producing countries have brought strong pressure to control production within their boundaries. Like the OPEC countries, all four major

TABLE 12–3   MAJOR COPPER PRODUCERS, 1974

(*Copper in Ore in Thousands of Metric Tons*)

Country	Copper Production	Country	Copper Production
United States	1,445	Peru	39
USSR (est.)	740	South Africa (est.)	153
Zambia	698	Philippines	221
Chile	906	Australia	162
Canada	840	Total	5,653
Zaire	449	World total	7,350

SOURCE: *Encyclopedia Britannica, Book of the Year, 1976, Chicago, 1976.*

Third World copper-producing countries have nationalized their production and have also formed an OPEC-like cartel to control production and prices.

Control of production facilities has not meant control over the copper market, however, for this depends on consumption in the industrial countries, and copper is notorious for tremendous price fluctuations. Because it is mainly used by industry, its consumption varies sharply with shifts in industrial production. In 1970, at the very time that the Freeport Sulphur Company was planning to develop deposits in West Irian, copper supply was running far ahead of demand (see box on p. 403). The stockpile in the London Metal Exchange had increased more than fourfold, and U.S. inventory supplies were even higher. The price of copper on the London exchange hit a six-year low of forty-seven cents per pound, having decreased 40 percent in nine months. The director of one of the largest copper companies predicted a surplus of more than 850,000 metric tons by 1975 at current demand rates.

The effects of the slump in copper demand and prices on Zambia, Chile, and Zaire can be readily imagined. Loss of income can be a serious blow to development plans and cause extensive unemployment in mining districts. In times of depression, an underdeveloped country learns the price of national control as it attempts to maintain production in a market it cannot manage.

The producing countries have been meeting to discuss their common problem of the copper price slide and have discussed various solutions, including limits on copper production and a program of massive buying to minimize the price drop. Neither method offers much hope, however, because production cuts would increase unemployment and would work only if all major producing countries agreed. Large-scale stockpile buying to support prices would be possible only if huge loans of international capital could be obtained.

PRODUCTION WITHOUT CONTROL

Variations in the flow and price of copper, oil, or other basic commodities produce serious short-run difficulties for the interconnected system and the industrial countries that are a part of it. But for the underdeveloped countries involved, such variations are little less than disaster. Although

THE COMPLEXITIES OF RESOURCE DEVELOPMENT IN AN UNDERDEVELOPED COUNTRY: THE CASE OF COPPER IN WEST IRIAN

The complexities of an international resource development operation can be seen in the exploitation of a rich copper deposit in the mountains of West Irian, a remote and little developed part of Indonesia on the island of New Guinea.

The deposit, the largest known to exist in the world, was discovered in 1936 by Dutch explorers when New Guinea was still part of the Dutch East Indies. But the field party's report was not acted upon until 1958 when another Dutch firm discovered it in the files. Lacking adequate financial and research resources, the new company asked Freeport Sulphur, a large U.S. mining-chemical firm, for help.

In 1960 engineers supported by the Dutch government (which was still nominally in control of West Irian even after the Indonesian republic was established) surveyed the deposit. Two years later the Dutch ceded West Irian to Indonesia. The result was a complete hiatus in development, for the Indonesian government of the time was very nationalistic and strongly anti-Western. For five years no development took place, but when the regime was overthrown in 1967, a new government, much more receptive to resource development by foreigners, came to power. Within three months Freeport Sul-phur had reached an agreement with the government and plans were reactivated. Development costs were set at $7.4 million, but it was estimated that another $120 million would be needed to start production.

The result was a complex bit of international money raising. Since U.S. copper smelters had adequate sources of copper, eight Japanese companies contracted to buy two-thirds of the mine's output, and a German company agreed to take the rest. On the basis of these agreements, Freeport Sulphur obtained $20 million from the Japanese firms and $22 million from the West German development bank, guaranteed by the government. The remaining funds were raised in the United States. Seven commercial banks put up $18 million based on Export-Import Bank guarantees. Forty million dollars came from five life insurance companies using Agency for International Development guarantees. The agency provides risk insurance protecting Freeport Sulphur against war, expropriation, and the inability to convert money from Indonesian accounts.

The actual cost of the on-site undertaking has been tremendous. The company had to build a port that could accommodate 15,000-ton ore ships and cut a sixty-nine-mile road through tropical rain forests into the mountains. Then an aerial tramway had to be built to the base of the ore body. The ore is brought down by conveyor belt, crushed at the base of the mountain, concentrated into a liquid mixture called slurry, and then pumped through a pipeline to the coast.

Now that a market is assured, the results of this complex process make the development worthwhile. The deposit contains an estimated 33 million tons of ore of exceptionally high quality. The ore contains 2.5 percent copper, whereas U.S. mines are working copper at less than 1 percent. This means 825,000 pounds of pure copper valued at more than $900 million at 1970 prices. In addition, other minerals can be extracted from the ore as by-products, including an estimated $28 million of gold and $13 million of silver.

To encourage the development, the Indonesian government agreed to a three-year tax-free period even though the mine may be in operation. After the first three years, the tax will be 35 percent for seven years and 41.75 percent thereafter. Freeport Sulphur expects to have exhausted the ore in slightly more than thirteen years.

PROBLEMS OF REGIONAL DEVELOPMENT IN THE UNITED STATES: A CASE STUDY OF AN AREA RICH IN RESOURCES BUT DEPRESSED IN INCOME

To understand more fully the problems of underdeveloped countries in managing their own affairs, let us look at the problems of economic development in an area in the middle of the United States. The area is centered roughly on the confluence of the Mississippi and Ohio rivers and extends approximately 500 miles (800 km) from northeast to southwest and 300 miles (480 km) from north to south (the area is outlined in Figure 12–3).

This area has a diverse resource base and, even by world standards, ranks as a major producer of several primary resources. Its mines produce more than 90 million tons of coal; only five countries in the world produce more. Only twenty-one countries produce as much oil. The area produces over 70 percent of U.S. output of lead and is a major source area for high-grade iron ore. It also contains major deposits of fluorspar, glass sand, and ceramic clay, and ranks as one of the important cement-producing areas of the United States. Total agricultural output is equal to that of such leading agricultural states as Indiana and Florida. This region is the leading cotton-producing area in the South, a major rice-producing area, and important in tobacco production; it also grows large quantities of soybeans. As the focal point of the country's largest river network, it is exceptionally well endowed with water resources.

In terms of income, however, the area is one of the economically depressed parts of the United States. As of 1969, median family income in most counties was well below the national average of $9,600, and over half the counties had median family incomes below $6,000. More than half the counties were designated as "depressed" in the 1960s, and as a result they were eligible for special government assistance.

The area's problem is not so much the lack of development as the nature of the development. It is dominated by resource extraction and primary production, and the output is moved to major manufacturing centers outside the area for processing and use. For example, the coal from one area was developed to serve markets such as industries and power stations in northern Illinois and the upper Mississippi valley. Moreover, investment capital has come largely from outside the area, primarily corporate interests located in the large metropolitan centers of the Middle West and Northeast, and profits return to those areas. In agriculture the story is similar. Most of the cotton and soybeans are shipped out for processing and to be used in manufacturing. No major industrial complex has developed within the area based on any of its primary resources.

Because of such an exploitative economic base, the area has only one metropolitan area, Evansville, Indiana, and it has less than 200,000 people. Only five other cities have over 25,000 people. As might be expected, the transportation system is primarily designed to make connections from the separate productive parts of the area to peripheral cities rather than to focus on any point within the area.

The lack of economic focus has its counterpart in political and cultural fragmentation. The area is divided among six states; political action and loyalties are therefore fragmented. Culturally, the backgrounds of the people of hill farms in southern Indiana or the Ozarks of Missouri are different from those of the people of the Mississippi lowlands in Arkansas, Missouri, and Tennessee, many of whom are of Southern origins.

This case illustrates that natural resources alone do not determine regional success. Even though, through its mines, farms, towns, and transportation, this region of the United States is a functioning part of the modern interconnected system, its people obtain below average shares of the national economic wealth. The reason is that employment, prices, and profits in such activities as mining and farming are dependent on decisions made outside the area. Consequently, the economy of an area such as this is peripheral to the national system, and no obvious means exist by which the people within the area can overcome their subordinate economic position.

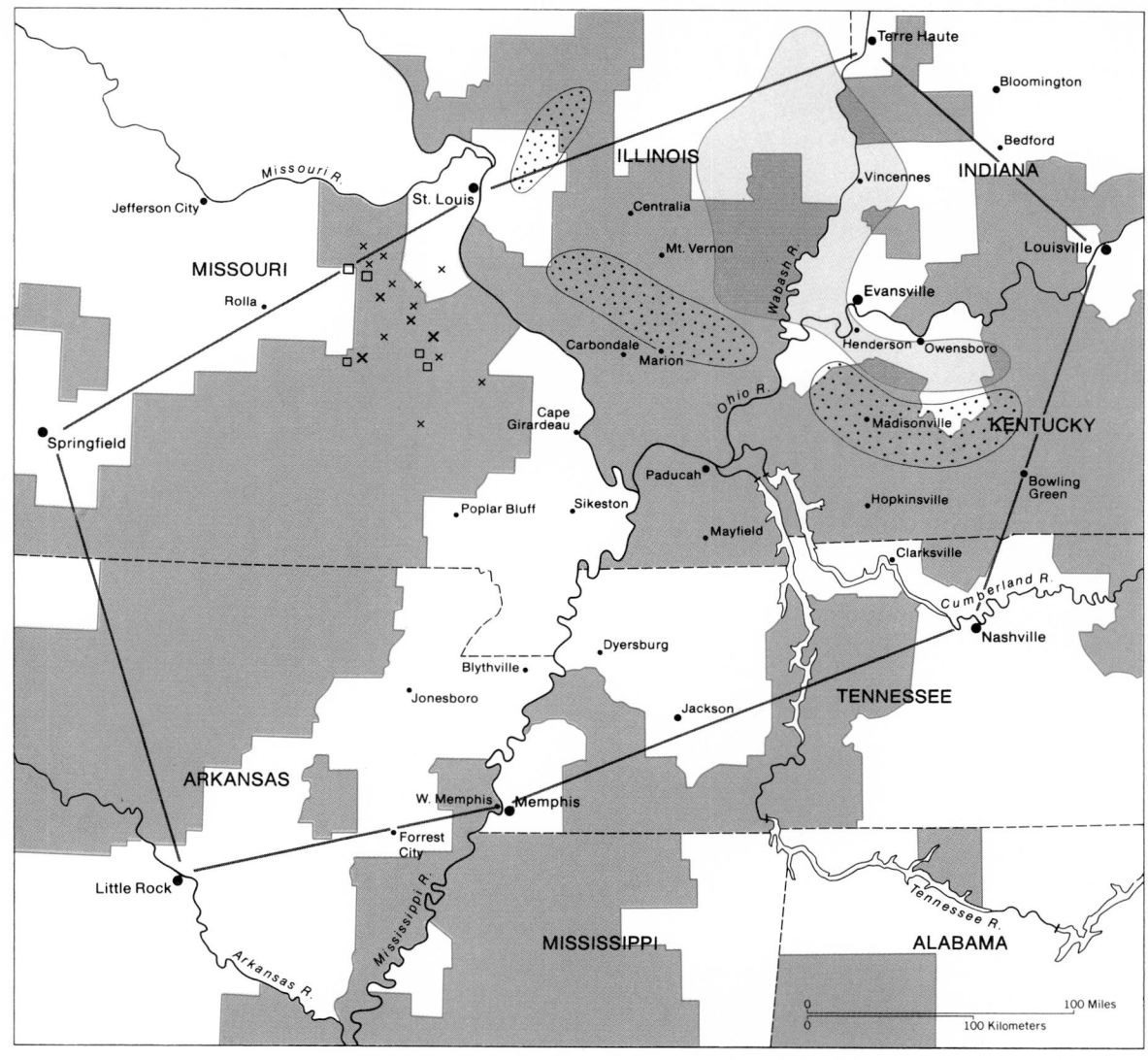

FIGURE 12-3 REGION SURROUNDING THE CONFLUENCE OF THE MISSISSIPPI AND OHIO RIVERS IN THE MID-1960s

Coal mining area

Oil fields

× Lead ore

□ Iron ore

Eligible redevelopment area

• 10,000 to 25,000 population

• 25,000 to 100,000 population

● Over 100,000 population

these countries have recently achieved political independence, this has done almost nothing to eliminate the subservient economic position in which they still find themselves.

It is striking how underdeveloped countries compete with one another for markets for their relatively few major products (Table 12-4). Look at the wide range of countries producing these tropical products: coffee, cocoa, peanuts, bananas, rubber, tea, palm products, and tropical woods. The same pattern also exists for subtropical products—sugar, cotton, jute, and rice—and for petroleum, copper, diamonds, bauxite, tin, iron ore, and phosphates. Note, too, how often neighboring countries have the same specialty economies and thus must compete with each other.

The difficult position of the underdeveloped countries in the world trade structure can also be seen. Not only are these countries not in control of the market and movement of their products, they are also in the weak posture so characteristic of farmers and other primary producers throughout the world. They produce raw materials that sell at low prices, and they need the manufactured goods and services of the industrial world that sell at much higher rates. The foreign exchange gained in the sale of coffee, rubber, or bauxite does not go far when one needs to buy machinery, transportation equipment, and other sophisticated technical supplies from industrial areas.

Moreover, financial control often is not in the hands of the local people and, as a result, neither is the profit-making component of the transaction. Though payments are made to the government of the country, and though workers in those countries are paid wages, the profit from the operation goes to the foreign company that made the investment. Commonly, total investment in facilities in a country is much less than the profits taken out. This problem is compounded by the fact that in the modern interconnected system money is made not only in producing but also in financing, managing, transporting, and selling the product. Both New England and Japan suc-

ceeded well as primary producers and simple manufacturers because they were able to gain control over the business operations. Soon they were their own entrepreneurs, managing and financing the whole undertaking from the production of raw materials through the sale of the finished product. Few underdeveloped countries have managed that.

THE CHANGING POSITION OF
THE MULTINATIONAL COMPANIES

Traditionally, underdeveloped countries have been viewed by the large multinational companies that dominate the modern, worldwide interconnected system only as sources of raw materials. With only a fraction of their populations in the modern interconnected system, these countries, even those with large populations such as India and Indonesia, offer only minor markets of little importance compared to those of the large industrial countries. Their workers have been viewed as too unskilled and poorly educated to provide a reliable pool of manufacturing labor. Thus little interest has been generated in locating modern factories in such countries.

In the last decade, in particular, the situation has changed. An important segment of the population of countries such as Brazil, Argentina, Mexico, Kenya, and Taiwan now lives in cities and works for wages. Although the wages are well below those of industrial countries, they are sufficient to buy modern products such as motorcycles, radios, televisions, and home appliances as well as Western clothes. These countries, with populations of 10 to 100 million people—even if only 10 to 15 percent of the population are in the money economy—offer a significant market to the multinational companies. Moreover, city people are now learning to read and write. Thus the multinational companies have begun to locate factories employing local workers in large underdeveloped countries such as Brazil and Mexico to serve local markets. Such countries, in which a

TABLE 12-4 IMPORTANCE OF PARTICULAR COMMODITIES IN EXPORTS
OF UNDERDEVELOPED COUNTRIES, 1974   *(Percent of Country's Exports)*

Sugar		Nicaragua	12	Honduras	27	Swaziland	28
Mauritius	89	Brazil	11	Somalia	26	Burma	22
Cuba	74	Guinea	11	Panama	25	Ghana	21
Fiji	67	Papua New Guinea	10	Costa Rica	22	Ivory Coast	18
Dominican Republic	53	Dominican Republic	7	Ecuador	12	Honduras	16
Belize	48	Zaire	7			Malaysia	15
Guadaloupe	43	Ecuador	6	*Citrus Fruit*		Paraguay	15
Guyana	31	Mexico	5	Cyprus	21	Equatorial Guinea	9
Barbados	30					Philippines	9
Philippines	27	*Cocoa*		*Peanuts*		Gabon	8
Swaziland	18	Equatorial Guinea	66	Gambia	94		
Togo	14	Ghana	63	Guinea-Bissau	46	*Palm Oil or Nut Copra*	
Brazil	12	Cameroun	27	Senegal	34	Tonga	79
Jamaica	12	Togo	26	Sudan	15	Benin (Dahomey)	34
India	10	Benin (Dahomey)	24	Niger	14	Philippines	22
Mozambique	10	Ivory Coast	21	Mali	10	Malaysia	11
Peru	10	Ecuador	10	Malawi	5	Sri Lanka	11
Haiti	9	Dominican Republic	8			Guinea	6
Mexico	7			*Fish*		Sierra Leone	6
Thailand	7	*Cotton*		French Guiana	76		
Taiwan	6	Chad	63	Peru	13	*Tobacco*	
		Egypt	56	Mauritania	11	Rhodesia	51
Coffee		Mali	36	Panama	7	Malawi	39
Portuguese Timor	90	Nicaragua	36			Turkey	13
Burundi	84	Sudan	35	*Rubber*			
Uganda	71	Central African		Cambodia	93	*Meat and Cattle*	
Rwanda	64	Republic	31	Malaysia	28	Somalia	54
Colombia	47	Pakistan	31	Sri Lanka	21	Botswana	42
El Salvador	45	Syria	26	Liberia	16	Uruguay	38
Haiti	33	Mozambique	20	Singapore	14	Argentina	24
Central African		Benin (Dahomey)	19	Thailand	10	Paraguay	21
Republic	30	Tanzania	17	Indonesia	7	Lesotho	18
Malagasy	30	Turkey	16			Costa Rica	9
Costa Rica	29	Afghanistan	15	*Tea*		Nicaragua	6
Guatemala	28	Uganda	12	Sri Lanka	39		
Ethiopia	27	Guatemala	11	Malawi	17	*Rice*	
Angola	26	El Salvador	10	Kenya	9	Burma	39
Cameroun	25	Paraguay	10	India	7	Pakistan	21
Equatorial Guinea	24	Colombia	7	Rwanda	6	Thailand	19
Ivory Coast	22	Peru	7	Uganda	5		
Kenya	18	Mexico	6			*Jute*	
Honduras	16			*Timber*		Bangladesh	88
Tanzania	13	*Bananas*		Congo (Brazzaville)	47	Nepal	15
		Guadeloupe	36	Laos	36	India	9

(cont.)

SOURCE: *Encyclopaedia Britannica, Book of the Year, 1976,* Chicago, 1976.

TABLE 12–4 (*cont.*)

Petroleum

Saudi Arabia	100
Brunei	99
Libya	99
Oman	99
Qatar	98
United Arab Emirates	98
Iran	96
Iraq	96
Kuwait	96
Venezuela	96
Nigeria	91
Trinidad	89
Algeria	88
Gabon	87
Indonesia	68
Southern Yemen	62
Ecuador	58
Syria	55
Congo (Brazzaville)	53
Panama	42
Tunisia	34
Angola	29
Bolivia	29
Singapore	26
China	20
Kenya	19

Copper

Zambia	93
Chile	86
Zaire	62
Papua New Guinea	55
Peru	23
Philippines	15
Cyprus	11
Mauritania	10
Uganda	5

Diamonds

Zaire	62
Sierra Leone	60
Namibia	40
Central African Republic	32
Angola	10
Lesotha	7
Rhodesia	7
Ghana	5

Bauxite–Aluminum

Jamaica	72
Guinea	65
Guyana	57
Ghana	7

Tin

Laos	57
Bolivia	42
Malaysia	15
Rwanda	12
Cyprus	11
Thailand	6

Zaire	6

Iron Ore

Mauritania	73
Liberia	65
Swaziland	11
Sierra Leone	10
Chile	7

Nickel

New Caledonia	95

Phosphates

Morocco	55
Togo	46
Jordan	21
Tunisia	12
Senegal	11

Uranium

Niger	39

significant proportion of the population has entered the money economy, might better be called "developing countries" rather than underdeveloped countries.

The multinational companies have also discovered that urban laborers in developing countries can do assembly-line jobs as well as workers in industrial countries can, and at lower wages. Moreover, since the cities are part of the worldwide interconnected system, they have surface transport to bring in needed production materials and to ship finished products anywhere in the worldwide system. They also have long-distance telephones and jet airplanes to allow management to continuously monitor operations. The result has been the development of **export platforms**—factories in developing countries built, not primarily to serve national markets, but to export products for sale in the developed countries. Thus factories in cities in developing countries can produce clothing, radios, sewing machines, televisions, cameras, and other consumer goods for sale in the United States and Europe. Engines for Ford Pinto cars, for example, are made in Brazil. Such production is often encouraged by the government of the developing country. Governments provide special tax benefits to multinational firms locating there. Some make the export of a proportion of plant output a condition for location of the industry in their country. Mexico, for example, requires foreign auto manufacturers to export a proportion of their output. Companies that want to enter the market of such a country must be prepared to accept such conditions.

As with the oil industry, underdeveloped and developing countries are increasingly requiring that all foreign investment involve local partners in ownership. Sometimes the local share in the investment is only a small proportion. But often the underdeveloped or developing country requires majority ownership by its own nationals. Some countries also call for the transfer of technology into their countries. Brazil has recently signed an agreement with West Germany to create a nuclear power industry that will over the next fifteen years make Brazil virtually self-sufficient in nuclear power—from designing and building reactors to producing the fuel.

THE DUAL ECONOMIES OF DEVELOPING COUNTRIES AND PROBLEMS OF NATIONAL DEVELOPMENT

Although the underdeveloped countries are regarded by the industrial countries primarily as sources of raw materials and foods, and although they depend on other countries for the capital for their economic development, the people in most of these countries are not primarily producers or supporting workers of the interconnected system. Rather, they follow a completely different way of life. They are still part of the traditional, locally based society typified by the village of Ramkheri. Moreover, the traditional society is older and has roots deep in the people's culture. The people not only depend almost entirely upon the resource base of their tiny corner of the world, but they have their primary cultural and political allegiances within that same local area.

TWO WORLDS IN JUXTAPOSITION

In the developing world, what we really find in each country are **dual economies** in juxtaposition. Mines in Africa depend on workers drawn from the traditional tribal society, and peasant farmers in various parts of the underdeveloped world sell some of their production whenever they can in exchange for a few manufactured products from the outside world. Surely, the two systems are in contact with one another in the local area. Such contacts are more than economic meetings; more often they are also cultural and political. This happens when government officials make efforts to enlist the local population by educating them, taxing them, indoctrinating them, or looking after their health.

TIES WITH INDUSTRIAL COUNTRIES BUT NOT OTHER DEVELOPING COUNTRIES

Traditionally, the people of underdeveloped countries have not had contact with other underdeveloped areas even on their own continent, not to mention other parts of the world. They form separate cultural and economic units. The ties that have grown between them have come with the modern interconnected system. They follow the routes of transportation and communication established by the Westerners. Often European ties, such as the English or French language, provide the communication link between one culture and another. Long-distance telephone calls between African countries often must be routed through London. Other contacts arise when people of different cultures are brought together at a work site, as when workers are recruited from different regions.

At the national level the native leaders of the new countries have been trained in the ways of the modern world through contacts with Westerners in Western schools, colleges, or training programs in the industrial countries. Until the recent entry of the Soviet Union and the Japanese into the field, the contacts that native leaders made were generally with the industrial countries that had the closest economic and political ties to their homelands.

It is not surprising that India and the African countries that are members of the British Commonwealth of Nations have emerged with governmental structures and even protocol manners very like those of their former colonial masters or that, indeed, they have even joined the Commonwealth. Former territories of France, whether in Africa or Asia, tend to show a similar French orientation.

Politically, the most important embassies of underdeveloped countries have been those in the capitals of the leading industrial countries. Until recently governments had little basis for discussion with those nearby. Neighboring countries have been more rivals for economic and political favors of industrial countries than trade partners. However, the United Nations has brought diplomatic representatives of the underdeveloped world together and enabled them to discover common problems and destinies. In the U.N. General Assembly and in many of its supporting agencies, the underdeveloped countries have emerged as a special-interest group. Under the U.N.'s rules giving each country one vote in the General Assembly, the Third World countries have a clear majority; they have found that voting solidarity means that they have a very large voice in U.N. affairs despite the veto power of the major industrial countries: the United States, Soviet Union, United Kingdom, and France.

As the result of the new political contacts in the United Nations, underdeveloped countries regularly meet to discuss their problems and develop strategies for dealing with industrial countries. Having seen the strength that the OPEC countries have shown in world oil matters, other producing countries have shown interest in establishing similar production cartels.

Moreover, the Third World bloc has begun to clamor strongly for a reordering of the modern interconnected system. The Third World countries have accused the industrial countries of exploiting their resources and undervaluing the raw materials that still provide their main tie with the modern interconnected system. They call for a "New International Economic Order" that will overhaul economic relations between the industrial countries and the Third World. In the name of equity and justice Third World spokespeople call for a restructured order to eliminate those disparities. They see the present affluence of the developed countries as based initially on colonial exploitation and subsequently exploitation by multinational corporations. Now compensatory restitution is demanded. They feel they have new leverage because of both their General Assembly majority and the strong bargaining power of OPEC.

CULTURAL FRAGMENTATION RESULTING
FROM THE IMPACT OF THE MODERN SYSTEM

To appreciate the problems of superimposing the modern interconnected system on traditional life in underdeveloped and developing countries, we must remember that new ways add yet another dimension to the already fragmented pattern of separate local cultures. The modern system is not limited to the economic sphere. It is a whole new way of life, offering a different culture with its own values, customs, and lifestyle that serve to break up traditional national patterns and often divide families. Some people in developing countries, commonly the young people and the city dwellers, heartily accept the modern ways. Others, usually those older and more conservative, reject anything new and try to maintain traditional values. Still others are often confused. New values make them question old ways that were once accepted, but they are afraid of anything new. Thus a society that was once accepted by most of its people has become divided into factions. We saw this revealed in the changes that were occurring in Ramkheri. On the scale of a whole country, the conflict between old and new appears in political as well as cultural form and tears the fundamental fabric of society.

THE POPULATION EXPLOSION AND THE RESOURCE BASE

The introduction of the modern interconnected system into areas of traditional culture has led to rapid population growth in underdeveloped countries. The population of Africa, estimated at 120 million in 1900, was 413 million in 1976, and similar growth has occurred in Asia and Latin America. Some of this increase, particularly in Latin America, was the result of the immigration of Europeans (in the early part of the century accompanied by their slaves), but, in large part, the increases represent natural growth by more births than deaths.

Much of the rapid increase can be attributed to the introduction of modern health and sanitation measures. Inoculation against diseases, water purification, better diets resulting from higher living standards, and the use of insecticides have all contributed to lower infant mortality and increased life expectancy. Airlifts of food supplies and medicines have cut the toll taken by famines following floods, droughts, earthquakes, and plagues. Birth rates have not declined, however. In rural areas, where children are regarded as economic assets because they can work in the fields, the tradition favoring large families has resulted in a population explosion.

In the modern interconnected system the population explosion means congestion, air pollution, and problems of waste disposal. In traditional areas where the population is sparse or the resource base particularly rich, population growth does not appear significant, but in those where the existing economic system is producing near the upper limits of its capacity for providing food and other basic needs, it will produce a new manifestation of the Malthusian problem. Without major improvement in production, sooner or later one of the population controls that Malthus described—famine or war—will appear. In-

creased productivity depends upon technological change, and that can come only by shifting from the old closed system to longer-range connections.

CHANGES IN THE CLOSED SOCIETY

Changes have, of course, taken place in the traditional world. Some segments of all countries have become productive components of the interconnected system. In addition, scientific knowledge has spread to the locally based society from the modern world in such forms as programs for agricultural education, schemes for improving drainage and irrigation, plans to implement the use of fertilizers, and improvements in simple tools. In the 1960s major breakthroughs were made in the development of "miracle" rice and wheat strains said to increase yields 50 percent or better. Even with these improvements the traditional agricultural pattern is still the way of life for most people.

Although some romanticists would like to see a return to the "good, simple, primitive life of the past," this is impossible. In most places population densities are too high for complete support from the local base. Furthermore, social and economic institutions have been so modified that societies have become dependent on the modern world and are no longer self-sufficient.

Members of the Kikuyu tribe of Africa, while still residing in villages and maintaining their traditional social life, now live on poorer lands as a result of the settlement of their areas by Europeans. Money has replaced traditional barter, and cattle and goats are no longer a measure of a person's wealth. Children go to school to read and write, not to become hunters or warriors. The government has encouraged the Kikuyu to abandon their tiny family fields and to work together on larger properties. Coffee and tea are planted for sale. Dairy herds are kept within fenced pastures and no longer allowed to wander wild. The Kikuyu are still primitive by Western standards, but their life has been changed fundamentally. Clearly, they could never return to their old way of life.

Governments in underdeveloped countries hoping to improve the standard of living are faced with dual developmental problems. On the one hand, they want to see the modern interconnected component grow, for with it come both a rising living standard and increasing political prestige. On the other, they must solve the problem of the population explosion within the traditional segment of the population. In countries where the population pressure on the traditional component is high, the government must often tap the money gained from exports in the modern component to purchase the food necessary to support the overpopulated traditional society.

The problem of development in the modern interconnected segment of a developing country centers upon control. The industrialized countries of the world have always been interested in development within underdeveloped areas, but they plan for their own ends, not for those of the country in which they invest. Thus companies from the modern industrial countries are anxious to tap rich, new mining areas and potential plantation regions. They hire local labor to build the new facilities and to help in production. However, as we have shown in the discussions of the oil and copper industries, the foreign industrialists make the basic decisions, control the movement and marketing of products, and reap the profits. Governments of underdeveloped countries get only an indirect share of the operation—royalties and other payments on the amount of product produced. Today their leaders, trained in universities in the industrial countries, know economics and the power of their political position. They are now working to get a larger share of the return.

Although underdeveloped countries might like to develop and control their own mines and plantations that export to the world, they have little capital to invest in such costly enterprises and far less leadership and management experience than the companies with which they must deal. Nor do they have any prospect of raising the necessary capital from taxes. Poor people in these countries have little monetary income and little agricultural surplus to be taken as taxes. Indeed, if population pressure is great, the government will barely be able to provide for its people's needs.

Like the American republic in the eighteenth century, underdeveloped countries commonly have no adequate transportation and communication systems connecting the different regions of the country. The railroads, highways, and ports have been developed by foreign investors to get the products of mines and factories to the coasts, not to provide internal integration for the country. Usually, no transport network connects the capital to all settled parts of the country. This would offer the necessary base for economic and political integration. Instead, the productive parts of the country are tied to overseas markets, not to the largely subsistence areas elsewhere in their own country.

WHAT TO DO WITH THE CLOSED SYSTEM

In addition to the problem of control, underdeveloped countries may have a severe problem of population pressure on the resource base. In a country where there has been a population explosion, the pressure of the traditional economy on its productive base may be almost critical. The problem may call for both short-term programs to provide emergency food and longer-term programs of education to try to promote birth control. In the long run, an increase in the productivity of the system is needed.

The long-term solution is to increase the output of the subsistence component of the population, whether that increase continues to be used simply to support a locally based system or to begin to provide some cash sales to the subsistence producers. This will require nothing less than an industrial revolution, which will result in a complete upheaval.

There is, however, a shortage of both capital and management leadership. People living in a traditional society are conservative with regard to change. Often religious beliefs, built around trying to find meaning in their lives of work and privation, lead them to revere traditional ways. Moreover, leadership and power are most often vested in the elders—the most experienced, the most conservative, and the ones with the greatest stake in the status quo. In addition, living on the brink of starvation makes people extremely reluctant to try something different. Inefficient and unproductive as old ways may be, they have been tested through time, and new ways are regarded as risky.

The dual societies of the developing countries of Africa, Latin America, and Asia vary greatly. Although each of these countries in various regions of the world is unique, they are roughly similar. In eastern Asia the problem of population pressure on the land is most acute. In the Arab countries and in northern Africa, basically feudal societies have begun to modernize. In tropical Africa some of the newest and smallest countries have recently emerged from colonialism. In Latin America, European immigrants are part of the dual economies there. We shall explore each submodel in succeeding chapters.

THE PROBLEMS OF THE
DUAL ECONOMY IN BRAZIL

The problems of the dual society can be sketched by looking at Brazil, one of the largest of the developing countries. Brazil has essentially had five different producing areas that have fed goods into the interconnected system (Figure 12–4): first, the old sugar-producing area of the northeast coast; second, the interior frontier cattle lands in the south; third, the old rubber lands of the Amazon where, at the close of the nineteenth century, a tremendous rubber boom took place; fourth, the scattered mining communities northwest of Rio de Janeiro where first gold and diamonds and now iron have been developed; finally, the rich coffee-growing area centered on São Paulo near the coast along the Tropic of Capricorn.

With the exception, perhaps, of the western cattle frontier, which had connections with both São Paulo and Rio de Janeiro, each of these different producing regions has had its own transportation system to get products to overseas markets. Ocean ships traveled up the Amazon to the rubber-producing centers, and the ports of Recife and Salvador served the sugar coast. Rio was the focal point for a rail network into the mining areas, and São Paulo was the point for another one into the coffee country. From colonial days each region has fared differently economically. Rubber and cattle booms, at least for overseas shipment, are long gone. The sugar coast has never regained the prominent position it held from the colonial period until the early nineteenth century. On the other hand, mining has become profitable, and the São Paulo hinterland has become both the leading coffee-producing district in the world and the economic core of modern-day Brazil. This area has the most modern industrial development in Latin America.

Outside of these areas and in nonexporting communities within them, the population of Brazil lives in the traditional closed system. Some areas are frontier regions occupied by Indians, traders, and missionaries; others are peasant-farming areas where the local marketplace is the destination of most goods sold.

How does a government put all the disparate and essentially independent components of Brazil into a single functioning country? The question has not yet been answered. If giant Brazil, one of the more advanced countries of the Third World, has not been able to answer it, how can others, smaller and weaker, accomplish it?

The problem is compounded when one recalls that the boundaries of many underdeveloped countries were set by European powers in colonial days and reflected European interests and compromises, not native interests. Thus borders cut through the territories of a single native group or surround tribal peoples who have traditionally been rivals. Furthermore, since colonies were organized to exploit the resources of an area, Europeans often set up separate colonial units wherever an individual exporting system focused on a port.

The result is that the new countries are commonly both small and fragmented. Over half the

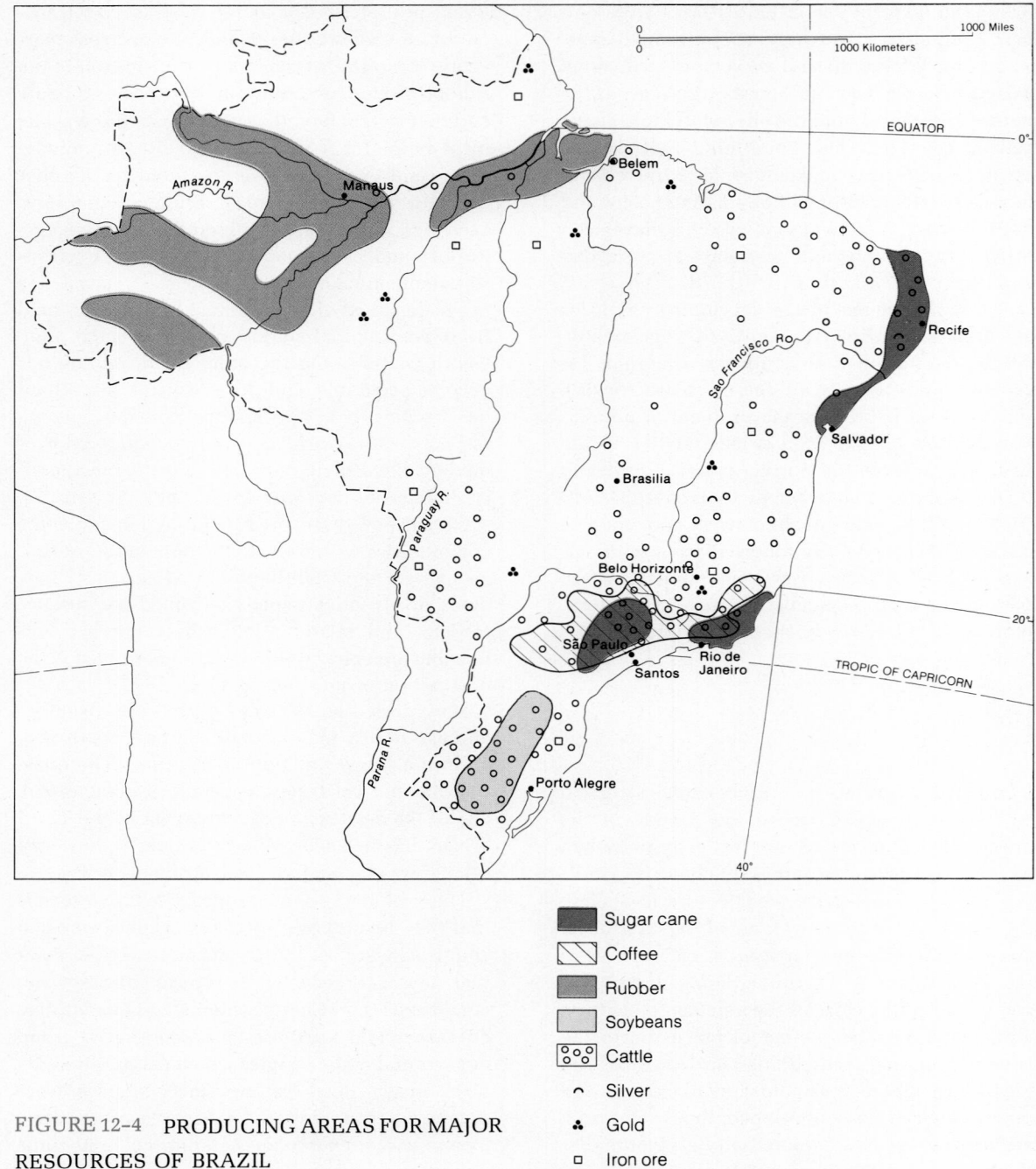

FIGURE 12–4   PRODUCING AREAS FOR MAJOR
RESOURCES OF BRAZIL

Sugar cane
Coffee
Rubber
Soybeans
Cattle
Silver
Gold
Iron ore

102 underdeveloped and developing countries have populations of 10 million people or less, and only 14 have more than 25 million. Eleven countries have fewer than 1 million people. The typical underdeveloped country is therefore in an extremely disadvantageous position in the world when one considers that it must compete economically, politically, and militarily with the industrial world. Yet, even underdeveloped countries with their small populations face the problem of integrating peoples of different local backgrounds, even different languages, into a functioning nation.

Individually, these underdeveloped countries and their problems do not loom large in the press and other media of the industrialized countries. But Latin America, North Africa and the Middle East, Africa south of the Sahara, and South and East Asia other than Japan have more than two-thirds of the people of the world. Their problems cannot help but be world problems, and their future is of major importance to the whole world. In the next part of this book we will examine four of the most important areas of the developing world: Latin America; Black Africa; the Middle East–North Africa; and South and East Asia, including India and China, the two most populous countries in the world. Each of these regions shares the problem of the dual economies described in the Brazilian example, but, as with the different industrial areas, each has its own distinctive problems and potential.

SELECTED REFERENCES

Keyfitz, Nathan. "World Resources and the World Middle Class." *Scientific American,* July 1976, pp. 28–35. An effective extension and elaboration of many of the points made in this chapter. A demographer and sociologist discusses the problems of raising living standards in the world's poor countries under conditions of rapidly rising populations and limited resources and capital assets for development. Only those countries with small populations and abundant exploitable resources, such as Saudi Arabia, are likely to succeed.

Lloyd, Peter S. *International Trade: Problems of Small Nations.* Duke University Press, Durham, N.C., 1968. An economist's view of the position of small underdeveloped countries in a world dominated by the industrial nations.

O'Dell, Peter. *Oil and World Power: Background to the Oil Crisis,* 3rd ed. Penguin, Baltimore, 1974 (paper). Basically, an overview of the geography of world oil production and consumption. The author discusses how the present geography of oil supply and demand affects economic conditions and the political power of nations.

Rostow, Walter W. *The Stages of Economic Growth.* Harvard University Press, Cambridge, Mass., 1960. A classic statement that argues a case for economic development as a process following a definite sequence of steps or stages. Though the proposed sequence for development is questionable, the descriptions of different stages of development provide a useful framework from which to compare areas.

Thoman, Richard, and Edgar Conklin. *Geography of International Trade.* Prentice-Hall, Englewood Cliffs, N.J., 1967 (paper). Provides a basic treatment of international trade from the viewpoint of a geographer.

Vernon, Raymond, ed. *The Oil Crisis.* W. W. Norton, New York, 1967 (paper). An excellent review of the background, effects, and major actors in the 1973 oil crisis. The new realities, for both the producer and the consumer nations, stemming from that crisis are discussed as well as the changing role of the international oil companies. An appendix provides data relevant to the geography of oil today.

PART THREE

Developing Areas: Modern and Traditional Worlds in Conflict

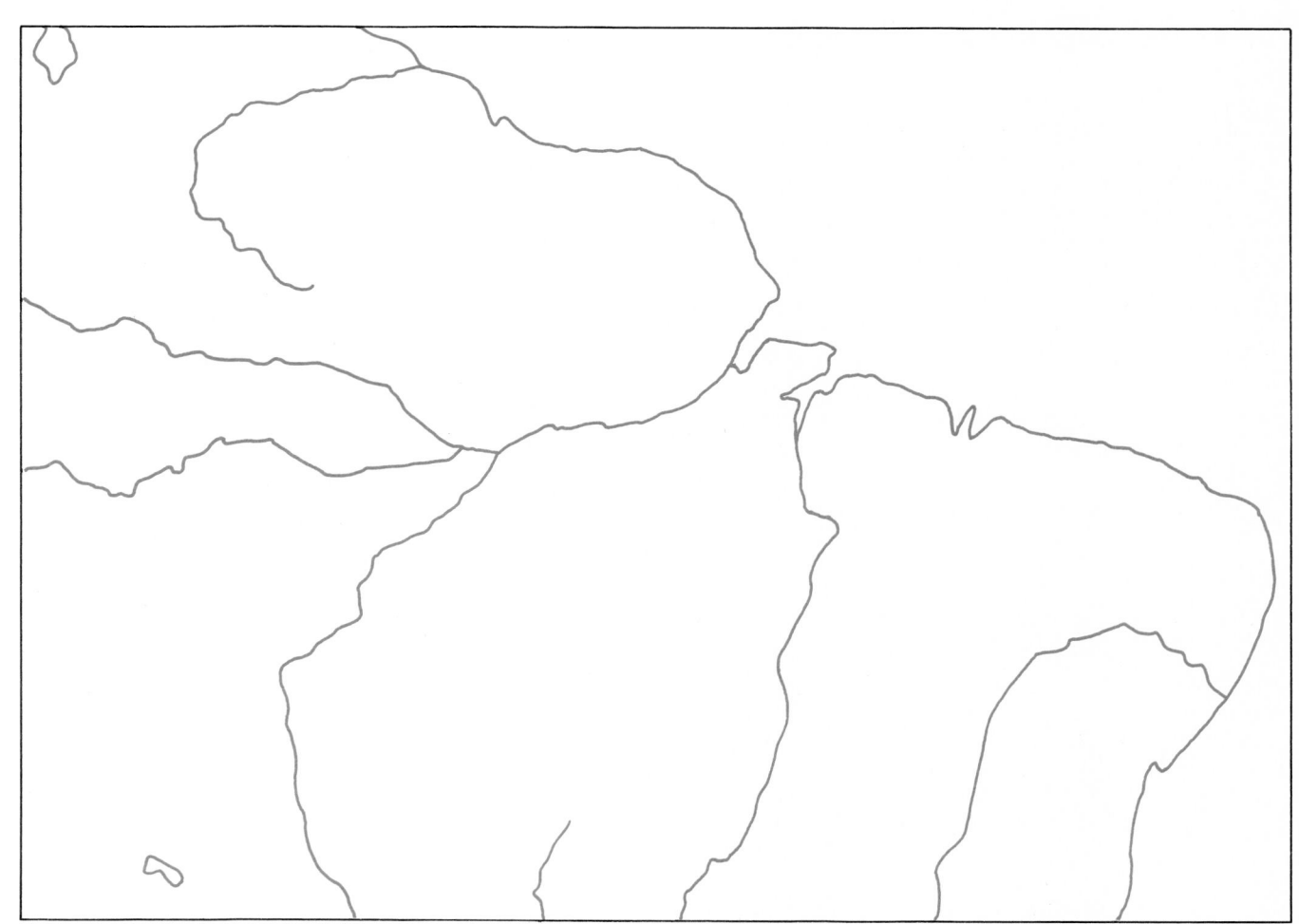

CHAPTER 13

Latin America: A Developing Offshoot of European Culture

A basic Latin culture dominates this part of the world despite ethnic diversity and the physical separation of population concentrations. Some coastal areas and parts of the Andean highlands are densely populated, while the extensive tropical lowlands are for the most part sparsely settled. Mineral resources, mostly in the areas of Indian populations, have been important historically in the pattern of development, beginning with the Spanish conquests. The rise of urban-industrial populations in the twentieth century, in contrast to the large numbers of rural peasants, has given rise to the dual societies that mark the contemporary scene in most countries.

Left: Cultivation on a small farm in the highlands of Ecuador. (FAO photo by J. G. Thirion)   *Above:* São Paulo, the largest city in Brazil, with a metropolitan population of approximately 8 million. (A. Greenberg/Photo Trends)

Above: Automobile engine factory in Brazil. (Pictorial Parade) *Below:* Settlement in an upland valley in Colombia. (United Nations)

Above: Construction of the Amazon highway, which runs across Brazil through the Amazon rain forest to the Peruvian border. (Edward Leahy/Photo Trends)

419

In contemporary Latin America, extending from the northern border of Mexico to the southern tip of South America and including the Caribbean islands, we find the full range of ways of life, from primitive Stone Age Indians to some of the largest, most modern metropolitan centers in the world. Aspects of the two different worlds represented by Washington, D.C., and Ramkheri are intermingled. No country is all modern or all traditional. Instead, the governments face problems of managing the affairs of two different ways of life, each with its own problems, as shown by the illustration of Brazil in Chapter 12.

Stone Age Indians are found in the remote tropical forests of the upper Amazon Basin, an area that until recently has been of little interest to outsiders other than explorers, anthropologists, and missionaries. In the twentieth century these people live tribal lives very much like those of prehistoric societies, but their traditional lifestyles are being threatened as Brazil, Peru, and Ecuador seek to develop these remote areas.

As much as half the total population in Central America and the Andes live in more advanced Indian cultures (vestiges of the great Mayan, Aztec, and Incan cultures). Many still speak their own native dialects as a first language and are not even nominally a part of the Latin culture. Some have been "converted" to Christianity, but in many cases, traditional religious beliefs have been incorporated in the Christianity that is practiced. The lives of these people depend mainly on the produce they can extract from the local region.

Latinos are more widespread than the Indians. They are descendants of Iberian immigrants and may also be of partly Indian or African parentage. They were the workers on colonial haciendas, the large rural estates established under European landlords. The hacienda workers traditionally worked part time for the landlord, but they had to depend on their own plot for much of their subsistence. Today some are squatters on the fringe of settled regions. As long as they just subsist, the landlord lets them stay, but if they show signs of developing more products to sell and of functioning successfully in the modern system, they may be asked to leave or to share their production with the landlord. These people remain in the traditional, locally based system almost as completely as do the Indians.

The modern interconnected system dominates in the cities, mines, and commercial farming areas of all countries (Figure 13–1). Parts of Latin America have been tied to the modern system from the first colonial days in the sixteenth century. Like the British colonies in Anglo-America, Latin America's colonial ties were to Europe. The colonies provided tropical specialties or other products that Europeans wanted. Sugar cane and other tropical crops were developed for export, as was the output of the mines and the plantations. Where the modern interconnected system has been developed on a large scale, as in the coffee lands around São Paulo in Brazil or in the farmlands near Buenos Aires in Argentina, a full-fledged, regional network of the modern system has developed. It has its own transportation and communication systems, financial organization, and specialty food producers supplying fruits, vegetables, meat, and dairy products. São Paulo and Buenos Aires, the centers of these regions, are among the largest, most modern cities in the world.

Brazil also contains traditional areas, and the Brazilian government must deal with the whole spectrum of its populace. Some other regions of Brazil in the interconnected system are not so successful, and there the people are virtually locked into the traditional, locally based system. The same is true in all Latin American countries. Elements of both systems exist side by side, often within the same region. In general, countries with high per capita GNP or energy consumption are closer to the modern interconnected system; those with low per capita figures are more traditional. Nowhere in Latin America is there a country as developed as Japan, Canada, or the other examples of modern industrialization we have studied.

Gulf of Mexico

ATLANTIC OCEAN

MEXICO

Mexico City

Havana
CUBA

TROPIC OF CANCER

BELIZE
HONDURAS

Port-au-Prince
HAITI

PUERTO RICO (U.S.A.)

Kingston
JAMAICA

Santo Domingo
DOMINICAN REPUBLIC

Guatemala City
GUATEMALA

Tegucigalpa

Caribbean
Sea

WEST INDIES

Bridgetown
BARBADOS

San Salvador
EL SALVADOR

Managua
NICARAGUA

Barranquilla
Cartagena

Port-of-Spain
TRINIDAD AND TOBAGO

San José
COSTA RICA

Panama City
PANAMA

Maracaibo

Caracas

VENEZUELA

Georgetown

Paramaribo

Cayenne

Bogotá

GUYANA

SURINAM

FR. GUIANA

COLOMBIA

EQUATOR

Quito

ECUADOR

Manaus

Amazon R.

Belém

PERU

BRAZIL

Recife

Lima

La Paz

BOLIVIA

Sucre

Salvador

Brasília

Belo Horizonte

PARAGUAY

Rio de Janeiro

Asunción

São Paulo

PACIFIC OCEAN

TROPIC OF CAPRICORN

CHILE

ARGENTINA

Santiago

URUGUAY

Buenos Aires

Montevideo

ATLANTIC OCEAN

Caracas National capital

• Cities over 1 million

• Other important cities

FIGURE 13–1 POLITICAL MAP OF LATIN
AMERICA

The mix of the modern interconnected system and the traditional, locally based system throughout Latin America results in a "dual society" or "dual economy." The two different systems exist side by side within each country, often in a single area, but the ties are loose and direct contacts are few. Although government officials visit the rural villages bringing health programs and exacting taxes, the village social structure and cultural values may be those that have existed locally for centuries. The products of most farms in such areas are destined for local marketplaces. People may not fully comprehend that their local world is part of a larger political unit. This is a whole different world from life in São Paulo, Lima, or Mexico City, where businesspeople and scholars are part of a world community, shops stock products similar to those in Europe and the United States, and streets are clogged with automobile traffic. Different as these two worlds are, a single national government must deal with both.

THE ENVIRONMENTAL BASE OF LATIN AMERICA

About three-fourths of Latin America is within the tropical zone, characterized by high year-round temperatures and climatic environments that are either high-work or seasonal-work with a dry dormant season (see Figure 3–7). No part of Latin America extends into high latitudes comparable to those of northern Canada, northern Europe, or the Soviet Union. Only the southern tips of Argentina and Chile lie beyond the 49th parallel, which, in the Northern Hemisphere, is the southern boundary of Canada. Thus the frost-free season is typically year round for most of Latin America; only southern Argentina and Chile lie at latitudes where frost is likely at all elevations during the winter months of the year.

The climatic environments of the tropical latitudes of Latin America are complicated by the presence of high mountains. Because atmospheric temperatures decrease as elevation increases, high mountain areas stand out as environmental anomalies at all latitudes. Their climates do not conform to the pattern one would expect from a consideration of latitude and potential evapotranspiration, and for this reason highland areas are usually delineated on climate maps. The high Andes and the highlands of northern Mexico experience several months during the year when the danger of frost is present; and even close to the equator, glaciers and permanent snow cover the highest peaks.

LANDFORMS

High mountains extend the full length of Latin America (Figure 13–2). In the north they appear as extensions of the mountain system of the western United States into Mexico and Central America. In South America the higher mountains in the Andes form an S-shaped curve from southwest Venezuela through Colombia, Ecuador, and Peru, and then extend all the way down the western margin of South America to Tierra del Fuego at the southernmost tip. The highest peaks of the Andes are south of Ecuador, and many are higher than the tallest North American peaks. Mount McKinley in Alaska, which rises to 20,320 ft (6,100 m), is surpassed by thirteen mountains in South America, the highest of which, in Argentina, is 23,034 ft (6,910 m).

A notable feature of the Andes and the Mexican highlands is the presence of high mountain basins that support large populations. Mexico City is located in such a basin, and three South American capitals, Bogotá, Quito, and La Paz, are in basins high in the Andes. The largest of these plateau-like areas in South America is the Bolivian **Altiplano,** which contains two large lakes and a large proportion of the Bolivian population.

Most of the rest of Latin America's surface area consists of plains, low mountains, and plateaus. A great expanse of lowland plains is located

0 [____|____] 1000 Miles
0 [____|____] 1000 Kilometers

TROPIC OF CANCER

20°

Orinoco R.

3

EQUATOR 0°

Amazon R.

5

4

Tocantins R.

6

São Francisco R.

3

Paraguay R.

7

Paraná R.

6

20°

TROPIC OF CAPRICORN

Salado R.

8

3

9

40°

120°   100°   80°   60°   40°   20°

Note: Desert areas are shown here for convenience, although *desert* is an environmental rather than a landform term.

SOURCE: Adapted from *A Geography of Man,* 3rd ed., by P. E. James, copyright © 1966 by John Wiley & Sons, Inc. Reprinted by permission of John Wiley & Sons, Inc.

FIGURE 13-2   LAND SURFACE FEATURES OF
LATIN AMERICA

☐ Plain
⋰ Desert
▨ Hilly upland and plateau
▨ Mountains and basins
■ Mountain
〰 Escarpment

1 Mexican Highlands
2 Central American Ranges
3 Andes Mountains
4 Guiana Highlands
5 Amazon Basin
6 Interior Uplands
7 Paraguay-Paraná Plain
8 Pampas
9 Patagonian Desert

around the Amazon Basin. The Amazon River drains a large portion of South American lowland, flowing from west to east into the Atlantic Ocean roughly parallel to the equator. A less extensive lowland area extends to the north coast of Venezuela along the Orinoco River. To the southeast of the Amazon Basin is the Paraguay-Paraná Plain, which extends through Paraguay and northern Argentina. This lowland and that of southern Brazil are drained by the Paraguay, Paraná, and Uruguay rivers, which flow into the estuary known as the Río de la Plata, on which Buenos Aires is situated. South of the Río de la Plata is the **pampa** of Argentina, a fertile, humid, productive agricultural region that supports the bulk of Argentina's population. Other extensive lowland areas in Latin America include the Yucatán Peninsula, most of Cuba, and the Caribbean coastal regions of Honduras and Nicaragua.

The largest highland region outside the Mexican highlands and the Andes is the highland of eastern Brazil. It rises gradually from the Amazon Basin, culminating in a plateau with a steep Atlantic-facing escarpment that drops 2,000 to 8,000 ft (600 to 2,500 m) in a short distance to the coast. Much of the plateau's drainage follows its gentler slope toward the Amazon Basin and the Río de la Plata. Several rivers do, however, descend from the Brazilian plateau to the Atlantic, providing a source of hydroelectric power as they fall from the escarpment.

CLIMATIC VARIATIONS WITH ALTITUDE

The presence of mountains in all latitudes adds an important variable to the climate of Latin America. The decrease in atmospheric temperature that accompanies increased elevation causes cooler weather and lower potential evapotranspiration as one ascends to the higher elevations in the mountains. Latin Americans, accustomed to this **vertical zonation** in climate, have developed terms for the different living zones (Figure 13-3): the *tierra caliente,* the hot lowland zone

from approximately 2,000 to 3,000 ft (610 to 915 m); the *tierra templada,* the lower upland zones to 6,000 ft (1,830 m), where it is warm rather than hot; and the *tierra fría,* or the cold country, from 6,000 to about 10,000 ft (3,050 m), where temperatures are cool. Quito, the capital of Ecuador, lies almost on the equator. It is over 9,000 ft (2,750 m) above sea level and has an average annual temperature of 56°F (13°C). As one would expect at the equator, there is very little seasonal variation in temperature. La Paz, at an elevation of about 12,000 ft (3,655 m), has an average annual temperature of 50°F (10°C) with only slight seasonal variation.

When considering vertical zonation, remember that latitude affects the elevations at which the different zones begin and end. Figure 13-3 shows a typical cross section, but at a higher latitude, where there is less radiant energy from the sun, the *tierra templada* and the *tierra fría* would begin at elevations lower than 3,000 and 6,000 ft. At a lower latitude than the one represented by Figure 13-3 they would begin at higher elevations.

GROWING POTENTIAL

Recall from Chapter 3 that potential evapotranspiration (PE) varies with latitude. Thus tropical areas have much greater potential for plant growth than do the midlatitudes. The tropical areas of Latin America receive up to 50 percent more radiant heat per year than is received in southern Argentina or Chile (see Figure 3-3). The highest PE, however, does not occur at the equator but in the subtropical dry areas of Mexico, because of less cloudiness and consequent greater solar intensity.

Since PE conditions are generally favorable over much of Latin America, the limiting factor in growing potential is the availability of moisture. As Figure 3-7 shows, the Amazon Basin is the largest expanse of high-work environment in the world. Here deficits are small and surpluses of 20

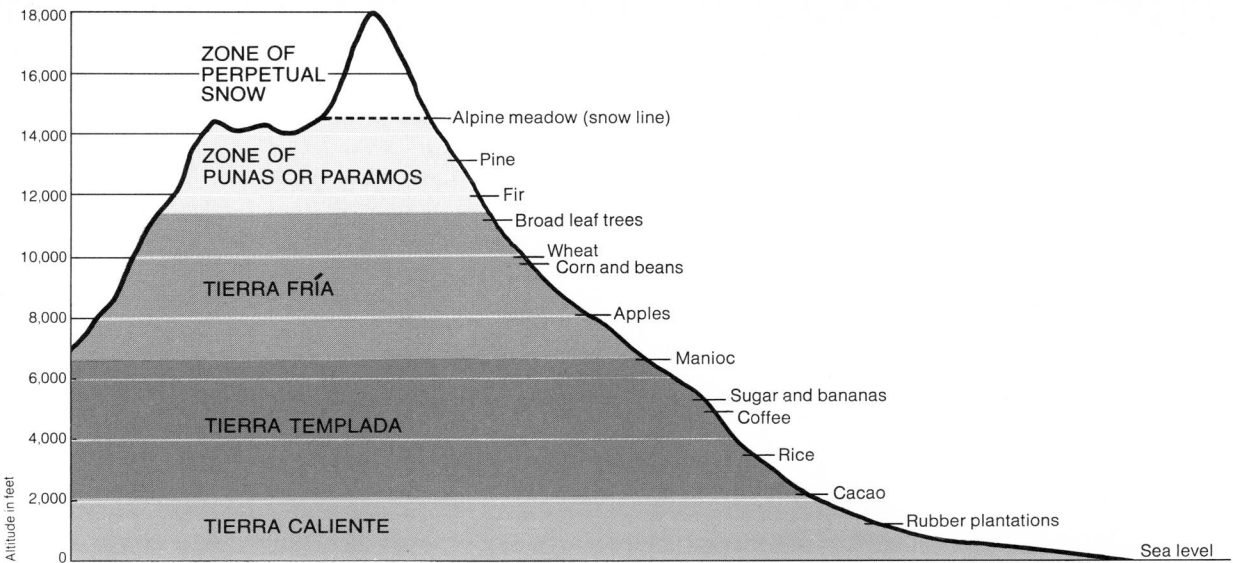

FIGURE 13-3 CROSS SECTION OF *TIERRA FRÍA, TIERRA TEMPLADA,* AND *TIERRA CALI- ENTE* IN LATIN AMERICA

to 40 in. (50 to 100 cm) are broadly distributed (see Figures 3–5 and 3–6). High-work environments also occur on some of the Caribbean Islands, along the eastern side of Central America, and along part of the east coast of Brazil. These areas have little or no moisture deficit during their dry seasons and show moderate surpluses during their wet seasons. North and south of the Amazon Basin, in Venezuela, central Brazil, and Paraguay, are the extensive seasonally wet and dry environments, where periods of high and low work alternate and the low-work seasons are caused by dryness. Central Chile has a Mediterranean type of climate, a midlatitude seasonal-work environment with a dry summer. A seasonal-work environment with sequencing seasons, in which vegetative growth occurs year round, is found in a large area comprised of southern Brazil, Uruguay, the Argentine portion of the Paraguay-Paraná Plain, and the pampas.

The southern coast of Chile, which is at a latitude comparable to the Pacific Northwest of the United States, has a similar seasonal-work environment with sequencing seasons.

Low-work environments in Latin America are mostly dry, rather than cold, climates, except in the high mountains. These environments occur in northeastern Brazil, along the coasts of Peru and northern Chile, and in northern coastal Venezuela. More extensive arid and semiarid environments are found in northern Mexico and southern Argentina. Irrigated agriculture has made the northwest coast an important Mexican agricultural region, but northeastern Mexico is mostly poor farmland. The Patagonian Desert of Argentina and the adjacent semiarid grasslands are used mainly for livestock ranching.

Low-work environments, however, are not a major feature in Latin America. When compared with areas of the world of similar size, Latin

America has a small percentage of dry areas (see Table 5–1). In addition, only 1 percent of the land in Latin America is so cold that agriculture is not feasible. In terms of moisture availability and growing potential, therefore, Latin America is among the best-endowed areas of the world.

ENVIRONMENTAL LIMITATIONS ON AGRICULTURE

The favorable growing potential of Latin America is not a good measure of its agricultural productivity. The high growing potential of the vast Amazon Basin, for example, belies poor soil and other conditions unfavorable for current agricultural systems. Soils in the hot, wet tropics are subject to intense chemical weathering and leaching year round, which leads to thorough depletion of plant nutrients. Natural vegetation in such areas conserves nutrients by recycling them through the decay of its own organic debris. When land is cleared for agriculture, however, the abundant standing vegetation, which represents the major store of nutrients, is removed. Crops planted on the nutrient-poor soils cannot be productive for very long under these circumstances. Conventional agriculture in places like the Amazon Basin faces the additional problems of competition from diseases, insects, and other plants.

Outside the Amazon Basin agriculture is quite widely practiced, except in areas of dry climates; yet there are few agricultural areas of any great productivity or commercial importance. The most favorable area is the region that extends from southern Brazil through Uruguay to the pampas of eastern Argentina. In this area commercial livestock ranches and crop and dairy farms are successfully operated. In addition, there are commercial plantations in some areas of eastern Brazil, along the coasts of Venezuela, Guyana, and Surinam, and in interior Colombia. Elsewhere, strips of coastal lowland, major river lowlands, and areas of suitable topography and soil in the highlands make up the agricultural resources for most of the people of Latin America.

Variations in growing potential with elevation have a significant effect on the kinds of crops that are grown in the highlands of Latin America. In more than half the Latin American countries vertical zonation affects the pattern of agricultural activity. The *tierra caliente* is suitable for tropical crops such as sugar cane, cacao, and bananas. In the *tierra templada* corn, coffee, and oranges reach the limits of their temperature tolerance. In the *tierra fría* only hardy livestock, such as sheep and llamas, and durable crops, such as potatoes and barley, can be raised successfully.

At present Latin America has few areas of well-developed agriculture comparable in productivity to those of the U.S. Middle West or of Europe or India. In part, this reflects environmental limitations on conventional agriculture. Modern agricultural methods similar to those used in Europe and North America are most successful in the midlatitude regions, such as southern Brazil and the pampas of Argentina. Modern irrigation has been applied successfully in Mexico and Peru. Elsewhere, as in the Amazon Basin, most attempts to introduce modern agriculture have failed. Over much of Latin America, traditional agricultural practices prevail.

VEGETATION

The wet tropics of the Amazon Basin, the coastal regions of Colombia and Ecuador, and much of Central America are covered with evergreen tropical **rain forests.** Rain forests, which receive abundant moisture and have a year-round growing season, contain hundreds of plant species, and their vegetation is more dense than that in any other environment. Eastern and southern Brazil also had extensive natural forests, now partially cleared for commercial development.

Tropical grasslands separate the Brazilian coastal forests from the Amazon forests and are also found in the lowland stretching along the north shore of the Orinoco River in Venezuela and continuing into eastern Colombia.

In the midlatitudes of Latin America, extensive grasslands, **scrub forests,** and deserts coincide with areas of moderate to severe moisture shortages. The pampas of Argentina resemble the Great Plains of the United States except that the pampas do not suffer from dryness. The vegetation of central Chile is like that of southern California, southern Europe, the Mediterranean coast of northwest Africa, and the tip of South Africa, and like all these areas, it is an important region of wine production. In the mountains, vegetation varies with elevation, ranging from tropical forests on the lower slopes to tundralike growth at the highest levels.

MINERAL RESOURCES

Minerals have been one of the key resources in the development of Latin America since Indian times. The Aztecs and Incas mined gold, silver, and emeralds, and early Spanish explorers sought the sources of these valuable minerals. These minerals are still mined in many Latin American countries. The Western Hemisphere's largest gold mine has only recently begun producing in the Dominican Republic, and Mexico and Peru remain among the world's leading silver producers. However, these precious minerals are now less important economically than industrial minerals—petroleum, coal, iron ore, bauxite, copper, tin, lead, and zinc. Latin America has major deposits of all these, but production is limited to only a few of the largest deposits.

Figure 13–4 shows the major Latin American areas of mineral deposits that are producing today. Venezuela, Colombia, Trinidad and Tobago, Argentina, and Mexico have major oil reserves. Brazil has discovered some offshore oil fields, but they are unlikely to make the country self-sufficient in petroleum, at least in the foreseeable future. Argentina has some petroleum production but little other mineral output. Iron ore is mined in the eastern highlands of Brazil, in Venezuela, and in the West Indies. Bauxite is an important resource of Guyana, Surinam, the Amazon lowland of Brazil, and some Caribbean islands. The high mountain chain from Mexico to southern South America produces nonferrous metals—copper, tin, lead, and zinc—as well as iron ore. The most important mineral districts in the mountains and immediately adjacent lowlands are in Peru, Chile, Bolivia, and northern Mexico. Cuba is a producer of nickel. Little production has yet come from the vast lowland of the Amazon or the east flank of the Andes, but this is still frontier country. With new roads into the Amazon Basin, new projects are expected.

EVALUATIONS OF THE LATIN AMERICAN ENVIRONMENT

Traditional societies and modern interconnected society have exploited the climate, soil, landforms, and minerals of Latin America in different ways. Each group has had a different technological base and thus has perceived the resource possibilities differently. As a result, each group has developed different patterns of settlement and economic activity.

Traditional Societies

Except for the deserts and highest mountain elevations, the environments of pre-Columbian Latin America had plant life and animal populations sufficient to support hunters, gatherers, and, in places, some form of agriculture. For this reason Indians were widely spread over the area at the time of Columbus. In the areas of advanced Indian culture—the highlands of Mexico and Peru—where some produce was exchanged with other areas, the economy of the Indians rested largely on a well-established local food-producing base.

Mineral deposits

□ Iron ore	✕ Lead
▲ Petroleum area	▼ Zinc
○ Coal	△ Tin
▽ Copper	♣ Gold
◇ Bauxite	∩ Silver

FIGURE 13-4 MAJOR AREAS OF MINERAL DEPOSITS IN LATIN AMERICA

The hot, wet tropical forests have always had sparse population. Although plant and animal life was plentiful in such areas, there were major environmental problems, particularly of disease from the abundant insect and microbe populations. The thick forest vegetation was difficult to clear for agriculture, and when it was cleared, the soil rapidly lost its meager supply of nutrients. Although Indian populations were spread over the wetter margins of both tropical and midlatitude grasslands, the densest populations prior to European settlement were in central Mexico and in the Andean highlands of what are now Peru, Bolivia, and Ecuador. Highland valleys were removed from the hot, wet climate of the tropics, the poor soil, and the insects, and agriculturally they were productive enough to support the Indian populations. In the lower highlands it was possible to grow subtropical crops such as corn, tobacco, and bananas, and at higher elevations to grow small grains and potatoes.

Traditional societies were hampered not only by inefficient methods of production, but also by their inability to store the fruits of the environment effectively. This was particularly true in the tropics where, although some products were harvested every month, it was impossible to keep them more than a few weeks because of animal pests and the rapidity of decay.

Modern Interconnected Society

Like the Indians, the early European settlers selected the midlatitude areas of Latin America or the highlands of tropical latitudes. As commercial development of particular resources expanded, however, it went into any area where it was economically feasible to exploit resources. There are mines in mountain valleys over 15,000 ft (4,575 m) above sea level and in the driest deserts; and there are plantations in some of the hottest, wettest tropical forests. If the financial returns are great enough, the modern interconnected system, with its technology and sophisticated methods of transportation, will bring together in remote places all the workers and materials necessary to support development.

Thus the crucial question for modern development is not what the environment is like but rather what its productive possibilities are. Potential for profit must be weighed against the cost of overcoming environmental obstacles. In this respect the wilderness areas of Latin America, such as the Amazon Basin and the high Andes, are similar to the Canadian North. Because development costs are so great, these areas will be opened only as their very rich deposits of raw materials are needed in the modern world.

Through much of the period of European settlement, important decisions about what to develop and where to live were made in terms of the larger interconnected system, centered first in Spain or Portugal. More recently, these decisions have been made in the United States, in European countries, and in Japan. Decisions in the past were often made by merchants and companies based in the industrial countries, but in the past twenty years, more and more of Latin America's development has been centered in its own metropolitan centers, particularly Buenos Aires, São Paulo, and Mexico City. These are developing subcenters of the modern interconnected system that increasingly are built around their own markets, farm and industrial producing areas, and financial institutions. They are the largest metropolitan areas of the largest Latin American countries.

The geographical location of Latin America, especially South America, has had an important effect on its functioning in the modern interconnected system. To the east, the nearest neighbors are the countries of Africa, which are themselves developing and therefore in most cases have little to contribute to development in South America. To the west lies the huge expanse of the Pacific Ocean. The major centers of the modern interconnected system lie far to the north—Europe to the northeast and the United States to the northwest. Distance has tended to isolate South America, but in recent years contacts with other countries—both developed and developing, in Europe, Africa, the Middle East, and the Far East (Japan)—have increased.

THE LATIN AMERICAN CULTURE

Latin culture, transported mainly from the Iberian Peninsula (Spain and Portugal) over several hundred years, can be found in modified versions from Argentina and Chile in the south to Mexico in the north. Although Latin America is a large geographic expanse, its cultural coherence is the primary reason for treating it as a region. Spain and Portugal dominated this entire region for over 300 years. Most of the countries have been independent for almost 150 years, but vestiges of earlier Spanish and Portuguese domination are still present in the government, the economy, and the social life of the region.

Except on a few islands once held by the French, Dutch, and English in the Caribbean, Spanish and Portuguese are the dominant languages. Roman Catholicism is recognized in most countries as the national church. Roots of an essentially feudal economic and political structure are found in most Latin American countries.

The Latin culture, so called because it is associated with languages derived from Latin, exists despite the great ethnic diversity among the people (Figure 5-6). In some areas native Indians are predominant, particularly in some Central American countries and in the Andean highlands from Colombia down through Peru. In others, such as the West Indies and the northeastern coast of Brazil, descendants of African slaves form a major component of the population. In Argentina, parts of Brazil, and Venezuela, English, German, and Italian roots are significant. In Guyana are areas that are dominantly East Indian. These variations in ethnic character do add diversity to Latin America; nevertheless, these peoples have found themselves dominated by the Latin American culture of Iberian origin. The Indians were conquered by the Spanish and Portuguese, and other ethnic groups were brought in to serve particular functions in an already established system. Centuries have passed since the arrival of other groups, yet the Indians have largely continued their traditional ways in the midst of the system brought from Spain and Portugal.

The Latin American countries and their peoples lack close ties, yet the Latin culture persists. The map of population density (Figure 5-4) implies that there is little contiguity between the populations of the various countries. The centers of population of adjoining countries are usually islands separated from one another by vast thinly populated stretches. Moreover, the map of world surface transportation (Figure 5-16) shows how few surface connections exist between Latin American countries. Like the former colonies of Africa and Asia, Latin American countries have their primary ties to Europe and the United States rather than to their neighbors. Each country stands largely in cultural and economic isolation from its neighbors.

MANY POLITICAL ENTITIES

Latin America is divided into twenty-five sovereign countries and several dependent territories (Figure 13-1). Most of these states are small in both area and population (Table 13-1). Nineteen out of the twenty-five had populations of fewer than 10 million in 1976—fewer people than the state of Illinois—and fourteen are smaller in area as well. Only four countries had populations larger than California. Brazil, exceptional in Latin America, accounts for 46 percent of the total land area and 33 percent of the total population, and is almost 50 percent larger in area than Mexico and Argentina combined.

CROWDED "POPULATION ISLANDS" IN A VAST SEA OF LAND

The average population density of Latin America is very low, but as Figure 13-5 indicates, population distribution is uneven. Most of Latin America has fewer than two persons per square mile and is, in fact, largely frontier.

TABLE 13-1　POPULATION CHARACTERISTICS OF LATIN AMERICAN
COUNTRIES, 1976

Country[a]	Area (Thousands of Square Miles)	Population (Millions)	Current Rate of Growth (Percent)	Population Under Age 15 (Percent)	Per Capita GNP (Dollars)	Density (per Square Mile)
Central America						
Mexico	762	62.3	3.5	46	1,000	82
Guatemala	42	5.7	2.8	44	570	135
El Salvador	8	4.2	3.2	46	390	525
Honduras	43	2.8	3.5	47	340	65
Nicaragua	50	2.2	3.3	48	650	44
Costa Rica	20	2.0	2.3	42	790	100
Panama	29	1.7	2.6	43	1,010	59
West Indies						
Cuba	44	9.4	1.8	37	640	213
Jamaica	4	2.1	1.9	46	1,140	525
Haiti	11	4.6	1.6	41	140	418
Dominican Republic	19	4.8	3.0	48	590	253
Puerto Rico[b]	3	3.2	2.4	37	2,400	1,066
Barbados	0.2	0.2	0.8	34	1,110	1,000
Trinidad and Tobago	2	1.1	1.5	40	1,490	550
Tropical South America						
Venezuela	352	12.3	2.9	44	1,710	35
Colombia	440	23.0	3.2	46	510	52
Ecuador	109	6.9	3.2	47	460	63
Peru	496	16.0	2.9	44	710	32
Bolivia	424	5.8	2.6	43	250	14
Paraguay	157	2.6	2.7	45	480	17
Brazil	3,286	110.2	2.8	42	900	34
Guyana	83	0.8	2.2	44	470	10
Surinam	55	0.4	3.2	50	870	7
Mid-latitude South America						
Uruguay	72	2.8	1.1	28	1,060	39
Argentina	1,072	25.7	1.4	29	1,900	24
Chile	292	10.8	1.7	39	820	37

SOURCE: Population Reference Bureau, Inc., *World Population Data Sheet, 1976*, Washington, D.C., 1976.
[a]Data not available for French Guiana or British Honduras.
[b]Nonsovereign country.

Persons per square mile (square km)

0-2 (0-0.8)

2-25 (0.8-10)

25-60 (10-23)

60-250 (23-96)

Over 250 (Over 96)

• Cities over 1 million

FIGURE 13-5 POPULATION DENSITY OF LATIN AMERICA

The majority of the population lives close to the seacoasts, and even there some areas are sparsely populated. Nowhere in Latin America does dense settlement extend as far inland as it does in the eastern United States and in the core triangle of the Soviet Union. Only in southern Mexico does an important population concentration extend from coast to coast, and there the distance between coasts is less than 300 miles (480 km).

Each population island forms the core of a different country except in Brazil, where there are several nuclei of population along the Atlantic coast. Population density varies sharply, with some of the highest densities occurring in the smallest countries. Moreover, relatively high densities appear in areas that suffer from arid environments and rugged, mountainous terrain, as in parts of Bolivia, Peru, and Colombia. Where the country is small and the population island occupies all, or almost all, of the territory, as in the West Indies, densities average 200 or more per square mile. Where the countries are large, the population islands occupy only a part of the total land area, and overall population densities are low. The highest density in Table 13–1 is for the tiny island of Barbados, which has an area of less than 200 square miles (320 square kilometers). Barbados is exceptional in that it has no empty, virtually unpopulated land.

MIGRANTS, NATIVES, AND THEIR OFFSPRING

The composition of the population in Latin America is very different from that in any of the other underdeveloped areas of the world. Here, as in Anglo-America, peoples of European origin outnumber the native Indian. Unlike Africa and Asia, Latin America is an underdeveloped region where the largest group is not the native peoples, but the descendants of European settlers. However, there has been much interbreeding between Europeans, Indians, and blacks.

Indians and Blacks

Indians are more prominent in present-day Latin America than in Anglo-America. The relatively pure Indian stock, however, is highly localized. The largest Indian populations are in the former centers of advanced Indian civilizations—the highlands of Mexico, Guatemala, and the Andes from Ecuador to northern Chile (Figure 13–6). Indians are still in the majority in Bolivia, Paraguay, and Guatemala and constitute almost half the population of Peru. The Indians of these countries form the one large group not closely integrated into Latin American culture. Most speak native Indian languages.

The black population of Latin America is also considerably larger than it is in Anglo-America. Like the Indians, this group is also sharply localized. The blacks today are descendants of slaves brought from Africa to work on plantations and in the mines in the days before modern machinery. Their distribution correlates highly with the areas of important early sugar plantations along the northeastern coast of Brazil and in the West Indies. Blacks are in the majority in Haiti, in Jamaica, and on most of the small sugar-producing islands such as Guadeloupe and Martinique. They account for almost half the population of Trinidad and Tobago, and a significant portion of the population of the Dominican Republic, Cuba, Puerto Rico, and Brazil. **Mulattos** (descendants of both black and white ancestors) total 60 percent of the population of the Dominican Republic and constitute large groups in Cuba, Puerto Rico, Trinidad, and Brazil.

Europeans

Whereas Indians and blacks are sharply localized in the Latin American population, Europeans are prominent everywhere. In Argentina, Costa Rica, and Uruguay Europeans have not mixed with other groups, but elsewhere they have contributed to the racial intermix. As mulattos are the result of interbreeding between whites and blacks, **mestizos** are the offspring of mixed white and Indian parentage.

Predominantly Indian

Predominantly European

Predominantly Negro

Mixed European and Indian

Mixed, with large proportion Negro

SOURCE: Adapted from *The World Today,* 3rd., ed., by D. F. Kohn and D. W. Drummond, copyright © 1971 by Webster/McGraw-Hill, Inc. Used with permission of Webster/McGraw-Hill, Inc.

FIGURE 13-6 ETHNIC ORIGINS FOR LATIN AMERICAN POPULATION

The origins of the European population in Latin America are very different from the origins of Europeans in Anglo-America. The settlers of the two Americas came not only from different parts of Europe, but at different times with different purposes.

During the colonial period in the sixteenth to eighteenth centuries Latin America was valued more highly than Anglo-America for its resources. The tropical environment provided sugar and other crops that could not be produced in midlatitude Europe. In addition, the Spanish conquered the two great Indian civilizations, the Aztecs in Mexico and the Incas in Peru, seizing ornate gold and silver art objects and leading to their further development of gold and silver deposits.

Latin American settlement also began much earlier than in Anglo-America. Europeans explored and settled these areas in the sixteenth century before any permanent colonies were established farther north. Mexico City, Lima, and Havana were major cities when the Jamestown (Virginia) colony was founded in 1608. Moreover, European migration to Latin America was more than twice as large as that to North America. An estimated 14 million Europeans entered Latin America from the fifteenth to the eighteenth centuries, whereas only 6 million Europeans migrated to Anglo-America in the same period.

In the nineteenth and early twentieth centuries, however, the situation was reversed. Anglo-America, which offered productive midlatitude agricultural lands and a rising industrial economy, became the land of opportunity for European immigrants. Between 1821 and 1932 an estimated 40 million Europeans entered Anglo-America; only 16 million went to Latin America. Moreover, immigration to the southern areas was sharply localized. Well over half the Europeans went to either Argentina or Brazil, particularly to the areas with midlatitude environments.

Since World War II moderate migration into Latin America has continued, again primarily to Argentina and Brazil but also to Venezuela. To an important degree these immigrants have chosen centers of economic development. Latin Americans in turn have emigrated to the industrialized parts of the world. Large numbers of Cubans and Puerto Ricans have entered the United States, and Jamaicans and other West Indians emigrated to the United Kingdom until limitations were established in 1968.

THE EVOLUTION OF THE PRESENT DUAL SOCIETIES

From the beginning the interest of the Spanish and Portuguese in Latin America was primarily commercial. There is little evidence of settlement groups seeking freedom of religion or trying to establish a utopian society. The expeditions of the sixteenth century were after riches—gold and silver, and exotic forest products such as cacao and dyewood. Like Columbus, the first explorers were content to trade with the Indians and attempted no permanent settlements. The lure of gold, however, brought settlements to the islands of Hispaniola, Cuba, and Puerto Rico in the Caribbean, and these were followed by the conquests of the Aztec Empire of Mexico and the Inca Empire in Peru and Bolivia. These centers of Indian civilization offered accumulated treasures and wealth, which could be looted, and mineral deposits whose locations were already known. They became the focal points of Spanish interest and the chief sources of the wealth of the empire (Figure 13–7).

THE COLONIAL PERIOD: MERCANTILISM

In the colonial system the commercial connections between Spain and its Latin American colonies were closely controlled. All trade had to be with the mother country and, in fact, could only pass through the Spanish ports of Sevilla and

NEW
SPAIN

Havana 1519

CUBA

HISPANIOLA

Mexico City 1521 • • Veracruz 1519

WEST INDIES

CENTRAL
AMERICA

Cartagena
Portobelo 1533
1597

• Caracas
1567

1519
Panama City

• 1538
Bogotá

NEW GRANADA

GUIANA

EQUATOR

PERU

• Lima
1535

BRAZIL

• Potosí
1545

TROPIC OF CAPRICORN

LA PLATA

• Rio de Janeiro 1567
São Paulo 1554

PERU

Santiago
1541

Buenos
Aires • Montevideo 1726
1580

PATAGONIA

TROPIC OF CANCER

1000 Miles
1000 Kilometers

40°

20°

20°

40°

120°   100°   80°   60°   40°   20°

SOURCE: R. A. Harper et al., *Learning About Latin America,* Silver Burdett, Morristown,
N.J., 1964

Spanish lands

Portuguese lands

FIGURE 13–7 COLONIAL MAP OF LATIN

AMERICA, 1800

Dates are founding dates of cities

Cádiz. Moreover, all trade in the New World had to move through one of several seaports: Veracruz in Mexico and Portobelo in Panama for goods from the Incan lands; Cartagena in Colombia, San Juan and Santo Domingo for cargoes from Spain to the West Indies. Havana was a shipping port to Spain.

The Portuguese found little wealth in Brazil for there was no Indian civilization and there were no gold deposits close to the coast. Sugar plantations were established along the coastal lowland around what are now the port cities of Recife and Salvador, and by 1600 Brazil was the world's chief supplier of sugar. This was the first settlement in Latin America based on plantations and large-scale agricultural exports. Portugal was not able to control the sugar trade, and it was handled by Dutch and English merchants. Soon sugar plantations were developed in the Caribbean islands as well.

Although Spanish **mercantilism** attempted not only to restrict all trade to its new colonies but to supply all manufacturing goods as well, the need for tools, clothing, and buildings resulted in the development of important manufacturing trades in the busy colonial centers. Still, imports from Spain remained important.

The growth of the export centers also provided a market for other parts of the colonies. Food staples normally were produced in the immediate lands around most commercial centers, but in the case of the miners in the highlands of Bolivia and Peru, food and textiles were carried overland from what is now Argentina. In the same way, cattle raisers moved into the grasslands of northeastern Brazil from the coast to supply meat to the coastal sugar plantations. Hides and skins from the operation were shipped to Europe.

Labor Supply in the Early Colonial Period

Natives, rather than Europeans, were recruited for work on the plantations and in the mines. Those in the mines were commonly Indian slaves, while agricultural laborers worked under the **encomienda,** the antecedent of the hacienda. An encomienda gave the owner not only a vast tract of land but a certain number of Indians as well. Under the encomienda the Indians were not slaves. Rather, they were entrusted to the landowner, who was supposed to teach them the Catholic faith and provide various forms of protection and help. Essentially, no wages were paid, and the bulk of the population lived at or near subsistence level. In the sugar lands the supply of native labor proved to be inadequate, and soon African slaves were being imported. In all, 14 or 15 million Africans were brought to these areas, far more than the million brought to the United States.

The Hacienda

The Latin American heritage came from preindustrial Europe. Modern mining, plantation, and industrial developments in Latin America that are a part of the interconnected system were superimposed on an essentially rural society in which political and economic power was in the hands of the landowners. At the time of early settlement, land was given out in large tracts to Europeans in favor at the courts of Spain and Portugal or in governmental circles of the colonies themselves. Many of these lands and the wealth they have generated remain under the control of the descendants of these first seigniors.

The basis of Latin American life has been the **hacienda** (or estancia) in Spanish-speaking lands and the **fazenda** in Brazil. The hacienda owner and family lived in luxury in a large dwelling while the workers were virtually bound to the land in a self-contained community that included meager living quarters and also a store and a chapel. Nominally free, the workers were in fact tied to the land by indebtedness to the owner. They were given supplies and food on credit by the hacienda owner, but they could never make enough money to pay off their debts and leave. They paid their obligations by providing labor. They received land and possibly a shack, and tried to eke out a subsistence living on the side after they finished their work on the hacienda.

The hacienda had much the appearance of Anglo-American plantations, but in fact, it was more like a feudal fiefdom. It was never organized for efficient commercial production. Like the feudal landowner, the hacienda owner lived off the workers' production. The hacienda was geared to produce staples for local consumption rather than specialty production for export.

Officially, the hacienda has been done away with in many Latin American countries, but, like caste in India and racial discrimination in the United States, it continues as an important way of life in the culture. Like the estates of feudal Europe, Japan, or czarist Russia, the hacienda was part of a system in which the ruling minority lived well and enjoyed power, while the masses who did the work remained very poor and politically impotent. Governments were drawn from the landed class; sons of the landed gentry joined the Catholic clergy or the army. In this way the wealthy controlled all facets of society.

Developmental Centers

The export centers—Cuba, Mexico, Panama, Venezuela, and Peru in Spanish-speaking Latin America, and the Portuguese colony of Brazil— were the focuses of European settlement in the New World of Latin America. Other areas, such as the present-day countries Guatemala, Chile, and Argentina, were the poorly developed fringes of colonial times.

Large cities emerged in the export centers. Mexico City, Lima, Havana, and Potosí, in the Bolivian mining district, were cities with tens of thousands of people before the first settlements in Anglo-America. Here the modern interconnected system first touched the Americas. These cities were the focuses of administration and trade and were important points in the links between the colonies and the mother countries. They contained the palaces of the governing viceroys and the homes of the wealthy colonialists.

INDEPENDENCE WITHOUT DEMOCRACY OR INDUSTRIALIZATION

Most of the Latin American countries gained independence before 1850. The revolts from Spain came after the U.S. Revolutionary War, but generally they were not attempts by the people as a whole to gain political control democratically. Rather, they were revolts by the landed gentry, the hacienda owners, who wanted freedom to manage their own lands and to govern themselves. This wealthy, powerful minority has maintained control of most countries ever since.

During the period from independence until World War I, Latin America developed differently from Anglo-America. In the United States this was a time of regional integration and specialization. New transportation and communication developments were used to tie the parts of the country into a series of specialized components, while considerable political and cultural unity was building. Latin America responded to the changes brought about by the Industrial Revolution in Europe in a very different way, probably because the former colonies emerged as politically and socially weak entities. Instead of becoming economically independent of European markets, Latin American countries grew more closely tied to those markets than they had ever been before.

As we have seen, the specialization of European countries in industry, trade, and finance increased their dependence on raw materials and foodstuffs from other parts of the world. Growing affluence and reduced transport costs lowered prices on European markets and increased demand. As a result world trade increased fivefold between 1820 and 1870, and fivefold again between 1870 and World War I. The classic pattern of European imports of raw materials and food and exports of manufactured goods was established.

In response, Latin American population centers, which already had come to depend on European markets during colonial days, sharply expanded their role as suppliers of raw materials. Colonial restrictions on trade were gone, and the new landlords who ruled the countries were anxious to satisfy their desires for European consumer goods through imports. New plantations and mines and even new producing regions came into being. In the latter part of the nineteenth century the major exporting centers of Latin America were established—the coffee lands near São Paulo, the sugar plantations of Cuba, the banana plantations in Central America, the nitrate and copper mines in northern Chile, and the cattle and grain-producing areas of Argentina. Oil production in Venezuela began early in the twentieth century. These developments, on a scale never before seen in Latin America, brought a new demand for labor and capital.

In the nineteenth century slavery was rejected in Latin America as it was in the United States. After it was outlawed, a new supply of labor came from European immigrants. In Brazil and Argentina the new wave of immigration came not just from Spain and Portugal, but also from Italy, Germany, and other European countries. As in the days of industrial expansion in the United States, European commercial interests invested large amounts of capital in these underdeveloped areas in order to establish vital trade links.

In Brazil and Argentina, European capital was not put to the plantations and estates held by the wealthy landowners. Rather, investments were made in railroads, public utilities, and other aspects of the infrastructure necessary to get products to ports for shipment overseas. Complete control was established only where major organization was needed, such as in the tropical banana plantations or in the desert wastes of northern Chile where large-scale mining operations were started. Such areas lacked local entrepreneurs who could handle the task. Investments were made by early multinational corporations seeking raw materials to supply manufacturing in the industrial countries.

Like the United States, Latin America was developing regional specialties that fitted the new long-range, interconnected system made possible by the Industrial Revolution. The organizational core of the system, however, remained in Europe. Thus while local entrepreneurs benefited, often handsomely, by organizing coffee, sugar, and livestock production, and while regional cities like São Paulo, Havana, and Buenos Aires emerged as the business centers of production, the decision makers—shippers, financiers, and dealers at commodity markets—were in Europe.

Moreover, each of the specialized producing regions was in a different country. Where export development was great and local production remained in the hands of the landowners, the population island grew and became prosperous, as in the coffee area of Brazil. Countries where this occurred emerged as the more advanced in Latin America. But where developments were smaller and all phases remained in the hands of foreign capitalists, the countries remained weak.

The Ruling Oligarchy and the People

In the nineteenth century economic and political control in most Latin American countries was similar to that in Japan in the same period. In both, power was in the hands of the small landowning class, which controlled not only the export-producing units but also the government. The results in Latin America, however, were very different. Unlike the Japanese oligarchy, Latin American leaders in the various countries were content to be a part of an externally controlled system. They fed their specialties into the system and purchased the desired manufactures and luxury goods from overseas; the payoff for them was handsome, and they found the arrangement satisfactory. In Latin America wealth was seen in preindustrial terms—one became rich and gained power through controlling land.

As a result, little was invested in consumer manufacturing in Latin America. Most of the industry that did develop was concerned with processing export products: refining sugar, processing meat, and smelting and refining metal. Moreover, even these industries were largely in the hands of foreign investors. As in Japan, the mass of the population received little return from the new economic developments. They were Indians, freed slaves, or just subsistence squatters on the land. Or, if they worked on haciendas or plantations, they received little more than a shack and a plot of land. Since they were poor and uneducated, they did not provide a market for the products of industry. The hacienda system lacked the Japanese tradition of industriousness, worker loyalty to the system, and the strong work incentive that was found in Japanese feudalism. Hacienda labor had been inefficiently utilized, and there was a tradition of underemployment.

In any case, Latin American leaders largely chose to ignore the industrial sector of the economy; they were satisfied to depend on exports of raw materials and food specialties. Instead of encouraging industry through financial support and protective tariffs as Japan did, Latin American countries were content to encourage greater and greater importation of manufactured goods. Thus although the growth of export markets in most parts of the region during this period provided money that might well have been invested in establishing an industrial base, such an undertaking was not considered. Of course, most countries had much smaller populations and capital bases than Japan, but even in Mexico, Brazil, and Argentina, the industrial sector lagged.

Centers of Economic Activity

The size of export development was reflected in the size of the large cities in Latin America. In 1920 two cities, Buenos Aires in Argentina and Rio de Janeiro in Brazil, had populations of over 1 million people. These were not only national capitals, but also the headquarters of the two countries most closely tied to the new economic developments in Latin America. Three other cities had populations over one-half million: Mexico City, Santiago, and São Paulo. Other large centers were Montevideo, Uruguay; Havana, Cuba; and Rosario, in the agricultural region of Argentina.

The shift in the major centers of export production from colonial times was striking. Not only did the coffee area of Brazil and the meat and grain sections of Argentina and Uruguay gain major stature, but two major centers of colonial times declined in importance. In Brazil the sugar plantation owners were slow to convert to modern milling techniques developed in the nineteenth century, and there was also a shift from primary interest in gold and silver toward industrial minerals. Peru and Bolivia participated in the new mining with production of copper and tin, but these developments did not restore to these areas the relative importance they had during the colonial period.

THE TWENTIETH CENTURY: THE BEGINNINGS OF INDUSTRIALIZATION

World War I, the depression of the 1930s, and World War II produced major breakdowns in the long-range, interconnected systems of Europe and the United States, thus greatly affecting Latin American producers of raw materials and their governments. During the wars, as the European countries undertook all-out war efforts, the demand for Latin American raw materials increased. But because the economies of the European nations were so preoccupied with war material, the supply of imports to Latin America broke down. The depression that followed the 1929 stock market crash brought a severe drop in the value of export trade. The total trade of Chile declined 85 percent in value between 1929 and 1932, and seven other countries in Latin America experienced drops of more than 65 percent.

In World War I the need for domestic supplies of goods normally imported had caused some of the larger countries, particularly Argentina and Chile, to embark on increased domestic manufacturing. But prior to World War II no Latin American country had a really significant base of heavy industry in steel, metalworking, chemicals, or cement production.

The long drought of imports from 1930 through 1945 changed the thinking of political and economic leaders in Latin America. First because the export trade was depressed during the 1930s, and then because U.S. and European manufacturing centers were disrupted during World War II, for some fifteen years Latin American countries had to cut back their desires for consumer goods and the equipment needed to expand their production facilities.

As a result, since then the governments of the largest Latin American countries have had programs encouraging the development of modern industry in their countries. Brazil, Argentina, Mexico, Colombia, Chile, and even Peru and Venezuela have their own steel mills and basic metalworking plants. Many countries, including Brazil, Mexico, and Argentina, have encouraged automakers from industrial countries to establish assembly plants and have been urging (in some cases, requiring) companies to produce more and more of the component parts in their countries. North American retailers such as Sears Roebuck have been allowed to enter Latin American national markets with the agreement that an increasing share of the products sold will be made domestically.

The new manufacturing has been designed almost exclusively to provide goods for the growing domestic market, not to provide new export products. Thus it cuts down the demand for imports rather than increases exports. The result is that the export trade of Latin American countries is essentially the same as in colonial times—the export of primary products of the resource base, manufactured only to the degree necessary for shipment.

Actually, except for petroleum, raw material exports from Latin American countries in the mid-1960s were slightly less than in 1929. Former colonial territories receive preferential treatment from the European Economic Community countries. In large measure Latin America has suffered from not having been recently colonial, and African exporters have prospered in the European markets at the expense of Latin Americans. Also, most Latin American countries have provided little in the way of research assistance and long-range governmental planning to raw material producers.

Changes in the Trade Structure

As Table 13-2 shows, considerable shifts in the relative importance of particular export commodities have, in turn, affected particular producing areas. Today Europeans are less interested in midlatitude staple foods—wheat, corn, and meat. They have higher per acre yields, depend on more intra-European trade, and draw imports from Canada and Australia. At the same time, with the population increase at home, Argentina now consumes three times as much beef as it exports. Exports of agricultural raw materials, particularly wool and hides, are only half the totals of the early 1930s. Lead, zinc, tin, and nitrate exports are also lower today than forty years ago.

The greatest gains by far have come in petroleum exports. Crude oil production in Venezuela in 1971 was ten times that of the early 1930s. Over 90 percent of Venezuelan oil was exported. Iron ore exports, though much less in total, also increased sharply in the past forty years. Over 80 percent of this output comes from Brazil, Venezuela, and Chile.

One remarkable new export industry in the period since World War II has been the fish meal and fish oil industry of Peru and, to a lesser degree, Chile. The industry is based on the demand for high-protein livestock feed in the United States and Western Europe and the presence of vast quantities of a single variety of fish,

TABLE 13-2 RELATIVE IMPORTANCE OF MAJOR EXPORT
COMMODITIES OF LATIN AMERICA (*Percent*)

Commodity	1934–38	1946–51	1963–64	1972
Temperate foodstuffs	17.2	10.8	8.2	15.1
Wheat and flour	5.1	4.2	1.7	1.3
Maize	6.3	2.0	2.0	2.3
Meat	5.4	4.0	4.0	10.0
Butter	0.1	0.2	0.1	—
Cattle	0.3	0.4	0.4	—
Soybeans	—	—		1.5
Tropical foodstuffs	21.3	30.0	21.2	30.2
Coffee	12.8	17.4	15.0	15.0
Cocoa	1.2	1.6	0.8	1.3
Sugar[a]	6.1	10.2	3.3	9.4
Bananas	1.2	0.8	2.1	4.5
Agricultural raw materials	12.6	11.8	7.6	6.8
Cotton	4.6	4.7	4.3	5.4
Wool	4.3	3.7	2.0	1.4
Hides	3.5	3.2	0.5	—
Oils and oilseeds	0.3	0.2	0.8	—
Forest products	1.0	2.3	1.0	—
Timber manufactures	0.3	1.2	0.8	—
Quebracho	0.7	1.1	0.2	—
Fishery products	0.0	0.1	2.4	2.9
Petroleum	18.2	17.3	26.4	30.1
Iron and steel	0.0	0.2	2.9	5.5
Copper	4.7	3.4	4.9	0.1
Lead	1.5	1.3	0.7	0.04
Zinc	1.0	0.6	0.6	0.05
Tin, predominantly ore	1.6	1.0	0.5	0.1
Nitrates	1.4	0.8	0.3	—
Total of above products	80.5	79.6	76.7	90.9

Sources: Grunwald and Musgrove, *Natural Resources in Latin American Development*, p. 21.
Published for Resources for the Future by Johns Hopkins University Press. Copyright © 1970 by the
Johns Hopkins Press. All rights reserved. *World Trade Annual, 1973.*
[a]Excludes Cuba; this affects most of the data for sugar.

the anchovy, in the cold waters off the Pacific coast. The anchovy fishing industry and processing plants have risen to serve the export market. Oil is extracted in processing plants along the coast, and it is exported, along with the resulting fish products, particularly to the Netherlands, West Germany, and the United States. In recent years, however, the fish population of the Pacific fishing grounds has dropped rapidly, either because of environmental changes or because of overfishing. The result has been disastrous for the Peruvian economy.

Through the years the total export of tropical agricultural specialties has stayed about the same, but coffee and banana exports have increased and sugar shipments have declined. Coffee production in Brazil remained steady until 1975, when it dropped sharply because of a severe frost that killed a large percentage of the coffee bushes. Colombia has doubled its coffee output, and Mexico has quadrupled its production. The small Central American producing countries have had increases of 50 to 300 percent, but the most significant expansion has come in Peru and Ecuador, insignificant coffee producers forty years ago. Brazil, Colombia, and Mexico have important domestic markets, but these account for only a tiny fraction compared to exports. Banana exports, too, have doubled, because of a growth in demand, especially from Japan. Ecuador now provides over one-third of the total Latin American banana exports.

Localization of Trade

The major export trade of Latin America is concentrated in a few centers. Table 13-3 indicates that almost 75 percent of total exports come from just eight of the twenty-four major producing countries: Venezuela, Brazil, Argentina, Mexico, Chile, Peru, Cuba, and Colombia. Venezuela, a leading source of petroleum in the world, is by far the most important exporting country, with almost 18 percent of the total exports for the twenty-four major producing countries of Latin America. It is followed by Brazil, Argentina, and Mexico, each exporting approximately one-half to two-thirds the amount that Venezuela does. Together, these three countries total 35 percent of the major countries' exports.

The large Latin American countries, particularly those in South America, dominate the area's exports. The eight largest exporters are the eight largest countries in population, and all but Mexico and Cuba are in South America. However, the sizes of their export trade are not directly correlated with the sizes of their populations. Venezuela, the leading exporter, is sixth in population. Argentina, with less than half the population of Mexico, is a larger exporter. Colombia, with twice the population of Cuba, Chile, or Venezuela, has the fewest exports of the eight. All this indicates the existence of economies with varying degrees of dependence on export trade, and hence varying degrees of involvement in the modern interconnected system. Export trade appears to be much more important to Venezuela and Argentina than to Peru or Colombia, and similar variations can be seen in the smaller countries.

Most countries of Latin America depend on one, two, or three commodities for virtually all their exports (Table 13–3). This means that only a few parts of the country are tied into the export system. A few producing areas have rail or highway links to a port or a handful of ports. This is the case not only with the small countries but also with major exporting countries including Venezuela, Chile, Cuba, and Colombia. These producing districts and their connecting infrastructure are separate islands within each country. The whole producing network is designed not to pull the resources of different parts of the country together, but to deliver particular, specialized products to ports for shipment overseas. This is a classic example of the commercial export component of the dual economy.

TABLE 13-3   EXPORT CENTERS OF LATIN AMERICA

Total Value of Exports, 1971 ($ millions)

Venezuela	3,130	Trinidad and Tobago	563	Honduras	188
Brazil	2,904	Jamaica	343	Nicaragua	183
Argentina	1,740	Guatemala	290	Guyana	135
Mexico	1,502	Dominican Republic	243	Panama	120
Chile	1,247	El Salvador	228	Paraguay	65
Peru	892	Costa Rica	225	Haiti	46
Cuba	859	Bolivia	222	Total	17,770
Colombia	732	Uruguay	206		

Leading Exports, 1974 (Percentages of Total Exports for Each Country)

Venezuela		*Cuba*		Bananas	22
Crude oil	64	Sugar	74	Meat	9
Petroleum products	32	Nonferrous metals	16	Chemicals	7
Brazil		*Colombia*		Sugar	6
Sugar	12	Coffee	47	*El Salvador*	
Coffee	11	Emeralds	7	Coffee	45
Iron ore	7	Cotton	6	Cotton	10
Mexico		*Trinidad and Tobago*		*Bolivia*	
Nonferrous metals	13	Petroleum products	61	Tin	42
Textiles	9	Crude oil	29	Crude oil	19
Chemicals	9	*Jamaica*		Zinc	7
Sugar	7	Alumina	52	Antimony	5
Machinery	6	Bauxite	20	Silver	5
Cotton	6	Sugar	12	*Uruguay*	
Coffee	6	*Guatemala*		Meat	38
Argentina		Coffee	28	Wool	23
Meat	24	Cotton	11	Hides and skins	6
Corn	11	Sugar	11	*Honduras*	
Wheat	8	Bananas	6	Bananas	27
Wool	6	*Dominican Republic*		Coffee	16
Animal feed	5	Sugar	53	Timber	16
Chile (1973)		Cocoa	8	Meat	9
Copper	86	Coffee	7	*Nicaragua*	
Iron ore	7	Tobacco	6	Cotton	36
Peru		*Ecuador*		Coffee	12
Copper	23	Crude oil	58	Meat	6
Fish meal	13	Bananas	12	*Guyana*	
Silver	11	Cocoa	10	Bauxite	48
Zinc	11	Coffee	6	Sugar	31
Sugar	10	*Costa Rica*		Alumina	9
Cotton	6	Coffee	29	Rice	9

SOURCE: *Encyclopaedia Britannica, Book of the Year, 1976,* Chicago, 1976

THE MEASURE OF ECONOMIC DEVELOPMENT

Use the figures given in Table 13–4 to determine the relative economic development of Latin American countries. Determine which countries are most integrated into the modern interconnected system. What countries are least involved? Is the range great?

1. We might hypothesize that the higher the relative importance of export trade to a country, the more a part of the modern interconnected system it is. Does this hypothesis hold up? For what countries? If it does not work in some countries, which are they? Why might they not fit?

2. If export trade is not a satisfactory measure, what measure of involvement in the modern interconnected system would you suggest?

3. Consider the eight largest countries in population and per capita GNP (Table 13–1 and the Appendix). Is the modern interconnected system well developed in each of them? What variations do you see? Do all of them function primarily as exporters to other countries, or do some seem to have internal industrial development?

4. Brazil and Mexico each have between one-third and one-half of their work force engaged in agriculture,

yet they are the two leading Latin American countries in total industrial energy consumption. How do you explain that?

5. Classify the countries as (a) those that are predominantly traditional, (b) those that are important suppliers of exports to other countries, and (c) those with important beginnings of industry of their own.

TABLE 13-4 INDEXES OF ECONOMIC DEVELOPMENT FOR THE LARGEST LATIN AMERICAN COUNTRIES, 1972

Country	Per Capita Value of Exports (Dollars)	Per Capita GNP (Dollars)	Per Capita Industrial Energy Consumption (Pounds)	Percent of Work Force Employed in Agriculture	Percent of National Income Derived from Agriculture
High exports
Venezuela	345	980	5,452	22	7
Moderate exports
Cuba	92	530	2,574	42	25[a]
Chile	82	720	3,342	21	9
Argentina	81	1,160	3,809	15	12
Peru	65	450	1,371	43	16
Low exports
Brazil	40	420	1,173	44	16
Mexico	35	670	2,905	39	11
Colombia	33	340	1,345	44	27

SOURCE: Inter-American Development Bank, *Economic and Social Progress in Latin America, 1973*, Washington, D.C., 1973.
[a] Estimate.

Not all countries are tied to one or two export specialties. The exports of Brazil, Mexico, Argentina, and Peru include a half-dozen or more major types, not all of which are listed in Table 13–3. In large countries such as these with diverse environments, there is a series of different, independent export systems. Brazil is a case in point. The coffee district close to the Tropic of Capricorn has traditionally been the most important, but the sugar, iron, forest-products, and banana areas are all in different parts of the country. The products of each of these districts move out of the country along separate routes. The producing regions are not tied together to contribute to an integrated national economy, as in the industrial countries discussed in previous chapters.

Tourism and the External System

A new sort of dependence on external markets that does not appear in the export totals has emerged in Latin America since World War II. Tourism represents a new evaluation of the environmental resources of the region by developers in the United States and Europe. With increasing affluence, shorter working hours, paid vacations, and jet air transportation, the market for long-range vacation travel has burgeoned. Latin America, especially Mexico and the Caribbean, has become the major tropical center of the U.S. travel industry.

Cuba and Mexico have long been favored by tourists from the United States. Ideological differences between Cuba and the United States have closed that country to tourists but jet travel has opened up the more distant islands of the Caribbean. Not surprisingly, foreign investors have speculated in tropical island real estate and developed hotels and resorts throughout the Caribbean. The influx of large numbers of tourists with money to spend has given an important boost to local economies, but a large share of the profits are siphoned off by external owners of major facilities, making this just one more example of resource-oriented activity controlled by

outside interests. In this case, however, instead of the products of the environment being transported to other countries, tourists are taking over sections of the Latin American landscape from the local people and making them their own.

THE DUAL SOCIETIES TODAY

Although each Latin American country is a combination of traditional, locally based societies and the modern interconnected system, various degrees of modernization and traditionalism can be seen in the region. Some countries are greatly involved in exporting specialized products from mines and plantations to Europe and Anglo-America; some have large populations that are predominantly traditional; and some have established significant bases of modern industrial production.

EMERGING INDUSTRIAL CENTERS

Mexico and Argentina have relatively well developed economies, having moved from the colonial dependence on outside markets toward regional integration within the respective countries. Both rank high in indexes of development (see Appendix). High energy consumption is a measure of industrial development, and high GNP indicates the affluence that goes with it.

Table 13–5 gives further indication that Mexico, Argentina, Venezuela, Brazil, Colombia, Chile, and Peru have the beginnings of modern industry. Not only do they have the largest electrical consumption in Latin America, but they also have steel production and, except for Colombia, automobile assembly. This contrasts sharply with other Latin American countries, which may produce cement and have textile mills and agricultural processing plants, but depend on overseas sources for all the products of modern industry. These other countries also have much

TABLE 13-5 INDUSTRIAL PRODUCTION, 1972

Product	Major Industrial Countries							
	Mexico	Brazil	Argentina	Venezuela	Colombia	Chile	Peru	Cuba
Coal (thousands of tons)	2,170	2,479	675	40	2,800	1,332	—	99
Crude oil (thousands of tons)	22,163	8,197	22,130	168,066	10,134	1,613	3,194	117
Electricity (millions of kilowatt-hours)	34,457	53,767	25,319	14,656	10,300	8,934	5,949	5,208[a]
Pig iron (thousands of tons)	2,778	5,300	849	537	286	489	171	140
Steel (thousands of tons)	4,396	6,518	2,151	1,127	275	581	192	800
Cement (thousands of tons)	8,753	11,381	5,454	2,508[a]	3,006	1,404	1,428	1,474
Autos and trucks (thousands of vehicles)	233	613	268	52[a,c]	25[c]	26[c]	24[c]	—

Product	Other Countries										
	Guatemala	Ecuador	Dominican Republic	El Salvador	Panama	Bolivia	Nicaragua	Costa Rica	Honduras	Paraguay	Haiti
Electricity (millions of kilowatt-hours)	830[a]	1,117	1,201	820	951[b]	872	649[a]	1,266	310[b]	273	114
Cement (thousands of tons)	235[a]	482	678	218	181[b]	65	150[a]	261	134[b]	75	89

Source: *Encyclopaedia Britannica, Book of the Year, 1976*, Chicago, 1976.
[a] Figures are for 1971.
[b] Figures are for 1970.
[c] Assembly of autos and trucks.

BRAZIL'S ECONOMIC GROWTH IN THE 1970S

Brazil in the 1970s has been referred to as the "Japan of Latin America." Its government is intent on industrial development and a rapidly expanding industry producing both for a growing internal market and for export. The industrial growth rate was 11 to 15 percent per year between 1970 and 1973, and fixed industrial investment almost doubled between 1970 and 1974. Brazil now accounts for over a quarter (28 percent) of the total output of manufactures for Latin America, and exports of manufactured goods have more than tripled since 1970.

As in Japan, this growth has been by design of the ruling powers. In Brazil the ruling elite, consisting of a mix of military and civilian personalities, has encouraged industrial development while repressing political freedom. The limitations on political freedom are regarded as a necessary evil to accomplish economic and social development goals. The regime has protected Brazilian manufactures with tariff barriers and encouraged foreign investment with a carefully thought out battery of tax incentives. Particularly important, Brazil attracted large multinational corporations such as the major auto producers, first by requiring the establishment of a local assembly plant for the sale of cars in the Brazilian market, then by asking the manufacturers to produce more and more of the sophisticated components used in assembly in Brazil, and finally by encouraging firms to export products from Brazil.

As a still largely underdeveloped country, Brazil promoted its advantage as an export platform for Brazilian goods—such as shoes, soluble coffee, clothing, and furniture—produced by cheap Brazilian labor. Soon Brazilian auto plants were exporting engines to West Germany and the United States.

Brazilian industry still has a long way to go. Modern industry is largely concentrated in the south around Rio de Janeiro and especially São Paulo. Despite more than fifteen years of efforts to bring industry to the drought-plagued Northeast, the region lags farther behind the core area than it did before. Tens of millions of Brazilians still are largely outside the modern interconnected money economy and the rich resources of the North and West remain largely undeveloped. Vertical integration of industry even in the major industrial region is not complete; not all stages of manufacture and marketing, from raw material to end use, are in Brazilian hands. Brazilian iron ore still moves to Japan to be made into steel and then is returned to São Paulo to be rolled into sheets for the country's auto industry. Wages remain low, and even in São Paulo almost three out of four people live on inadequate diets.

Brazil is investing heavily to overcome its problems. In addition to the development program for the North-

east, it has invested heavily in roads and other infrastructure to open the Amazon North and the far western parts of the country. Before the 1973–1974 petroleum crisis, highways rather than railroads were built because the government is required to build only the roads themselves; the transportation equipment is supplied by the private sector of the economy. If railroads had been built, the government would have had to finance both the railroads themselves and the rolling stock and service facilities. Since the crisis, however, the government has continually reassessed its decision and has given railroad investment revitalized attention. Steel production increased more than three times between 1970 and 1974 and immediately ahead lies a $10 billion investment program to expand and develop the industry further. Also, a massive program of social investment in schools, houses, medical services, sanitation, and basic food production is under way.

It could be said that Brazil is the United States minus one hundred years in terms of development. It has a population of over 100 million people and a large territory, but it has not yet developed an internal functioning system of interrelations between its regions. Do you think this is a valid analogy? What do you see as important differences in the two cases? What do you think Brazil's prospects are?

Where do the workers needed in

the booming factories of the São Paulo area of Brazil come from? Migrants from the poorer rural regions arrive in São Paulo at the rate of one every eight minutes, more than 60,000 per year. Most are untrained, even illiterate; thus a shortage of skilled workers still remains. Qualified personnel from the executive level downward are lacking. University professors, boiler workers, nuclear physicists, cooks, and salespeople and other skilled workers are in short supply. As a result, the government has marked $20 million per year for a special intensive training program to prepare 450,000 people for skilled and semiskilled jobs. In addition, persons with inadequate skills find in the military a training institution that prepares them for factory work and business administration.

Brazilian workers are still very poorly paid. Most work at the minimum wage, which, despite a 41 percent increase in 1975, is $67 per month in São Paulo. Workers on an auto assembly line in São Paulo receive about $100 per month, less than half what they would receive for the same job per week in the United States. Yet these are good wages for Brazil, where it is estimated that as many as 6 million people either earn no money from farming or work for money less than fifteen hours per week. They total more than all the industrial workers in the country.

The sudden increases in petroleum prices enacted by the OPEC countries in 1973–1974 had an adverse impact on Brazil's economic prospects.[1] Despite some oil production within the country, Brazil relies heavily on imported oil. Petroleum is especially important to a country that is expanding its transportation infrastructure and its industrial capacity. It is needed not only as a fuel, but also as a raw material in petrochemical industries.

Brazil had been quite successful in managing its petroleum needs. A state enterprise created in 1954, Petróleo Brasileiro (generally known as Petrobrás), controls all petroleum exploration and production. Its staff includes many Brazilian-educated engineers, geologists, and geophysicists. Between 1954 and 1972 Petrobrás increased Brazil's domestic refining capacity from less than 9,000 cubic meters per day, in a single refinery, to more than 100,000 cubic meters per day, mostly in six refineries operated by Petrobrás. The objective was, of course, to reduce the net cost of petroleum to the country by reducing the amount imported in refined form; imported crude oil is significantly cheaper than imported refined petroleum.

In the course of developing its petroleum program, Brazil expanded its contacts with Africa and the Middle East. A subsidiary of Petrobrás, formed in 1972, has been exporting Brazilian technology to other developing countries. Arrangements have been made with Colombia, Iraq, the Malagasy Republic, Egypt, and Algeria for Brazilian participation in petroleum exploration and development in those countries. Another subsidiary of Petrobrás has established relations with oil-rich Nigeria.

Despite an effective and well-managed development program, Brazil's economy was hit hard by the price increases imposed by the OPEC countries. Its trade deficit was worsened, and the inflation rate, which had begun to decline in the years preceding 1973, rose again. Inflation had been 22 percent in 1969, was down to 15 percent in 1973, but rose to about 28 percent in 1974. To sustain its program of economic development, Brazil intends to increase its production of manufactured and semimanufactured goods for the export market, thus reducing further its reliance on agricultural exports. Now, however, it faces not only decreased demand for its manufactures in other countries as a result of worldwide recession, but also the risk of tariffs or other trade barriers erected by countries with similar industries. Brazil has made remarkable progress, but it has only a small chance of becoming another Japan overnight.

[1] See Thomas F. Kelsey, "The Impact of the Energy Crisis in Brazil," in *The Energy Crisis and the Environment: A Comparative Perspective,* ed. Donald R. Kelley, Praeger, New York, 1977.

lower indexes of development and lack the population, and hence market size, necessary for large-scale production.

The aim of internal industrialization is to reduce dependence on overseas sources of manufactured goods. Theoretically, a region might expect to specialize in the things it can produce best and to depend on other areas for the rest of its needs. However, this system does not work very well for producers of primary goods, as farmers in the United States have learned. They sell farm products at low prices and must buy manufactured goods at higher prices. Countries such as those in Latin America that export food and raw materials face this problem.

A difficult task confronts Latin American countries that have been able to build up their own industrial base to the point where they can produce both consumer goods and essentials such as steel and cement for further economic growth. Large amounts of capital are needed to build and equip factories. Since most Latin American countries have had a shortage of available capital, they have encouraged industrial firms from Europe, Anglo-America, and even Japan to build plants in their countries. American automobile companies and Volkswagen, for example, have built automobile plants in Brazil and elsewhere.

Foreign companies, however, are not allowed simply to produce goods within the country and then take the profits out. Usually, new industry is a joint undertaking in which the government of the Latin American country or some of its own businesspeople have an important share in the ownership of the corporation. In recent years, Latin Americans have been asking for an increasing share in such companies. In many cases they have the controlling influence, owning at least 50 percent of the stock.

Moreover, Latin American countries are no longer content to let foreign manufacturers import component parts from overseas and then simply assemble automobiles or other products in Latin America. Agreements now call for some manufacturing to be done within the country. In Brazil automobiles are increasingly being assembled from Brazilian-made components. Similarly, manufacturers are being asked to train more Latin Americans for supervisory and management jobs as well as for factory jobs. In the past decade the Brazilian and Mexican governments have begun to pressure foreign companies in their countries to export part of their production. Both countries export clothing and sophisticated products such as automobiles and engines.

Some countries have taken over all or large parts of the foreign investment within their jurisdiction. Mexico expropriated all oil production in the 1930s. When Castro's regime took over the government of Cuba, it outlawed all private property. In recent years Chile and Peru have nationalized such major foreign industrial developments as oil fields, sugar plantations, and mines. Venezuelans have followed other OPEC countries in taking control of oil company operations. Notice in all these cases that the industries taken over were those developed primarily to serve export markets. Latin Americans now prefer to control profits and production levels themselves rather than depend on taxes, duties, and royalty payments.

Despite the emphasis on industrialization in Latin America, little attention has been given to integrating the industrial growth of one country with that of another. As in Europe, each country is a separate decision maker, anxious to enhance its own development. Each has fostered its own industry instead of working cooperatively with other countries to serve a combined market. Neighboring Peru and Chile, for example, have each developed separate steel industries even though the domestic market in each is marginal. Moreover, the mills are not even complementary to the extent that one might specialize in heavy rods, rails, and construction components and the other in sheet metal for roofing and siding for buildings. Instead, each produces the same products. As long as countries develop industry in isolation from their neighbors, market size will be the key factor inhibiting Latin American industrialization except in the few large countries. Even in the large and populous countries, people in the traditional society still have very low incomes. For them, the prices of domestic manufactures are likely to be too high, so they will continue to rely on handmade products. Small plants with inefficient production techniques simply cannot produce goods cheaply enough for such a market.

METROPOLITAN NODES AS INDEXES OF THE MODERN SYSTEM

In the industrial countries studied in previous chapters, the large metropolitan centers were the focuses of the modern interconnected system. We might logically expect the large urban centers of Latin America to be associated not only with the largest countries but also with the most advanced industrialization and trade. Evaluate this hypothesis in terms of Figure 13–8.

If the presence of large metropolitan areas is a good index of the modern interconnected system in Latin America, their absence in parts of the region can be taken as a measure of the lack of modern development. Thus Central America south of Mexico appears the least developed area, and the islands of the Caribbean (except Cuba) and the South American countries of Paraguay and Bolivia fall into the same category.

Even in the larger, more modern countries the distribution of metropolitan areas may indicate uneven development of the country. São Paulo and Rio de Janeiro lie close to one another in one small area of Brazil, the coffee region, which is a very successful but highly localized development. There are no cities of comparable size elsewhere in Brazil, although there are large regional centers such as Recife, Salvador, Belo Horizonte, and Pôrto Alegre.

RURAL POPULATION AND SUBSISTENCE LIVING

In most Latin American countries the majority of the population is still rural and agricultural. Only in the more highly developed countries of Venezuela, Chile, Argentina, and Uruguay is the proportion of people employed in agriculture well below 50 percent. Even large developing countries like Mexico, Brazil, Peru, and Colombia have close to half their labor force employed in agriculture. In Haiti, Guatemala, Nicaragua, and Honduras, two-thirds or more of the labor force work the land. Yet despite the large population involved in agriculture, most countries derive less than 25 percent of their national income from it, which suggests an inefficient use of labor in agriculture and the lack of alternatives.

In all countries except perhaps Argentina and Chile, a large proportion of the agricultural population is still engaged in subsistence farming, especially countries that have large Indian populations. Guatemala, Nicaragua, Mexico, Peru, Bolivia, Colombia, Ecuador, and Paraguay all have large Indian groups and predominantly agricultural populations. In Guatemala, Peru, Bolivia, and Ecuador, in particular, large numbers of Indians live essentially as they have for centuries. Their homelands in highland valleys far from the coast were of little interest to Europeans seeking accessible plantation crops, so they were left in semi-isolation. They farm tiny plots of land with a digging stick and hand tools, using most of their production for family consumption. Occasional surpluses are traded in the local market for essential goods from other areas. In most of these countries, the Indians live in highland valleys of *tierra fría* above the productive agricultural lands of *tierra templada* (see Figure 13–3). Others live in tropical forests in the lowlands beyond the frontier of European settlement. Speaking native languages and centering their activities on traditional social and political organizations, these people live essentially within their own separate, local worlds.

Raw materials
Industrial area
Railroad

Metropolitan population
· 250,000-1,000,000
● 1,000,000-4,000,000
● Over 4,000,000

FIGURE 13-8   CENTERS OF THE MODERN IN-
TERCONNECTED SYSTEM IN LATIN AMERICA

THE TWO WORLDS OF MEXICO

In November 1976, less than two weeks before he left office, the president of Mexico expropriated almost 250,000 acres of farmland in the Yaqui Valley in northwest Mexico. This valley produces almost half of Mexico's wheat and cotton crops. The expropriated land, which had been held by seventy-four wealthy families, was turned over to more than 9,000 previously landless peasants and migrant workers.

The expropriation came almost on the eve of Mexico's celebration of the 66th anniversary of the 1910 revolution, in which the country was freed from the dictatorial rule of Porfirio Díaz. Although the expropriation was greeted with great enthusiasm by Mexico's 28 million peasants, it involved an infinitesimal amount of the farmland of Mexico, indicating that the ideals of the 1910 revolution have yet to be fully achieved. Under Díaz, 90 percent of the farmers were landless; most were serfs on vast haciendas. Since the revolution, the Mexican government has expropriated more than 130 million acres of land and redistributed it to more than 11 million people, less than half the peasant population.

In accordance with the Constitution of 1917, the expropriated land is divided into about 18,000 **ejidos,** which are properties owned by entire communities rather than individuals. (*Ejidos* originated in the practice of communal farming in Mexico before the Spanish Conquest.) A small proportion of them, concentrated in the arid north and in the Yucatán Peninsula, are worked communally, and

the whole community shares the crops. Most *ejidos,* however, are assigned to individuals, who work them and are entitled to the produce. They can pass the land on to their children, but they cannot rent it or sell it. The individual plots are small, and much of the land in *ejidos* is relatively unproductive.

In 1960, the time of the last census, large properties of more than 6,000 acres accounted for more than 78 percent of all farmland in Mexico, and small farms of fewer than thirteen acres constituted less than 1 percent of farmland. In terms of numbers of farm properties, large farms of more than 6,000 acres accounted for less than 2 percent of the total number of farms, and about two-thirds of the farm properties were smaller than thirteen acres.

What this means is that a few landowners still control most of the farmland in Mexico, and that the remainder is divided among a very large number of owners. The most productive land is irrigated, and although the amount of irrigated land held by any individual is legally limited to 250 acres, the wealthy landowners circumvent this restriction by registering property in the names of other family members. The result is that most Mexican farmers live in harsh poverty, trying to feed their families from tiny slivers of land. With little to sell, they can obtain little capital to invest in farming. They are largely a part of the traditional, locally based world with the village as the center of life.

At the same time, Mexico's large farms produce important commercial

items. Mexico is second only to Brazil among Latin American countries in the value of its exports, of which agricultural products, especially cotton, coffee, and sugar, are most important. Fruit and vegetables from the modern irrigated farms of northern Mexico are marketed in the United States.

Mexico is also one of the leading industrial countries of Latin America. Like Brazil, it has a well-established base of modern industry: steel mills, chemical plants, oil refineries, and modern assembly industries making household appliances, automobiles, and television. Industry is concentrated in Monterrey—the steel center and a city of more than 1 million people—and in Mexico City.

Mexico City is among the half-dozen largest metropolitan areas in the world. As modern as any city in Europe or the United States, it is not only the capital and center of the Mexican economy, but also part of the worldwide system of cities. Scheduled flights link it to U.S. and European cities, to Tokyo and Sydney, Australia, as well as to other Latin American cities. One of Mexico City's modern research hospitals has gained international attention for its research on birth defects; the head of the study team was a doctor trained in Yugoslavia.

Mexico's ties to the world bring problems as well as benefits. Mexico has borrowed money for its industrialization and modern growth, and now its economy is burdened with the necessity of paying back those loans. It is estimated that over one-third the value of Mexico's exports goes toward interest on that debt. Moreover, events elsewhere in the world affect Mexico's economy. For example, the recession of 1975–1976 reduced the demand for Mexico's exports, and the result was widespread unemployment, with 600,000 workers unemployed in Mexico City alone.

Like Brazil and other developing countries, Mexico still has not tied its different regions into a single functioning economic system. However, Mexico does have an extensive system of railways and highways that link the different regions to Mexico City and the major ports. Cotton from the irrigated farms of northern Mexico, oil from the Gulf Coast, and manufactured goods such as automobiles move into export markets. The domestic market for Mexico's production remains limited because a large proportion of the population cannot afford to buy the products of industrialization.

Because of its proximity to the United States, Mexico's economic ties to the United States are closer than its ties to any other Latin American country. Over half of Mexico's exports go to, and over 60 percent of imports come from, the United States. Mexico also heavily depends on U.S. tourists; it has the largest influx of tourists of any Latin American area. Mexico City, the resort city of Acapulco on the west coast, and the ancient Mayan area of Yucatán are the major centers of tourism. Tijuana, Mexicali, and Ciudad Juarez along the border are visited by large numbers of Americans who cross the border just for a day's recreation and shopping.

In the same way, descendants of African slaves and hacienda dwellers with mixed European-native backgrounds live off the land of their local areas. Many are simply squatters on the fringe or part of a landlord's vast property holdings. Farmers in such positions have neither the educational training, the capital, nor the land to operate efficiently. In contrast, in the coffee area of Brazil, the Argentine pampas, and around Mexico City and near big cities, farming has shifted to a commercial base to supply the urban population.

THE RULING MINORITY

Although most of Latin America has been independent for more than 100 years, the Spanish and Portuguese descendants of the former colonial rulers have found it difficult to change the deeply rooted colonial patterns. The backward and inefficient hacienda system provides the landowners with a good life. The vast landholdings, combined with a ready supply of cheap labor, have enabled hacienda owners to have luxurious homes in the capital cities, European schooling for their children, and, in the twentieth century, vacations and buying trips to the United States and Europe.

With these advantages, hacienda-owning families were not attracted to popularly supported political movements such as those led by Fidel Castro in Cuba, Juan Perón in Argentina, and Salvador Allende in Chile. In fact, the landowners controlled the government, and for them "revolution" meant overthrowing a particular member of their ruling clique if that leader did not protect the landlords' interests. They expected the new leader to come from the same ruling class that has held power since independence.

A large percentage of the Catholic clergy continued to come from wealthy families. Moreover, the church controlled great wealth and often vast estates of land as well. While it ministered to the spiritual needs of the poor majority, it did little to improve their social and economic plight.

Political power was supported by the military, whose leaders commonly came from the same landholding families that dominated politics and the church. The ruler was commonly a military officer supported by a junta of other officers.

Today the hacienda system has been weakened by the rise of the modern interconnected system with its emphasis on city life. Many of the landowning families have shifted their interests and become part of a new urban elite consisting of professionals, bankers, industrialists, and businesspeople. The mass of the population in most countries remains as it was in the past: poor, uneducated, and politically weak.

In some countries like Mexico and Chile, the old haciendas are being broken up and parceled out in small units to formerly landless peasants. However, this is more a measure of the declining value of the hacienda than a real social revolution. Plots are often so small that the newly landed peasant has little hope of getting ahead. While the peasants have gained land, hacienda owners have moved their families and their commercial interests into the city.

Income distribution reflects the continued control of the ruling minority. A recent study by the United Nations Economic Commission for Latin America estimated that the upper 5 percent of Latin America's income groups received one-third of the total income, and the upper 20 percent of the population had almost two-thirds (Figure 13–9). By contrast, the lowest 50 percent, the rural and the poor, had less than 15 percent of total income, and the lowest 20 percent received only 3 percent. Compare these figures with those of the United States.

Major attempts to change the control of power and money in Latin America have not been widespread. Castro's successful revolution in Cuba is the exception. There a centralized system of state control similar to that of the Soviet Union has been instituted. Both the plans and the goals are in the hands of the government, but, as in the Soviet Union, one of the goals is to redistribute the results of production more evenly among the people than ever before. In Chile socialist attempts at pushing land reforms and the nationalization of the country's major industries resulted in a military coup supported by U.S. secret services. In Mexico a real peasant

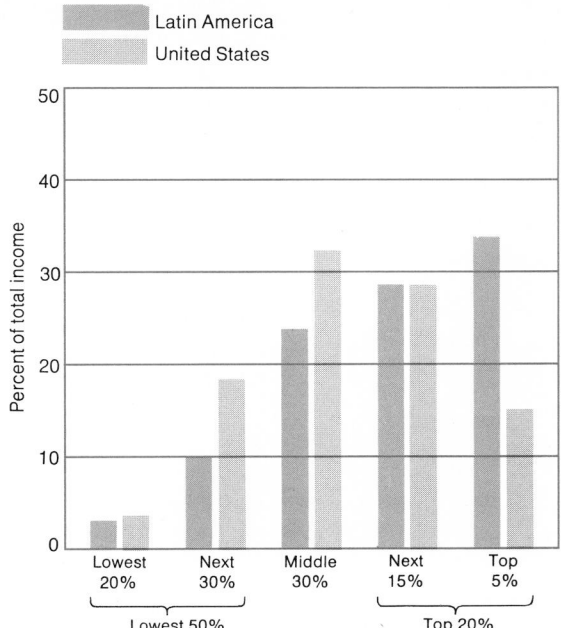

FIGURE 13–9 INCOME DISTRIBUTION IN LATIN AMERICA AND THE UNITED STATES

revolt took place early in the twentieth century. Since that time the Mexicans have developed a degree of democracy that is remarkable among Latin American republics for its effectiveness and stability. Other countries such as Venezuela and Peru are also attempting political and cultural change.

The Catholic church, too, has begun to change. Since World War II the number of native Latin American priests has declined. About a quarter of the priests are foreigners, mostly from Spain and Italy, but increasingly they are from other countries, including the United States and Canada. With the growth of cities, the traditional ties to the church that were a part of rural life have been broken. Middle-class city families, who once financially supported the church in addition to supplying it with nuns and priests, have drifted away. The number of entrants to the ministry and attendance at Mass have dropped sharply. The "new church" has turned away from its traditional support of established power to become the ally of the poor. The result has not been a great success. The increasingly secular governments have been hostile to the church's interest in social change, and persecution has resulted.

WILDERNESS AND THE PROBLEM
OF FRONTIER SETTLEMENT

Despite the large agricultural populations and vast tracts of land beyond the settlement frontier, in most countries there has been no mass movement to the frontier. In recent years the poor of Latin America, like the poor in rural parts of the United States, have found the lure of city life and its potential riches far more attractive. They flock to the cities without adequate education, capital, or knowledge of urban life, and thus in most countries squalid slums now surround most large urban places and rural population has declined. This type of metropolitan growth contrasts with the growth that occurred in European and U.S. cities as a consequence of industrialization. There the new city dwellers were needed as workers. In Latin America today the large numbers of newcomers are a burden on the cities' infrastructure of transportation, communication, and sanitation services.

Latin American slums are mostly on the edges of the cities rather than in the "inner" city, although Mexico City and Rio de Janeiro have slums in both locations. In many cases rural people have simply settled as squatters on the urban fringes and built shacks. Although they do not own the land or pay rent, they think of their dwellings as their property and constantly strive to improve them. What might start as a cardboard cover on a wooden frame with a bit of tarpaper becomes in time a hut of cement blocks, perhaps with flooring and a corrugated iron roof. Usually, the dwellings have no water or plumbing, but the inhabitants see their shacks in the city as better than adobe huts in a village without hope. These are the pioneers who, not satisfied with what their villages offered, are striving to better themselves economically and socially. To them the city offers much greater possibilities than the rural frontier.

Faced with the problem of trying to develop frontier areas to counteract the urban migration, governments have devised incentive programs to encourage settlement. Particularly active in frontier settlements are the countries with lands in the upper Amazon lowland: Brazil, with its remote western lands; and Colombia, Ecuador, Bolivia, and Peru, with forested lowlands east of the Andes that have never been a functional part of these countries. But such development seems to attract more immigrants from other countries than natives. Mennonites have settled land in Paraguay and Mexico, Japanese have gone into the Amazon country, and various land companies and colonies have been established by North Americans and other outsiders in Peru and Colombia.

In evaluating the laggard development of the rural frontier during the more than 450 years of European settlement in Latin America, it is important to recall that the great frontier areas are in tropical environments, not in the midlatitudes. The potential for management is very different. Moreover, most of the frontier areas are hot, wet rain forests that provide a satisfactory life with a minimum of effort. Human needs for shelter and clothing are not great. Hunters and gatherers, with almost no capacity for storing perishable food, and farmers have learned to manage the tropical environment. They plant a variety of crops together in a clearing and then shift to new clearings when unwanted weeds inhibit growth. They plant root crops, grains, vegetables, and small trees (such as bananas) in a seemingly helter-skelter fashion in the clearing. This many-tiered field replicates the great variety of the forest itself and thrives for several seasons before becoming overgrown.

The rain forest area is totally unsuited to the midlatitude agricultural system, in which an area is cleared, one crop is planted, and one or two crops are harvested each year. Such a system draws on plant nutrients stored in the soil, nutrients that are replenished regularly by artificially fertilizing the soil. But in the high-work environments of tropics the growth system depends on the continuous recycling of nutrients through the plant system rather than on storage in the ground, as was discussed in Chapter 3. Tropical soils are storage places for moisture but not for the other essentials for plant growth. Thus a system that breaks up the variety of plant life in the tropical forest system, that depends on nutrients stored in the soil, and that disrupts the surface soil layer through heavy plowing (all characteristics of midlatitude agriculture) faces severe problems in the wet tropics.

In general, European settlers have avoided the forested tropics, and so the major frontier of Latin America in the Amazon lowland of Brazil, Colombia, Ecuador, Peru, and Bolivia remains semiwilderness. Even major schemes by industrial corporations have generally been unsuccessful. An attempt by the Ford Motor Company to establish rubber plantations failed. A new private scheme to develop 2.4 million acres of timberland by clearing the forest and planting a single species has got off to a slow start. Early plantings have been hindered by floods, unusual rains, and the low-spreading growth of the plants. However, the Brazilian government is gambling that new accessibility to the Amazon Basin will open the frontier to farming.

Frontier growth has been further handicapped by the lack of institutional development in the back country. Development depends on roads and railroads, the availability of land, the incentive of markets. But, except for Brazil's recent efforts in moving the capital to Brasília and undertaking a new road network into the Amazon Basin, the amount of investment by Latin American governments in their tropical frontiers has been minimal. Such areas are far from the major population islands of each country; and in the case of the western countries, they lie across the rugged Andes from the population centers. Frontier development has been an almost overwhelming undertaking for the small countries of Latin America, faced at the same time with the need for further investment within their major population centers.

ATTEMPTS AT INTERREGIONAL COOPERATION

Despite the continuing separation of the social, political, and economic structures of individual national states in Latin America, attempts have been made in recent years to establish institutions among Latin American countries similar to the European Economic Community. The first formal organization was the Central American Common Market, which bound together the five weak Central American republics of Guatemala, Honduras, El Salvador, Nicaragua, and Costa Rica. This pact called for the elimination of tariffs between member countries and established a single set of tariffs for the rest of the world, like the EEC. Both external trade and internal trade within the Central American Common Market have increased, and the availability of a single market for the five has encouraged foreign investment in large-scale enterprises. There was at one time talk of possible political union.

Simultaneously, a more ambitious project was the Latin American Free Trade Association (LAFTA), which involved all the largest and most advanced Latin American countries. This organization includes countries that comprise 80 percent of the population and produce 50 percent of the output of all Latin America. LAFTA aimed at reducing trade barriers and at encouraging trade among its members. It has been much less successful than expected, however, probably because it is overambitious in scope at this stage in

Latin American development. Its membership is too widely scattered and diverse, and subregional organizations are emerging in its place. Most successful is the Andean Group, which includes all the Andean countries including Venezuela. The Andean Group is working toward the "common market" goals of internal free trade and a set of common external tariffs by 1990. Duties have already been reduced on thousands of items. The result has been increased trade between members, but the primary trade of all countries is still with the industrial countries outside Latin America. The members have also established a joint policy on foreign industrial investment. However, the Andean Group suffered a setback in 1976 when Chile withdrew.

Most recently, small Caribbean countries that are former or present dependencies of Britain have formed the Caribbean Economic Community in order to lower internal tariffs and other barriers between them. Included in this group are Guyana, Belize, Jamaica, Trinidad and Tobago, and various small island units.

The major problem with attempts at breaking down economic barriers between countries in Latin America is that different countries are at different stages of development. Brazil, with a growing industrial base, benefits greatly from free trade within Latin America. Its goods can flow freely into largely nonindustrial countries such as Bolivia and Paraguay. Those countries, however, want their own industrialization, and they think it is endangered by the influx of Brazilian goods.

A new approach was taken in 1975 when eighteen Latin American countries formed Sistema Economico Latinoamericano (SELA), an economic compact designed to replace the Organization of American States, which had been dominated by the United States. SELA sees itself as a problem-solving, coordinating organization. It has already shared information concerning food, fertilizer, housing, world prices for raw materials, markets, and national capacities to meet demands. To meet the needs of its member countries, SELA plans to set up corporations financed by private or state funds or a combination of private and state funds. Each nation is to invest what it can in projects that will further its own interests while SELA provides coordination and technical aid. Membership includes not only the Spanish and Portuguese countries, but English-speaking Caribbean countries as well. However, the midlatitude countries of South America—Argentina, Chile, Uruguay, and Paraguay—have not joined SELA. Since Argentina and Chile are among the most economically advanced countries in Latin America, their absence limits SELA's potential.

CHANGE BUT NOT SOLUTIONS

In view of the problems still to be faced, these attempts at economic, political, and social change are moving at an extremely slow pace. Industrialization has just begun, agricultural methods in most places are still largely traditional, and the entire economy remains oriented toward the production of raw materials for industrial Europe and Anglo-America. Political stability has come from power wielded by the military, which represents the traditional landed class and international interests rather than the mass of Latin Americans, who have no political voice, live on the brink of starvation, are uneducated, and remain largely ignored by the governments that purport to represent them.

Some governments in Latin America have reacted sharply against attempts at social change. In Argentina and Chile, where governments dedicated to change were recently in power, military juntas now govern, and individual freedoms and democratic elections have been curtailed. The military governments often are more representative of the ruling minority of the past than of the people as a whole. Military governments in Brazil and Peru, however, have been just as dedicated to change as previous governments have been. Change comes most easily in the large cities, where it is powered by economic forces.

Mexico, Brazil, Venezuela, and Argentina have better internal regional integration than most other Third World countries. Mexico is cited as the one Latin American country that has achieved durable political stability and steady economic growth. This is largely the result of the revolution of 1910, which broke up the estates of the oligarchy, gave land to the peasants, and brought social reforms. Mexico is ruled by a benevolent, one-party dictatorship of the Institutional Revolutionary party, and the system works well because it provides some voice within the party for all sectors of the population. But Mexico still has a host of other problems. Population is expanding at a rate of 3.5 percent per year, and the economy is hard pressed to absorb the growing youthful population into the work force. Nearly half the people live in family units that earn less than $1,000 annually. Thus even the most successful of the countries of Latin America has a long way to go.

SELECTED REFERENCES

Cole, John P. *Latin America: An Economic and Social Geography*, 2nd ed. Butterworth, London, 1975. A basic geography of Latin America. Chapters 1 through 10 systematically treat such topics as historical background, commodity production, population, and poverty. Chapters 11 through 17 discuss each of the seven largest countries in turn.

Furtado, Celso. *Economic Development of Latin America*. Cambridge University Press, Cambridge, England, 1970. A treatment of economic development in Latin America since colonial times.

Gilbert, Alan. *Latin American Development: A Geographical Perspective*. Pelican, Baltimore, 1974 (paper). A study of the processes and problems of modernization in Latin America. Chapter 2 (historical setting), Chapter 3 (industrial change), Chapter 4 (pattern of urbanization), and Chapter 5 (rural sector) support and augment the theme of this chapter.

Odell, Peter R., and David A. Preston. *Economies and Societies in Latin America: A Geographical Interpretation*. John Wiley, New York, 1973. Focuses on the spatial organization of the economies and societies of Latin America. Chapters 6, 7, and 8 are particularly pertinent to the ideas discussed in this chapter.

Taylor, Alice, ed. *Focus on South America*. Praeger, New York, 1973 (paper). A collection of articles by different authors. Part I gives an overview of such matters as population, unemployment, and rural development, and Part II consists of short geographical reviews for each of the South American countries. A very worthwhile supplementary source.

U.N. Economic Commission for Latin America, *Economic Bulletin for Latin America*. New York (published twice yearly). A periodical of substantial reviews of population, income, employment, trade, and resource conditions in relation to economic development in Latin America. Useful statistical summaries of current conditions or recent trends are usually an important part of most articles.

Wilkie, James W., ed. *Statistical Abstracts of Latin America*, vol. 17. UCLA Latin American Center Publications, Los Angeles, 1976. A compilation of the most recent data for countries of Latin America on population, economic indicators, and international statistics. Data for the civil subdivisions—such as provinces or states—or individual nations are included. The publication is revised frequently, usually annually.

CHAPTER 14

Black Africa: The Struggle with a Colonial Heritage

Present-day Africa south of the Sahara is the product of two layers of modernization on a traditional base. The first is colonialism, the result of Europe's industrialization in the nineteenth century. The second is the twentieth-century modern interconnected system. Today the boundaries of African countries do not fit the realities of Black African life, and the existing countries experience difficulties in developing their environment and in competing economically with larger, wealthier nations. Much is still unknown about how to manage the fragile tropical environment, with its two distinct areas: the higher, drier, and cooler eastern and southern portion with a seasonal-work environment; and the lower, wetter, and hotter western and northern portion with mostly a year-round high-work environment.

Humid tropical forest in Ghana in West Africa. (Owen Franken/Stock, Boston)

Above: Cargo being unloaded from small boats, with modern mechanized port in background, in Dar es Salaam, Tanzania. (Lynn McLaren)

The governor's palace in Brazzaville, Republic of the Congo. (S.E.F./Editorial Photocolor Archives)

Two gemsbok (or South African oryx), inhabitants of the dry grassy plains of southern Africa. (Ira Kirschenbaum/Stock, Boston)

The problems that are being faced in Africa south of the Sahara are not new. Many of today's industrialized countries faced them in the past, as the following description of England in the seventeenth century suggests:

> [The] . . . economy . . . was one in which the methods of production were simple and the units of production were small; in which middlemen . . . were both hated and indispensable; in which agricultural progress was seriously impeded by the perpetuation of communal rights over land. The chronic underemployment of labour was one of its basic problems and, despite moral exhortations, among the mass of the people the propensity to save was low. . . . It was an economy heavily dependent on foreign sources for improved industrial and agricultural methods and to some extent for capital, but in which foreign labor and businessmen were met with bitter hostility. In it ambitious young men often preferred careers in the professions and government service. . . . Men increasingly pinned their hopes on industrialization and economic nationalism to absorb its growing population; but industrialization was slow to come and the blessings of economic nationalism proved to be mixed.[1]

Today a similar situation exists in Africa as the native population emerges from a heritage of tribalism, subsistence living, and many years of outside colonial domination. Most of its people have been living within local traditional societies that formed tightly knit culture groups. Most economic development in Africa has come as a result of, first, colonial activity and, later, modern exploitation of the mineral and agricultural resources of the region for external markets in Europe and the United States. Colonialism is

[1]From "The Sixteenth and Seventeenth Centuries: The Dark Ages in English Economic History?" *Economica* (London School of Economics), new series 24, no. 93 (February 1957), pp. 17–18. Quoted in A. M. Kamarck, *The Economics of African Development,* Praeger, New York, 1967, pp. 47–48.

gone, but the modern components of the African economies still depend on exports to industrial countries.

In this chapter we focus on the countries south of the Sahara Desert. The term "Black Africa" is commonly applied to this region because of the predominance of black (negroid) cultures, even though Rhodesia and the Republic of South Africa are ruled by white minority governments.

THE PEOPLES OF AFRICA

In contrast to Latin America, and despite the years of colonial domination, the African countries emerging today are inhabited by an overwhelming majority of indigenous people. Africa never experienced the magnitude of European migration that occurred in the Americas. Probably 5 million people in the total African population of 310 million are of European origin, most of them in South Africa. And outside of South Africa no more than 2 percent of the population is European; for the most part the Europeans are temporarily employed only by the government or foreign business interests on a contract basis. Asian immigration has been even smaller; the Asians in the area total only somewhat over 1 million persons, half of whom live in South Africa.

INDEPENDENCE FOR TRIBAL PEOPLES

The countries themselves, almost without exception, have been independent for fewer than twenty-five years (Table 14–1). As late as 1956 there were only three independent countries in the entire region—Ethiopia, South Africa, and Liberia. Today only one important area, the territory of Southwest Africa (Namibia), is not independent, and it is controlled by South Africa, not a European colonial power. Since 1957 more than thirty-five new states have gained political independence in the region. The new countries' governments are in the hands of native peoples

Country and Territory	1976 Population (Millions)	Population per Square Mile	Rate of Population Growth (Percent)	Per Capita GNP (Dollars)	Energy Consumption (Million Metric Tons of Coal Equivalent) 1972	Date of Independence
Western Africa						
Benin	3.2	74	2.7	120	.09	1960
Gambia	0.5	114	1.9	170	.03	1965
Ghana	10.1	110	2.7	350	1.38	1957
Guinea	4.5	47	2.4	120	.4	1958
Guinea-Bissau	0.5	36	1.5	330	.04	1974
Ivory Coast	6.8	55	2.5	420	1.40	1960
Liberia	1.6	37	2.9	330	.6	1847
Mali	5.8	12	2.4	70	.12	1960
Mauritania	1.3	3	1.4	230	.12	1960
Niger	4.7	10	2.7	100	.12	1960
Nigeria	64.7	181	2.7	240	3.83	1960
Senegal	4.5	59	2.4	320	.64	1960
Sierra Leone	3.1	111	2.4	180	.35	1961
Togo	2.3	106	2.7	210	.13	1960
Upper Volta	6.2	59	2.3	80	.07	1960
Middle Africa						
Angola	6.4	13	1.6	580	1.19	1975
Cameroon	6.5	35	1.8	260	.6	1960
Central African Empire	1.8	7	2.1	200	.09	1960
Chad	4.1	8	2.0	90	.07	1960
Congo (Brazzaville)	1.4	11	2.4	380	.22	1960
Zaire	25.6	28	2.8	150	1.57	1960
Equatorial Guinea	0.3	27	1.7	260	.05	1968
Gabon	0.5	5	1.0	1560	.45	1960
Eastern Africa						
Burundi	3.9	362	2.4	80	.03	1962
Ethiopia	28.6	61	2.6	90	.9	1941
Kenya	13.8	61	3.4	200	1.99	1963
Malagasy Republic	7.7	34	2.9	170	.49	1960
Malawi	5.1	111	2.4	130	.24	1964
Mozambique	9.3	31	2.3	420	1.22	1975
Rwanda	4.4	433	2.8	80	.05	1962
Somalia	3.2	13	2.5	80	.10	1960
Rhodesia	6.5	43	3.4	480	3.31	1965
Tanzania	15.6	43	2.7	140	.98	1961
Uganda	11.9	130	3.3	160	.69	1962
Zambia	5.1	18	3.1	480	2.24	1964

(cont.)

TABLE 14–1 (*Cont.*)

Country and Territory	1976 Population (Milllions)	Population per Square Mile	Rate of Population Growth (Percent)	Per Capita GNP (Dollars)	Energy Consumption (Million Metric Tons of Coal Equivalent) 1972	Date of Independence
Southern Africa						
Botswana	0.7	3	2.3	270	—	1966
Lesotho	1.1	93	2.1	120	—	1966
South Africa	25.6	54	2.7	1,200	71.30	1910
Swaziland	0.5	75	3.2	400	—	1968
Southwest Africa (Namibia)	0.9	3	2.2	—	—	South Africa

SOURCE: Population Reference Bureau, Inc., *World Population Data Sheet, 1976,* Washington, D.C., 1976.

except for South Africa and Rhodesia, where European minorities, descendants of the settlers who colonized the countries, are in control.

We can get some idea of the problems surrounding African independence if we think of European colonization in the Americas and then imagine that the Europeans pulled out, leaving the Indians to rule not their own ancestral lands, but the colonial territories established by the Europeans. African life at the time of European colonization was not very different from the life of the American Indians when the U.S. colonies were established. The Africans lived in locally based tribes and nations, each with its own cultural heritage, language, customs, political organization, economic system, and prescribed territory. Some African tribes lived highly organized lives like the Sioux in North America; a few approached the sophisticated civilizations of the Incas and Aztecs.

European **colonialism** superimposed a new layer of political and economic power on the traditional system. Boundaries were laid out by European political powers with little regard for the real patterns of African life. Europeans governed through their own institutions; Christian missionaries introduced different cultural and moral values; and foreign business enterprises established transport facilities to tap valuable resources for their home markets.

European development had a tremendous impact on Africa, but it was sharply localized in plantation and mining areas and in capital cities. Its effect on most of the rural population was primarily indirect. The majority of Africans not only continued to support themselves from their traditional local base but also maintained their ancient tribal and village loyalties. The social structure remained that of the closely knit **extended family.** In some places class or caste structures were almost nonexistent.

This pattern has continued after independence. A small fraction of the African population has left the rural communities to be educated by the Europeans or to work in the government or business community, but most continue their old lives unchanged. New basic crops such as manioc, beans, maize, and peanuts were introduced into local native economies, but traditional farming practices continued.

COUNTRIES THAT DON'T FIT THE REALITIES OF LIFE

The African countries that have emerged after independence are European creations that do not fit the realities of African life (Figure 14–1). Independence has come separately to each former colony, which was defined in European

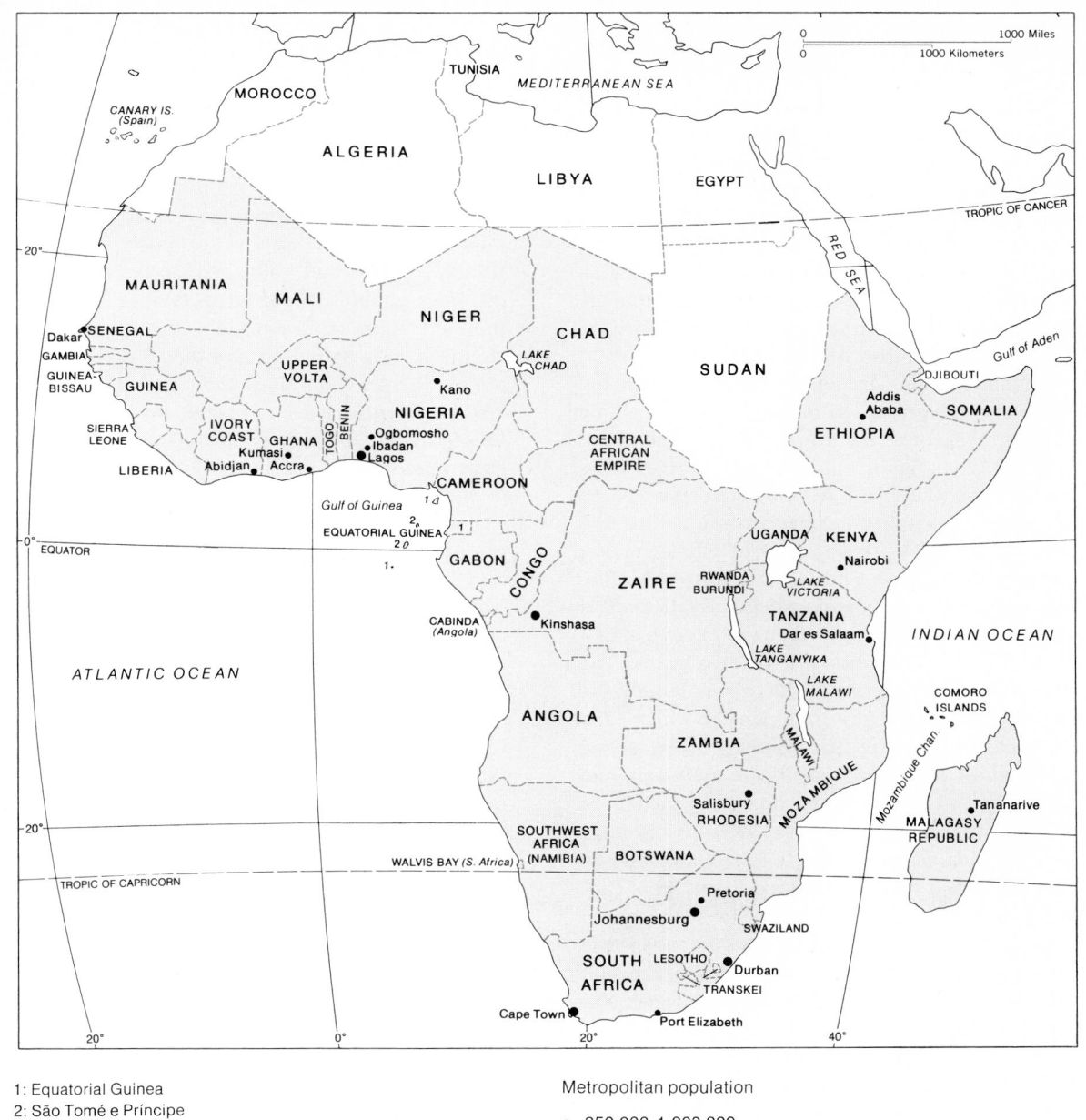

Metropolitan population

- • 250,000–1,000,000

- ● Over 1,000,000

   Tinted countries are those comprising
   Africa south of the Sahara

1: Equatorial Guinea
2: São Tomé e Príncipe

FIGURE 14–1   POLITICAL MAP OF AFRICA

terms, not African. Europeans established the boundaries of their colonies not in response to patterns of African life, but to gain control over Africa in the late nineteenth century. Negotiations among competing European colonial powers established boundaries that often cut right through the territories occupied by individual tribes or across the migration routes of nomadic herders. A given colony commonly consisted of the lands of several tribes, some of them fierce rivals. Thus today the new African governments must cope with political territories that are, in terms of the realities of life, artificial.

In many ways African countries are both too large and too small. In terms of African ethnic groups, the countries are too large. The tribal map of Africa (Figure 14–2) shows far more fragments than the forty African countries. More than 100 important language variations are recognized in Black Africa, most of which have a variety of dialects. The result is that new governments are split by serious tribal rivalries. It is common for a dominant tribe to gain control of a government and develop it in its own terms. The civil war in Nigeria in the 1960s was largely the result of tribal warfare in which the Ibos of the southeastern part of the country broke away. Ibos, whose homeland was in the southeast, had gone into northern areas dominated by the Hausa tribe to work and do business. Tribal friction led first to massacres of Ibos who had moved to northern areas, then to a mass evacuation of Ibos to their homeland, and finally to secession in 1967. The independent state of Biafra engaged in civil war with the federal forces of Nigeria until it was defeated in January 1970 and reabsorbed into Nigeria. The struggle for power in Angola in the 1970s, seen in the Western world as a conflict between groups backed by the Soviet Union and the United States, was first of all tribally based. The Eritreans in Ethiopia are another group trying to break away from the country of which they are politically a part.

At the same time, many African countries are too small to be viable in the modern intercon-

nected world. Black Africa has almost twice as many countries as Latin America, even though the land area is only 10 percent larger and the population 4 percent less. Most Black African countries are small in size and in population (Table 14–1). Only eight have populations of more than 10 million people; of these, only Nigeria, with 65 million people, has either the size or population of the largest European countries. Ethiopia, Zaire, and South Africa, the next largest, have populations of just a few million more than the state of California. Of the remaining countries only ten have more than 5 million people and five have no more than one-half million. The largest of these small countries has a population similar to that of Michigan; the smallest have numbers similar to those of Alaska, Wyoming, and Vermont.

The large number of small African states has produced one strength for the region. Black Africa has the largest voting bloc in the United Nations. In that body, where each member country has one vote, the 40 African countries account for more than one-fourth of all 143 voting members. This means that not only do Black African countries have a strong voice in the United Nations debates, but they have the voting power to push through the programs that will benefit them. Together with the countries of Latin America, the Middle East–North Africa, and South and East Asia, Black African countries give the developing world a majority in the United Nations and in its important economic and social developmental agencies.

THE MEANING OF GNP FOR
UNDERDEVELOPED AREAS

Not only are Black African countries small, but by world standards they are poor as well. Table 14–1 shows the very low living standard of most people of this region. South Africa approaches the world average of $1,360 in per capita GNP, and the small, oil-producing country of Gabon

uninhabited

uninhabited

TROPIC OF CANCER

EQUATOR

ZAMBIA

TROPIC OF CAPRICORN

TRIBES OF ZAMBIA

1 Tabwa	9 Senga	17 Kaonde	25 Mashasha	
2 Lungu	10 Mombera	18 Lamba	26 Ila	
3 Iwa	11 Unga	19 Lala	27 Lenje	
4 Lambya	12 Bisa	20 Kunda	28 Mashi	
5 Tumbuka	13 Chewa	21 Mpezeni	29 Totela	
6 Luapula	14 Luvale	22 Mbunda	30 Subia	
7 Aushi	15 Ndembu	23 Lozi	31 Tonga	
8 Bemba	16 Lukolwe	24 Nkoya	32 Nsenga	

Inhabited by several tribes
with small territories.
Each dot represents one tribe.

SOURCE: G. P. Murdock, *Africa*, McGraw-Hill, New York, 1959

FIGURE 14-2   TERRITORIAL BOUNDARIES OF
AFRICAN TRIBES

TABLE 14-2 COMPARATIVE PER CAPITA GNP, 1975 (*Number of Political Units in Each Category*)

Region	Over $1,000	$750–$1,000	$500–$750	$250–$500	$100–$250	Under $100
Africa south of the Sahara[a]	2	0	1	14	16	7
Latin America	13	4	6	7	1	0
Middle East–North Africa	10	1	3	4	3	0
South and East Asia	3	0	3	6	10	3

SOURCE: Population Reference Bureau, Inc., *World Population Data Sheet, 1976,* Washington, D.C., 1976.
[a] Namibia included with South Africa.

exceeds it; however, twenty-three of the other countries have per capita GNP below $250 per year (Table 14–2). This is in striking contrast to the U.S. 1976 per capita GNP of $6,640.

Although we have been using per capita GNP as a measure of wealth in the modern interconnected world, it is really a very different sort of measure in underdeveloped areas. GNP is a monetary measure, designed as a useful index for a **money economy,** but much of the production of traditional society is never translated into monetary terms. A family may support itself almost entirely from the subsistence production from its own lands. It consumes almost all it produces and has little surplus to sell. Its total commercial sales in the marketplace during the year may be $100 or less, and much of that may be transacted by bartering without using money at all.

Thus the question of measuring subsistence production remains. If a subsistence family's living standard is very low, its total assets, or its total production in money terms, may be very low when compared with a city worker's annual wages. If the family continues to live from year to year on a given standard, life is supported, and the threshold necessary for life has been reached, regardless of the level of actual production. One could ask whether a threshold should be considered in money terms, or measured in terms of hours of contentment, units of family togetherness, or degrees of personal satisfaction. There is

no meaningful way to "cost out" the input and output of a subsistence farm as one might cost out the production of a modern commercial farm or of a factory worker.

Although GNP cannot be used to measure the standard of living of people in underdeveloped areas such as Africa, it can measure the degree to which the country has a money economy and hence the extent to which its population is a part of the modern interconnected system. GNP may indicate the relative importance of the modern interconnected (money) economy as compared to the traditional, locally based (nonmoney) economy. If a country has a per capita GNP of $200 per year, one can assume that a much larger segment of its population is in the modern interconnected system than a country with a GNP of $50 per capita per year. In the second case we would presume that most of the population is in the traditional system.

Using GNP in this way, it becomes obvious that Black Africa is one of the least developed parts of the world. Compare it with Latin America in Table 13–1. The modern interconnected system is prominent in certain areas that produce goods for the rest of the world, such as South Africa or Gabon. In most African countries, however, this type of production is very small when compared to the total population.

When used in the same way, figures on the consumption of modern energy sources support

TABLE 14-3 COMPARATIVE PER CAPITA CONSUMPTION
OF INDUSTRIAL ENERGY SOURCES, 1972 (*Number of*
Political Units in Each Category)

Region	Million Metric Tons of Coal Equivalent					
	Above 5	2.5–5	1–2.5	0.5–1	0.1–0.5	Under 0.1
Africa south of the Sahara	1	1		2	15	20
Latin America	9	1	7	6	9	1
Middle East–North Africa	2	2	1	6	8	1
South and East Asia	9	1	1	2	8	5

SOURCE: *U.N. Statistical Yearbook, 1974.*

the thesis developed with GNP data. The greater the use of modern industrial energy sources, the greater a country's role in the modern interconnected system. Conversely, countries that rely on animate energy are presumably still traditional societies.

When one considers the lack of modern energy consumption (Table 14–3), the lack of involvement in the money economy, and the small population of most African states, one can understand their weak position in the modern world. If they are to participate fully in it, they would seem to need a major technological revolution and the development of at least a considerable degree of interstate economic cooperation.

DIFFUSE RURAL POPULATION
AND A FEW URBAN NODES

As we can see from the world map of population distribution (Figure 5–2), the population distribution in Black Africa is very different from that in Latin America. Africa has fewer areas with very low densities, indicating less wilderness area beyond the frontier of settlement. The wider spread of the population also suggests that African population distribution is much more the result of indigenous traditional settlement than of

outside immigration. Concentrations of population can be found well back from the coast, but as in Latin America, high densities generally stand as separate islands.

In Table 14–1 we can see that few of the countries of Africa south of the Sahara have dense populations. The highest densities occur in some of the smallest countries, such as Rwanda and Burundi, but most African countries have densities lower than those of Latin American countries (see Table 13–1) and far lower than those of Europe. In measuring the meaning of these densities, we must consider three factors: (1) the degree to which a population depends on its local resource base rather than on outside supplies, (2) the degree to which the people can use the resource base, and (3) the degree to which the base can support the economy. With so much of its population still in the traditional, locally based system, Africa still depends on its own resource base. Such people have not had the technical capability to use the environment fully, and in many respects the environment of tropical Africa presents major obstacles to any development.

The Pattern of Traditional Peoples

Basically, the population distribution of Black Africa today is the result of the evaluation of the environment by each local tribal group in the traditional, locally based system over an extensive period of time. Under these circumstances, areas of dense population represent either areas where the environmental base is better suited to traditional use or areas where a particularly perceptive and industrious group has developed its local economy above the level of most groups.

In general, African economic systems use the land extensively, and population densities remain low. Traditionally, most African cultures have been dependent on **shifting cultivation,** herding, hunting, fishing, and gathering. These activities primarily depend on harvesting wild plants in the natural environment rather than on expending labor to make the land produce particular products. Tribes dependent only on gathering, fishing, and hunting in rain forests and those that are primarily livestock herders in drier areas require many square miles to support a small population. An economy based on shifting cultivation can support a slightly larger population, but it still requires a large area. Even in the few places where Africans have developed more intensive agriculture, the land is not nearly as densely settled as in the agricultural areas of Asia.

The two dominant forms of traditional living in Africa have been subsistence agriculture and pastoralism. Agriculture has been centered in the wetter forest lands that have been cleared for crop cultivation. In this process, called **slash-and-burn agriculture,** clearings are burned out of the forest during the dry season. Smaller trees and debris are cut back and a variety of crops are planted in the clearings among the stumps of trees too large to remove. The burning leaves a thin layer of plant nutrients at the surface and turns the acid soil slightly alkaline and hence more suitable for crops. A single clearing may be used for "multistory" agriculture: root crops such as taros, manioc, and sweet potatoes; surface vines; tall standing grains such as corn and millet; and tree crops such as plantains. After a few years, when soil nutrients have been depleted, a clearing is abandoned. The people then move on to a new clearing, where they repeat the process.

On the drier margins in the north and east of Africa the raising of cattle, sheep, and goats is traditional. Herding, or **pastoralism,** is carried on under extremely difficult conditions in areas of almost constant aridity. Pastoralists are highly skilled at managing their animals, keeping them alive in periods of drought, ministering to them when they are sick, sharing scarce pasture and water with other herders, and finding sources of water. Nevertheless, these herds show the results of malnutrition and poor breeding. Frequently, the herds are too large for the dry environment to support, and large areas show the results of centuries of overgrazing.

Today most of the herders are, in fact, farmers as well. They raise millet and maize, root crops, and bananas to supplement their traditional diet of milk, butter, blood, and meat. In some areas of the grasslands pastoral tribes and farming tribes have developed a **symbiosis** based on local trade of surpluses. On the areas along the margin of the forests and the grasslands, where the soils are better and the droughts not too severe, farming communities based on permanent farmland are found. Traditionally, they have raised the same grains and root crops found in the slash-and-burn clearings, but this more intensive use of the land supports denser populations. Permanent agriculture is increasing as farming shifts increasingly from subsistence to the production of cash crops suited to the local environment.

A Few Modern Population Nodes

Westerners, with their greater technology and long-range interconnected system, perceived the same African environments used by native peoples in very different ways. They superimposed their pattern of use of African resources on the old system in particular places. They perceived the region not in terms of its potential to provide basic support of a local population, but as a source of particular commodities produced in quantity for consumption within the long-range system centered in the industrial countries. As a result, they established mines and plantations in only a relatively few, most advantageous locations. In addition, they established political capitals and seaports with their overseas connections in mind as much as their relation to the local area. Europeans rejected most of Africa's tropical environment as unsuited to supporting large numbers of Western people. Instead, they came only in sufficient numbers to govern and manage mines and plantations. The only areas that they seriously tried to settle were those they saw as similar to their home midlatitude regions. Thus the only important European migration was to midlatitude South Africa and to higher, drier Rhodesia and Kenya in the tropics.

The centers that were linked to the Western-centered modern interconnected system became focuses for African settlement. Many Africans abandoned their traditional homelands and migrated to sites where Europeans offered jobs. The large concentrations of population in South Africa, in the Copper Belt of Zaire, Zambia, and Rhodesia, and in the interior of East Africa are examples of European-imposed development.

Mining and even plantation agriculture in the modern system call for intensive use of labor to extract, process, and ship the products to market. Population size is proportional not only to the intensity of labor use but also to the external demand for the product. Large-scale mining operations, such as those in the Copper Belt or in the diamond-mining district of Witwatersrand in South Africa, attracted large numbers of Africans from far away. The concentrations of production and population are at focal points in transportation systems designed to get the resources to external markets. Since they are nodal in nature, they stand out as tiny points on the map of population. In turn, these areas draw on a wide network of agricultural producing regions for the food supply necessary to support the working population. Food producers who supply the mining and plantation developments no longer farm their land on a subsistence basis. Instead, they use it to produce **cash crops** for the African urban market.

Rapid Population Growth After Centuries of Stagnation

One other aspect of Africa's population map should be noted. Although the population of most African areas is growing very rapidly, growth is largely a recent phenomenon. Between 1650 and 1850 the subsistence economy was tapped by slave traders who carried off an important segment of the youthful population. This exodus, coupled with a high death rate, caused a decline in Africa's total population. Later, Europeans exploring and occupying land not only exposed the local populations to European diseases but also spread native ones. Henry Stanley, an American who explored part of Africa on behalf of the King of Belgium, in 1887 carried the sleeping sickness across Africa. The Italians invading Ethiopia in the 1890s are thought to have brought in cattle with rinderpest. Within a few years, the disease swept East Africa, killing millions of cattle, which had been the support of herding tribes. Subsequently, large grazing areas returned to bush, the native environment of the tsetse fly that carries the sleeping sickness. The effect of these diseases was that until the 1930s most local governments were concerned with depopulation, not population growth.

Today public health measures have eliminated much of the danger of disease, and death rates have dropped sharply. With little out-migration from Black Africa, populations are growing rapidly (Table 14–1). They are expected to continue to rise because a large proportion of the population is young. Nearly half the African population is under the age of fifteen, as compared with less than one-third of the population in most industrial countries. The large proportion of young people means a large population coming into the childbearing ages in the future and promises explosive population growth unless birth control is encouraged. Moreover, the need to provide necessary schools, hospitals, and jobs for the young is a heavy burden on the new African governments.

A TROPICAL ENVIRONMENT

The African environment is almost exclusively tropical. The continent extends southward only to 32° south latitude, and only South Africa is primarily in the midlatitudes. The Cape of Good Hope at the tip of South Africa is closer to the equator than is Savannah, Georgia. In the north, only small portions of Mauritania and Mali lie beyond the Tropic of Cancer. We would expect tropical Africa to be a year-round high-work area with a high total potential evapotranspiration (PE) comparable to tropical Latin America (see Figure 3–3).

In terms of growing potential, Black Africa may be divided into two principal areas: (1) a higher, drier, and cooler eastern and southern portion dominated by seasonal-work environments and (2) a lower, wetter, and hotter western and northern portion with mostly year-round high-work areas. This division is reflected in the pattern of settlment by Europeans: no significant settlement has occurred in West Africa, but East and South Africa have been the sites of settlement in agricultural as well as in urban and mining areas.

A LAND OF GRASSES AND SCRUB FORESTS

Contrary to the impressions of early explorers, who saw only the wetter coastal fringes of the continent, tropical rain forests cover less land area in Africa than in either Latin America or Asia. A major difference between Latin America and Africa is the pattern of elevations. Most of Africa south of the Sahara stands 800 ft (245 m) or more above the sea and, like eastern Brazil, is separated from the coastal lowland by a sharp rise. This is plateau rather than plain. The top of the plateau is level or rolling, not hilly or mountainous (Figure 14–3), although it is broken by two major **rift valleys** in East Africa. There are no mountains comparable to the Andes of Latin America in either altitude or extent. Probably fifty peaks in the Andes are higher than Mount

Kilimanjaro, Africa's highest mountain, and only nine mountains in all of Africa rise to heights of over 13,000 ft (3,960 m).

An increase in elevation has an environmental effect similar to that of an increase in latitude, as we saw in Chapter 13. Much of the African environment has closer similarities to midlatitude than to other tropical environments. Tropical rain forests can be found in Africa, particularly in the Zaire Basin and on the Guinea Coast, but most of tropical Africa is covered by a variety of natural grasses and open woodlands. In wetter areas the woodlands predominate; in drier places trees are less numerous and grasses are most prominent. In most of Africa the combination of grasses and trees produces a parklike **savanna** landscape different from either the forests or the grasslands found in the midlatitudes.

The presence of open woodlands and tree-strewn grasslands rather than forests suggests an insufficient moisture supply for full forest growth; actual evapotranspiration (AE) is probably a good deal less than potential evapotranspiration. Indeed, the problem of soil-moisture shortage is chronic to most of tropical Africa. While there are year-round temperatures conducive to rapid plant growth over most of tropical Africa, the areas of open forests and grasses suffer from pronounced dry seasons each year and from great fluctuations in rainfall from year to year. The greatest potential evapotranspiration is along the fringes of the tropics rather than at the equator, and the drier areas, where there is less cloudiness, have greater PE than the rain forests. But all areas, except a small portion of the rain forest on the equator, have water deficits producing a dry season, and the deficiency rises with increasing distance from the equator. The only areas of moisture surplus during the wettest season occur in the forest areas of West Africa near the equator.

Tropical Africa is therefore less a high-work environment than its earth location would indicate. Most of it is a seasonal-work environment because of a long dry season. Extensive areas are also so dry as to be low-work environments. The

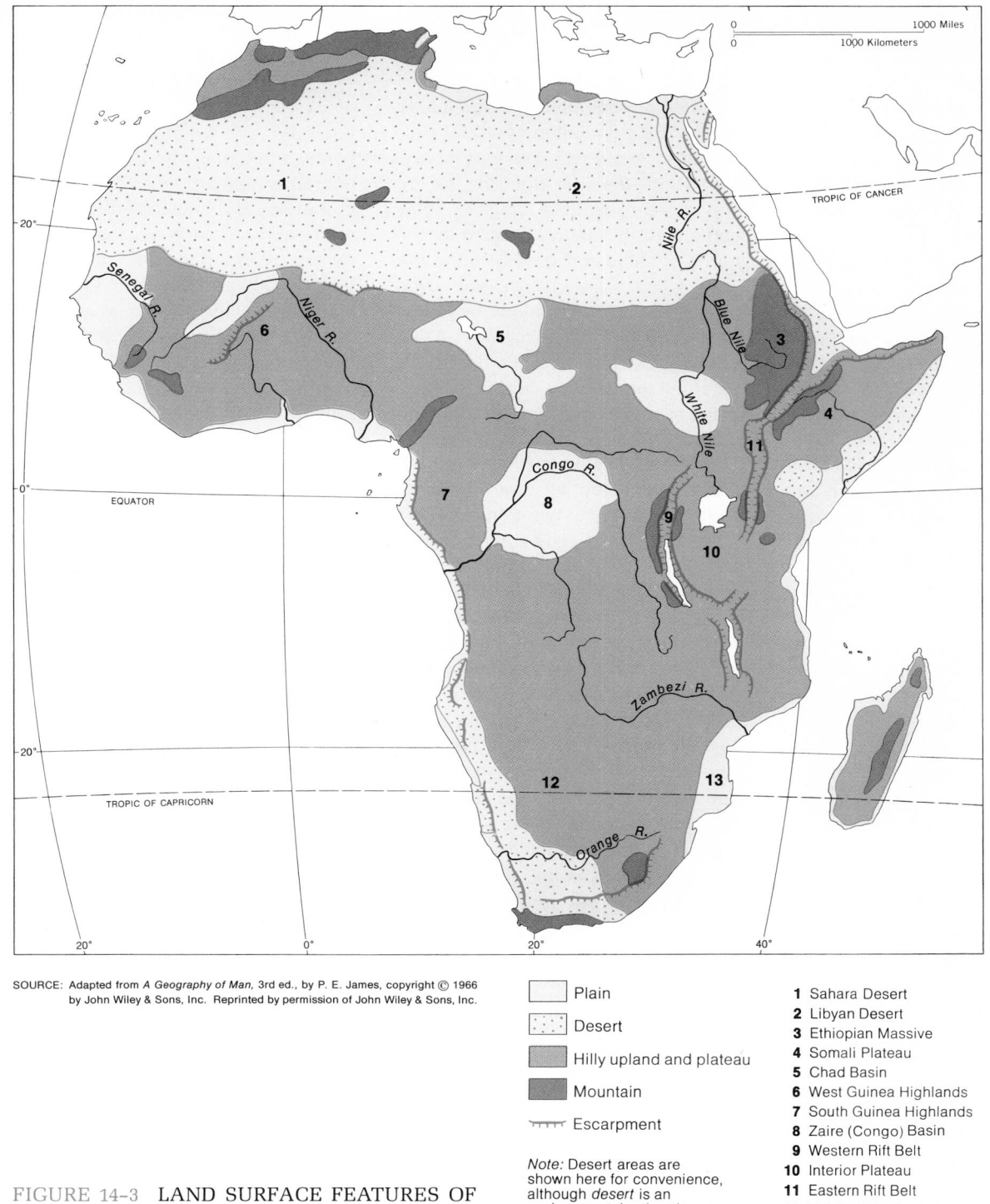

SOURCE: Adapted from *A Geography of Man,* 3rd ed., by P. E. James, copyright © 1966
by John Wiley & Sons, Inc. Reprinted by permission of John Wiley & Sons, Inc.

Plain

Desert

Hilly upland and plateau

Mountain

Escarpment

Note: Desert areas are
shown here for convenience,
although *desert* is an
environmental rather than
a landform term.

1 Sahara Desert
2 Libyan Desert
3 Ethiopian Massive
4 Somali Plateau
5 Chad Basin
6 West Guinea Highlands
7 South Guinea Highlands
8 Zaire (Congo) Basin
9 Western Rift Belt
10 Interior Plateau
11 Eastern Rift Belt
12 Kalahari Region
13 Mozambique Plain

FIGURE 14-3 LAND SURFACE FEATURES OF
AFRICA

THE SAHEL DROUGHT

The tragic drought of the Sahel along the northernmost fringe of Black Africa shows the dangers that can result from basic changes in traditional economic patterns. Although natural conditions caused the drought, the human misery was in large part the result of human errors.

Traditionally, this dry world has been inhabited by Tuareg and Fulani nomads supported by herds of camels and cattle; on its southern fringe were thousands of villages dependent on subsistence crops of millet, maize, and sorghum. In 1960 Mauritania, Mali, Upper Volta, Niger, and Chad, some of the weakest and least prepared of the new African countries, became independent. All had been part of the French empire and maintained the traditional French style of strong centralized government in a region where poor communications made it difficult for the government to know what was going on in remote parts of the country. No tax rolls or census were available to provide guideposts for development. But development did take place, aided by a series of wetter than usual years. International aid was used to provide new wells and to vaccinate cattle and nomads. The growing economies of nearby Nigeria and the Ivory Coast provided markets for all the cattle the nomads could raise. With water and pasture both the cattle population and the nomad population that depended on the cattle grew. This population could be supported in wet years, but not in dry.

In 1968—the first dry year of the decade—the trouble began. Few cattle died, but heavy grazing by the herds of cattle and goats around the wells killed grass, brush, and even trees. By 1972 conditions began to be disastrous. Almost no rain fell that year and by December the herds had eaten all the pasture, forcing the nomads to move south into the farming communities in search of fodder. There they discovered that the farm crops had failed as well. By June 1973 one-half to two-thirds of the herds were dead. The survivors were far south of their normal feeding grounds, and the families left in the north were in desperate need of food. Peasants from the villages fled to cities, living in unsanitary makeshift encampments. Worst of all, the Sahel vegetation on which the whole system had been built had been destroyed. The desert had expanded southward as much as 60 miles (95 km). Next the weakened people were afflicted by epidemics—measles, influenza, chickenpox, and diphtheria—adding to the death toll. Those who survived showed severe effects of malnutrition.

This tragedy occurred in countries with the lowest per capita GNP in Africa. In Niger the loss of cattle deprived the government of 35 percent of its tax base, which came from a head tax on animals. Aid had to come from the modern interconnected system based in the industrial countries. But since there were few roads and railroads available, it was necessary to airlift relief supplies, an expensive operation for long-term aid. It is estimated that in Mali one ton of airplane fuel was required for every ton of food delivered. By 1974 over 400 trucks were moving 50,000 tons of food each month. Trucks wore out quickly under severe desert conditions. When the rains came again, they washed out the few railroads and truck roads. The political effects of the disaster hit in 1974, and the government of Niger was overthrown.

What of the future? Studies indicate that much of the Sahel has been destroyed, transformed into desert. There are no plans to irrigate and restock the area. The traditional life of the nomads appears at an end. The problem began with a program of economic development, the expansion of herds tied to new commercial markets. The disaster showed what can happen when rural development is allowed to disrupt the local ecological balance. Much of the northern and eastern fringe of Black Africa, however, has a delicate fringe environment of dry land, and economic development is needed.

Based on "The Sahel Drought," *Encyclopaedia Britannica, Book of the Year, 1975.*

extent of these environments gives the continent relatively low potential for agricultural productivity, particularly for the traditional native subsistence farmer or herder who does not have access to modern expertise in agricultural production. Over the centuries the natives have worked out agricultural practices by trial and error. Even where Europeans work land for agriculture, they find it does not always respond to available scientific practices.

GRASSLANDS: THE PROBLEM OF DROUGHT

In central Africa rainfall is concentrated during a rainy season, with almost no rainfall during the rest of the year. The severe moisture deficits of the dry season turn the landscape yellow and brown and make this a difficult season for the traditional, locally based system. During the dry season herders search the landscape for pasture, animals become bags of bones, and farmers watch their crops wither in the hot sun. Traditional subsistence herding and farming can get by in normal or wet years, but the problem is that much of the area is close to the lower moisture limit for most plant growth. Thus excessively dry years create terrible problems to traditional societies. There is little surplus food and no means to store water for later use. Successive drought years require relief supplies from other areas to alleviate suffering and starvation. In 1973–1974 the culmination of a series of dry years across the southern fringe of the Sahara, a region called the Sahel, affected as many as 25 million people. An estimated 50,000 to 100,000 people died. Nomadic herders in the region lost as many as 70 and 80 percent of their herds. But dry years affected much of East Africa, too.

Soils in Africa are often a problem, as they are in other tropical areas. Even in the dry tropical areas most soils are thin and low in organic matter. When land is cleared of natural vegetation, the heat of the sun kills essential soil bacteria, and heavy rainstorms destroy the vital soil struc-

ture and leach away mineral components necessary for plant growth. Native agriculture has adjusted to these problems by planting a profusion of different crops in clearings to provide soil cover and by abandoning clearings when productivity decreases. Only the best soils have supported continuous cultivation year after year, since native agriculture generally uses no fertilization and depletes the fertility of the soil.

RAIN FORESTS:
THE WET WORLD OF AFRICA

West Africa and the coastal lowlands have tropical forests similar to those in Latin America. Too much moisture—not too little—is the problem in converting these areas to agricultural uses.

Moisture in an area with high temperatures throughout the year results in rapid vegetative growth. The hothouse conditions produce a tremendous variety of plants, but they also result in the rapid proliferation of insects and bacteria, which foster rapid decay of plants and cause disease. Malaria and yellow fever are carried by mosquitoes. Hookworm, amoebic dysentery, and schistosomiasis are also problems. The greatest scourge is sleeping sickness, which is spread by the bloodsucking tsetse flies native to Africa. In its different forms sleeping sickness is fatal both to humans and to domestic livestock. It has held down the native population and discouraged European settlement; more important, it has eliminated livestock as a part of the local farming systems.

Since the rapid-decay cycle goes on throughout the year, it is difficult to store produce without controlled storage facilities. But because the rain forest continues to produce fruits throughout the year, it will support a limited subsistence population. The native people in these areas have acquired some immunity to the worst diseases.

From the start Europeans considered the climate unhealthy. West Africa was known as the "white man's grave," and the wet tropical areas

FARMING IN A RAIN FOREST: A PROBLEM NOT YET SOLVED

In most underdeveloped countries one of the great problems is the increasing pressure of population on the traditional, locally based economies. Such pressure of increased numbers of people creates difficulties for traditional shifting cultivation in the tropical forests. Under that system farmers essentially rotate the land used for crops rather than rotate crops in continuously cultivated fields as is common in the midlatitudes.

This system, common among peoples in the tropical forests where both degree of technology and population density are low, depends on "slash-and-burn" clearing or **bush fallow,** which involves using a field for two or three years and then allowing it to return to wild vegetation for several times that many years. When the fields are planted, a great variety of crops are intermixed.

Farmers depending on trial and error and using only hand tools such as the hoe and digging stick make as much use of the environmental systems as possible. Clearing is done during the dry season, and the debris is burned off to leave the ash for important plant nutrients. The variety of crops planted not only keeps rapid-growing weeds from taking over but has the advantage of providing the farm family with the essential variety of food needed during the year. Different crops mature at different times, so some harvesting takes place during most of the year. Root crops also can be left in the ground as long as two years before they deteriorate;

thus there is a natural storage system. Most important, the new clearing depends on the natural soil regeneration process; green fertilizer is provided by abandoning the field to forest growth for the years between the clearing of the area.

In Ghana, Nigeria, and other African countries, where population growth is as much as 3 percent per year, increasing pressure is being placed on the rain-forest lands for use in agriculture. Moreover, forest lands are in demand for commercial plantation production of tree crops such as cacao, rubber, and oil palm. Such pressure has meant that the bush-fallow time between use and reuse of clearing has been reduced to two or three years. The result has been inadequate refertilization from fallowing and inadequate yields in the recleared areas.

Given this problem, what solutions might be tried? Two possibilities are apparent: (1) the addition of fertilizer to the farming system to reduce or eliminate the fallow period needed and (2) the planting of special high-nutrient fallow cover or of fallow crops that would have some food value.

In either case the present agricultural system calls for fundamental changes in the perspective and work practices of the people using shifting cultivation. The use of fertilizer not only requires technology and knowledge that the farmers do not have, but also may not work well on these soils. Moreover, such farmers will

either have to be subsidized by the government for fertilizer purchases, or they will have to enter further into the money economy. If they go more deeply into the money economy, several things are needed: a guaranteed market for cash-crop sales, some price guarantees, and adequate storage and shipping facilities—in short, a whole new structure for agriculture.

In the case of more valuable fallow growth, there is the problem of planting and maintaining desired crops in the presence of wild vegetation and its inevitable competition. Clearing land for fallow planting does not fit into present work practices. Moreover, the brush in the fallow areas presently is used as a major source of wood supply for cooking and other fuel needs. One possibility would be the addition of mechanized clearing and planting equipment, but this would probably mean government subsidization of the program.

Accompanying any of the solutions is the problem that individual clearings are small and each farmer works several fields interspersed among neighbors' fields. The use of either fertilizer or mechanization depends almost surely on the redistribution and consolidation of land, which in turn calls for fundamentally altering the present farming system. But the existing traditional system is no longer capable of supporting the population. Something must be done. What?

of West Africa have never received any appreciable influx of Europeans. Government and company officials and their families who have served tours of duty there have viewed the assignments as temporary. West Africa has the lowest percentage of Europeans of any part of Africa.

AFRICA'S EMERGENCE ON THE WORLD SCENE

The African coastline was explored by Europeans before much was known of the coast of the Americas. The ancient empires of Ghana and Mali are supposed to have been the chief sources of gold for Europe before the discovery of America. Nevertheless, most of Africa beyond the coast was out of contact with the rest of the world until less than a hundred years ago. Because of the steep rise in the land surface near the coast, it was impossible to navigate upriver over the waterfalls.

The first important contacts were across the Sahara, and the most sophisticated African societies in political and economic terms were those of Ghana and Mali along the northern desert fringe in West Africa. There was a steady trade between the contrasting lands on either side of the Sahara. In many ways these cultures were similar to early societies in Europe and Japan. Political control was built on an economic system that included both traditional and subsistence agriculture and trade connections with other areas. Most of the population depended on the local agricultural base, but the wealth of the empire, in the hands of the nobility, came from the profitable cross-desert trade. Merchants shipped gold, ivory, spices, hides, forest products, and slaves northward across the desert and received salt, copper, cloth, and dried fruits in return.

This trade represented an earlier model of the interconnected system, and nodal points were necessary in the transport and business management of such trade. Trading cities such as Timbuktu became centers of commerce and learning, and political kingdoms rose above the level of the subsistence tribe. The Moslem religion was introduced over the trade routes, and to this day much of the population of the desert fringe remains Moslem.

By the time Europe established trade contacts with Africa, much of the trans-Saharan trade had broken down, and the once prosperous empires had declined. Invasions by nomadic peoples with firearms disrupted political control. The newcomers took over the trade and brought herds of livestock that broke up the delicately balanced agricultural system of the area.

For centuries afterward the Sahara Desert proved a formidable barrier to overland trade, and the coastal zone, particularly the wet tropical lowland along the Atlantic, was forbidding to Europeans. The African coasts offer few good harbors, and for centuries ocean-going vessels were forced to anchor offshore and transfer cargo by means of small boats. Only in the twentieth century have artificial harbors been built to allow ships to reach dockside.

Because vast stretches of the West African coast are backed by dense tangles of **mangrove** swamp, Europeans were content to let native merchants bring wanted goods out to points along the coast for trade. The same was true, in general, of the east coast, where Arab and Indian traders had contacts before the arrival of the Europeans. The natives controlled the trading system and brought gold, ivory, and slaves to the coast from great distances, but they did not develop a real empire. Instead, they tapped each local trading system in the tribal village or at markets and periodic fairs. They also benefited from the booty taken as a result of tribal warfare. Trade routes were commonly paths rather than roads, and human carriers were used for transport. The connections were from tribe to tribe and nation to nation; these local interchanges added up to regional systems that brought goods to the coast.

Despite such contacts, most of Africa remained within the tribal group and the traditional, locally based system. There was a multitude of different local variations in the subsistence economy throughout Africa. In the drier areas nomadic herding predominated; in areas with somewhat better moisture conditions livestock herding was combined with slash-and-burn cultivation. In wetter areas, particularly near the equator, raising livestock was impossible because of the tsetse fly, and shifting cultivation was combined with forest gathering, hunting, and fishing. In areas with better soil and moisture conditions or where population pressure was great, communities maintained permanent fields. There farmers terraced hillsides with stone walls, rotated crops, kept cattle for manure, collected weeds to make compost, and used wood ash or night soil to maintain the fertility of their fields in an agricultural system somewhat like the intensive farming system of Asia.

THE SLAVE TRADE: FIRST EXPLOITATION OF AN AFRICAN RESOURCE

European traders along the coasts and on the desert fringe brought some changes to Africa. The traders were interested in African goods, but since transport was difficult, trade was limited to commodities of great value and little bulk—gold, ivory, pepper, and gum—and to slaves, which were mobile. Sections of the West African coast were known as the Gold Coast, Ivory Coast, and Slave Coast, but trade in each of these commodities took place at trading outlets throughout Africa.

European and Arab traders established trading posts at or near the coasts and depended largely on native-controlled trade to gather what they wanted from the continent. Sales allowed the Africans to purchase European guns, powder, rum, cloth, and hardware. For the most part they had little surplus from their subsistence life with which to trade. Slaves provided the one ready surplus commodity, and tribe fought with tribe for hostages. Within the tribal structure the chief might sell his subjects, or in dire circumstances the parent a child. Over much of Africa personal

security was lost and the area fell into anarchy. Rulers along the coast built up kingdoms as intermediaries in the slave trade, and their power blocked any plans of European traders to move inland.

While there is no good estimate of the total number of slaves taken from Africa, it is assumed to have been at least 15 and perhaps 25 million. West Africans were taken to Brazil, to the West Indies, and North America, where the demand for labor on plantations, in days before farm mechanization, was great. East Africans were marketed in Arabia and the Ottoman Empire. Whatever the numbers, the slave trade drew off the strong young people, a resource that any region could ill afford to lose.

COLONIALISM: PRODUCT OF EUROPE'S NEW INDUSTRIALIZATION

European colonialism burgeoned in the latter part of the nineteenth century. New industries in Europe brought an increasing demand for raw materials, and the developing urban society and its increasing wealth opened a new market for tropical specialty products. At the same time, steam-powered, steel-hulled ships made possible the movement of vast quantities of goods by sea, and railroads could reach the interior of continental regions.

Europeans also had political and cultural interests in Africa. Claiming new lands brought political prestige, and explorers were folk heroes to European society. The tales that they brought back concerning the primitive conditions of the native life they had seen inspired a wave of missionaries to convert the "heathen" and to better the lives of the "downtrodden."

Whereas in 1884 European claims in Africa consisted only of coastal footholds plus South Africa (Figure 14–4), by 1895 virtually the whole of Africa south of the Sahara was claimed by European powers (Figure 14–5). At first, development had been turned over to private trading companies, which soon were raising capital to build railroads and ports.

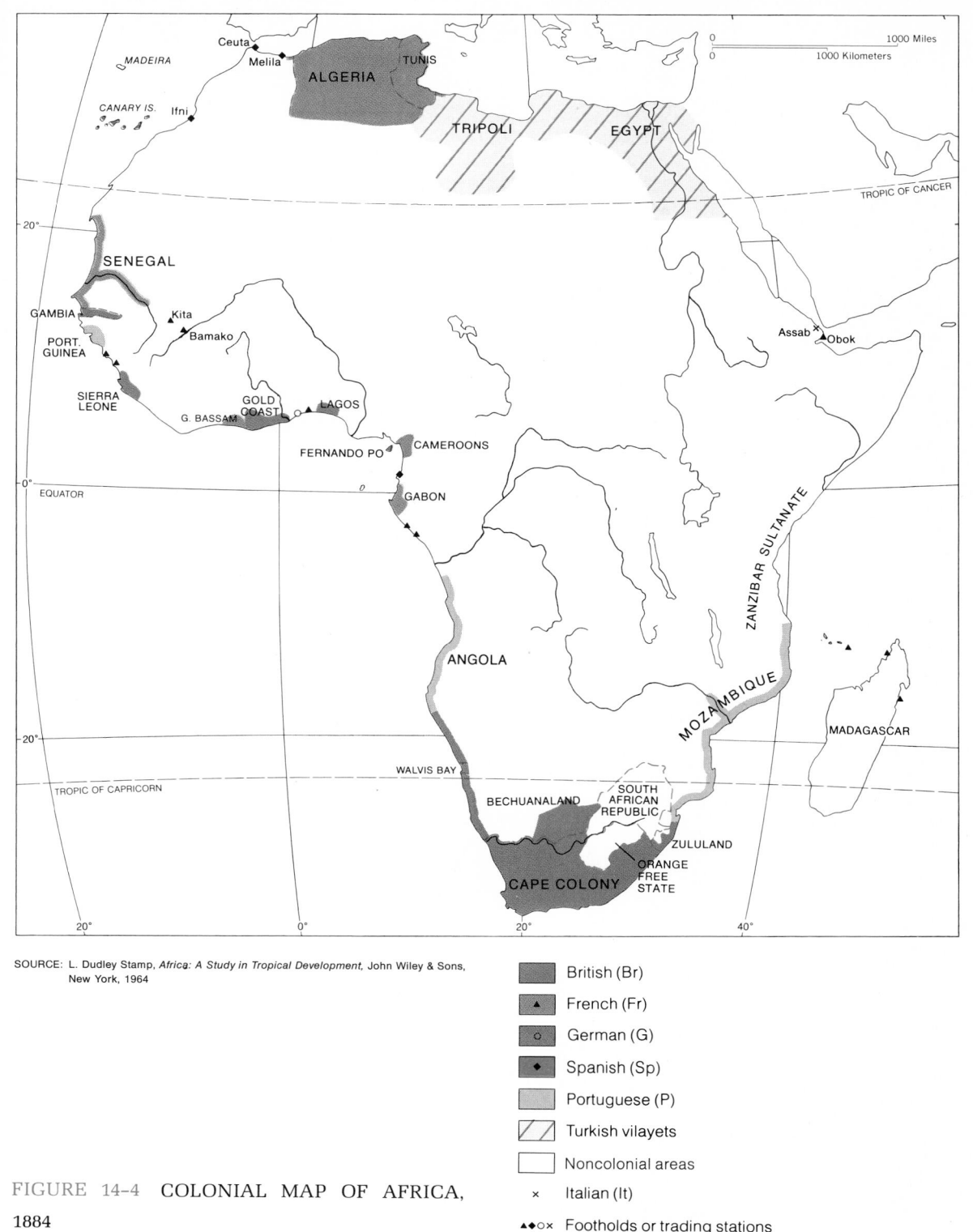

SOURCE: L. Dudley Stamp, *Africa: A Study in Tropical Development*, John Wiley & Sons,
New York, 1964

British (Br)

▲ French (Fr)

○ German (G)

◆ Spanish (Sp)

Portuguese (P)

/// Turkish vilayets

Noncolonial areas

× Italian (It)

▲◆○× Footholds or trading stations

FIGURE 14-4 COLONIAL MAP OF AFRICA,
1884

Map labels:

MADEIRA
Ceuta
Melila
ALGERIA
TUNIS
CANARY IS.
Ifni
TRIPOLI
EGYPT
TROPIC OF CANCER
20°
SENEGAL
GAMBIA
Kita
Bamako
PORT. GUINEA
SIERRA LEONE
GOLD COAST
G. BASSAM
LAGOS
FERNANDO PO
CAMEROONS
Assab
Obok
0°
EQUATOR
GABON
ANGOLA
ZANZIBAR SULTANATE
MOZAMBIQUE
20°
MADAGASCAR
WALVIS BAY
TROPIC OF CAPRICORN
BECHUANALAND
SOUTH AFRICAN REPUBLIC
ZULULAND
ORANGE FREE STATE
CAPE COLONY
20°
0°
20°
40°

1000 Miles
1000 Kilometers

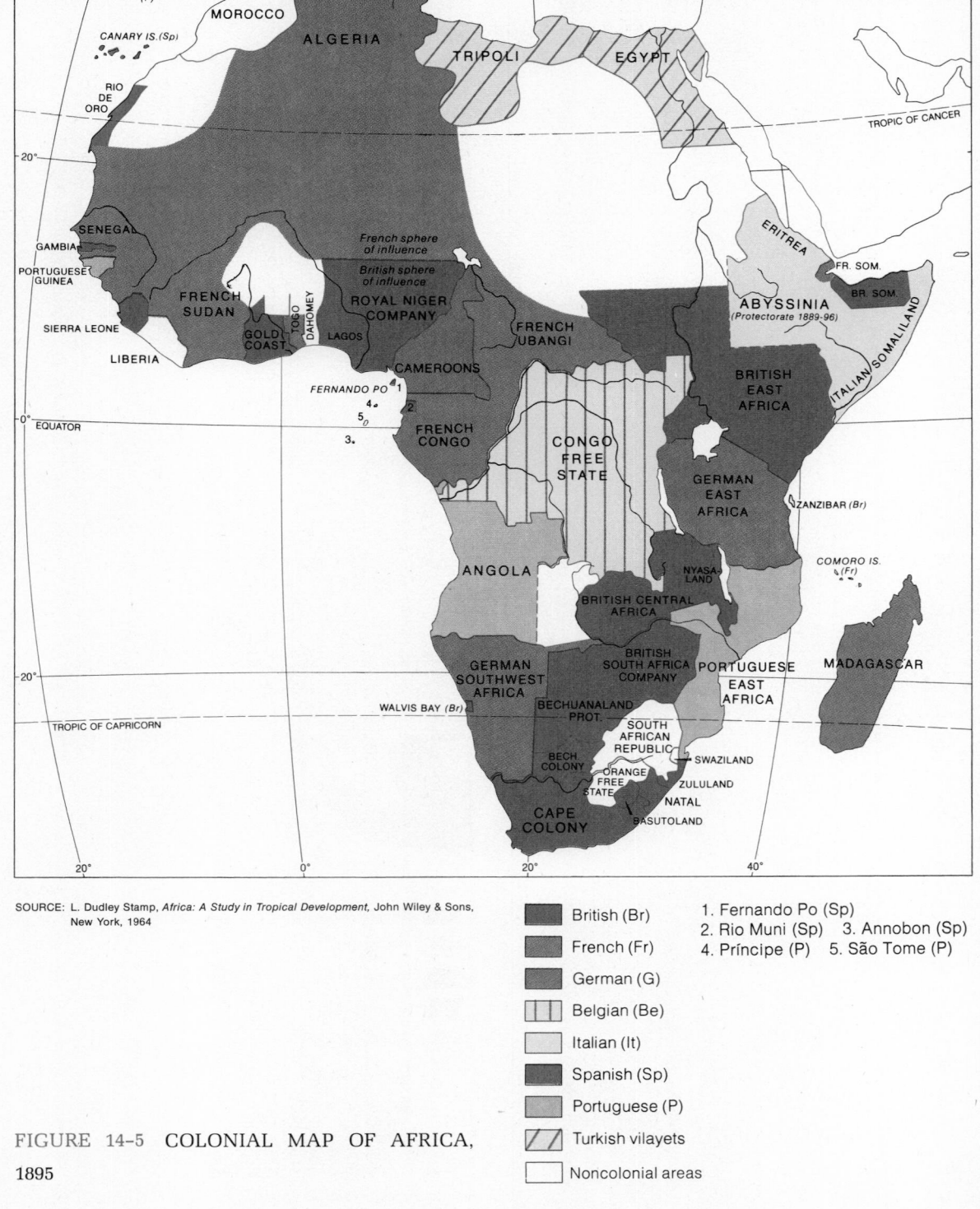

MADEIRA (P)

CANARY IS.(Sp)

RIO DE ORO

MOROCCO

ALGERIA

TUNIS

TRIPOLI

EGYPT

TROPIC OF CANCER

20°

SENEGAL

GAMBIA

PORTUGUESE GUINEA

SIERRA LEONE

LIBERIA

FRENCH SUDAN

GOLD COAST

TOGO

DAHOMEY

LAGOS

French sphere of influence

British sphere of influence

ROYAL NIGER COMPANY

CAMEROONS

FERNANDO PO

FRENCH UBANGI

FRENCH CONGO

ERITREA

FR. SOM.

BR. SOM.

ABYSSINIA (Protectorate 1889-96)

ITALIAN SOMALILAND

BRITISH EAST AFRICA

1

4

5

2

0

3.

EQUATOR 0°

CONGO FREE STATE

GERMAN EAST AFRICA

ZANZIBAR (Br)

COMORO IS. (Fr)

ANGOLA

NYASA-LAND

BRITISH CENTRAL AFRICA

PORTUGUESE EAST AFRICA

MADAGASCAR

20°

GERMAN SOUTHWEST AFRICA

BRITISH SOUTH AFRICA COMPANY

WALVIS BAY (Br)

BECHUANALAND PROT.

SOUTH AFRICAN REPUBLIC

SWAZILAND

TROPIC OF CAPRICORN

BECH. COLONY

ORANGE FREE STATE

ZULULAND

NATAL

BASUTOLAND

CAPE COLONY

20°

0°

20°

40°

1000 Miles

1000 Kilometers

SOURCE: L. Dudley Stamp, *Africa: A Study in Tropical Development,* John Wiley & Sons, New York, 1964

British (Br)

French (Fr)

German (G)

Belgian (Be)

Italian (It)

Spanish (Sp)

Portuguese (P)

Turkish vilayets

Noncolonial areas

1. Fernando Po (Sp)
2. Rio Muni (Sp)   3. Annobon (Sp)
4. Príncipe (P)   5. São Tome (P)

FIGURE 14-5 COLONIAL MAP OF AFRICA, 1895

The first important commodity from the tropical African coast was palm oil for soap, candles, and lubricants for machinery. Unilever, the British-Dutch parent company of Lever Brothers, the well-known soap company, has trading interests in West Africa that date from this period.

Most trading companies were not profitable, however, and soon African development was largely in the hands of European governments. Although native peoples struggled on several occasions to remain free of colonial control, in most cases land was taken by treaties containing promises of trade, protection, or both. Boundaries were arbitrarily established, often by compromises between European powers and usually with no relation to traditional boundaries. The land of the Somalis was divided five ways; there were British, French, and Italian Somalilands, a Somali enclave in Kenya, and another in Ethiopia. The Somalis are still disputing with the latter two states.

European investment in Africa was made both by colonial governments and private investors interested in resource development projects. By 1936 about $6 billion had been invested in Africa, slightly less than half of it by the colonial governments and the rest by private investors. Most government money went to build railroads, ports, and public works, whereas the greatest part of private investment was used to develop the mines.

Development as Counterpoint to Latin American Tropical Trade

Originally, Europeans saw tropical Africa in terms of its potential for producing agricultural specialties. However, with industrialization the development of mineral resources attracted much more attention. Perhaps because Africa did not come into the functioning European colonial system on a large scale until the twentieth century, agricultural specialization was less concentrated there than it was in Latin America. By the time African areas were under European control, Latin American plantations had already been well organized to produce cotton, coffee, sugar, and bananas, and Asian colonies were producing rubber, tea, and jute. Africa has produced all these crops, but not as a primary source of supply for any of them.

Africa south of the Sahara, not Latin America, has emerged as a treasurehouse of precious metals. Africa produces more than two-thirds of the world's gold, most of the world's diamonds, and more than one-fifth of its copper. A very large share of European investment went into production facilities for these three minerals alone.

Thus although European political control in colonial times covered almost the whole of Africa, European economic investment was centered on the three major minerals and the few localities where they were produced. By far the largest amount of economic investment went into South Africa, where first gold and then diamonds were discovered. Most of the rest of European investment went into the copper-mining districts of Zambia, Zaire, and Rhodesia. Investment in other colonies was minor by comparison, and because much of it was in plantation agriculture, it was widely scattered.

Islands of European Development in a Sea of Rural Africa

With investment in mines and plantations, the European colonial powers and the favored private investors established localized "islands" of the modern interconnected system within the vast native subsistence system. The nature of this localized system can readily be seen by examining Figure 14–6.

Notice the areas of commercial production, particularly the mineral areas. You can see that railroads reach back from ports to make connection with these productive areas. These railroads are part of schemes by European investors to obtain specific products needed by the worldwide modern system, primarily in Europe, Anglo-America, and Japan. They can move products to export markets but do not form an extensive rail network within any country, much less within Black Africa as a whole. The most important railroads are those that bring out the

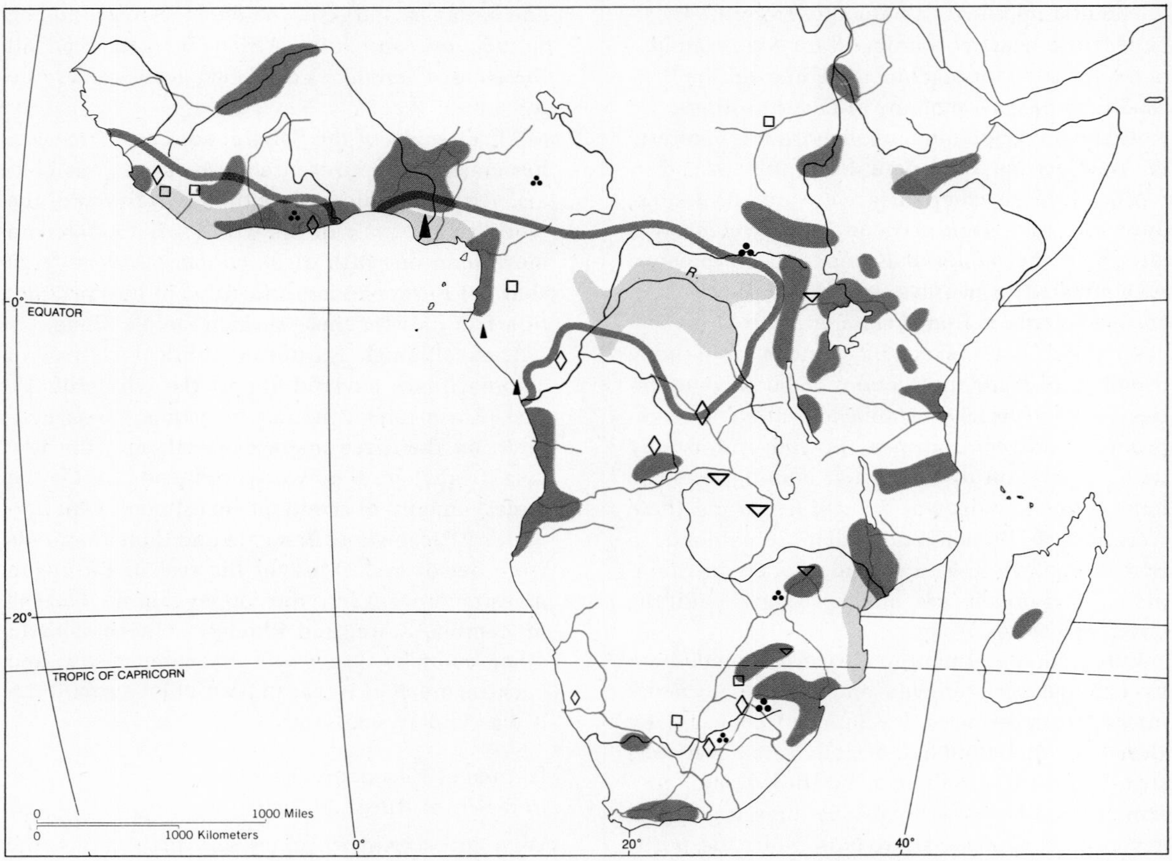

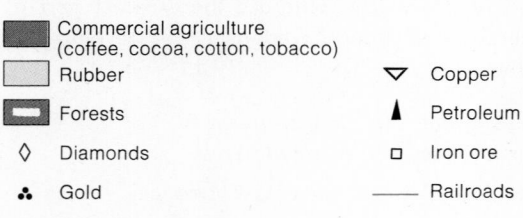

FIGURE 14-6 **MAJOR AREAS OF COMMER-
CIAL PRODUCTION IN AFRICA SOUTH OF
THE SAHARA**

copper of Zaire and Zambia. In the case of Zambia, the railroads that carry the copper to the coast must pass through neighboring Angola or through Rhodesia and Mozambique. Copper movements were handicapped in the mid-1970s by the struggle for power in Angola; shipments through Rhodesia and Mozambique were disrupted by the black-white confrontation centered in Rhodesia.

*Migratory Labor: A Lifestyle
Bridging Both Worlds*

The commercial islands and their transport links have formed new centers of employment for indigenous Africans. The mines and plantations need workers, and since slavery has long been outlawed, most work is done by people recruited from rural tribal units. Since rural densities are low, individuals often travel hundreds of miles to work in the mines; there they live in workers' dormitories. Typically, the worker's family remains behind, continuing in its traditional lifestyle. This is possible because much of the work in agriculture has always been handled by women. Males were traditionally hunters and warriors and only incidentally concerned with farming. They cleared new fields and watched after livestock but did not participate in the daily farm routine. The men who now work in the commercial mines are transients in the interconnected system; neither they nor the companies see them as permanent miners. They are recruited and assigned for a contract period of three years. They live in bachelor compounds and are prohibited from bringing their families. A worker stays until he feels that he has acquired enough money to meet his family's needs or until some pressing tribal or family matter calls him back. If he does not return for another contract period, the mining company will hire another young man from the countryside to replace him.

Life in the mines is dangerous—over 500 men were killed and 23,000 injured in South African mines in 1974. The mining companies believe that the work force is inefficient because of its transient nature, and it is hard to imagine that the workers benefit from such a lifestyle.

These migratory workers, more than any other people we have seen, are members of both systems. On the average they spend over 60 percent of their working life from age sixteen to forty-seven outside their tribal area. Nonetheless, they are still a part of the tribal family, and the pay they send home supports traditional village life at home.

Remarkably, the migratory workers cross national boundaries regularly as they move between work and village. The low densities in rural areas make it necessary to draw workers from a very large area. Mines in South Africa utilize some 400,000 workers from Lesotho, Botswana, Swaziland, Mozambique, and Malawi. Malawians also work in Rhodesia, Rwandans work in Uganda, and people from Niger and Chad work in the cotton fields in the Sudan. Some people from Nigeria and Upper Volta are employed in Ghana and the Ivory Coast. In this way the nodes of production in the modern interconnected system serve as growth points that pump money into widespread rural areas.

People from countries such as Malawi depend on employment in other countries for money. To a degree, the emerging modern interconnected system of one country depends on the expansion of resource production in another. This is a particularly difficult situation for African countries whose citizens work in South Africa and Rhodesia, where European minorities control the governments. How can such countries enforce economic sanctions against apartheid policies when their citizens depend on jobs in racist countries?

Obviously, the presence of persons who migrate between cultural-economic systems results in modifications in both systems. The workers of mines and plantations come from various rural areas and bring their own customs and languages with them. Training programs and living facilities must take into account such ethnic diversity. Workers are exposed to a whole set of new standards and values, and they must learn to work under supervisors who speak a different language. When these workers return to their villages, they inevitably transmit some of this new culture to their families.

LARGE CITIES: INDEXES OF THE
MODERN INTERCONNECTED SYSTEM

The system of transient workers can operate well only where masses of unskilled workers are needed. It is not an answer to the industrial and business needs of mechanized mines and modern cities, where people trained to handle complex tasks are increasingly needed and masses of unskilled workers are decreasingly needed. Industry calls for a stable work force committed to continuous labor within the modern interconnected system. In South Africa, where industrialization has moved ahead with production of steel, machinery, automobiles, and appliances, migratory labor has been largely replaced by a permanent population of indigenous Africans who have given up tribal life. Thus in large cities throughout Africa, a trend of accelerated urban growth and rural depopulation has begun, similar to trends in the industrial countries and Latin America, but on a much smaller scale.

Nevertheless, Africa south of the Sahara remains the most rural of all regions of the world. In most African nations over 80 percent of the population is rural. South of the Sahara, just two metropolitan areas have over 1 million persons: Johannesburg, center of mining and industrial development in South Africa; and Kinshasa, capital of Zaire. Only seven other cities have over one-half million people, and two of them are in South Africa (Figure 14–1). A dozen African countries have capital cities with populations of less than 100,000 persons, and seven of these are below 50,000. As a rule, the capital of an underdeveloped country is the largest city, and city sizes follow the primate pattern described in Chapter 9. Only South Africa and Nigeria have more than two cities of 100,000 or more people.

The size and number of large cities correlate more closely with a country's degree of commercial development than with its total population size. South Africa, with less than half the population of Nigeria, has eleven cities of over 100,000 as compared to seven in Nigeria. Ethiopia, which is second in terms of population, has only one city of more than 100,000.

Political as well as Economic Links

Large cities indicate the degree to which a country has become involved in the modern interconnected system. The capitals are not only the political centers around which the internal affairs of the country develop, but are also the links with the modern, international political scene. Even the smallest, least developed country has links with the United Nations and with the major political powers of today's world. Such contacts are vital not only in political terms but also for economic and cultural programs. Through such contacts comes most of the investment capital for economic development. Africa receives assistance from individual industrial countries, the United Nations, and the World Bank, and private investments in transportation and resource development. The People's Republic of China has given aid to Tanzania. These sources also provide a stream of technical assistance for development programs and offer ties to institutions of higher education through which the new leadership of African countries can be trained. All aid is normally administered through the large city.

Political Capitals: Heritages of Colonialism

Almost without exception political capitals were created by the European colonial governments, which built streets, government buildings, and official residences after the style of their homelands. Governmental form also reflected European views. Since the leaders of new native governments were trained in the industrial countries, even after independence, the governmental institutions still follow Western culture. The business cores of almost all African cities seem very familiar to visitors from industrial countries. Moreover, these cities are very modern, since they have grown tremendously in the last few years.

Capital cities may be the center of life in African countries, but they also impose major burdens on the new governments. Paternalistic colonial governments developed elaborate, free public services for cities including free schools, free medical services, free roads, free water, free sewerage, free or subsidized electricity, and subsidized housing, often at the expense of the agricultural population in the rest of the country. Today, weak, underfinanced national governments believe that they must not only continue such services, but must also expand them to make their capitals worthy showplaces of national life. Moreover, city populations are insistent in their demands for such services. Thus in countries with very limited financial resources, money spent to subsidize city life leaves less for other essential national investment.

Moreover, city people dominate Africa's new governments. Government is in the hands of the educated elite and military officers. Since independence no country has held democratic elections. Unable to reach consensus among the variety of tribal groups that inhabit their territories, African countries have resorted to one-party and military regimes. Not one of Black Africa's leaders has lost power through elections. All have been deposed in military coups, and the soldiers who have taken power have generally kept it. Several African countries, including Senegal, Ivory Coast, Guinea, Kenya, Tanzania, and Zambia, still have the same leaders today they had more than fifteen years ago. Thus, like most Latin American countries, African countries are run by a tiny power group, usually the military, and the masses of the population have no voice in the government. Leaders remain in power as long as they carry out policies acceptable to the elite. Then they are deposed in a coup and new military officers take over. This ruling clique, centered in the capital, sees itself increasingly as a part of the modern interconnected system of the world and loses its identity with the rural masses.

Governments as an Economic Force

Government domination in a country's economic affairs was typical of colonial regimes and continues today. The government is the single largest enterprise, and its expenditures usually amount to about 20 percent of gross national expenditures. If publicly owned railways, electricity, ports, and other basic industries are included, governmental expenditures may amount to one-third of the total, and the private investment sector may account for less than one-third of the economy.[2]

[2]For comparison, in the United States private business production accounts for over 80 percent of total production, whereas the government total is about 11 percent.

The government is also the chief employer. It hires from one-third to half the wage earners of the country, and most government employees live in the capital city. However, the number of paid workers in the labor force of the largely subsistence African countries is very low, usually from 5 to 10 percent.

Government jobs traditionally carry the highest prestige and pay the highest wages. Most educated Africans aspire to such positions. Even the relatively massive government system cannot absorb all the Africans looking for clerical and other low-level government jobs, and large numbers of urban Africans with educations are unemployed. Once their families have made sacrifices to send them to school, they are regarded as failures if they become farmers like the uneducated, so they do not take laboring jobs.

As the locus of government, the big city is the decision-making center for the entire country. The city-based government sets fiscal priorities, decides on rural land programs, and underwrites the development of transportation and public utilities for the whole country. It approves resource concessions for foreign investors and trade programs, and also determines who will be educated, what the cultural image of the country will be, and even what the national language will be.

Metropolitan Centers in the Modern Interconnected System

The largest metropolitan centers are also the major commercial centers and the headquarters for the modern business organizations in the country. Here are found the national management functions of the large international companies that are developing the resources of the country as part of the modern interconnected system centered in industrial areas outside Africa. This system of international business, begun in colonial times, remains important to the foreign investor and essential to the new government as a source of foreign exchange, taxes, and native jobs.

SOUTH AFRICA

South Africa has developed differently from the rest of Africa south of the Sahara. Although more than 80 percent of its population is native black or of mixed black-white ancestry, its political and economic system has been controlled by Europeans since the first colony of the Dutch East India Company was established there more than 300 years ago. Today there are more than 4 million descendants of European settlers in South Africa. Most are Afrikaners, of Dutch ancestry, who speak their own version of Dutch, but there are a large number of British people as well.

European settlers were attracted to this region, the only midlatitude area of Black Africa, by the farming possibilities. Most of South Africa is high plateau over 2,000 ft (610 m) above sea level. Thus the climate is more moderate than that of Florida, which lies in approximately the same latitudes (though in the Northern Hemisphere). Much of the middle and high **veldt** (grassland) of the interior consists of farmland and ranches producing corn, wheat, cattle, and sheep. These were the lands of the original Dutch settlers, but today many of the European settlers have moved to the cities as absentee landlords and their rural properties are worked by black Africans. In fact, from the start the European settlers imported African and Asian slaves rather than seek white laborers from Europe. Thus European settlement has developed a different pattern than in other areas of European settlement, such as Canada, Australia, and New Zealand.

Farmers also occupied the lowlands around Cape Town at the southern tip of the continent. With their subtropical dry-summer climate, these lands were suitable for growing grapes and citrus fruit. European farmers also settled the wetter subtropical eastern coast, where sugar cane and cotton could be grown.

Much of South Africa's agricultural output is exported to the United Kingdom and Japan. The European population of South Africa is too small to provide an adequate market, and the black majority is mostly too poor to be important consumers. Many of the blacks are settled on native reserves where they carry on traditional, largely subsistence agriculture typical of other parts of Black Africa.

South Africa's economic success has come from its minerals, not its agriculture. South Africa produces about 60 percent of the world's gold and is the third leading diamond producer, after Zaire and the Soviet Union. In recent years, however, the revenues of the South African government have been substantially lower than anticipated as the result of a depression in the world price of gold.

Large coal deposits were found to provide power to run the mining machinery and the railroads. Coal also formed the base for an industrial revolution. Iron ore, manganese, asbestos, platinum, and copper have been found, and uranium is produced as a by-product of gold mining. Steel mills using local coal and iron produce more than 12 million tons of steel per year for use in making automobiles, trucks, buses, machinery, and household appliances. More than a million persons are employed in manufacturing in South Africa.

As a result of mining wealth and industry, South African whites live in an affluent modern society. Johannesburg, located among the gold fields, is a modern metropolis of over 1 million people, and Cape Town is a city of about 800,000. Both are centers in a national rail and road network and their airports are part of the worldwide jet system. South Africa participates in many ways in the modern interconnected system of the world. One such way is medical research; the first heart transplant was accomplished by a South African medical team.

As with agriculture, South African mining and industry and many of its city services depend on black labor. Black Africans provide most of the labor in the mines and factories, on the buses and trains, and in the shops. Yet these black Africans have had very little personal share in South Africa's modern economy. They hold only the more menial jobs. More important, they must live apart in black compounds separate from the white cities and carry work permits so they can travel outside their compounds. They go to separate schools and churches and have separate medical facilities.

A conscious policy of **apartheid** has been followed since 1948. Apartheid calls for the complete segregation of whites from blacks and coloreds (people from other racial stocks, such as Orientals). In addition to the urban black residential compounds, native reserves are set aside in rural areas. One aspect of apartheid is the intention to grant "independence" eventually to the native reserves, even though most reserves are geographically disunited and could not survive economically. In 1976 the native reserve of the Transkei along the east coast was granted "independence." However, this new "nation" has not yet been recognized by any of the major powers of the world.

The suppression of the political, social, and economic rights of South African blacks under apartheid is under increasing pressure from the world community, but particularly from black leaders within the country. Since the early 1960s there have been sporadic riots, and they have increased in frequency in the mid 1970s. It remains to be seen if the white South Africans can continue to rule their black majority in a culture area where most other countries are governed by blacks.

It is logical to have the business management operations of a company located in the largest city of the country. The local corporate management in the capital has direct air and communications connections to the corporate headquarters elsewhere. At the same time it is at the node of national transport and communications and at the seat of government, where economic decisions are made. Thus in the capitals of the proud new African countries are found the management operations of the international corporations whose chief concern is the export of the country's resources for use elsewhere. The impact of the modern interconnected world is very great, and to an important degree capital size is proportional to the importance of the country in the external, interconnected system. In the largest capitals private office buildings housing local headquarters of these corporations and trading companies are becoming as prominent as government buildings.

THE PROBLEMS OF A COLONIAL PAST

Each of the European powers colonized Africa with varying degrees of political independence, social concern, and economic development. Understandably, the newly emergent independent states reflected the differing styles of the colonial governments. Zaire, for example, was known as the Belgian Free State under the direct control of King Leopold and was operated as a gigantic plantation with an absentee landlord. When the territory was taken over by the government of Belgium in 1908, the natives were regarded as childlike wards to be managed by paternalism. Profit was of first importance to the king, who had little concern for the people. The natives had duties, and some rights; their "place" was determined by the Belgians. They were given no preparation for self-government; thus none of the people were trained to be leaders, and when independence came there was virtually no one to take charge.

France, which ruled its colonies in a very authoritarian way at first, later espoused the principle that all French territories were a part of France and all residents were French people. They made an attempt to educate as many of the natives as possible to become responsible and intelligent French citizens.

The Portuguese also strove for assimilation of all colonies and all colonial peoples as Portuguese. Interbreeding and intermarriage between natives and Europeans was encouraged. Theoretically, natives who could read Portuguese and who accepted Christianity were entitled to full Portuguese citizenship, but only if they put aside tribal loyalties, proved that they were of good character, and demonstrated that they could earn enough to support a family. Obviously, few natives could achieve that status. The remaining natives were subject to another set of rules and could be sentenced to forced labor by authorities.

British colonial rule varied greatly from colony to colony, partly depending on the interest of Europeans in settlement. In hot, wet West Africa, where the British showed little interest, they were not allowed to own land. But in cooler, drier East Africa, European immigrants not only owned land but forced natives from their traditional homelands to make way for their settlements. Because of this, independence in British colonies brought different problems in different places. In some the question was how to arrange native rule; in others it was whether the native majority or a white minority should rule.

Where European populations were present in former British colonies, they remained in control. For example, in Rhodesia the European minority constitutes less than 6 percent of the total population, but it controls the government. But in Kenya the black population controls the government even though a European population of less than 1 percent remained dominant in business and commercial agriculture after independence. Europeans who have stayed in Kenya have retained their economic position.

Today twofold pressure, from black majorities within the countries and from the other countries of Black Africa, is forcing changes, especially in Rhodesia. In 1976–1977 the white government of Rhodesia was pressured to agree to a conference with black leaders to negotiate the transfer of power to the black majority in the country. Guerrilla activity along the border with Mozambique raised the possibility of full-scale war. Most other African countries had no significant European populations and are now largely in the hands of native peoples, although their leaders have been trained in European institutions. Government forms follow European models, and the boundaries and cultural patterns established by Europeans also remain. Most new governments have emerged as one-party dictatorships rather than democracies. The leaders tend to be charismatic figures whose support comes as much from military power as from widespread popular endorsement.

Most governments are strongly nationalistic—a nationalism resulting from an attempt to weld many tribal cultural heritages into a single, national culture. However, the models of government that they must follow are those created in the industrial countries. African leaders learned about government from colonial rule and from university education in Europe, the United States, or the Soviet Union. Moreover, in the modern interconnected system of the United Nations and the international business world, a national language that reflects native culture does not do much good. Leaders in politics and business operating on the international scene must be fluent in at least one of the recognized international languages—English, French, Spanish, German, or Russian. In countries where there are several tribal groups, it is often easier to speak the former colonial language that everyone in the country learned in mission schools. An unhappy result is that not only the governmental body of the country but also the modern interconnected system uses a language unknown to the masses of people in the traditional system. The rural tribes who speak only their local dialects are cut off from the modern elements of national life, and the gulf between the traditional system and the modern system of the country is maintained.

CONTACTS WITH THE REST OF THE WORLD

As in Latin America, most of the exports from Black Africa originate in a few countries (Table 14–4). The Republic of South Africa accounts for

TABLE 14-4 GROWTH OF EXPORT TRADE IN COUNTRIES
OF AFRICA SOUTH OF THE SAHARA: 1938, 1948, 1958,
1968, 1972 ($ millions)

Country	1938	1948	1958	1968	1972
South Africa	496	946	2,018	3,235	2,602
Nigeria	47	253	475	587	2,180
Zaire	64	273	501	516	1,004
Gabon	—	—	48	125	905
Zambia	49	140	362	759	758
Angola	15	75	124	276	511
Rhodesia	52	135	193	275[a]	499
Ivory Coast	11	79	151	425	443
Ghana	76	216	325	336	393
Tanzania	22	82	174	241	319
Kenya	40	58	113	176	267
Uganda	4	81	120	186	261
Liberia	2	28	83	169	244
Cameroon	7	47	97	189	218
Senegal	22	72	113	151	216
Mauritania	—	—	—	72	193
Mozambique	9	37	78	154	175
Malagasy Republic	24	69	75	116	164
Sierra Leone	12	22	83	96	117
Swaziland	—	—	—	42	86
Malawi	—	21	24	58	81
Guinea	3	11	55	55	78
Niger	1	3	13	38	54
Togo	2	9	15	39	50
Congo (Brazzaville)	7	43	18	49	46
Somalia	5	6[a]	25[a]	30	43
Benin	3	13	18	22	42
Central African Empire	7	43	14	36	36
Chad	7	43	13	28	36
Mali	—	—	—	17	34
Upper Volta	[b]	[b]	4	21	20
Gambia	2	6	8	15	19
Lesotho	—	—	—	7	9
Total	966	2,753	5,417	8,647	12,380

SOURCES: Andrew M. Kamarck, *The Economics of African Development*, revised edition (New York: Praeger, 1971), pp. 115–116. Copyright © 1971 by Andrew M. Kamarck.
[a] Estimate.
[b] Included with Ivory Coast.

21 percent of all exports, Nigeria for 17 percent, and together the top eight countries account for more than 70 percent of the total. Thus less than 30 percent of all Black African export trade is shared among thirty-two countries.

To a large degree the concentration of African output in a few places reflects the production of minerals in just a few countries. The value of mineral exports from Black Africa is much greater than that of agricultural shipments. Petroleum ranks first, and Nigeria is by far the leading oil exporter. Gabon and Angola, less important oil exporters, are nevertheless among the leading exporting countries. The copper exporters, Zaire and Zambia, which share the African Copper Belt, are also among the leading exporting countries. South Africa owes its leading position to diamonds, uranium, and copper.

Agricultural exports are lower in total value, widely scattered among countries, and divided between the products of the hot, wet tropics and the higher, drier areas. None of the top half-dozen African exporting countries is primarily an agricultural producing nation. The Ivory Coast is a major producer of both coffee and cocoa, the leading agricultural export crops of Black Africa, and Rhodesia has important tobacco exports. Cacao, a crop grown in the hot, wet region of Africa, is exported mainly from Ghana, Nigeria, and the Ivory Coast. Coffee is most important in highland East Africa, particularly Uganda, Kenya, and Ethiopia, but it is found in the western highlands too, in the Ivory Coast, Angola, Cameroon, and Zaire. Cotton is an important crop of the highlands and the grasslands of the fringe of the Sahara Desert; Tanzania, Uganda, Nigeria, and Mozambique are important cotton producers.

RAPID GROWTH IN EXPORTS
IN RECENT YEARS

In contrast to Latin America, where exports have not increased greatly since 1929, Black Africa has seen the value of its exports increase over tenfold between 1938 and 1972 (Table 14–4). Most of this growth has come since World War II. In 1948 the value of trade was almost four times what it had been in 1938, and between 1948 and 1972 exports increased more than five times. Remarkably, the growth has occurred in almost every country. As a result of the general growth everywhere, the major exporting countries have not changed much in relative importance. Kenya and the Malagasy Republic are comparatively less important today, and although the leaders such as South Africa, Nigeria, Zambia, Zaire, Ivory Coast, and Ghana have grown rapidly in absolute terms, there has been little change in their ranking.

A Competitor of Latin America

Remember that colonial development occurred in Latin America more than a hundred years earlier than it did in Africa, and that commercial development in Africa is largely a phenomenon of the twentieth century. Because Latin America and Africa have similar resources, Africa really could offer little that was not already being developed in Latin America. The early exploitation of Africa emphasized gold, diamonds, and palm oil.

For the most part, exports from Black Africa compete with similar products from Latin America and South and East Asia. They include the basic tropical and subtropical export products: coffee, cocoa, cotton, tobacco, copper, petroleum, iron ore, and timber. Of these, only copper, cocoa, and timber are produced in greater quantity in Africa than in Latin America (Table 14–5).

Of the tropical agricultural export products, only palm oil is indigenous to Africa. All the other products were introduced by Europeans during colonial days to provide revenue for the new colonies and to compete with products from the independent Western Hemisphere countries. Except for the cocoa market, however, Africa has not been able to make up for its late start, and it trails Latin America in every other major agricultural export.

TABLE 14-5 COMPARISON BETWEEN AFRICAN AND LATIN AMERICAN
PRODUCTION OF EXPORT PRODUCTS, 1925–1973

	Avg. 1925–29	Avg. 1940–44	1950	1960	1967	1973
Copper (thousands of metric tons)						
Africa	120.7	429.2	519.1	975.8	1,128.3	1,580.0
Latin America	379.5	564.2	481.5	796.9	922.1	1,177.2
Petroleum (thousands of cubic meters)						
Africa	237	1,331	2,656	16,744	181,370	119.7
Latin America	28,327	47,232	110,669	210,401	277,217	265.3
Iron ore (thousands of metric tons)					Black Africa	
Africa (incl. North Africa)	3,751	2,666	7,004	15,464	43,001	44,100
Latin America	1,183	1,526	5,628	43,643	62,890	68,400
	1929–30 1933–34	1939–40 1943–44	1950–51	1960	1967	1973
FAO coffee (thousands of metric tons)						
Africa	72.9	172.0	266.5	817.5	1,145.0	1,382
Latin America	1,988.7	1,623.6	1,784.0	2,945.0	2,750.0	2,502
	1930–34	1940–44	1950	1960	1967	1973
FAO cocoa (thousands of metric tons)						
Africa	376.0	407.2	485.4	661.3	968.2	962
Latin America	162.1	208.9	260.0	367.4	337.1	356
	1925–29	1940–44	1950–51	1960–61	1966–67	1973
FAO cotton (thousands of metric tons)						
Africa	438	479	685	904	1,078	587
Latin America	259	749	864	1,305	1,462	1,690

(cont.)

TABLE 14–5 (Cont.)

	1925–29	1946–49	1950	1960	1966	1973
Forest products (thousands of cubic meters roundwood)						
Africa	—	90,550	89,780	190,073	238,278	297
Latin America	—	177,910	191,978	209,347	274,268	238

	1948	1958	1968	1972
Value of total exports				
Africa	430	1,640	3,040	5,510	8,800
Latin America	2,010	7,460	9,600	13,900	20,370

SOURCES: Grunwald and Musgrove, *Natural Resources in Latin American Development*. Published for Resources for the Future by Johns Hopkins University Press. Copyright © 1970 by the Johns Hopkins Press. All rights reserved. *U.N. Statistical Yearbook, 1974.*

The recent increase in African production reflects the growing demand for its products during the last half-century, particularly since World War II. The new affluence of the industrial countries, particularly in Europe, which has been Africa's traditional market, has been a vital factor in improving the African export base. World consumption of copper tripled between 1929 and 1972; coffee and cocoa consumption almost doubled; and cotton use rose over 40 percent.

African countries have also benefited from old colonial trade ties with European countries that were retained after independence. Former British colonies that became members of the Commonwealth have had privileged access to markets in the United Kingdom, and later to the European Economic Community; and former French colonies were given associate status in the EEC with preferential treatment. These advantages over the independent Latin American countries have helped Africa's international position greatly. In 1975 all of Black Africa plus former EEC colonies in the West Indies and the Pacific were covered in an agreement that allows almost all their exports to enter EEC countries duty free. It provides guaranteed annual payments in case of crop failure, yet allows associated countries to trade outside the EEC Association and it provides monies for economic development.

Africans have also benefited from attempts by other producing areas to raise the price for raw materials. Coffee production in Africa was encouraged by efforts of Latin American countries to establish a floor on coffee prices, though the agreement was ended in 1972. With a guaranteed minimum price in world markets, much of the risk of entering coffee production was gone. Likewise the efforts of the United States, the world's leading cotton exporter, to support domestic prices through acreage control and price subsidy on domestic sales have raised cotton prices throughout the world and encouraged other countries to increase production. Now African countries are joining international producers' groups in order to support the price of such key commodities as coffee, cocoa, copper, and bauxite.

Problems in Trading Raw Materials for Fuel and Manufactures

For most countries in Black Africa the pattern of trade remains as it was in colonial days: agricultural and raw material exports are exchanged for manufactured products from the industrial countries. Most countries are also dependent on the outside world for basic industrial fuels, particularly oil. General inflation and increased demand have resulted in a long-range upward trend in the price of such commodities as coffee, cacao, tobacco, and cotton, but there has been an even more rapid increase in the cost of the manufactured goods that African countries must buy. Development in most Black African countries has been particularly hard hit by the rapid increase in oil prices as the result of actions taken by OPEC in 1973–1974. This has been a great boost to the African OPEC members, Nigeria and Gabon, but other countries have found themselves paying four times as much as before for essential energy sources.

African countries are more dependent than most developing countries on trading markets in the industrial countries. Most Black African countries are only beginning the process of industrialization. They must depend on imports both for manufactures used in current consumption and for those needed to produce dams, railroads, and other requirements of modern development. Imports are paid for by exports, and the market for those exports is in the industrial countries—mostly Western Europe, the United States, and Japan. As a result, Africa feels the repercussions of economic events in these industrial regions. In 1974–1975, as a result of recession in the industrial nations, the prices of major African commodities except petroleum dropped sharply while, at the same time, as a result of inflation, the prices of industrial goods that Black African countries wanted to buy continued to rise. The purchasing power of exports dropped sharply, leading to large trade deficits. For most countries it was a time of zero economic growth at best, and in some cases the progress made in previous years was partially reversed. As a result, developing countries, led by the Black African bloc, have been agitating for international agreements that will bring both greater stability and parity to commodity trade.

MEASURING AFRICA'S MODERNIZATION

The relatively high per capita value of exports in Africa as compared to Latin America would suggest considerable potential purchasing power in African countries. However, there are a number of problems, particularly the small populations of most African countries. Although the per capita figure may appear significant, the total income from exports remains small. Moreover, income distribution is a problem. Money may be spent on governmental institutions, national symbols, or national defense, rather than on the economic and social well-being of a nation's people.

In Latin America we used per capita exports as an initial index of the degree of industrialization. Using this measure, we would expect Zambia, among the countries with large populations, to be the most highly industrialized country south of the Sahara, with Rhodesia and South Africa well behind and with Ghana, Zaire, and Nigeria lagging badly (Table 14–6). But in fact, only South Africa has any significant steel production, and if electrical energy production is used as a measure, all the other African exporting countries show poorly in comparison with Latin American exporters, as we can see in Table 14–7. Except for South Africa, few nations have much development of home-based modern industry.

Also apparent from Table 14–6 is the dependence of many of the African economies on one major export. In five of the countries listed—Burundi, Gabon, Gambia, Nigeria, and Zambia—one commodity accounts for more than 80 percent of total exports. Price fluctuations in a particular commodity can therefore have a great impact on the economic fortunes of one of these countries. The recent price increases in petroleum have added to the revenues of Gabon and Nigeria. On the other hand, a steady decline in copper prices, discussed in Chapter 12, has caused difficulties for Zambia.

Most of Africa Remains Underdeveloped

In terms of per capita electricity consumption, South Africa is far ahead of any of the other

TABLE 14-6 VALUE OF EXPORTS OF AFRICAN COUNTRIES SOUTH OF THE SAHARA, 1972

Country	Total Value ($ Millions)	Per Capita Value ($)	Leading Commodities (Percent of Country's Total Exports)	
Western Africa				
Benin	42[a]	15	Palm oil	34
Gambia	19	48	Peanuts	94
Ghana	393	45	Cocoa	63
Guinea	78	19	Bauxite	65
Ivory Coast	553	123	Coffee	22
			Cocoa	21
Liberia	244	325	Iron ore	65
Mali	34	6	Cotton	36
Mauritania	193	148	Iron ore	73
Niger	54	13	Uranium	39
Nigeria	2,180	31	Petroleum	91
Senegal	216	53	Peanut products	53
Sierra Leone	117	45	Diamonds	60
Togo	50	25	Phosphates	46
Upper Volta	20	4	Meat	38
Guinea-Bissau	3	5	Peanuts	46
Total	4,196			
Central Africa				
Angola	511	85	Crude oil	29
Cameroon	218	36	Coffee	25
Central African Empire	36	23	Diamonds	32
Chad	36	9	Cotton	63
Congo (Brazzaville)	46	42	Crude oil	53
Zaire	1,004	44	Copper	62
Gabon	905	905	Petroleum	87
Total	2,756			
Eastern Africa				
Burundi	16	7	Coffee	84
Ethiopia	167	7	Coffee	27
Kenya	267	22	Coffee	18
Malagasy Republic	164	21	Coffee	30
Malawi	81	17	Tobacco	39
Mozambique	175	20	Cashew nuts	22
			Cotton	20
Rwanda	19	5	Coffee	64
Somalia	43	14	Livestock	54
Rhodesia	499	88	Tobacco	51

TABLE 14-6   (Cont.)

Country	Total Value ($ Millions)	Per Capita Value ($)	Leading Commodities (Percent of Country's Total Exports)	
Tanzania	319	23	Cotton	17
Uganda	261	25	Cotton	16
			Tobacco	13
			Hazelnuts	11
Zambia	758	168	Copper	93
Total	2,778			
Southern Africa				
Botswana	59	91	Minerals	44
Lesotho	9	8	Wool	36
South Africa	2,602	119	Diamonds	10
			Gold coins	10
Swaziland	94	196	Wood pulp	20
Total	2,764			

SOURCE: *Encyclopaedia Britannica, Book of the Year, 1975*, Chicago, 1975.
[a] 1971.

TABLE 14-7   COMPARISON OF ELECTRICITY PRODUCTION IN LEADING AFRICAN COUNTRIES SOUTH OF THE SAHARA AND LATIN AMERICA, 1973

Country	Aggregate Electricity (Billions of Kilowatt-Hours)	Population (Millions)	Per Capita Electricity (Thousands of Kilowatt-Hours)
South Africa[a]	64.9	24.0	2,704
Brazil	61.4	101.6	604
Mexico	37.1	55.4	670
Argentina	26.7	24.5	1,090
Venezuela	16.4	11.3	1,451
Chile	8.8	10.0	880
Colombia	10.3	23.2	444
Rhodesia	7.3	5.9	1,237
Cuba	4.2	8.9	471
Zaire	3.9	23.6	165
Ghana	3.3	9.4	351
Nigeria	2.6	59.6	44
Zambia	3.4	4.6	739

SOURCE: *Encyclopaedia Britannica, Book of the Year, 1976*, Chicago, 1976.
[a] Includes Southwest Africa (Namibia).

countries, with figures comparable to those of some European countries. In terms of this measure of industrialization, South African development is greater per capita than in any of the Latin American countries including rapidly developing Brazil, Mexico, and Venezuela.

Aside from South Africa, which might well have been considered an example of the modern, industrialized world as described in Part Two of this book, and which is run by its European immigrant population rather than by indigenous Africans, no country south of the Sahara has taken more than the first steps leading to industrialization. The major manufacturing countries are those processing export products—Zambia, Rhodesia, and Zaire have copper smelters and refineries, and Ghana has an aluminum plant.

The Beginnings of Industry

Most African countries have food-processing plants for local markets and produce simple building materials, textiles, and other consumer goods that require little capital investment or skilled labor. Some automobile parts are produced, tires are retreaded, and trucks are made from parts that are shipped from European countries.

Almost all African nations have an oil refinery built by one of the international oil companies in order to give that company a dominant position in the local energy market. Rhodesia, Kenya, Uganda, and Nigeria even have small iron and steel mills. But this is just a beginning. Most African countries have such small proportions of their populations in the money economy that there is neither the market to attract foreign investors nor local capital available for native initiative. Only where the government sees a particular industry as part of its development plan or of particular prestige value is an industry established. Outside of South Africa, Black Africa has no industrial developments comparable to those in parts of Mexico, Argentina, and Brazil.

In per capita terms two small African nations, Gabon and Liberia, place great emphasis on exports (see Table 14-6). In Gabon, oil has been developed by foreign oil companies; in Liberia, important iron ore production has been financed by foreign industrial interests. For their size, these two countries rank among the highest of any nonindustrialized countries in the world with regard to their involvement in foreign trade. Ethiopia, Mali, Chad, Guinea-Bissau, Burundi, Rwanda, Lesotho, and Upper Volta are at the other extreme, among the lowest countries in the world in per capita exports. They are only marginally connected to the worldwide economic system. In fact, Mali and Upper Volta's exports are primarily geared to markets in Africa; they provide livestock to neighboring countries, much in the way that Ireland sells food to Britain. These countries are in an extremely weak position because they depend on an African market that, in turn, depends on markets external to Africa. In the same way, Lesotho and Swaziland, members of a customs union with South Africa, provide that country with food and raw materials.

Notice that, except for Ethiopia and Guinea-Bissau, the countries ranking low in foreign trade are in the interior of Africa. Most are in the north, along the fringe of the Sahara Desert. These are the countries that suffered in the Sahel drought. They are also among the least developed, with per capita GNP of $100 or less.

THE AFRICAN WAY OF LIFE: LAYERS OF MODERNIZATION ON A TRADITIONAL BASE

Black Africa clearly lives with a dual economy. Two layers of the modern interconnected system are superimposed on Africa's traditional, locally based way of life. The first of these layers was established during the colonial period, which began when Europe exploited African resources and then carved political realms out of the continent, ignoring the cultural groupings of the native peoples. Under colonial control Europeans not only exploited the countries' resources through mines and plantations, but they tried to impose European values on the people. To be successful under the colonial system natives had to adopt European dress, manners, and religion. Those who were employed in important government or business positions were usually educated in European-style schools.

When independence came to political units whose boundaries had been set by Europeans, not Black Africans, the leadership of these new countries was derived from the cadre of Western-trained natives who carried on European-style political and social institutions. As a result, the new states controlled by Africans follow models inherited from colonial days, often modified by modern political-economic theories such as socialism or communism that also had their origins outside Africa. Each native government, then, develops its own particular blend of the ideas derived from the Western world with native customs and values.

The second layer of the interconnected system imposed on Black Africa has been the economic system powered by foreign investors attracted to Africa's resources. This system had its roots in colonialism, but it has grown since independence with the increasing demand of the industrial world for raw materials and tropical specialties. Resource exploitation by the modern system has required an infrastructure of transportation and communication routes, cities, and ports in order to get products to the coast for shipment overseas. But the big multinational corporations were interested only in getting the products to the coast and to overseas markets in industrial countries. As a result, Africa has almost no interlacing infrastructure uniting even the region's modern interconnected system into a functioning whole. Instead, each area of commodity development has its own individual transport corridor to the coast and, aside from air travel and radio communication, one African country has little direct interaction with its neighbors.

Until now the interconnected system has significantly touched only a tiny fraction of Africa's population. It reaches a small, modernized elite, the particular export-producing areas, and the big cities. Some development of commercial agriculture has occurred in traditional, tribal subsistence areas, but its expansion depends on transport and marketing facilities that are simply not yet available in most of Africa. Tribal peoples affected by migratory labor opportunities are only partly involved in the modern system. Working in the large mines provides a way to pay taxes and buy desired staples of the modern system, but it offers little incentive to break with traditional social and political patterns.

THE HANDICAPS OF COUNTRIES
THAT HAVE EMERGED FROM COLONIES

The political divisions of Africa, made arbitrarily by Europeans, are both too large and too small. The countries are larger than local cultural groups, yet most people regard themselves first as members of their own local world and only secondarily as citizens of the new country. Often large groups not well represented in a government still regard themselves as subject peoples. This was the case when the Ibo of Nigeria revolted in the late 1960s. Somalis are seeking to break away from both Ethiopia and Kenya.

Viewed from an external economic perspective, however, most of the countries of Africa are far too small in population. Nigeria, with its 65 million people, is the only country with a population size comparable to the European nations, which, in turn, are faced with the need for economic integration to provide a modern-scale market. Most African countries have fewer than 5 million people, numbers that hardly provide an adequate base for modern industry and commerce, assuming that most of the people were part of a modern interconnected system. But with most of the population still primarily at the subsistence level, such countries have little hope

LATIN AMERICA AND AFRICA COMPARED

We have now examined two under-developed areas: Latin America and Africa. Each has a dominantly tropical environment; each has traditional and modern societies juxtaposed. The countries of each area have problems in taking their place in today's political and economic world. The people of each area are faced with making choices between old ways and new.

Do you think that the two areas are fundamentally alike in terms of their developmental problems? Do they have the same potential for development? What do you see as the major obstacles to the development of each?

Both areas consist in large part of countries with very small populations. What value would there be in developing closer economic ties with their neighbors or with countries in other culture realms?

Underdeveloped peoples are very jealous of their independence. To what degree are the countries of these areas free? Do you see different degrees of independence in the two areas?

Do you see any analogies between the developmental problems of Africa and Latin America today and those of the United States as it emerged from colonialism? What prospects do these areas have for developing along the lines of the United States or Japan? Should this be their aim?

---

of generating either the capital or the labor force needed for industrialization. Most African countries will continue to be dependent on external sources for capital and technical help for a long time to come.

Plans for the political and economic integration of Africa are much less developed than in Latin America. The Organization of African Unity is largely a political forum. Four small former French territories in West Africa have joined in a customs and economic union to work out common external tariffs and internal economic policies. An East African Community was formed by Uganda, Kenya, and Tanzania, countries that share common rail, post, and telecommunications systems, but it has fallen apart as a result of rivalries between these countries. In 1975 fifteen West African countries from Mauritania to Nigeria set up the Economic Community of West Africa. Most African countries, however, are more concerned with their dealings with industrial countries and international organizations such as the United Nations, the World Bank, or multinational corporations than with their neighbors.

PROBLEMS OF REDEVELOPING A FRAGILE ENVIRONMENT

Too little is known of the African environment to be able to guarantee the success of any economic development plans that call for radical changes in farming and herding practices or for the establishment of major facilities that produce change in a local environment. Too much still needs to be learned about the impact of agriculture on the fragile tropical environment with its high potential evapotranspiration. Scientific stations have been set up, but as Africans begin to learn about the environment, they remain confronted by the problem of translating ideas into practice on land that is now mostly held by very conservative subsistence farmers.

Modern facilities often produce environmental changes that result in unplanned side effects. In 1971 a huge new dam was completed in the Ivory Coast. It formed a lake that is 660 square miles (1,700 square kilometers) in area and provides enough water to increase hydroelectric power generation by about 20 percent. In twenty years it should be able to irrigate 25,000 acres of land, but it also inundated 178 villages, acres of existing farmland, and giant teak forests, and it displaced great numbers of wildlife. Nearly 100,000 people had to be moved into new settlements. The greatest danger is that the new dam may result in a serious threat of disease. A much smaller dam in 1964 led to raging epidemics of river blindness and sleeping sickness as insect carriers found the new environments around the lake to their liking.

The experts from advanced countries as well as the African leaders do not really know how best to plan for African development. The initial steps have only recently been undertaken and the area has only recently begun to be scientifically studied. One can wonder whether it will be possible to quickly produce the dramatic changes that in England, for example, took 300 years to accomplish. Africans in the traditional system may not want life to change. But, as we saw in the discussion of the Sahel, it has changed. Traditional life can no longer support people adequately using traditional methods. Change is inevitable. What form will it take? Will Africans be able to direct that change or will it be directed by outside forces?

SELECTED REFERENCES

Hance, William A. *Population, Migration, and Urbanization in Africa.* Columbia University Press, New York, 1970. A detailed examination of the key issues of population pressure, growth, population movements, and urbanization in Africa. The maps are especially useful.

Hance, William A. *The Geography of Modern Africa,* 2nd ed. Columbia University Press, New York, 1975. An excellent text and source on the geography of Africa. The first three chapters and the conclusion provide a concise interpretation of Africa's economic setting, physical assets and limitations, population, and developmental potentials. The rest of the book is a regional geographical study of its major regions and countries.

O'Connor, A. M. *The Geography of Tropical African Development.* Pergamon Press, New York, 1971. A focus on the changes in geographical patterns of measures of development that have taken place in recent years. The book includes chapters on agricultural change, mining, industrial development, transportation, power supply, and urbanization. Collectively, the maps offer a comprehensive view of the basic pattern and structure of economic activity and development in tropical Africa.

Prothero, R. Mansell, ed. *People and Land in Africa South of the Sahara: Readings in Social Geography.* Oxford University Press, New York, 1972. A collection of articles drawn from professional journals. Most address topical issues in specific countries or areas of Africa, but all deal in some degree with the basic theme of population-land relationships.

Uppal, J. S., and Louis R. Salkever. *Africa: Problems in Economic Development.* Free Press, New York, 1972. A collection containing several articles relevant to the discussion in this chapter. Part I is an overview of recent trends in economic development, and Part III treats agricultural and industrial development in Africa.

DUAL SOCIETIES

Tractor and oxen work side by side in Tunisia. (Paul Conklin)

The modern interconnected system, with its expanding networks of transportation and communication, has spread its influence into many developing countries. The most readily visible effects of the modern influence are in urban architecture; the capitals of the developing countries look much like other cities throughout the world. More important is the spread of higher education and the technology that is necessary for development. The universities serve as points of entry for individuals into international scholarship and science and as training grounds for national leaders. The universities, in turn, reach out to the rural areas, bringing aspects of modern education to the villages. In most developing countries, however, only a small percentage of the population is directly affected by such changes; the rest of the people maintain their traditional lifestyles. Any country that has distinctly different sectors, modern and traditional, is said to be a dual society.

The University College of Ghana, where most of the country's contemporary leaders were educated. (Anthony Howarth/Photo Trends)

Government Road in Nairobi, the capital of Kenya. (Mohamed Amin/Nancy Palmer)

Television being used for educational purposes in Niger. (Marc Riboud/Magnum Photos, Inc.)

European-style opera house in Oran, Algeria. (Klaus Francke/Peter Arnold)

Dual societies are evident in the cities, towns, and villages of developing countries. Tall modern buildings dominate the larger cities, while within the cities—often on their outskirts—are overcrowded slums or shantytowns built from waste materials by migrants from the rural areas. These people living on the fringes of the modern system see greater hope for advancement here than in the poverty-stricken villages. For those who have stayed in the villages, traditional ways remain relatively unchanged. Despite some agricultural innovations and the introduction of

Above: Grass roofs characterize the homes of the inhabitants of this African village. (FAO photo by A. Kannangara)   *Right:* A section of Medellín, a city with a population of approximately 1 million in northern Colombia. (Paul Conklin)

some consumer goods produced in the cities, village life is dominated by the need to gain subsistence from the local earth environment. A few villagers are able to produce surplus goods to sell in the marketplaces of the larger towns, and thus to a limited extent enter the cash economy of the modern system. But the links between the modern and traditional sectors in most developing countries are tenuous. For the most part, these countries have two economies existing side by side, rather than an integrated national economic system.

Above: Goods for sale in Sullana, a town in northwestern Peru not far from the coast. (Standard Oil of N.J. Collection)　*Left:* A street in Calcutta. (Hiroji Kubota/Magnum Photos, Inc.)

Manufacturing and a high level of energy consumption characterize the modern interconnected system with its factories and use of inanimate energy. Traditional societies, too, require energy and manufacture products, the latter often of finer quality than the output of modern factories. Contrasting energy sources and methods of manufacture are frequently found within the same developing region or country. In cloth making, for example, workers may be seen finishing dyed cloth in the traditional manner, using hand methods, while elsewhere a factory employee tends a modern weaving machine. Here we see the contrast in energy sources: in traditional cloth making the power comes entirely

Pounding newly dyed cloth in Kano in northern Nigeria. (George Gerster, Rapho/Photo Researchers, Inc.)

Cotton mill in Lagos, the costal capital city of Nigeria. (Bruno Barbey/Magnum Photos, Inc.)

from human muscles, while in the factory the banks of machines are powered by electricity, which is commonly derived from fossil fuels. In traditional villages, domestic energy needs are supplied mainly by firewood, but a shortage of firewood has forced reliance on less desirable sources of energy for cooking and heating. In a village where dried cow-dung cakes are the chief fuel, the total energy supply would probably be less than that needed to run one of the large trucks that deliver refinery products. Despite differences in the sources and uses of energy between the traditional and the modern systems, energy is a serious concern in both systems.

Cow dung being prepared for use in place of firewood in Varanasi in the Indian state of Uttar Pradesh. (United Nations/J. P. Lafonte)

Refinery for petroleum products in Saudi Arabia. (Marc Riboud/Magnum Photos, Inc.)

CHAPTER 15

The Middle East-North Africa: Can Oil Bring a Great Culture Region Power in the Modern System?

Here is the world's largest expanse of dry climate, with severe moisture deficits, very little forest vegetation, and mostly unproductive soils. Population is concentrated, as it has been for centuries, in the valleys of the Nile, Tigris, and Euphrates rivers. The Islamic way of life unifies much of the region. Rich reserves of petroleum and natural gas have greatly increased the economic power of this region, but major issues confront the oil-producing countries today. They need to establish a long-term economic base and to manage their enormous foreign earnings to bring about internal development. The Arab–Israeli conflict is significant not only to this region but to the world.

Electrical power station at the Aswan High Dam in Egypt. (Jacques Pasloviski/Sygma)

Israeli tanks on the Golan Heights, October 17, 1973. (Stern/Black Star)

Above: Rest stop at an oasis in Libya. (Robert Mottar/Photography for Industry) *Below:* The Nile River near its delta. (FAO photo by Patrick Morin)

Entrance to a mosque in Iran. (Inge Morath)

509

The Middle East–North Africa has been a major center of human culture, wealth, and power since the dawn of history. During the past two centuries it has found itself left behind by the new technology of the modern interconnected system. A dramatic change has occurred during the past few decades, however, for the region has been proven to contain the world's largest known petroleum reserves. Thus the countries of this part of the world control much of the key resource that powers today's modern interconnected system. This control has brought wealth and power to a few countries within a few years. Today the oil-producing countries of this region seem likely to join the United States, Europe, Japan, and the Soviet Union as centers of the modern interconnected world.

The oil was discovered and developed by corporations with headquarters in the United States and Europe, as we saw in Chapter 12. These companies invested huge sums in a system of wells, pipelines, ports and tankers, and even refineries, to get the oil to markets in Europe and Japan. And they reaped huge profits because they were tapping the world's richest oil fields at shallow depths, usually as holders of monopoly concessions granted by the host country. With these advantages production costs were as little as twelve cents per barrel (42 U.S. gallons to the barrel) compared to over three dollars in oil fields elsewhere. Royalties and taxes were paid to the host government but great profits went to stockholders in the modern world.

In the early 1970s all this changed. The major oil-producing countries of the region, along with key oil-exporting countries elsewhere, decided to take advantage of the possibilities offered by OPEC. The organization, formed in 1960, includes Algeria, Iran, Iraq, Kuwait, Libya, Qatar, Saudi Arabia, and the United Arab Emirates from the Middle East–North Africa; Ecuador and Venezuela from Latin America; Gabon and Nigeria from Black Africa; and Indonesia from South and East Asia. Together these countries form a cartel that controls the price of crude oil in the international market. Using their control over 60

TABLE 15–1 ESTIMATED OIL REVENUES OF MAJOR OPEC COUNTRIES OF THE MIDDLE EAST–NORTH AFRICA, 1974

	Total Revenues ($ U.S. Millions)	Revenue per Barrel of Oil Exported ($ U.S.)
Saudi Arabia	20,000	6.54
Iran	17,400	8.40
Libya	7,600	13.89
Kuwait	7,000	7.73
Iraq	6,800	10.15
United Arab Emirates	4,100	6.68
Algeria	3,700	10.57
Qatar	1,600	8.46

SOURCE: *The Petroleum Economist*, March 1975.

percent of world oil production, a much higher percentage of the oil moving in world trade, and an even greater share of known oil reserves still in the ground, in 1973–1974 they more than tripled the price of oil from their fields. Moreover, several countries in OPEC took over (or had earlier taken over) the controlling share in the operations of the major international oil companies within their borders.

All this sent a shock wave through the modern interconnected system, particularly Japan and the industrial countries of Europe, which are largely dependent on Middle Eastern oil for fuel. Some of the less developed countries, such as India and Brazil, were also severely affected by the price increases. Most important to the Middle East–North Africa, however, was the fact that some of the poorest countries in the world had been transformed into some of the richest. With tremendous revenues from oil (Table 15–1) and yet relatively small populations, the per capita incomes of some oil-producing countries rose almost overnight to levels higher than those of

TABLE 15-2 POPULATION AND PER CAPITA GROSS
NATIONAL PRODUCT FOR COUNTRIES OF THE
MIDDLE EAST–NORTH AFRICA

Country	Population Estimate, 1976 (Millions)	Per Capita GNP, 1974 ($ U.S.)
Algeria	17.3	650
Bahrain	0.2	2,250
Egypt	38.1	280
Iran	34.1	1,060
Iraq	11.4	970
Israel	3.5	3,380
Jordan	2.8	400
Kuwait	1.1	11,640
Lebanon	2.7	1,080
Libya	2.5	3,360
Morocco	17.9	430
Oman	0.8	1,250
Qatar	0.1	5,830
Saudi Arabia	6.4	2,080
Sudan	18.2	150
Syria	7.6	490
Tunisia	5.9	550
Turkey	40.2	690
United Arab Emirates	0.2	13,500
Yemen Arab Republic	6.9	120
Yemen (People's Republic)	1.7	120

SOURCE: Population Reference Bureau, Inc., *World Population Data Sheet, 1976,* Washington,
D.C., 1976.

the United States and other industrial countries
(Table 15–2).

ABSOLUTE GOVERNMENTS, OIL WEALTH, AND DEVELOPMENT PROGRAMS

In most of the oil-producing countries, the government controls all oil fields and the citizens
have little, if any, voice in government affairs.
The shah of Iran, the king of Saudi Arabia, and
the rulers of the tiny Persian Gulf sheikdoms are
absolute monarchs, vestiges of traditional tribal
control in the region. Other countries such as
Iraq, Libya, and Algeria have centralized one-
party socialist governments. In all cases, oil
money goes into treasuries of the central government, which decides how it is to be spent.

The oil-rich sheiks have lived very well personally. The Saud family in Arabia was long
known for its air-conditioned limousines and jet
aircraft. The shah of Iran has five palaces, a
personal fleet of jet aircraft and helicopters, and
a villa in Switzerland. In 1971, to celebrate 2,500

years of Persian civilization the shah decided to hold a celebration the world would notice. It was not a world's fair open to tourists; instead, it was a $100 million entertainment for world rulers, attended by ten kings and nineteen other heads of state.

Most of the oil-rich governments are concerned about raising living standards and establishing a modern industrial base. They have embarked on major development programs that will modernize education, health care, and other social services. They are attempting to improve transport and communications, develop new irrigation sources, and establish modern industry. Traditional unrest in the region, in particular the hostilities with Israel, and the quest for world prestige, have led these nations to purchase vast amounts of the world's most modern military equipment—jet planes, tanks, and missiles.

Even so, a few of these countries are earning more money from oil than they can spend in their own development. Like the Texas oil millionaires before them, they must find ways to invest their money surpluses. The best investments are in the modern interconnected system of the world—in banks, industrial corporations, and real estate in the largest centers of the modern system. Middle Eastern countries own parts of Daimler-Benz, the manufacturer of Mercedes-Benz automobiles, and Krupp steel of Germany; wealthy Middle Easterners have invested in banks in Washington, D.C., Detroit, and California. They own important shares in a new real estate development in downtown Atlanta, control valuable real estate in London and Boston, and are developing a fashionable resort-retirement enterprise in South Carolina.

Immediately after the oil price rise in October 1973 it was feared that there was so much oil money coming into the world financial system that the traditional institutions and their operations were endangered. There was fear that Middle Easterners would make such large deposits in major U.S. and British banks that they could threaten the security of loans made by those banks and cause financial panic by withdrawing their deposits at any time. But the dangers appear to have been overstated. First, Middle Eastern countries have emerged as conservative investors. They have put their money in the hands of money managers from the industrial countries. More important, the majority of the oil-rich countries have found that development projects at home and military purchases in a world of inflating prices have taken most of the money.

Actually, the surpluses depend on the balance between oil production and population. Each of the four major oil-producing countries—Saudi Arabia, Iran, Kuwait, and Libya—has outputs of more than 100 million metric tons per year. Saudi Arabia and Iran each produce twice as much as the other two. But the four countries differ greatly in population. Iran has 34 million people, whereas Saudi Arabia has fewer than 7 million. Iran is in the midst of a $70 billion, six-and-a-half-year development plan intended to double the present per capita GNP, bringing it to $1,570, still less than half that of most industrial countries. The task of development seems far easier in Saudi Arabia, which has more oil money and fewer people. It seems even easier in Libya, with half as much oil production and fewer than 3 million people. The country with the largest surpluses of money to invest is Kuwait, the third ranking oil producer in the Middle East–North Africa. Even now, after more than ten years of rapid immigration, Kuwait has only 1.1 million people.

The other Middle Eastern oil-producing countries have only modest oil production. There is limited oil money for development in Syria with over 7 million people or Algeria with about 17 million. But tiny Oman, Qatar, and the United Arab Emirates, each with fewer than a million people, have money for ambitious small-scale developments. In 1974 per capita GNP in Qatar was $5,830 and in UAE, $13,500. Thus because of the localized nature of oil production, the problems of the Middle East–North Africa have not been solved by oil wealth. Indeed, more than half the countries of the Middle East–North Africa are *not* important oil producers. The majority of their populations carry on the traditional occupations: agriculture, nomadic herding, and trade with urban centers. The per capita GNP figures for these countries are much less than half those of modern industrial countries.

"THE CRADLE OF CIVILIZATION"

The Middle East–North Africa is larger than Black Africa and extends from the Atlantic coast of North Africa eastward along the south shore of the Mediterranean Sea, then across the neck of land that connects Africa and Asia to the Persian Gulf (Figure 15–1). It is separated from Black Africa by the largely uninhabited Sahara Desert and from Pakistan on the east by other dry areas. Seas are also important boundaries of the region; the Mediterranean and Black seas separate it from Europe to the north. Despite its large area, the Middle East–North Africa is sparsely populated. It has fewer people than the United States and less than two-thirds the population of Black Africa.

The eastern part of the region is known as the "cradle of civilization," for here agriculture and urban life apparently had their origins long ago. Even though most of the region is a low-work environment where actual evapotranspiration (AE) is far below potential evapotranspiration (PE), agriculture began in the basins and valleys of the highlands east of the Mediterranean Sea as many as 10,000 years ago. The region where agriculture originated, called the **Fertile Crescent,** includes the valley of the Tigris and Euphrates rivers, the wet highlands of what is now Syria, Lebanon, Israel, and the Nile Valley in Egypt. Rainfall in the Fertile Crescent was sufficient to support a variety of wild plants along with wild goats, sheep, and cattle. Early peoples cultivated barley, wheat, and peas, tended goats and sheep, and developed systems of irrigation.

The developments of worldwide significance took place along the great rivers of the region—the Tigris, Euphrates, and Nile—and in the Indus Valley of Pakistan on the eastern margin of the region. In these areas potential evapotranspiration was great, and annual flooding of the rivers provided moisture for crops and replenished soil fertility each year from the plant nutrients in the silt deposited by the floods. Key sites in the Middle East were at Jarno in what is now western Iraq and Jericho along the Jordan River in present-day Israel. Cultivation of wild wheat and barley has been dated as early as 7,000 B.C. Evidence of the domestication of pigs can also be found at these sites. By 5,000 B.C. permanent agriculture based on irrigation, the plow, and oxen to do heavy work had spread throughout the Fertile Crescent. The system was very productive, and by 4,000 B.C. there were enough agricultural surpluses to support a number of cities.

THE MIDDLE EAST–NORTH AFRICA:
A CENTER OF WORLD POWER UNTIL
THE TWENTIETH CENTURY

The Middle East–North Africa has played a major role in world history ever since it was the "cradle of civilization," but the contrast between wealth and poverty within the area has also continued to be a part of its history. The wealth of agricultural societies such as Egypt and Mesopotamia (today Iraq) contrasted with the life of nomadic herders including the early tribes of Israel. Even today the nomadic life of the Bedouins remains much the same as in early times.

Great wealth concentrated early in trading cities such as Carthage, Babylon, Alexandria, Nineveh, and Persepolis. Moreover, the region has been an important center of science and the arts. The Arabs made major advances in chemistry, mathematics, and astronomy, and the followers of Islam passed Greek science to Christians in Europe during the Islamic conquest there. Arab commentaries were used as texts in European

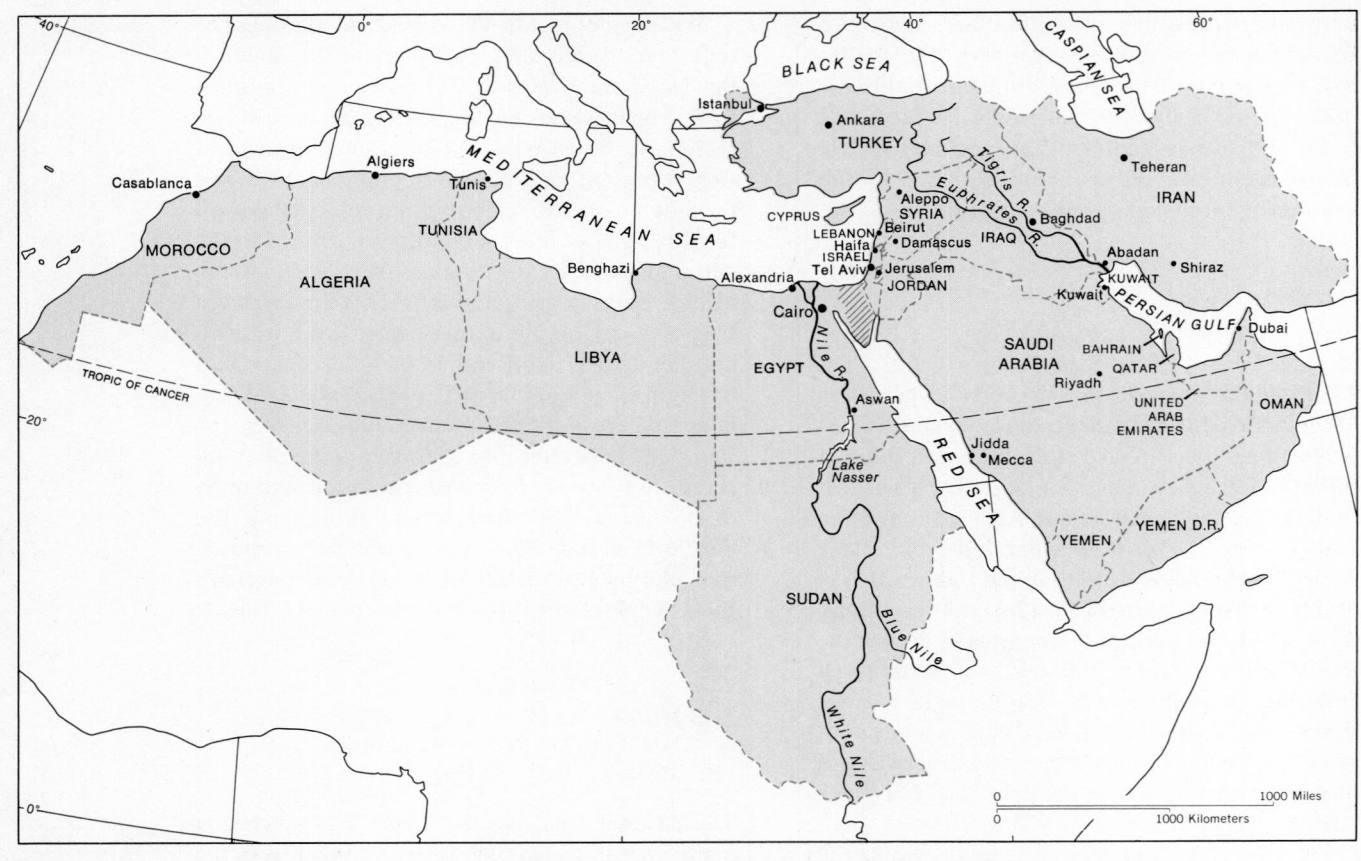

FIGURE 15–1 POLITICAL MAP OF THE MIDDLE EAST–NORTH AFRICA

Metropolitan population

•   Less than 1,000,000

•   1,000,000 - 4,000,000

●   More than 4,000,000

▨   Occupied by Israel

universities through the eighteenth century. During medieval times steel, silverware, cottons, leather, and damask and muslin cloth from the region were highly prized in world trade. Moslem culture is known for its abstract poetry and literature, calligraphy, and music.

Cities of the Middle East–North Africa have been trading centers since early times. There was busy trade across the Mediterranean and into the Black Sea and in the Red Sea and the Persian Gulf as well. Moreover, the Middle East lay between four other key civilizations—China and India to the east and Europe and Africa to the northwest and southwest. Goods flowed in and out of the Middle East to those centers, and Middle Easterners served as the intermediaries for long-distance trade between the Orient and Europe. Not only did North African cities benefit from the wealth of the trade within the Mediterranean and, later, the Arab world, but they also controlled the northern ends of the trade links across the Sahara to Black Africa. Damascus, Baghdad, Cairo, and the religious cities of Jerusalem, Mecca, and Medina were the focuses of important caravan routes, and Alexandria, Muscat, Tripoli, and Tangier were ports. All were among the large cities of the world before 1800. Their trade involved more than just the movement of the luxury products that dominated trade before the Industrial Revolution. In addition, foodstuffs had to be provided for ships and caravans, and camels, tents, and guides were needed for the long desert journeys. Bankers and other trade specialists emerged in the trade centers.

From the seventh century to the fifteenth century, Islamic forces invaded Europe. Spain and Portugal were overrun, and the invaders reached as far north as Tours in central France in A.D. 732. Later in the fourteenth and seventeenth centuries, the Ottoman Empire centered in Turkey spread the Moslem religion northward into southeastern Europe and the southern Ukraine.

OUTSIDE CONTROL: EUROPEAN MANDATES
AND OIL EXPLORATION

After World War I the Ottoman Empire broke up and most countries of the region came under the control of British or French empires. The French and British brought great change to this part of the world—what had been a world of great wealth and power was now under the control of alien European powers. Although Islam had considered Christianity inferior, now Christian nations were in control of this center of the Moslem world. The Middle East–North Africa, once famous for its science and learning, was now overwhelmed by the European industrial technology. Under these circumstances the large multinational oil companies first began to enter the region in search of petroleum. Iraq was under the control of the French, and the other Persian Gulf countries had weak governments supported largely by British military power. Such conditions made it easy for the companies, exploiting a resource that had no direct use in the nonindustrial economy of the region, to obtain very favorable terms for the right to produce oil. The local governments saw oil output as providing an important new source of income, even if their return per barrel was very low.

The governments soon learned the importance of the tremendous oil pool in their part of the world, and in the past two decades have increased their demands on the companies for a greater return from oil resources. For them the recent price increases and takeovers of companies have been nothing more than the return to their rightful place of importance in the world.

THE MIDDLE EASTERN ENVIRONMENT

In terms of the geographic equation, the distinctive Moslem culture and the large expanse of arid environment in the region—not oil—identify the Middle East–North Africa as a first-order world region separate from the rest of Asia and Africa. Its land area is more than 25 percent greater than that of the United States (Figure 15–2), yet the majority of this vast region receives less than 16 in. (40 cm) of rainfall per year (Figure 15–3)—it is the world's largest expanse of dry climate. Only small coastal areas of Turkey, the mountains of northern Iran, and the southern Sudan have an average rainfall above 40 in. (100 cm) a year. In the rest of the region, only the fringes of the Mediterranean Sea, the highlands of western Iran, the Fertile Crescent and central Turkey, and much of the Sudan have more than 16 in. (40 cm) of precipitation annually.

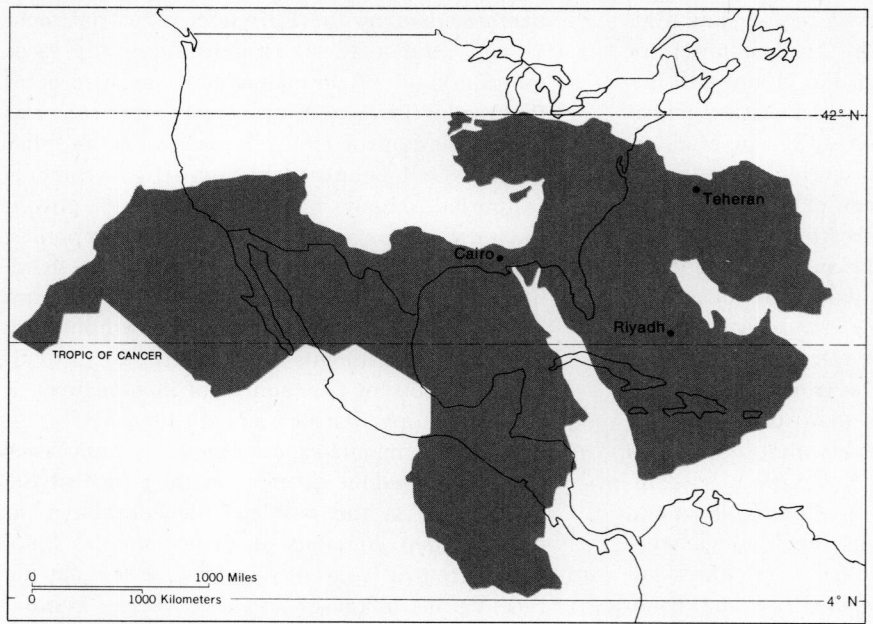

FIGURE 15–2 THE MIDDLE EAST–NORTH AFRICA
AND NORTH AMERICA COMPARED IN AREA AND
LATITUDE

HIGH POTENTIAL EVAPOTRANSPIRATION BUT MAJOR MOISTURE DEFICITS

The region has some of the highest potential evapotranspiration figures in the world (see Figure 3–3). When water is brought to the fields, the region is very productive agriculturally. The Middle East–North Africa lies along the fringe of the northern tropical zone, and as can be seen from Figure 3–3, the world's highest PE figures normally are found on the margins of the tropics. Potential evapotranspiration here is 25 to 50 percent greater than in most parts of the United States. The largest expanse of this high PE is found across the Sahara Desert and into the Arabian Peninsula. Potential evapotranspiration decreases northward in the region, and in southern Iran, southern Israel and Jordan, northern Egypt,

and much of the southern coast of the Mediterranean it is similar to that of southern Florida and the southwestern United States. In the rest of Iran, most of Iraq, Turkey, the eastern end of the Mediterranean Sea, northern Algeria, and Morocco, PE is about like that in the U.S. South, southern Middle West, and California.

Offsetting the high PE are some of the most severe moisture deficits in the world (see Figure 3–6). From the northern tip of the Red Sea and the northernmost part of the Persian Gulf southward, the water deficit is over 80 in. (200 cm)—more than the total PE for most parts of the United States. Only the eastern coast of the Mediterranean Sea south of Turkey normally has any surplus moisture during any part of the year.

The moisture problem is complicated by the fact that even in the wettest areas around the

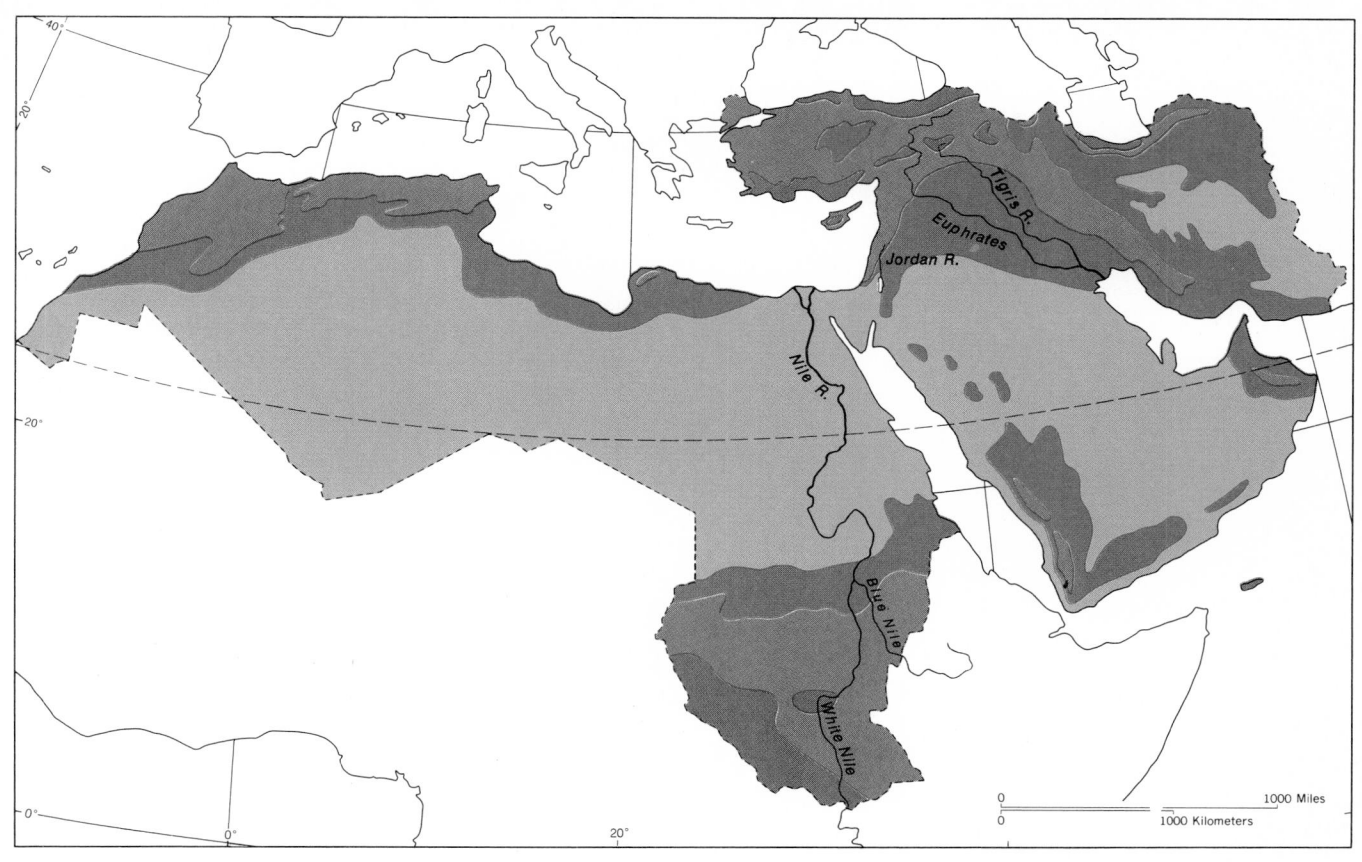

SOURCE: Adapted from *The Middle East and North Africa*, Europa Publications, 1974, London, pp. 8–9

Average annual rainfall

Under 4 inches (under 10 cm)

4-16 inches (10-40 cm)

16-40 inches (40-100 cm)

Over 40 inches (over 100 cm)

FIGURE 15–3   AVERAGE ANNUAL RAINFALL
IN THE MIDDLE EAST–NORTH AFRICA

Mediterranean Sea, most precipitation comes during the cool season of the year, as in California, rather than in the period of active plant growth. Winters are short and mild, and winter crops can be grown, but the moisture deficiency during the hot summer is severe. Even the wetter regions receive less than 5 in. (12.7 cm) of precipitation during the May–October half of the year, when the temperature is most suitable for growing crops. As in California, even these wetter areas require irrigation for a large farming output.

July daytime temperatures in Cairo are over 100°F (38°C) most days, but may fall to 70°F

(21°C) or below at night. In the deserts daytime temperatures reach more than 125°F (52°C). Although winters are mild around the Mediterranean Sea and in the Arabian Peninsula, snow occurs in the mountains of North Africa and Yemen. Farther north, in the plateaus of Iran and Turkey, winters are severe.

SURFACE LIMITATIONS OF THE WETTER AREAS

Most of the coastal fringes of the Mediterranean and other water bodies, where moisture is adequate for agriculture, are hilly and mountainous rather than broad plains. The Atlas Mountains, with heights to 13,000 ft (3,965 m), lie back from the coast in Morocco, Algeria, and Tunisia in North Africa. Much of Turkey is hilly and mountainous, and there are mountains along the southern coast of the Caspian Sea in Iran. Mountains also rim the southeast coast of the Red Sea, the Gulf of Aden, and the Gulf of Oman. Even where the southeastern corner of the Mediterranean bounds Israel, Lebanon, and Syria, the land quickly rises to a series of hills. Within a few miles of the coast just beyond the hills, the land drops off into a rift valley that forms the trough through which the Jordan River flows to the Dead Sea. The continuation of this valley is submerged in the Gulf of Aqaba, an arm of the Red Sea. The Dead Sea lies more than 1,300 ft (396 m) below sea level. The best agricultural areas are the alluvial lowlands along the Nile and the Tigris–Euphrates. This land is level and suitable for agriculture, and with sufficient irrigation water high yields can be produced.

DRYLAND VEGETATION AND ARID SOILS

In so dry a world there is very little forest vegetation. This is the result partly of the dryness, and partly of deforestation by humans throughout history. In this dry region demand on forests for fuel and building materials has been greater than supply. Crops and trees can be planted in the Mediterranean dry-summer climate, but the natural vegetation consists mostly of evergreen shrubs and herbs. Oak, olive, walnut, and poplar trees grow well, and there are thickets of dense thornbushes and shrubs topped by scrubby oaks and myrtles that are difficult to penetrate.

In the grassland steppes of Turkey and much of Iran drought-resistant grasses, green only after rain in winter and spring, dominate in the lowlands. On the slopes juniper and other trees and shrubs give the area a parklike appearance. The major areas of forests are on the mountain slopes. Evergreen oaks and Mediterranean pine grow on the lower slopes, but give way to other oak and pine and, east of the Mediterranean, to the famous cedars of Lebanon (now protected like the sequoia of California). At the highest elevations, above the tree line, are alpine pastures. In the deserts vegetation is sparse and grows only during the few weeks following rain. Most vegetation is found along the dry stream valleys where the water rushes down slopes after a rain.

Most soils in the area show the effects of aridity. Organic content is low, and texture is often sandy. Desert soils have sand and often gravel interspersed with masses of stone of various sizes. The best soils by far are those derived from the deposits of flooding rivers. Thus, again, the alluvial lands along the Nile and the Tigris–Euphrates stand out as having the greatest use potential.

Minerals such as copper, iron, gold, and silver have been known and used in the area since prehistoric times, but today few deposits there measure up to commercial standards elsewhere in the world. Iron is mined in Algeria and mineral salts in Israel and other desert areas. Much of the world's phosphate comes from Morocco, and Egypt produces phosphate fertilizers, but aside from the production of petroleum and natural gas, mineral resources are not currently of much importance in this region.

ISLAM: UNITY AMONG CULTURAL DIVERSITY

The region has held a great mixture of peoples and ideas throughout its history. Today six major peoples and cultures are recognized within the region: Iranians, Turks, Greeks, Arabs, Berbers, and Jews. Each has its own language and customs. For most people throughout the history of the region, one's basic loyalty has been to one's family and to one's tribe, village, or town. Thus although broad cultural patterns have spread over the region, each community has its own distinctive dialect and local customs. As in medieval Europe, czarist Russia, or Japan before the twentieth century, most people have lived in traditional, locally based communities even though the nobility and merchants in cities have benefited from long-range links and cultural ideas from other parts of the world.

Despite the cultural diversity, major forces give this region strong cultural unity. Most important is the Islamic religion. There are other religious groups—the Jews and various Christian sects around the eastern end of the Mediterranean Sea—but Islam dominates in almost all countries. This region is often called the Moslem world, but there are major Moslem countries—most notably Pakistan, Bangladesh, and Indonesia—in South and East Asia, and Islam is practiced in parts of Black Africa.

Moslem practices that have formed the basis of life in this region for more than 1,000 years are still important today. From Morocco in the west to Iran and Afghanistan in the east, five times every day faithful Moslems throughout the Middle East–North Africa fall on their knees to pray to Allah. Each person faces toward the Saudi Arabian city of Mecca, the birthplace of Mohammed, the founding prophet of Islam. Mecca is the sacred place of all Moslems and, if possible, each male will make a pilgrimage to Mecca at least once in his lifetime. The pilgrims come on foot or by bus, ship, or plane, and all change from their modern clothes or desert robes into two lengths of seamless white cloth. Then they walk seven times around a cube-shaped, black building, the most sacred shrine of Islam.

Islam, like Christianity, has several different sects. Most important are the Sunnis and the Shi'as. Most Moslems in this region are Sunnis, but the Shi'a sect is the state religion of Iran and is found in parts of Iraq and Yemen. There are several minor sects found as minorities in cities and as locally important groups in some rural areas. Saudi Arabia is dominated by a puritanical sect, the Wahhabites, which arose in the nineteenth century calling for a return to the simple faith preached by Mohammed. The conservative Wahhabite traditions have had an important influence on the way that Saudi Arabia has responded to its oil riches.

Some of the governments today have broken their traditional ties with the Moslem religion and are following political lines not patterned after the Koran. Turkey cut itself off from direct ties to religion, and Egypt has outlawed the Muslim Brotherhood, which seeks to reinstate closer church-state ties, but in Saudi Arabia the ties of religion and state remain. In countries where these ties have been broken, religious roots are strong among the people. In Turkey there is religious support for conservative policies. Thus the issue of the proper sphere of religious influence remains.

Six major language groups are found in the region: Turkish; Persian; Greek, the language of Cyprus and parts of western Turkey; north Semitic languages, including Hebrew and Aramaic found in the eastern Mediterranean and Syria; south Semitic, primarily Arabic from Mecca and surrounding desert areas; and Berber, the language of western North Africa.

One of these languages, Arabic, from the home of the prophet Mohammed, has become the written language of all Islam. Classical Arabic is the language of religious instruction throughout the Moslem world. It is used in literature and broadcasts from Morocco to Afghanistan, and it forms a common bond of communication for educated

SYRIA

Syria is in many ways a microcosm of life in the Middle East–North Africa today—its problems and hopes. Its 7 million people occupy a strategic location near the center of the Fertile Crescent and at the crossroads of sealanes on the Mediterranean and former caravan routes to Iran and the Persian Gulf. The people form a diverse mix. Most are Sunni Moslems, but there are other Moslems, Druzes (a fundamentalist offshoot of the Moslem faith), and various Christian groups. City and town dwellers, on the one hand, and agricultural villagers, on the other, have contrasting lifestyles. Mountain people have still other ways. People of the Jebel Druz in the southwest have remained socially and politically apart. Levels of sophistication range from primitive, illiterate herders to well-educated merchants with long traditions of trading with the world.

Syria contains both coastal Mediterranean lands where irrigation forms only a part of the agricultural system and larger expanses of steppes and deserts where irrigation is essential. The two largest cities, Damascus, the capital, and Aleppo, once the second most powerful city in the Ottoman Empire, vie for economic and political power. While the country itself is just a minor producer of oil, two pipelines—from Iraq and Saudi Arabia—pass through the country. Syria also has two oil shipping ports on the Mediterranean.

Syria is both modern and traditional, advanced and backward, nationalistic and pan-Arab. Syria has been struggling with the problems of government since achieving independence in 1946; there have been a series of coups and governments of various leanings. Since the early 1960s the influence of pan-Arabism and anti-Israeli sentiment have played the major part in political affairs. Syria has had its own national party of the larger Ba'ath Middle Eastern party that calls for the unity of Arab countries, the introduction of socialism and the more equitable distribution of wealth, and the removal of the last vestiges of Western imperialism. In Syria the party has been split into two factions. One is a progressive faction, strongly Marxist-oriented and supported by the Soviet Union, concerned most of all with a one-party socialist state. The other is a nationalist faction, less concerned with ideology, more pragmatic in its socialism, and concerned with good relations with its neighbors, including Israel. Syria has been a principal combatant in the wars with Israel, and its ports and Damascus have been bombed by air and sea. As a result, it has been the recipient of aid and loans from oil-rich Arab countries. The Israeli problem can be better understood when one recalls that Syrians consider theirs the most profoundly Arab country and the remnants of a larger Syrian domain that once included Palestine.

The government that took power in Syria in 1969 has tried both to remove one-party government and to lead Syria more into the world scene. National elections were held and a legislative body was formed in 1971. Government came into the hands of a coalition of the five major political parties instead of being a product of the single Ba'ath party. Cabinet members are drawn from several parties. Syria, a nation that had been closely identified with the Soviet Union, has reopened diplomatic relations with the United States and West Germany and has gained recognition as a leader of the Arab world.

Syria's economic development plans are built around expanding its agricultural base, both to be able to feed its own people and to build industry linked to that farm economy. The industry is to focus on processing of cotton and wool into textiles, and on the manufacturing needs of agriculture, such as farm machinery and fertilizer. Most important is the Euphrates Dam, which is planned to double the country's irrigated land in twenty years.

The Euphrates Dam, one of the world's largest earth dams, was built across the Euphrates River in northeastern Syria. It is ultimately to irrigate 1.5 million acres and revitalize agriculture on the dry eastern fringe of Syria's fertile plain. With a capacity of 800 megawatts, the hydroelectricity it will produce is enough to provide the basic energy for Syria's planned industrial expansion during coming decades. The basic financial aid and technical supervision for the project came from the Soviet Union.

Construction began in 1968, and the first turbines started to operate in 1974. By 1975 the reservoir behind the dam was one-third filled, and 250,000 acres of land were ready for cultivation. It is expected that the region around the dam will have a population of 1.5 million when the irrigation project and its supporting towns are completed near the end of the century. In late 1974 a new town, Thawra, already had a population of 60,000 and sixteen large villages had been constructed. A planned industrial complex located more than 100 miles (160 km) away at the city of Homs will use electricity from the dam as well as oil from fields to the northeast beyond the dam.

The project does not lack problems, however. The drainage basin for the artificial lake is mostly in an area of Turkey where soil erosion is severe and persistent. The water entering the lake carries a heavy load of sediment that will probably result in a siltation problem in the reservoir. The reservoir's lifetime will be short; it may be measured in decades rather than centuries.

Syria has a greater area of land in crops than most countries of the Middle East. The new agricultural program is designed to turn the country once more into "the granary of the Eastern Mediterranean." Syria hopes to export wheat and other grains in the near future.

Previous socialistic governments reformed landholding. Whereas once 1 percent of the landowners held 30 percent of the land and hence the ownership units were large in area, most holdings are now less than twenty-five acres. Still, over 80 percent of the land is in private units with most of the rest in cooperatives. There are currently more than 1,700 cooperatives. Some are simply for marketing, but others have developed cooperative farming practices. The government has been working to improve the infrastructure for farmers—this has included health facilities, housing, sanitation, and roads. Efforts are being made to encourage farmers growing wheat and barley to replace a year or two of fallow between crops with legume crops that enrich soil nitrogen and provide livestock pasturage. In drier areas farmers are being taught to use new seeds and better sowing methods and to adjust their plowing to the weather cycle.

The major program of development is based on government investment, since Syria nationalized most industries as part of its socialist goals. Investment by foreign private firms is not ruled out, but the government would prefer to contract with private companies to construct and manage plants under government ownership. Still, Syrians are being encouraged to invest in businesses within the country. Syrian traders are being allowed a freer hand to import goods, and strict controls over money flows have been relaxed. The government is also trying to lure back rich Syrians who left the country during troubled times since independence, taking large sums of capital with them. Since the government controls most industry, it is expected that private money will go mostly into real estate and commerce.

One of the important problems in Syria, as in the rest of the Middle East–North Africa, is the shortage of skilled labor. Even now one-third of boys aged ten to fourteen are illiterate; and girls, in a society influenced by traditional Islam, generally receive less education than boys. Moreover, many of Syria's intelligent, educated professionals have moved to other countries that offer them better prospects. Professionally trained people are being encouraged to remain in the country and appeals to return are being made to skilled Syrians living abroad. The government is emphasizing improvements in the educational system, particularly technical training. In the short term it appears that Syria, like its oil-rich neighbors, will have to turn to Egypt or other countries for skilled technicians, although these countries, engaged in their own development programs, cannot afford such a "brain drain."

There is a shortage of skilled workers, but no shortage of population. Syria's population has been growing at a rate of 3 percent per year. Half the population is less than fifteen years old, and so continued growth can be expected in the future. The first business of economic development, then, is simply to keep the economy growing fast enough to keep up with the new population demands.

Although Syria is not one of the oil-rich countries, annual oil production does exceed 10 million metric tons. This is enough to supply the bulk of the country's exports. Moreover, oil exploration has been haphazard up to now, since all operations have been restricted to a national company and concessions to Soviet bloc companies in Eastern

Europe. Syria now plans to offer oil concessions to Western oil companies.

Most trade still is with Soviet bloc countries and with other Arab countries, but Italy, West Germany, and France are major customers of Syria now. Exports of petroleum and cotton to the world and of wheat and barley to Lebanon and other neighbors are encouraged. Despite plans for self-sufficiency in agriculture, Syria still must import part of its food needs. Much of its industrial equipment must come from more technically advanced countries.

Questions

1. Syria has been presented as a microcosm of the Middle East–North Africa. In what ways is it so? In what ways may it not be representative?

2. Do you think the Euphrates Dam will make a major change in the Syrian economy, increasing the per capita GNP from the present level of $490?

3. It is expected that the area around the Euphrates Dam will support a population of 1.5 million in a country that in 1976 had a population of approximately 7 million. With a population growth rate of 3 percent per year, how far will the dam go toward solving Syria's population problem?

4. Syria is a tiny country by world standards; its population is smaller than that of the state of New Jersey. What do you think its prospects are in the modern interconnected system?

Moslems throughout the Middle East–North Africa. Thus the leadership of the whole region is linked by the common bond of written Arabic. However, strong governments interested in breaking religious ties have pushed for modernization of language. Turkey has expunged Arabic words from the vocabulary of its language.

POPULATION ISLANDS SEPARATED BY EMPTY LAND

The population of the Middle East–North Africa is about 90 percent of that of the United States. The area of the region is larger than that of the United States, however, and the population density is only about two-thirds that of the United States. Such figures tell very little about the distribution of people in the region; that can better be understood by looking at the photograph on page 509. This aerial view of the Nile Valley is vivid evidence of the contrasts in population typical of most of the region. In this arid world population densities are high wherever it has been possible to find water for agriculture. Densities of over 6,000 persons per square mile make the Nile Valley one of the most densely populated agricultural areas in the world. Considering that over half the population still depends on agriculture for its living, the pressure on the land is tremendous. As the photograph shows, where water is not available the land is almost deserted in an arid country such as Egypt.

Ninety-nine percent of the population of Egypt lives in the irrigated Nile Valley, a narrow ribbon only 2 to 10 miles (3.2 to 16 km) wide for 500 miles (800 km) along the river. Only north of Cairo, in the vast **delta** of the Nile River, 150 miles by 100 miles (240 km by 160 km), does the population spread widely over the land. There are only 30,000 to 40,000 nomads living in Egypt's deserts, even though the deserts account for 95 percent of the land area. Only five major **oases** have significant population outside the Nile Valley.

Not all situations are as extreme as Egypt, but most of the people in the Middle East–North Africa live in population "islands" separated by empty arid lands. The population centers mostly hug the coast of the Mediterranean Sea, the Red Sea, the Gulf of Aden, and the Persian Gulf (Figure 15-4). Only in the more humid areas of Turkey and western Iran and along the irrigable river valleys of the Nile and the Tigris–Euphrates is

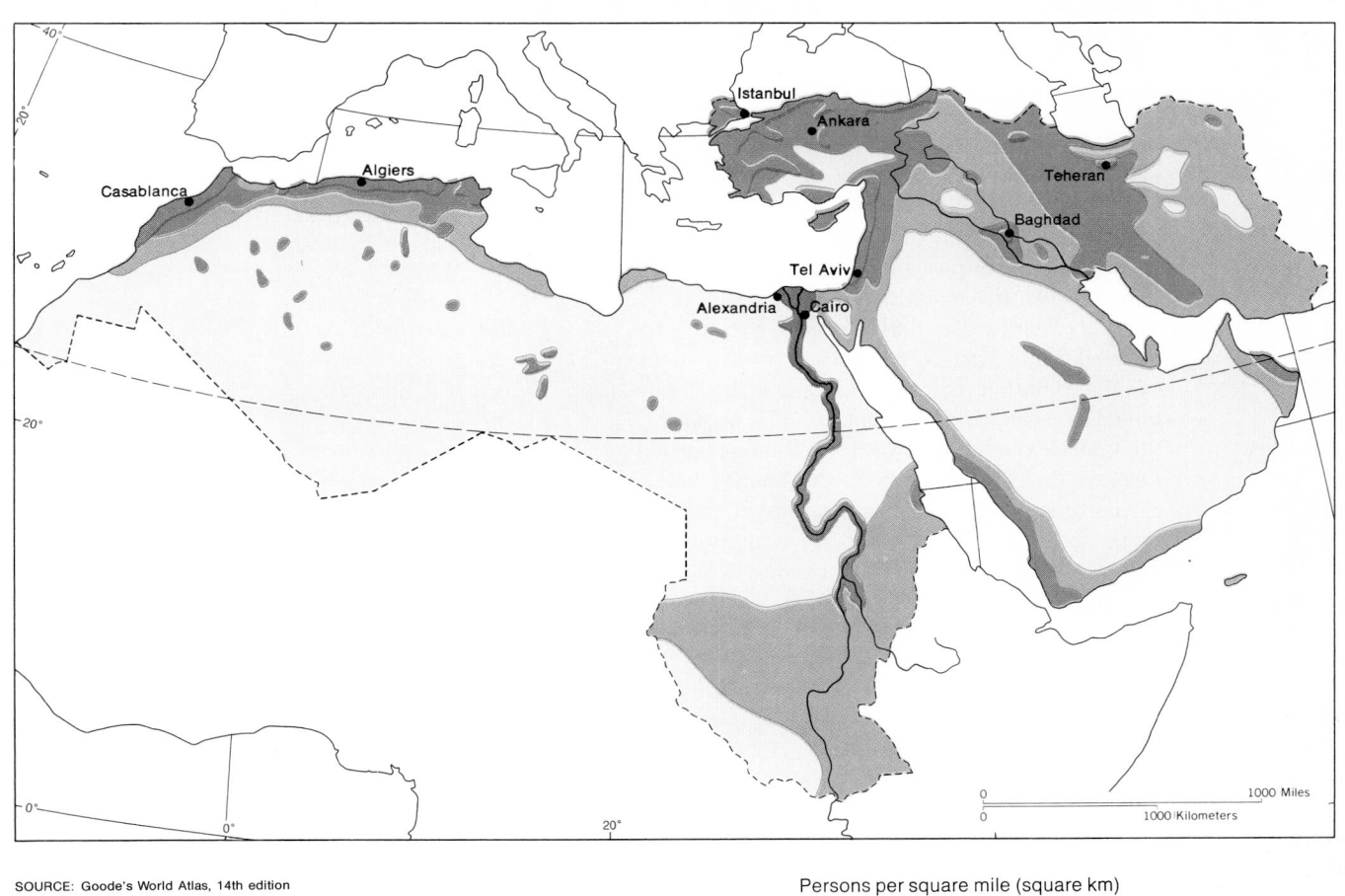

Istanbul
Ankara
Teheran
Baghdad
Casablanca
Algiers
Tel Aviv
Alexandria   Cairo

0   1000 Miles
0   1000 Kilometers

SOURCE: Goode's World Atlas, 14th edition

Persons per square mile (square km)

☐ 0-2 (0-0.8)

▢ 2-25 (0.8-10)

▢ 25-60 (10-23)

▢ 60-250 (23-96)

▢ Over 250 (over 96)

● Cities over 1 million

FIGURE 15–4   POPULATION DENSITY OF THE
MIDDLE EAST–NORTH AFRICA

there dense settlement more than 200 to 250 miles (320 to 400 km) from the coast. The desert even extends to the coast of the Mediterranean Sea in Libya and Egypt, along the shores of the Red Sea and the Gulf of Aden, and on much of the coast of the Persian Gulf. Much of the Sahara across North Africa, the interior of the Arabian Peninsula, and some areas in the interior of Iran have so few water sources that they are scarcely inhabited even by nomadic tribes.

Turkey, which occupies one of the wetter areas of the Middle East–North Africa, has about the same population as Egypt. Turkey's population, however, is supported largely by dry farming in the interior plateau or by Mediterranean agriculture like that of Italy and Greece. **Mediterranean agriculture** involves a combination of irrigated crops in the lowlands; dry farming of grain, grapes, and olives on the slopes; and grazing of sheep on the hillsides. Turkey is, in fact, the only country in the Middle East–North Africa with a diverse environmental base typical of other regions of the world. Its population is therefore more evenly distributed than that of most other countries of the Middle East–North Africa. In Iran, which also has a large population (about 34 million), most people live within a number of separate irrigation-based settlements, and there are, perhaps, a half-million nomads scattered widely over the rest of the country.

The population of countries in the Middle East–North Africa is closely related to the amount of irrigable land. There are large countries—Egypt, Turkey, and Iran—with important water sources and large populations, each over 30 million. There are large countries with fewer water resources and thus fewer people: Morocco and Algeria, each with about 17 million people; Saudi Arabia, with 7 million; and Libya, with fewer than 3 million. At the other extreme, there are heavily populated small countries such as Syria, Yemen, and Tunisia, each with over 5 million people; and other countries such as Israel, Lebanon, and Jordan in the wetter eastern Mediterranean, each with a population of approxi-

mately 3 million. There are also the dry sheikdoms of the Persian Gulf—Kuwait, Bahrain, Oman, Qatar, and the United Arab Emirates—of which only Kuwait has as many as a million people.

DUAL ROOTS: FARM REGIONS AND CITIES

From early times the population centers of the Middle East–North Africa have had a dual base: successful agriculture and important urban development. Agriculture, as we have seen, has been centered in the Nile Valley, the Tigris–Euphrates Valley, and the rest of the Fertile Crescent. Important, too, have been the farmlands of Turkey and the Mediterranean shores of Algeria, Tunisia, and Morocco. The agricultural system of the region has been feudal, with tenant farmers working the large estates of wealthy landlords. Thus life in the farming villages has been difficult and the peasants are very poor.

The efforts of the farm population have provided wealth for cities, but the cities of the Middle East have also been centers of political and religious power and long-distance trade. Rural landlords preferred the sophisticated life of cities to the isolated, backward ways of farm villages. Their wealth, created by the work of peasants in rural areas, was transferred to the cities. Cities also were major political centers: they could be easily defended, there were people and wealth to support armies, and they were linked to other places by overland routes and seaways. Moreover, in many places, cities controlled the only major water sources. Conquerors seized control of farmlands and other territory simply by capturing the regional center served by those areas. Religion, first Christianity and later Islam, spread from city to city with major shrines and large numbers of priests and administrators centered in the cities.

CITIES TODAY:
PART OF A LARGER NETWORK

Today many of the major cities of the past remain important centers of the Middle East–North Africa. Cairo, with more than 6 million people in its metropolitan area, is the largest. Istanbul, formerly Constantinople, has almost 3 million. Alexandria has more than 2 million, and Baghdad about 2 million. In contrast, Teheran, with more than 3 million people, is a relatively new center. Its rise dates from the eighteenth century when the site, which had previously served as a watering spot for animals, was chosen as capital of what is now Iran. Beirut and Casablanca are modern cities now with about a million people each; Algiers is about the same size.

All these cities are tied into the international jet air system of the world with links to major European centers and to those in South and East Asia or Black Africa. They function as regional centers in much the same way as Atlanta and Denver do in the United States. With the rise of demand for petroleum, new centers have been tied into the air network. Now there is service to Kuwait, Manama in Bahrain, Abu Dhabi in the United Arab Emirates, and Dhahran and Riyadh in Saudi Arabia. Mecca and Jerusalem and to a lesser extent Medina still remain religious centers. More than a half-million pilgrims come to Mecca each year, but since only Moslems may live there or even visit, much of the commercial development of Saudi Arabia is centered in Riyadh and the port of Jiddah on the Red Sea and Dhahran on the Persian Gulf. Jerusalem, a city sacred to Jews, Moslems, and Christians, remains the subject of a dispute over jurisdiction between Israel and the Arab world.

FARMING: A WAY OF LIFE FOR MORE
THAN HALF THE POPULATION

Arable land accounts for less than 20 percent of the land area of most countries of the Middle East–North Africa. The major farming areas today are those of antiquity: the Fertile Crescent, the oases of Iran, the coastal lowlands and interior plateau of Turkey, and the coastal lowlands between the Atlas Mountains and the Mediterranean Sea in northwest Africa. Despite the relative scarcity of arable land, farmers make up over half the working population in most countries (Table 15–3). Most farmers still follow traditional practices that have been carried down from generation to generation.

A Village World

The farm village is home to most people, and in many ways it shows striking similarities to Ramkheri (Chapter 1). The typical village is a cluster of crowded sun-baked mud houses, most with only two rooms, one for the family and the other where animals are kept at night. In most houses cooking is done in a central courtyard. Usually, houses do not have electricity, and water is obtained from village wells or water taps. There are a few shops and artisans' workshops, but these are farm villages and most of the families work nearby fields. Each morning farmers walk to their fields following an ox, camel, or bullock and during planting or harvesting the women and children come, too. Workers in the fields are barefoot, and even today they use hoes and sickles or scythes.

For most people the village is their world. Few people have traveled beyond the nearby town more than a few times in their lives, and only a handful in any village have been able to afford a pilgrimage to Mecca. Members of the village form a large extended family, including brothers, sisters, cousins, uncles, and aunts. Intermarriage is common, and family ties are close.

The Beginnings of Change

Although much of village life goes on as it has for centuries, there are important signs of change. The imprint of the government is widespread. In many villages the house of the village "headman" has a telephone, and he is in frequent contact

TABLE 15-3 POPULATION STATISTICS FOR COUNTRIES OF THE
MIDDLE EAST–NORTH AFRICA

Country	Population Estimate, 1976 (Millions)	Annual Population Growth (Percent)	Years to Double Population	Population Under Age Fifteen (Percent)	Urban Population (Percent)
Algeria	17.3	3.2	22	48	50
Bahrain	0.2	2.9	24	44	78
Egypt	38.1	2.3	30	41	43
Iran	34.1	3.0	23	47	43
Iraq	11.4	3.3	21	48	61
Israel	3.5	2.9	24	33	86
Jordan	2.8	3.3	21	48	43
Kuwait	1.1	5.9	12	43	22
Lebanon	2.7	3.0	23	43	61
Libya	2.5	3.7	19	44	29
Morocco	17.9	2.9	24	44	37
Oman	0.8	3.1	22	—	—
Qatar	0.1	3.1	22	—	—
Saudi Arabia	6.4	2.9	24	45	18
Sudan	18.2	2.5	28	45	13
Syria	7.6	3.0	23	49	44
Tunisia	5.9	2.4	29	44	40
Turkey	40.2	2.6	27	42	39
United Arab Emirates	0.2	3.1	22	34	65
Yemen Arab Republic	6.9	2.9	24	45	7
Yemen (People's Republic)	1.7	2.9	24	45	26

SOURCE: Population Reference Bureau, Inc., *World Population Data Sheet, 1976*, Washington, D.C., 1976.

with government officials in the regional capital. Government buildings may include both a school and a clinic. Formal schooling has replaced the traditional informal classes where children were taught to memorize passages from the Koran. The clinic probably has a doctor, nurses, and midwives. Fewer babies die, and adults live longer.

There may also be an office of the local government agricultural specialist who makes suggestions on how to improve crop yields and recommends new crossbreeds of cattle that yield four times as much milk as the native cattle of the village. Farmers are encouraged to plant crops for sale in the growing cities, and money is replacing traditional barter. The small coffee shop where men play cards at night probably has a radio that gives news of the outside world as interpreted by the government station. It also gives suggestions on farming and public health. Young people leave the village to seek their fortunes in cities, and the share of their pay sent back helps villagers to pay taxes and buy important "extras" they couldn't afford before.

Traditionally, farmers have been tenants of wealthy landlords or religious orders based in the cities. A single landlord has often owned not just all the land of one village, but of several others as well. In early-twentieth-century Egypt, wealthy landowners made up only 0.1 percent of all landowners, but they controlled 20 percent of the cultivated land. Sixty percent of the farmers did not own the land they worked. Farmers worked a number of tiny plots, slivers of fields whose total land area added up to less than ten acres per farm. Tenants could not be sure that they would be able to work the same land more than one year. Thus they had little reason to work to improve the land. The absentee landlords commonly saw land merely as a safe outlet for surplus capital, and thus had little interest in farming or in investing in new methods or machines. As long as they received the same level of return each year, they were satisfied.

Here, too, the government has stepped in. In Iran, Iraq, Syria, and Egypt governments have undertaken land reform. The large estates—often more than 100 acres in size—have been confiscated, and their lands divided into units of two to six acres and distributed to the landless farm workers. Farmers are organized into cooperatives that work out credit, plan cultivation, and handle crop marketing. Often the small units owned by single farmers are pooled and worked in large fields. New laws also have introduced reforms in rental agreements. Maximum rents in Egypt have been reduced about one-third and minimum leases are for three years. In Syria rents are one-sixteenth of former levels.

The result of these measures is that farmers are better able to invest in fertilizers, insecticides, and small rototillers, even tractors. The government often supplies threshing machines. Government aid helps with improvements in irrigation systems and drainage facilities.

Government development plans also are bringing the villages into the modern world. New roads and trucks provide access to city markets, and farmers are being encouraged to shift from subsistence to money crops. Farmers in southern Turkey are changing from age-old patterns of goat herds and grain fields to citrus orchards with production going to cities in Turkey and Europe. Electricity and diesel pumps are replacing ancient systems of lifting water in the irrigated lands of Egypt, and there are even some tractors and powered garden cultivators. At the same time, government schools and health dispensaries are providing villagers with improved life prospects. Some of these changes are occurring in all countries, but they are being accelerated in the oil-rich countries, particularly those with small populations.

Crop yields in the area have traditionally been low, except in the Nile Valley (Table 15–4). Returns per unit area for the Middle Eastern farmer are sometimes only one-eighth or one-fourth those for farmers in Western Europe or the United States. This is partly the result of handicaps of aridity and excessively high temperatures. Soil temperature during the summer reaches 130°F to 180°F (54°C to 82°C). This destroys needed organic material and makes it difficult to successfully apply fertilizer. Moreover, much of the crop is lost to pests and disease. In this dry world dust storms and seasonal desiccating hot winds result in severe crop losses. Locust plagues are still serious in the region, as are scale diseases, rusts, and mildew. Workers suffer from malaria, plagues, tuberculosis, typhoid, and eye diseases such as trachoma. Barefoot workers in the field are subjected to parasitic infections, schistosomiasis, and dysentery.

In spite of the many problems, new farm methods have produced higher yields. At the same time better diets and health care have improved the farmers' ability to work. Improved health has also resulted in rapid population growth, however. In many villages half the population is under fifteen, and adults are living longer. Thus each year there are more mouths to be fed in the village although the meager diet does not improve either in quantity or quality. Most farmers eat meat only once or twice a month, and the poor may have meat only twice a year at the two great Islamic feasts.

TABLE 15-4 AGRICULTURAL LAND USE IN THE MIDDLE EAST–
NORTH AFRICA, WITH COMPARISONS WITH THE UNITED STATES
AND THE WORLD

	Total Area	Area of Permanent Pastures and Meadows	Area of Cropland	Percentage of Cropland in Cereals	Cereals Yield
	Thousands of hectares[a]				Kilograms[a] per hectare
Middle East– North Africa	1,439,388	230,493	90,926	45.1[b]	1,174
Algeria	238,174	37,416	6,792	39	506
Egypt	100,145	—	2,852	69	4,004
Iran	164,800	11,000	16,153	42	937
Iraq	43,492	39	4,999	45	960
Israel	2,070	818	417	37	2,287
Jordan	9,774	100	1,300	22	928
Kuwait	1,782	134	1	—	—
Lebanon	1,040	10	345	18	1,171
Libya	175,954	7,000	2,521	17	413
Morocco	44,655	12,500	7,437	61	1,070
Oman	21,246	1,000	36	—	—
Qatar	2,201	50	2	—	—
Saudi Arabia	214,969	85,000	878	48	1,325
Sudan	250.581	24,000	42	10	2,482
Syria	18,541	6,497	5,874	39	1,023
Tunisia	16,361	3,250	4,510	31	812
Turkey	78,058	26,135	28,196	46	1,303
United Arab Emirates	8,360	200	20	—	—
Yemen Arab Republic	19,500	7,000	1,200	98	936
Yemen (People's Republic)	28,768	9,065	252	23	1,643
United States	936,312	244,277	190,053	36.2	2,975
World	13,399,313	2,992,308	1,472,929	49.8	1,818

SOURCE: U.N. Food and Agriculture Organization, *Production Yearbook*, 1974.
[a] 1 hectare = 2.47 acres; 1 kilogram = 2.2 pounds.
[b] Of the total cropland in cereals, wheat comprises 56 percent, barley 21 percent, and sorghum 8 percent.

Fortunately for the village, many young people leave each year to go to the large cities. These youngsters are poorly educated and have little prospect of anything but the most menial city jobs. Compared to village life, however, many think that even life in shantytowns of mud and tin on the fringes of the city is exciting. Poorly paid as they are in the city, they commonly send portions of their wages back to the village, where even these small amounts are important to relatives whose incomes may be less than $100 per year.

Crop and Livestock Choices

Crop choices vary from place to place over this huge region, on the basis of both environmental variations and cultural choices. Farming areas have traditionally been more concerned with feeding their own populations and those of cities and towns in their countries than with export. The chief crops are cereals: wheat, barley, and rye in the north; millet, maize, and rye in the south. Hard summer wheat is the chief crop of Turkey and the other countries at the eastern end of the Mediterranean Sea, and of Algeria and Tunisia. Barley is the major crop in the drier areas of Libya, Iraq, Morocco, and parts of Iran because it is more resistant to drought and insects. Rye is grown in the colder mountain areas of Turkey and Iran. Rice is highly prized because of its high yields under irrigation and is considered a luxury food. It requires both abundant water and heat, and so is limited to warmer areas. It also requires great amounts of irrigation water, and the flooded fields present a malaria danger. Therefore, the amount of rice that can be planted is restricted by some governments. Maize is the chief grain crop of Egypt and is of increasing importance in Israel. Various millets are important grains in dry Saudi Arabia. Sugar beets are an important basic crop in Iran and Turkey.

Cotton, tobacco, citrus fruits, wine, and narcotics have been the chief export crops, but production is sharply localized. Fine, long staple cotton has long been grown in Egypt, and it ac-counts for about 20 percent of cropland along the Nile. Cotton and cotton manufactures represent two-thirds of Egypt's exports. The crop is also important in Turkey and Iran. Middle Eastern tobacco, known as Turkish, has long been exported. It is grown along the northern and eastern coasts of Turkey, in Lebanon, in Iran, and even in Egypt. Grapes are widely grown for wine that is drunk locally by non-Moslems, but Algeria exports cheap table wine to France. Fruits and olives are important in local diets, and citrus exports are particularly important to Israel. Iraq leads the world in date exports. Opium and hashish are grown legally in Turkey to supply the world pharmaceutical industry, and its production is a great source of concern to drug control officers of the world.

Livestock is more important as an input into the farming system than as an output. Bullocks, oxen, donkeys, and camels are important work animals in farming, but animals are not a major part of the farm products produced. Irrigated land is too precious to be used as livestock pasture, and the abhorrence of pork by Moslems and Jews limits the presence of hogs that serve as scavengers in other intensive farming operations. Most villages still have a shepherd or two that look after all animals in common while the rest of the farmers devote themselves to their crops. The only important areas of mixed farming that include livestock are in parts of the Nile Valley, in Lebanon, and in the new agricultural settlements of Israel.

THE BEDOUIN MINORITY

About 7 percent of the population consists of nomadic herders, or **Bedouins,** who occupy most of the arid and mountainous lands of the region. They are found in the Arabian Peninsula, where there has been little agriculture, along the northern fringe of the Sahara, in Iran, and in the highlands of Iraq. In these arid lands it is necessary to move flocks or herds over vast areas in search of pasturage and water. Some groups move through wide districts in the desert throughout the year, staying closely within recognized tribal boundaries. Others move from lowland pastures in the desert during the winter up into mountain pastures in the summer. Some sheep and camel nomads, who move long distances—often hundreds of miles—during the year in search of pasture, and other shepherds of sheep and goats, occupy much more limited territory. Nomads often have some cultivated crops as part of their system. They may feed sheep or cattle on planted pasture during the winter, and then move the animals to wild pastures in the mountains in summer. All nomadic groups commonly trade with farm communities and towns for grain, water, and other supplies. The animals of Bedouins depend on scanty arid vegetation; thus they are often small, and their hides are of indifferent quality. Wool from sheep tends to be "hairy" and tough, so it cannot be used in weaving fine textiles. Instead, it is best used in making carpets, for which parts of this region, such as Iran and Iraq, are world famous. Livestock does provide the basic diet of meat and milk products.

Bedouin groups are limited in size by the availability of food and water for their flocks and herds. Groups are organized in tribal fashion under the leadership of a sheik who is often supported by a tribal council of elder men. Since decisions often must be made quickly, the sheik has a great deal of personal power. The presence of absolute monarchies and strong military leaders in Saudi Arabia, Kuwait, and the tiny states of the Persian Gulf is tied to their Bedouin backgrounds. The Saud clan that rules Saudi Arabia is descended from Ibn Saud, a sheik who consolidated the tribes of Arabia; and Kuwait and Qatar are sheikdoms. Tribal customs are very important and include not only Moslem practices but also traditions that date back to before the conversion to Islam.

Bedouin life is nevertheless changing. Governments see the independent tribal life as a threat to their authority, and nomads have been blocked from following traditional herding patterns by modern political boundaries. As a result, many Bedouins have left their nomadic life and have settled down in cities or farming communities. The national army of Jordan has been based on the use of Bedouin soldiers. The nomadic Kurds who occupy the mountains of western Iraq are one of the few tribes to resist the authority of the national government and have been fighting for many years for independence from Iraq. Only in 1975 was peace apparently established.

The Tuareg nomads of southeastern Algeria have almost completely given up their traditional ways. Once they were the nobles of a complex nomadic system. Tuaregs raised camels and raided villages and caravans. They were supported by vassal tribes that guarded the camps and herded goats that provided meat, milk, and leather. Moreover, slaves were captured and brought back by the caravans to act as domestic servants to the nobles and their vassals.

Years of French and now Algerian rule changed all that. First peoples from other parts of the Sahara were allowed to move into the region and establish agricultural land under contract to the landowning Tuaregs. The slaves were freed and the farming peoples awarded title to the lands they worked. The French established the town of Tamanrasset, which now has an international airport to serve an experimental atomic base, and traditional caravan trade has been replaced by vehicular travel over an asphalt road that stretches from Tamanrasset across the Sahara. Tourists are now encouraged to come into the area.

Many Tuaregs now work in town as wage earners, despite the fact that they once refused to do any manual labor. Children go to boarding school in Tamanrasset, then take jobs in town. Despite traditions that scorn farming, some Tuaregs have settled in small villages and have begun to cultivate their own gardens. Others produce traditional leather goods for sale to tourists, while some use their experience with camels to provide animals and guide tourist parties traveling through the wild mountain area. Most have given up their nomadic ways. Some Tuaregs, however, prefer life in the old tent camps. They settle on the edge of town and try to maintain tribal social customs, but the possibility that Tuareg traditions may soon be completely altered is very real.

THE PRESSURE OF PEOPLE ON FARMLAND

The harshness of the physical environment precludes settlement of population in areas that are now sparsely populated. Today's rapid population growth in the Middle East–North Africa therefore raises the Malthusian problem of population growth outstripping the productive base. Population growth in most countries in the region has been very rapid—over 2.5 percent per year in most countries. In many countries more than 40 percent of the population consists of children under fifteen years of age who will soon be coming into the childbearing age. Many Moslems marry early, still practice polygamy and easy divorce, and place a high premium on large families. Population growth is already threatening Egypt, with its dependence on Nile water, and some other populous countries. Thus governments are advocating programs of population control.

Such population growth puts pressure on national economies. However, although only a small proportion of land in most countries is arable, that land has a high potential for production by world standards because it is basically irrigated land. With irrigation, actual evapotranspiration can be made to equal potential evapotranspiration, and in this part of the world, as we have seen, PE is very high. With an almost year-round growing season in many areas, two to three crops a year can be harvested from the same land. It would seem that with water control and scientific farm practices the prospect of increasing agricultural output is good.

Egypt: Population Pressure on Irrigated Land

Egypt's population has increased tremendously in the last two centuries. In 1800 there were only 2.5 million Egyptians; by 1900 the figure was 10 million. Today the population is almost 40 million, and with a growth rate of about 2.5 percent a year, it will be more than 70 million by the end of the century. Almost all Egypt's population lives on a narrow strip of irrigated land along the Nile and its delta, with densities that are among the highest in the world. Over half the population of Egypt is supported directly by farming.

The Egyptian government has worked hard to increase agricultural production with some success. We have noted its land reform and the cooperatives that encourage farmers to use commercial fertilizers, new crop varieties, and improved systems of crop rotation. Most important, the irrigation system dependent on the Nile has been improved. From the time of the Pharaohs until a hundred years ago farming along the Nile depended on the river's annual flood from August through November. Fields were cultivated and seeds planted in the mud left by the subsiding floodwaters. In the nineteenth century dams were built across the river to hold back floodwaters for use during the dry season and thus allow a second crop. Most important was the Aswan Dam built in 1903 more than 500 miles (800 km) from the mouth of the river. Water stored behind Aswan was released into an elaborate system of irrigation channels that allowed irrigation along the valley from the dam to the delta.

Since then the amount of land irrigated has depended on the amount of floodwater stored behind Nile River dams. Aswan was raised in 1912 and again in 1934 and another dam was built in the Sudan to the south. The reservoirs could not store the full flow of the largest floods on the river, so the amount of irrigation possible in any year still depended on the size of the flood. When the flood was below normal not all land could be irrigated throughout the year. In the 1960s this was changed by the construction of a

ECOLOGICAL IMPACT OF THE NEW EGYPTIAN HIGH DAM

The new Aswan High Dam is expected to increase Egypt's cultivated land by 50 percent. However, it has major negative aspects.

1. The Nile floodwaters brought not only moisture to the parched earth, but also nutrient-rich silt eroded from the Ethiopian highlands thousands of miles away. That silt built up the alluvial valley and delta over thousands of years, and each year provided a new supply of nitrates and phosphates needed for plant growth. With the dam, the silts from Ethiopia are deposited in the water behind the dam, not on the fields. Thus Egyptian farmers now need to apply fertilizers to the land, which is both costly and requires understanding of proper farm practices. Fortunately, Egypt has important deposits of phosphates, mineral salts common to desert areas, and nitrates can be produced from the air using electricity generated at the dam.

2. Lake Nasser, the reservoir behind the dam, is 6 miles (9.6 km) wide and more than 90 miles (144 km) long. This huge area of water increases the breeding grounds for the parasitic blood fluke that causes schistosomiasis, mosquitoes carrying malaria, and black flies carrying trachoma. Egyptians already suffer from these diseases and the number of cases is expected to increase markedly. For example, almost half the Egyptian population is already infected with schistosomiasis. This disease, involving a fluke that must incubate in snails, is the result of the infestation of the intestine. It results in a swollen abdomen, fever, diarrhea, and, in chronic cases, increasing emaciation and weakness. Egyptians who use the Nile and irrigation canals as a place to bathe, dump wastes, wash clothes, and swim expose themselves to the incubated flukes carried in the water. Unregulated floods formerly washed many of the host snails away, but now 55 to 70 percent of persons in the newly reclaimed farmlands—2.6 million or more people—are expected to become infected, bringing the total of Egyptians with the disease to more than 17 million.

3. The nutrient-laden floodwaters also used to reach the Mediterranean Sea, where they provided food for phytoplankton, the lowest order in the food chain of the eastern Mediterranean, an important fishing ground between the mouth of the Nile and the coast of Lebanon. Since the floodwaters have been cut off, the number of phytoplankton cells has decreased from 2.4 million per liter to 35,000 per liter. Egyptian sardine harvests have decreased from 18,000 tons to 500 tons since the construction of the dam. The result is a severe blow to the Egyptian fishing industry as well as a drop in essential protein in the Egyptian diet. On the other hand, Lake Nasser should increase freshwater fish production, which is expected to rise from 2,000 tons to 12,000 tons per year.

Questions

1. The dam has significantly increased the amount of irrigated land in Egypt and has eliminated the uncertainty of year-to-year variations in the flow of the Nile. Do you think it can be considered a success?

2. Most major development projects have positive and negative aspects. How can a country decide whether or not to undertake a particular development? What factors should be considered in the decision-making process?

3. Irrigation is considered the key to agricultural expansion in the Middle East–North Africa. What else can be done to increase agricultural productivity in this region of the world, where there is a shortage of agricultural land and the population is increasing at a rate of almost 3 percent per year?

new high dam four miles upstream from the earlier Aswan Dam. This high dam eliminates flooding completely in the lower valley by storing all floodwater in any year. Now water can be distributed by a system of control gates to all lands as needed during the year. The new dam together with other irrigation projects is planned to expand the irrigated farmland of Egypt 50 percent by 1980. However, during this period it is expected that the rural population of Egypt will have nearly doubled, and thus per capita land in agriculture will continue to drop.

Finding Water Sources

The problem in other countries is much the same as in Egypt: agricultural output from existing land must be increased and new farmland must be found if possible. Finding new farmland means developing new water resources, which is expensive. Dams are major undertakings that require vast sums of money and take years to build. Although other countries do not have rivers the size of the Nile or the Tigris–Euphrates to tap, they can seek new water sources under their arid lands. Major potential sources of water for new farmland have been found under parts of the Sahara and in other dry areas thousands of feet below the surface, but drilling for this water is as expensive as drilling for oil, and additional money must be spent to build the surface irrigation system. To tap these deepwater sources requires modern technology. Moreover, this water, where tapped, is being "mined" from deposits that may have accumulated over centuries. Thus it is being used faster than it is being replenished and is only a temporary source.

Some countries have begun distilling fresh water from seawater. Such plants may be practical for city and industrial water, where there are no other sources, but generally they are too expensive for most farming operations. The cost of financing whatever irrigation system is used to obtain, store, and distribute water to the farms must be borne by either the farmers or the government in this region. Humid farming regions of the world are spared this financial burden.

Population Shifts

One of the solutions to the population problem is out-migration from the most populous areas. Population has, in fact, been moving both from one population island to others within the region and from the Middle East–North Africa to other parts of the world.

Since much of the oil development of this region has occurred in the sparsely populated regions of the Persian Gulf where there is little native technological expertise, Kuwait, Qatar, the United Arab Emirates, and even Saudi Arabia need engineers and other professional personnel, as well as oil field workers. Many have been recruited throughout the Middle East–North Africa, and some professionals have come from the Western industrial countries.

Other people from the region have gone to other areas—Syrians and Lebanese to West Africa, Brazil, Argentina, and the United States. Nearly 1 million inhabitants of Brazil and Argentina are either new immigrants or first-generation Middle Easterners, and they still send part of their wages back to the area. Recently, Turks, Algerians, and Tunisians have found employment in West Germany and other industrial European countries.

One crowded country still has immigration. Jews throughout the world consider Israel as their homeland. Thus almost 2 million Jews migrated into Israel in the twentieth century, mostly from Europe. As recently as 1970 over half the Jewish population of Israel had been born abroad.

The Importance of Trade

The Middle East–North African countries have long known another solution to the dilemma of population pressure: trade. The most populous centers have for centuries been ports and trade centers on caravan routes. Today the volume of transactions in the modern, long-range interconnected system centered in Europe, the United States, and Japan is much greater than in the past. As the oil-rich nations have found, if a country has a product important in this system, it can buy the food and manufactures its growing population needs and still raise living standards.

Middle East–North African countries without oil do not have major output of key resources in demand by the worldwide modern interconnected system. They have few manufactures, other than artisans' products, to export. Agricultural products such as cotton, tobacco, olive oil, and Mediterranean fruits or mineral phosphates are not as essential as oil and they are produced in modest amounts. Thus the non–oil-producing countries are only minor parts of the worldwide system. With the exception of Israel, their exports do not provide the funds for large-scale imports of food and manufactured products.

OIL: THE WORLD'S CHIEF SOURCE OF ENERGY AND A NEW SOURCE OF WEALTH

As we saw in Chapter 12, the Middle East–North Africa is today the world's most important oil-producing area (Figure 15–5). Some of the oil is used within the region to power modern motor vehicles, trains, airplanes, ships, factory machines, and electric generating equipment. However, farmers practicing traditional farming methods in villages without electricity and Bedouin herders are not important consumers of modern energy. With 40 percent of world petroleum production but only 5 percent of total world population, the region generates enormous oil surpluses. Oil-short Europe and Japan and most of the Third World countries depend almost entirely on the Middle East–North Africa for this vital fuel and raw material for petrochemicals.

In recent years the United States has become a net oil importer, buying oil not only from nearby Canada and Venezuela, but from the Middle East as well.

THE CASE OF TINY KUWAIT

In Kuwait the new oil money has drastically changed the country. Before World War II Kuwait was an old-fashioned port where Arabic *dhow* ships traded with India, ports of the Persian Gulf, and the eastern coast of Africa. The port city was surrounded by barren desert occupied by Bedouins. Most of the people were fishers, pearl divers, and boat builders. With the discovery of oil after World War II, the old walled city has been replaced by a modern city of wide boulevards, new apartment blocks, and modern suburbs. Free schools and medical care for all citizens are now provided. Large sums have been paid to landowners for property within the city to be used for the new construction. Many of the old landlords and merchants themselves have become wealthy. Large desalinization plants have been built. The government has become the largest employer and many other Kuwaitis are employed, not just in the petroleum industry, but in the construction industry as well. Supermarkets, shopping centers, and office buildings have turned Kuwait into a modern, Western-style city. In the new industrial park, factories produce items demanded by the newly wealthy city dwellers such as bricks and cement building materials. An oil refinery and a fertilizer plant use Kuwait's natural gas as the primary raw material.

Still, money is left over. Kuwait has been buying jet fighters and other military equipment, and it has begun investing in real estate in Europe and the United States. Kuwaiti investors have taken significant shares in the ownership of European businesses. Because of its oil wealth the

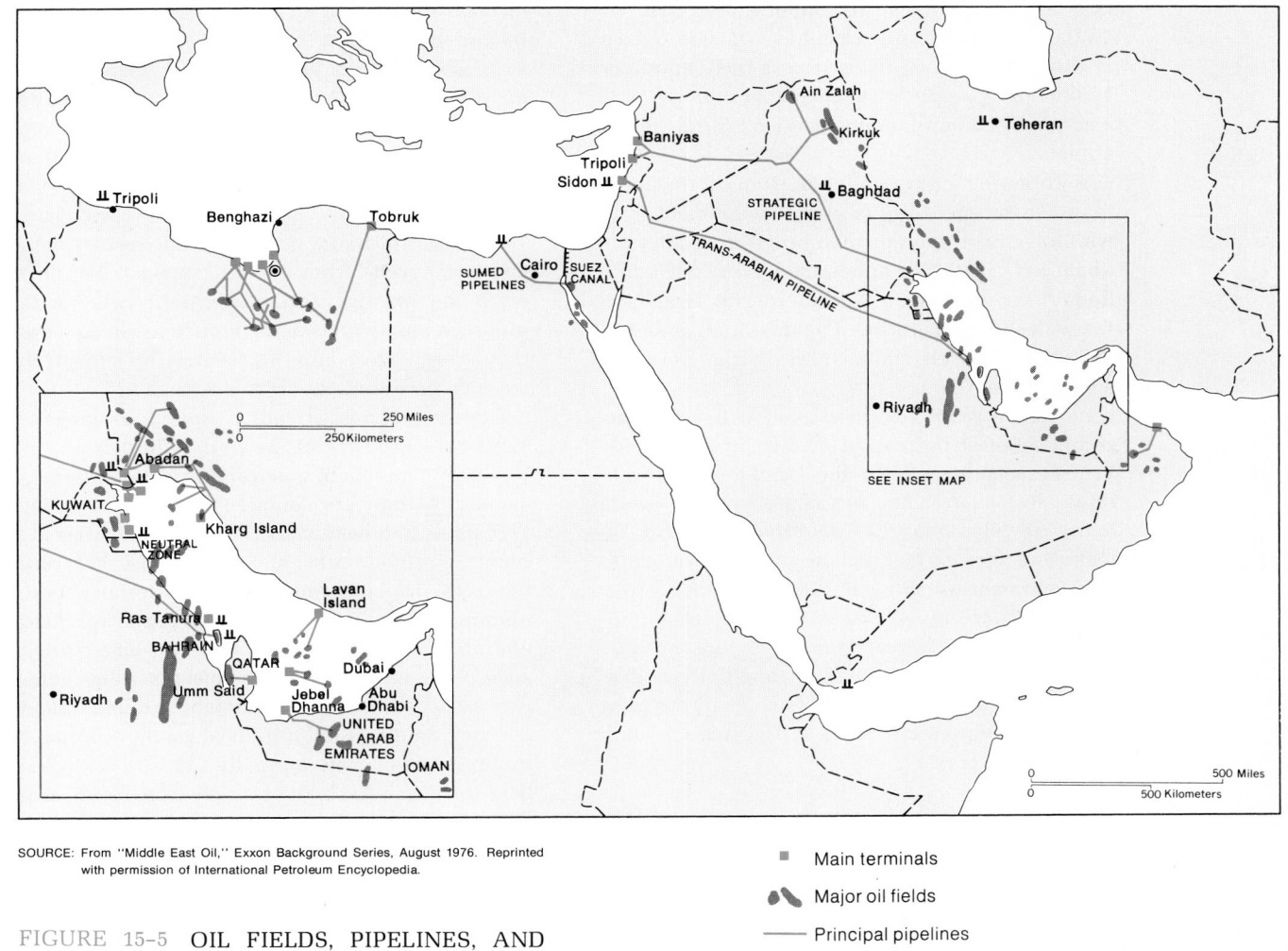

SOURCE: From "Middle East Oil," Exxon Background Series, August 1976. Reprinted with permission of International Petroleum Encyclopedia.

FIGURE 15-5 OIL FIELDS, PIPELINES, AND DISTRIBUTION CENTERS IN THE MIDDLE EAST–NORTH AFRICA

- ■ Main terminals
- Major oil fields
- — Principal pipelines
- ⚓ Large refineries
- ◉ Liquefied natural gas plant

city is emerging as an important center of finance. Major world banks have opened offices in Kuwait, and Kuwait Airport is busy handling businesspeople and financiers drawn to this emerging center of money and national development.

With more money coming in from oil revenues than can be spent on its own development, Kuwait has undertaken its own program of international aid. In 1975 it announced aid of $415 million for the three Arab states bordering Israel, for the Palestine Liberation Organization, and for four poorer Arab states—Mauritania, Somalia, and the two Yemens.

In 1976 Kuwaitis were estimated to have the second highest per capita income in the world, $11,640, second only to the United Arab Emirates. But change has not come without problems. Development has required the help of masses of foreign technicians and professionals. A recent census showed that native Kuwaitis accounted for only 47 percent of the total population. This is of great concern to the government. Citizenship is being withheld from foreigners and, along with it, the benefits of the new Kuwaiti welfare state, such as free schools and free medical care.

THE CASE OF OIL-RICH SAUDI ARABIA

Saudi Arabia, the largest Middle Eastern oil producer, which has a population of about 7 million people, has been accumulating vast amounts of "petrodollars" since the OPEC price increases. In 1975 it was estimated that its reserves of gold and currencies had surpassed those of the U.S. government. Saudi Arabia has begun a program of investing oil money to improve its economy and the living standards of its people. Its second Five-Year Plan, begun in 1975, calls for ten times the investment of the first. The program calls for heavy industry in an area that at present seems

an unlikely location for industry. As one Western businessman put it, "If you looked at the world as a whole, you'd put nothing [in terms of industry] in Saudia Arabia." He might have added that, even if there are 7 million people, and even if they all were to live at a standard equal to that in the United States, the local market for goods would only be the size of the state of Michigan. The leaders of Saudi Arabia do not look at their industrial scene from a global perspective; they want the prestige and power that come with industry, and they have the money to afford what they want. They hope to become an industrial country before oil reserves are depleted.

Oil money has already resulted in government-built industry at the port of Jiddah on the Red Sea. There both a petrochemical plant and a small steel mill are completed, and as of 1974–1975 plans had been made for sixteen small-scale plants to provide consumer goods. Included were plans to produce construction materials, food, aluminum, plastics, spare auto parts, oil field equipment, assembled automobiles, and refined sugar. The most ambitious project is a steel complex to be located on the Persian Gulf and aimed at using Saudi Arabian natural gas to produce 3 million tons of steel annually. This is far more than the Saudi Arabian economy can absorb, and the surplus is to be sold to Europe and Japan. Iron ore will be brought from Brazil, half a world away. It will be mixed with water so that it can be carried as a slurry by crude oil tankers on their return voyages and loaded and unloaded by pipeline. The steel will supply a new pipe mill, which in turn will supply the Saudi Arabian oil industry. There are also plans to build a shipyard to produce tankers to carry gasoline and other refined products from Arabian refineries.

Saudi Arabia also aims at increasing agricultural production through an ambitious program to develop water resources. Plans call for drilling 1,000 wells, some of them to tap water sources as far as 7,000 ft (2,135 m) below the surface of the

OIL-RICH COUNTRIES APPLY THE MOST ADVANCED TECHNOLOGY

In using their wealth to buy the technology of the modern interconnected system based in the United States, Western Europe, the Soviet Union, and Japan, the oil-rich countries are not satisfied to simply copy existing facilities. Most often they ask technicians to produce new systems designed specifically to fit their needs and to incorporate the newest technology.

King Faisal Specialist Hospital in Riyadh is planned as the most modern hospital in the world. Not only will it serve patients from throughout the Arab world, but it will attract the outstanding specialists and surgeons of the world. Computerized services will automate basic medical services, thus freeing doctors and nurses to carry out advanced porfessional treatment and care. A communications system will allow continuous monitoring of patients and instantaneous retrieval of patient records. At the same time, it will provide computer printouts of work schedules for nurses in the wards and technicians in laboratories, as well as a continuous inventory of supplies and medicines. Computer communications will also provide the operating theaters with access to patient records and all laboratory facilities. In addition, the latest nuclear treatment and laser facilities will be available.

There are plans for a regional satellite communications system throughout the Arab world. Stations will be set up in all major capitals and more than 1,500 stations in communities throughout the region. It is expected that an Arab-financed satellite will be placed in stationary orbit over the region. Such a system will provide both long-distance telephone communication and television.

Even now some of the smallest, most traditional countries are making use of the latest communications systems. Oman is installing an advanced telecommunications system tied to satellites, and the United Arab Emirates are building a microwave network.

Questions

1. If, in fact, the oil-producing countries of the Middle East carry out these ambitious plans and develop a modern technology, how do you see these countries fitting into the worldwide modern interconnected system centered in the United States, Western Europe, Japan, and the Soviet Union?

2. Compare the development problems of Saudi Arabia with 7 million people, Iran with 34 million, and the United Arab Emirates with 200,000. Which of these countries has the most favorable prospects for development?

desert. The new program also calls for increasing the number of dams from about a dozen to at least fifty within the next few years. Under the scheme settlers on the farmlands created by the use of the new irrigation water will receive free land. Plans are being implemented to settle many of the country's Beduoin tribes as farmers in the newly developed land.

THE BIG QUESTIONS FOR
THE OIL-RICH COUNTRIES

Several questions in the development of oil countries remain unanswered:

1. Is there enough water?
2. Are there enough workers?

3. Are there enough expert managers?

4. Will oil last long enough to provide funds for national development plans?

We have seen that there are serious questions about water supply. Oil-rich countries around the Persian Gulf are in some of the driest environments on earth. This is true of Libya, too, which might be regarded as an Egypt without a Nile River. Dams are expensive and bring environmental problems. Deepwater drilling in arid regions is really another form of mining and will soon deplete resources. The most likely source of water for the new industry is the sea. Saudi Arabia's new industries will depend on the **desalinization** of seawater and Kuwait uses the same source. Such plants not only supply fresh water but produce electricity at the same time. The problem is that desalinization of seawater remains one of the most expensive means of obtaining water in the world.

Finding Enough Workers

Labor supply is a more critical question. Some of the small countries do not have enough people to meet industrialization plans. There has been major immigration from elsewhere in the Middle East. The population of Kuwait, for example, more than trebled in the decade between 1963 and 1972 as workers were attracted by high wages.

Throughout the Middle East–North Africa the problem of recruiting a new industrial labor force involves more than mere numbers of people. Illiterate farmers and Bedouins will not be suited to the semiskilled jobs needed in modern metallurgical and petrochemical industries until they are both educated and given specialized training. More important, tradition among the leading tribes and clans of the region scorns manual labor. Bedouins have always felt that they should not have to work with their hands. Thus although education is partly designed to supply workers for the new industry, many young people attending universities view education as preparing them for office jobs, not factory work. Most seek careers in government. In Saudi Arabia it is estimated that 300,000 Arabians work for the government, the largest number in office jobs.

The Shortage of Trained Professionals

Most serious of all is the shortage of scientists, engineers, and plant managers. Modern universities are new to the oil countries, and most of their nationals with university educations have received them overseas in the United States, Europe, Japan, or the Soviet Union. Many have found markets for their new skills in the industrial countries where they were educated, and they have never returned. Many, because of traditions of despotism and corruption in their countries, felt that they would be unable to find well-paid jobs at home. Thus although the producing countries have gained majority interests in the oil operations, they still depend on foreign managers who work under contract. Some of the government officials in the oil countries fear that leaving operations and administration in the hands of foreigners will lead to gradual loss of both control and the ability to direct their reserves in accordance with their best interests. A veteran Arab oil man put it bluntly, "We must develop know-how or we will kill ourselves." The building of technical colleges and universities has high priority, and efforts have been made to encourage professionals living overseas to return. Many overseas Arabs, however, already live very well, have changed from traditional Moslem ways, and have even married foreign non-Moslems.

How Long Will Oil Last?

Although new oil fields may yet be discovered in the Middle East and North Africa, the oil in this, the richest oil region known in the world, is limited. Even Saudi Arabia, with the greatest reserves, is expected to have oil for only sixty-five years if production continues at the 1972 rate. Other Persian Gulf countries—Iraq, Kuwait, Qatar, and Oman—have similar prospects. Iran, the most populous oil country, where the demands for development monies are greatest among the oil countries, will be out of oil in less than thirty-five years at 1972 output. Libya and Qatar have similar prospects for their oil; in Algeria the wells may run dry in less than twenty-five years, in Bahrain in twenty years.

The tiny oil countries with small populations may be able to invest enough of their petrodollars in the modern interconnected system of the world so that, like wealthy retirees, they can live off the income of their investments. But Saudi Arabia, Iraq, and particularly Iran will not have that luxury. They are investing in their economy while the oil money lasts in hopes of developing as industrialized countries that will be able to sustain themselves after the oil has been depleted.

TIES TO INDUSTRIAL COUNTRIES

Even though the oil countries of the Middle East–North Africa have gained wealth and control over oil operations, they remain very dependent on the modern interconnected system centered in the industrial countries, which provide the bulk of the market for their oil. The greatly increased price of oil that has brought prosperity to the Middle East–North African oil countries has called forth oil conservation policies and a search for alternative power sources in the industrial countries and oil-poor Third World nations. The result has been a decrease in the demand for Middle Eastern–North Africian oil, and thus a slowing of the flow of oil monies into the region coupled with mounting pressure for lower prices. However, Japan and Western Europe continue to rely in large part on oil from the Middle East–North Africa; and, with the decline in domestic production of crude oil, the United States increasingly depends on foreign oil.

As we have seen, the oil countries are tied to the modern system through more than oil and money. The technology that has made large-scale oil production possible and that is behind the new industrialization in oil countries comes from the industrial countries. So does management. So, too, does the modern lifestyle. Wealthy Middle Easterners drive Cadillacs and European luxury cars and live in homes furnished with Western products. New television networks beam reruns of American and European television programs, and Middle Eastern cities and resorts look like those elsewhere in the modern world. Cairo, Teheran, and the other large cities have jet airports, modern hotels, and restaurants. Office blocks house branches of international corporations and banks, and headquarters of new Middle Eastern firms that themselves do business throughout the world. There are modern suburbs occupied by nationals and foreign managers and technicians, and thousands of high-rise apartments of workers employed in the offices, stores, transport facilities, oil fields, and new factories.

Still, the old ways remain, not only in the farming areas, but in the shanty towns around the fringes of the cities that attract the poor and uneducated villagers. These people eke out a living doing the jobs no one else wants in the city. Even in the oil countries they live in great poverty on the edge of great wealth.

ISRAEL: A JEWISH STATE
IN A MOSLEM WORLD

The state of Israel appears insignificant on the map of the Middle East–North Africa. Although Israel is only slightly larger than New Jersey and has only 1.4 percent of the total population of the

Middle East–North Africa, it has presented one of the largest problems in this part of the world.

Israel was created in the twentieth century not simply by Jewish immigration into the area, but by the displacement of some 400,000 Arabs through force. Palestine, which as late as the outbreak of World War II was still overwhelmingly Arab, was the scene of immigration of Jews after 1895 under authorization of the United Kingdom, which held mandate over the territory. By 1946 there were 650,000 Jews in Palestine—two-fifths of the population. The Jewish Zionist movement throughout the world was calling for the partition of Palestine into separate Jewish and Arab states. The matter was brought before the newly formed United Nations, and a partition plan was approved that included Arab control of the west bank of the Jordan River and the setting aside of Jerusalem, a city sacred to Jews, Christians, and Moslems alike, as an international city. The plan was not implemented. The Arabs attacked the Jews, who carried out the first of two military campaigns that have pushed back the Arabs. In 1948 the state of Israel was established. The new Israeli state included the territories proposed by the U.N. partition plan; the rest of Palestine west of the Jordan River was occupied by Israel in 1967.

The Palestinians who fled from Israel have various attitudes and strategies, but all retain the hope of returning to at least the lands along the west bank of the Jordan River occupied by the Israelis in 1967. Israel has continued to place new Jewish settlements in that territory. Palestinians have formed the Palestine Liberation Organization (PLO), recognized by the Arab states as the sole legitimate representative of Palestinians, and the PLO leader has spoken before the United Nations. The PLO has a national assembly, an army, and guerrilla forces. It has successfully mobilized support of many countries in the world for the return of Palestinians to Palestine.

The conflict between the Arabs and Israel has resulted in three bloody wars, the most recent in 1973. In the thirty years since the formation of Israel there have been only cease-fire agreements; no peace treaty has ever been signed.

A MODERN INDUSTRIAL COUNTRY

With aid from the outside world and leadership educated in modern ways in Europe, the United States, and the Soviet Union, Israel has emerged as a modern country (Table 15–5). Israel stands as an example of what modern scientific expertise can accomplish in the arid environment of the Middle East–North Africa.

Agriculture accounts for less than 8 percent of total employment, but this is in large part the result of application of modern technology to farming. A major effort has been made to establish Jewish agricultural communities, to irrigate more land, and to intensify all farm production by using modern scientific methods and machinery. The farm communities are cooperatives, either individually owned and operated farms or collectives—the *kibbutzim* (singular *kibbutz*). Major investment has gone into a national irrigation policy, based to a large extent on the use of Jordan River water diverted from the north into the southern parts of the country and the Negev Desert. This has created controversy in the Arab world because the sources of the Jordan River are in Syria and Lebanon and the river forms the border with Jordan.

Israel wants to be as agriculturally self-sufficient as possible. Ample food production helps reduce its critical import needs. Israel has managed to become largely self-sufficient in foodstuffs except for cereals and vegetable oils and fats. Citrus fruits marketed in Western Europe have become a major export.

Most of all, Israel is a modern industrial economy, the first in the Middle East–North Africa. It reached the stage of industrialization twenty years ago that Iran and Saudi Arabia seek today.

TABLE 15–5 SELECTED ECONOMIC INDICATORS FOR ISRAEL AND EGYPT

Indicator	Unit of Measure	Israel	Egypt
Per capita energy consumption, 1974	Kilograms of coal equivalent	2,914	322
Per capita GNP, 1973	U.S. dollars	3,010	257
Total GNP, 1973	Millions of U.S. dollars	9,932	9,361

GNP by economic sector, 1973 (percent)[a]

		Israel	Egypt
Agriculture		5	29
Industry		18	19
Construction		10	3
Wholesale and retail trade		8	9
Transportation and communication		6	4
Other		32	24

Employment by economic sector, 1973	Number (Thousands)	Percent	Number (Thousands)	Percent
Agriculture	81.4	7.5	4,179	46.6
Industry	269.8	24.8	1,160	12.9
Construction	96.1	8.8	359	4.0
Trade and finance	206.9	19.0	860	9.6
Transportation and communication	78.9	7.2	409	4.8
Utilities	10.4	1.0	82	0.9
Other	344.9	31.7	1,927	22.5
Total	1,088.4		8,976	

SOURCE: Data reprinted from *The Europa Year Book 1976*, vol. 2 (Europa Publications Limited, 18 Bedford Square, London W.C.1); and from *U.N. Statistical Yearbook, 1975*.
[a]These percentages do not add to 100 because of import duties on industrial materials.

Most production goes to provide essential consumer and construction goods for the country, but there is an important export trade with Europe and the United States. Israel is the world's leading diamond cutting and polishing center and has important exports of textiles, chemicals, paper, tires, and electrical goods. Israel even produces airplanes. The products of Israel's industries could well be used by other countries in the Middle East–North Africa, but there is virtually no trade because of the enmity between Israel and the Arab governments. Only Iran and Turkey, non-Arab countries, trade with Israel. Several Arab governments officially boycott multinational corporations that do business with Israel, and they will not invest their oil money in banks known to have major Jewish ownership, wherever they may be located.

THE COST OF CONFLICT

Arab governments vow that they will not relent until the almost 1 million Arabs who have fled from Israeli territory since 1948 have been repatriated. There is strong feeling that an Arab state must be established in the West Bank lands occupied by Israel. To achieve these goals, Arab countries, rich and poor, have been spending large amounts of their national budgets that might have gone for new irrigation facilities, new roads, and better housing, on arms to overthrow the Israelis. The Israelis, too, have spent an important share of their national production on arms. The Middle East–North Africa has a larger concentration of modern military equipment than any other region in the Third World. The modern industrial countries that have supplied most of these arms now spend a great deal of effort trying to maintain peace in this part of the world.

The conflict between Jews and Arabs is not the only one in the Middle East–North Africa. There is enmity between Moslems and Christians in some areas, particularly Lebanon, where armed hostilities raged between 1975 and 1977. In that country both Christians and Moslems make up major components of the population. The Moslems, however, estimated to be 60 percent of the population, attempted to change the governmental system established in 1932, which called for the president and the commander of the armed forces to be Christian, the prime minister to be Moslem, and the legislature to have a six-to-five ratio of Christians to Moslems. Bitter guerrilla fighting broke out between Moslems and Christians, especially in the capital and largest city, Beirut. The commercial core of Beirut, one of the most important banking and business centers in the Middle East–North Africa, was almost totally destroyed. It took intervention by the Syrian army and an Arab peace-keeping force to re-establish order.

THE FUTURE OF
THE OIL COUNTRIES

The countries of the Middle East–North Africa, particularly those with oil resources, have become important parts of the worldwide modern interconnected system, even if many of their citizens who are farmers and nomadic herders have not. But today it is difficult to see what the future role of the oil countries in the worldwide system will be. Will they be rich raw-material producers and consumers of goods produced by industrial countries, but still play a minor role in comparison to the industrial countries? Or will they be able to create a viable industrial base and develop the professional management expertise that will enable them to emulate Japan, and to join the fraternity of wealthy industrialized countries? Can Arab-Israeli conflict be peacefully resolved or will there be more war?

SELECTED REFERENCES

Fenelon, K. G. *The United Arab Emirates: An Economic and Social Survey.* Longmans, London, 1973. A British scholar's look at these little-known shiekdoms.

Fisher, W. B. *The Middle East: A Physical, Social, and Regional Geography,* 6th ed. Methuen, London, 1971. The standard geographic text on the area by a distinguished geographer.

The Middle East and North Africa, 1976–77. Europa Publications, Ltd., London, 1976. A basic handbook on the area with articles on major area problems and individual countries by recognized authorities.

Powell, G., M. Gelb, and A. Spencer. *Atlas of the Middle East.* Kendall-Hunt, Dubuque, Ia., 1975. A well-illustrated basic atlas of the Middle East with accompanying text dealing with problems and regions.

Sampson, A. *The Seven Sisters: The Great Oil Companies and the World They Shaped.* Viking Press, New York, 1975. A popular book dealing with the impact that oil has had on this area and the rest of the developing world.

CHAPTER 16

South and East Asia: Can Half the World's People Attain Greater Economic Development?

The problem of development is posed in its most extreme form in South and East Asia, where increases in food supply and economic development must keep pace with a huge and rapidly growing population. India and China are our major case studies, illustrating two different approaches to modernization—one following Western economic experience, the other more ideological. Southeast Asia has centers that have become major nodes in today's business world, especially Singapore and Hong Kong. In most of Asia, however, dual societies remain, and further development depends on greatly increased productivity, both agricultural and industrial, to meet the vast demands of nearly half the world's people.

Left: Rice seedlings, of one of the improved hybrid varieties, being transplanted into the rice fields in the Philippines. (CARE photo)   *Above:* Sorting and drying fish in India for storage. (Lynn McLaren)

Above: Commune workers in the People's Republic of China. (Rene Burri/Magnum Photos, Inc.)

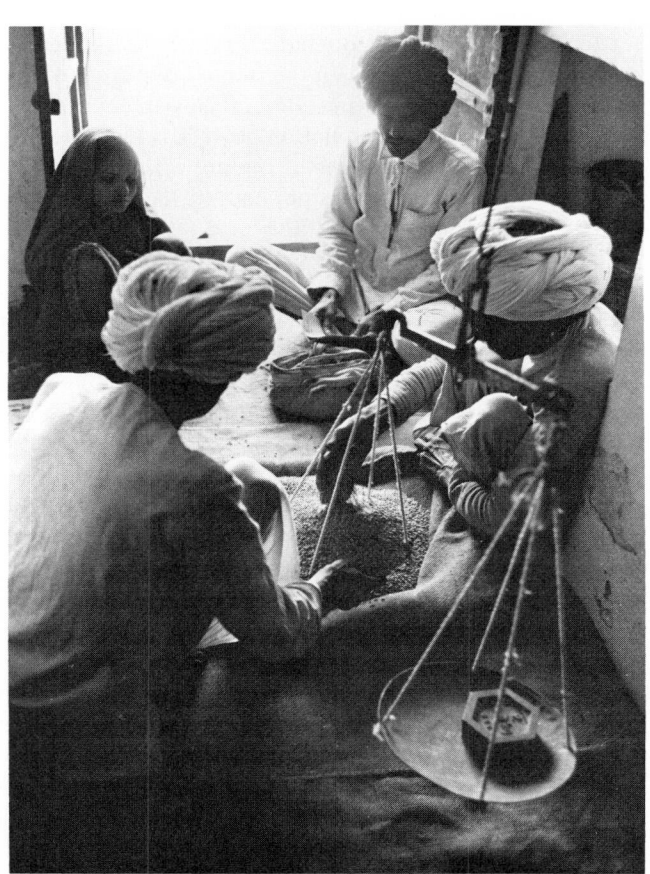

Left: Thai students of agriculture examining an experimental strain of corn. (Ted Spiegel)   *Above:* Indian villagers in the state of Rajasthan buying grain. (FAO photo)

South and East Asia is in many ways the most crucial region of the developing world. The challenges this region presents to the modern interconnected system, especially to its capacity to improve the quality of human life, are of a dimension never faced before. No experts—individual, corporate, or government—can confidently put forward plans to solve its developmental problems.

The problems themselves are basically the same as those in the other developing regions we have studied: Latin America, the Middle East–North Africa, and Black Africa. The majority of the population live in a predominantly traditional, locally based system, struggling to support themselves on the limited resources of their own particular portion of the earth environment. The rate of population growth is high in most countries, almost half the population is under fifteen years old and per capita GNP figures are among the lowest in the world.

The overwhelming problem facing development programs in South and East Asia is that of scale. South Asia, dominated by India, and East Asia, dominated by China,[1] contain the two most populous countries in the world. Southeast Asia, comprising the Indochinese peninsula (Laos, Thailand, Khmer Republic, and Vietnam), the Malay Peninsula, and the islands of Indonesia and the Philippines, is also noted for major population concentrations. If we exclude Japan, which we studied separately in Chapter 10, we find that South and East Asia contains more people than all the regions studied in previous chapters combined. Here are more than half the people on earth, over 2 billion in 1975. Not only is the scale of the development problem greater than that in any of the other developing regions of the world, it is greater than that faced by any of today's industrial regions in the past. At the beginning of the Industrial Revolution, Europe

[1] By China we mean the People's Republic of China, or Communist China. We refer to Nationalist China as Taiwan.

probably had less than one-tenth as many people as South and East Asia have today. Industrialization began in the United States with less than one-twentieth the population of this part of the world. Even now, Europe has less than one-fourth as many people and the United States just over one-tenth as many.

In South and East Asia the race between population, on the one hand, and food supply and other aspects of economic development, on the other hand, involves half the population of the world. The degree of success in overcoming the population problem here will have a major impact not only on the people of this region, but on the people of the whole world. Half the world's population cannot be ignored.

THE MALTHUSIAN PROBLEM

Population densities are much higher in South and East Asia than in either Latin America or Africa (Table 16–1). Only five of the twenty-three countries listed in the table have population densities of less than 100 people per square mile (40 per square kilometer), and over half have more than 200 per square mile. These densities are three or more times those in Latin America or Africa. However, with the exception of Bangladesh, Taiwan, South Korea, and the independent cities of Singapore and Hong Kong, these densities are lower than those of such countries as the United Kingdom, the Netherlands, Belgium, West Germany, or Japan. Those countries are centers of the modern interconnected system with most of their population centered in large metropolitan agglomerations, not spread over the rural landscape; and the European and Japanese systems are supported by imports from throughout the world. In South and East Asia, as the low per capita GNP implies, most people are still subsistence farmers depending on their own local environmental base.

TABLE 16-1 POPULATION DATA FOR COUNTRIES OF SOUTH AND
EAST ASIA

Country	Population Estimate (Millions)	Persons per Square Mile (Square Km)[a]		Rate of Growth (Percent per Year)	Population under Age Fifteen (Percent)	Per Capita GNP ($ U.S.) 1974
South Asia						
Afghanistan	19.5	78	(30)	2.2	44	100
Bangladesh	76.1	1,381	(533)	2.7	46	100
Bhutan	1.2	66	(25)	2.3	42	70
India	620.7	506	(195)	2.0	40	130
Nepal	12.9	237	(91)	2.3	40	110
Pakistan	72.5	210	(81)	2.9	46	130
Sri Lanka	14.0	553	(213)	2.0	39	130
East Asia						
China	836.8	227	(88)	1.7	33	300
Hong Kong	4.4	11,000	(4,250)	2.1	36	1,540
Japan	112.3	781	(301)	1.2	24	3,880
North Korea	16.3	351	(135)	2.7	42	390
South Korea	34.8	916	(354)	2.0	40	470
Mongolia	1.5	2.5	(1.0)	3.0	44	620
Taiwan	16.3	1,173	(453)	1.9	43	720
Southeast Asia						
Burma	31.2	119	(46)	2.4	41	90
Indonesia	134.7	179	(69)	2.1	44	150
Cambodia (Khmer Rep.)	8.3	119	(46)	2.8	45	—
Laos	3.4	37	(14)	2.4	42	—
Malaysia	12.4	97	(37)	2.9	44	660
Philippines	44.0	380	(147)	3.0	43	310
Singapore	2.3	11,500	(4,450)	1.6	39	2,120
Thailand	43.3	218	(84)	2.5	45	300
Vietnam	46.4	361	(139)	2.2	41	150

SOURCE: Population Reference Bureau, Inc., *World Population Data Sheet, 1976,* Washington, D.C., 1976.
[a]Population density calculated from land area data in Edward Espenshade and Joel Morrison, eds., *Goode's World Atlas,* 14th ed., Rand McNally, Chicago, 1974.

High population densities in an area still dominated by the traditional, locally based system mean that the pressure of population on the environmental base is approaching the critical stage in terms of the Malthusian principle. As we saw in Chapter 4, population, if unchecked, will increase in a geometric progression, while the means of subsistence will increase in only an arithmetic progression. Therefore population will tend to outrun the growth in the food supply. It will, however, expand only to the number that can be supported at the subsistence level by the local environment; at that point, famine, illness, and wars resulting from competition for the scarce food supplies will tend to keep the population stable. Stabilizing factors such as these are known as **Malthusian checks.** The critical point in population growth is the one at which the population demands begin to exceed the carrying capacity of the resource base.

In Europe at the time of the Industrial Revolution, achievements in technology based on the use of inanimate energy were able to increase the productiveness of the resource base, and thus the number of people that it could support. Most Asian countries have yet to feel the impact of the energy revolution. People depend largely on their own labor and the work of domesticated

THE DEMOGRAPHIC TRANSITION

Demography is the study of the characteristics of human populations—their distributions, densities, and changes that occur over time. Most countries of South and East Asia are in one of the middle stages of the process of demographic transition. Demographers have observed and described a sequence of changes in birth rates and death rates that roughly parallels the development of a country from the traditional to the modern model. **Demographic transition** has four stages:

Stage 1. Birth rate and death rate are both high. A culture group in this stage is almost completely dependent on its local environment. When environmental disasters such as floods and droughts occur, the results are famines, illness, and wars. Population size fluctuates in response to such events, but it remains approximately the same over the long term.

Stage 2. The death rate begins to decrease, but the birth rate remains the same. This change in the birth and death rates has only one possible result: a steady increase in population. The most populous countries of South and East Asia—India, China, and Indonesia—are in this stage of the demographic transition. Improvements in medicine, public health, and sanitation have reduced the death rate. Increasing ties with other countries reduce their dependence on the local environment, and when natural disaster does strike, aid from more developed countries usually moderates its effects.

Stage 3. The birth rate begins to decline, while the death rate remains stable at a low level. The decrease in birth rate is usually the result of urbanization accompanied by a rising standard of living. People in urban-centered industrial economies tend to have fewer children because children are seen as economic expenses rather than as assets, as they are viewed in rural areas. Urban families have the incentive and educational opportunities to learn contraceptive methods. In this stage of the demographic transition, the population continues to increase because the birth rate is still higher than the death rate, but the rate of increase is slower. Japan and South Korea may be considered examples of this stage.

Stage 4. Birth rate and death rate are both low. This stage has been reached by most countries of Anglo-America and Western Europe, where industrialization and urbanization have led to a decrease in the birth rate to the point where it is approximately the same as the death rate. The result is little if any population growth.

Consider the following list of birth and death rates (taken from *U.N. Demographic Yearbook, 1974*):

Country	Birth Rate	Death Rate	Country	Birth Rate	Death Rate
	(per 1,000 population)			(per 1,000 population)	
Group A			*Group B (cont.)*		
Angola	50.1	30.2	Singapore	19.9	5.3
Benin	50.9	25.5	South Africa	40.3	16.6
Burundi	48.1	25.2	South Korea	35.6	11.0
Mali	49.8	26.6	Soviet Union	18.2	8.7
Group B			Sri Lanka	29.5	7.7
China	33.1	15.3	*Group C*		
Egypt	35.4	12.9	Austria	12.8	12.5
Hong Kong	19.3	5.2	Canada	15.5	7.4
India	42.8	16.7	Denmark	14.2	10.2
Indonesia	48.3	19.4	France	15.2	10.4
Ireland	22.5	11.0	West Germany	10.1	11.7
Japan	19.4	6.6	Sweden	13.4	10.6
Mexico	43.2	8.9	United Kingdom	13.1	12.1
Pakistan	36.0	12.0			

Questions

1. How would you characterize each group of countries in terms of demographic transition?

2. Hong Kong and Singapore have the lowest death rates. What factors do you think might account for this?

3. Austria has one of the lowest rates of population growth in the world, 0.1 percent. What variables in Austria's geographic equation would explain this?

4. West Germany's death rate is even higher than its birth rate. However, the rate of population growth in this country is 0.2 percent, greater than that of Austria. What might account for this difference?

5. What part of the world has the largest number of countries in the early stage of the demographic transition?

6. How would you rank the countries in Group B in terms of the demographic transition? Which of these countries have the most serious population problems, and why?

7. Try to represent the demographic transition by a simple graph based on the written descriptions of the four stages. Plot time (stages 1 through 4) along the horizontal axis and rates (birth rate and death rate) on the vertical axis. Then, using a different color pencil, draw a line representing roughly the pattern of population growth during the demographic transition.

animals such as oxen, bullocks, and horses to provide energy. They have only simple tools and crude farm implements. Birth rates remain high. Most people are uneducated in contraceptive methods, and children are prized as an additional source of energy. Thus population growth, increasing geometrically, puts great pressure on an economic system that has expanded, but only sluggishly, in an arithmetic progression.

The checks on population that Malthus described have been at work in Asia. Illness and early death have long been common; infant mortality has been high; and large numbers of older children die of diseases and malnutrition. Life expectancy in India, Pakistan, Indonesia, and most of Southeast Asia is less than fifty years. The ever-present danger of famine is increased by natural disasters such as flood, drought, or insect plague. A million and a half people died of famine in India in 1943, and only imports of grain from Canada and Australia and emergency subsistence rations halted widespread famine in China in the winter of 1960–1961. **Typhoons,** the Asian counterpart to hurricanes, are a threat each spring and fall particularly to the Philippines, coastal Vietnam, and Taiwan. Typhoons contributed to the twenty-five days of heavy rain in the Philippines in 1972 that killed more than 400 persons and destroyed most of the country's rice crop. The storms left 2.4 million people without food or employment. Floods in Bangladesh in 1974 covered half the country, producing famine that went largely unchecked despite foreign aid.

The population problem of South and East Asia has been compounded by the introduction of modern medical techniques and public health to the area. Now fewer children die, pregnant mothers are better cared for, and adults are inoculated against diseases. Consequently, the death rate has decreased, but there has been little decrease in births in traditional rural areas. These two factors have worked to create a population explosion: Asian population has more than tripled since 1800, and it has more than doubled since 1900. Even more telling is that this doubling has added about 1 billion people, one-fourth of all the people on earth today.

As noted in Chapter 10, population pressure on a particular environment can only be assessed in terms of the economy under which people exist. Europe, for instance, was overpopulated before industrialization in the eighteenth century, when most of the population was engaged in subsistence agriculture. The modern industrial economy now supports four times the population of the eighteenth century at much higher standards of living. Economic development brought about a new system and an entirely different resource base. South and East Asia is the major area of the underdeveloped world where the existing traditional economic system cannot provide for the basic food needs of the large and growing population. Most nations of South and East Asia must change their economic system and reduce population growth—or face catastrophes brought about by huge population growth. A temporary alternative might be for the rest of the world to agree to a continuing, and probably increasing, subsidy of food to support Asia's tremendous population. This, however, is politically and economically unlikely. Furthermore, it would soon become impossible for surpluses from other parts of the world to sustain the large absolute additions of population in South and East Asia.

A certain degree of change has already occurred in the economic systems in most Asian nations, but the pressing question is how to move economic development fast enough to avert serious human problems and economic disasters. Fortunately, as Table 16–1 shows, for many of the countries of the region the scale of the problem appears to be within reasonable bounds. Twelve of the twenty-two countries (excluding Japan) have populations of fewer than 25 million people; thus they are comparable in size to countries in Africa and Latin America. Moreover, all but four of the remaining countries have populations of fewer than 50 million people, and so the scale of their economic development needs is less than those of the largest Latin American and African countries: Brazil, Mexico, and Nigeria.

However, three countries of South and East Asia—India, China, and Indonesia—have populations greater than any other underdeveloped countries in the world. India's population in 1976 was almost 150 million more than the combined populations of the United States and the Soviet Union. China's population was more than 100 million larger than that of Europe and the Soviet Union combined. Together India and China account for 70 percent of the total population of South and East Asia. Moreover, 21 out of every 100 people in the world live in China and 15 out of every 100 live in India. Thus developments in these two countries alone directly affect 36 percent of all the people of the world. The population of Indonesia is considerably smaller than that of either the United States or the Soviet Union, but so is its area. Furthermore, it is a nation of islands comprising an archipelago extending more than 3,000 miles (4,800 km).

In India and China we can see the meaning of the population explosion. India has a relatively high growth rate of 2 percent per year, although this is not high by Latin American or African standards. When translated into absolute numbers the problem of India's population growth appears more formidable. In 1976, for example, the population increased by 12.4 million people. This is like adding the total population of Venezuela or of the whole of the New York City and Jersey City metropolitan areas. The Indian government is faced with the need for major economic development just to hold its own against the continuing wave of population that increases by one person almost every two seconds throughout the year. In China the growth rate is less (1.7 percent), but the number of people added in 1976 was greater than in India: 14.2 million, almost the total combined population of Ohio and Indiana. In contrast, the population growth for the United States in 1976 was estimated to be 1.7 million.

In the remainder of this chapter we shall ex-amine the dual system as it operates in South and East Asia today as well as present developmental problems. We shall sketch the environmental and cultural background for understanding the whole of South and East Asia, but will focus attention on the two population giants, India and China. Each has not only its own cultural and environmental mix, but also its own distinctive strategy and plan for moving into the modern world.

THE GEOGRAPHIC EQUATION

Despite its large population, the land area of South and East Asia is not very different from that of other regions we have been studying. It is slightly larger than Anglo-America, slightly smaller than South America, and about 10 percent smaller than Africa south of the Sahara or the Soviet Union.

TROPICAL AND MIDLATITUDE ENVIRONMENTS

Like Latin America, the region includes both tropical and midlatitude areas; thus there is great climatic variation. South and East Asia extends from just south of 10° south latitude at the southernmost island of Indonesia across the equator to about 53° north latitude along the northern border of Chinese Manchuria. The southern tip of the Asian mainland reaches almost to the equator in the Malay Peninsula, where the city-state of Singapore lies slightly more than 1° north of the equator.

South and East Asia has more land area in the midlatitudes than in the tropics (Figure 16–1). Nearly all China is outside the tropics, and even most of India, Pakistan, and Bangladesh lie in the midlatitudes; Nepal and Bhutan are midlatitude; and a portion of Burma extends north of the

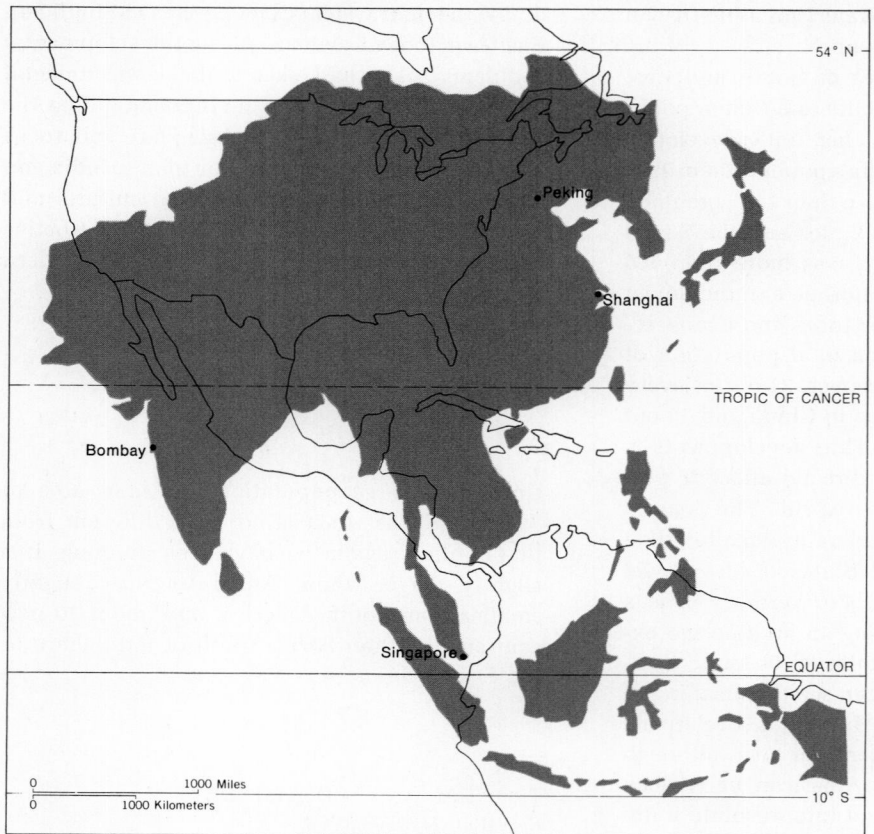

FIGURE 16–1   SOUTH AND EAST ASIA COM-
PARED WITH NORTH AMERICA IN AREA AND
LATITUDE

Tropic of Cancer. The tropical parts of the region
consist mostly of peninsulas and islands.

Growing Potential

The tropical areas have a year-round growing
season. Given adequate rainfall, year-round high
temperature makes it possible to grow two and
sometimes three crops a year in the tropics. Two
crops are also grown in the better lands of southern China, but along the northern margins and in

highland China the growing season is minimal.
On the northern frontier of Manchuria the growing season is only a little more than ninety days,
often considered the minimum for successful agriculture; and in the highlands of Tibet it is generally less than ninety days.

The growth potential of tropical areas, in terms
of potential evapotranspiration (PE), is more than
150 cm (over 60 in.). In parts of the Indian peninsula, PE reaches as much as 200 cm (over 80 in.).

In general, all tropical areas have PE totals greater than does any part of the United States. PE drops sharply away from the tropical peninsular parts of Asia, except over eastern China, where the decline is more gradual. Growing seasons are generally more than 180 days. Parts of southeastern China, in fact, have growing seasons long enough for two crops per year. These conditions in eastern China are quite similar to those of the eastern United States. Interior and northern China have less than 100 cm (40 in.) of PE, which is comparable to conditions along the Canadian–U.S. border and farther north in North America.

Moisture Availability

The availability of moisture varies widely over South and East Asia, and moisture deficiency is a common limitation on growing season and agricultural potential for much of this region (see Figures 3–5 and 3–6). Only Indonesia and the Malay Peninsula in tropical Asia have no deficits throughout the year. Elsewhere in Southeast Asia, between India and China, a short season of deficit normally occurs, but for the year as a whole surpluses exceed deficits, often by substantial amounts.

The Indian peninsula has both seasonally dry and year-round dry conditions. Seasonally wet-and-dry environments prevail over all but northwestern India and Pakistan, which have mostly dry environments. The west coast and northeast of India receive exceptionally heavy precipitation during the wet season, which normally begins in May. For these two areas precipitation far exceeds PE during the wet season and often for the year as a whole. The dry season generally begins in November and the highest temperature and greatest dryness for the year come in April and early May. The contrast between the two seasons is so striking that the term "monsoon" is usually applied to this particular version of seasonal wet-and-dry climates.

Monsoon refers to the shift in winds between the seasons: moist winds from the Indian Ocean during the summer and dry winds from the continent during the winter. The shift from dry to wet in May is often abrupt and signals the onset of a new year for Indian farmers. Many of their activities are timed according to the monsoon.

Precipitation during the wet season in the interior of India is approximately equal to PE; here moisture deficits may even occur during the growing season. In northwest India and over most of Pakistan desert conditions prevail, with year-round moisture shortages.

In eastern China moisture deficiencies are generally small, and summer is the wettest season, with maximum energy present for vegetative growth. Western and northern China are much drier, with moisture deficits occurring even in the summer, which is the wettest season, and with cold and very dry winters.

The effect of these PE and moisture conditions is that relatively little of tropical South Asia has a year-round high-work environment (see Figure 3–7). High-work environments are limited to the coastal areas of Southeast Asia and most of Indonesia. Most of the Indian peninsula and Indochina—the mainland area east of India and south of China—have seasonal environments with a wet summer and a dry winter or cool season. Northwestern India and most of Pakistan have year-round moisture deficiencies and are semi-arid to arid. The settled areas of southeastern China have sequencing seasonal-work environments similar to those in the southeastern United States. Northeastern China and Korea have a seasonal-work environment because of cold, dry winters. Western China and Mongolia have low-work environments due to dryness, whereas the Tibetan highlands have a short growing season because of low temperatures at such high elevations.

Vegetation, Soil, and Terrain

Because much of South and East Asia is warm year round and is wet seasonally or year round, vegetative growth is abundant. Forest cover is especially lush from eastern India across Burma,

Thailand, the other Indochinese countries (Laos, Vietnam, and the Khmer Republic), and on to the Indonesian and Philippine islands. It is hard to speak of "natural" vegetation in a region that was settled so long and has so many rural people today. Not only have large areas been cleared for intensive farming, but other areas have been burned repeatedly by slash-and-burn agriculture, in which forests are cleared, used for crops for a few years, then abandoned to grow wild again. The forests have been harvested of fruits, nuts, and useful woods, but the warm tropical climate and the adequate moisture supply leave little doubt that almost all of Southeast Asia must once have been covered with lush tropical forest. Today forests remain only in those areas where terrain, soils, and drainage conditions are unsuited to intensive subsistence agriculture, or where people still live directly from the forest environment by some combination of hunting and fishing, gathering, and slash-and-burn agriculture.

Forest cover in the tropics, however, is generally misleading as an index of agricultural potential. As discussed in Chapter 3, the soils under tropical forests are often quite infertile and difficult to manage for productive crop agriculture. Consequently, much of the upland areas of South and East Asia are not important agriculturally. The best soils are those of the **floodplains** and deltas of the major rivers, such as the Ganges, Brahmaputra, Irrawaddy, Nam Chao Phraya, Mekong, and Red, and those developed from the mineral-rich basaltic bedrocks of the Deccan Plateau in India and on the island of Java in Indonesia. On these lands agriculture is intensively practiced and population densities are especially high.

The alluvial lands are particularly suitable for rice cultivation, not only because they contain more soil nutrients than the uplands, but water for **paddy cultivation** is readily accessible from rivers and easily distributed over the flat lands. In fact, availability of water for irrigation may be more important than natural soil fertility for agriculture here. Supplemental irrigation is widely practiced in South and East Asia, but especially in China, where much more agricultural land is under irrigation than in any other country of the world.

The terrain of South and East Asia further complicates the use of land for agriculture (Figure 16–2). Comparatively little of South and East Asia is level land, and there are no extensive plains areas like those in the United States, the Soviet Union, parts of South America, or even the African plateau. The most extensive plains extend from the Ganges Delta in Bangladesh across northern India and down the Indus Plain in Pakistan. Northeast China and Manchuria have important plains areas but elsewhere most of the flat land is of limited extent, mostly confined to the deltas and floodplains of large rivers. Most of Bangladesh is floodplain, with a recurring threat of devastating floods. Most of South China, the peninsular mainland of Southeast Asia, the islands of Indonesia and the Philippines, and much of the Indian peninsula are roughlands consisting of mostly mountains and hills.

In southwest China and along the border between China and India rise the highest mountains in the world. These reach their peak in the Himalayas immediately north of the Indus-Ganges lowland. Extensions of these high mountains reach southward into Burma and the Indochinese peninsula. As in Latin America, these highland areas offer only *tierra fría* possibilities for agriculture, and their ruggedness serves to isolate places within them from other parts of the region. Mountain valleys and basins are inhabited by peoples of distinctive traditions whose isolation is reflected by the fact that separate political entities such as Nepal and Bhutan still exist (Sikkim, formerly independent, was absorbed by India in 1975). Tibet, inhabited by peoples different from those in the heavily populated parts of China, was seized earlier by the People's Republic of China.

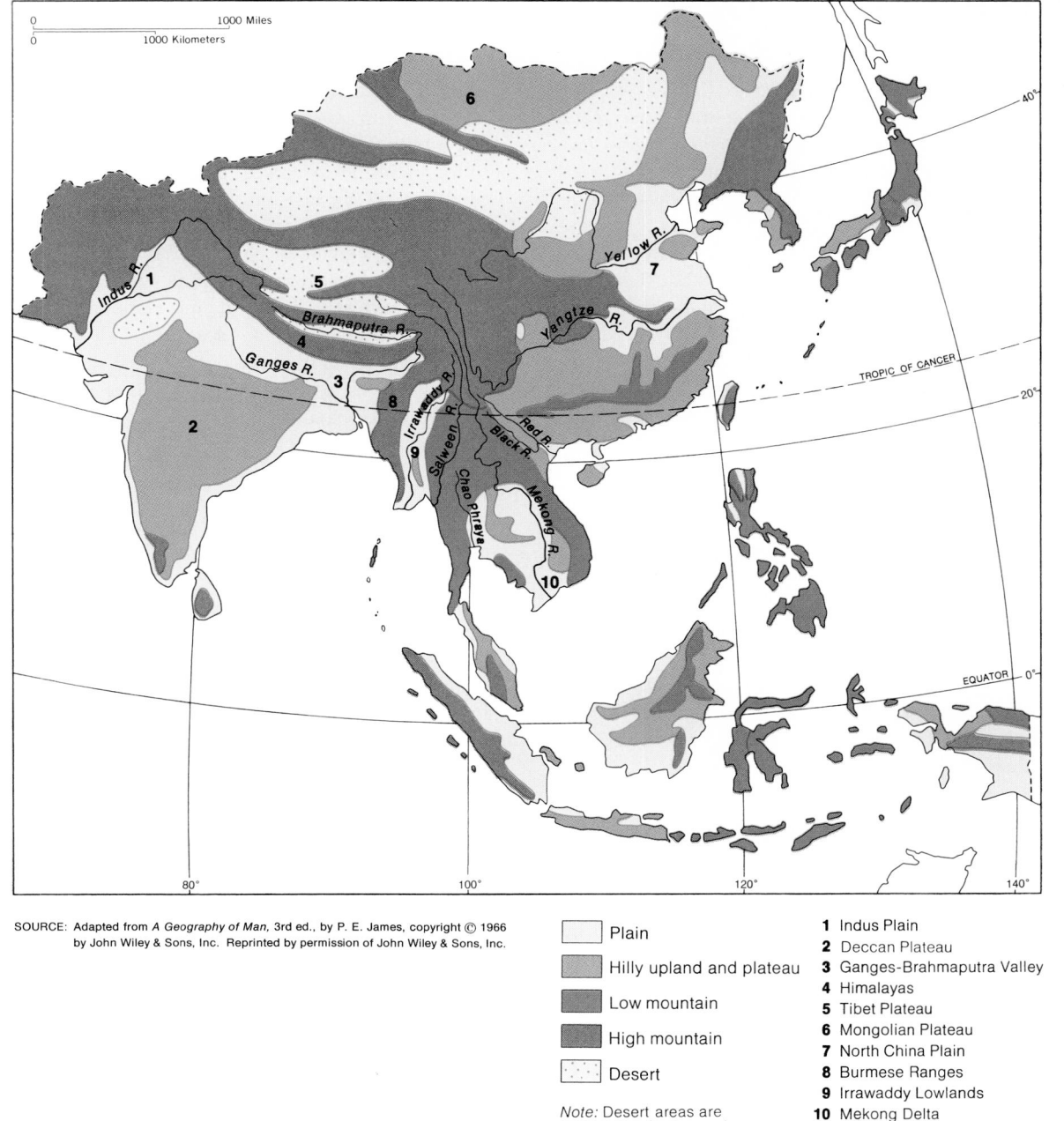

SOURCE: Adapted from *A Geography of Man,* 3rd ed., by P. E. James, copyright © 1966 by John Wiley & Sons, Inc. Reprinted by permission of John Wiley & Sons, Inc.

☐ Plain

▨ Hilly upland and plateau

▨ Low mountain

■ High mountain

⋯ Desert

Note: Desert areas are shown here for convenience, although *desert* is an environmental rather than a landform term.

1 Indus Plain
2 Deccan Plateau
3 Ganges-Brahmaputra Valley
4 Himalayas
5 Tibet Plateau
6 Mongolian Plateau
7 North China Plain
8 Burmese Ranges
9 Irrawaddy Lowlands
10 Mekong Delta

FIGURE 16–2 LAND SURFACE FEATURES OF SOUTH AND EAST ASIA

A MIX OF PEOPLES AND CULTURES

The peoples of South and East Asia present a tremendous ethnic diversity. Three of the world's major racial groups live in this area, each in its own distinctive portion of the region (see Figure 4–1). **Mongoloid** peoples spread over eastern China and into the peninsular areas of southeastern Asia. Today they continue as the dominant group in countries from Korea to Burma, Thailand, and the fragmented states that once made up French Indochina. The majority of people in India, western China, and parts of Pakistan, though dark-skinned, are of the same **Caucasian** racial stock as Europeans. In the Malay Peninsula, Indonesia, and the Philippines, peoples of **Polynesian** stock are predominant.

Unlike Latin America, where Europeans and blacks have been added to the population through recent migration, each of the Asiatic peoples has occupied its own particular portion of Asia for many centuries. In more recent times Chinese and Indians moved into most parts of Southeast Asia and now form minorities within the indigenous population. Malaysia and Indonesia have large Chinese or Indian minorities. They commonly live in the cities and larger towns and have an influence greater than their numbers; many are successful businesspeople who have gained money, property, and power. Singapore withdrew from the Malaysian Federation because it was a Chinese city in a Malay country. As in Africa, there has been no large-scale settlement of Europeans, even though most of the area outside China was once occupied by one European power or another.

Site of Two of the World's Greatest Civilizations

South and East Asia is the site of two of the world's earliest and most advanced civilizations. The origins of Indian and Chinese civilization have been traced back beyond 2,000 B.C. When the Europeans first established direct contact with them through the travels of Marco Polo in the thirteenth century and early voyages of Por-

tuguese traders, these civilizations had far surpassed Europe in artistic, scientific, and political achievements. The European explorers who reached the Americas in the fifteenth and sixteenth centuries were, in fact, searching for new sea routes to China and the Indies to tap the lucrative trade in spices, porcelains, and silk—products of the artisans of sophisticated societies. As late as 1773, when an envoy of the British king was received in the court of the Chinese emperor, he was brushed aside with the comment: "I have already noted your respectful spirit of submission. . . . I do not forget the lonely remoteness of your island cut off from the world by intervening waters of sea. . . . Our Celestial Empire posesses all things in prolific abundance and lacks no product within its borders. We do not need to import the manufactures of outside barbarians in exchange for our own produce." European traders to India were dazzled by the great palaces and other displays of wealth of the maharajas and upper castes.

Primarily Agriculturally Based Cultures

Despite wealthy and literate upper classes and skilled and productive artisans, both China and India were agriculturally based civilizations. Farm practices today throughout South and East Asia are essentially those developed thousands of years ago through the use of simple tools and human and animal power. They are not, however, primitive tribal systems. South and East Asian agriculture has depended to a high degree on the development of sophisticated irrigation systems, which over the centuries have created one of the most efficient means people have yet devised to harvest crops from tropical and subtropical wet environments. But that system is the result of the application of massive amounts of human and animal energy. The result has been high rural population densities—people and work animals that must be fed—that largely negate the relatively high returns from the land.

Wherever it can be planted, the major crop is rice. Although wheat is the staple crop of North China and the major grain of India is sorghum, rice is predominant in most of South and East Asia. It is well suited to this densely populated region because it produces the highest food value (in calories) of any grain crop. The most productive rice varieties grow in standing water, which seems to be an especially effective agricultural adaptation for the tropics. The standing water reduces weed growth, insulates the soil, and harbors nitrogen-producing organisms. Irrigation is required, and its development in South and East Asia is on a scale unknown anywhere else. The river valleys of South and East Asia have been altered by thousands of years of effort on the part of generations of peasant farmers. The lowlands have been almost completely transformed in order to fit the rice crop and its need for irrigation. Dams, diked fields leveled to grade so irrigation water will flow onto the land when needed and off when necessary, and terraced hillsides are all evidence of the tremendous effort of people using only domesticated animals and simple tools. The system has produced not only the most labor-intensive way of farming in the world, but it has been aimed at perpetuating itself from generation to generation. Crops have been carefully rotated where there is a year-round growing season, and, in an area with few large domesticated animals, human waste has often been returned to the soil as fertilizer.

Eighty to 90 percent of the huge population of this area remains agricultural, living in farm villages of five to a hundred families or more. Farm properties have traditionally been separate private holdings of ten acres or less. Any individual farm may consist of as many as five or more separate pieces of property. Such fragmentation is the result of hundreds of years of land inheritance, but it also provides a way of adjusting to differences in soils, slope, and water availability. Farming is a family enterprise, with women and children helping with the transplanting of seedlings into the fields and with the harvesting, and often working a vegetable patch close to the house. This system produces almost 80 percent of the rice crop of the world, and large quantities of wheat, sorghum, and other grains, yams and other vegetables, yet it appears doubtful that it can expand output fast enough to keep up with population growth. It needs the input of modern capital and scientific expertise.

Cities in a Village World

In such a world, most life is centered around the local village. As we saw in Ramkheri, very little in the way of goods, people, or ideas has traditionally moved into or out of the village. The Oriental system, like other traditional systems, has been based primarily on the subsistence use of the local environment.

Nevertheless, there have been and are large cities in South and East Asia (Figure 16–3). As elsewhere, cities here are a measure of advanced civilizations. In 1800 China had eight cities with more than 250,000 population as part of the "Celestial Empire," while at the same time the Indian cities of Calcutta, Madras, and Bombay had populations of 600,000, 300,000, and 200,000, respectively. But many of the large cities that have emerged since then were established by European colonial powers. Today China has nineteen metropolitan areas containing over 1 million people, including seven of the eight large cities of 1800. India has nine metropolitan centers of over 1 million, including the three large cities of 1800. In addition, there are eleven other million-plus metropolitan areas elsewhere in South and East Asia.

Most of the large metropolitan areas of Asia continue to be political centers. The largest metropolitan areas in the region—Shanghai, with 7.9 million people, and Calcutta, with 9.1 million— are not the capitals of their countries. Instead, they are commercial and industrial centers whose origins are related to European influence rather than to indigenous history. Peking, the Chinese capital, has 4.8 million people; and Delhi–New Delhi in India, 4.5 million.

The number of large metropolitan areas is still small in comparison to the United States or Western Europe, and their relative importance is much less than in Japan or the Soviet Union. Only 5 percent of the Chinese population lives in

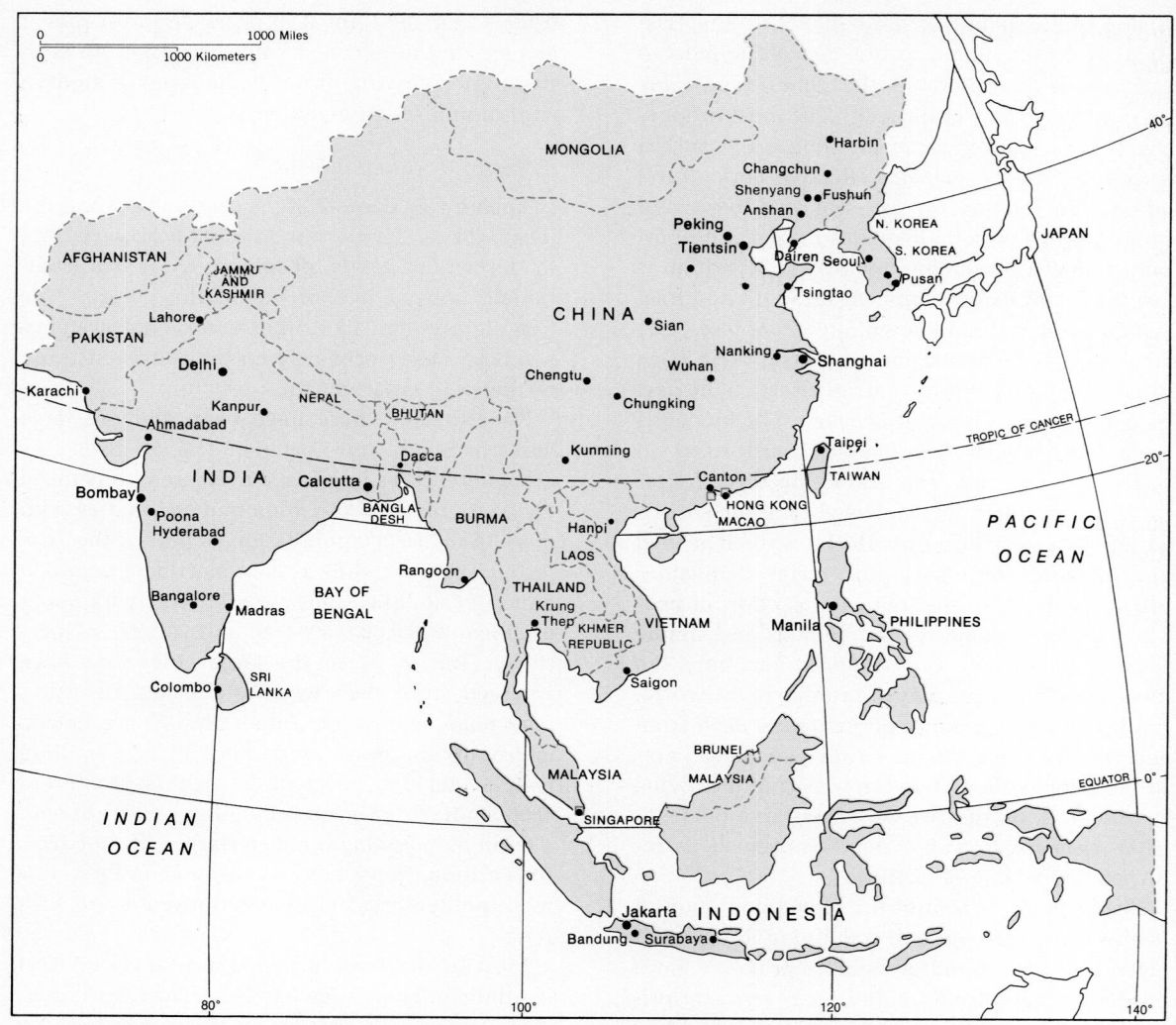

FIGURE 16-3 POLITICAL MAP OF SOUTH AND EAST ASIA

Metropolitan population
• Less than 1,000,000
• 1,000,000–4,000,000
• More than 4,000,000

the large metropolitan areas of 1 million or more, and in India the proportion is only 4 percent. In some of the smaller countries, such as South Korea and Taiwan, more than 10 percent of their populations lives in large metropolitan areas, as is the case in some Latin American countries. Their urbanization in part reflects their contact with outside industrial powers as a result of Western involvement in Asiatic defense and war since World War II. The presence of foreign military operations or military aid missions has resulted in the rapid growth of the largest city in each of these countries, and of Saigon in South Vietnam until 1975. These cities served as the chief management and communications bases for the complicated logistics networks needed to carry on modern military and economic development. Calcutta, Shanghai, and Rangoon had served the same purpose in the nineteenth century.

CONTACTS WITH THE REST OF THE WORLD

South and East Asia has had unique relations with the present-day industrial countries. All countries of the region were visited by European explorers and traders during the Age of Exploration in the fifteenth, sixteenth, and seventeenth centuries. In fact, the desire to trade with the Indians and Chinese and with the tropical lands of this part of the world stimulated European sea exploration. Blocked by Moslem power in the Middle East–North Africa from overland trade with the Orient, the Europeans turned instead to the seas. They wanted Asia's tea and spices, its silk, and the products of its skilled artisans.

At first the Europeans were content to establish trading posts—called "factories," as in Africa—in port cities. Soon industrialization brought greater demand for trade. Europeans wanted markets for textiles and other manufactures, as well as sources of supply for tropical and subtropical products such as jute, cotton, coffee, and rubber. The result was the colonial occupation of almost all of South and East Asia except China (including Mongolia), Japan, and Thailand, and even these countries found themselves under trade and economic domination by the Europeans. Their cultures were also threatened by the impact of Western values brought by missionaries and traders. As in Africa, European colonial governments were established to protect the commercial interests and to maintain order, and inland transportation and communication systems were built to bring the products of the area to ports for export.

The Europeans and later the Japanese and Americans gained control over almost all of South and East Asia except China. Pakistan and Bangladesh were once part of British India. Burma, Malaysia, Singapore, and Sri Lanka (formerly Ceylon) were also British colonies. Indonesia was ruled by the Dutch, and Vietnam, Laos, and Cambodia were ruled collectively as French Indochina. Spain and later the United States ruled the Philippines. Japan made Korea the first important addition to its empire. Only Thailand remained nominally independent as a **buffer zone** between British and French colonies. Mongolia was a buffer zone between China and Russia, and Afghanistan between Russia and British India.

The Search for Resources

The chief interest of the Europeans was to acquire resources to meet their growing demands. With the opening of the Suez Canal in 1869, South and East Asia became readily accessible by sea. Europeans saw the region as a productive alternative to the tropical areas of the Caribbean and turned it into the most important plantation area in the world. Tea was grown in Sri Lanka, jute in what is now Bangladesh, rubber in Malaysia, Indonesia, and Indochina, and coconut products in the Philippines. In Malaya, tin was drawn from placer stream deposits by dredges, and in what is now Indonesia petroleum was extracted from oil fields.

The United States, too, developed both a trading and a colonial interest in Asia. It opened Japan to the world when Admiral Perry sailed into Tokyo Bay with four ships on a summer day in 1853. The United States, faced with exclusive trading rights by the British in China, negotiated equal privileges. It also gained control over the Philippines as one of the results of the Spanish-American War.

European and U.S. colonial power in the area was challenged in World War II not by the peoples of South and East Asia, but by the Japanese, the new industrial trading country. Japan, once embarked on the road to modern industrialization, sought control of the rich resources of the rest of South and East Asia since its own physical resource base was so inadequate. Japanese influence developed in Korea and Manchuria early in the twentieth century as a result of victory in the Russo-Japanese War (1904–1905). By the 1930s both areas were part of the Japanese Empire, and the Japanese invaded China proper. By 1940 the Japanese held the Chinese capital of Nanking and all the coastal cities. During World War II Japanese armies moved over most of Southeast Asia while the Europeans were engaged in battle on the European continent. Japan captured the Philippines, the Dutch East Indies (now Indonesia), French Indochina, Thailand, and most of British Burma.

Independence and Its Consequences

Independence movements had arisen before World War II in South and East Asia. The Philippines had been scheduled to gain independence when the war broke out, and during the Japanese occupation the Allies encouraged and financed guerrilla movements aimed at breaking Japanese control. With the defeat of the Japanese in 1945 both local independence movements and the world political climate called for independence for South and East Asia. European countries recovering from the war had little money or energy to recover their colonies or, later, to assist lost Asian colonies in reorganizing.

The two Koreas emerged from Japanese control after the war, representing the zones of Soviet and U.S. influence. India, Pakistan, Ceylon (Sri Lanka), and Burma became independent from the United Kingdom in the 1940s, Indonesia from the Netherlands, and the Philippines from the United States. Thailand regained its prewar independence. French Indochina was broken up into Laos, Cambodia, and North and South Vietnam in the early 1950s, leaving a struggle for power just recently concluded, and Malaysia gained independence in 1957, only to have the city of Singapore secede from it in 1963. Thus all the major political units of South and East Asia are politically independent today. Hong Kong, the British island colony on the southern Chinese coast, is the major dependency in the area, although Bhutan has a dependency relation with India, and Macao, another island colony near Hong Kong, is still Portuguese.

In much of South and East Asia the story of the newly independent countries is similar to that in Africa. Diverse local peoples struggle for control of government, and learning to govern is a problem. These countries depend on revenues from export commodities, and each struggles to control foreign investment and carry out a program of economic development. The particular set of circumstances in each region is unique, but the problem of dual economic systems in juxtaposition, a problem we have seen in other underdeveloped areas, is common here also.

A problem for some new countries of South and East Asia is that their colonial boundaries were not consistent with the traditional territories of ethnic groups. Thus each new government inherited boundaries that encompassed a variety of different ethnic groups and rival claims to leadership in the new state. For example, the French in 1948 divided the previous five Indochinese states into three states—Laos, Cambodia, and Vietnam—none of which represented a cultural entity. The Vietnamese and the Khmers of Cambodia are lowland peoples, yet their territory includes important upland areas occupied by very different peoples. Only the northwest section of Laos has historic continuity as an entity. The rest of Laos contains diverse upland peoples. Segments of each of the colonies designated by the French were occupied by peoples very different from those who formed the plurality of the population.

Vietnam was further divided between the northern people, who were very closely identified with Chinese culture, and those in the center and south, who were not. It is not surprising that the agreement for independence in 1954 called for the division of Vietnam into north and south zones, each with a separate government. This did not satisfy the North Vietnamese and the South Vietnamese dissidents (the Vietcong). Hostilities during the 1950s eventually led to full-scale war in South Vietnam in the 1960s. The United States was involved, at first in an "advisory" capacity, but more and more U.S. military personnel and equipment were sent to Vietnam in the late 1960s and early 1970s. Despite U.S. military assistance to South Vietnam, in 1975 the north invaded the south and once more unified the Vietnamese state originally established by the French. It is important to remember that this move through the power of arms does not change the cultural distinctions between North and South Vietnamese, nor does it solve the problems of the different, less advanced minorities in the highlands.

INDIA AND CHINA: TWO VARIATIONS ON THE DEVELOPMENT THEME

In the late 1940s new governments in both China and India inherited the problem of a national economic system unable to supply its population adequately—the Indian government of Mahatma Gandhi achieved independence from the United Kingdom in 1946, and the Chinese regime of Mao Tse-tung defeated the Chinese Nationalists in 1949.

Each government has had the same aim: to build its rural economic base so that it can adequately support its population. At the same time both governments want to develop a modern economy that will provide industrial and military strength. But their development plans are very different. India, though strongly nationalistic, has built its new political-economic system firmly on British colonial inheritance and the Western approach to economic development. It has had an elected parliament and a socialist economic system that relies heavily on private enterprise. The Indian government makes financial plans and sets economic priorities, but a large sector of the economy remains in private hands, including farmlands that support rural India and many of the major industrial and business concerns that support city life. As a result the government has no direct control of total economic development. It tries to present goals, provide loans, offer assistance, levy taxes, and make regulations to direct the private sector in following its plans.

China, by contrast, has chosen a communist model politically similar to that of the Soviet Union. The central government not only maintains control, it also makes all basic developmental decisions and then sees that they are carried out. It sets priorities and has the power and investment support to push them forward. The priorities of economic development, however, have not followed the Soviet model. The Chinese have emphasized revitalizing peasant agriculture and have attempted to organize and apply the huge labor supply for economic advancement.

These two different forms of development programs have led to very different degrees of support from the industrial countries in a position to provide economic and technical assistance. India has received important amounts of assistance from the United Nations, the United Kingdom, the United States, West Germany, and the Soviet Union. China, on the other hand, was refused admission to the United Nations until 1971 because the United States continued to recognize the Nationalist regime of Taiwan as the legitimate government for China.

The primary concern of Mao's regime has been with internal development. There has been virtually no trade with capitalistic industrial countries. For more than a dozen years the Soviet Union provided both capital goods and technical assistance, but after an ideological break in the 1960s, China refused all further aid from the Soviet Union. Thus the Chinese have been developing their economy with a minimum of outside assistance and foreign trade with the industrial countries. Only since its admission to the United Nations and President Nixon's visit in 1972 have normal trade relations between China and other industrial countries begun.

The push of each government toward economic development must also be seen in terms of the great cultural heritages of these countries. Each government sees itself re-establishing the proud position the country once held in precolonial times. In particular, much of China's appeal to its people is the pledge, presented continually in government propaganda, to re-establish China as the world's greatest cultural and political power.

THE TWOFOLD DEVELOPMENT TASK

India's and China's problems are in many ways similar to those of the United States and the Soviet Union rather than to those of Japan. Although the population density of China and India is much higher than that of the United States and the Soviet Union, it is not greater than that of Japan or the small Western European countries. The large territories of both China and India provide them with relatively large resource bases

in terms of both agricultural and mineral resources. The problem for each is to develop the varied resources to meet the needs of the internal market, not to produce specialty products for export and foreign earnings. They must change a largely locally based system into one that will function on a national scale.

MODIFICATIONS ON THE BRITISH AND SOVIET MODELS

At first glance, the two countries seem to have attacked their problem in the same way: the governments have embarked on a series of long-range economic plans, like the five-year plans of the Soviet Union.

India has followed a plan of **socialization** very much like that in the United Kingdom. The state finances essential new projects—hydroelectric irrigation works, transportation, communication, and even new industries such as steel mills. Moreover, it has taken over from private firms banking and major insurance companies. It is planned that more than half of each five-year plan will be based on public investment, but private investment is still strongly supported. Many industries are privately owned, including the country's largest steel operation, and agriculture remains in private hands.

The Indian government attempted to carry out land reform. The vast hereditary estates of the upper castes and those created under the British regime were broken up. Land was given to former tenant farmers. Not all land was distributed in this manner, however, and large landowners are being encouraged to prosper by the government. They tend to be the most productive farmers. The Indian government, recognizing that the best way to cope with the continuing pressure for farm production is to focus resources on the most successful farms, has given these farms subsidized credit, fertilizer, and the new high-yielding "miracle" seeds. Thus India with its socialistic goals is creating a group of successful capitalistic farmers. Big farm operations often return more than $100 per acre on farms of 200 acres.

In China the central government has become the major decision maker and manager of the whole economy. Central planners determine the country's needs in terms of industrial development and the supporting food base. Priorities are set among all the myriad facets of the economy of this huge country and goals are established in terms of the availability of production, labor and personnel, and supporting facilities such as transportation. Production quotas are set for rice production, steel output, and every other commodity needed in the economy, and then the needed resources are allocated to achieve those goals. Moreover, national quotas are allocated regionally over the country until each county, commune, and factory has been presented with its expected share of the national plan.

Once allotments have been made to the local producing units, whether factory or commune, the local working unit must plan how it will achieve its quota. Local councils then muster their labor, resources, and capital to do the task assigned. In this structure the whole economy from the largest to the smallest item is under the review of the central government.

DIFFERENT GEOGRAPHIC EQUATIONS

Although the development problems facing India and China are essentially the same, the geographic equations of the two countries are very different. In brief summary, India is located mostly in tropical latitudes and is dominated by a seasonal-work climatic environment with a dry dormant season, although northwest India has a low-work dry environment. The most populous part of China is in the midlatitudes, with a sequencing seasonal-work environment. Although China is three times the size of India in area, it has less than one-half as much arable land. Thus China, with one-third more population than India, has to supply its food needs from much less agricultural land.

Over the past two decades India and China have been influenced by differing degrees of interaction with the rest of the world. China has been isolated, while India has sought ties with and economic aid from the industrialized countries. India has concentrated on industrial development, while China has focused on agricultural productivity and harnessing the energies of its huge rural labor force.

INDIA IN CLOSER PERSPECTIVE

In area, India is one of the ten largest countries in the world. Environmental variations within its borders present differing production possibilities. Variations in agricultural production from place to place are a key to its problems. Most Indians are rural and depend on production from their own local area to provide the essentials of life. Variations in agricultural output suggest differences in the growing potential and environmental base from one part of India to another and, as we shall see later, differences in population density.

Since grain rather than meat is the staple Indian diet, the distribution of its production (Figures 16-4a, b, and c) provides a measure of growing potential. Rice is the favored grain. Wheat is grown only where rice cannot grow; it usually does not do well in tropical environments. Millet and sorghum, rough grains that are used as livestock feed in other countries, are raised as human food where it is not possible to raise other grains.

The rice lands of India are concentrated in the lower Ganges River valley in the northeast and along the coastal fringes of the Indian peninsula. Rice will grow anywhere in the tropics and sub-

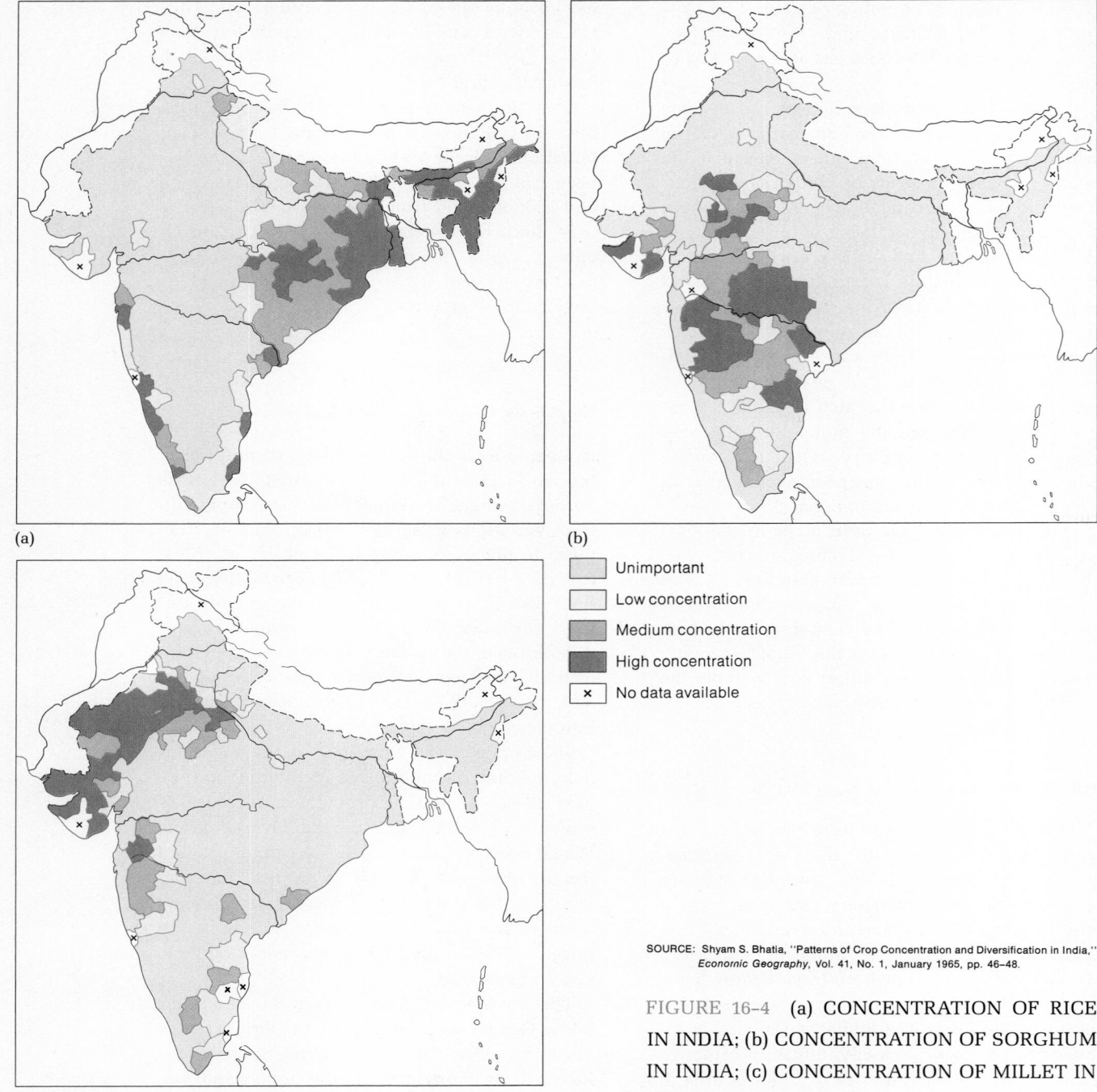

(a)

(b)

Unimportant

Low concentration

Medium concentration

High concentration

✕ No data available

(c)

SOURCE: Shyam S. Bhatia, "Patterns of Crop Concentration and Diversification in India," *Economic Geography*, Vol. 41, No. 1, January 1965, pp. 46–48.

FIGURE 16–4 (a) CONCENTRATION OF RICE IN INDIA; (b) CONCENTRATION OF SORGHUM IN INDIA; (c) CONCENTRATION OF MILLET IN INDIA

tropics, but the major varieties are wet crops requiring standing water on the field during much of the growing period. Hence rice is grown in areas of high growing potential where AE is almost as high as PE or where there is access to irrigation water.

Over most of the peninsula where moisture deficits are great, millet and sorghum are the dominant grains. This reflects environmental conditions more than human choice. Millet and sorghum are often planted in the same fields. If summer rains are good, the millet dominates, but if the year is dry, sorghum is the chief harvest. As we saw in Ramkheri, both can be used to produce a porridge or ground into a meal for bread or cakes.

The relative importance of these different grain regions can be seen by comparing their areas of production (Figures 16–4a, b, and c) with population densities (Figure 16–5). Since most of India's rural population still lives like the villagers of Ramkheri, very little grain moves from one rural agricultural area to another. Little surplus is available for sale, and most villagers would not have the money to buy it anyway. Some grain is sold to markets in cities, but major movements occur only during times of crop failure, when emergency arrangements are made to redistribute surpluses to areas in need. Most cities usually must depend on imported grain, which accounts in part for the fact that India is a leading importer. The primary function of India's agricultural system is simply to feed the people of the country, and the agricultural program will succeed or fail on that basis. India, however, cannot afford to shift areas that are producing export or important industrial crops into food production.

In contrast to Latin America or Africa, agricultural exports are a secondary matter to India. Cotton and other crops are used as raw materials in the growing industrial programs, not for exports. Exports of tea, jute, and cotton provide important foreign exchange, but increasing amounts of jute and cotton are being processed within the country, not only to clothe Indians but also for export. Cotton manufacture is one of India's largest industries, and tea accounts for 40 percent of agricultural exports.

AGRICULTURAL PROBLEMS

The map of major cereal regions indicates one of India's leading agricultural problems: a shortage of moisture over much of the country. Except along coasts in rice-growing areas and along the southern flank of the Himalayas in the north, the deficits in the dry season are comparable to those in the Great Plains and desert areas of the United States (see Figure 3–6). Over most of the cereal-producing area there are no surpluses after the rainy season.

Dryness

Since water shortages are endemic to most Indian agricultural areas, development of available water resources for irrigation is needed if agricultural productivity is to be improved substantially beyond current levels. Most irrigation now depends on ground water from shallow, hand-dug wells. The Indian government has undertaken drilling of deeper wells with greater volume, but such endeavors face rising costs for pumping. Another source of water has been storage ponds, called "tanks," which collect runoff during the wet season for use during the dry months. Just as the U.S. Department of Agriculture's program of pond construction has been an important factor in improving farming in some parts of the United States, the tank-digging program of the Indian government is being counted on to improve vital water supplies in areas of the Indian peninsula where there is a shortage of water for all purposes during the dry season.

One of the major efforts of India's agricultural program under the five-year plans has been to provide irrigation. Large-scale modern projects using the runoff from the mountains in the northwest have also been undertaken to bring water to dry areas never cultivated before. Three major

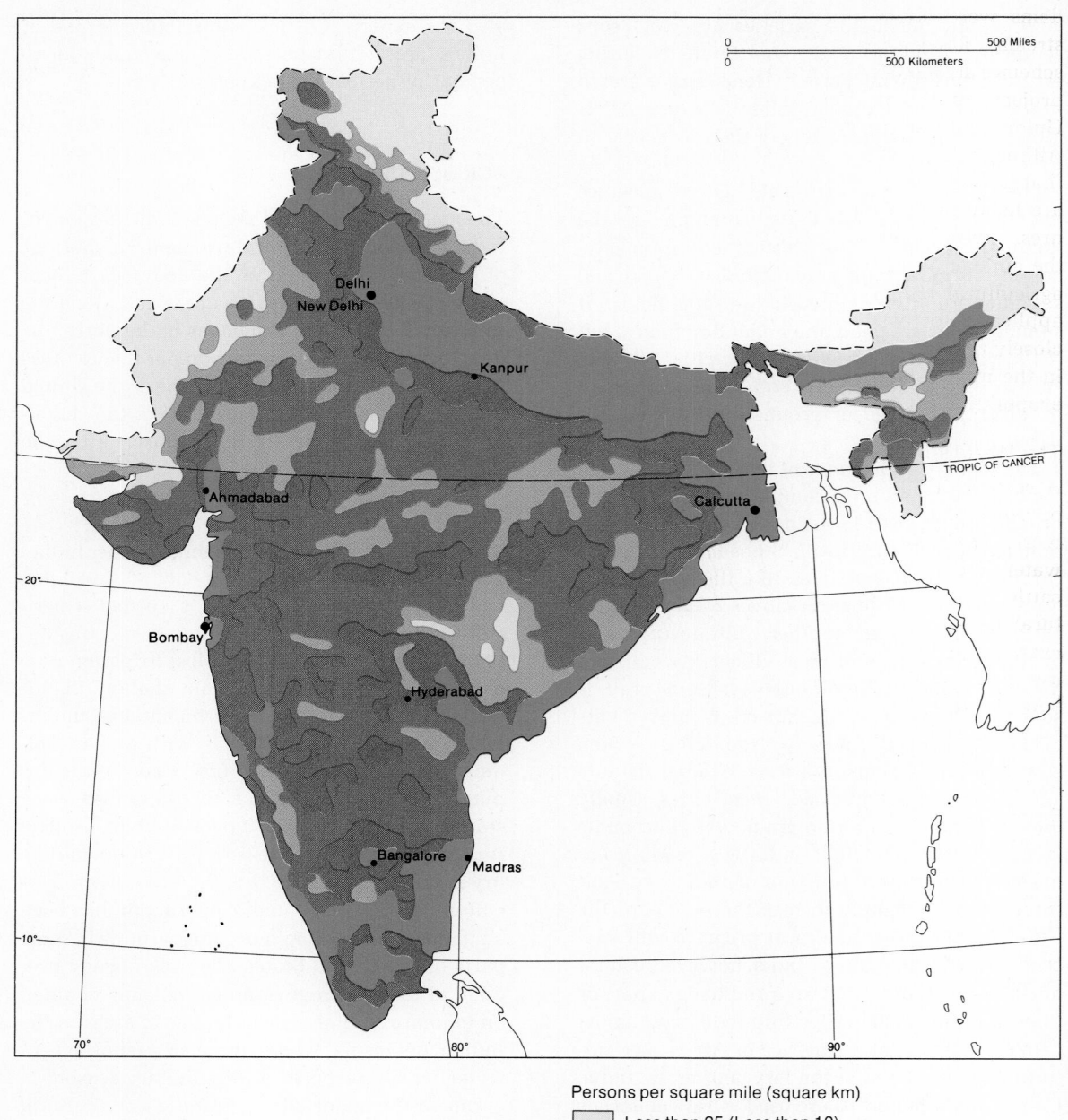

Persons per square mile (square km)

  Less than 25 (Less than 10)

  25-125 (10-48)

  125-250 (48-96)

  250-520 (96-200)

  More than 520 (More than 200)

Metropolitan area population

  • 1-4 million

  ● Over 4 million

FIGURE 16–5 POPULATION DENSITY OF INDIA

dams were under construction in 1970 along streams flowing from the Himalayas. These schemes are modeled after the huge multipurpose projects of the United States and the Soviet Union and require large capital investments. Although they will open new acreage in dry areas that have hitherto been impossible to farm, they are likely to be little more than stopgap measures. Irrigation will increase productivity, but the number of acres being added to India's total agricultural land is very small. Moreover, the application of water and soil drainage must be closely managed to keep salt from accumulating in the upper soil levels as a result of excessive evaporation.

The World's Largest Cattle Population

A second problem of Indian agriculture is the presence of the largest cattle population in the world, estimated at 177 million plus 55 million water buffalo in 1974. In most countries such cattle numbers would be seen as a major agricultural asset. In predominantly Hindu India, however, cattle are considered sacred. The bull is associated with the Hindu god Shiva, and the cow is recognized as the representative of "mother earth." All cattle are inviolate, and even castes that eat meat do not eat beef. Except for dairy products, this huge cattle population is not a direct source of food. Instead, it competes with the human population for the available food supply.

The cattle have important uses in the agricultural system, however. They and the buffalo are the chief source of energy in agriculture other than human muscles. They plow fields, power threshing machines, and pull carts. Since woodland is scarce in most parts of India, animal dung mixed with straw and dried in the sun is often the chief fuel for cooking and heating. Nearly half the cattle manure that might be used as fertilizer on the land is used as fuel.

Since cattle in India are essentially beasts of burden, they are most populous in areas where crop production is greatest. However, the Indian agricultural system is designed to produce food crops for humans, not animals. There is little room in the system for growing fodder, and pasture land is limited. Work animals are fed "hay" of rice and other grain stalks that are left after the harvest. Without seeds it is not very nutritious, and the supply is woefully inadequate. Such animals are too undernourished to provide good power, dairy products, fuel, or fertilizer.

The biggest problem is the indiscriminate, largely uncontrolled breeding of cattle. Government plans call for gradual reduction in the number of cattle and improvement in breeding. Implementing such plans, however, is very difficult in Hindu India.

Low Productivity

In contrast with other parts of Asia where the intensive agricultural system produced some of the highest per acre crop yields in the world even before the introduction of modern scientific techniques, India's yields are among the lowest. An important component in India's problem is drought, but much of the problem is the result of Indian farming practices. The land has been worked continuously for thousands of years without adequate nutrient replenishment. Little fertilizer was used, tools and cultivation practices were poor, and the seed was of low quality. Wooden plows were ten times as common in 1972 as iron plows, and many cultivators did not own plows at all.

Overcoming India's agricultural problem will require more than new wells, water tanks, and better breeds of cattle. Farmers will have to accept innovations. A new infrastructure of roads and markets must be developed that will bring them into the cash-crop economy. A better system of credit and a readjustment of farm prices will be needed, as will better storage facilities and more processing plants for farm produce.

India has two agricultural economies. In addition to traditional farming, a small modern sector is organized to produce commercial crops for export. Management is in the hands of entrepreneurs who have applied modern techniques. Much of the work is mechanized and fertilizers are used generously—an important share of India's meager fertilizer supply goes into these operations. Unfortunately, this modern agricultural sector is a world apart from that of the overwhelmingly predominant traditional agriculture. Mechanization does not increase employment opportunities and these farms bring little wealth to the masses where they exist. They are essentially a closed system within Indian agriculture.

PLANS FOR AGRICULTURAL DEVELOPMENT

Indian agricultural plans must rest largely on expanding the output on existing farmlands, since there is no significant frontier land that can be developed. It is not mechanization that is called for, but rather increased productivity within the traditional farming system based on human and animal power. The essentials are education, markets for surplus production, credit on reasonable terms, and agricultural research. Since none of these factors is independent of the others, many changes are needed simultaneously. Farmers who learn about new methods must be assured that they can market their products and that they will have the money to undertake necessary changes such as purchases of new seeds or fertilizer. All this requires remodeling the rural way of life, a task that, as we saw in Ramkheri, cannot be accomplished in a short time.

The goals for India may seem modest, but the task has been overwhelming. The new program must reach more than 750,000 Indian villages and about 100 million farm families. With the help of agricultural specialists and community development experts, the Indians have been working on the problem for more than twenty years. Early strategies called for schools and programs to train one influential leader from a given village with the idea that that person would return and spread the word. Farmers have been encouraged to organize into **cooperatives,** to provide credit and to share know-how. Village councils were started both to educate people and to allow them to participate in change.

Yet even when the new ideas are presented, it is hard to change tradition, particularly those things that have been much the same as far back as anyone in the village can remember. The Indian government, in contrast to the Chinese or Soviet government, relies on mutual cooperation rather than assigned goals and coercion to achieve its goals. There are no collective farms or communes that manage the productive lands. Land is in the hands of a vast number of independent farmers. The farmer is probably illiterate, in debt, and without ready access to working capital or better tools and seeds. Under these circumstances India faces the biggest educational and developmental program in the world. Moreover, the Indian government itself is far from efficient. It has an elaborate bureaucracy burdened with a mass of directives and cumbersome communication channels.

In recent years development efforts have been concentrated in just a few areas of the country. Great effort has been expended to find the key places where agricultural investment can be expected to yield high returns. Areas with rich soil, adequate water, and adequate connections to regional markets are chosen. Most success has occurred in the new irrigated areas of the northwest, near the Ganges Delta, and in the states of Tamil Nadu and Kerala in the south. Farmers are provided with supplies, storage facilities, transport, credit, and price incentives.

The most significant need is probably fertilizer. In response to this need, India has expanded fertilizer production. Between 1950 and 1972, fertilizer production increased 375 percent, and by 1975 plans call for increasing that output twofold. Increased use can be seen as an index of progress made in intensifying agriculture. It does not pay, however, to use heavy doses of fertilizer unless other factors are also improved, particularly water availability and the quality of seed.

THE GREEN REVOLUTION

Over the past fifteen years there have been times of both optimism and pessimism concerning the food supply of the large populations of South and East Asia. After the pessimism of the early and mid-1960s, when India had a succession of poor crops, great optimism accompanied what was called the **Green Revolution** in the late 1960s. Despite the technological accomplishments of the Green Revolution, the early 1970s was another period of pessimistic predictions for world food supplies, as the reserve stocks of grain became extremely low because of poor harvests in the Soviet Union and India. These events raise questions about the significance of the Green Revolution, especially for South and East Asia.

The Green Revolution is part of the technological evolution in agricultural development that began several decades ago. Its origins lie in the development of hybrid corn in the United States in the 1930s. The crops resulting from the Green Revolution are mostly improved varieties of wheat and rice. Genetic modification, like the **hybridization** of corn, increased the potential yields when fertilizers are used in the presence of adequate soil moisture.

The genetic improvements of wheat were initiated under the auspices of the Rockefeller Foundation in Mexico in 1948. Research and development that produced the higher-yielding varieties of rice came later through the

work of the International Rice Research Institute in the Philippines financed by the Ford and Rockefeller foundations. The accomplishments of these programs have led to the more recent establishment of similar research centers for other crops, especially those of tropical and subtropical areas (Table 16–2).

By 1966 high-yielding varieties of both rice and wheat were ready for field use. These varieties had three major advantages over traditional seeds: (1) better responsiveness to fertilizer, especially nitrogen, without **lodging** (a condition in which the stem grows too tall and the plant falls over); (2) greater tolerance of the shorter day length characteristic of the lower latitudes; and (3) a shorter maturation period, which permits more crop cycles each year. When adequate fertilizer is supplied, particularly nitrogen, and soil moisture conditions are favorable, high-yielding varieties can produce double the yields of traditional varieties. When rainfall is below normal or irrigation water is not available, however, the high-yielding varieties may not do any better than, or even as well as, traditional varieties. Moreover, farmers who cannot afford fertilizers cannot obtain the full benefits of the higher yield potential.

The adoption of high-yielding varieties has been rapid. In 1967 only 4 million acres were planted to high-yielding varieties of wheat and rice;

by 1973 the figure had risen to 80 million acres. The use of the high-yielding varieties is concentrated in a few countries in Asia. In 1972–1973 India and Pakistan together accounted for about 81 percent of the total acreage planted to high-yielding wheat in Asia. Rice was somewhat less concentrated; India, the Philippines, Indonesia, and Bangladesh accounted for about 83 percent of the total. India alone represented 61 percent of the wheat and 55 percent of the rice area in Asia outside communist countries.[1]

The yields could be even higher than those now obtained in India or the Philippines if the support and management systems were geared to modern agricultural production. In countries such as Mexico and Taiwan, which have developed further toward the modern interconnected model than have India and Pakistan, the increases in yields due to the new strains have been much more dramatic than in the more traditional countries. To realize the full potential of these new varieties of wheat and rice requires changes in traditional farming practices, a transportation infrastructure that can supply the necessary inputs and collect the output, and stable supplies of manufactured fertilizers. The need for such changes in the supporting facilities and institutions points to the overall need for a general modernization of a nation's transportation

[1] U.S. Department of Agriculture, Economic Research Service, *The World Food Situation and Prospects to 1985,* Foreign Agricultural Economic Report No. 98, 1975, pp. 66–67.

systems and manufacturing, in addition to new management practices by farmers and basic support for agricultural research and development. Thus to sustain the full benefits of the Green Revolution requires more than the introduction of new seed varieties. Far-reaching changes are needed toward modernization of a nation's economy and related social institutions.

The Green Revolution has spread very unevenly over developing countries such as India. Only farmers who have sufficient capital can benefit. The many small farmers who cannot afford **canal** or **tube-well irrigation,** fertilizers, and pesticides cannot participate in the new high-yield crop programs. The programs have instead attracted many educated and wealthy people to farming through such advantages as high food-grain prices, tax-free agricultural income, price supports, and low-interest agricultural loans. Thus where traditionally farming was considered a menial job practiced mostly by under-educated and poor villagers while the landlord lived in town, now the wealthy are prospering in new agricultural ventures. The gap between rich and poor in developing countries often has been widened rather than narrowed by the introduction of the new varieties of food grains. The poor farmer is comparatively worse off than ever.

In most respects the Green Revolution is less a revolution and more an evolutionary introduction of new cropping systems based on higher-yielding seeds into areas of traditional agriculture. Its full success depends on fundamental changes toward modernization, not just on using seed with higher potential yields. Farmers must use manufactured fertilizer and soil moisture must be adequate, which often means using irrigation. There must also be the capability to harvest, store, transport, and market the surplus output not needed locally for food and feed. These requirements of the production system demand fundamental restructuring of traditional agricultural societies, which takes time. Thus the new higher-yielding varieties should not be seen as a quick solution to the problem of food supply in South and East Asia, but as a stimulus to change in agriculture and other aspects of the traditional societies where the new strains are introduced. If these changes do not come about, neither will the potential increase in yields that the new seeds promise.

India has been keeping fertilizer demand, supply, and distribution roughly coordinated over this vast agricultural country.

India's agricultural effort has a high cost. In 1968 alone, over a third of India's $1.4 billion foreign-exchange deficit resulted from imports of food, fertilizer, phosphate, and petroleum, all primarily involved in the agricultural program. The increases in petroleum prices in 1973–1974 exacerbated India's economic difficulties and endangered the program of agricultural development. Fertilizers and pesticides are petroleum products, and oil and gas run the irrigation pumps. Thousands of acres were put out of production for want of fuel for these pumps, and fertilizer production was hampered by the increased cost of the energy to make it.

By some measures India's agricultural program has had remarkable success. Rice output from 1973 to 1977 was more than double that in the early years of Indian independence. Wheat production in the same period almost tripled. By 1970, with the addition of "miracle" grain varieties, the agricultural problem shifted to questions of adequate storage, and shipment and delivery before spoiling. Restrictions on interstate movement of grain were lifted, and with the help of imports there was a reserve of more than 6 million tons. The 1969–1974 five-year plan called for self-sufficiency in food production by 1974, but a series of dry years upset those plans.

In spite of remarkable progress, India's central problem remains one of improving the food-to-population ratio. Cutting imports is one thing;

TABLE 16–2 AGRICULTURAL RESEARCH CENTERS IN
DEVELOPING COUNTRIES

Institute	Areas of Research	Funded	Location
International Rice Research Institute (IRRI)	Rice	1960	Philippines
International Maize and Wheat Improvement Center (CIMMYT)	Wheat, maize, barley, triticale	1966	Mexico
International Institute of Tropical Agriculture (OOTA)	Corn, rice, cowpeas, soybeans, lima beans, roots and tuber crops	1968	Nigeria
International Center of Tropical Agriculture (CIAT)	Field beans, cassava, rice, corn, forages, beef production	1969	Colombia
International Potato Center (CIP)	Potatoes	1972	Peru
International Crops Research Institute for the Semiarid Tropics (ICRISAT)	Sorghum, millets, chickpeas, pigeon peas, peanuts	1972	India
International Laboratory for Research on Animal Diseases (ILRAS)	Livestock disease	1973	Kenya
International Livestock Center for Africa (ILCA)	African livestock	1974	Ethiopia
International Center for Agricultural Research in Dry Areas (ICARDA)	Wheat, barley, lentils, broad beans, oil seeds, cotton	(planned)	Lebanon

SOURCE: From "The Amplification of Agricultural Production" by Peter R. Jennings. Copyright © September 1976 by Scientific American, Inc. All rights reserved. (Vol. 235, p. 188.)

improving the diet of the people is another. The average daily diet contains less than 2,000 calories, 400 less than the generally recognized requirement.

To succeed, India will have to stabilize its population growth, or it will lose any gains by the Malthusian principle that population will outgrow food production. The government has been making an effort to reduce the rate of population growth in absolute terms. An elaborate birth control educational campaign has been instituted. There are almost 2,000 urban family planning centers and over 2,000 rural units, but there are more than 6,000 villages and the program is voluntary. Vasectomies have been legalized, and a government program encourages men to have them. But the vastness of the population problem—in terms of capital, educational efforts, and experts needed to execute the program—is overwhelming. India estimates that 14 percent of its population is participating in family planning, but even if the effort was multiplied many times, the rate of growth would continue to rise in the immediate future because of the large proportion of the population now in or soon to reach the reproductive age. India has worked successfully during the past twenty years to eliminate epidemic and endemic diseases. Millions of children carried through their early years by new disease-control measures are now reaching childbearing age.

INDUSTRIAL DEVELOPMENT

When the Indian government achieved independence from the United Kingdom, it inherited the beginnings of modern industry. In the nineteenth century the British viewed industry mostly in terms of export products. British capitalists took advantage of India's agricultural exports and cheap labor. Jute mills developed in and around Calcutta, cotton mills sprang up in the Bombay area, and the large Indian population constituted a ready market for domestic textiles. Coal was produced for use in factories, and a steel mill was built near Calcutta, close to coal and iron deposits. In the 1920s and 1930s this plant became the largest steel-producing unit in the British Empire and one of the world's lowest-cost producers of steel. The large ports were connected to all major parts of the country by a road and railway network. Originally designed for military and political purposes, the network now enables goods from the large port cities to be marketed over most of the country. Not surprisingly, the widespread marketing of factory-made goods has hurt traditional, small-scale handicraft industries located in the villages.

Industrialization after Independence

India's industry at the time of independence in 1946 was inadequate for the home market in the quantity of goods produced and lacked sophisticated production systems. Moreover, India could not afford to spend the money necessary to import its industrial needs. Then, as now, the basic need is to build up domestic industry to supply the national market.

Industrialization has had top priority in Indian economic plans. The government has used duties on imports to protect newly developing industries and to limit foreign purchases. It has encouraged foreign investors to enter the market in partnership with Indian investors if they are willing to train Indians to do supervisory as well as menial jobs. As in most plans for industrialization, emphasis falls on heavy industry—iron, steel, aluminum, and finished metal products such as bicycles, sewing machines, automobiles, and trucks. The second emphasis is on fertilizers and other chemicals needed to serve the Indian market.

Early results were significant. The establishment of three government-owned plants in the northwest, central India, and the south built with foreign assistance from the United Kingdom, the Soviet Union, and West Germany helped increase steel production three and one-half times between 1950 and 1969, but there has been little increase since then. Coal output increased almost two and one-half times in twenty years. India makes the trucks and buses that are important to its political and cultural pride as well as to its economy. It has an expanding chemical industry, exports textiles, and makes its own telephone equipment. It also makes a variety of machine tools of sufficient quality and quantity to export to industrialized countries.

India is among the ten leading industrial countries in the world, but produces less than 2 percent of the world's industrial output. That places it roughly at the level of Sweden, a country with fewer than 10 million people, or Yugoslavia, one of the least developed European countries. In per capita terms, industrial production has risen from nine pounds of goods per person in 1950 to more than twenty. In the United States industrial production per capita is more than half a ton, and in Japan it is almost one ton per person. Furthermore, India's industrial growth is still below the world average. In the past twenty years Japan's industrial production has more than doubled, whereas India's has increased less than 30 percent. Thus although India's industry is proceeding a bit more rapidly than its population growth, it is not keeping pace with the major industrial powers of the world.

The Problem of Scale

India's problem is not primarily a shortage of natural resources. Although it does lack certain key resources such as petroleum, timber, and some of the nonferrous metals, India has adequate supplies of both coal and iron ore and excellent possibilities for hydroelectric development. The real problem is one of organizing human resources and institutions to cope with the scale of development needed. Although it has far better physical resources than Japan, India lacks the experienced leadership and organizational skills of the Japanese. The imbalance is between the physical industrial plant and the human component needed to make the plant operate at full capacity and efficiency.

As did most modern industrial countries, India located its factory-industry complex in the large metropolitan areas that are traditional centers of skilled workers and capital, and in new urban centers built near basic industrial resources. Seven of the eight metropolitan areas with 1 million or more people are major manufacturing centers. By far the most important centers are the two largest cities, Calcutta and Bombay, ports originally developed by the British as junctions of empire trade. Each is surrounded by supporting industrial towns. Calcutta has an iron and steel district to its west. Bombay is near both an old complex of textile mills and agricultural processing towns, and a new area for chemical, engineering, and modern light industry. The third and fourth industrial districts are centered around the cities of Ahmadabad, north of Bombay, and Bangalore in the southern interior. New Delhi and Madras, two of the largest cities in the country, are also major industrial centers in their own right. Both have shifted from traditional textile mills to more modern technical industry. Madras, a port, concentrates on heavy industry, whereas New Delhi, the political center, is more orientated toward consumer goods.

Industry in Domestic Development

Even though the first five-year plans gave top priority to industrial development, agriculture has increasingly become India's first order of business. In the long run, industrial development in India depends on agricultural change to overcome the poverty of the hundreds of millions in largely subsistence villages. Most of these people have cash incomes of less than $100 per year. Their needs are simple—a few condiments, pots, some kerosene, a few hand tools, and a simple plow—which limits the domestic market.

Moreover, Indian policy since independence has encouraged local village artisans and cottage industry, which uses the artisans' homes as opposed to factories or even shops. Cottage industries using human power alone cannot be as productive as factories using machine and inanimate energy. Nevertheless, as we saw in Ramkheri, cottage industry has important advantages to the villager. Artisans can repair or design products to fit individual needs, and they can be paid with a portion of the farmer's crop rather than cash. They also usually offer easy credit terms until the harvest, and they create employment where the people are living. Indian planners had hoped cottage industry would expand to take care of the demands of villagers so that early industrial development could concentrate on heavy industry.

Industrial development in India is hampered by other problems as well. The country has always had a trained intellectual elite, but there is a severe shortage of managerial leadership. Economic controls can cause delays. Labor unions have developed along political, not just economic, lines, and thus work is often disrupted for political reasons. There are often shortages of particular raw materials and replacement parts for industrial machines.

THE SPATIAL PATTERN OF FUNCTIONING INDIA

As the population density map indicates, the core area of India is along the Ganges Valley in the north (Figure 16–5). The states from Delhi on the west to Calcutta on the east account for 38 percent of India's population. In that zone the state of Uttar Pradesh (Figure 16–6) has a population of more than 70 million, which would make it one of the ten largest countries in the world, and the state of Bihar has almost as many people. The other densely populated areas are along the coasts of the peninsula. The interior of the peninsula, the desertlike northwest, and the northeast frontier are least densely settled.

Although the Ganges area does have both major agricultural output and some of the most industrialized centers in the country, it is much less a focus than the cores of major industrial countries are. Calcutta, the largest and most important metropolitan center of India, anchors one end of the Ganges core, and New Delhi, capital of the country, anchors the other. The languages are different, with Hindi spoken in the west and Bengali in the east around Calcutta. Bombay, the second most important commercial-industrial area, is not part of this core area, nor are all but one of the other metropolitan centers with a million or more population. The first-order flows in India are between the three largest centers: Calcutta, Bombay, and New Delhi. The second-order connections are between those places and other metropolitan centers scattered throughout the country.

Despite the local orientation both economically and culturally of the majority of its people, India does function as a national entity more than most of the underdeveloped areas we have examined thus far. Even with diversity of languages and local cultures, India is a political entity. Its member states are larger in size and population than most of the underdeveloped countries. Most, in fact, have nationalistic movements. But, because the United Kingdom organized India into a single colonial entity instead of a series of separate colonies, most Indian leaders have regarded the entire subcontinent as an entity. Gandhi, who led the fight for independence, urged a unified India, and it was only the fear of Hindu dominance on the part of the Moslem minority that resulted in the creation of Pakistan, a country consisting of the areas of Moslem majority in British India.

To govern so large and diverse an area is not easy. The country is held together by a nationally oriented and organized civil service administration—a Western-educated, strongly nationalistic elite—spread over the country, and by an integrated transportation and communication system that is probably the best of any underdeveloped country.

INDIA IN THE WORLD TODAY

Modernization in India since independence has not been spectacularly successful. The one-sixth of the human race that lives there has reaped little benefit from the modern world. In fact, industrial countries have shown surprisingly little concern for India's problems. United States policy has been tuned to the strategic "hot spots" of the world, particularly in Latin America and the smaller countries of Southeast Asia. The largest annual loan from the United States to India was $435 million in 1966, equivalent to about eighty-seven cents per capita. The size of loans has declined in recent years. The European Economic Community countries have been more concerned with African development, and the

Scale bar: 0 ... 500 Miles / 0 ... 500 Kilometers

JAMMU AND KASHMIR

HIMACHAL PRADESH

PUNJABI SUBA

HARIANA

Delhi 2
New Delhi

RAJASTHAN

UTTAR PRADESH

Kanpur

ARUNACHAL PRADESH

SIKKIM

ASSAM

NAGA LAND

MEGHALAYA

MANIPUR

BIHAR

TRIPURA

MIZORAM

TROPIC OF CANCER

MADHYA PRADESH

Ahmadabad

GUJARAT

3

3

4

Bombay

WEST BENGAL
Calcutta

20°

MAHARASHTRA

ORISSA

BAY OF BENGAL

ARABIAN SEA

Hyderabad

5

GOA 3

ANDHRA PRADESH

MYSORE

Bangalore

Madras

ANDAMAN AND NICOBAR IS.

5°

KERALA

5°

TAMIL NADU

5°

PONDICHERRY

5°

10°

LACCADIVE, MINICOY AND AMINDIVI IS.

INDIAN OCEAN

70°

80°

90°

1. Chandigarh   2. Delhi   3. Goa, Daman and Diu
4. Dadra and Nagar Haveli   5. Pondicherry

FIGURE 16-6   THE STATES OF INDIA

United Kingdom has not been in a financial position to underwrite development on a large scale. In recent years the Soviet Union has provided some economic and military support.

Large foreign industrial corporations have shown an interest in India. They are lured by the huge internal market and the Indian government's need for private investment to complement its own expansion of transportation and key industries. The government prefers to have Indian participation in the control of foreign industrial operations in the country and calls for the use of Indian nationals in management roles. These rules, however, have been flexible for industries the government badly needs. Foreign investment is greatest in oil, chemical, and metal production.

In 1975 the democratic processes that had been operating in India since independence received a severe setback. Prime Minister Indira Gandhi called a state of emergency, and used emergency powers to round up a large number of opposition leaders and to impose censorship on the press. There was little opposition to the moves, and popular economic measures were proclaimed— fixed bonuses for industrial workers, equal pay for men and women, new regional banks, and the abolition of bonded labor. Parliamentary elections due in March 1976 were postponed, the first such postponement since independence. The prime minister justified these actions as within the framework of the constitution and as needed to save the country from anarchy and to restore democracy. Critics saw them as an interruption, perhaps the end, of Indian democracy. Indira Gandhi's government was decisively defeated, however, in elections held in March 1977.

CHINA IN CLOSER PERSPECTIVE

Decision making in China is highly centralized in the national government. The political rulers in Peking control the economic and moral life of the entire country. Such control is not only characteristic of current Communist political systems, but is also a Chinese cultural tradition. Throughout Chinese history a highly developed and complex civil service bureaucracy enforced the power of the emperor and established the rules of life for the people. These rules included a code that placed service to the state above family loyalty and family loyalty above personal gain.

Making such a centralized system work in the world's most populous country requires a tremendous organizational structure. Centralization has great efficiency in carrying out decisions, both good and bad. Correct decisions produce great benefits, but poor decisions produce horrendous disasters. The Chinese found this out during the Great Leap Forward in the late 1950s and early 1960s. At that time the government's emphasis on industrial priorities and the collectivization of agriculture disrupted agricultural production, and the problem was exacerbated by bad weather. The result was a severe food crisis for the country that required emergency imports and forced a re-evaluation of national economic priorities.

AN ENVIRONMENT LIKE THAT OF THE UNITED STATES

In its environmental setting, China is probably more like the United States than any other country in the world. It is almost exactly the same size and has the same midlatitude position. Eastern China has somewhat greater latitudinal range than the continental United States (see Figure 16–1). Both countries have ocean frontage along their entire eastern extremity and extend almost the same distance from east to west. Of course, China, as part of the huge Eurasian landmass, has no western seacoast.

Climatic Diversity
Like the United States, China has great climatic diversity. South China has a year-round growing

season, whereas most of North China has less than half of that, and northern Manchuria has only ninety days. Most of southeastern China gets more than 60 in. (150 cm) of rainfall a year, approximately the same as most of the southeastern United States (see Figure 3–4). This rainfall, coupled with year-round growing temperatures, offers great growing potential. At the same time, most of northeastern China receives less than 40 in. (100 cm) of precipitation annually, less than that of the northeastern seaboard of the United States. To the west more than half of China is arid like the western United States. Rainfall ranges from less than 20 in. (50 cm) annually to less than 10 in. (25 cm) in the Gobi Desert of Sinkiang Uighur and in parts of Tibet. Even in the midlatitudes this small amount of rainfall is inadequate for agriculture without irrigation.

Landforms

Although China has climatic conditions comparable to those of the United States, its land surface is very different (Figure 16–2). China has more mountains and highlands. Most of South China is hill land similar to Appalachia. Western China is similar to the mountainous area of the western United States, but on a grander scale. The highest mountains rim the country along the frontiers of the Soviet Union and India, while Tibet is mostly high plateau about 15,000 ft (4,500 m) above sea level.

Despite the significant differences between the terrain of China and that of the United States, growing potentials are about the same (see Figures 3–3 and 3–6). The Chinese culture, however, has made very different evaluations of the rather similar environment. Floodplains are highly prized by the Chinese for rice culture, but such land in the United States was the last to be occupied by European settlers. By contrast, grasslands similar to those of the Great Plains were among the last lands developed by the Chinese. The important distinctions between Chinese and American life reflect culture more than natural environment.

POPULATION DISTRIBUTION

In rough outline Chinese population distribution is similar to that of the United States. The eastern third of each country is heavily populated, while most of the west is sparsely populated (Figure 16–7). In China the distinctions between east and west are even more pronounced, however. Over three-fourths of the Chinese population is found in the east on less than one-third of the land.

Population density in the crowded areas is about 400 persons per square mile (155 per square kilometer), and more than two-thirds of these people live directly from agriculture. Population per square mile of cultivated land is more than 1,600 (620 per square kilometer). Such figures are comparable to those of Japan, India, the Netherlands, and Belgium, but they are exceptional when applied to a country the size of China. They indicate the tremendous burden placed on agricultural land resources.

SEPARATE, LOCAL AGRICULTURAL SYSTEMS

China's regional crop specialization provides a measure of environmental growing conditions perceived by the traditional agricultural system. Here, as in India, the population depends on grain as the staple food, and the different crop possibilities result in varied diets from place to place. In China the order of grain preference is the same as in India. Rice is the first choice, wheat is second, and millet and sorghum are grown on poorer lands (Figure 16–8).

China's humid tropical and subtropical south is well suited to rice, and two crops a year are possible in many places. As a result, production in terms of calories per acre is very high, and it is possible to support dense rural populations. In the Yangtze Valley of central China rice can still be grown as the chief summer crop, but winter wheat replaces the second rice crop each year.

In most of northern China summers are too dry for rice or spring wheat. Instead, kaoliang, a

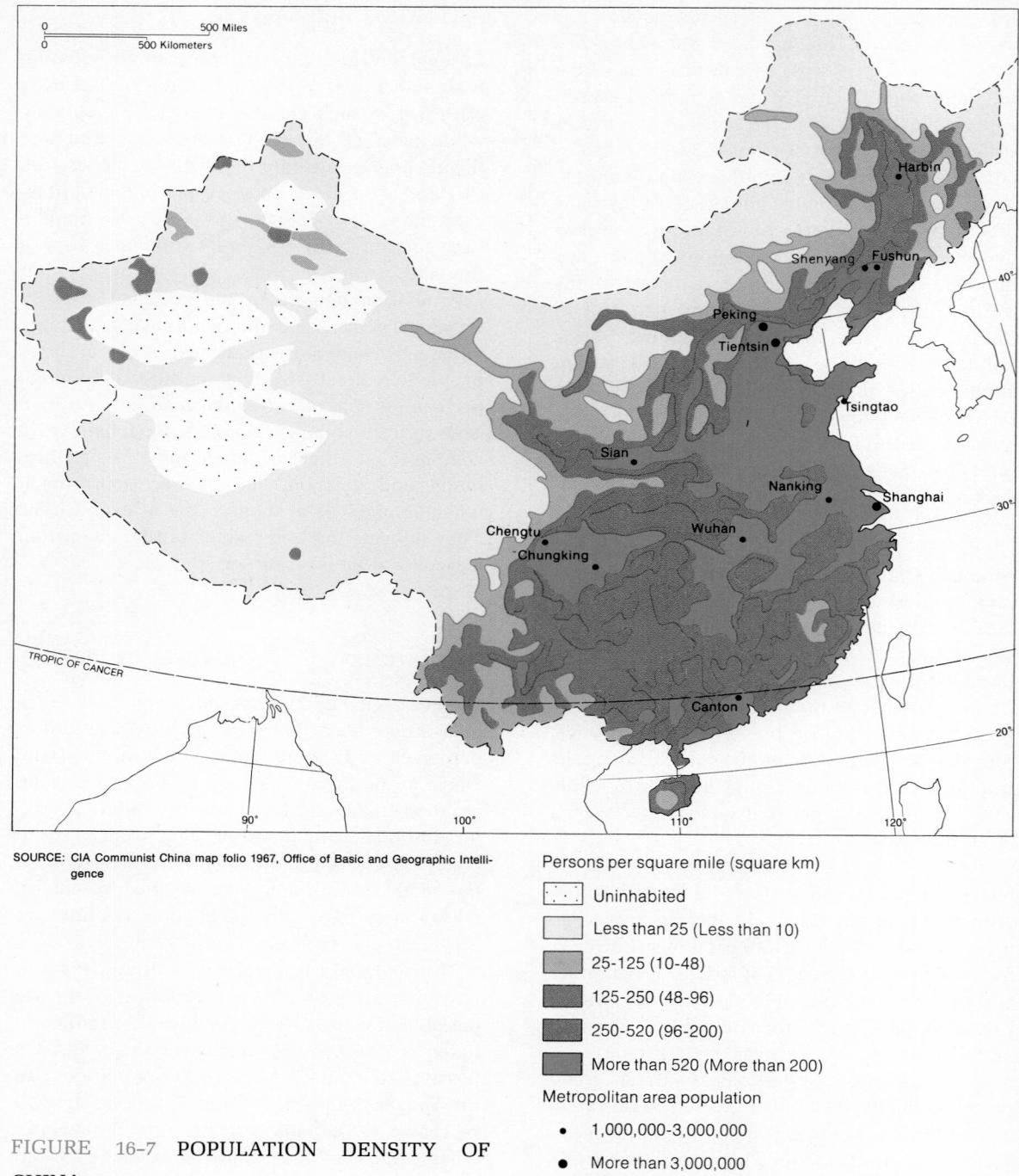

SOURCE: CIA Communist China map folio 1967, Office of Basic and Geographic Intelligence

Persons per square mile (square km)

- Uninhabited
- Less than 25 (Less than 10)
- 25-125 (10-48)
- 125-250 (48-96)
- 250-520 (96-200)
- More than 520 (More than 200)

Metropolitan area population

- • 1,000,000-3,000,000
- ● More than 3,000,000

FIGURE 16-7 POPULATION DENSITY OF CHINA

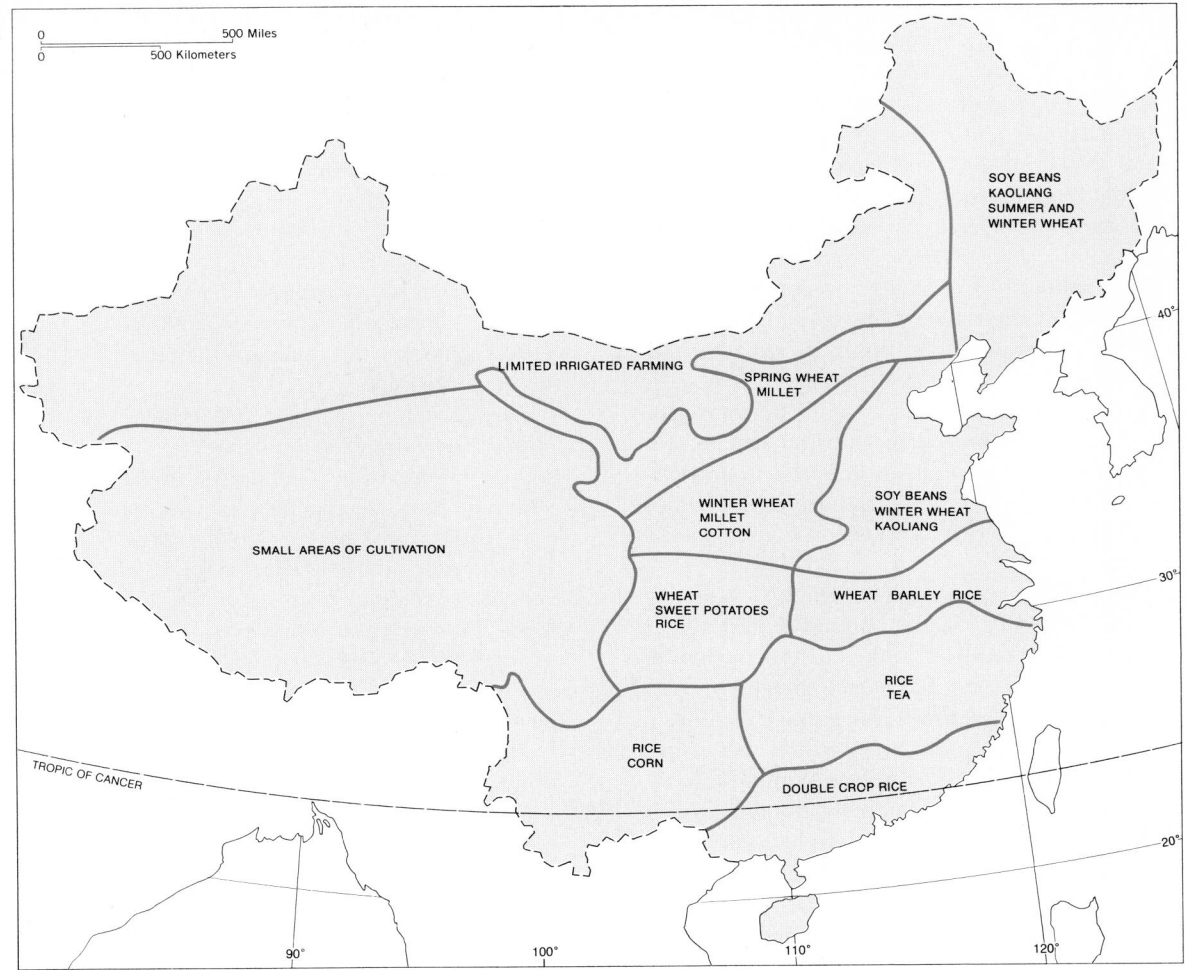

SOURCE: From *Asia, East by South,* 2nd ed., by J. E. Spencer and W. L. Thomas, John
   Wiley & Sons, New York, 1971.

FIGURE 16-8 AGRICULTURAL REGIONS OF
CHINA

giant sorghum, is the summer crop. These crops are supplemented with soybeans, grown not only for oil but also as a bread substitute. On the drier western margin of populated China, the crop choice shows adaptation to more arid conditions. North of the Yangtze, winter wheat is the leading crop, and farther northwest, where winters are more severe, millet and spring wheat dominate.

Root crops such as yams and potatoes are the basic supporting crops. Cotton, the major industrial crop, is grown in northern China rather than in the south. In all these areas livestock are raised either to serve as beasts of burden in the largely nonmechanized agricultural system or as scavengers that provide an important additional food source. Cultivation is intensive, and in most of China the land is worked as a giant garden patch.

Through the centuries this farming system primarily has fed the farm population, particularly the individual family and the clusters of families that occupy the myriad farm villages of rural China. It has also produced enough surplus to supply the cities and towns, which make up a small fraction of total Chinese population. Although China has nineteen metropolitan areas of over 1 million whose populations total 40 million people, this figure is only slightly more than 5 percent of the total population of the country. China also contains many cities with populations less than 1 million, but the urban population has not represented a major burden on either production or transportation. Almost every farm contributes a small amount of production in this commercial system and, in turn, receives some cash payment.

THE COMMUNIST DEVELOPMENT PLAN

China had an industrial base even in pre-Communist times. To a large degree foreign investors, not Chinese, had built that industry either to process Chinese raw materials or to provide clothing and supplies to the population, particularly the city workers most involved in the money economy. Manufacturing developed by European interests was concentrated in the major ports such as Shanghai, Canton, and Tientsin. During the 1930s and early 1940s the Japanese developed a heavy industrial complex based on local deposits of coal and iron ore in occupied Manchuria. This development had no relationship to the Chinese economy.

The Chinese economy meanwhile suffered through almost constant warfare. First, the Japanese occupied China in the 1930s; and then, following World War II, the Communists finally took over the mainland in 1949. Except in Manchuria, investment in industry was neglected in the 1930s and 1940s. The modest railroad system suffered from war damage, and national control was lost in most regions of the country. Local landlords dominated both the political and the economic scenes in many areas, and tenants were forced to make exorbitant payments for the use of land. In the postwar period the Soviets, who occupied industrial Manchuria when the Japanese surrendered in World War II, systematically confiscated machinery and railroad cars to rebuild their own war-devastated economy.

Once in power in 1949, the Chinese Communists began to consolidate the nation within the framework of the Communist political and economic philosophy. They promised **agrarian reform** to the peasantry, and in the 1950s the holdings of the landlords were redistributed among the landless families in the villages.

Initial emphasis and major investment priority was placed on modern heavy industry as a base for expanding industrial activity and military power. The first five-year plan (1952–1957) assigned almost half the state's investment to industry, over 80 percent of this amount for heavy industry. Only 3 percent of the investment went directly to agriculture. The Soviets provided the basic industrial support for the new Communist regime by offering trade, foreign assistance, technical advisers, and the opportunity for Chinese students to train in Soviet universities.

Heavy Industry

Chinese strategy called for developing a full, modern industrial base along the Soviet model to give independence from foreign sources of supply. The program resulted in successful completion of modern metalworking, chemical, and petroleum plants. The design and construction of jet aircraft, the detonation of nuclear bombs, and the launching of space vehicles attest to Chinese progress.

To accomplish this in a country beleaguered by a shortage of modern overland transportation, the Chinese relied primarily on the expansion of industry in the traditional industrial centers of the major cities (Figure 16-9). Most important was the heavy industrial complex in Manchuria centered in Shenyang (formerly Mukden), Fushun, and Anshan. Second were the great port of Shanghai, center of industry under Nationalist China, and the Yangtze Valley centers of Hangchow, Nanking, and Wuhan. Third was the Peking–Tientsin area. All other major metropolitan areas were centers of regional industry.

In addition, new centers emerged at sites of new raw-material development. Examples include the iron and steel industry on the upper Yellow River and the Yangtze, and oil refining and petrochemicals at Lanchow on a new railroad line built by the Communists to connect with Sinkiang Uighur, the westernmost province.

A Structural Agricultural Revolution

The traditional Chinese agricultural system, with its intensive methods concentrated on the best land resources of the country, has been a productive one. Yields have always been much higher than in India. Still, the Chinese have increased agricultural productivity appreciably. A small amount of additional land on the drier margins of agricultural China has been put into cultivation by use of mechanization and by irrigation and other means of land reclamation. Essentially, however, the increase has come through planting improved crop strains, improving supplemental irrigation, practicing double and even triple cropping, and developing adequate storage facilities.

All agricultural growth in China since 1949 has been achieved without fundamentally changing the intensive garden-patch agricultural system. The major change has been in control. The Communists brought about land redistribution by dividing large farms among the peasants. In 1953, however, farmers were urged to pool their work as part of mutual-aid teams, and by 1956 all farmers were being encouraged or coerced into participating in producer cooperatives where they pooled their land and worked the combined lands in work teams. Returns were a share of the harvest proportionate to the amount of land contributed. Equipment was cooperatively owned.

Gradually, the basis for sharing in the cooperative became the amount of work done by members of the cooperative. Soon "advanced cooperatives" were established in which farmers pooled their land and several small cooperatives were united. The basis of compensation became the time a person worked for the cooperative. This change eliminated the advantages of the large and middle-sized landowner. In time the cooperatives became **collectives,** which held land rights from the government, which had taken over all farmland. Equipment and livestock were owned collectively rather than individually, and the designated leaders of the collective made management decisions. Farmers owned only their homes, small garden patches, and a few animals.

In 1958 the central government moved one step further. In what was called the Great Leap Forward, **communes** were formed by consolidating collective groups into units with 18,000 to 20,000 people, bringing the inhabitants of numerous nearby villages together. These communes were planned to be far more than agricultural entities. They were to be social and cultural units with supporting educational and social services. They were also to function as the local units of government. Moreover, they provided a means to mobilize the mass of labor to undertake industrial

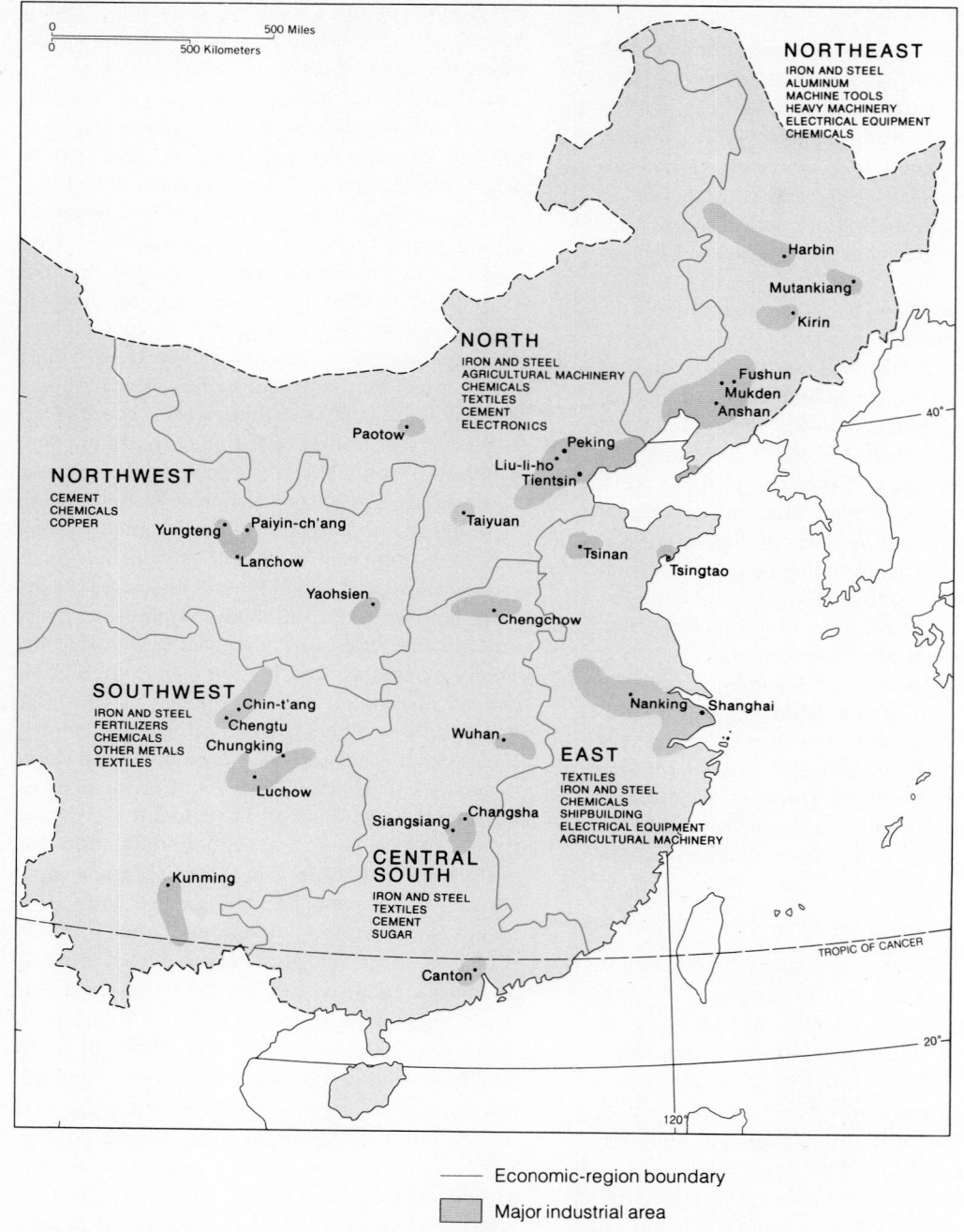

FIGURE 16-9   INDUSTRIAL AREAS OF CHINA

Legend for the map:

—— Economic-region boundary

[shaded] Major industrial area

NORTHEAST
IRON AND STEEL
ALUMINUM
MACHINE TOOLS
HEAVY MACHINERY
ELECTRICAL EQUIPMENT
CHEMICALS

NORTH
IRON AND STEEL
AGRICULTURAL MACHINERY
CHEMICALS
TEXTILES
CEMENT
ELECTRONICS

NORTHWEST
CEMENT
CHEMICALS
COPPER

SOUTHWEST
IRON AND STEEL
FERTILIZERS
CHEMICALS
OTHER METALS
TEXTILES

EAST
TEXTILES
IRON AND STEEL
CHEMICALS
SHIPBUILDING
ELECTRICAL EQUIPMENT
AGRICULTURAL MACHINERY

CENTRAL SOUTH
IRON AND STEEL
TEXTILES
CEMENT
SUGAR

Cities labeled: Harbin, Mutankiang, Kirin, Fushun, Mukden, Anshan, Paotow, Peking, Liu-li-ho, Tientsin, Taiyuan, Tsinan, Tsingtao, Yungteng, Paiyin-ch'ang, Lanchow, Yaohsien, Chengchow, Nanking, Shanghai, Chin-t'ang, Chengtu, Chungking, Wuhan, Luchow, Changsha, Siangsiang, Kunming, Canton

TROPIC OF CANCER

projects, build roads and dams, and carry out such developmental operations as reforestation and land reclamation. As part of the Great Leap Forward, even traditional family groups were reorganized, with separate dormitories for men and women and communal dining halls. Special boarding schools for children were established. Workers received free food, clothing, and a small salary. Women were expected to do much of the field work in agriculture, in order to free men for work on irrigation facilities, roads, and factories.

This program, however, was far too great a departure from traditional Chinese life to be fully accepted by the people, and it was too hastily implemented to function properly. Many of the rural people resisted the changes. Moreover, shortages occasioned by droughts in some areas and floods in others required China to import wheat and rice.

Faced with the failure of the Great Leap Forward, the government was forced to backtrack on its ambitious schemes. It abandoned communal living and allowed families to have private plots of vegetables, other cash crops, and livestock. Traditional trade fairs were revived for selling the surplus. Much of the working management shifted from the larger commune level to more local production brigades and teams that really represented the old collective and cooperative organizations. Ownership was returned to the cooperative level, and each team was guaranteed the fruits of its production. Thus individual production teams within the same commune could have differences in living standards.

The Commune as the Primary Institution

The objectives of the multipurpose commune are fundamental to Chinese Communism. Long-term aims include elimination of individual incentives and encouragement of people to work for the goals of the state and commune. Chinese Communism aims to help those people without political power, a high level of education, or material wealth—the rural majority of China.

In contrast to most other countries, China does not see cities as the high points of civilization. The Chinese are striving to minimize the level of urbanism and to move many essentially urban functions, including industrialization and government control, to rural areas. The commune is designed to serve as the basic organizational unit.

In carrying out this rural program, China, unlike India, has geared its price structure to the needs of the rural people rather than the city population. The creation of locally based commune industry is an attempt to provide inexpensive industrial goods for rural areas. Communes are to be as independent of city factories as possible. Local industries are seen as a part of the policy of integration of industry and agriculture and of town and countryside, and communes are expected to supply their own basic nonagricultural needs as well as food. City factories supply national needs.

Communes are recognized as local governmental units. They are also social and cultural units that operate their own educational and medical facilities and their own militias. Each local area—that is, each county within a province—is to build a self-contained economic, cultural, and political unit with the communes as the bases. Ideally, the local area should provide for most of its own economic needs and most of its own management and decision making within the guidelines and authority of the strong central government.

Counties are encouraged to establish integrated local industrial systems with their own iron and steel plant, coal mines or other power sources, agricultural processing facilities, metalworking and cement plants, fertilizer works, and insecticide and consumer goods factories. Many of these supplement commune-run operations, and others depend on commune workers as sources of labor.

During the mid-1960s Mao Tse-tung inspired a return to revolutionary values and reinforcement of basic Communist principles. This Cultural

Revolution aimed at modernizing China not according to Western models but within the value structure of traditional Chinese culture. Chinese leaders planned not just to bring China into the modern world, but to develop an alternative form of modernization. The focus of life was to be in the rural areas, not in the cities. Urban centers were to serve the people in rural areas, not vice versa, as has been the rule in other industrialized societies, even the Soviet Union. As a result, the Chinese experiment has become significant to other underdeveloped areas of the world that, up to now, have had only the Western urban-centered model to follow. It is also of interest to industrialized countries, which have all followed a basically similar path to modernization and are now concerned about the future of their large metropolitan areas.

As part of the Cultural Revolution the central government tried to replace elite education with universal education of the people. People trained in universities and technical schools, including doctors and technicians, were expected to return to rural areas to work. The government wanted to end the traditional Chinese reverence for scholarship and professional pursuits and the emphasis on higher education. Practical skills, not classical studies, were now stressed. The Chinese tradition of an intellectual elite was reoriented to encourage functional education for all. Communes had public schools for all children, and more and more students in higher education came from the communes rather than from the cities. Management cadres worked one-half day in the fields and one-half day in the office, so as not to lose touch with the masses. All this fit the plan to modernize the life of the largely rural population without a massive movement of people to urban areas.

The blueprint for the modernization of China under Communism was set down by Mao Tsetung. His vision and his power as a leader kept the country united behind the plan and dissidents in check. In 1976 Mao died, and the struggle for power after his death that had been predicted by experts on China soon emerged as one centered on the very question of development strategies. One group strongly supported the Maoist view of a country centered on its great rural masses; the other called for the expansion of China's urban-industrial base along largely Western-style lines.

THE FUNCTIONING CHINESE SYSTEM

As in India, the problem in China is not so much knowing what to do as how to carry it out on a vast scale. China has major cultural and historical advantages. Chinese culture had placed a premium on service to the state and on the development of organizational and communication systems designed to operate over the whole country. In contrast to India, the government can take advantage of a largely homogeneous culture. Although there are many different languages and dialects, there is one written language that literate persons in all parts of the country can read. Most important, the people have great national pride in China's heritage and destiny. The Communist regime has taken advantage of these valuable assets.

The Chinese system operates on two different levels, loosely interconnected. The lives and support of most of the people are on a local level, that of the county, the commune, or even the work team. At the local level, traditional life within China has been centered on the individual family, the village, and the local market.

The other level is that of the national government. The governmental system reaches out from Peking, the capital, to the twenty-two provinces, the five autonomous regions, and the municipal districts of the three largest cities—Shanghai, Peking, and Tientsin (Figure 16–10). The provinces were a part of traditional China, in contrast to the autonomous regions, such as Sinkiang Uighur and Tibet, which have a non-Chinese

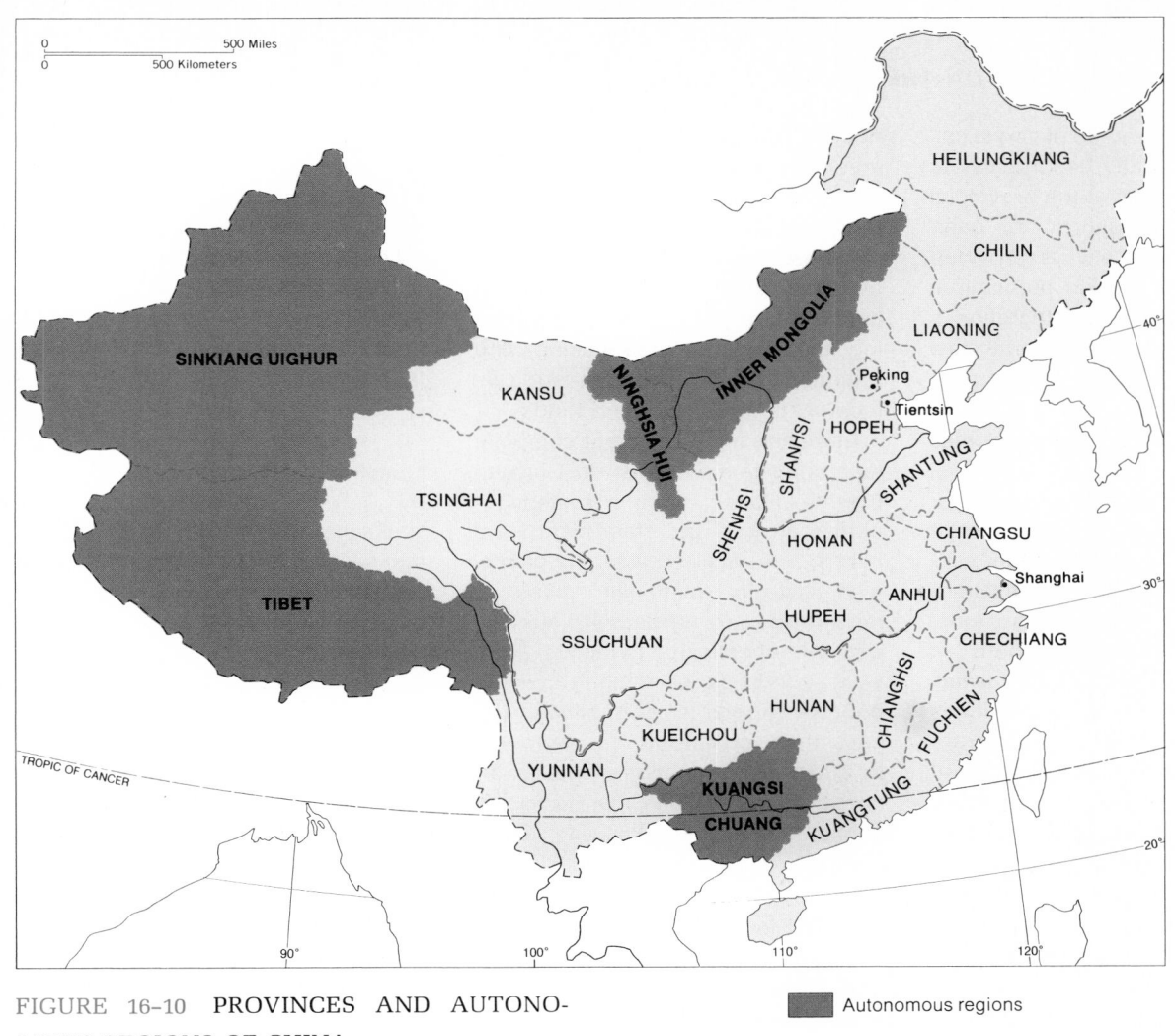

FIGURE 16-10 PROVINCES AND AUTONO-
MOUS REGIONS OF CHINA

Autonomous regions

SMALL-SCALE PRODUCTION ON THE BYWAYS OF CHINA

Plans for the rural areas of China call for integrated regional economies at the local rather than at the provincial level. Since agriculture has been dominant and has always provided local support to its own population, the problem of regional integration is to provide essential industrial raw materials, fuel sources, and manufacturing facilities.

Fuel for the local economies comes from two primary sources: local coal mines and small hydroelectric plants. In Yunnan Province in southern China in 1970 there were an estimated 2,100 small coal mines with a total production of 1.8 million tons, and Chekiang Province south of Shanghai had over 1,200 small mines. Several Chinese provinces each claim over 2,000 small hydroelectric power stations, with Kansu at the headwaters of the Yellow River having more than 4,400. Electric lights have replaced the pine and bamboo torches used for lighting in remote rural areas, and radio networks tie together out-of-the-way villages in the highlands.

Manufacturing is centered around four small industries essential to local needs: iron and steel manufacture, machine building (especially farm machinery), chemicals (particularly fertilizers and insecticides), and cement works. Other small-scale industries include agricultural and timber processing, textile and clothing manufacture, papermaking, and pharmaceuticals. The idea is to use local materials, labor, and funds.

Emphasis is on small but complete local industrial systems. In Chiyuan County, Honan, such a system includes an iron and steel plant, coal mines, sulfur mills, machinery plants, a metallurgical plant, and plants or mills producing refractory materials, cement, chemicals, ceramics, knitwear, and paper among twenty county-run plants and mines and seventeen commune-run factories. In addition, production brigades on communes have their own farm-tool shops, small coal pits, and workshops processing agricultural and farm by-products. In Fencheng County, Kiangsi, a system comprising iron and coal mines, a coke oven, iron-smelting furnace, steel-making furnaces, and rolling mills produces several hundred tons of rolled steel.

In Kiangsu Province two-thirds of the counties have small nitrogenous fertilizer plants, and half have small phosphate works.

Output for Fencheng County in 1969 included 4,000 tons of pig iron, 99 tons of steel, 6,500 pieces of farm machinery, and some 1,300,000 small farm tools.

Questions

1. Given the wide range of environmental conditions over China, what problems would you anticipate in developing real local economic self-sufficiency?

2. Are there problems of small-scale production? Can steel be made as effectively in small plants as in large? If so, what are the advantages of the large-scale steel plants that have developed into some of the largest industrial mass-producing facilities in countries such as the United States, the Soviet Union, and Japan? What problems will China have to face as it moves in this direction?

---

cultural base and a greater degree of self government. Plans for the political, economic, and social functioning of the entire Chinese state are made in Peking. Shanghai and Tientsin play special roles, too. Shanghai, the largest port and the largest metropolitan area, is China's most important international center and the major focus of manufacturing and internal commerce. It functions in relation to Peking somewhat as New York City does in relation to Washington, D.C. Tientsin, the combined port and industrial base closest to Peking, is the country's leading chemical center and has a wide range of both heavy and light industry. Most of the government programs filter down through the capital cities of the provinces and the five autonomous regions.

Peking's ties to the major provincial political capitals are provided by the country's transportation and communication links, the most important of which are within the populous areas of China. Thus in contrast with the United States, the busiest links are those on a north-south rather than an east-west axis. The first order of interaction is between Peking and centers within heavily populated eastern China. Next, connections are made from Peking to the outlying areas in the west—Sinkiang, Tibet, and the others.

For the present, China seems to have the population problem under control. There is no famine or serious food shortage. Birth control clinics in the communes appear to have the rate of population growth under greater control than in India. Under centralized regimes it is easier to enforce birth control measures. Programs are required, not optional as in India, and there are fewer units to deal with. But still the population growth is tremendous in absolute terms—over 14 million more people each year. As in India, the absence of famine and starvation is one thing; raising the living standard of such a huge population is another. This is the ultimate test that will determine the success of the Chinese Communist model of development.

CHINA IN THE WORLD TODAY

In its development, China more than India has depended on both its own natural resources and its own sources of investment capital. Since the United States supported the anti-Communist Nationalist regime during the civil war that followed World War II, it did not formally recognize the Chinese Communist government and for many years blocked China's entry into the United Nations. Moreover, during the Great Leap Forward and the Cultural Revolution, China made an ideological break with the Soviet Union. Thus China has been cut off in terms of economic aid and technical assistance from the two greatest powers of the world today. Although most European countries and underdeveloped countries recognize China and carry on trade and diplomatic relations, they have contributed little to the nation's vast capital and technical needs, and China has been forced into becoming self-sufficient.

However, through trade and aid with underdeveloped areas, particularly African countries, China has worked to develop its own sphere of influence in the world. It built a 1,100-mile (1,760-km) railroad for Tanzania and Zambia. Its greatest trade is with Japan, its nearby industrial neighbor. China exports textiles and clothing, metallic ores, rice, tea, and coal, and imports grains and machinery. China has been the protector of tiny Albania, a European Communist country that broke with the Soviet Union. China accounts for two-thirds of Albania's trade.

In the 1970s the isolation of China from the power circles of the world began to break down. In 1971 China was admitted to the United Nations as a permanent member of the Security Council where it sits along with the United States, the Soviet Union, France, and the United Kingdom. Before this date the Chinese seat at the United Nations had been held by the Chinese Nationalist government-in-exile on Taiwan. In 1972 the president of the United States visited China in an effort to open up channels for communications and trade.

Relations with the industrialized countries, however, are still tenuous. The United States and China still have not agreed to full diplomatic recognition and thus do not yet exchange ambassadors. Trade with the United States and Western Europe is still just a tiny fraction of China's total trade and is unimportant to the industrialized countries. The chief trade exchanges are with the industrialized areas of Asia, particularly Japan and Hong Kong. Japan is eager to tap China's rich mineral resources, especially oil, and Hong Kong, which is leased from China, must depend on China for food and even water. China finds Japanese and Hong Kong industrial products useful in the drive toward full industrialization.

China has achieved a position of political prestige in the world far above India's, and its political influence will probably continue to grow. In economic terms it has also made considerable progress since the end of World War II, but the road to modernization is difficult and many challenges still lie ahead.

THE REST OF SOUTH AND EAST ASIA

The rest of South and East Asia outside India, China, and Japan seems at first glance to be an extension of the human system and environmental base that we have just studied. The setting and daily routine of most people seem much the same: most live in villages; most are farmers; and farming methods are basically the same as those in China and India. This is primarily a world of intensive subsistence farming. Tremendous effort is expended to get the best return possible, but much depends on the physical environment. If the rains come too early or too late, if they last too long or are particularly heavy, crops may wither away or be washed out. Drought and flood are the twin terrors of all Asian farmers, and of all residents of the hundreds of thousands of villages.

FOUR VARIATIONS ON ASIA'S GEOGRAPHIC EQUATION

The smaller states of South and East Asia, despite their superficial similarity to India and China, have their own geographic equations and their own particular variations on the development problem faced by India and China. These countries occupy four strikingly different envi-

ronments, each of which differs from the physical environment of most of India and China. Pakistan, Mongolia, Korea, and Bangladesh illustrate four distinctive variations on the basic geographic equation for South and East Asia.

Pakistan

Pakistan is a great lowland lying astride the Indus River and extends from the Himalayan Mountains to the Indian Ocean. It is a dry world with less than 10 in. (25.4 cm) of annual rainfall in most places. The Indus River system is its lifeline and is fed by rains and snows in the Himalayas to the north. The waters of the Indus have been used for irrigation for more than 5,000 years, and today they support most Pakistani agriculture. An elaborate system of canals connects the tributaries of the Indus with the main river, redistributing water southeastward across the lowland. Wheat is the chief irrigated crop, with rice limited to lands close to rivers and canals. Barley replaces wheat on the poorer soils and along drier margins.

This dryness is rather like that of Texas and Oklahoma in the United States. It happens that Pakistan is slightly smaller in area than those two states combined. It has, however, more than 72 million people, most of them in rural areas, whereas Texas and Oklahoma together have fewer than 15 million.

Pakistan, along with the area around the delta of the Ganges River 1,200 miles away (which has become Bangladesh), was a part of British India. These were the predominantly Moslem areas of British India and their leaders feared that the Moslems would have minority status in Hindu-dominated free India. As a result, the independent state of Pakistan, including both Moslem areas, was established in 1947. It soon became clear that the two parts of Pakistan were separated by more than just distance. The people of East Pakistan lived in a very different environment, spoke a different language, faced very different problems, and felt that they were not receiving their proper share in the new state of

Pakistan. In 1971 they fought for secession and, after a war in which they were supported by India, became the independent country of Bangladesh in 1972.

Pakistan faces many of the same developmental questions as India: how to improve crop yields in a traditional farming system dependent on human and animal power; how to control population; and how to develop a national industrial base. As in India, the chief question is how to increase actual evapotranspiration (AE) in a dry environment, a question complicated by occasional floods and earthquakes. One part of the answer has been the construction of large dams on the upper Indus River to store floodwaters and distribute them as needed for irrigated agriculture.

Manufacturing is less advanced than in India, with production concentrated on textiles made from Pakistani cotton and on fertilizer materials used in agriculture. The two largest industrial centers and metropolitan areas are Karachi, which is also the capital and chief port and has a population of about 3.5 million people, and Lahore, which has almost 2 million.

Pakistan's goal, like India's, is to develop its internal economy into a functioning system that can support the whole population rather than to focus on exports and imports with the rest of the world. But Pakistan is a smaller, less diverse country with fewer mineral resources, and it therefore depends on imports, particularly grain and industrial products from the United States and oil from the Middle East. In 1975 Pakistan had a large trade deficit. The value of its imports was over 50 percent greater than that of its exports: floods and earthquakes had destroyed an important part of the agricultural output, thus requiring the import of grain.

Mongolia

Mongolia is part of the semiarid world of central Asia, and its population lives in much the same way as do people in the Middle East–North Africa: by raising sheep, goats, horses, cattle, and camels, and by oasis agriculture. But the short-grass-shrub vegetation of the Mongolian Plateau does not support a dense rural population; Mongolia's population of 1.5 million is the smallest of all countries of South and East Asia. In most circumstances the small, poor nation of Mongolia could not maintain political independence from its neighbors, the Soviet Union and China. To a certain extent Mongolia has done so by serving as a buffer between the two large states.

Most of its population still leads a seminomadic life, following flocks of sheep, cattle, goats, and horses. With Soviet help, hay and grains are being produced for winter feed for the livestock. Most farmland is state-operated and is concentrated in a few river valleys. Ulan Bator, the capital, has slightly more than 300,000 people. Small amounts of lignite are mined, mostly to supply electric power to the capital. Mongolia, like some of the small countries of Black Africa, seems too small and poor to be viable in today's world. It has little that the outside world might buy, and thus has little chance of altering its traditional ways without subsidy.

Korea

The Korean environment is in many ways like that of Japan: it is mountainous and, being a peninsula, has a climate much moderated by the sea, although the dry winter monsoon that sweeps from Manchuria affects the northern part of the peninsula. Korea's highlands are not very high—the highest peak reaches only 9,000 ft (2,700 m)—but the peninsula is rugged, laced with river gorges and canyons, particularly in the north. Rivers are short and flow through narrow valleys, and so there are no extensive lowlands. Most Koreans are farmers in the traditional Asian style. Because of the predominantly maritime climate, it is possible to grow rice in most Korean valleys. Barley is the second grain crop, for it can be grown during the winter in fields used for rice and other crops in summer. Wheat and millet are also grown, particularly in North Korea, where the climate is drier and colder than in the south.

Korean culture is very ancient, and it reflects the position of this small country between the dominant cultures of China and Japan. Korean customs show Chinese influence, the language is closely related to Japanese, and the dominant religion is Buddhism, with both Chinese and Japanese sects present. The importance of the family in everyday life reflects both Chinese and Japanese practice.

Japan dominated Korea from 1904 until the end of World War II. The Japanese built a railroad network and, as in Manchuria, systematically exploited iron ore, coal, and other minerals as part of their own industrial expansion. By the 1930s the Japanese had established an iron and steel industry; chemical, cement, and machinery plants; and hydroelectric plants. Most of this industry was in the northern part of the peninsula, where the mineral resources were found. Japan also supplied fertilizers and agricultural advisers to Korea, so that Korea became a source of rice supplies needed to make up the Japanese deficits in the 1930s.

When Japan surrendered in World War II, Korea was divided at the 38th parallel into a Soviet occupation zone in the north and a U.S. zone in the south. Under the influence of the two great powers, two very different political systems were established. Since then, all efforts at peaceful unification of the country have failed. In 1950 North Korean troops invaded South Korea, and war involving U.S. and other United Nations forces continued for three years. At the end of it the boundary between the two separate countries remained in the vicinity of the 38th parallel. Today the two Koreas continue to develop according to two very different philosophies, North Korea following Communism and South Korea developing under capitalism.

North Korea has expanded steel production; in 1975 it produced 2.6 million tons. It has also increased coal and other mining output, and cement and hydroelectric production. South Korea has also increased its steel output (to 1.9 million tons in 1975), but its manufacturing has been geared primarily to export markets. South Korea serves as an export platform, which means that its textiles, clothing, and small consumer goods are produced primarily for sale in Japan and the United States, not for the local market. Various U.S. and Japanese corporations have established production in South Korea to take advantage of the cheap, easily trained labor force and government tax incentives. As a result of South Korea's economic advances, its capital, Seoul, has become one of the largest and most modern of Asia's cities.

Bangladesh

Bangladesh is in the hot, wet monsoon deltalands of the Ganges and Brahmaputra river system. The area receives over 80 in. (over 200 cm) of rainfall a year. Its rivers also bring abundant run-off from areas in the Himalayas that receive 40 or more inches (100 or more cm) a year. As elsewhere in the Indian subcontinent, the rain is concentrated during the summer monsoon. The danger of floods in low-lying Bangladesh is therefore very great. The problem is compounded frequently by northward-moving tropical typhoons (equivalent to hurricanes in the Atlantic). Devastating floods in July and August 1974 following a typhoon inundated more than half the area of the country. Almost all the standing crops were destroyed, and widespread famine was averted only through international assistance. The official death count in the month-long floods was 1,500, and nearly one-third of the population was directly affected.

Bangladesh is a small country, slightly smaller than the state of Illinois, but it is one of the most densely populated areas in the world with almost 1,400 people per square mile (540 per square kilometer). Farming follows the traditional methods of neighboring India. Jute (used for making burlap sacks) is the most important cash crop and is the country's major source of foreign exchange. Bangladesh accounts for 54 percent of world jute production. Rice, however, is the principal crop in acreage and in the support of the population.

Bangladesh ranks fourth, after China, India, and Japan, in world rice production, yet does not produce enough to feed its huge population of 76 million adequately.

Agricultural productivity could be improved, but the fragmentation of landholdings is an obstacle. By tradition, each male member of a family inherits several parcels of land of differing quality—a patch of good irrigated land and a patch of less desirable irrigated land; a patch of good dry land and a patch of dry land of lesser quality; and a patch of poor land. Another obstacle to increasing Bangladesh's agricultural productivity is the government's lack of the necessary capital to launch a successful program of rural rehabilitation.

SOUTHEAST ASIA

Southeast Asia, lying between India and China, includes nine countries with a total population of 326 million in 1976. This is one and one-half times the population of the United States. Lying east of India and south of China, the area is one of the most extensive hot, wet tropical regions in the world. As in Bangladesh, much of this area has more than 80 in. (200 cm) of rainfall per year. Large moisture surpluses during the wet monsoon usually dwarf the small deficits of the dry monsoon. Only in limited areas of Burma, Thailand, and Laos is drought a regular problem.

Actual evapotranspiration equals potential evapotranspiration in most of Southeast Asia. Since this is predominantly a tropical region, growing potential in most areas is greater than it is anywhere in the conterminous United States. There is a year-round growing season with no danger of frost. Unfortunately, year-to-year variability in weather frequently is great enough to bring disaster. The monsoon rains are often late or early, too heavy or too light. As in Bangladesh, floods are a serious threat, because the best farmland and most densely populated farming areas are in the lowlands of the major rivers such

as the Irrawaddy, Mekong, and Nam Chao Phraya.

Except for Indonesia, the countries of Southeast Asia do not have the resources or the population to embark on programs of self-contained development similar to those of India and China. Like states in the United States, or like Canada and Australia, if they are to participate in the modern world they must do so as parts of a larger system. They must contribute specialty products—raw materials, food, or even specialized manufacturing and financial services such as Singapore and Hong Kong have developed. Economic plans of most countries of Southeast Asia must be more like those of African and Latin American countries than like those of India and China.

Despite political independence, the countries of Southeast Asia still depend on their economic ties to the modern industrial countries. Over half the imports of most of these countries come from trade with the major industrial countries of Europe, the United States, and Japan. Vietnam and some other countries trade with the Soviet Union and China. Exports are critical in providing income for government programs and economic development. Export value is more than one-third of total gross national product in Malaysia, Taiwan (formerly Nationalist China), and Sri Lanka.[2] In most cases, about half the value of exports is accounted for by one or two tropical products or other raw materials. These are usually still some combination of the key exports of colonial days—rubber, tea, jute, coconut products, timber, tin, or petroleum.

The Impact of Outsiders

The impact of outsiders on the peoples and cultures of Southeast Asia has been great since early times. The native populations are predominantly

[2]Strictly speaking. Taiwan is in East Asia and Sri Lanka is in South Asia. We discuss these countries along with Southeast Asia for convenience.

Malayan, Tibeto-Burman, and Thai (Figure 16–11). In about A.D. 100 Indian traders started arriving, bringing their own religious ideas (first Hinduism, then Buddhism) and agriculture. They pushed into Southeast Asia in search of minerals and spices. Chinese influence in Southeast Asia began about the same time but was less extensive, concentrated mainly in Vietnam. Many people of Chinese origin moved southward with the spread of the Chinese Empire, and later the empire itself established loose control over part of Southeast Asia and called for tribute. Confucian ideas of government and administration came into the area, as well as Chinese agricultural practices. Another important influence on Southeast Asia was that of Moslem traders from India, who came by sea during the thirteenth, fourteenth, and fifteenth centuries, before the Europeans arrived. The Moslems were most influential in Sumatra, the Malay Peninsula, and the Indonesian islands.

As elsewhere in the Third World, the most far-reaching outside influences have come in the past 500 years of contact with the Europeans. The Europeans came to organize trade and production of tropical products and minerals, not to settle. In the process they established political control over most of the area, built new cities, and attempted to influence the ways of most peoples through the work of their missionaries and public health specialists. Rival European countries established sharp boundaries between parts of the region to protect their interests from each other, and they built separate systems of transport and communication within each country. New rail and water connections focused on the new colonial capital of each colony. Thus Southeast Asia was carved into different functioning territories, each under the control of a separate European country.

The new commercial possibilities offered by the Europeans brought a second influx of Chinese and Indians. Southern Chinese migrated to every part of Southeast Asia, first to set up businesses in the new commercial cities and then into small towns and villages. There they became important in the organization of local production for export. Many Chinese came as single men and married into local societies, but before long Chinese women came as well. Practically every city of Southeast Asia has a Chinese community made up of families who view themselves as permanent residents. Some families have been there 150 years. Most Indians came to Burma and Malaysia where they, too, became either commercial agents and shopkeepers or contract laborers on commercial plantations. Indians were expelled from Burma in the mid-1960s. Malaysia today illustrates the complexity of ethnic distribution in this part of the world. Half the people are Malays, over a third are Chinese, and more than 10 percent are Indians or Pakistanis.

New Business Centers

Singapore, Hong Kong, Taiwan, and to a lesser extent Thailand have found that Asian countries can play a distinctive role in the modern interconnected economy. Singapore and Hong Kong have emerged as *entrepôts* and business centers of Asia outside China and Japan. From the beginning these were European cities, created as centers of trade with this part of the world. Not only are both major ports that have long served as intermediaries for goods entering and leaving other Asian countries, but they have also evolved into major points in the international air network. They have both the business experience and the accessibility required of major business nodes, and they perform much the same function for Asia that London, New York, and Tokyo do for the modern industrial world. Here one finds major locally based banks and businesses—often in the hands of the Chinese community—and also branches of the major international banks and business corporations. The skylines of both cities gleam with modern office buildings.

The economies of Hong Kong, Singapore, and Taiwan now include major industrial firms as well. Like Japan in earlier days, these areas have pools of hard-working, low-paid, semiskilled

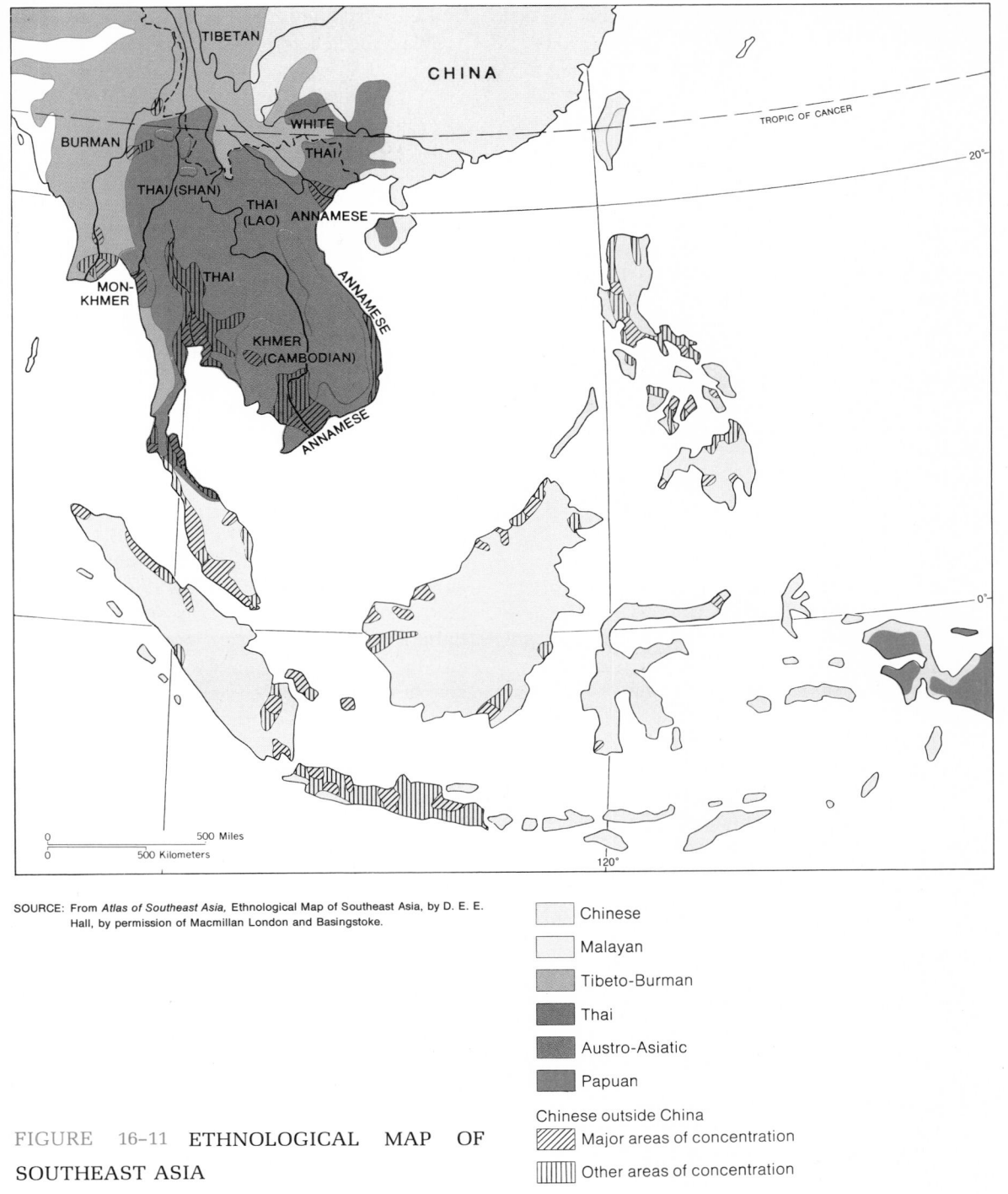

SOURCE: From *Atlas of Southeast Asia*, Ethnological Map of Southeast Asia, by D. E. E. Hall, by permission of Macmillan London and Basingstoke.

Chinese

Malayan

Tibeto-Burman

Thai

Austro-Asiatic

Papuan

Chinese outside China

Major areas of concentration

Other areas of concentration

FIGURE 16–11 ETHNOLOGICAL MAP OF SOUTHEAST ASIA

workers who, with the help of modern machinery, can produce electronics and optical goods, clothing, and other consumer goods more cheaply than the major industrial countries can. In those countries labor costs are much higher. The major multinational companies from the United States, Europe, and Japan have established factories in these Asian countries not to produce goods for native markets, but as export platforms, producing goods to be imported into the United States and other major industrial markets. The labels on electronic goods and moderate and low-priced clothing will reveal that many of these items are made or assembled in Hong Kong, Singapore, or Taiwan. These countries demonstrate that it is possible for Third World centers to be more than raw material producers in the modern interconnected world.

DUAL SOCIETIES TODAY AND PROBLEMS OF DEVELOPMENT

In most countries of South and East Asia, dual economies or societies remain. The modern component of the dual society is based on the production of raw materials from mines or plantations controlled by foreign capital. Workers are paid wages that are good by national standards, but very low by world standards. The government gets much of its revenues from taxes and duties on the exports, even though most producing countries in the Third World have not been as able to control the production or the prices of the materials they produce as have the oil-producing countries.

Export production from mines and plantations means jobs for only a small percentage of the total work force. In the countries of South and East Asia the majority of families remain subsistence farmers; some still farm by primitive slash-and-burn techniques, but most work at intensive farming on their tiny "garden plots." They live precarious lives very much as subsistence farmers do elsewhere. Their lands do not produce an adequate diet, and thus they face malnutrition. This makes them susceptible to disease, which means that they cannot farm at peak efficiency and thus cannot overcome their malnutrition. Densities of population per square mile of cropland average more than 900 in Southeast Asia (350 per square kilometer) and more than 3,600 (1,390 per square kilometer) in Indonesia, whereas in the United States the figure is less than 50 (20 per square kilometer). Densities in the crowded alluvial lowlands of Southeast Asia may reach more than 2,500 per square mile (965 per square kilometer).

The problem may be as much underproduction as overpopulation. With few alternative sources of labor and without modern energy application to agriculture, the labor of South and East Asia remains inefficient. So many family members must work to produce from the tiny family plot that, even with higher returns per acre than in India and Pakistan, there is not enough food to give the family an adequate diet—much less provide a surplus for sale. Recent agricultural improvements have increased output. They include particularly the use of "miracle" rice and wheat—hybrid seeds that can more than double yields from a given field. At least one-sixth of the rice in Southeast Asia consists of the new varieties. The new varieties, however, require high inputs of fertilizer and complete irrigation control. These add to the cost of farming and make it difficult for poor farmers to use the new seeds. Much of the high-yield rice comes from lands owned by relatively prosperous commercial farmers. It has been difficult for the countries of South and East Asia to obtain adequate supplies of fertilizer, particularly since the price of oil-based fertilizers has risen in recent years. The new genetic technology has not yet been applied to millets, oil seeds, and other crops basic to the diets of people in drier areas or to commercial crops such as sugar cane and cotton.

For China, India, and Southeast Asia the Malthusian problem of the race between people and economic production is reaching a critical stage. Much has been written of the need to control population growth in this region. Despite more than a decade of birth control information in India and other parts of South and East Asia, rates of population growth in most countries remain high. But increasing economic productivity is a greater need. The net growth in the economies of most of the countries has been slow. It has been difficult for agriculture following largely traditional farming practices to increase output as rapidly as necessary. Modern farming technology, such as the miracle grains and chemical fertilizers, has been added too slowly. The problem is largely one of human organization—the education of farmers, the generation of farm capital, and the opening of markets. Land reform that has resulted in small farms owned by poor farm workers has only compounded the problem, because only the wealthy have capital to invest in seed and fertilizer and education in modern production methods.

Industrialization has brought progress in modern manufacturing to only a few places such as Taiwan, Hong Kong, South Korea, and Singapore. Although living standards in these places are still low by Western standards, major changes have occurred in the last decade. A problem persists, however, in that much of the capital has been invested by the multinational corporations and international banks, and the region shows few signs of being able to control the management function itself as Japan did in its move into the developed world. Only China, which has largely excluded outside investment, is in complete charge of its own affairs. It remains to be seen if its centrally planned economy can maintain the rapid progress it has made since 1949.

It is appropriate to conclude our examination of the two contrasting human systems with South and East Asia. This part of the world is more than just another example of the juxtaposition of the traditional and modern systems in particular environments in the developing world—it also demonstrates the problems and conflicts involved in the shift from predominantly traditional ways to the modern system. South and East Asia is the key area for the future of human life on this planet. Here are more than half the people of the world, many of them struggling with the process of change from old ways to new. Major questions for the next few decades concern the degree of success of that changeover. Will it be peaceful? What will be the form of the new Asian lifestyle? What will be the impact of a modern Asia on life elsewhere in the world? Can the earth environment support mass consumption, and perhaps mass pollution, by an additional 2 billion or more people? Will production be able to outpace population growth in this area?

Undoubtedly, change will come in this stronghold of the traditional system. China, India, and other governments are committed to taking their places among the modern countries of the world. The question is how. India and the smaller countries of the area have been following closely the blueprint for industrialization that has come from Europe, the United States, and the Soviet Union, while at the same time trying to maintain traditional values as much as possible. The modern system as it emerges in these countries must pass through the filter of each culture, being modified by the people's perceptions of the modern system and fine-tuned by their value systems.

China has tried to go one step further, to bend the framework of the modern system to fit traditional Chinese ways. Under Mao Tse-tung industrialization was made to fit into China's predominantly rural, family-centered lifestyle. Manufacturing, medical services, education, and various functions that in other modern systems have been focused in large metropolitan areas were dispersed in the rural communes. Specialists from the cities were sent to the farms, rather than farm workers migrating to city jobs. Mao is now dead, and there seems to be a major controversy among China's leaders concerning the future course toward modernization. Strong voices call for Western-style, urban-centered industrial development as the most rapid way to give China parity with other great powers.

South and East Asia exemplifies the transitional situation of the human race today. We stand, indeed, between two worlds. The transition to the modern world is almost complete in Anglo-America, Europe, and the Soviet Union. Japan, too, is a predominantly modern country. The footholds of the modern system in the rest of the world are still relatively small. With the huge population of South and East Asia still largely in the traditional system, the human race is a long way from being completely integrated into the modern interconnected system.

SELECTED REFERENCES

Adams, John, and Walter C. Neale. *India: The Search for Unity, Democracy, and Progress.* New Searchlight Series, D. Van Nostrand, New York, 1976 (paper). Two economists view India's society, political system, economic system and economic development, and foreign policy. The book contains nine worthwhile maps, especially those on population, crop regions, and major mineral resources.

Buchanan, Keith. *The Transformation of the Chinese Earth.* Praeger, New York, 1970. An interpretative overview of China today as an extension of its long past interaction of people and place. Chapters 8 through 15 are most appropriate for those interested primarily in China today. They deal with agriculture, industry, population, and an assessment of the achievements under the Communist regime.

Fryer, Donald W. *Emerging Southeast Asia: A Study in Growth and Stagnation.* McGraw-Hill, New York, 1970. A study of the countries east of India and south of China. The focus is on current economic development and the basic character of the economies of individual countries. Part I deals with land resources, industrialization, and modernization for the region as a whole. Part II studies Thailand, the Philippines, Malaysia, and Singapore as the more progressive economies. The remaining countries are taken up in Part III, which discusses their economic stagnation. Part IV discusses the economic prospects for the region. The book is a useful supplement for the material in this chapter.

Mellor, John W. "The Agriculture of India," *Scientific American,* September 1976, pp. 154–163. Reviews the present state of agriculture in India and discusses the basic problems facing the development needed to meet rising food demands.

Myrdal, Gunnar. *Asian Drama: An Inquiry Into the Poverty of Nations,* abridged by Seth S. King. Pantheon, New York, 1971. The prospects for economic development in Asia examined by a well-known Swedish economist. Myrdal selects and evaluates important circumstances that have produced the current trends of development in South and East Asia, and argues an interpretation of conditions shaping future economic progress in Asia.

Petrov, Victor P. *China: Emerging World Power,* 2nd ed. New Searchlight Series, D. Van Nostrand, New York, 1976 (paper). A short but useful overview of the current geography of China. The current state of agricultural development, mineral resource reserves, and industrial capacity are emphasized. Comparative figures on Chinese, Soviet, and U.S. production offer an effective measure of the economic power of China today.

Spate, O. H. K., and A. T. A. Learmonth. *India and Pakistan: Land, People, and Economy.* Methuen, London, 1967. (Paperback, University Paperback, Harper & Row, 1972.) Uses a topical organization and is most appropriate as a supplement to the discussion of this chapter. Chapters on population, agriculture, industrial activity, and transportation are particularly useful. A series of statistical tables provides a fund of data on the topics included. Though somewhat dated now, this is among the best available texts on India and Pakistan in English.

Spencer, J. E., and William L. Thomas. *Asia, East by South: A Cultural Geography,* 2nd ed. John Wiley, New York, 1971. A basic text that deals with South and East Asia from Pakistan to Japan and includes the insular areas of Indonesia and the Philippines. Part I is a systematic treatment for the whole area on such topics as population, language, religion, health and disease, and the physical environmental base. Part II looks at the cultural geographic development of individual countries. The historical-cultural emphasis offers a different context from that presented in this chapter for seeing the current economic development of countries of South and East Asia.

Wortman, Sterling. "Agriculture in China," *Scientific American,* June 1975, pp. 13–21. A review of the current state of Chinese agriculture, with special emphasis on how China has achieved virtual self-sufficiency in agriculture in this decade. The article evaluates the prospects for maintaining self-sufficiency under the demands of population growth.

Appendix

TABLE A-1 POPULATION, ENERGY, AND GNP DATA FOR COUNTRIES

Country	Population estimate, mid-1977 (millions)	Population projection to 2000 (millions)	Urban population (percent)	Dietary energy supply, 1974 (kilocalories per person per day)	Per capita gross national product, 1975 ($ U.S.)	Per capita commercial energy consumption, 1974–75 (kilograms of coal equivalent)
Afghanistan	20.0	36.1	15	1,970	130	67
Albania	2.5	4.1	34	2,390	600	725
Algeria	17.8	36.5	52	1,730	780	531
Angola	6.3	11.7	18	2,000	680	191
Argentina	26.1	32.9	80	3,060	1,590	1,861
Australia	13.9	19.6	86	3,280	5,640	5,997
Austria	7.5	8.1	52	3,310	4,720	3,883
Bahamas	0.2	0.3	58	—	2,600	7,985
Bahrain	0.3	0.6	78	—	2,440	11,819
Bangladesh	83.3	154.9	9	1,840	110	31
Barbados	0.2	0.3	44	—	1,260	1,175
Belgium	9.9	10.7	87	3,380	6,070	6,709
Benin (Dahomey)	3.3	6.0	14	2,260	140	42
Bhutan	1.2	2.1	3	—	70	—
Bolivia	4.8	8.7	34	1,900	320	283
Botswana	0.7	1.4	12	2,040	330	—a
Brazil	112.0	205.0	59	2,620	1,010	652
Bulgaria	8.8	9.9	59	3,290	2,040	4,195
Burma	31.8	53.3	22	2,210	110	56
Burundi	3.9	7.2	2	2,040	100	13
Cameroon, United Republic of	6.7	11.6	20	2,410	270	86
Canada	23.5	31.6	76	3,180	6,650	9,816
Cape Verde	0.3	0.4	8	—	470	90

TABLE A-1 *(Cont.)*

Country	Population estimate, mid-1977 (millions)	Population projection to 2000 (millions)	Urban population (percent)	Dietary energy supply, 1974 (kilocalories per person per day)	Per capita gross national product, 1975 ($ U.S.)	Per capita commercial energy consumption, 1974–75 (kilograms of coal equivalent)
Central African Empire	1.9	3.3	36	2,200	230	57
Chad	4.2	6.9	14	2,110	120	17
Chile	11.0	15.8	76	2,670	760	1,361
China, People's Republic of	850	1,126.0	24	2,170	350	650
Colombia	25.2	47.1	64	2,200	550	636
Comoros	0.3	0.5	10	—	260	47
Congo, People's Republic of	1.4	2.7	40	2,260	500	216
Costa Rica	2.1	3.6	41	2,610	910	491
Cuba	9.6	14.9	60	2,700	800	1,178
Cyprus	0.6	0.8	42	2,670	1,180	1,419
Czechoslovakia	15.0	16.9	67	3,180	3,710	6,826
Democratic Kampuchea (Cambodia)	8.0	14.7	12	2,430	70	17
Denmark	5.1	5.4	80	3,240	6,920	5,114
Dominican Republic	5.0	10.7	44	2,120	720	433
East Timor	0.7	1.2	11	—	150	16
Ecuador	7.5	14.7	41	2,010	550	363
Egypt	38.9	63.9	45	2,500	310	322
El Salvador	4.3	8.6	39	1,930	450	248
Equatorial Guinea	0.3	0.5	45	—	320	87
Ethiopia	29.4	53.8	12	2,160	100	31

Country	Population estimate, mid-1977 (millions)	Population projection to 2000 (millions)	Urban population (percent)	Dietary energy supply, 1974 (kilocalories per person per day)	Per capita gross national product, 1975 ($ U.S.)	Per capita commercial energy consumption, 1974–75 (kilograms of coal equivalent)
Fiji	0.6	0.8	38	—	920	487
Finland	4.8	4.8	58	3,050	5,100	4,636
France	53.4	61.7	70	3,210	5,760	4,342
Gabon	0.5	0.7	32	2,220	2,240	1,070
Gambia	0.6	0.9	14	2,490	190	73
Gaza	0.4	0.9	87	—	—	—
German Democratic Republic	16.7	17.7	75	3,290	4,230	6,946
Germany, Federal Republic of	61.2	65.5	88	3,220	6,610	5,698
Ghana	10.4	21.1	31	2,320	460	184
Greece	9.1	9.8	65	3,190	2,360	2,048
Grenada	0.1	0.1	15	—	370	340
Guadeloupe	0.3	0.5	48	—	1,240	550
Guatemala	6.4	12.2	34	2,130	650	252
Guinea	4.7	8.5	20	2,200	130	94
Guinea-Bissau	0.5	0.8	23	—	390	41
Guyana	0.8	1.3	40	2,390	560	931
Haiti	5.3	7.9	20	1,730	180	31
Honduras	3.3	6.9	31	2,140	350	224
Hong Kong	4.5	5.9	92	—	1,720	1,232
Hungary	10.7	11.1	50	3,280	2,480	3,557
Iceland	0.2	0.3	86	—	5,620	5,138
India	622.7	1,023.7	21	2,070	150	201
Indonesia	136.9	226.9	18	1,790	180	158
Iran	34.8	66.1	44	2,300	1,440	1,268
Iraq	11.8	24.3	64	2,160	1,280	906
Ireland	3.2	4.0	52	3,410	2,420	3,296
Israel	3.6	5.5	82	2,960	3,580	2,914
Italy	56.5	61.9	53	3,180	2,940	3,224
Ivory Coast	7.0	13.2	20	2,430	500	370
Jamaica	2.1	2.8	37	2,360	1,290	1,439
Japan	114.2	133.4	72	2,510	4,460	3,839
Jordan	2.9	5.9	42	2,430	460	388
Kenya	14.4	31.5	10	2,360	220	179
Korea, Democratic People's Republic of	16.7	27.5	43	2,240	430	2,721
Korea, Republic of	35.9	52.9	48	2,520	550	961
Kuwait	1.1	2.9	56	—	11,510	9,913

TABLE A-1 (Cont.)

Country	Population estimate, mid-1977 (millions)	Population projection to 2000 (millions)	Urban population (percent)	Dietary energy supply, 1974 (kilocalories per person per day)	Per capita gross national product, 1975 ($ U.S.)	Per capita commercial energy consumption, 1974–75 (kilograms of coal equivalent)
Lao People's Republic (Laos)	3.5	5.8	15	2,110	70	65
Lebanon	2.8	5.6	60	2,280	1,070	1,073
Lesotho	1.1	1.8	3	—	180	—[a]
Liberia	1.7	3.0	28	2,170	410	432
Libya	2.7	5.2	30	2,570	5,080	975
Luxembourg	0.4	0.4	69	3,380	6,050	19,539
Macao	0.3	0.4	97	—	310	263
Madagascar	7.9	16.4	14	2,530	200	71
Malawi	5.3	9.9	10	2,210	150	56
Malaysia	12.6	21.6	27	2,460	720	566
Maldives	0.1	0.2	11	—	100	—
Mali	5.9	11.0	13	2,060	90	24
Malta	0.3	0.3	94	2,820	1,220	1,156
Martinique	0.4	0.5	50	—	1,540	1,077
Mauritania	1.4	2.4	23	1,970	310	112
Mauritius	0.9	1.2	44	2,360	580	275
Mexico	64.4	134.6	62	2,580	1,190	1,269
Mongolia	1.5	2.7	46	2,380	700	3,046
Morocco	18.3	35.5	38	2,220	470	257
Mozambique	9.5	17.4	6	2,050	310	141
Namibia	0.9	1.7	32	—	800	—[a]
Nepal	13.2	23.2	4	2,080	110	12
Netherlands	13.9	16.1	77	3,320	5,590	6,191
Netherlands Antilles	0.2	0.4	48	—	1,590	14,284
New Zealand	3.2	4.4	81	3,200	4,680	3,444
Nicaragua	2.3	4.8	49	2,450	720	453
Niger	4.9	9.6	9	2,080	130	31
Nigeria	66.6	134.9	18	2,270	310	94
Norway	4.0	4.5	45	2,960	6,540	4,926
Oman	0.8	1.6	5	—	2,070	250
Pakistan	74.5	145.5	26	2,160	140	192
Panama	1.8	3.2	50	2,580	1,060	846
Papua New Guinea	2.9	5.1	11	—	450	250
Paraguay	2.8	5.3	37	2,740	570	173
Peru	16.6	31.2	55	2,320	810	650
Philippines	44.3	83.7	32	1,940	370	309
Poland	34.7	40.2	54	3,280	2,910	4,687
Portugal	9.2	10.4	26	2,900	1,610	1,026

TABLE A-1 (Cont.)

Country	Population estimate, mid-1977 (millions)	Population projection to 2000 (millions)	Urban population (percent)	Dietary energy supply, 1974 (kilocalories per person per day)	Per capita gross national product, 1975 ($ U.S.)	Per capita commercial energy consumption, 1974–75 (kilograms of coal equivalent)
Puerto Rico	3.2	4.1	58	—	2,300	4,280
Qatar	0.1	0.2	69	—	8,320	18,423
Réunion	0.5	0.7	51	—	1,550	484
Rhodesia	6.8	15.2	20	2,660	540	805
Rumania	21.7	25.9	43	3,140	1,300	3,543
Rwanda	4.5	8.7	3	1,960	90	13
Samoa, Western	0.2	0.3	21	—	320	126
São Tome and Principe	0.1	0.1	16	—	570	104
Saudi Arabia	7.6	14.8	21	2,270	3,010	976
Senegal	5.3	9.3	32	2,370	370	184
Seychelles	0.1	0.1	26	—	520	317
Sierra Leone	3.2	5.8	15	2,280	200	123
Singapore	2.3	3.1	100	—	2,510	2,060
Solomon Islands	0.2	0.4	9	—	310	223
Somalia	3.4	6.5	28	1,830	100	40
South Africa	26.1	51.2	48	2,740	1,320	2,754
Soviet Union	259	314	60	3,280	2,620	5,252
Spain	36.5	45.4	61	2,600	2,700	2,063
Sri Lanka	14.1	20.7	22	2,170	150	140
Sudan	16.3	32.7	13	2,160	290	125
Surinam	0.4	0.9	50	2,450	1,180	2,834
Swaziland	0.5	1.0	8	—	470	—[a]
Sweden	8.2	9.2	81	2,810	7,880	5,804
Switzerland	6.2	6.9	55	3,190	8,050	3,616
Syria	7.8	16.0	46	2,650	660	590
Taiwan (Republic of China)	16.6	22.0	63	—	890	—
Tanzania, United Republic of	16.0	33.1	7	2,260	170	75
Thailand	44.4	84.6	13	2,560	350	300
Togo	2.3	4.6	15	2,330	270	70
Trinidad and Tobago	1.0	1.3	12	2,380	1,900	3,885
Tunisia	6.0	10.8	47	2,250	760	416
Turkey	41.9	72.4	43	3,250	860	628
Uganda	12.4	24.7	7	2,130	250	51
United Arab Emirates	0.2	0.5	52	—	10,480	13,503
United Kingdom	56.0	61.9	78	3,190	3,840	5,464
United States	216.7	262.5	74	3,330	7,060	11,485
Upper Volta	6.4	11.0	4	1,710	90	14

TABLE A-1 (*Cont.*)

Country	Population estimate, mid-1977 (millions)	Population projection to 2000 (millions)	Urban population (percent)	Dietary energy supply, 1974 (kilocalories per person per day)	Per capita gross national product, 1975 ($ U.S.)	Per capita commercial energy consumption, 1974–75 (kilograms of coal equivalent)
Uruguay	2.8	3.4	81	2,880	1,330	900
Venezuela	12.7	23.2	74	2,430	2,220	2,895
Vietnam, Socialist Republic of	47.3	78.9	22	2,340	160	194
Yemen	5.6	11.0	9	2,040	210	30
Yemen, Democratic	1.8	3.5	33	2,070	240	360
Yugoslavia	21.8	25.7	39	3,190	1,480	1,883
Zaire	26.3	50.5	26	2,060	150	76
Zambia	5.2	11.3	36	2,590	540	557

SOURCE: *World Population Data Sheet 1977* for population and GNP data; *World Population Data Sheet 1975* for dietary energy data; *World Energy Supplies, 1950–1974*, U. N. Statistical Papers, Series J. No. 19, Table 2, for commercial energy consumption data.
[a] Included in data for South Africa.

TABLE A-2 CONVERSION TABLE

English to metric		Metric to English	
1 inch	= 2.54 centimeters	1 centimeter	= 0.39 inches
1 mile	= 1.61 kilometers	1 kilometer	= 0.62 miles
1 acre	= 0.40 hectares	1 hectare	= 2.47 acres
1 mile^2	= 2.6 kilometers^2	1 kilometer^2	= 0.38 mile^2
1 mile^3	= 4.19 kilometers^3	1 kilometer^3	= 0.24 mile^3
1 ounce	= 28.3 grams	1 gram	= 0.04 ounces
1 pound	= 0.45 kilogram	1 kilogram	= 2.20 pounds
1 ton	= 0.9 metric ton	1 metric ton	= 1.1 tons
1 quart	= 0.95 liter	1 liter	= 1.06 quarts

Energy Units

1 barrel of crude oil	= 43 gallons		
7 barrels of crude oil	= 1 metric ton	=	40 million Btu's
1 metric ton of coal	= 28 million Btu's		
1 Btu	= 252 calories		

Glossary

accessibility A measure of the ease of access to resources, ideas, and people of other places. Traditional places have a low accessibility, whereas modern places with a high degree of connectivity have a high accessibility.

actual evapotranspiration (AE) The amount of water actually lost by evapotranspiration into the atmosphere. Desert surfaces, for example, have low AE even though PE (potential evapotranspiration) may be high.

administrative center An urban area in which employment in government management and decision making dominates other sectors of employment.

agglomeration economies Savings resulting from reduced transportation costs in a mass market of a metropolitan area. Because of the high population densities of such compact geographic areas, markets for goods are large and economy of scale becomes important.

agrarian reform Reallocation or redistribution of land in a country for agricultural purposes. Generally, this means breaking up large landholdings into smaller plots.

alluvial soils Soils developed from recently deposited sediment in floodplains and deltas. They are generally deep and flat, and hence attractive for agriculture unless drainage is poor.

Altiplano A string of high intermontane basins in Bolivia. Lake Titicaca occupies the northern most of these basins at an elevation of 12,507 ft (3,752 m) above sea level.

apartheid The official South African governmental policy of separation of the races, whites from blacks and coloreds (racial mixtures as well as other races such as Orientals).

arithmetic progression A sequence of numbers (terms) that increases by a constant added to each, such as 1, 3, 5, 7, 9, . . .

atmosphere The gaseous envelope above the earth's surface. The principal gases are (by volume): nitrogen (78.08 percent), oxygen (20.95 percent), argon (0.93 percent), and carbon dioxide (0.03 percent). Small amounts of other gases are constant except water vapor, which varies from 0 to 4 percent. The lower part of the atmosphere, where most of the "weather" occurs, is a layer called the troposphere. The second, higher layer is the stratosphere. The gases are held close to the earth by gravity.

atmospheric forces Dynamic aspects of the solar energy system that result in pressure, wind, temperature, and moisture variabilities from place to place and season to season. The net result on the landforms is a gradual change through weathering and erosion of the surfaces.

balance of trade The difference in value between the imports and exports of a country. A favorable trade balance occurs when the value of exports is greater than the value of imports.

biodegradable Decomposable by bacterial action; a term applied to waste materials.

biological forces The activities of living organisms interacting with the atmosphere, water bodies, and landforms to change the soil and surface conditions over time. An example is the effect of plant cover in retarding surface erosion due to heavy rainfall and runoff.

biosphere The total of the earth's living things, from the smallest forms of life to the largest. It is commonly broken into three zones: salt water, fresh water, and land.

Bedouins Nomadic herders of the arid and mountainous areas of the Middle East and North Africa.

birth rate The number of births per unit of time for a given country or geographical area divided by the average population during that time. For example, the U.S. birth rate in 1973 was 15 per 1,000 population. This is often expressed as a percentage—in this case, 1.5 percent. (Rate of population growth is the difference between birth rate and death rate.)

break-in-bulk point The place where goods in shipment are transferred from one type of transportation to another, usually from a low-cost water mode to rail or truck transport.

buffer zone An independent area situated between two powerful countries and serving as a deterrent to conflict. Thailand was a buffer state between French and British colonies in Southeast Asia.

bush fallow A slash-and-burn clearing that is unused and allowed to return to its natural vegetation for a few years before being recleared.

Canadian Shield (Laurentian Shield) A large area of exposed metamorphosed Precambrian rocks centered on Hudson Bay. Glaciers stripped the shield of its soil cover and created a landscape of exposed rock, lakes, and bogs. Where soil remains, extensive forests have developed.

capital Supply of wealth created by people. It can be in many forms ranging from machinery to clothing to money.

cash crop A crop grown for sale and consumption by others.

Caucasian One of the groups of humans displaying certain identifiable traits such as an aquiline nose and light features. Many Caucasians, however, have dark skins, as for example, the people of India.

centralization Complete control of a national economy by the central government, as in the Soviet Union.

clay content Amount of soil in a given sample that is in the form of small particles (less than 2 micrometers in diameter). Clay affects movement of water through the soil and also chemical activity and thus weathering. The higher the clay content, the more impervious the soil is to water movement. Clay attracts and stores plant nutrients in the form of positively charged ions.

collective An agrarian landholding by a group, with rights to the land remaining with the government. These farms are managed by leaders elected from among the collective members.

collectivization Reorganization of agricultural lands into large collective farms, as in the Soviet Union. During Soviet collectivization, all private estates and peasant farms were eliminated, except for small private plots in the villages.

colonialism System whereby colonies are held by foreign countries for the purpose of economic exploitation. A superimposition of foreign institutions and cultural values on the native peoples in the colonies is a by-product of this system.

commune A small group of people living together and sharing in work, culture, and social activities. The basic organizational unit of China, where it is recognized as the local governmental unit.

complementary region A region that compensates for deficiencies in another region. Some of the smaller European countries are complementary regions to the major industrial countries and serve as markets and also as suppliers of goods for interregional trade.

concession A privilege granted by a government to a company to use land for a specific purpose, for instance, drilling for oil.

confederation The uniting of separate colonies, states, or nations into a single political entity. Canada, for example, was formed in 1867 through the confederation of Ontario, Quebec, Nova Scotia, and New Brunswick.

connectivity Transportation and communication links between places or between a place and the rest of the world. A high degree of connectivity is necessary for the modern interconnected system to function.

consortium An association or alliance of two or more countries for a common purpose. The Organization of Petroleum Exporting Countries (OPEC) is a consortium of oil producers.

consumer market A market for consumer goods produced to satisfy people's needs rather than to be used for producing other goods. Goods produced for a consumer market include such items as food, clothing, appliances, and automobiles.

consumption The use of a product. The product may be used up immediately, such as a crop, or its use may extend over a long period of time, such as a building.

cooperative A collectively owned organization for the purchase and marketing of goods. Agricultural cooperatives are the most common.

core region Area of concentration of nodes of population centers and economic activity. The core region in the United States consists of the Middle West and the Northeast.

cottage industries Industries in which subcontractors employ a small number of workers, usually with low pay and long hours, to make component parts for larger industries. Most cottage industries specialize in manufacturing one or two parts.

craft industry Making of goods by hand; handicrafts. This type of manufacturing preceded the Industrial Revolution and remains in many of the traditional areas of the world.

crop rotation The growing of two or more crops in succession on the same field to improve soil fertility. Some crops add nitrogen to the soil, and others remove it.

cultural dualism Existence of two cultures side by side in the same society, for instance, Western influences and traditional Japanese culture in Japan.

cultural homogeneity Existence of the same characteristics throughout a culture region. For instance, in Japan, because of isolation and protective laws, there was little contact with foreign culture until late in the nineteenth century.

cultural pluralism Existence side by side of several cultures having different technologies and traditions.

culture The total of the learned behavior of a human group, including its language, religion, politics, and technology.

culture group The people who comprise a particular culture in a particular place at a particular time.

culture hearth The area where a distinctive culture originated and developed.

culture realm A large part of the world that has developed broad but distinctive cultural patterns that differ from other parts of the world, for example, Africa versus Europe.

death rate The number of deaths per unit of time for a given country or geographical area divided by the average population during that time. For example, the U.S. death rate in 1973 was 9.4 per 1,000 population, or 0.94 percent. (Rate of population growth is the difference between birth rate and death rate.)

delta Alluvial deposits at the mouth of a river where its capacity to carry sediment is lowered as it enters the sea or ocean. The term comes from the Nile River, whose mouth is shaped like the Greek letter Δ.

demographic transition Changes in birth rates and death rates that normally parallel the development of a country from traditional to modern. Both birth rate and death rate move in stages from a high to a low level.

demography The study of the characteristics of human populations—distributions, densities, and changes occurring over time.

desalinization Removal of salt from seawater.

detritus Residue of dead vegetation; also, an accumulation of rock fragments from mechanical weathering.

diastrophic forces Internal processes of the earth that cause deformation in the crust. Such deformations are manifested on the landscape as faults, folds, continents, and ocean basins.

dormant season A period when little or no environmental work occurs in a seasonal-work environment. It may occur as a result of dryness or low temperature.

drainage Loss of water in response to gravity. The drainage system of an area includes surface streams and ground-water discharge. Poor drainage may result in waterlogged soil, bogs, and swamps.

dry farming Cultivation and mulching of fields so that soil-moisture loss from evapotranspiration is reduced. It is practiced in some semiarid areas of the world.

dual economies In developing countries, the existence of two separate economic systems side by side. The dualism often includes cultural and political factors as well. An example is the urban areas of modern economic activity in nations that are mostly occupied by native peoples following traditional lifestyles.

earth environment The natural environment of the earth, which includes the crust and its resources as well as the atmosphere; the sum total of conditions where people must live.

economies of scale Savings resulting from lower cost per unit in large operations.

ejido Communally owned lands in the Mexican land-tenure system established by the Constitution of 1917. *Ejido* land can be farmed privately or communally, with each member doing a share of work and receiving a share of the profits.

elevation Altitude of a point above mean sea level, or above some other reference plane.

encomienda In the colonial Spanish agricultural system in the Americas, a large tract of land that was given to a landowner along with a number of Indians to work it. In turn, the landowner was to teach the Indians the Roman Catholic faith and offer them protection and help.

erosion The reshaping of the land surface by the agents of moving ice, running water, wind, and mass movement.

escarpment A line of steep slopes or cliffs forming the edge of a surface feature such as a plateau or the exposed edge of a resistant rock formation.

evaporation Conversion of liquid water into a vapor (gaseous) form by heat; also called vaporization.

evapotranspiration Combination of water evaporated from land and water surfaces and transpired from vegetation. Together, these processes account for the water lost from the earth's surface to the atmosphere.

export platform A developing country in which factories are built to manufacture products for export to the developed countries. In South Korea, for instance, clothing is made for the U.S. market.

extended family A closely knit social unit having the characteristics of a family but including a large number of related persons; in contrast to a family consisting of one pair of parents and their offspring.

feedback Mutual interaction between the human and natural systems, whereby people modify the environment and the environment in turn modifies people's ideas.

Fertile Crescent An area of irrigated land in the Middle East stretching in an arc from the valleys of the Tigris and Euphrates rivers to the Nile Valley of Egypt, including much of Syria, Lebanon, and Israel.

fertilization Enrichment of the soil to supply it with needed nutrients or organic matter for continued crop production.

feudal system (feudalism) The medieval European economic, political, and social system of serfs, vassals, and overlords. Serfs worked the land owned by vassals, who exchanged military service to overlords for the land.

first-order trade Trade between industrialized areas of the world.

floodplain Land area along a stream subject to flooding.

foreign exchange Transfer of credit to another country for payment of accounts; currency or bills of credit in a foreign country for purchases in that country.

fossil fuels Hydrocarbon deposits found in sedimentary rock formations that can be burned or otherwise used for fuel.

geometric progression A sequence of numbers (terms) each of which has a constant ratio to the preceding number, such as 2, 4, 8, 16, 32, . . .

Green Revolution Recent efforts to rapidly improve agricultural yields through education of farmers, improved seeds, and use of artificial fertilizers in areas mostly outside the modern world. The development of higher-yielding seeds for wheat and particularly rice in the 1960s gave rise to this designation.

gross national product (GNP) The total production of goods and services in a country in a given period of time, usually a year. The per capita GNP is often used as a yardstick in comparing productivity of countries and thereby evaluating the living standard of their people.

ground water Water stored in underground reservoirs or conduits (aquifers) of porous rock. Sandstone is normally an excellent aquifer, whereas shale is impervious to the flow of water.

hacienda (fazenda in Brazil) A large estate, virtually a self-contained community with a chapel, store, and living quarters. Haciendas were granted by the Spanish and Portuguese to officers, soldiers, and others for service to the Crown. Workers on the hacienda raised staple crops and livestock for the wealthy landowners, but seldom rose above the subsistence level themselves.

hierarchy of places Relationship of places according to their importance in a functioning system. For example, villages, towns, small cities, and metropolitan areas form a hierarchy.

high-work environment An environment where there is a high rate of biological and chemical activity throughout the year. These environments are typically regions with a hot, moist climate.

hill Terrain of moderate local relief lacking in extensive gentle slopes.

hinterland Area surrounding and serving an urban center as a source of labor resources and as a market for goods and services.

hybridization Genetic modification (cross breeding) of grains such as corn, rice, and wheat that increases yields per acre under good conditions of soil, climate, and fertilizer application.

hydrologic cycle A constant cycle of evapotranspiration of water from the earth's surface, condensation of the vapor into clouds and fog by cooling, and finally the return of the water to the surface by precipitation. The last results in overland runoff, replenishing the water in lakes, streams, underground storage, and oceans.

hydrosphere The portion of the earth covered by water, including oceans, seas, lakes, rivers, and underground water.

hydrospheric forces Actions of the earth's water systems, which transfer heat in association with the atmosphere.

igneous rock Rock formed from the cooling and solidification of molten material. It is largely crystalline when cooled slowly, as in granite, or it may erupt as lava.

inanimate energy Energy derived from sources other than humans or animals. Fossil fuels, electricity, and atomic reactors are inanimate energy sources.

industrial center An urban area in which manufacturing employment dominates other sectors of employment.

Industrial Revolution A development, beginning in the eighteenth century in Great Britain and continuing into the nineteenth, that involved harnessing inanimate energy through the use of machines to produce manufactured goods. The Industrial Revolution spread to other European countries and the United States quite rapidly, changing the social and economic structure of these societies.

infrastructure The framework of structures and facilities supporting the complex, modern interconnected system. This involves the systems of communication, transportation, buildings, and services.

intensive farming Application of large amounts of labor, fertilizer, and capital to small parcels of land to gain high crop yields.

intermediary An organization or institution that serves as an agent between persons or things. The government is an intermediary in that it collects taxes and distributes services.

interregional system A linkage of trade, transportation, and communication between regions. An interregional system developed in the United States as a stage in the emergence of the modern interconnected system.

interregional trade Exchange of goods between regions of a country, for instance, between the Middle West and the Northeast of the United States.

irrigation Addition of water to soils for agricultural use.

Latino In Latin America, a person descendant from Iberian immigrants; may have Indian or African parentage as well.

leaching The removal of chemical constituents in the soil by solution (dissolving) and percolation of water from the surface downward.

lithosphere The solid crust of the earth, which varies in thickness and underlies both the continents and the oceans.

lodging A condition in plants that causes the stem to grow too tall and fall over.

low-work environment A dry or cold climate where biological growth, chemical change, leaching, and erosion are confined to short periods of time.

macronutrient Any of the seven elements needed by all plants in large amounts for growth: oxygen, hydrogen, carbon, nitrogen, calcium, potassium, and phosphorus.

Malthusian checks Stabilizing factors in the geometric progression of population growth, such as wars, famine, and illness resulting from competition for scarce food supplies in the local environment.

Malthusian principle A doctrine of the English economist T. R. Malthus (1766–1834) that population increases geometrically while primary production increases arithmetically. According to this principle, population would outstrip food supply if unchecked by war, famine, or natural disasters.

management The control and direction of others in the organization and production of resources and their distribution.

mangrove A number of species of trees that grow in coastal mudflats in the tropics. They create swamps and make river deltas and coasts difficult to penetrate.

marginal areas Areas that are relatively unsuccessful in the modern interconnected system, for instance, Appalachia in the United States.

maritime climate A moderate climate influenced by location close to the sea or ocean and in line with prevailing winds. This maritime influence tends to make such places milder in winter and cooler in summer than landlocked places at the same latitudes.

market-oriented Refers to an industry that benefits from being located near its market, for example, the electronics and meat-packing industries.

mass consumption The large-scale purchase and use of goods and services. Assembly-line production and easy credit have increased the demand for goods and services and thus encouraged mass consumption.

mass production Production of an item in large numbers, usually manufactured on an assembly-line.

massif (massive) A general term for a compact mountainous region (as opposed to an extended mountain range) or plateau, such as the Massif Central of France or the Deccan Plateau of India (sometimes called the Indian massive).

Mediterranean agriculture A combination of dry farming of grains, irrigated farming, and specialty tree crops such as olives. The peculiar nature of the Mediterranean climate requires special techniques because the rainfall comes in the cool season and the summers are hot and dry. Most crops require both moisture and heat in the growing season.

Mediterranean climate A climate noted for dry summers and winter rainfall; characteristically mild.

mercantilism An economic system used by the European colonial powers, which protected and strengthened home industries by government tariffs, increased foreign trade, monopolies, and a favorable balance of trade.

mestizo In Latin America, a person of both white and Indian parentage.

metallic minerals Minerals containing metals; more specifically, ore, a profitably mined mineral.

metamorphic rock Rock created by heat and pressure or chemical reaction upon other metamorphic, igneous, or sedimentary rocks.

metropolitan area A large central city and its surrounding suburbs and smaller satellite cities.

metropolitan population Population living in a metropolitan area, including the city and its surrounding suburbs.

micronutrients Elements required in small amounts by plants for growth, such as iron.

mind-set A preconceived notion in a culture that may hinder its development and that of its environment; a "cultural blinder."

model In the geographic sense, a translation of space that can be structured, as in an equation or a hypothesis.

modern, long-range, interconnected system A worldwide network tapping widely scattered resources and ideas through highly developed transportation and communication systems; the result of highly developed technology. Manufactured products, as well as decisions and ideas, are disseminated to the culture groups worldwide that have the knowledge or need to use them and the resources or capital to purchase them. The modern interconnected system has a worldwide environmental base and is not limited to the local environment.

money economy An economy that uses money as a basis for barter, trade, and wealth measurable in terms of gross national product (GNP). Developing nations, lacking a well-developed money economy, may be based upon a subsistence economy, where very little surplus is produced and therefore little money is exchanged.

Mongoloid Member of a group of humans displaying certain identifiable characteristics such as straight black hair, slanted eyes, broad face, and brownish-yellow skin; includes Chinese, Koreans, Japanese, and peoples of Southeast Asia.

monsoon Seasonal shift in wind direction. In Asia, particularly India, there is a "monsoon season," which normally means the wet season; however, there is a dry monsoon season during the winter. A daily monsoon occurs along coasts, where there are sea breezes during the day and a reversal to land breezes at night due to differential heating and cooling of land with respect to the ocean.

mountain Terrain of high local relief and many steep slopes. The distance between hills and mountains is relative. A reasonable, though arbitrary, minimum of 1,000 ft (approximately 300 m) of local relief is often used to define mountainous terrain.

mulatto A person of both black and white parentage.

multinational corporation A business organization with interests in several different countries, usually managed from headquarters in a large metropolitan area; made possible by advanced transportation and communication technology.

nation-state A political entity that exercises sovereign power over its people and territory. It is an extension of the idea of the culture group into the modern era. Loyalty of the people within the boundaries of the territory are expected by the nation-state.

nested hierarchy A hierarchy of places in which there are simultaneous multiple interconnections, for example, trade and communications, at all levels.

nesting In the context of market areas, a situation in which smaller market areas lie entirely within larger ones; for instance, the market area of a village within the market area of a town; that of a town within that of a small city, and so on.

network The interconnectivity of functional nodal areas or core regions through communication and transportation services.

nodal Refers to a core area or node; a city or metropolitan area is nodal in character in relation to its surrounding area.

node Central point or nucleus of a network or system; for instance, a metropolitan area in the functioning of a national system, or the inner city in the functioning of a metropolitan area.

nonmetropolitan population The people who live outside metropolitan areas.

nutrient Nourishment or food required by all living organisms to build tissues and carry on physiological processes.

oasis A desert area where water is available for settlement and growth of vegetation. Springs are usually the source of water.

oil shale A fine-grained sedimentary rock (shale) comprised of silt or mud that contains hydrocarbons from which petroleum can be extracted. In the United States thick deposits of oil shale occur in northwestern Colorado, southeastern Wyoming, and northeastern Utah.

organic debris Matter in the soil derived from living organisms; decayed organic matter is humus. Organic materials supply part of the nutrients required by plants.

paddy cultivation Cultivation of irrigated crops in flooded fields, for instance, rice.

pampa A humid, fertile agricultural district of Argentina that supports the bulk of the country's population. There are four divisions of the humid pampas: (1) pastoral; (2) alfalfa-wheat; (3) maize (corn); and (4) intensive truck farming, dairying, and fruit growing.

pastoralism The practice of herding animals, usually in semiarid to arid environments, by nomadic peoples.

peninsula An elongated projection of land into an ocean, sea, or lake, for example, the Iberian Peninsula comprising Spain and Portugal.

peripheral region A region that is outside the core region of a country or functioning system.

permafrost Permanently frozen subsoil.

plain A land surface with low relief and much of its surface in gentle slopes. Plains may occur at any elevation.

plantation system An agricultural system based on a specialty crop, managed by outsiders who infuse capital, and using local labor to produce for an overseas market. Historically, imported slaves supplied the labor when the native population was sparse or nonexistent. This system is associated with colonialism.

plateau Relatively flat surface elevated above its surroundings. Plateau areas are also called tablelands.

Polynesian One of the human groups inhabiting part of the Pacific Ocean islands known as Polynesia, including Hawaii, Indonesia, and the Philippines.

population density An expression of population distribution in persons per unit area, usually per square kilometer or square mile. This idea is useful for comparing the relation of people to the area they occupy.

population distribution The pattern of where people live on the earth, with concentrations in some places and sparcity in others.

population explosion Rapid growth in the world's population, especially during the twentieth century. The world population growth rate is close to 2 percent per year but is nearly 3 percent in Latin America and parts of Africa.

potential evapotranspiration (PE) The amount of water that can be lost by evapotranspiration at a particular place if there is unlimited availability of water. Under this condition, the loss of water is directly proportional to the energy received from the sun.

prairies An extensive area of grassland. This term is applied particularly to the interior plains area of the United States and Canada.

primary land use The principal economic activity associated with a particular unit of land, such as grazing, farming, or manufacturing.

primary producing region A region noted for primary production, which includes agriculture, fishing, forestry, and mining.

primary production Creation or gathering of products directly from the environmental base by hunting, gathering of fruit and nuts, fishing, herding of animals, agriculture, forestry, or mining.

primate city The largest and dominant city in a hierarchy, for instance, Moscow among cities of the Soviet Union.

private sector That part of the economy represented by those interested in the profit motive. This includes all private enterprises where economic self-interest predominates.

production The creation of something of value, such as turning environmental resources into food, clothing, shelter, or manufactured products.

promontory A projection of cliffs or high headlands into a sea or lake.

public sector That part of the economy represented by governmental agencies at all levels that are supported by taxes; the community at large and its supporting services.

race A group of people distinguished by particular physical traits such as hair, color of skin, shape of head or nose, or other features.

rainforest A tropical forest of evergreen and deciduous broadleaf trees and associated vines and parasite plants. Rainfall is well distributed throughout the year, so the rainforest remains luxuriant. Hundreds of species of trees abound, but large stands of a single species are rare. Trees generally are arranged by height in three layers or canopies of leaves, shielding the rainforest floor from light. A smaller, midlatitude rainforest also exists, but it differs in species of trees and the occurrence of pure stands.

region A grouping of places having similar qualities or an organizational cohesiveness. A region could be based on aridity (a desert), continental affiliation (Latin America), or sense of common purpose as in a nation-state (United States).

regional self-sufficiency Ability of a region to meet the needs of its population from its own resource base. Regions remote from the major population and industrial centers of a country must strive for this ideal. In the United States, California is a nearly self-sufficient region.

regional specialization Geographic division of labor, in the sense that each region produces the crops or commodities for which it is best suited.

regionalization Geographic generalization in which places having like characteristics are grouped and categorized.

relief Difference between the highest and lowest points in a local area, for instance, valley bottom to hilltop.

resource base The portion of the earth environment on which people in a particular area depend for the support of life.

resource-oriented Refers to an industry that locates close to its source of raw material, usually because of the bulky nature and consequent high cost of shipment. An example is the location of iron and steel industries near iron ore and coal or adjacent to low-cost water shipping.

rift valley A downfaulted valley between two scarps. Rift valleys are often the sites of elongated lakes, especially prominent in eastern Africa.

royalty payments In the context of extraction of mineral resources, a share of the proceeds paid to the owner or lessor of land that has been leased for oil or mineral production.

runoff water Water that exceeds the storage capacity of the soil and flows into creeks, streams, and rivers. Before runoff occurs, the soil-moisture storage zone becomes saturated.

savanna Tropical grasslands with interspersed clumps of trees that give a parklike landscape. The savanna grass ranges from one to twelve feet high and is generally tough. At the arid margins, the grass is sparse and grows in clumps. The wetter margins contain numerous forest growths and the tallest grasses (elephant grass).

scrub forest Vegetation consisting of bushes, stunted trees, brush, and spring grasses; common in semiarid climates or on sandy or stony soils.

seasonal-work environment An environment with distinct climatic seasons, including an extended period of high environmental work. The seasons may be sequencing, in which case environmental work goes on in all seasons but changes in nature from season to season; or there may be a dormant season, in which there is little environmental work.

secondary production Creation of something of value out of the products of primary production; includes all processes of manufacturing and construction.

second-order trade Trade between industrialized and nonindustrialized areas.

sedimentary rock Rock resulting from cementation of particles of weathered, transported, and deposited rock, or from the chemical precipitation of calcium carbonate from solution. Examples are sandstone (cemented sand), shale (cemented silt or mud), and limestone (calcium carbonate).

sequencing seasons Seasons in which environmental work continues throughout the year but the type of work varies from one season to another.

shifting cultivation Type of agriculture practiced in tropical forest, scrubland, and savanna regions, where patches of land are cleared and cultivated until the natural fertility of the soil declines and new fields are prepared. Clearing is usually accomplished by slashing the bark of trees, allowing the trees to die, and then burning them. This slash-and-burn method provides nutrients from the ashes of the burned vegetation. Whole villages may follow the shift in cultivation sites, or the sites may rotate around a central village site.

shogun A military minister of the Japanese emperor and the holder of the real power in the country during feudal times. The Tokugawa Shogunate was the last before the country began to modernize under Emperor Meiji in 1868.

slash-and-burn agriculture The practice of clearing portions of tropical forests and scrublands by slashing and burning trees. Nutrients are provided by the ashes, and crops are grown until the fertility of the soil declines and the cultivators move on to new fields. See also shifting cultivation.

socialization Process of altering social structure so that the producers possess more political power and more of the means of producing and distributing goods; usually involves nationalization of certain institutions and industries.

soil-moisture storage Water held in the soil against the pull of gravity and available to plant roots. The moisture transpired by plants originates with the roots in the soil-moisture storage zone. Water moves into this zone under gravity. All moisture is never completely removed from the zone by plants or capillary action because of a thin film of water held to each soil particle by surface tension.

spatial interaction The connections and flows between places, ranging in scale from adjoining streets to continents.

spatial variable Any phenomenon that varies from one section of space on earth to another, such as people, vegetation, or crops.

specialization of labor Division of complex work into a large number of individual tasks performed by persons possessing highly specific skills. Modern manufacturing and production depend on specialized labor skills and supporting services.

specialty crops Crops particularly suited to be grown in a particular area or for a particular market.

Standard Metropolitan Statistical Area (SMSA) An area made up of an entire county or counties and centered on a city of 50,000 population or larger. It is presumed to represent the area dominated by the economic functions and employment of the central city. The U.S. Bureau of the Census determines the definition of SMSAs.

subculture A distinct human group having its own interests and goals within the larger framework of a culture, for instance, the Scandinavian subculture within the Germanic culture.

subsidiary A branch plant or company whose stock is controlled by the parent company.

supplemental region A region that serves a "colonial" function, supplying raw materials to a core region. A supplemental region contrasts with a complementary region.

symbiosis The mutual interdependence of two groups, such as pastoral tribes and farming tribes that depend on each other's surpluses for existence.

taiga Extensive subarctic forests consisting of coniferous evergreens.

technology Accumulated expertise that allows a society to overcome problems posed by its environment and to advance or maintain its standard of living.

tertiary activities The service functions of a society, such as banking, education, and government, as opposed to primary production (farming, fishing, mining, forestry) and secondary activities (manufacturing, construction).

third-order trade Trade between nonindustrialized areas.

Third World Countries with a developing market economy (that is, money economy) as distinguished from (1) countries with a developed market economy, especially the affluent Western industrialized nations and (2) countries with centrally planned economies, including the Soviet Union, China, and other Communist countries. A large proportion of the population in Third World countries usually lives at the subsistence level.

tidewater plantation A colonial plantation that could be reached directly by ship. Thus goods such as cotton or tobacco could be shipped directly to Europe from the field docksites and manufactured goods could be unloaded directly from the ships to the plantation.

tierra caliente A hot, lowland zone of settlement and agriculture in Latin America. It generally ranges in altitude from sea level to 3,000 ft (1,000 m), although the upper limit is dependent upon latitude. Tropical crops and fruit, for instance, bananas, are products of this zone.

tierra fría A cold, highland zone of settlement and agriculture in Latin America. It generally ranges in altitude from 6,500 ft (1,950 m) to more than 10,000 ft (3,000 m), depending upon latitude. Potatoes and hardy grain crops are cultivated in this zone.

tierra templada A warm upland zone of settlement and agriculture in Latin America. It generally ranges in altitude between 2,000 and 6,500 ft (600 and 1,950 m), depending upon latitude. It is sometimes referred to as the "zone of coffee"; corn is also grown.

trade deficit A negative balance of trade. This situation occurs when the value of a country's imports exceeds the value of its exports.

trade surplus A positive balance of trade. This situation occurs when the value of a country's exports exceeds the value of its imports.

traditional, locally based system A way of life based on a limited local environment. Only local resources are available to support the traditional culture group, and limited technology does not allow extensive contacts with other groups outside the local environment.

transformational processes Weathering of minerals and rocks, and decay of organic materials.

translocational processes Dislocation or transport of materials from one place to another through erosion, stream flow, leaching, and ground-water movement.

transpiration Release of water vapor into the air through the stomata of plant leaves.

tube-well (canal) irrigation Irrigation system in which a tubular channel is used to distribute the water (in contrast to paddy cultivation). Source of water is often a well dug into the ground-water table.

tundra Generally treeless areas of Eurasia and North America except where stunted growth occurs in protected hollows. Dominant vegetation is a groundcover of mosses and lichens, with some grasses and flowering plants.

typhoon A violent tropical storm in the western Pacific, identical to a hurricane in the Atlantic Ocean.

urban Refers to places where people live in close proximity to one another, as in towns and cities.

value added by manufacture The money value of all costs incurred in the manufacturing process; the difference between the value of the manufactured product and the raw materials. This index measures the economic importance of manufacturing industries.

veldt In southern Africa, the plateau areas of open grassland, occupied and farmed by Europeans as well as Africans.

vertical integration Structure of a manufacturing company whereby the sources of raw materials, in addition to factories and distribution facilities, are owned or controlled by a single company.

vertical zonation In mountainous areas, especially in Latin America, division of environmental (climatic) zones according to altitude. Because of decreasing temperatures with increasing altitude, zones exist from sea level to the highest basins and mountains in the Andes. The boundaries between zones vary with latitude, but generally range from *tierra caliente* (hot lowland) to *tierra templada* (warm, temperate) to *tierra fría* (cold highland). The *páramos* (alpine meadows) and *tierra helada* (zone of permanent snow and ice) are sometimes included in this vertical zonation of climates in Latin America. Vertical zonation exists, but is less important, in other mountainous areas.

water balance On a local scale, a water balance consists of two primary factors: (1) supply of moisture available to plants and (2) demand for moisture as expressed by evaporation from the soil and transpiration by plants. On a global scale, water balance is equivalent to the hydrologic cycle, the exchange of water among oceans, atmosphere, and land surfaces.

water budget analysis Balance of water supply and demand for an entire year at a particular place. It takes into account monthly precipitation, potential evapotranspiration (PE), and water surpluses or deficits.

water deficit Potential evapotranspiration (PE) minus actual evapotranspiration (AE) in the water budget for a particular place: deficit = PE − AE. This amounts to moisture demand exceeding supply and takes into account both precipitation and soil moisture storage.

water surplus Precipitation minus the actual evapotranspiration (AE) in the water budget for a particular place: surplus = precipitation − AE. This amounts to the supply of moisture exceeding the demand at that place.

weathering Disintegration and decomposition of rock and minerals through chemical or physical (mechanical) action or both. Mechanical weathering results from freeze-thaw activity of moisture in rock or differential expansion and contraction of constituent minerals caused by heating and cooling. Chemical weathering results when the bonds between molecules are broken, as in the process of oxidation, in which oxygen combines with a mineral in the rock and weakens the rock.

Index

Accessibility, 150–151. *See also*
  Communications; Transportation
Actual evapotranspiration (AE), 60, 66, 67,
  68, 73, 77, 79
  in Asia, 565, 589, 591
  in Black Africa, 472
  in Canada, 386
  in Middle East–North Africa, 513, 531
Afghanistan, 559
Africa, Black, 352, 413, 429, 460–499, 514,
  533. *See also* North Africa; South
  Africa
  agriculture in, 42, 142, 144, 157–158, 427,
    464, 470–499 *passim*
  communications in, 409, 487, 497
  cultures in, 248, 466, 477, 478, 483, 488,
    496–497
  energy consumption in, 117, 148,
    468–469, 493–496
  environments in, 66–84 *passim,* 460,
    469–470, 472–476, 498–499
  and Europe, 153, 234, 244, 263–264, 279,
    410, 413, 441, 460–497 *passim*
  GNPs in, 147, 466–469, 474, 496
  governments in, 410, 462–464, 466, 474,
    481, 484–485, 488, 496, 497
  and Latin America, 429, 449, 490–492, 498
  metropolitan areas in, 142, 483–487
  and minerals, 395, 399, 401–402, 409, 449,
    470–471, 477–496 *passim,* 508, 510
  politics in, 50, 462–466, 474, 477, 478,
    484–485, 487, 497, 498
  population distribution in, 129, 134, 135,
    466, 469–471, 484, 497
  population growth in, 120, 411, 471, 476
  religions in, 108, 464, 477, 488, 519
  slaves from, 181, 393, 430, 433, 437, 454,
    471, 478, 486
  and Soviet Union, 320, 466
  trade of, 153, 178, 180, 265, 392, 401,
    477–481, 488–493, 496
  traditional system in, 38–40, 157–158,
    409–412, 462–470 *passim,* 476–478, 488,
    496–499
  transportation in, 150, 464, 474, 478,
    481–482, 486, 487, 497, 587

wages in, 406, 485
Afrikaners, 486
Age of Exploration, 102, 234, 559
Agrarian reform, 580
Agriculture, 34, 35, 54, 69, 72, 79–87, 117.
    See also Livestock; individual crops
  in Asia (general), 142, 144–145, 481,
    552–553, 554, 556–557
  in Australia, 142–143, 264, 383
  in Bangladesh, 559, 569, 590–591
  in Black Africa, 42, 142, 144, 157–158,
    427, 464, 470–499 *passim*
  in Canada, 142–143, 264, 362–363, 365,
    366, 376, 382–387
  in China, 81, 144, 393, 554, 556, 563,
    577–580, 581–583
  in Europe, 109, 142–143, 145, 238–240,
    244, 249–254, 278, 299, 427
  in India (general), 81, 119–120, 135–136,
    144, 393, 556, 562–570, 573
  in Japan, 332, 334, 339–341, 343, 344,
    356–357
  in Korea, 589
  in Latin America, 42, 116, 180, 264, 393,
    420–457 *passim,* 481, 569
  Mediterranean, 524
  in Middle East–North Africa, 134, 136,
    513, 518–533 *passim,* 536–537, 540
  in modern system, 42, 45, 87, 109,
    142–143
  in New Zealand, 264
  in Pakistan, 569, 588, 589
  in Ramkheri, 19, 20–22, 25, 26–27, 36, 40,
    144–145
  slash-and-burn, 470, 476, 478, 554, 594
  in Southeast Asia, 81, 393, 591, 594–595
  in Soviet Union, 143, 292, 293–295, 303,
    304–305, 308–312, 317
  in traditional systems, 86, 87, 109, 142,
    409, 411
  in United States. *See* breakdown under
    United States
Ahmadabad, 573
Air travel, 11–13, 44, 46, 150, 156, 408
  Asia and, 28, 150, 592
  Black Africa and, 150, 486

Canada and, 13, 150, 380
Latin America and, 13, 150, 446
in Middle East–North Africa, 525, 530,
  536
in Soviet Union, 319
in United States, 7, 10, 11–13, 15, 150,
  206–207, 380
Akron, 188
Alabama, 181, 227
Alaska, 164, 312
  environments in, 83, 84, 167, 171, 422
  oil in, 83, 173, 308
Albania, 234, 242, 258, 587
Alberta, 363, 365, 368, 369, 379
Aleppo, 520
Alexandria, 513, 515, 525
Algeria, 320, 449, 511, 530–531
  agriculture in, 524, 529
  environments in, 516, 518
  European workers from, 279, 533
  and oil, 510, 512, 539
  population in, 512, 524
Algiers, 525
Allende, Salvador, 454
Alps, 57, 237, 240, 248
Altiplano, 422
Amazon Basin, 420, 427
  agriculture in, 393, 426, 457
  environments in, 67, 72, 79, 84, 135, 424,
    426, 429
Amazon River, 413, 424
American Indians, 38, 39
  in Canada, 365, 379, 380
  in Latin America, 413, 418, 420, 427–429,
    430, 433, 435, 437, 440, 451
  in United States, 40, 157, 183, 226, 227,
    464
Americas. *See* Canada; Latin America;
  United States
Amsterdam, 258, 271
Andean Group, 458
Andes Mountains, 422, 427, 429, 430, 433,
  472
Anglo-America. *See* Canada; United
  States
Angola, 466, 482, 490

617

Factories, 110. *See also* Manufacturing
  assembly-line, 116, 408
  energy for, 177, 180, 187, 256, 293
  in India, 572, 573
  Japanese, 351–352, 358
  in Kuwait, 534
  in Soviet Union, 291, 293
Families, extended, 464, 525
Famines, 48, 411, 550
Far East (Soviet Union), 292, 293, 295, 300,
  307
Farming. *See* Agriculture
Fazendas, 437
Fencheng County, 586
Fertile Crescent, 513, 515, 520, 524, 525
Fertilization, 79, 85, 87, 569
  in Africa, 475, 476
  in Asia, 568–570, 586, 589, 594
  in Europe, 239
Feudal system, 392
  in Africa, 413
  in Arab countries, 413, 524
  in Europe, 248, 251
  in Japan, 339, 341–342, 440
  in Latin America, 438, 440
Finance, 45, 406. *See also* Banks; Capital;
  Foreign investment
  in Canada, 366, 372–375, 376–378, 380,
    388
  in Europe, 234, 264, 512
  in Japan, 358
  in Kuwait, 536
  in United States, 178, 188, 216, 222, 404,
    512
Finland, 239, 265, 268, 269
Fire, 39, 75, 112, 117
Fishing
  in Africa, 478
  in Canada, 365, 366, 372, 376
  Japanese, 339, 344, 357
  in Latin America, 441–443
  in Middle East–North Africa, 532
  in traditional systems, 40, 86, 365
  in United States, 171, 178
Floodplains, 81, 83–84, 532, 554, 577
Floods, 75, 81
  in Asia, 550, 554, 583, 588, 589, 590, 591
  in Soviet Union, 304
Florida, 164, 174, 175, 188, 194, 204, 216
Food, 145–146, 411, 412
  in Africa, 145–146, 471, 474
  in Asia (general), 145–146, 546, 550, 557,
    569, 594
  in Canada, 146, 368, 372, 379
  in China, 119, 136, 550, 563, 576, 577, 587
  and environment, 34, 54, 88
  in Europe, 117–118, 120, 270, 278
  in Iceland, 84
  in India, 119–120, 136, 550, 563, 570–571
  in Japan, 120, 339, 344, 354, 356, 357
  in Latin America, 180, 427, 429, 437, 441
  in Middle East–North Africa, 119, 522,
    540
  and modern system, 48, 87, 146, 392,
    393–394
  in Soviet Union, 293, 308, 312
  in traditional systems, 40, 86

in United States, 16, 42, 171, 175, 178,
  180, 183, 225
Forces, environmental, 55–56, 57–60, 171
Ford Foundation, 569
Foreign aid, 320, 332, 484, 536, 561–562,
  572, 574–576
Foreign exchange, 400, 401, 406, 485, 565,
  570, 590
Foreign investment, 406, 409, 412, 439. *See
  also* Finance; Multinational
  corporations
  in Africa, 481, 484, 485–487, 496
  in Anglo-America, 372–375, 376–378, 380,
    388, 512
  in Asia, 403, 572, 576, 580, 594
  in Europe, 512
  in Latin America, 418, 439–440, 446, 448,
    450, 457, 458
Foreign Investment Review Act, 376–378
Forestry, 34, 35, 40, 45, 117
  in Africa, 490
  in Canada, 364–365, 366, 368, 372,
    375–376, 379
  in Europe, 239, 269
  in Latin America, 446, 457, 490
  in Lebanon, 518
  in United States, 167, 169, 171, 178,
    217–218, 229–230
Forests, 76
  in Africa, 472, 475–477
  in Asia, 553–554
  in Canada, 363, 364
  in Europe, 239
  in Japan, 336
  in Latin America, 426–427, 429, 456–457
  in Lebanon, 518
  rain, 426–427, 456–457, 472, 475–477
  scrub, 427
  in Soviet Union, 304, 311, 364
  in United States, 167, 171, 178, 181
Fort Worth, 213
Fossil fuels, 34, 35, 57, 171–173, 256. *See
  also* Coal; Gas, natural; Oil
  limited supplies of, 47, 89
  for transportation, 43, 111, 256
France, 234, 242, 256, 271–272, 395
  and Africa, 244, 264, 410, 474, 487, 492,
    529, 530
  agriculture in, 251, 252
  and Americas, 264, 365, 371, 430
  and Bretons, 40, 248, 281
  and economics, 244, 275–278, 279, 280
  environments of, 235, 237, 238, 239
  and Indochina, 556, 559, 560–561
  manufacturing in, 271, 280
  metropolitan areas in, 9, 224, 255, 263,
    264, 271
  and Middle East, 515, 522
  minerals of, 240, 269
  trade of, 265, 270, 271, 274, 320, 522
Frankfurt, 258, 280
Freeport Sulphur Company, 402, 403
French Canadians, 40, 362, 365, 366, 370,
  371–372
French language, 365, 372, 309, 488
Fruits
  in Black Africa, 475, 486

citrus, 216, 251, 264, 269, 334, 356, 426,
  486, 529, 540
  in Europe, 251, 252, 264, 269
  in Japan, 334, 356
  in Latin America, 426, 429, 439, 443, 446,
    453, 481
  in Middle East–North Africa, 524, 529,
    540
  in United States, 188, 216
Fuels, 40, 117. *See also* Fossil fuels
Fukuoka, 347
Fulani nomads, 474
Fur trade, 365, 366
Fushun, 581

Gabon, 466–468, 490, 493, 496, 510
Gaelic, 105
Gambia, 493
Gandhi, Indira, 576
Gandhi, Mahatma, 561, 574
Ganges River, 554, 588, 590
Ganges Valley, 574
Gas, natural, 57, 394–395
  in Anglo-America, 173, 376
  in Europe, 240, 265, 278
  in Middle East–North Africa, 508, 518,
    534, 536
  in Soviet Union, 305, 308, 314
Gases, atmospheric, 78
Gasoline, 113, 399
Gathering, 40, 86, 365, 427, 478
General Motors, 10, 218–222, 349, 351, 362
Genoa, 271
Geographic equation, 30, 33, 36–37, 39–40,
  49
  and Asia, 563, 588
  and Canada, 362
  and Japan, 357
  and Middle East–North Africa, 515
Geography, defined, 32, 33
Georgia, 178, 334. *See also* Atlanta
Germany, 248, 256, 264, 292. *See also* West
  Germany
  agriculture in, 251, 252
  East, 238, 242, 259–262, 272, 279, 320
  Latin American immigrants from, 430,
    439
Ghana, 476, 477, 483, 490, 493, 496
Ghettos, urban, 157
Glasgow, 258, 263, 271
Glass, 188, 343, 404
GNP. *See* Gross national product
Goats, 513, 530
Gold, 393, 403
  in Africa, 477, 478, 481, 486, 490
  in Canada, 365
  in Latin America, 413, 427, 435, 440
  in Soviet Union, 305
Gold Coast, 478
Governments, 40, 138, 408, 412. *See also*
  Management; Politics
  in Asia (general), 403, 560–563, 594
  in Black Africa, 410, 462–464, 466, 474,
    481, 484–485, 488, 496, 497
  in Canada, 368, 372, 379, 388
  in China, 561, 562, 563, 576, 581, 583–587

Mount McKinley, 422
Mozambique, 482, 483, 488, 490
Mulattos, 433
Multinational corporations, 43, 44, 46–47,
    406–410, 412, 439
    in Asia, 399, 403, 576, 592, 594, 595
    Black Africa and, 395, 399, 485–487, 497,
    498
    in Canada, 375, 376
    in Europe, 271–272, 278, 395–400, 510
    in Latin America, 394, 395, 397, 398, 399,
    400–401, 448
    and Middle East–North Africa, 395,
    397–398, 400–401, 515, 539, 542
    in United States, 229, 264, 275–278,
    395–400, 510
Munich, 280
Muscat, 515
Muslim Brotherhood, 519

Nagasaki, 339
Nagoya, 346, 355, 356
Nam Chao Phraya River, 554, 591
Namibja, 462
Nanking, 560, 581
Narcotics, 529
Nationalism, 488, 520
National Road, 181
Nation-states, 39, 44–45, 109, 240, 248
Negev Desert, 540
Nepal, 117, 551, 554
Nesting, 223
Netherlands, 242, 252, 256, 272, 334
    colonies of, 264, 403, 430, 486, 559, 560
    environments of, 237, 238
    metropolitan areas in, 258, 263, 271
    and minerals, 269, 279, 403, 443
    and multinational corporations, 271–272,
    395–400
    trade of, 265, 268–269, 270, 271, 274, 339,
    437
Network, metropolitan, 206
New Brunswick, 365, 366, 368, 369, 372
Newcastle-upon-Tyne, 258
New Delhi, 557, 573, 574
New England. See Northeast
Newfoundland, 366, 368, 376
New France, 365, 371
New Guinea, 135, 403
New Jersey, 180, 182, 216, 222
New Orleans, 181, 186, 365
New York Central Railroad, 192
New York City, 10, 14, 189, 212, 222–223,
    227
    air travel in, 11–13, 207, 380
    climate in, 237
    colonial, 180
    communications in, 222, 225
    management in, 43, 216, 222–223, 224,
    263, 264, 351, 378
    population in, 186, 318, 349, 366
    trade of, 181, 192
New York State, 187, 188, 190, 204, 227,
    363, 369
New Zealand, 139, 142–143, 147, 148, 234,
    264, 362
Nicaragua, 424, 451, 457

Nickel, 305, 344, 365, 376, 427
Niger, 117, 474, 483
Nigeria, 474, 483, 484, 493, 496
    agriculture in, 476, 490
    civil war in, 466, 497
    and minerals, 395, 399, 449, 490, 493, 510
Nile River, 508, 522, 531
Nile Valley, 513, 518, 522, 524, 529
Nitrogen, 79, 80, 82
Nixon, Richard M., 562
Nodes, 44
    in Africa, 471, 477
    in Japan, 349
    in Soviet Union, 317
    in United States, 196, 205–206, 223–224,
    230
Norfolk, Virginia, 13
Norilsk, 300, 305
North Africa, 279, 320, 413, 415, 449,
    503–542. See also Sahara Desert
    and oil, 395, 399, 400, 508, 510, 512–513,
    534, 538–539
    religion in, 106–107, 508, 513, 514, 515,
    519, 524, 530, 531
North America. See Canada; United
    States
North Atlantic Drift, 237
North Dakota, 227
Northeast (U.S.), 177–180, 227, 312, 404, 406
    agriculture in, 183, 187, 226
    and Canada, 366, 368, 369, 370, 375, 388
    environments in, 167, 238
    fishing in, 171, 178
    manufacturing in, 180–181, 182, 185,
    188–189, 210, 213, 215, 343, 378, 406
    metropolitan areas in, 202, 203, 204, 205,
    210, 212, 213
    minerals in, 173, 174
    population in, 174, 202, 203, 204, 205, 256
    transportation in, 175, 178, 179, 181, 183,
    192, 343
    unions in, 188–189
North Sea, 240, 264, 278, 395, 399
North Slope, Alaska, 83, 173
Northwest Passage, 365
Northwest Territories, (Canada), 376, 379
Norway, 235, 242, 258, 259, 268, 275
Nova Scotia, 365, 366, 368, 379
Nuclear energy, 112, 115, 117, 278, 409
Nutrients, plant, 78–81, 87, 239, 532
    in tropics, 84–85, 426, 429, 457, 470
Nylon, 343

Oakland, 212
Oases, 522
Oats, 251
Ocean environment, 34, 61, 83, 171, 357,
    399
Oceania, 146, 153, 392
Ocean shipping, 150, 177, 178, 183, 413
Ohio, 188, 227
Ohio River, 81, 181
Ohio Valley, 167, 181
Oil, 47, 57, 84, 113, 390, 394–401
    and Black Africa, 395, 399, 449, 490, 493,
    496, 508
    in Canada, 365, 372, 376, 379, 395

in China, 399, 581
in Europe, 240, 264, 265, 271, 395, 399
India and, 399, 510, 570
in Indonesia, 399, 559
and Japan, 332–333, 344, 400, 510, 534,
    539, 587
and Latin America, 394, 397–400, 427,
    439, 441, 443, 449, 450, 454
and Middle East–North Africa, 150, 264,
    278–279, 308, 352, 395, 397–400,
    508–539 passim
prices of, 48, 174, 398, 401, 449, 493, 510,
    512, 515, 536, 570
in Soviet Union, 300, 305, 307, 308, 314,
    315, 395, 399
in United States, 119, 173–174, 186, 188,
    216, 264, 379, 395–399, 404
Oil shale, 89
Oklahoma, 173, 174, 188, 227
Olives, 251, 524, 529
Oman, 512, 524, 537, 539
Ontario, 366, 369, 376, 379, 388
    agriculture in, 362, 382, 386
    culture in, 371, 372
    manufacturing in, 378, 381
    population in, 369, 372
    transportation to, 363, 368, 379, 380
OPEC. See Organization of Petroleum
    Exporting Countries
Opium, 529
Oranges, 216, 334, 356, 426
Oregon, 167
Organic debris, 79, 304
Organization of African Unity, 498
Organization of American States, 458
Organization of Petroleum Exporting
    Countries (OPEC), 400, 410, 449, 493,
    510, 536
Orinoco River, 424, 427
Osaka, 330, 341, 344, 351, 356
    manufacturing in, 343, 355
    population in, 339, 346, 349
Ottawa, 380
Ottawa River, 372
Ottoman Empire, 478, 515, 520
Ozone layer, 47

Pacific Coast (U.S.). See West
Pacific islands, 106, 164, 167, 206, 218
Pacific Ocean, 171, 308, 375, 429, 443
Pakistan, 117, 550, 556, 588–589
    agriculture in, 569, 588, 589
    in British India, 559, 588
    Canadian immigrants from, 369, 370–371,
    372
    environments in, 134, 513, 551, 553, 554,
    588, 589
    European immigrants from, 279, 280
    independence of, 560, 574
    Islam in, 519, 574, 588
    population in, 129, 132, 134, 588
Palestine, 520, 540
Palestine Liberation Organization (PLO),
    536, 540
Palm oil, 476, 481, 490
Pampas, 80, 424, 425, 426, 427, 454
Panama, 437, 438
Paraguay, 424, 425, 433, 451, 456, 458

Paraguay-Paraná Plain, 424, 425
Paris, 9, 224, 263, 264, 271
Pastoralism, 470. *See also* Livestock
PE. *See* Potential evapotranspiration
Peace River settlement, 369
Peking, 557, 581, 584, 586, 587
Peninsulas, 235, 237, 430, 589
Pennsylvania, 180, 182, 190, 192, 218, 227
Pennsylvania Dutch, 251
Permafrost, 304, 363, 379
Perón, Juan, 454
Persian Gulf sheikdoms, 511, 525, 530. *See also* Bahrain; Kuwait; Oman; Qatar; United Arab Emirates
Persian language, 519
Peru, 158, 420, 456, 457
  agriculture in, 426, 427 443, 450, 451
  environments in, 422, 425
  Europeans in, 435, 437, 438
  fishing of, 441, 443
  industry in, 441, 446, 450
  minerals in, 401, 427, 440, 450
  politics in, 455, 458–459
  population in, 429, 433
  trade of, 443
  traditional system in, 158, 420, 429
Pesticides, 88, 570
Petróleo Brasileiro (Petrobrás), 449
Petroleum. *See* Oil
Philadelphia, 14, 180, 181, 186, 189, 207, 318
Philippines, 40, 556
  agriculture in, 559, 569
  environments in, 550, 554
  population in, 8, 546
  and United States, 559, 560
Phoenix, Arizona, 175, 204, 222
Phosphate, 174, 518
Pigs, 244, 251, 310, 356, 513, 529
Pindus Mountains, 237
Pittsburgh, 188, 207, 216, 218
Places, 32, 33, 54. *See also* Urban places
  hierarchy of, 223, 300, 317–318
  local, 124–125, 409, 583, 584, 586
Plains, 56, 57, 237, 239, 303, 422–424, 554. *See also* Great Plains
Plantations, 393, 394
  in Africa, 42, 470–471, 481, 482
  in Asia, 393, 559, 594
  in Latin America, 42, 116, 180, 393, 426, 429, 433, 437, 439, 440, 478, 481
  tidewater, 179, 181
  in United States, 116, 179, 180, 181, 186, 227, 478
Plateaus, 56, 237, 363, 424, 472, 486
Platinum, 486
Pluralism, cultural, 290
Poland, 238, 242, 251, 263
  coal in, 240, 269
  trade of, 269, 270, 272, 320
Polar areas, 77, 106, 138, 357
Political parties, 205, 242, 287, 290, 372, 459, 520
Politics, 109, 138, 409, 410, 412. *See also* Governments; Independence, political
  in Asia, 560–561, 574, 576, 588, 589, 590, 591
  in Black Africa, 50, 462–466, 474, 477, 478, 484–485, 487, 497, 498

in Europe, 234, 240–248, 267, 270, 275, 281
in Japan, 341, 343
in Latin America, 430, 438, 439, 441, 448, 453, 454, 455, 458–459
in Middle East–North Africa, 519, 520, 524
of Soviet Union, 234, 287–290, 292, 300, 319, 320
in United States, 404
Pollution, environmental, 15, 16, 47–48, 89, 411
Polo, Marco, 556
Polynesian racial stock, 556
Population density. *See also* Population distribution
  defined, 132–134
  and environment, 50, 54, 72, 83, 84, 86, 87, 90, 134–135
Population distribution, 41, 129–136, 138–139, 406, 415. *See also* Metropolitan populations
  in Asia (general), 129–132, 135, 415, 546–551, 556
  in Australia, 129, 135, 136, 139, 383
  in Bangladesh, 129, 132, 546, 590
  in Black Africa, 129, 134, 135, 466, 469–471, 484, 497
  in Canada, 129, 136, 139, 360, 362, 365, 369, 372, 383
  in China, 119, 120, 129, 132, 135–136, 144, 546, 562, 577
  in Europe, 129, 132, 134, 135, 136, 234, 242, 249–256, 297, 334
  in India, 120, 129, 132, 135–136, 144, 546, 562
  in Japan, 129, 132, 135, 334, 339–341, 344, 345–349
  in Latin America, 120, 134, 135, 415, 418, 429, 430–433, 440, 443, 451
  in Middle East–North Africa, 134, 415, 508, 512, 513, 522–524, 533, 536, 539–540
  in Mongolia, 589
  in Pakistan, 129, 132, 134, 588
  in Southeast Asia, 132, 134, 546, 551, 591, 594
  in Soviet Union, 132, 234, 284, 289, 293, 295–303, 317, 318–319
  in United States, 47, 129, 133, 136, 174–175, 234, 256, 334, 341, 362, 594
Population explosion, 117, 411, 412
Population growth, 117, 118–120, 411, 412
  in Asia (general), 117, 411
  in Black Africa, 120, 411, 471, 476
  in Canada, 366, 368, 388, 548
  in China, 548, 551, 587, 595
  in Europe, 117–118, 239, 546, 548
  in India, 119–120, 548, 551, 571, 595
  in Japan, 332, 345, 548
  in Latin America, 120, 411, 459
  in Middle East–North Africa, 521, 531, 538
  in Southeast Asia, 119, 548, 595
  in United States, 118–119, 169, 175, 189, 204, 546, 548, 551
Portland, Maine, 366, 368, 379
Portugal, 235, 237, 254, 515

and Africa, 487–488
and Americas, 234, 429, 430, 435–438, 439
and Asia, 556, 560
and trade, 265, 275
workers from, 279, 280
Potatoes
  in Anglo-America, 188, 366
  in China, 580
  in Europe, 251
  in Latin America, 426, 429
  in Soviet Union, 308, 310, 312
Potential evapotranspiration (PE), 60, 66, 67–69, 74, 75, 84
  in Asia, 66, 552–553, 565, 591
  in Black Africa, 66, 472, 498
  in Canada, 362, 363, 364
  in Europe, 238
  in Japan, 334
  in Latin America, 66, 424
  in Middle East–North Africa, 513, 516, 531
  in Soviet Union, 304
  in United States, 66, 164, 165–167, 304
Potlatch Corporation, 217–218
Potosí, 438
Poverty, 227, 453, 539, 573
Power. *See* Energy
Prairie Provinces (Canada), 380, 386, 388
  minerals in, 365, 376, 379
  population in, 369, 372
  transportation in, 363, 368, 376, 379
Precipitation. *See* Rainfall
Prices
  of agricultural products, 492, 493
  of copper, 402, 403, 492, 493
  of oil, 48, 174, 398, 401, 449, 493, 510, 512, 515, 536, 570
  in Soviet Union, 290–291
Primary production, 40, 392, 406, 569–570. *See also* Agriculture; Fishing; Forestry; Gathering; Hunting; Livestock; Minerals
  in Asia, 563–565, 567–568, 569–570, 580, 581, 591
  in Australia, 383
  in Black Africa, 475
  in Canada, 372, 376, 382, 383
  in Europe, 234, 239, 254, 264
  in Japan, 344, 349, 356, 406
  in Latin America, 443
  in Middle East–North Africa, 527
  in modern system, 45, 120, 169
  in Soviet Union, 293, 312
  in United States, 164, 169, 174–175, 177, 183, 192, 198, 201–202, 404
Primate city, 318
Prince Edward Island, 366, 368
Prince George, British Columbia, 369
Private sectors, 223, 320
Production, 40–41, 117–118, 120, 147, 512. *See also* Primary production
  in Africa, 492
  in Asia, 572–573, 586, 594, 595
  automobile. *See* Automobile production
  in Canada, 368, 379, 380
  of copper, 402–403
  in Europe, 270, 275–278, 393, 572
  in Israel, 542